COMPUTATIONAL METHODS IN SCIENCE AND TECHNOLOGY

This book contains the proceedings of the 4TH International Conference on Computational Methods in Science and Technology (ICCMST 2024).

The proceedings explores research and innovation in the field of Internet of things, Cloud Computing, Machine Learning, Networks, System Design and Methodologies, Big Data Analytics and Applications, ICT for Sustainable Environment, Artificial Intelligence and it provides real time assistance and security for advanced stage learners, researchers and academicians has been presented.

This will be a valuable read to researchers, academicians, undergraduate students, postgraduate students, and professionals within the fields of Computer Science, Sustainability and Artificial Intelligence.

PROCEEDINGS OF THE 4TH INTERNATIONAL CONFERENCE ON COMPUTATIONAL METHODS IN SCIENCE AND TECHNOLOGY (ICCMST 2024), 2–3 MAY 2024, MOHALI, INDIA

Computational Methods in Science and Technology

Volume 1

Edited by

Sukhpreet Kaur, Sushil Kamboj and Manish Kumar

Chandigarh Engineering College – CGC Landran, Mohali, Punjab

Arvind Dagur and Dhirendra Kumar Shukla

Galgotias University, Gr. Noida

CRC Press
Taylor & Francis Group
Boca Raton London New York Leiden

CRC Press is an imprint of the
Taylor & Francis Group, an **informa** business
A BALKEMA BOOK

First published 2025
by CRC Press/Balkema
4 Park Square, Milton Park, Abingdon, Oxon, OX14 4RN

and by CRC Press/Balkema
2385 NW Executive Center Drive, Suite 320, Boca Raton FL 33431

CRC Press/Balkema is an imprint of the Taylor & Francis Group, an informa business

British Library Cataloguing-in-Publication Data
A catalogue record for this book is available from the British Library

Library of Congress Cataloging-in-Publication Data
A catalog record has been requested for this book

Set

ISBN: 9781032911571 (hbk)
ISBN: 9781032911588 (pbk)

Volume 1
ISBN: 978-1-032-81760-6 (hbk)
ISBN: 978-1-032-81768-2 (pbk)
ISBN: 978-1-003-50124-4 (ebk)

DOI: 10.1201/9781003501244

Volume 2
ISBN: 978-1-032-91160-1 (hbk)
ISBN: 978-1-032-91161-8 (pbk)
ISBN: 978-1-003-56165-1 (ebk)

DOI: 10.1201/9781003561651

Typeset in Times New Roman
by MPS Limited, Chennai, India

Computational Methods in Science and Technology Volume 1 – Kaur et al. (Eds)
© 2025 the Editor(s), ISBN: 978-1-032-81760-6

Table of Contents

Computational Methods in Science and Technology Volume 1 – Kaur et al. (Eds)
© 2025 the Editor(s), ISBN: 978-1-032-81760-6

Preface

This book contains the proceedings of a non-profit conference with the objective of providing a platform for academicians, researchers, scholars and students from various institutions, universities and industries in India and abroad, and exchanging their research and innovative ideas in the field of Computer Science and its emerging fields. This conference provides opportunities to our college students & faculty members to interact with the external researchers and delegates to exchange new ideas and application experiences. It also helps to find global partners for establishing educational and research relations and for future collaboration. The conference holistically aims to promote translation of basic research into applied investigation and convert applied investigation into practice. This conference will also create awareness about the importance of basic scientific research in related fields and synchronizing with the product market.

Acknowledgements

The conference chairs would really like to appreciate the efforts of all the volunteers and helpers who have devoted their time and endless patience for this conference. CGC is a powerful and ever-growing learning association of many enthusiastic people who organize conferences and events from time to time and we are very grateful to be a small part of it. In particular, we have to thank the chairs, who were working on a voluntary basis for a whole year to make this conference a success.

We would also like to thank all the members of the international program committee, who provided timely and insightful suggestions and reviews without complaint and little credit. Finally, we would like to thank the advisory board members and organizing committee members of the conference for their support in this amazing endeavour. These are the people who have worked incredibly hard behind the scenes to guide all the aspects of the conference. Special thanks go to Hon'ble Chairman sir, S. Satnam Singh Sandhu and Hon'ble President sir, S. Rashpal Singh Dhaliwal for believing in us and giving nod to organize such an event. We would also like to take this as an opportunity to thank all those who have actively taken part as local organizers of the conference. We would like to thank the administration of CGC group, technical partners, publication partners and our sponsors for their timely support. We especially welcome conference delegates who are attending ICCMST – 2024 and hope it will be a great learning experience. Last but not least, we would like to acknowledge Taylor & Francis for providing us the opportunity to publish our conference proceedings.

© 2025 the Editor(s), ISBN: 978-1-032-81760-6

Editor Biography

Prof.(Dr.) Sukhpreet Kaur is working as a Professor in CSE department at Chandigarh Engineering College, Landran, Mohali. She has in total of 15 years of vast experience in teaching and research. She has done Ph. D in CSE from I K Gujral Punjab Technical University, Jalandhar and has done her Masters in Technology in CSE from GNDEC, Ludhiana. The various research areas in which she worked includes Image Processing, Artificial Intelligence and Computer Vision.

Prof.(Dr.) Sushil Kamoj is a distinguished Professor and Head of the Computer Science and Engineering (CSE) Department, with over 20 years of experience in academia and research. Renowned for his expertise in Computer Networks, Dr. Kamboj has made significant contributions to the field through his innovative research, teaching excellence, and leadership. He is dedicated to fostering academic growth and developing cutting-edge solutions to complex technological problems.

Prof.(Dr.) Manish Kumar is working as a Professor in the Department of Computer Science and Engineering at Chandigarh Engineering College, Chandigarh Group of College, Landran, Mohali. He has a career academician with about 20 years of experience in teaching, research, academic and administrative outreach activities. As far as academic qualification is concerned he did B.Tech., M.Tech. and Ph.D. degree in Computer Science and Engineering. He has primarily specialization research in the area of Wireless sensor network, data mining and machine learning. To his credit he has more than 40 publications in widely circulated journals of national and international repute. He has actively organized several national and international conferences. He has organized the various conference and proceedings editor of ICMST 2020, 2021 and 2022.

Prof.(Dr.) Arvind Dagur is currently working as Professor and Program Chair in school of Computing Science and Engineering, Galgotias University, Greater Noida, India. He worked as Professor & Dy. Head, in Department of Computer Science and Engineering, KEC (KIET Group) Ghaziabad. He has more than 18 Years of experience in Teaching and research. He completed Ph.D. (Computer Engineering) from Jamia Millia Islamia, New Delhi, India (A Central University), M.Tech. (Computer Science and Engineering) from MNNIT, Allahabad, India and B.Tech. (CSE) from UPTU, Lucknow. He has published more than 30 research papers in reputed international Journals / Conferences. His research papers are indexed in SCI, SCOPUS, DBLP, and ESCI, etc. He also published 5 book chapters in Springer Book series such as LNCS, AICS Series, and LNICST. He has chaired many sessions in various conferences and been a reviewer of many conferences and journals. He is the conference chair and proceedings editor of ICABCS 2023 and ICAIIT 2023.

Dr. Dhirendra Kumar Shukla is currently working as Associate Professor in School of Computer Science & Engineering Department, He has total experience of more than 11 years of teaching Computer science and Engineering Department and has been associated with Galgotias University since 2022. He has published 11 Journal research Paper area of Cloud computing.

Advisory Committee

International Technical Program Committee

Dr. Ranbir Singh Batth, *Western Sydney University-ATMC Campus, Melbourne, Australia*
(Prof.) Dr. Rajkumar Buyya, *The University of Melbourne, Australia*
(Prof.) Dr. Jemal Hussein Abawajy, *Deakin University, Australia*
Dr. Marek Bolanowski, *Rzeszow University of Technology, Poland*
Dr. Daniel Dasig, Jr., *De La Salle University, Dasmariñas, Philippines*
Dr. Nguyen Ha Huy Cuong, *The University of Danang, Vietnam*
Dr. Ahmed A. Elngar, *Beni-Suef University, Egypt*
Dr. Ivan Perl, *ITMO University, Saint Petersburg, Russia*
Dr. Vishal Sharma, *Soonchunhyang University, South Korea*
Dr. Osama Mokhtar, *Obour Institutes, Cairo, Egypt*
Dr. Wei Cai, *The Chinese University of Hong Kong, Shenzhen, China*
Dr. Victor C.M. Leung, *The University of British Columbia, Canada*
(Prof.) Dr. Prihandoko, *Deputy Director, G. University, Indonesia*

National Technical Program Committee

Dr. Arvind Dhingra, *GNDEC, Ludhiana*
Dr. Virender Rihani, *PEC, Chandigarh*
Dr. Pramod Kumar, Prof. & Principal, *Swami Rama Himalayan University (SRHU), Dehradun*
Dr. Viney Sharma, *Sharda University, Agra*
Dr. Vijay Luxmi, *Guru Kashi University, Bhatinda*
Dr. Vinay Singh, *Lovely Professional University, Punjab*
Dr. Lalit Goyal, *BVCOE, New Delhi*
Dr. Deepak Garg, *Vice Chancellor – SR University*
Dr. Rajeev Tiwari, *IILM University, G.Noida*
Dr. Savita Gupta, *UIET, PU, Chandigarh*
Dr. Naveen Aggarwal, *UIET, PU, Chandigarh*
Dr. Lakhwinder Kaur, *UCoE, PU, Patiala*
Dr. Ashutosh Kumar Bhatt, *Computer Science and Information Technology, Uttarakhand Open University, Haldwani, Distt- Nainital*

© 2025 the Editor(s), ISBN: 978-1-032-81760-6

Organizing Committee

Chief Patrons
S. Satnam Singh Sandhu, Chairman, CGC, Mohali
S. Rashpal Singh Dhaliwal, President, CGC, Mohali

Patron
Dr. P N Hrisheekesha, Campus Director, CGC, Mohali

Executive Chair
Prof. (Dr.) Rajdeep Singh, Executive Director (Engg), Chandigarh Engineering College, CGC, Mohali
Dr. Ruchi Singla, Dean (R& D), CGC, Mohali

General Chair
Dr. Valentina Balas (Aurel Vlaicu University of Arad, Romania)

General Co Chair
Prof. (Dr.) Sukhpreet Kaur, Professor & Head, CSE, Chandigarh Engineering College-CGC Landran, CGC, Mohali
Prof. (Dr.) Amanpreet Kaur, Professor & Head, IT, Chandigarh Engineering College-CGC Landran, CGC, Mohali

Organizing Chair
Dr. Manish Kumar (Professor, CSE, CEC)

Publication Management Committee
Dr. Gagandeep Jindal (Professor, CSE, CEC)
Ms. Parneet Kaur (Assistant Professor, CSE, CEC)

Web Presence Management Committee
Mr. Pardeep Singh Tiwana (Assistant Professor, IT, CEC)
Mr. Anish Kumar (Student, IT, CEC)

Financial Affair Committee
Dr. Parveen Sharma (Associate Professor, CSE, CEC)
Mr. Amitabh Sharma (Assistant Professor, IT, CEC)

Public Relations and Sponsorship Committee
Mr. Sachin Majithia (Assistant Professor, IT, CEC)
Mr. Gaurav Goel (Assistant Professor, CSE, CEC)

Hospitality and Events Management Committee
Dr. Rohit (Assistant Professor, CSE, CEC)

Registration Committee
Dr. Ishpreet Singh (Associate Professor, CSE, CEC)
Ms. Lakhvinder Kaur (Assistant Professor, IT, CEC)
Ms. Navpreet Kaur (Assistant Professor, IT, CEC)
Ms. Sukhdeep Kaur (Assistant Professor, CSE, CEC)

Quality Assurance and Integrity Review Committee
Dr. Arvinder Kaur (Associate Professor, IT, CEC)
Dr. Renu Puri (Associate Professor, CSE, CEC)

Dr. Vikas Gupta (Associate Professor, CSE, CEC)
Dr. Ishpreet Singh (Assistant Professor, CSE, CEC)
Dr. Monika (Assistant Professor, IT, CEC)
Mr. Sachin Bhardwaj (Assistant Professor, CSE, CEC)
Ms. Kamalinder Kaur (Assistant Professor, CSE, CEC)
Ms. Anupam Bonkra (Assistant Professor, IT, CEC)
Ms. Megha Sharma (Assistant Professor, IT, CEC)
Ms. Saumya (Assistant Professor, CSE, CEC)

Computational Methods in Science and Technology Volume 1 – Kaur et al. (Eds)
© 2025 the Author(s), ISBN: 978-1-032-81760-6

A synergistic approach to enhancing teaching-learning processes through the integration of ICT and Machine Learning

Rajeshwari Sharma
UPES, Dehradun, India

Himanshi Sharma
Graphic Era University, Dehradun, India

Richa Gaur
Akashvani, Dehradun, India

ABSTRACT: In the landscape of education ICT is one of the fields that presents a promising avenue for transformative advancements in the teaching-learning process. In order to remain relevant in the present digital landscape, educators must integrate ICT into their regular lesson plans and swap out outdated teaching strategies for more contemporary ones. This study's main goal is to determine how well ICT integration works in the classroom. The study's specific objective is to evaluate the computer-based instruction and learning competency of secondary school instructors. Another goal of this investigation is to assess the degree of ICT integration in instruction. Another goal of this paper is to ascertain how much ICT secondary school instructors have included into their lesson plans and classroom instruction. For this quantitative study, a dataset is considered that consist of student's interest in the classroom when they indulge with the ICT. The findings indicate that most instructors are mediocre users and that many would rather use ICT for work-related purposes at their offices than for teaching and learning in the classroom. Additionally, research shows that to provide students with ICT-based learning opportunities to improve the quality of their education, teachers must always be ready, equipped with ICT skills, and have a positive attitude. Future studies should consider other aspects of ICT integration, especially when it comes to management issues like policymaking and strategic planning.

Keywords: Teaching Learning, ICT, Teaching Process, Learning Strategies

1 INTRODUCTION

Advances in information and communications technology (ICT) and societal revolution have brought about significant changes in people's ways of thinking, working, and living [15]. This calls for the curriculum integration of ICT into schools and other educational institutions to assist students prepare for life in "a knowledge society," according to Ghavifekr [2] instructors are considered to be the most crucial people when it comes to integrating ICT into traditional classroom settings and preparing students for the modern digital age. There's no denying that technology is being used more and more these days, especially in educational settings. This is so that teachers can employ a variety of tools provided by current technology to raise the standard of instruction in the classroom [3,4,5]. ICT can also help pupils get ready for life in the twenty-first century. Students that are proficient with ICT have the skills and knowledge needed to overcome obstacles in the future [6]. According to Bransford, Brown, and Cocking [7], ICT use can assist students in gaining the skills necessary for the modern global economy. The goal of the Indian educational system is to produce future workers that are creative, technologically literate, and proficient in their field. This is consistent with the information highway and globalization [8]. Therefore, one of the key components of the strategic planning for ICT integration with educational system and an effective ICT-based curriculum.

2 INCORPORATION OF ICT IN EDUCATION

Information and communications technologies (ICTs) have developed quickly to become an essential part of contemporary industrial society. These days, becoming proficient in information technology and understanding the core ideas and skills associated with ICT is highly valued in many countries [9,10]. The use of ICT in education has been rapidly expanding in classrooms. ICT is the collective term for computers, software, networks, satellite links, and other related technologies that allow people to access, analyze, create, exchange, and use data, information, and knowledge in previously unimaginable ways [11]. The roles of ICT in the field of education are depicted in Figure 1.

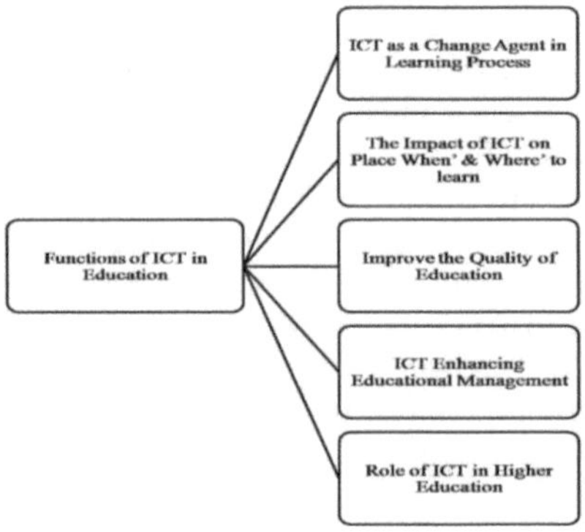

Figure 1. Functions of ICT in Education.

ICT plays a crucial role in transforming and enhancing the teaching-learning process. Some key functions of ICT in education are as follows:

- Access to Information: ICT facilitates access to a vast array of educational resources, including digital textbooks, online databases, and multimedia content, thereby expanding the availability of information.
- Interactive Learning: ICT enables the creation and use of multimedia presentations, interactive simulations, and educational videos, making learning more engaging and interactive.
- Virtual Laboratories: Virtual labs and simulations provide a safe and cost-effective environment for students to conduct experiments and explore concepts.
- Communication and cooperation: Regardless of geographical location, ICT tools that facilitate communication and cooperation between students and educators include discussion boards, video conferencing, and collaborative document editing.
- Personalized Learning: Information and communication technology (ICT), particularly when combined with machine learning, enables adaptive learning systems that adjust activities and content to each learner's preferences, learning styles, and pace.
- Assessment and Feedback: ICT makes it possible to design and administer online tests, quizzes, and assessments, which expedites the evaluation process.
- Data Analysis and Learning Analytics: Information and communication technology (ICT) makes it easier to gather and analyze data on behavior, performance, and

engagement of students. This data may then be used to influence decisions and provide individualized interventions.

- Predictive analytics: Algorithms based on machine learning can forecast student achievement and pinpoint areas that might require further help.
- Digital Citizenship Education: ICT offers chances to teach students about the ethical and responsible use of digital technology, encouraging awareness of online safety and digital citizenship.
- Flexibility in Learning: ICT supports blended learning models, combining traditional face-to-face instruction with online resources, offering flexibility in learning schedules and environments.

Haleem *et al.* [12] contend that instructors and students alike must come to trust technology to function as intended if increase adoption is selected and decrease opposition to it. Teachers who are proficient and self-assured in their use of a range of ICT tools can inspire more confidence in technology among their students. Without teachers who are skilled and adept at integrating ICT in accordance with their needs, it would be impossible to use technology to give teaching quickly and effectively. Teachers' ought to possess a broad variety of technical and communicative abilities in this regard. Hepp *et al.* [13] state that ICTs are used in the educational system for administrative, professional, social, cultural, and instructional goals.

3 THE NATIONAL ASPIRATION FOR ICT POLICY

The Indian Ministry of Education has placed a strong emphasis on ICT integration into the curricula of basic and secondary schools. In India's most recent new education policy, ICT is listed as one of the transformation shifts as the national education future development concentration. Three areas are given special attention in the ICT transformational shift "Leverage ICT To Scale Up Quality Learning Throughout India" [15]. The first wave of reform evaluates the present teacher-student ratio for ICT device allocations and developments in distant learning to find the best plan for providing Indian students with access to the whole curriculum and the best learning outcomes. As increase student-teacher interaction and provide culturally relevant content for indigenous students. The government wants more IT proficiency in schools, particularly in rural areas and less equipped areas. Moreover, it seeks to lessen the differences in ICT capacity across the nation in terms of quantity and quality in rural, urban, and distant locations. Raising the number and quality of ICT capabilities has the potential to improve teaching and learning. According to Kundu [16], ICT integration in schools is one of the most challenging projects due to potential obstacles and issues with stakeholder resistance to change, which could result in failure, especially in the early stages of implementation. ICT integration issues that remain unresolved in schools, especially in rural ones, have the potential to exacerbate the gap between students in urban and rural areas and lead to serious economic and social inequality nationwide.

4 RESEARCH OBJECTIVES

The following goals serve as the foundation for this research:

1. To assess secondary school teachers' computer competency and knowledge in relation to the teaching and learning process.
2. To gauge the extent to which secondary school instructors are incorporating Technology into their lessons.
3. To determine the connection between teachers' computer knowledge and skill levels and the use of ICT in the classroom for teaching and learning.

5 RESEARCH QUESTIONS

1. To what extent are secondary school instructors proficient in computers and cognizant of the teaching and learning process?

2. To what extent do secondary school instructors incorporate ICT into their lesson plans and instruction?
3. How does the teacher's degree of computer expertise and knowledge relate to the usage of ICT in the classroom for teaching and learning?

6 RESEARCH METHODOLOGY

The research design used for this study was quantitative. The purpose of this study is to determine the computer competency and knowledge levels of secondary teachers as well as their degree of ICT integration into the teaching and learning process in the classroom. This study also examines the connection between teachers' ICT use in the normal teaching and learning activities in the classroom and their computer knowledge and skills. Teachers at government schools and secondary schools are the sample population. The dataset[1] is available online that includes the students, computers, and learning with connection to examine the how students use ICT in the recent years. And evaluates the use of ICT for their learning purpose. The dataset is digitally dividing the education system, genders, geographical locations, and socio-economic status. The questionnaires were filled out voluntarily by teachers, students, and mentors. There is no formal differentiation between instructors' responsibilities and positions in schools because all of the samples were employed by government schools with the same organisational and hierarchical structure and geographic location.

7 RESULTS AND FINDINGS

How much knowledge about computers and how the teaching and learning process works do secondary school instructors possess?

Descriptive analysis was used to ascertain the frequency of instructors' competencies in the use of instructional resources and materials [17], as well as their ICT knowledge and skills, in order to answer the first study question. Additionally, a frequency analysis was conducted on the instructors' training scores about the use of computers in the curriculum. In order to address the first research question, percentages of teachers using computers and teachers responding to actions related to using ICT in schools were generated. Figure 2 shows the correlation among all attributes of the dataset.

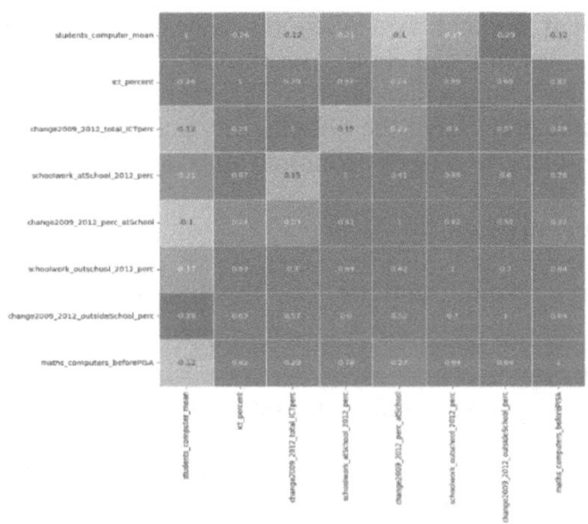

Figure 2. Correlation among the attributes.

[1]https://www.oecd.org/publications/students-computers-and-learning-9789264239555-en.htm

How frequently educators use resources and tools for education. According to the research, 78.7% of instructors reported using boards regularly, while 16.4% reported using them occasionally. This makes boards the most widely used tools and materials among teachers. In addition, printed resources like books or journals are preferred by teachers (60.7% frequently used them vs. 37.7% occasionally used them).

Figure 3 shows the mean distribution of computers with density of the end-users.

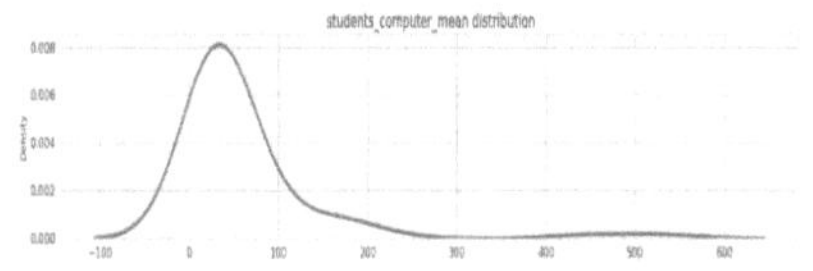

Mean: 467.98
Median: 604.5
Max: 940

Figure 3. Mean distribution of computers.

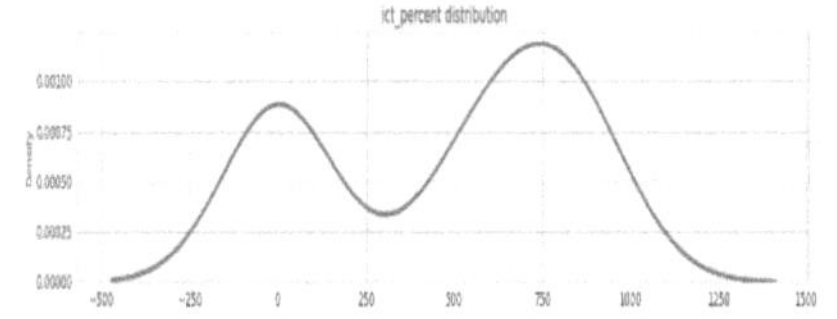

Mean: 14.63
Median: 0.0
Max: 138

Figure 4. ICT distribution in education sector.

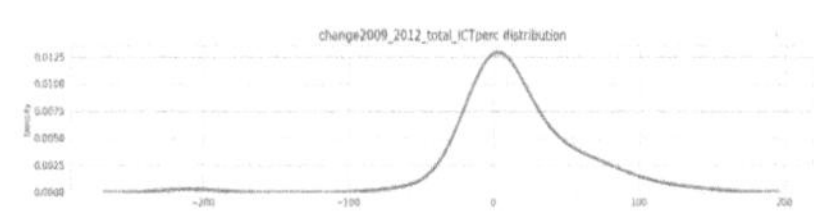

Mean: 246.34
Median: 287.0
Max: 808

Figure 5. Total percentage distribution.

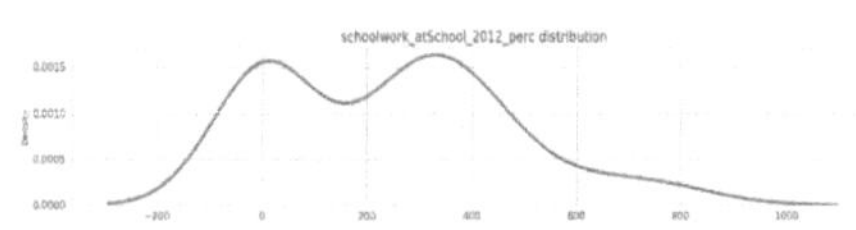

Mean: 17.66
Median: 0.0
Max: 158

Figure 6. ICT distribution in school.

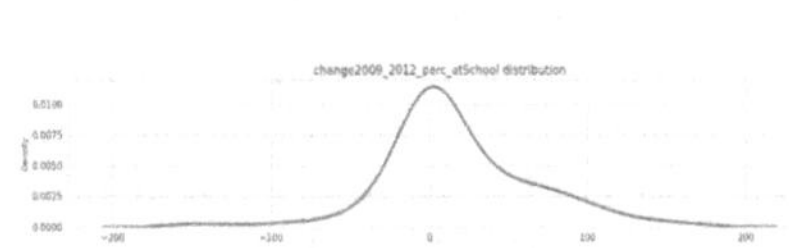

Mean: 350.13
Median: 457.0
Max: 756

Figure 7. Changes in distribution.

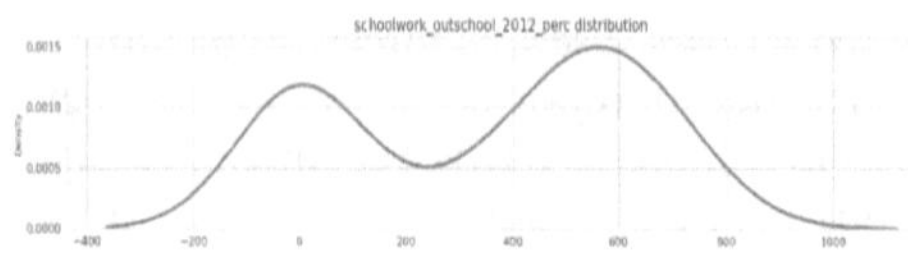

Mean: 65.89
Median: 58.0
Max: 294

Figure 8. Integrated ICT and School work distribution.

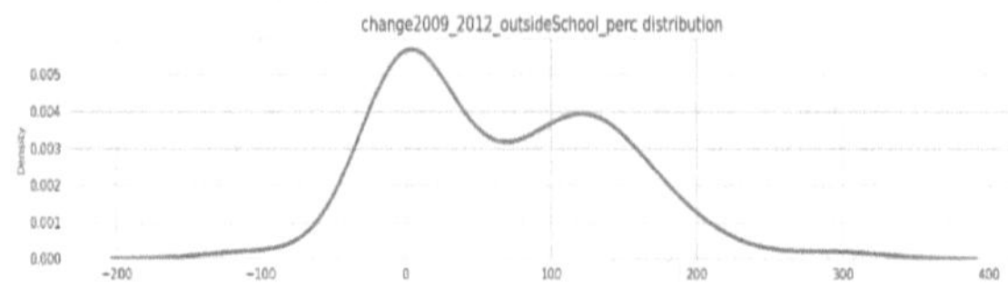

Mean: 206.51
Median: 237.5
Max: 731

Figure 9. Integrated ICT and outside cchool work distribution.

Figures 4 to 9 shows the distribution of ICT with the end-users and work distribution.

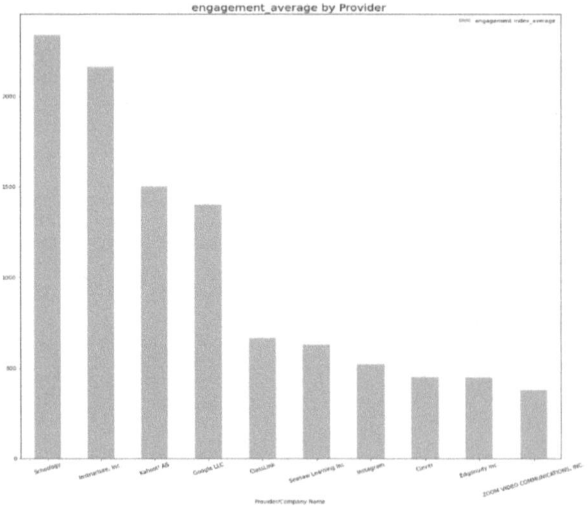

Figure 10. Distinct platform used for education.

Figure 10 shows the platform used for education purpose and its distribution. Most of the time, instructors used technological equipment to show the students the lesson materials. This addresses the initial research query. The frequency of instructors' proficiency with the use of instructional tools and materials has been found to be above average, with the majority of teachers utilising electronic tools for teaching and learning in at least some of their lessons and include those devices as teaching aids. When compared to training from schools, which had a rate of 63.9%, and training from family and friends, which had a rate of 49.2%, training from central office—whether through PPT—was far less common. Nonetheless, the majority of instructors who received training for using computers in the curriculum did so through their own research; in fact, 75.4% of teachers acquired this knowledge independently. The level of knowledge and abilities that each teacher has when it comes to using computers in the curriculum varies due to cultural and educational differences. No standard policy to provide any appropriate or organized training to the instructors. Furthermore, it could influence how much ICT is incorporated into the teaching and learning process and lead to nonstandard learning results for pupils.

8 CONCLUSION AND DISCUSSION

It is clear that ICT has a great deal of potential to enhance learning outcomes and students' critical thinking abilities by providing them with the most recent knowledge accessible without being constrained by distance or accessibility issues. ICT integration in the classroom will enable students to engage with a wider range of knowledge and information while they are studying [18]. Furthermore, how teachers use technology in the classroom will be influenced by their attitudes and beliefs [19]. The study found that most instructors use ICT resources and tools for teaching, like the Internet, multimedia PCs, projector systems, PowerPoint presentations, or word processing software, during the teaching and learning process. According to Hussain *et al.* [5], there doesn't seem to be much use of learning software, like instructional games, or more sophisticated ICT applications like learning websites when it comes to teaching and learning, or even when it comes to preparing for educational goals [20]. The results of this study indicate that teachers will be more likely to include ICT into their lessons when they possess the necessary knowledge and abilities. Together with IT expertise and knowledge, the attitudes of the teachers are a crucial element in the integration of ICT in education. Several research have revealed that the use of technology in the classroom will depend on the instructors' attitudes about it [21]. Incorporating ICT into the basic and secondary school curricula is important, according to India's Ministry of Education. In order to focus on the future growth of national education, the ICT component has been added as one of the transformative shifts in India. The results of this study are in line with those of other past studies and highlight the importance of teachers' attitudes and abilities with relation to ICT deployment in school systems. Peeraer and Petegem [22] argue that for effective ICT integration in secondary schools, researchers need to highlight that teachers' integration of ICT into the teaching and learning process is the key to fostering students' critical thinking skills and ultimately improving learning outcomes. Hence, to improve students' learning quality, teachers of today must always be prepared, well-versed in ICT competences, and full of a positive attitude [23]. Due to the world's information developing at a rapid speed, policymakers acknowledged that secondary school curricula should include the foundation and core skills of ICT to equip students with 21st century abilities in their early learning environments. As a result, students who receive ICT training in elementary school are better prepared to handle issues in the ICT networked world [24]. Technology helps teachers become more professional, but it also helps students become more competent with technology and more confident in their ability to flourish in the high-tech world of the future [25]. ICT also has a major impact on students' learning outcomes in science, math, and English language, according to several studies. To ensure successful ICT integration in schools, supportive policies, subsidies, and financial allocation from the government in education are also crucial.

REFERENCES

[1] Grabbe, M., & Grabe, C. (2007). *Integrating Technology on Meaningful Learning*. Boston NY: Houghton Mittlin.

[2] Ghavifekr, S., Afshari, M., & Amla, S. (2012). Management strategies for E-Learning system as the core component of systemic change: A qualitative analysis. *Life Science Journal*, 9(3), 2190–2196.

[3] Bingimlas, K. A. (2009). Barriers to the successful integration of ICT in teaching and learning environments: A review of the literature. *Eurasia Journal of Mathematics, Science and Technology Education*, 5(3), 235–245.

[4] Hamidi, F., Meshkat, M., Rezaee, M., & Jafari, M. (2011). Information technology in education. *Procedia Computer Science*, 3, 369–373.

[5] Hussain, A. J., Morgan, S., & Al-Jumeily, D. (2011, December). How does ICT affect teachings and learning within school education. In *2011 Developments in E-systems Engineering* (pp. 250–254). IEEE.

[6] Grimus, M. (2000, August). ICT and multimedia in the primary school. In *16th Conference on Educational Uses of Information and Communication Technologies*, Beijing, China (pp. 21–25).

[7] Bransford, J. D., Brown, A. L., & Cocking, R. R. (2000). *How people learn (Vol. 11)*. Washington, DC: National academy press.

[8] Ghavifekr, S., & Sufean, H. (2010). Management as visionary planning for dealing with systemic change: A case of malaysian open distance learning institution. *OIDA International Journal of Sustainable Development*, ISSN 1923-6654, Vol. 1, No. 4, pp. 79–85. The paper also presented at The Summer Congress /2010 . June 22–25. Laurentian University, Canada,Ontario.

[9] Daniels, J. S. (2002). Foreword. In *Information and Communication Technology in Education: A Curriculum for Schools and Programme for Teacher Development*. Paris, France: UNESCO

[10] Rampersad, C. A. (2011). Teachers perceptions of the contribution of information and communication technology to the teaching of modern studies, using an integrated system, In *An Urban Secondary School* (Doctoral dissertation), The University of the West Indies.

[11] Nepsha, O. (2021). Information technology and innovation for society development. *1.2. Use of Information and Communication Technologies in Education.*

[12] Haleem, A., Javaid, M., Qadri, M. A., & Suman, R. (2022). Understanding the role of digital technologies in education: A review. *Sustainable Operations and Computers*, 3, 275–285.

[13] Hepp, P., Hinostroza, J. E., Laval, E., & Rehbein, L. (2004). *Technology in schools: Education, ICT and the Knowledge Society* (pp. 30–47). Washington, DC, USA: World Bank, Distance & Open Learning and ICT in Education Thematic Group, Human Development Network, Education.

[14] Savage, C., Hindle, R., Meyer, L. H., Hynds, A., Penetito, W., & Sleeter, C. E. (2011). Culturally responsive pedagogies in the classroom: Indigenous student experiences across the curriculum. *Asia-Pacific Journal of Teacher Education*, 39(3), 183–198.

[15] Balli, O., & Singla, E. (2024). Dimensions of scale: Connected Learning Initiative (CLIx)—a case study of educational technology initiative in India. *Educational Technology Research and Development*, 1–23.

[16] Kundu, A., & Bej, T. (2021). Ingestion and integration of ICTs for pedagogy in Indian private high schools. *E-learning and Digital Media*, 18(2), 163–184.

[17] Pallathadka H., Wenda A., Ramirez-Asís E., Asís-López M., Flores-Albornoz J., and Phasinam K., Classification and prediction of student performance data using various machine learning algorithms, *Materials Today: Proceedings*, vol. 80. Elsevier BV, pp. 3782–3785, 2023. doi: 10.1016/j.matpr.2021.07.382.

[18] Buenaño-Fernández D., Gil D., and Luján-Mora S. (May 17, 2019). Application of machine learning in predicting performance for computer engineering students: A case study, *Sustainability*, vol. 11, no. 10. MDPI AG, p. 2833, doi:10.3390/su11102833.

[19] Hatlevik, O. E., & Arnseth, H. C. (2012). ICT, teaching and leadership: How do teachers experience the importance of ICT-supportive school leaders? *Nordic Journal of Digital Literacy*, 7(1), 55–69.

[20] Alam, A digital game based learning approach for effective curriculum transaction for teaching-learning of artificial intelligence and machine learning, *2022 International Conference on Sustainable Computing and Data Communication Systems (ICSCDS)*. IEEE, Apr. 07, 2022. doi:10.1109/icscds53736.2022.9760932.

[21] Seufert, S., Guggemos, J., & Sailer, M. (2021). Technology-related knowledge, skills, and attitudes of pre-and in-service teachers: The current situation and emerging trends. *Computers in Human Behavior*, 115, 106552.

[22] Peeraer J, & Petegem P. V. (2012). The limits of programmed professional development on integration of information and communication technology in education. *Australasian Journal of Educational Technology*. 28(Special issue, 6), 1039–1056.

[23] Dutta, A., & Nessa, N. (2022). Technological competencies of teachers in teaching learning process and the librarians role to enhance the technological skills: a study. *Library Philosophy and Practice (e-journal)*, 6706.

[24] Len, K. E., & Yuomeyse, K. ICT based instruction and the development of higher order learning skills in primary schools.

[25] Villegas-Ch, W., Román-Cañizares, M., & Palacios-Pacheco, X. (2020). Improvement of an online education model with the integration of machine learning and data analysis in an LMS. *Applied Sciences*, 10(15), 5371.

Computational Methods in Science and Technology Volume 1 – Kaur et al. (Eds)

Detection of brain tumor from MRI images using EfficientNet: A deep learning approach

Geetika Sharma, Moksh Sarda, Nikhil Malik, Harnoor Singh, Rahul Mishra & Somil Grover
Department of Computer Science and Engineering, Graphic Era Deemed to be University, Dehradun, India

ABSTRACT: Brain tumors are either cancerous or noncancerous masses of abnormal brain cells. Meningioma tumor, glioma tumor, and pituitary tumor are the three forms of brain cancers that are examined in this research. One essential diagnostic technique for the identification and diagnosis of brain tumors is brain tumor segmentation. Although manual brain tumor segmentation using MRI is possible, the accuracy and identification are not very good. The classification of abnormalities is a time-consuming task for physicians because it is not predictable and straightforward. However, tumor detection is difficult because tumors have complex appearances and boundary characteristics. In this work, we report on an automated method for detecting and diagnosing brain tumors from MRI images using EfficientNet. Specifically, we developed a preliminary report generation system that automatically provides the type of tumor, its corresponding ICD code, the tumor's diameter and area, and the original input image with the tumor highlighted for visualization. This will be accomplished through the classification and segmentation of MRI images, with classification using Dense EfficinetNet to identify the tumor type and ICD code and segmentation using EfficientNet UNet to visualize the tumor and determine its diameter and area. A preliminary report for our paper is being created, which will use one MRI image of a brain tumor to diagnose tumor diameter and tumor area.

Keywords: Brain Tumor, Detection, Diagnosis, EfficientNet, Classification, Segmentation

1 INTRODUCTION

An abnormal growth of brain cells, whether malignant or not, is called a brain tumor. A brain tumor can cause new or worsening headaches, impaired vision, loss of balance, disorientation, and seizures. There may be no symptoms in certain cases. Brain tumors can develop, as well as cancer that has spread to other regions of the body. Chemotherapy, radiation, and surgery are available as forms of treatment [1]. Brain tumors account for 85–90% of all primary malignancies of the central nervous system (CNS). It is anticipated that 308,102 primary brain or spinal cord tumor cases will be reported globally by 2020. In the US this year, about 4,170 children under the age of 15 will be diagnosed with brain or central nervous system tumors [2]. The three more common types of brain tumor which we can detected and diagnosed such as: ***Meningioma:*** The meninges, or membranes encircling the brain and spinal cord, are the site of a tumor known as a meningioma. Since it may compress or press on neighboring brains, nerves, and arteries, it is considered a brain tumor even if it is not exactly one. The most frequent kind of tumor to develop in the head is meningioma. The majority of meningiomas grow slowly, often going years without being seen. However, their effects on nearby arteries, nerves, or brain tissue can occasionally cause severe disability [3]. Most meningiomas do not require immediate treatment since they grow slowly and frequently without generating any symptoms. ***Glioma:*** One type of brain tumor that can harm both the brain and the spinal cord is called a glioma. Gliomas originate from the sticky, supportive cells known asglial cells that envelop and facilitate the function of nerve cells. Three factors can lead to glial cell tumorigenesis. The kind of glial cell that makes up the tumor and its genetic characteristics are used to categorize gliomas. This allows for the prediction of the tumor's

DOI: 10.1201/9781003501244-2

future behavior and the most successful course of therapy. Gliomas can cause profound impairments in brain function or possibly be fatal, depending on their location and rate of growth. Gliomas are an extremely prevalent kind of primary brain tumor [4]. ***Pitituary Tumor:*** Malignant growths in the pituitary gland that increase over time are called pituitary tumors. Certain pituitary tumors overproduce hormones that regulate vital bodily processes. In addition to producing less hormones, the pituitary gland may also develop pituitary tumors. Adenomas are benign (non-cancerous) growths that make up the majority of pituitary tumors. Benign tumors known as adenomas do not spread outside of the pituitary gland or the organs that surround it [5]. Figure 1 shows the MRI image of the three different brain tumors.

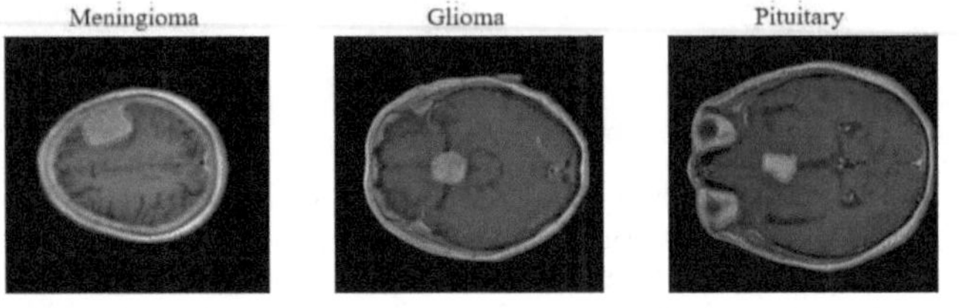

Figure 1. MRI image of the brain tumors.

Doctors employ imaging investigations to discover and diagnose brain tumors and to identify the type of tumor, such as cancer that has metastasized to the brain from another part of the body or a primary brain tumor. Imaging tests generate images of the interior of the body [6]. Doctors may take into account a variety of factors when selecting a diagnostic test, including the suspected tumor kind, symptoms and signs, age and overall health, results of past tests, and more. A preliminary report is being created in this research that uses one MRI scan of a brain tumor to determine tumor diameter and tumor area. To facilitate cross-national comparability, mortality data is gathered, analyzed, classified, and presented using the International Classification of Diseases (ICD) system [7]. For meningioma tumor, the corresponding ICD code is D32.9, For glioma tumor, the corresponding ICD code is C71.9, and for pituitary tumor, the corresponding ICD code is E23.7. Tumor diameter is defined as the maximum size of the invasive component of the primary tumor [8]. The definition of the tumor area is (maximum tumor depth) × (tumor length) [8]. On the other hand, because tumor appearance and border features are complicated, tumor detection is a challenging task [9].

EfficientNet is a more practical choice for real-world applications where resource constraints are a major concern such as brain tumor detection and diagnosis. Furthermore, EfficientNet has been shown to outperform other popular pretrained models such as ResNet, Inception, and VGG on a variety of image classification benchmarks in transfer learning which makes it an ideal choice for classification and segmentation of brain tumor detection and diagnosis.

In this paper, we present an Efficient Net based brain tumor detection and diagnosis using MRI images. We do this by building a system that automatically generates a preliminary report that includes the original input image with the tumor highlighted for visualization, along with the tumor's type, corresponding ICD code, diameter, and area. This will be achieved through the classification and segmentation of MRI images with efficientnet as the base pretrained model, where the classification is done to identify the tumor type and ICD code and segmentation is done to visualize the tumor and to find its size and area. For our paper, a preliminary report is being constructed which utilizes one MRI image of a brain tumor to diagnose tumor diameter and tumor area.

2 LITERATURE REVIEW

This section reviews EfficientNet-Based Brain Tumor Detection and Diagnosis from MRI Images. Meiyan Huang *et al.* [10] developed a content-based image retrieval (CBIR) technique for T1-weighted contrast-enhanced MRI (CE-MR) images of brain tumors. The system uses partition learning and the bag-of-visual-words (BoVW) model to extract informative features. Rank Error-based Metric Learning (REML) is introduced to bridge the gap between high-level semantic concepts and low-level visual data. Tested on a T1-weighted CE-MR dataset with meningioma, glioma, and pituitary tumor images, the method improves the CBIR system's mean average precision (mAP) to 91.8% and precision to 93.1%. These results demonstrate the method's effectiveness for locating brain tumors in T1-weighted CE-MR images.

Jun Cheng *et al.* (2015) [11] recognized three categories of brain malignancies using T1-weighted contrast-enhanced MRI (CE-MRI) images: meningioma, glioma, and pituitary tumor. Using spatial pyramid matching (SPM) to identify natural sceneries produces excellent results. SPM computes local data histograms from each subregion by dividing the image into progressively finer rectangular subregions. Because there is so much variation in the size and shape of brain tumors, this method is irrelevant. The authors of this study suggest a way to improve categorization performance. When compared to using the tumor region as ROI, using the enhanced tumor region increases the accuracy of the intensity histogram, GLCM, and BoW model from 71.39 percent to 82.31 percent, 84.75 percent to 78.18 percent, and 88.19 percent to 83.54 percent, respectively. Ring-form partition had improved accuracies by up to 87.54 percent, 89.72 percent, and 91.28 percent, in addition to region augmentation.

Using two publicly available datasets, Hossom H. Sultan *et al.* [12] proposed a DL model based on a convolutional neural network to classify several forms of brain cancers. The first classifies cancers into three groups: meningioma, glioma, and pituitary tumors. The other distinguishes between Grades A, B, and C of glioblastoma. The first and second datasets contain 233 and 73 patients, respectively, and total 3064 and 516 images on T1 weighted contrast-enhanced images. The input layer of the proposed network, which retains the previously processed images, is the first of sixteen layers. In the dataset considered in this work, their proposed architecture attained accuracy of 96.13 percent.

Ali Mohammad Alqudah *et al.* [13] utilized a Convolutional Neural Network (CNN) to classify brain cancers (glioma, meningioma, and pituitary tumor) using 3064 T1-weighted contrast-enhanced brain MR images. The study aimed to develop a novel CNN architecture for grading brain tumors with these images. Images were stored in a database, categorized into three folders based on tumor type. The CNN architecture comprised eighteen layers to enhance classification accuracy. The dataset was divided into training (70%), validation (15%), and testing (15%) subsets. The proposed method achieved accuracy rates of 97.4%, 99.0%, and 99.2% for cropped lesions, and 97.5%, 97.6%, and 98.4% for uncropped lesions. The design effectively classified brain tumors with high accuracy and sensitivity across all dataset cases: uncropped, cropped, and segmented.

According to Dillip Ranjan Nayak *et al.* [14], a CNN-based Dense EfficientNet was used to classify 3260 T1-weighted contrast-enhanced brain MR images into four groups: glioma, meningioma, pituitary tumor, and no tumor. The network, which includes drop-out and dense layers, applied min-max normalization and data augmentation to enhance tumor cell contrast. The dense CNN model accurately categorized a small database of images, achieving 97.78% accuracy and 99.97% accuracy after training. The proposed EfficientNet CNN architecture, with its high accuracy and positive F1 score, can be a valuable tool for brain tumor diagnostics.

3 PROPOSED METHODOLOGY

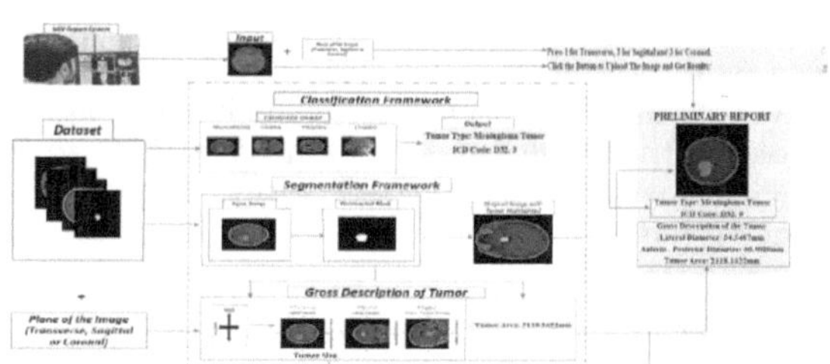

Figure 2. Overall proposed methodology for preliminary report generation.

Figure 2 shows the overall proposed methodology for preliminary report generation.

Jun Cheng *et al.* [15] proposed an open-source brain tumor dataset obtained from Nanfang Hospital, Guangzhou, and General Hospital, Tianjin Medical University, China (2005–2010).

Released online in 2015, the latest version came out in 2017. Approved by the Ethics Committees of both hospitals, the dataset contains 3064 T1-weighted contrast-enhanced MRI slices in transverse, sagittal, and coronal planes. It includes 708 meningioma, 1426 glioma, and 930 pituitary tumor slices, all 512 x 512 pixels with a 0.49mm x 0.49mm pixel size. Three experienced radiologists labeled the images based on patient pathology reports. Each patient is identified by their Patient ID. The data is in MATLAB (.mat) format. Adding 501 healthy control images from Kaggle [16], the total number of MRI images is now 3565. Table 1 shows the number of images per tumor class, their planes, and corresponding images.

Table 1. No. of images present in each category and views.

Tumor Category	No. of Slices	View	No. of Slices
Meningioma	708	Transverse	154
		Sagittal	119
		Coronal	435
Glioma	1426	Transverse	604
		Sagittal	491
		Coronal	331
Pituitary	930	Transverse	236
		Sagittal	416
		Coronal	278
Healthy Control	501	Transverse	357
		Sagittal	97
		Coronal	47

For Detection of Tumor Type and its corresponding ICD Code, we have developed a classification model, for which the proposed methodology is given below in Figure 3.

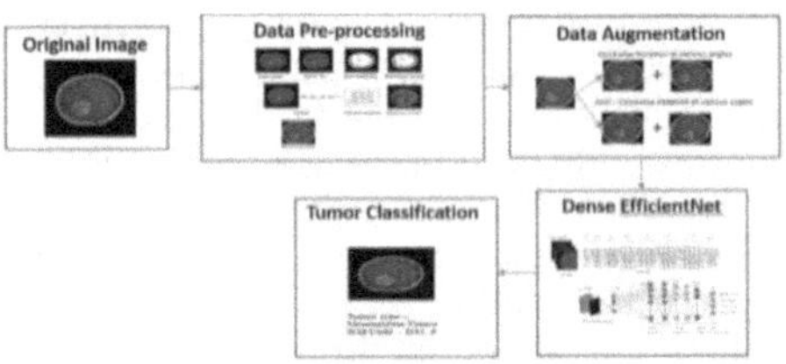

Figure 3. MRI image of the brain tumors.

3.1 *Data pre-processing*

The proposed method involves extracting the required image and mask based on their labels and cropping out excess background to retain only the Region of Interest (ROI) for efficient training. The process includes converting the image to greyscale, applying Gaussian Blur to reduce noise, and using binary thresholding (pixel intensity set from 45 to 255) to create a binary image. Contours are then found and extracted to generate a new image with only the tumor area, removing the excessive background.

3.2 *Data augumentation*

Large datasets are essential for deep neural networks, but our dataset is limited to 3565 brain images, with 80% for training and 20% for testing and validation. The dataset is significantly imbalanced, affecting training accuracy. To address this, data augmentation was applied. We rotated each glioma tumor image once by 5 degrees counterclockwise, each pituitary tumor image twice by 5 degrees clockwise and counterclockwise, and each meningioma tumor image three times by 5 degrees clockwise, counterclockwise, and 6 degrees counterclockwise. This increased the training data, improving learning capacity, generalization, and reducing overfitting. Data augmentation generates new samples to expand the dataset through oversampling or data warping.

Before discussing Dense EfficientNet, let's first understand EfficientNet, its modules, sub-blocks, and architecture. EfficientNet is a CNN architecture that uses a compound coefficient to uniformly scale depth, breadth, and resolution dimensions. This scaling method consistently increases network resolution, depth, and breadth using predefined scaling coefficients [17]. EfficientNet's compound scaling method scales each dimension equally, rather than randomly increasing width, depth, or resolution [17]. Figure 4 shows the architecture of an EfficientNet model. Now, let's look into Dense EfficientNet. This research introduces a novel dense CNN model that combines dense blocks with pre-trained EfficientNet. Dense blocks consist of convolution layers matching the size of EfficientNet input feature maps. They use the output feature maps of previous convolution layers to produce more feature maps with fewer convolution kernels. Dense EfficientNet includes drop-out and dense layers. The base layer, or dense layer, has neurons receiving all outputs from the layer above, and each neuron sends one output to the layer above. The drop-out layer thins the network during training, preventing overfitting and reducing capacity. The model includes a pooling layer, three dense layers with 180, 360, and 720 neurons respectively, and two drop-out layers with values of 0.25 each. A dense layer with four fully connected neurons and a Softmax output layer calculates and classifies the probability score for each class. Figure 5 shows the detailed structure of the proposed EfficientNet.

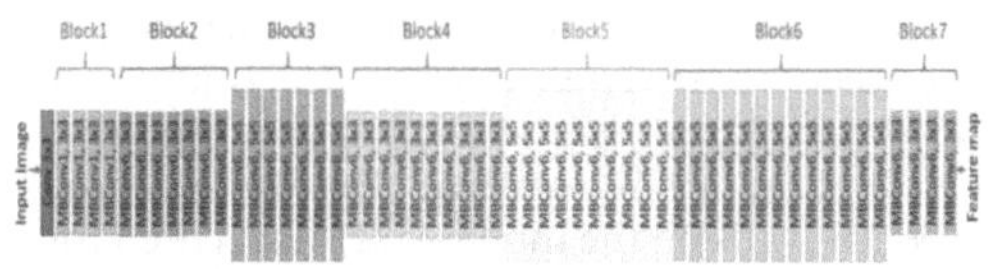

Figure 4. Efficient Net CNN model Architecture [18].

3.3 *Hyperparameters set and output generated*

Table 2 shows the hyperparameters set for our dense efficient net b2 to get the accuracy. The outputs generated are tumor type and the corresponding ICD code for that tumor type which will be visible on the report.

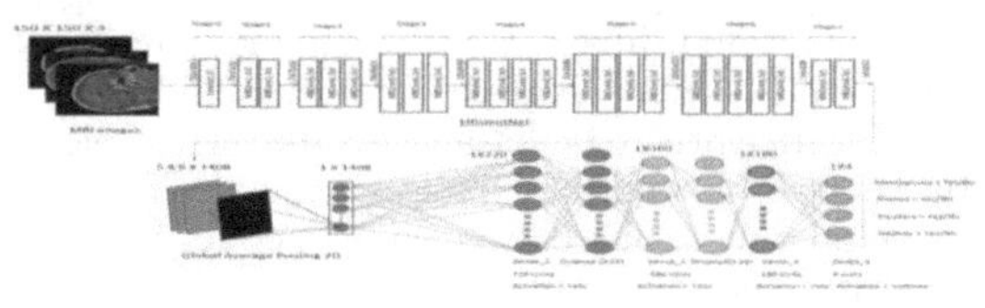

Figure 5. Proposed dense EfficientNet CNN model architecture.

Table 2. Hyperparameters for dense efficientnet

Hyperparameters	Values
Batch size	32
Epoch	20
Initial learning rate	0.0001
Reducing LOR at patience	2
Reducing LOR at factor	0.3
Optimizer	Adam

For obtaining tumor radiomic characteristics, which is nothing but the tumor diameter and area, along with the original image with tumor region highlighted, we have developed a segmentation model, for which the proposed methodology is given below in Figure 6.

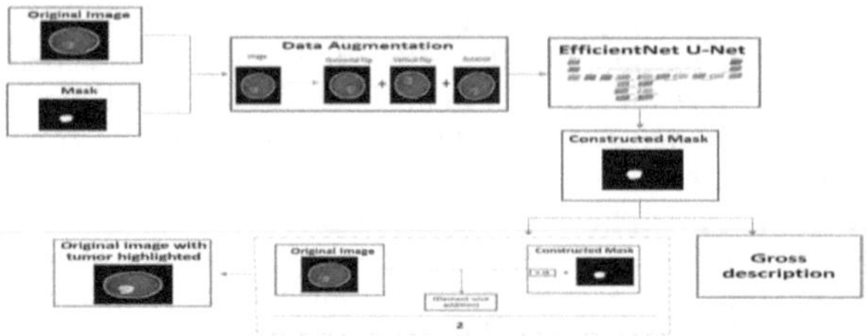

Figure 6. Overview of proposed methodology for brain tumor segmentation.

3.4 *Data augumentation*

To diversify the training data, the basic forms of data augmentation are used. The torchvision module of Pytorch is used for all augmentation methods. They are as follows: horizontally flip, vertically flip, and rotate between 75 and −15 degrees. Each augmentation method has a probability of 0.5, and the order in which they are applied is also random. The degree of rotation for rotation augmentation is chosen at random between 75 and −15 degrees. The dataset is split in the ratio 80:20, where 80% is used for training the model and 10% is used for validation and 10% is used for testing. A total of 18384 images (9192 images and mask respectively) were obtained which will be utilized for the brain tumor segmentation.

3.5 *EfficientNet UNet*

Mingxing Tan *et al.* [19] introduced EfficientNet UNet, which is a neural network architecture that combines the well-known UNet and EfficientNet architectures. UNet architecture is widely used for image segmentation tasks, in which each pixel in an image is classified into different classes. The architecture is made up of an encoder and a decoder, with the encoder extracting high-level features from the input image and the decoder generating a segmentation map by upsampling the features to the original image size. To balance the network's depth, width, and resolution, the architecture employs a compound scaling method. The strengths of both architectures are combined in the EfficientNet UNet architecture. It employs the EfficientNet as an encoder for efficient feature extraction and the UNet as a decoder for accurate segmentation. The architecture also includes skip connections between the encoder and decoder, which aid in the preservation of lowlevel features and the accuracy of segmentation.

3.6 *Hyperparameters set and outputs generated*

A mask of the tumor will be generated using our U – net which will be utilized to find the outputs required for the tumor radiomic characteristics which are all given below. Figure 7 shows the original image, mask in the dataset, and mask built by using EfficientNet model.

Figure 7. Original image, mask given in the dataset and the mask built by EfficientNet UNet.

4 RESULTS AND DISCUSSION

Several experimental tests have been performed to determine the validity of the proposed dense CNN model. All the experiments were carried out in a Python programming environment with GPU support. Pre-processing is performed first, in which the images are cropped for better training and then augmented for training. The evaluation of the distinct performance parameters such as accuracy, precision, recall, loss function, and validation accuracy can be evaluated for the evaluation of performance of the model. Table 3 shows the training and validation accuracy for dense efficientnet b0 to b7 value.

Table 3. Dense efficientnet models and their accuracies.

Dense EfficientNet Models	Training Accuracy	Validation Accuracy
Dense EfficientNet B0	99.81%	97.34%
Dense EfficientNet B1	99.89%	98.15%
Dense EfficientNet B2	99.93%	97.63%
Dense EfficientNet B3	99.91%	97.58%
Dense EfficientNet B4	99.90%	97.10%
Dense EfficientNet B5	99.84%	97.34%
Dense EfficientNet B6	99.85%	97.57%
Dense EfficientNet B7	99.84%	97.46%

4.1 *EfficientNet B2*

showed the highest training and validation accuracy, so it was selected for the Dense EfficientNet classification model, evaluated using loss, accuracy, precision, recall, and F1 score. To improve accuracy, augmented tumors were activated. The loss function used was "categorical cross-entropy." The model achieved 99.93% accuracy on training data and 97.63% on testing data over twenty epochs. The experiment used batch sizes of 32, image sizes of 150 × 150, and verbose 1. Initially, validation accuracy was 0.7013 and increased to 0.8670 after two epochs. Validation loss started at 1.1516 and dropped to 0.3502 after two epochs. The model's testing accuracy and loss were 97.63% and 0.1493, respectively, showing an overall trend of increasing accuracy and decreasing loss.

4.2 *Confusion matrix*

A number of performance metrics, including accuracy, precision, recall, and F1-score, were utilized to examine the effectiveness of the suggested model. The parameters were assessed using the confusion matrix. Figure 8 below illustrates the confusion matrix that was used to investigate the details. Three key metrics have been taken into consideration for technique comparisons: F1-score, precision, and recall.

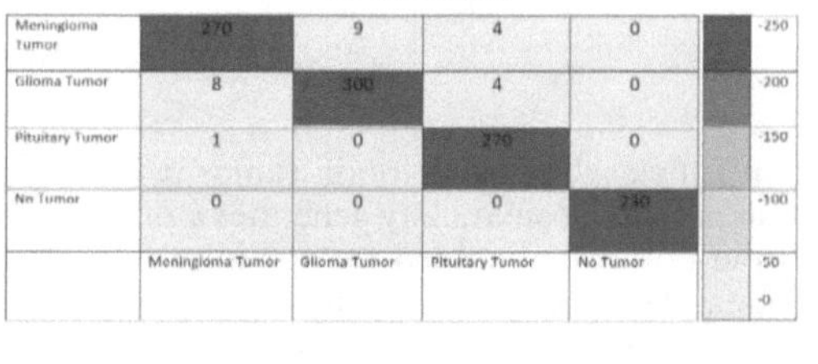

Figure 8. Confusion matrix for dense efficientNet B2.

4.3 *Classification report*

Three key metrics have been taken into consideration for technique comparisons: F1-score, precision, and recall. The dense efficientnet b2 classification report is displayed in Figure 9. Table 3 shows that when compared to other models, Dense EfficientNetB2 has the highest precision, recall, and F1-score. When compared to other tumor kinds, the no tumor performs the best across the board. Pituitary tumors have excellent precision, recall, and F1-score values. Overall, the dense EfficientNetB2 produces very good results. The suggested method has higher accuracy, precision, and F1-score (97.63%, 98%, and 98%, respectively) than existing comparison methods. Table 3 shows that dense EfficientNet B2 performs better than other methods due to the greater efficiency and vast data processing capacity of deep-learning-based approaches.

Table 4 shows the accuracy of our model at each plane and the class of the tumor.

Table 4. Accuracy of each planes and tumor class.

Image Plane	Accuracy	Tumor Class	Accuracy
Transverse	98.84%	Meningioma	96%
		Glioma	98%
		Pituitary	99%
		Healthy	100%
Sagittal	99.29%	Meningioma	98%
		Glioma	99%
		Pituitary	100%
		Healthy	100%
Coronal	99.42%	Meningioma	99%
		Glioma	99%
		Pituitary	99%
		Healthy	100%

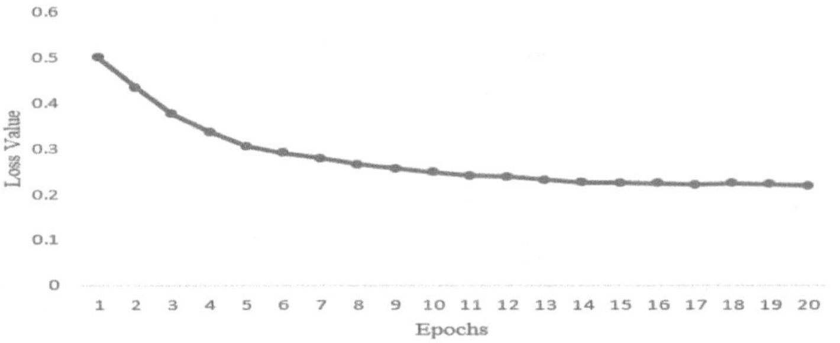

Figure 9. Loss plot for EfficientNet UNet B3.

The loss function here is "Dice Loss". The graph indicating Loss Value over 20 epochs are given below in Figure 9.

5 CONCLUSION

In this paper, we present an EfficientNet based brain tumor detection and diagnosis using MRI images. We developed a system that automatically generates a preliminary report that includes the original input image with the tumor highlighted for visualization, along with the tumor's type, ICD code, diameter, and area. The major limitation of this paper is in the accuracy of segmentation, which is only around 71%. Another limitation lies in the fact that, the dataset which we had utilized lacks demographics details of the patient. This work only utilizes .png images which are not the medical standard for detection and diagnosis. Using dicom images will be the medical standard. Nowadays, the issue of automated detection and diagnosis of brain tumors are major research area. Our work, which is automating the process of detection, diagnosis, and generation of a preliminary diagnosis report of the brain tumor will aid the radiologist and physician and will aid in more research work in the automation of the detection and diagnosis of brain tumor.

REFERENCES

[1] Brain Cancer Symptoms, Causes and Treatment in India – Apollo Hospitals, Apollo Hospitals. (2022). [Online]. Available: https://www.apollohospitals.com

[2] Brain Tumor – Statistics, Cancer.Net. (2022). [Online]. Available: https://www.cancer.net/cancer-types/brain-tumor/statistics.

[3] Meningioma – Symptoms and causes, Mayo Clinic. (2022). Available:https://www.mayoclinic.org/symptoms-causes/syc-20355643.

[4] Glioma – Symptoms and causes, Mayo Clinic. (2022). Available:https://www.mayoclinic.org/diseases/conditions/glioma/sy mptoms-causes/syc-20350251.

[5] Pituitary tumors – Symptoms and causes, Mayo Clinic. (2022). [Online]. Available: https://www.mayoclinic.org/

[6] Brain Tumor – Diagnosis, Cancer.Net. (2022). [Online]. Available: https://www.cancer.net/cancer-types/brain-tumor/diagnosis.

[7] International Classification of Diseases (ICD), *Who.int*. (2022). Available:https://www.who.int/standards/classifications/classification -of-diseases.

[8] Chen A., Ryu J., Donald P., and Colevas A., (2022). Cancer of the Nasal Cavity and Paranasal Sinuses.

[9] Shelke S. M. and Mohod S. W., Automated Segmentation and Detection of Brain Tumor from MRI

[10] Huang M. *et al.*, (2014). Content-based image retrieval using spatial layout information in brain tumor T1-weighted contrast-enhanced MR images, *PLoS ONE*, vol. 9, no. 7, p. e102754. Available: 10.1371/journal.pone.0102754.

[11] Cheng *et al.*, Correction: Enhanced performance of brain tumor classification via tumor region augmentation and partition 10.1371/journal.pone.0144479.

[12] Sultan H. H., Salem N. M., and Al-Atabany W., (2019). Multi-classification of brain tumor images using deep neural network, in *IEEE Access*, vol. 7, pp. 69215–69225, doi: 10.1109/ACCESS.2019.2919122.

[13] Alqudah A., (2019). Brain tumor classification using deep learning technique – A comparison between cropped, Uncropped, and segmented lesion images with different sizes, *International Journal of Advanced Trends in Computer Science and Engineering*, vol. 8, no. 6, pp. 3684–3691. Available: 10.30534/ijatcse/2019/155862019.

[14] Irmak E. (2021). Multi-classification of brain tumor mri images using deep convolutional neural network with fully optimized framework, *Iranian Journal of Science and Technology, Transactions of Electrical Engineering*, vol. 45, no. 3, pp. 1015–1036, Available: 10.1007/s40998-021-00426-9

[15] Brain tumor dataset, figshare. (2022). [Online]. Available: https://figshare.com/articles/dataset/brain_tumor_dataset/1512427.

[16] Brain Tumor MRI Dataset, Kaggle.com. (2022). [Online]. Available: https://www.kaggle.com/datasets/masoudnickparvar/brain-tumor-mridataset.

[17] Complete Architectural Details of all EfficientNet Models, Medium. (2022). [Online]. Available: https://towardsdatascience.com

[18] Baheti, B. *et al.* (2020) EFF-UNet: A novel architecture for semantic segmentation in unstructured environment. (2020). *IEEE/CVF Conference on Computer Vision and Pattern Recognition Workshops (CVPRW)* [Preprint]. Available at: https://doi.org/10.1109/cvprw50498.2020.00187.

[19] Tan M. and Le Q. V., EfficientNet-UNet: A Fast and Scalable Model for Medical Image Segmentation, in *International Conference on Medical Image Computing and Computer-Assisted Intervention*, 2020, pp. 405–415.

[20] Zhou M. *et al.*, Radiomics in brain tumor: image assessment, quantitative feature descriptors, and machine-learning approaches, *American Journal of Neuroradiology*, vol. 39, no. 2, pp. 208–216, 2017. doi: 10.3174/ajnr.a5391.

[21] tkinter — Python interface to Tcl/Tk — Python 3.10.4 documentation, Docs.python.org. (2022). [Online]. Available: https://docs.python.org/3/library/tkinter.html

[22] mAP (mean Average Precision) might confuse you!, Medium. (2022). [Online]. Available: https://towardsdatascience.com/map-meanaverage-precision-might-confuse-you-5956f1bfa9e2.

[23] Is the Dice coefficient the same as accuracy?, Cross Validated. (2022). Available: https://stats.stackexchange.com/questions/195006/

[24] Jaccard Index / Similarity Coefficient, Statistics How To. (2022). Available: https://www.statisticshowto.com/jaccardindex/

[25] Google Colab, Research.google.com. (2022). https://research.google.com/colaboratory/faq.html.

Computational Methods in Science and Technology Volume 1 – Kaur et al. (Eds)
 ISBN: 978-1-032-81760-6

The proposed framework for hybrid multilevel optimized machine learning classification in apple leaf disease detection

Anupam Bonkra & Sunil Pathak
Amity School of Engineering and Technology, Amity University, Kant Kalwar, Jaipur, Rajasthan, India
MMEC, MM(DU), Mullana, Ambala, Haryana, India

Amandeep Kaur
Chitkara University Institute of Engineering and Technology, Chitkara University, Punjab, India

ABSTRACT: Rust, scab, spot, and several infections are serious apple agricultural diseases. Expert disease detection from restricted areas costs time and money. To address the problem, Deep Learning (DL) is utilized to detect apple leaf illnesses early. This work diagnoses apple leaf diseases using image argumentation, preprocessing, feature extraction, and a hybrid multi-level optimum deep learning classification model. Comparing reliable feature methods with DL models for apple leaf disease detection. Research will be used to complete this analysis. The first shows the reliable feature set from the literature to develop an apple leaf disease classification model. A simulation study technique will be developed to determine the most precise illnesses utilizing DL. Apple leaf disease photographs will be identified using a hybrid multi-level optimum classifier algorithm. Classification affects detection system accuracy, precision, recall, etc. DL classifier model hybrid multi-level optimizer has better accuracy than ensemble CNN, DL model, and others.

Keywords: Apple Leave Disease Detection, Hybrid multi-level optimized Deep learning Classification model, feature extraction, ensemble CNN method

1 INTRODUCTION

China dominates fruit production with the most plant domain and apple output. Apple leaves are subject to ring-rot, scab, rust, early defoliation, and other pests. Verifying diseases with forestry and agricultural professionals is time-consuming, subjective, complex, and unproductive. Inexperienced farmers apply pesticides recklessly, which may cause infections, ruin the apple crop, and cost money. Detecting illnesses intelligently, rapidly, and accurately is critical [1]. Physical and professional procedures used to identify apple disease. Both take time, need fruit farmers, and are poorly dispersed [2]. Apples are the most costly fruit [3], and orchards are maintained everywhere. Apple production and consumption are rising, with 86 million tons expected by 2018. Standard National apple production is minimal for forecasting apple outcomes. Organic, post-harvest skills, socioeconomic constraints, etc. are the main causes of poor apple output [4]. Diseases lower apple efficiency, boosting plant life. Early and accurate issue categorization improves output and prevents illness. Automatic illness symptom identification will be cheaper and simpler than professional consulting . Therefore, a technology that spontaneously detects leaf illness early is crucial. Machine learning breakthroughs may enhance leaf disease diagnosis. Growers might save numerous losses by using ML to identify apple leaf diseases early. ML involves image preprocessing

 DOI: 10.1201/9781003501244-3

and feature extraction [5,6]. Bayesian, SVM, and DT are predictable ML algorithms. These methods are flat. The processes cannot be used on basic data. Figure 1 illustrates ML problem-solving. Essential data is sent directly to DL models. The main advantage of DL [7], is eliminating feature set extraction. Feature extraction is difficult and needs a thorough understanding. DL controls high-level information properties gradually. Figures 1 and 2 show DL problem stages. DL combines feature extraction with classification. Another virtue of DL is its ability to handle large data sets [8]. Apple leaf pictures are changed using R, G, and B to identify damage. RGB damaged apple leaf photos are pre-processed into HSV, HIS, LAB, etc [9,10]. To diagnose apple leaf diseases, image DIP retrieves texture, shape, edges, color, and other features from input photographs [11].

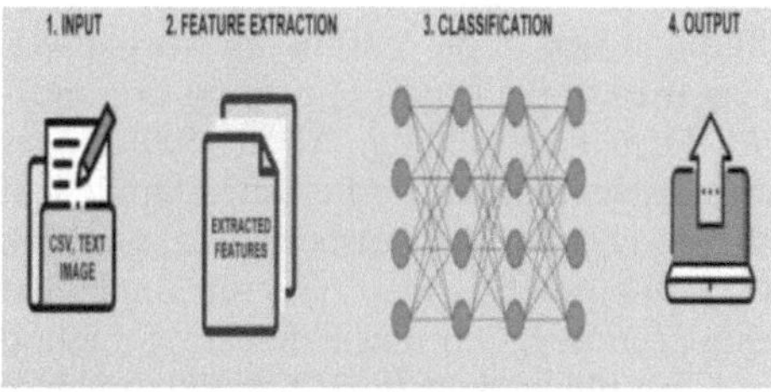

Figure 1. Generalized phases in ML.

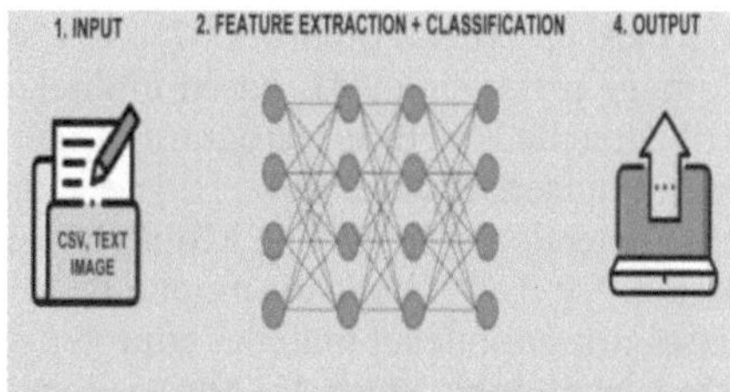

Figure 2. Generalized phases in DL.

1.1 *Problem statement*

Fruit diseases have devastated agriculture globally. Apple production—a popular fruit—remained steady at 86 million tons in 2018 despite increased demand. Post-sharvest technology issues, research gaps, and societal and economic restraints decrease productivity [12]. The most prevalent apple diseases are anthracnose, scab, and powdery mildew. Farmers need fast identification for safety. Outdated, non-automated detecting methods affect yields and consumption [13]. Poor picture quality, climate-related leaf changes, and limited datasets hinder disease identification, but agricultural experts are attempting to improve it.

1.2 *Aim and objectives*

I. To extract features and classify apple leaf diseases with a hybrid multi-level optimal deep learning classification system; (ii) to build and use this system to detect diseases in apple leaf image datasets; and (iii) to compare the proposed model to existing methods in accuracy, precision, and recall metrics. A hybrid multi-level optimal deep learning technique diagnoses apple leaf diseases quickly and cheaply. The recommended method enhances classification by modifying leaf colors and effectively handling overfitted training data. Results, disputes, and conclusions prove the technique works.

2 RELATED WORK

Researchers categorize apple illnesses using ML/DL deep learning. Some agricultural research used ML/DL to classify diseases. [14] suggested DL for apple disease categorization. Investigation databases require data collecting and classification. Automatic convolutional neural networks identified apple diseases. Because end-to-end training spontaneously derives feature sets and learns complex features from visual input, convolutional networks may recognize, categorize, and segment global objects. Transfer learning evaluates the algorithm. Conversion, rotation, reflections, and magnification avoided overfitting. CNN statistically increased to 97.18% dataset accuracy. [15] tested DeepNet and EfficientNet CNNs. Image collection, pre-processing, segmentation, augmentation, feature extraction, and classification were suggested. The four categories were healthy (standard), rust, scab, and others. Flipping, blurring, and clever edge detector helped classify. DenseNet was 99.75% accurate, EfficientNetB7 99.8%. [16] recommended apple leaf disease detection. Used Deep CNN. Many pre-trained classifiers performed well. The proposed system employs EfficientNetB7, Noisy Student, and DenseNet121 pre-trained classifiers. Healthy leaf, cedar rust, apple scab, etc. afflicted apple leaves. Successful dataset discoveries required data augmentation. Dataset validation produced 96.25% accuracy. Disease categorization was 90% correct. [17] classified apple leaf diseases using DL. Apple scab, cedar, and brown patches were discovered. TL used Plant Village healthy and sick photos to train Inception V3 deep learning. The novel variance-controlled down sample approach found features after TL. Modern classifiers pick up crucial qualities 97% of the time. [18] diagnosed apple leaf diseases using ML and image processing. ML-based multiclass SVM classified apple leaves as normal or abnormal. Segmented picture histogram equalization with Otsu thresholding. About 500 images properly judged effort at 96%. [19] developed an inexpensive, reliable, and effective apple leaf disease detection system. The MobileNet does this. It was cheaper than earlier neural network models since it was built on mobile devices. Second, non-specialists may pass our apple leaf diseases test. mobilenet matches complex deep learning accuracy. Our illness detecting method worked in multiple studies. [20] developed apple leaf fire blight detection. The Skierniewice Institute of Horticulture quarantine greenhouse produced photographs. Unhealthy apple plant statistics. The 1900 nm channel can identify all three leaves, whereas the 1450 nm SWIR spectrum can discriminate leaves. Pearson correlation tests indicated that illness development was mainly affected by ARI, MSR, and QFI. [21] classified apple leaf diseases using DL. CNN-based apple leaf disease diagnosis was faster and more accurate. Scanning healthy and unhealthy apple orchard leaves from growing sites and labs provides five regular apple leaf statuses and typical leaves. Instead of connected layers, our CNN apple leaf disease detection model uses DenseNet and Xception with average global pooling. After convolutional neural network feature extraction, SVM recognized apple leaf diseases.

3 METHODOLOGY

This research used an ALDD database with 9714 high-quality apple leaf photos accurately categorized into several illnesses. This work uses dataset collection, pre-processing, feature extraction, and a hybrid multi-level optimum classifier. Grayscale conversion, noise filtering, and attribute analysis are required before processing a picture. Classification uses a hybrid multi-level optimum deep learning technique to maximize feature sets and convolutional feature extraction. Apple leaf disease classification is more accurate than deep learning methods when thick layer vectors, which represent disease outcomes, are used in the classifier. The Figure 3 will show methodology of work.

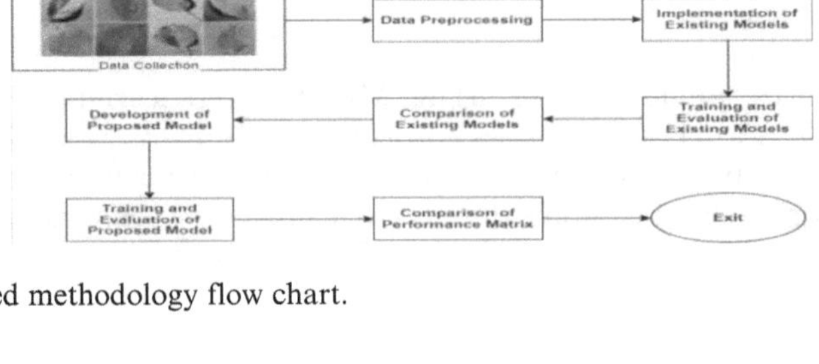

Figure 3. Proposed methodology flow chart.

4 RESULT AND DISCUSSIONS

The simulation setup is explained and surveyed using mathematical performance parameters and past results. The study will train a Python model. The research model was utilized with Python, GUIDE, and several graphical interfaces and instructions. It will open the UCI machine learning repository with KAGGLE [17]. Research and pre-trained models are assessed by accuracy, precision, recall, f1-score, and others [18]. Table 1 compares the hybrid multi-level optimized deep learning classification method, ResNet-152, shuffleNet, MobileNetV3, EfficientNet-B0, CNN models, and standard approach.

Table 1. Performance analysis.

Models	ShuffleNet [9]	MobileNetV3 [9]	Efficient Net-B0 [9]	Ensemble CNN [9]	Hybrid multi-level optimized DL method
Accuracy Rate (%)	62.3	93.4	94.3	96.25	98.3

On Kaggle Dataset, hybrid multi-level optimized deep learning beat ShuffleNet and other models with 98.3% accuracy.

5 CONCLUSION AND FUTURE SCOPE

The apple leaf disease detection DL model research proposal is described here. A hybrid multi-level optimum classifier model classifies apple leaf diseases using feature vectors or categories. Image argument, color transformation, RGB-to-grayscale conversion, feature extraction, selection, and classification grade illness symptoms. The study database includes high-quality photographs of ailments like spot, rust, and scab. The proposed model is accurate, simple, and fast. This model is 98.3% accurate compared to others. Apple illness is accurately detected using hybrid model with increased DL. More apple leaf pictures are available to improve future models. A new model will improve accuracy and other metrics. Apple leaf dataset pictures' superior texture feature sets will be recovered utilizing feature extraction.

REFERENCES

[1] Li, L., Zhang, S., & Wang, B. (2021). Apple leaf disease identification with a small and imbalanced dataset based on lightweight convolutional networks. *Sensors*, *22*(1), 173.

[2] Yan, Q., Yang, B., Wang, W., Wang, B., Chen, P., & Zhang, J. (2020). Apple leaf disease recognition based on an improved convolutional neural network. *Sensors*, *20*(12), 3535.

[3] Hossain, E., Hossain, M. F., & Rahaman, M. A. (2019, February). A color and texture based approach for the detection and classification of plant leaf disease using KNN classifier. In *2019 International Conference on Electrical, Computer and Communication Engineering (ECCE)* (pp. 1–6). IEEE.

[4] Singh, V., & Misra, A. K. (2017). Detection of plant leaf diseases using image segmentation and soft computing techniques. *Information processing in Agriculture*, *4*(1), 41–49.

[5] Dhiman, P., Kaur, A., Iwendi, C., & Mohan, S. K. (2023). A scientometric analysis of deep learning approaches for detecting fake news. *Electronics*, 12(4), 948.

[6] Bonkra, A., Pathak, S., Kaur, A., & Shah, M. A. (2024). Exploring the trend of recognizing apple leaf disease detection through machine learning: a comprehensive analysis using bibliometric techniques. *Artificial Intelligence Review*, 57(2), 1–26.

[7] Kaur, P., Harnal, S., Gautam, V. *et al.* (2023). A novel transfer deep learning method for detection and classification of plant leaf disease. *J Ambient Intell Human Comput* 14, 12407–12424 . https://doi.org/10.1007/s12652-022-04331-9

[8] Bonkra, A., Noonia, A., & Kaur, A. (2021, December). Apple leaf diseases detection system: A review of the different segmentation and deep learning methods. In *International Conference on Artificial Intelligence and Data Science* (pp. 263–278). Cham: Springer Nature Switzerland.

[9] Dhiman, P., Bonkra, A., Kaur, A., Gulzar, Y., Hamid, Y., Mir, M. S., ... & Elwasila, O. (2023). Healthcare trust evolution with explainable artificial intelligence: Bibliometric analysis. *Information*, 14 (10), 541.

[10] Rubbina, Pratibha Verma, and G. N. Verma. Brain tumor detection using hybrid model of DCT DWT and thresholding. *Brain 6*, no. 07 (2019): 855–860.

[11] Khairnar, K., & Dagade, R. (2014). Disease detection and diagnosis on plant using image processing–a review. *International Journal of Computer Applications*, 108(13), 36–38.

[12] Jiang, P., Chen, Y., Liu, B., He, D., & Liang, C. (2019). Real-time detection of apple leaf diseases using a deep learning approach based on improved convolutional neural networks. *IEEE Access*, 7, 59069–59080.

[13] Bonkra, A., kumar Bhatt, P., Kaur, A., & Shah, M. A. (2023). *Apple Leaf Disease Recognition with Segmentation Research Trend: A Systematic Analysis Using VoSViewer.*

[14] Khan, A. I., Quadri, S. M. K., & Banday, S. (2021). Deep learning for apple diseases: classification and identification. *International Journal of Computational Intelligence Studies*, 10(1), 1–12.

[15] Srinidhi, V. V., Sahay, A., & Deeba, K. (2021, April). Plant Pathology Disease Detection in Apple Leaves Using Deep Convolutional Neural Networks: Apple Leaves Disease Detection using EfficientNet and DenseNet. In *2021 5th International Conference on Computing Methodologies and Communication (ICCMC)* (pp. 1119–1127). IEEE.

[16] Bansal, P., Kumar, R., & Kumar, S. (2021). Disease Detection in Apple Leaves Using Deep Convolutional Neural Network. *Agriculture*, 11(7), 617.

[17] Tahir, M. B., Khan, M. A., Javed, K., Kadry, S., Zhang, Y. D., Akram, T., & Nazir, M. (2021). Recognition of apple leaf diseases using deep learning and variances-controlled features reduction. *Microprocessors and Microsystems*, 104027.

[18] Chakraborty, S., Paul, S., & Rahat-uz-Zaman, M. (2021, January). Prediction of Apple Leaf Diseases Using Multiclass Support Vector Machine. In *2021 2nd International Conference on Robotics, Electrical and Signal Processing Techniques (ICREST)* (pp. 147–151). IEEE.

[19] Bi, C., Wang, J., Duan, Y., Fu, B., Kang, J. R., & Shi, Y. (2020). MobileNet based apple leaf diseases identification. *Mobile Networks and Applications*, 1–9.

[20] Skoneczny, H., Kubiak, K., Spiralski, M., Kotlarz, J., Mikiciński, A., & Puławska, J. (2020). Fire blight disease detection for apple trees: Hyperspectral analysis of healthy, infected and dry leaves. *Remote Sensing*, 12(13), 2101.

[21] Chao, X., Sun, G., Zhao, H., Li, M., & He, D. (2020). Identification of apple tree leaf diseases based on deep learning models. *Symmetry*, 12(7), 1065.

Computational Methods in Science and Technology Volume 1 – Kaur et al. (Eds)
© 2025 the Author(s), ISBN: 978-1-032-81760-6

Estimation of sediment load in Mahanadi River, India using artificial neural networks

Aman Upadhyay & Ankita Sharma
Centre for Advanced Studies, AKTU Lucknow, India

Prateek Raj Gautam
School of Computer Science, UPES Dehradun, India

ABSTRACT: The Mahanadi River basin of India plays an important role in the water resources of the country. It provides benefits like irrigation, hydropower, etc. Sediment load is the amount of debris (rocks, sand, silt or any other type of waste). The problem of sediment load in the river can lead to problems like bad water quality, downstream issues like erosion, flood like situation, and much more. This requires proper management, estimation, and forecasting of sediment load so that proper steps can be taken to reduce the repercussions. Empirical methods available are very complicated and do not understand the underlying features of the date. This paper showcases the use of Artificial Neural Networks(ANN) to predict the sediment load in the river. The model is trained over 11 epochs using the data collected over the years through Basantpur gauge station on the river. The model showcases impressive results with 0.95 correlation value. Still, many Artificial Intelligence (AI) algorithms can be explored and combined with ANN to improve these results.

1 INTRODUCTION

The rivers are an integral part of our ecosystem which transports weathered material from one place to another. They are a major source of water on our planet and contribute to every aspect of our lives, be it environmental or for engineering purposes (U, P. *et al.*, 2019). The water that travels through a river system carries debris/sediment eroded with it. The total amount of erosional debris carried by a river basin is called the Sediment yield of that basin.

It is expressed in two ways, volume or weight per unit area of drainage basin. The sediment load of a stream is the total amount of sediment that is delivered or transported by it. Sediments are of three types, depending upon the size: the bigger ones with boulders, cobbles, and sand (Zanial, W. N. C. W. *et al.*, 2023). Smaller particles like clay, silt, and sand. Third type is the soluble sediments like chemical compounds.

The main source of sediments in the basin is the rocks in the bed of the stream. All the empirical methods include manual sampling of water which is prone to error. Many regression-based models are also used to estimate sediment yield, but they fail to understand the underline feature in the data (Ibrahim, K. S. M. H. *et al.*, 2022). The data on river systems is scarce and is collected over several years. These data are not linear in nature, so regression models cannot do full justice to it.

To solve this problem, we use AI and ML algorithms to estimate sediments yields. ANN model is used in this paper to estimate the sediment yield of the Mahanadi River from the data collected from the Basantpur gauge station (Fleming, S. W., Vesselinov, V. V. and Goodbody, A. G., 2021). It can understand underlying patterns from the non- linear data.

DOI: 10.1201/9781003501244-4

2 STUDY AREA AND DATA USED

Our investigation is centered on the Mahanadi River basin is in east-central India, as a demonstrative case for utilizing ANN for sediment load estimation. The Mahanadi basin, covering an area of 141,589 km^2, features a primary branch spanning a length of 851 km. The village of Pharsiya, situated in the Raipur district of Chhattisgarh, India, is known to be the source of the basin.

For this study, the Basantpur gauge station, which is located on the Mahanadi River is selected. Monthly data from January 1990 to December 2010 (20 years) is used for training the model (Yadav, A., Chatterjee, S. and Equeenuddin, S. M., 2018). It has water flow, rainfall and, temperature as variables. Data was normalized before feeding it to the model to ensure standardization of data.

3 METHODOLOGY

In scenarios where data is limited, ANN becomes an obvious choice for the researchers. Other features like, simplicity, flexibility, cost-effectiveness makes it more desirable (Karami, H. *et al.*, 2022).

ANN showcases outstanding performance when it comes to deriving intricate relationships between input and output values (Latif, S. D. *et al.*, 2023).

The count of neurons in input and output layers corresponds to the number of input and target parameters. Model is trained using Levenberg Marquardt (LM) training algorithm and determining the hidden layer's neuron count is based on the model's error statistics. Parameter selection for the ANN model is achieved through a trial- and-error approach.

While the methodology is not a novel study, in the field of water, prediction of SSY, is unique, particularly in the context of the Mahanadi River.

4 RESULTS AND DISCUSSIONS

The proposed model has a single hidden layer, output layer, inputs layer, and neurons. Levenberg Marquardt algorithm is used for the training with the learning rate (mu) in the range of 0.001 to 10×10^3. The learning rate is increased or decreased by a value of 10 and 0.1respectively. To reduce the complexity of the model, the number of nodes in the hidden layer ranges between 1–32. Best combination of mu and number of neurons were chosen. There are 18 neurons, and the model was trained for 11 epochs. Tan-sigmoid activation function is in in the hidden layer and pure linear transfer function in the output layer.

The model's effectiveness is assessed using various statistical indices, including Mean Absolute Error (MAE), Mean Squared Error (MSE), Root Mean Squared Error (RMSE), correlation coefficient (r), and error variance (VAR). It demonstrated a substantial correlation coefficient (r) of 0.96841. It's worth noting that a correlation coefficient (r) approaching 1 indicates a robust alignment between the observed and expected performance values. This analysis emphasizes the effectiveness and appropriateness of the proposed ANN model, supported by the high correlation coefficient (r) and consistently low values in other error statistics. Crucially, the model showcased well-balanced performance throughout the validation, testing, and training phases, alleviating concerns of underfitting or overfitting.

As illustrated in Figure 3, the estimated SSY values closely match the observed data. This alignment is further evident in Figure 4, a scatter plot illustrating the relationship between the SSY predicted by the ANN model and the observed SSY. Remarkably, a substantial number of points cluster closely around the bisector line, signifying consistency between predicted and observed values.

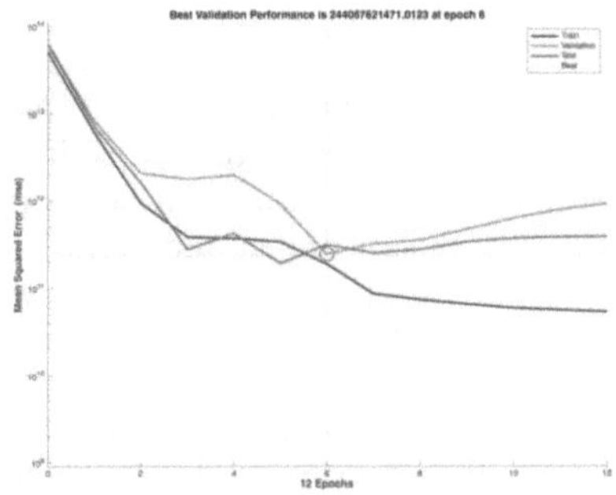

Figure 1. Performance variation with epochs in all phase.

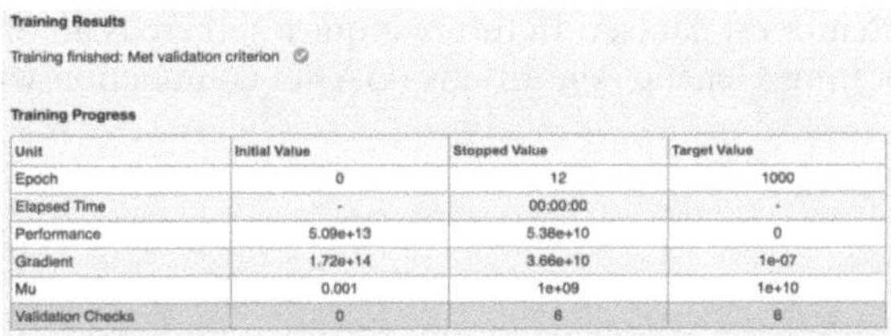

Training Results

Training finished: Met validation criterion

Training Progress

Unit	Initial Value	Stopped Value	Target Value
Epoch	0	12	1000
Elapsed Time	-	00:00:00	-
Performance	5.09e+13	5.38e+10	0
Gradient	1.72e+14	3.66e+10	1e-07
Mu	0.001	1e+09	1e+10
Validation Checks	0	6	6

Figure 2. Training result of the model.

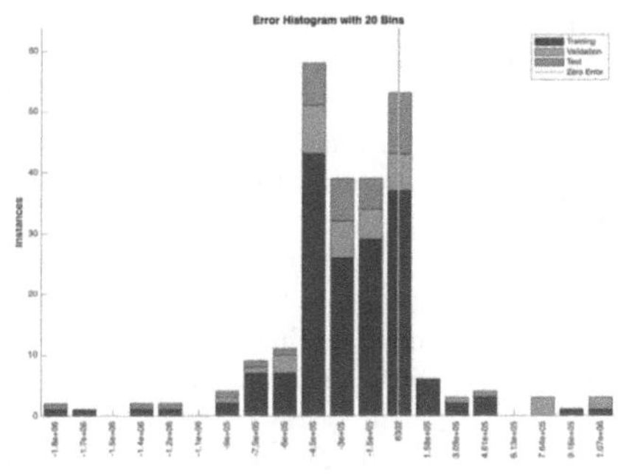

Figure 3. Comparison between actual and estimated water flow of ANN model in testing phase.

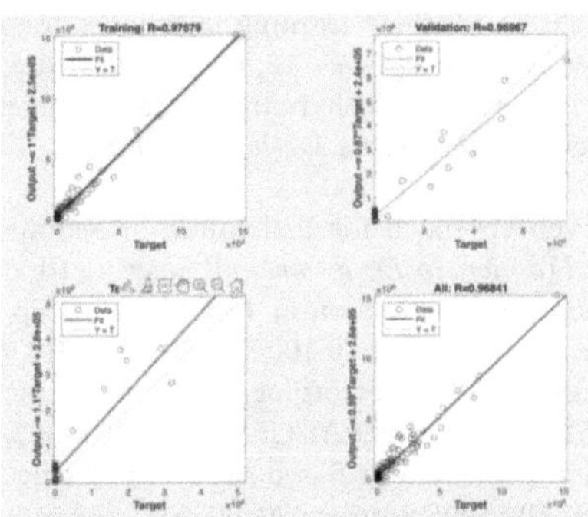

Figure 4. Scatter plot between actual and ANN estimated water flow based on training, validation and testing data.

5 CONCLUSIONS

The forecast of SSY involves utilizing an Artificial Neural Network (ANN)-based model that utilizes a dataset comprising water flow, rainfall, sediment yield, temperature, and water level information (Essam, Y. *et al.*, 2022). These factors have proven crucial for SSY prediction, particularly at the Basantpur gauge station within the Mahanadi River basin in India. It's noteworthy that, unlike some previous ANN models that produced negative values during periods of low SSY, all SSY predictions in this study remained positive.

Furthermore, the scatter plot depicting the predicted SSY versus the observed SSY demonstrates a significant agreement, indicating a close correspondence between the values predicted by the ANN model and the actual SSY. This agreement emphasizes the robust predictive capability of the ANN model, making it highly effective for estimating SSY, especially in situations where direct measurements are not feasible.

However, it's important to acknowledge that the current analysis focused on one gauge station and a limited temporal dataset. In future inquiries, there is an opportunity to expand the scope by incorporating Genetic Algorithms (GA) in conjunction with ANN.

REFERENCES

Aghelpour, P., Bahrami-Pichaghchi, H. and Varshavian, V. (8 2021) Hydrological drought forecasting using multi-scalar streamflow drought index, stochastic models and machine learning approaches, in northern Iran, *Stochastic Environmental Research and Risk Assessment*, 35, pp. 1615–1635. doi: 10.1007/s00477-020-01949-z.

AlDahoul, N. *et al.* (12 2022) A comparison of machine learning models for suspended sediment load classification, *Engineering Applications of Computational Fluid Mechanics*, 16, pp. 1211–1232. doi: 10.1080/19942060.2022.2073565.

Asadollah, S. B. H. S. *et al.* (2 2021) River water quality index prediction and uncertainty analysis: A comparative study of machine learning models, *Journal of Environmental Chemical Engineering*, 9, p. 104599. doi: 10.1016/j.jece.2020.104599.

Band, S. S. *et al.* (10 2020) Comparative Analysis of Artificial Intelligence Models for Accurate Estimation of Groundwater Nitrate Concentration, *Sensors*, 20, p. 5763. doi: 10.3390/s20205763.

Essam, Y. *et al.* (1 2022) Predicting suspended sediment load in Peninsular Malaysia using support vector machine and deep learning algorithms, *Scientific Reports*, 12, p. 302. doi: 10.1038/s41598-021-04419-w.

Fleming, S. W., Vesselinov, V. V. and Goodbody, A. G. (6 2021) Augmenting geophysical interpretation of data-driven operational water supply forecast modeling for a western US river using a hybrid machine learning approach, *Journal of Hydrology*, 597, p. 126327. doi: 10.1016/j.jhydrol.2021.126327.

Hussain, D. and Khan, A. A. (9 2020) Machine learning techniques for monthly river flow forecasting of Hunza River, Pakistan, *Earth Science Informatics*, 13, pp. 939–949. doi: 10.1007/s12145-020-00450-z.

Ibrahim, K. S. M. H. *et al.* (1 2022) A review of the hybrid artificial intelligence and optimization modelling of hydrological streamflow forecasting, *Alexandria Engineering Journal*, 61, pp. 279–303. doi: 10.1016/j.aej.2021.04.100.

Karami, H. *et al.* (3 2022) A Novel Approach for Estimation of Sediment Load in Dam Reservoir With Hybrid Intelligent Algorithms, *Frontiers in Environmental Science*, 10. doi: 10.3389/fenvs.2022.821079.

Latif, S. D. *et al.* (3 2023) Sediment load prediction in Johor river: deep learning versus machine learning models, *Applied Water Science*, 13, p. 79. doi: 10.1007/s13201-023-01874-w.

Mokhtar, A. *et al.* (2021) Estimation of SPEI Meteorological Drought Using Machine Learning Algorithms, *IEEE Access*, 9, pp. 65503–65523. doi: 10.1109/ACCESS.2021.3074305.

Noori, R. *et al.* (2 2022) An Efficient Data Driven-Based Model for Prediction of the Total Sediment Load in Rivers, *Hydrology*, 9, p. 36. doi: 10.3390/hydrology9020036.

Trach, R. *et al.* (5 2022) A Study of Assessment and Prediction of Water Quality Index Using Fuzzy Logic and ANN Models, *Sustainability*, 14, p. 5656. doi: 10.3390/su14095656.

U, P. *et al.* (6 2019) EODVGA: An Enhanced ODV Based Genetic Algorithm for Multi-Depot Vehicle Routing Problem, *ICST Transactions on Scalable Information Systems*, 6, p. 159099. doi: 10.4108/eai.10-6-2019.159099.

Yadav, A., Chatterjee, S. and Equeenuddin, S. M. (12 2018) Prediction of suspended sediment yield by artificial neural network and traditional mathematical model in Mahanadi river basin, India, *Sustainable Water Resources Management*, 4, pp. 745–759. doi: 10.1007/s40899-017-0160-1.

Zakaria, M. N. A. *et al.* (7 2023) Exploring machine learning algorithms for accurate water level forecasting in Muda river, Malaysia, *Heliyon*, p. e17689. doi: 10.1016/j.heliyon.2023.e17689.

Zanial, W. N. C. W. *et al.* (1 2023) River flow prediction based on improved machine learning method: Cuckoo search-artificial neural network, *Applied Water Science*, 13, p. 28. doi: 10.1007/s13201-022-01830-0.

Computational Methods in Science and Technology Volume 1 – Kaur et al. (Eds)
 ISBN: 978-1-032-81760-6

Prediction of water discharge in Mahanadi River basin, India using artificial neural networks

Ankita Sharma & Aman Upadhyay
Centre for Advanced Studies, AKTU Lucknow, India

Prateek Raj Gautam
School of Computer Science, UPES Dehradun, India

ABSTRACT: River water flow, also known as water discharge or streamflow of a river pertains to the volume of water passing through a point in a specified time frame. It is quantified in cubic meters per second or cubic feet per second. Maintaining and predicting the flow of water through a stream is essential yet a very difficult task. Increase or decrease in water levels can lead to flood or drought like conditions respectively. However, monitoring water behavior is tough due to lack of data, various hydrological factors, unmonitored terrains, and non-linear nature of water. The conventional approaches are not capable of handling non-linear, non-stationary nature of water. Therefore, this research focuses on developing an innovative hybrid artificial intelligence model. It involves training a neural network rigorously with monthly data collected for 10 years from Tikarpara gauge station over Mahanadi River for predicting the flow of water. This study produces notable results with correlation value of 0.94 between actual and predicted values. Artificial intelligent methods outperform any manual method and saves time, money, energy and chances of human errors.

1 INTRODUCTION

River water flow also known as water discharge or streamflow is an important aspect of nature, human civilizations, ecosystems and economies from decades. Rivers are a primary source of fresh water which is used for a lot of things like drinking, irrigation, industrial processes and other domestic uses (Zakaria, M. N. A. *et al.*, 2023).

Therefore, the sustainability and prediction of water flow is very important to ensure uninterrupted water supply to communities. Managing water flow is also important as it contributes to planning resources and taking precautionary measures in case of disasters such as flood or drought (Zanial, W. N. C. W. *et al.*, 2023). The level of water also effects transportation through water which can have effect on industrial and commercial sectors. To address all these issues, river basin planning, and resource management are very important, which in turn require proper water flow predictions.

Many important decisions, including strategic resource planning and management depend upon the ability to forecast water related events and phenomena with precision and in timely manner (Trach, R. *et al.*, 2022).

Neural networks, fuzzy logic, or a combination of both gained immense popularity amongst hydrologists for modeling hydrological processes. Artificial Neural Network (ANN) has garnered attention for its ability to handle the non-linear nature of the water (Mokhtar, A. *et al.*, 2021). A very important task for a hydraulic scientist is an accurate

 DOI: 10.1201/9781003501244-5

prediction of flow corresponding to a certain depth of the river, especially in the case of the overbank stage because it directly influences flood-risk mitigation.

Artificial Neural Network (ANN) is widely used by researchers to predict water flow as it is flexible and can find hidden relationships using scarce data (Ibrahim, K. S. M. H. *et al.*, 2022).

2 RESEARCH REGION AND UTILIZED DATASET

The Mahanadi basin, covering an area of 141,589 km^2, features a primary branch spanning a length of 851 km. Its source is traced back to Pharsiya village within the Raipur region in Chhattisgarh, India, positioned approximately 442m above the ocean's surface.

For the present investigation, the Tikarapara gauge station located in the Mahanadi River was selected. This station provided monthly data encompassing Rainfall, Water depth, Thermal conditions, and River flow. The dataset spans 1240 months (equivalent to 20 years), encompassing readings from January 1990 to December 2010 (Yadav, A. *et al.*, 2020). The study employed a systematic approach which involved conducting tests, training sessions, and validation processes to refine applied models. To ensure uniformity and comparability, data normalization was performed.

Normalization eliminates variable dimensions and preserves their distinctive characteristics during model training.

3 PROPOSED MODEL

Artificial Neural Networks (ANN) rank among the most frequently employed Artificial Intelligence (AI) technique due to various features like, cost-effectiveness, minimum data requirement, simplicity and, flexibility to name a few (Fleming, S. W., Vesselinov, V. V. and Goodbody, A. G., 2021).

The capability of ANN to model intricate, non-linear and dynamic systems makes it na easy choice in the field of water. Predicting and forecasting of complex processes, such as water flow, without the need of explicit mathematical formulation is extremely commendable.

The ANN model used in this study uses temporal data with variables like, rainfall, temperature, sediment load as inputs to estimate the flow of Mahanadi River. According to the estimated values, the model's effectiveness is assessed.

ANN also can derive hidden dependencies of input and output variables. In this study, a Multi-Layer Perceptron (MLP) model is employed. It is a forward propagating neural network which comprises an input layer, one or more intermediary layers called hidden layers and an output layer (Aghelpour, P., Bahrami-Pichaghchi, H. and Varshavian, V., 2021). The model uses Levenberg-Marquardt training algorithm. The count of neurons that will be present in the hidden layer and parameter selection for the model is done through hit-and-trial method. Using this method in water flow prediction imparts valuable perspective on using AI and machine learning within river systems.

4 RESULTS AND DISCUSSIONS

The proposed artificial neural network has multiple layers of perceptron (MLP-ANN) which uses Levenberg Marquardt algorithm that has a hidden layer, output layer, inputs layer, and nodes (neurons). The learning rate (mu) of the model varies between 0.001 to 10×10^9. The learning rate is adjusted by altering the factors containing values 10 and 0.1. Total nodes within hidden layer ranged from one to thirty-two to lower the complexity. An experimental trial was conducted to decide the number of neurons and the learning rate of the ANN model. The number of iterations for which the model ran came out to be 16 and the optimum number of neurons is 18. The hidden layer and output layer uses tan-sigmoid and pure linear transfer functions respectively.

Analysis of the error statistics unveiled the ANN model's favourable attributes, with a low RMSE value (0.044041) and a high correlation coefficient (r) of 0.9704. Notably, the correlation coefficient (r) close to 1 signifies a strong alignment between the observed and expected performance values. The model exhibited balanced performance across the validation, testing, and training phases, eliminating concerns of underfitting or overfitting.

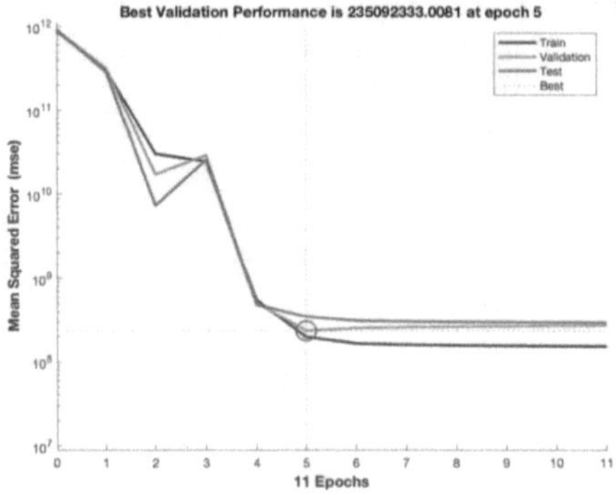

Figure 1. Changes in performance over epochs during different phases.

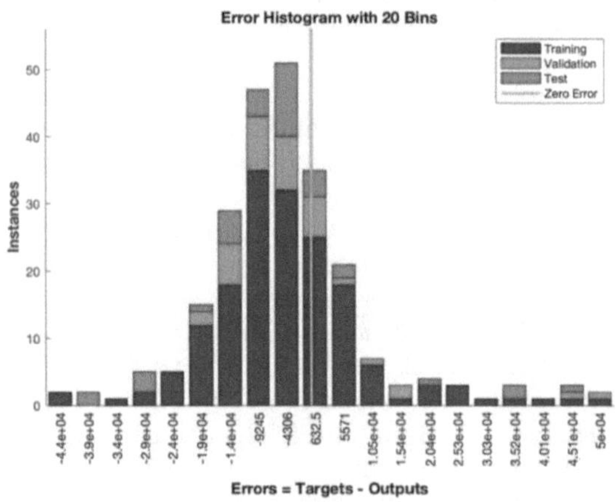

Figure 2. Variation in estimated and observed values.

The findings strongly indicate the successful development of the artificial neural network trained using the Levenberg-Marquardt algorithm, thereby showcasing the aptness of the selected model for the task at hand.

Evident from Figure 2, the estimated water flow values closely align with the observed water flow values. This proximity is equally apparent in Figure 3, a scatter plot portraying the association between predicted and observed values of water flow employing the ANN model. This outcome resonates with a satisfactory and broadly applicable result. The model did not generate any negative water flow value.

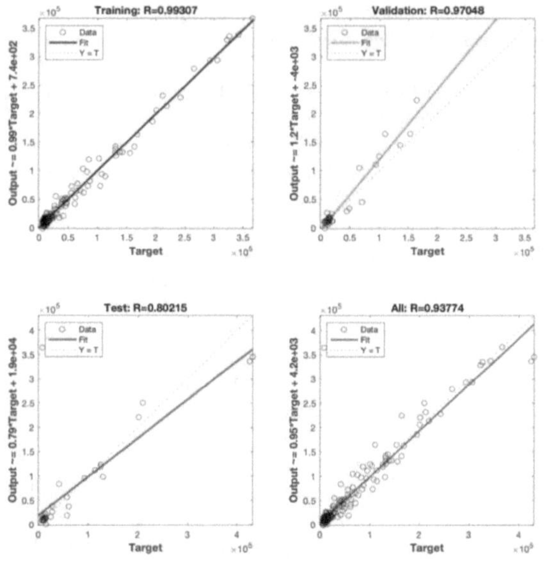

Figure 3. Scatter plot between actual and estimated water flow values.

5 SUMMARY OF RESULTS

Anticipating water movement entails the utilization of an ANN-based model, employing a dataset encompassing values for stream flow, precipitation, sediment yield, temperature, and water level (Band, S. S. *et al.*, 2020). These parameters have emerged as pivotal inputs for water flow prediction, particularly at the Tikarapara gauge station within the Mahanadi River basin in India. Noteworthy is the observation that while some prior ANN models yielded lower values during periods of reduced water flow, all water flow predictions in this study remained positive.

Furthermore, the scatter plot displaying the predicted water flow plotted against the observed water flow illustrates a remarkable congruence, signifying the close alignment between the ANN-predicted values and actual water flow. This consonance underscores the robust predictive capability of the ANN model, rendering it highly effective for water flow estimation, especially in cases where direct measurements are unattainable (Azad, A. S. *et al.*, 2022).

However, it's crucial to note that the current analysis was restricted to a singular gauge station and a limited temporal dataset. For future studies, there's potential to expand the scope by incorporating Genetic Algorithms (GA) in tandem with ANN, and subsequently comparing the performance of these two models across a broader dataset that includes spatial geological data from multiple gauge stations(Tikhamarine, Y. *et al.*, 2020). Furthermore, the exploration of diverse AI algorithms could yield valuable insights through comparative analyses and model variations.

REFERENCES

Aghelpour, P., Bahrami-Pichaghchi, H. and Varshavian, V. (8 2021) Hydrological drought forecasting using multi-scalar streamflow drought index, stochastic models and machine learning approaches, in northern Iran, *Stochastic Environmental Research and Risk Assessment*, 35, pp. 1615–1635. doi: 10.1007/s00477-020-01949-z.

Ali, S. and Shahbaz, M. (9 2020) Streamflow forecasting by modeling the rainfall–streamflow relationship using artificial neural networks, *Modeling Earth Systems and Environment*, 6, pp. 1645–1656. doi: 10.1007/s40808-020-00780-3.

Alvisi, S. and Franchini, M. (4 2011) Fuzzy neural networks for water level and discharge forecasting with uncertainty, *Environmental Modelling & Software*, 26, pp. 523–537. doi: 10.1016/j.envsoft.2010.10.016.

Asadollah, S. B. H. S. *et al.* (2 2021) River water quality index prediction and uncertainty analysis: A comparative study of machine learning models, *Journal of Environmental Chemical Engineering*, 9, p. 104599. doi: 10.1016/j.jece.2020.104599.

Azad, A. S. *et al.* (2 2022) Water level prediction through hybrid SARIMA and ANN models based on time series analysis: Red hills reservoir case study, *Sustainability*, 14, p. 1843. doi: 10.3390/su14031843.

Band, S. S. *et al.* (10 2020) Comparative analysis of artificial intelligence models for accurate estimation of groundwater nitrate concentration, *Sensors*, 20, p. 5763. doi: 10.3390/s20205763.

Fleming, S. W., Vesselinov, V. V. and Goodbody, A. G. (6 2021) Augmenting geophysical interpretation of data-driven operational water supply forecast modeling for a western US river using a hybrid machine learning approach, *Journal of Hydrology*, 597, p. 126327. doi: 10.1016/j.jhydrol.2021.126327.

Ghorbani, M. A. *et al.* (8 2020) Development and evaluation of the cascade correlation neural network and the random forest models for river stage and river flow prediction in Australia, *Soft Computing*, 24, pp. 12079–12090. doi: 10.1007/s00500-019-04648-2.

Hussain, D. and Khan, A. A. (9 2020) Machine learning techniques for monthly river flow forecasting of Hunza River, Pakistan, *Earth Science Informatics*, 13, pp. 939–949. doi: 10.1007/s12145-020-00450-z.

Ibrahim, K. S. M. H. *et al.* (1 2022) A review of the hybrid artificial intelligence and optimization modelling of hydrological streamflow forecasting, *Alexandria Engineering Journal*, 61, pp. 279–303. doi: 10.1016/j.aej.2021.04.100.

Mahmoodzadeh, A. *et al.* (7 2021) Presenting the best prediction model of water inflow into drill and blast tunnels among several machine learning techniques, *Automation in Construction*, 127, p. 103719. doi: 10.1016/j.autcon.2021.103719.

Mokhtar, A. *et al.* (2021) Estimation of SPEI meteorological drought using machine learning algorithms, *IEEE Access*, 9, pp. 65503–65523. doi: 10.1109/ACCESS.2021.3074305.

Tikhamarine, Y. *et al.* (3 2020) Improving artificial intelligence models accuracy for monthly streamflow forecasting using grey Wolf optimization (GWO) algorithm, *Journal of Hydrology*, 582, p. 124435. doi: 10.1016/j.jhydrol.2019.124435.

Trach, R. *et al.* (5 2022) A study of assessment and prediction of water quality index using Fuzzy Logic and ANN models, *Sustainability*, 14, p. 5656. doi: 10.3390/su14095656.

Yadav, A. *et al.* (12 2020) Application of artificial neural network and genetic algorithm based artificial neural network models for river flow prediction, *Revue d'Intelligence Artificielle*, 34, pp. 745–751. doi: 10.18280/ria.340608.

Yaseen, Z. M. *et al.* (2 2020) Hourly river flow forecasting: application of emotional neural network versus multiple machine learning paradigms, *Water Resources Management*, 34, pp. 1075–1091. doi: 10.1007/s11269-020-02484-w.

Zakaria, M. N. A. *et al.* (7 2023) Exploring machine learning algorithms for accurate water level forecasting in Muda river, Malaysia, *Heliyon*, p. e17689. doi: 10.1016/j.heliyon.2023.e17689.

Zanial, W. N. C. W. *et al.* (1 2023) River flow prediction based on improved machine learning method: Cuckoo Search-Artificial Neural Network, *Applied Water Science*, 13, p. 28. doi: 10.1007/s13201-022-01830-0.

Computational Methods in Science and Technology Volume 1 – Kaur et al. (Eds)

Online misinformation detection using deep learning on social media: A survey and analysis

Dharmaraj B. Vhanmore
M.E. Student, Department of CSE, D. Y. Patil College of Engineering & Technology, Kolhapur, India

Kishor T. Mane
Assistant Professor, Department of CSE, D. Y. Patil College of Engineering & Technology, Kolhapur, India

ABSTRACT: Social media platforms such as WhatsApp, Facebook, Instagram, Twitter and even others are being used more frequently nowadays. These platforms are considered a handy medium to share news, views, pictures, messages and videos by users. Though many individuals appreciate social media, false information can mislead users into believing propaganda through the transmission of false information like rumors and fake news. Furthermore, disseminating the vast quantity of misinformation on social media has turned into a worldwide threat. As a result, finding false material on social networking websites has garnered a lot of attention and is thought to be a potential field for future study. Numerous investigations on misinformation identification have been carried out in the literature in an effort to identify fresh study issues and methodologies. After doing the comprehensive survey of automated misinformation identification based on disinformation, rumors, false information, spam and fake news, there is a need for new techniques that will give more accurate information. A variety of machine learning approaches are utilized for the purpose of misinformation detection. The review paper provides a deep view of the current techniques and their limitations along with the research challenges in the field of deep learning algorithms-based detection of misinformation.

Keywords: Rumors, Fake news, deep learning, Machine Learning, Misinformation, Social Media

1 INTRODUCTION

Misinformation denotes a falsehood that aims to deceive individuals by concealing accurate information. It is also known as deception, ambiguity, falsehoods, and various other terms. The dissemination of misinformation fosters a sense of mistrust, subsequently undermining interpersonal relationships. This represents a negative deviation from expected behavior. More than that, people don't expect to be misled by strangers, relatives, or even close friends. Instead, people anticipate direct and honest communication. To illustrate, consider a Facebook conversation regarding a freshly released product, wherein both genuine and fake users participate. Sincere customers evaluate the attributes of the products in an open and sincere manner. On the other hand, the phony customers, regardless of their true views, lavish the goods with undeserved praise. People can express their thoughts, videos and news regarding their diverse activities on a variety of online social networks, including Facebook, Twitter, and WhatsApp. Even though people enjoy using these social networks, users may be misled into believing false information by dishonest tactics like rumors or fake news. Consequently, the identification of misinformation within social networking platforms has garnered significant attention and is presently regarded as a burgeoning area of research

DOI: 10.1201/9781003501244-6

interest. Accordingly, endeavor aims to present a technique founded on a deep-learning model to discover misinformation within online social network platforms.

The next segment provides a deep understanding of the existing algorithm along with its limitations and research challenges.

2 LITERATURE REVIEW

There are various studies and methodologies have been used in the detection process of misinformation on social media sites like Twitter, WhatsApp and many more.

In this (Mahbub *et al.*, 2022) put forth a groundbreaking feature space for identifying rumors on Twitter, specifically related to the COVID-19 pandemic. To evaluate their approach, the researchers employed various machine- learning models and utilized a publicly accessible Twitter dataset that had been labeled for COVID-19 rumors. Naive Bayes, Random Forest, J48 Decision Tree, and JRip Rule-based classifiers were among the machine learning models employed in this investigation. The authors effectively demonstrated improvements in performance metrics, including recall and precision, as well as detection accuracy for rumors related to COVID-19 on Twitter, using their experimental setups. Emphasizing the automated identification of rumors of COVID-19 on Twitter, the paper regarded whole information with uncertain veracity as potential rumors. The suggested hybrid feature space combined contextual characteristics unique to Twitter, psycho-linguistic features, and content-specific tweet attributes to improve COVID-19 rumor detection accuracy.

The article by (Kovács *et al.*, 2021) delves into the complexities associated with identifying hate speech on social media platforms. These difficulties arise from the uncontrolled proliferation of hateful content, the presence of paralinguistic cues, and inadequately written text. To solve these issues, the authors intend an advanced model built on natural language processing (NLP) which combines both recurrent and convolutional layers to detect hate speech. Remarkably, this model reached a 0.63 as F1 score when evaluated on the HASOC2019 corpus. The scarcity of training data, however, presents a significant risk of overfitting. The findings demonstrate that such strategies can substantially enhance classification accuracy. The authors investigate the utilization of an amalgamation of layers of Convolutional model and Long Short-Term Memory (LSTM) model (CNN-LSTM), transformer models such as Roberta, and the FastText classification model. Additionally, they conducted experiments using the OLID5 corpus, which bears resemblances to the HASOC corpus, as an external data source. Furthermore, the paper explores transfer learning techniques using RoBERTa and leveraging pre-trained word representations like FastText embeddings. These approaches aim to improve the generalization capabilities of the models and exploit additional data sources for hate speech classification.

In paper (Sansonetti *et al.*, 2020) the study examines characteristics that have the ability to forecast the identification of profiles on social network that are liable for disseminating fabricated information within the online sphere. The characteristics of monitored individuals were derived from Twitter, encompassing both social as well as personal data, in addition to their engagement with content and other users. The study showcases an analysis conducted offline through the utilization of the deep-learning based techniques, as well as an online analysis that involves actual users in the categorization of trustworthy and untrustworthy user profiles. The investigational findings, which have been statistically validated, shows which information is most effective in enabling both machines and humans to identify deceitful users. The main goal of this research work is to provide valuable insights for the development of more efficient tools to combat misinformation and those individuals who intentionally propagate it.

The study paper (Joshi *et al.*, 2023) presents a novel method for identifying misinformation on different social media platforms. With this method, a detector with universal applicability is developed by combining the both explainable AI approaches and the domain adaption. Explainability is important for AI models since it increases their credibility, as the paper emphasizes. For generalized detection, the authors use the Domain Adversarial

Neural Network (DANN), and for explanations of the DANN model's results, they use the Local Interpretable Model-Agnostic Explanations (LIME) model. The research shows improvements in classification performance and demonstrates through first-hand indication how effective this approach is in spotting COVID-19 disinformation. As a result, target labels are explained and the difficulties created by domain shift are effectively addressed by the suggested framework. In doing so, it enables the processing and extraction of trustworthy information, thereby effectively combating misinformation.

In the paper (Yadav *et al.*, 2022), the author proposes a hybrid deep learning method for detecting fake news, which involves the grouping of LSTM and CNN models. To train the model, a large consolidated dataset is utilized, which integrates publicly available two datasets: one consisting of fake as well as real news articles, and the other containing 64k+ labeled news articles referred to as "allData". In order to generate effective data representation, various techniques of word-embedding such as FastText, GloVe, and Word2Vec are used. For categorization purposes, deep-learning based models that include LSTM, CNN-BiLSTM, CNN-LSTM, and BiLSTM are utilized. The combination of the Word2Vec word embedding technique with the CNN-BiLSTM model demonstrates the lowest performance, as evidenced by inaccuracy, imprecision, non-recall, lowest F1 measure, and lowest AUC-ROC values of 0.025, 0.016, 0.030, 0.023, and 0.008, respectively. In the area of identifying real news from fake news using headlines, this suggested method works better than the most advanced alternatives at the moment. It is important to acknowledge that the study recognizes the limitation of not having a single large, standardized dataset solely dedicated to fake news detection. However, it does not delve into the potential limitations or challenges associated with the employed models and techniques. Moreover, the study does not provide detailed information concerning potential biases or limitations of the utilized datasets. Last but not least, the study makes no mention of the potential difficulties or restrictions that can result from applying word-embedding strategies and deep learning-based models to the identification of fake news.

Uma Sharma and colleagues (Sharma Uma *et al.*, 2021) aim to provide users with the ability to classify the news stories as authentic or falsified, while also confirming the reliability of the news source that is distributing the content. The authors use machine learning, natural language processing, and the power of intelligence to reach this aim. The main topic of this study is the widespread problem of spreading misleading information, which has serious consequences like the creation of twisted opinions and the altering of election outcomes. In order to address this challenging scenario, the authors suggest implementing machine learning algorithms to identify fake news, which will help to lessen the disinformation issue. As a result, this study advances the field of fake news identification by offering insightful information.

Based on the credibility of the domain names and the level of content understanding, the paper (Xu *et al.*, 2020) describes a wide range of widely read fake and real news pieces on Facebook. The websites of both fictional and real news publishers are examined by the writers for their registration practices, timing, rankings, and popularity. They also note that fake news tends to go off the internet after a long while. It explores several methods of fake news detection, including topic modeling using Latent Dirichlet Allocation (LDA) and TF-IDF. They also find that there is potential for using term and word vector similarity between publications to forecast both fake and real news. By carefully studying the reputations of domains and the characteristics of content in fictitious and genuine news, the authors offer valuable insights for well detecting fictitious news on social websites.

The paper by (Wang *et al.*, 2021) concentrates on the creation of a programmed method for identifying messages that oppose vaccination on social media sites, taking into account the presence of multiple forms of media in the posts. It introduces a deep-learning network that employs both textual and visual data, along with an innovative attention mechanism that prioritizes significant content. The model comprises three branches and produces comprehensive combined features for making predictions. The paper gathers and assesses a real-world dataset from Instagram, illustrating the model's capability to identify a significant number of antivaccine messages with a testing accuracy exceeding 97%.

The paper article authored by (Biradar *et al.*, 2023) examines the application of an early fusion-based technique to detect false information on social media websites pertaining to data of COVID-19. The proposed method (Biradar *et al.*, 2023) combines key features from context-based embedding's, like ELMo, XLNet, and BERT, to enhance the extraction of context and semantic data from social media posts. As a result, the accuracy of spotting fake news is increased. With an astounding 97% accuracy rate, the study shows that language and ensemble models are suitable for this use.

(Raza & Ding, 2022) endeavor to tackle the obstacles encountered in the detection of fake news by presenting an innovative framework that amalgamates data from news articles and social contexts through the utilization of a Transformer-based methodology. The suggested framework additionally integrates attributes derived from the news content & social contexts in order to enhance the precision of fake news categorization.

A BERT-based deep learning technique called FakeBERT is presented in a research paper by (Kaliyar *et al.*, 2021) with the goal of identifying deceitful content on social media. FakeBERT combines BERT with discrete parallel components of a single-layer deep CNN to handle ambiguity and improve classification performance. FakeBERT's bidirectional training methodology allows it to understand the semantic and long-range interdependencies present in sentences, which is a vital component in the revealing of fake news. The classification analysis's results confirm that FakeBERT outperforms other models and achieves a high accuracy rate.

Supply chain (SCD) disruptions may be lessened by the use of machine learning (ML) and artificial intelligence (AI) in fake news and disinformation (FNaD) recognition models, as described in this research (Akhtar *et al.*, 2023). Nevertheless, little research has been done on creating FNaD identification models based on AI and ML that are especially suited for SCDs. Practical evidence demonstrates the usefulness of the suggested FNaD identification model in guiding managerial decision-making, which is dependent on AI and ML. The lack of research on the combination of AI, SCDs, and FNaD has been noted, which emphasizes the importance of our work in filling this knowledge vacuum. By providing helpful insights and highlighting potential paths, this study signifies a substantial contribution to the body of knowledge on supply chain management and AI-ML.

The impact of false information on society and how social media algorithms influence its dissemination are the subjects of this paper written by (Siddhesh Chalke & Deepak Mishra, 2023). It emphasizes how vital it is to understand the characteristics of misleading information on social media. It uses a deep learning methodology to examine the content and emotions related to both kind of materials i.e. real and fake material. By examining real and fake information, this study work advances methods for assessing the emotional content of information. The study's conclusions point out that, a variety of emotions and the qualities of the content can have an impact on the spread of false information through social media platforms, leading to a range of changes. The objective of this paper is to make a valuable contribution to the regulation of false information and the preservation of network order, specifically about authentic information.

The scholars emphasize the extensive dissemination of misleading data on a number of social networking sites and other channels, underscoring the necessity for research that is specifically dedicated to identifying it. The scholarly paper (Khanam *et al.*, 2021) proposes a procedural method for establishing a framework that is capable of identifying and categorizing deceptive articles through the analysis of their lexical components, phrases, origins, and headings.

The author (Dadgar & Ghatee, 2021) presents a system for identifying inaccurate information that merges processes based on both network and content. In the process based on the network, the article concentrates on users, network characteristics, and social properties to identify instances of inaccurate information. In the content-based process, the article utilizes models for text classification and similarity to categorize inaccurate information based on the tweets' contents. The evaluation results show that in terms of incorrect information classification, the content-based models outperform the network-based models.

The study presently under consideration centers on the issue of identifying COVID-19 related misinformation on number of social networking sites. This topic is of great significance due to its widespread global impact and the resulting "Infodemic" phenomenon. a new framework called Contrastive Adversarial Domain Mixup (CADM) is introduced by (Zeng *et al.*, 2022) researchers for unsupervised domain adaptation, To create a link between the COVID-19 data domain and the training data source domain. Through the use of adversarial domain mixup and contrastive learning, CADM successfully transfers facts from the current source data to the domain of COVID-19-data. Empirical findings suggest that CADM successfully adjusts language models that have already been trained to the COVID-19 domain and consistently surpasses cutting-edge benchmarks in terms of detecting and addressing misinformation.

In (Sule *et al.*, 2023) examines the dissemination of inaccurate information regarding COVID-19 by American doctors across a range of social media sites and online sources after the availability of vaccines. Identifying the various categories of false information disseminated by physicians, encompassing misinformation about vaccines, pharmaceutical treatments, face coverings, and conspiracy theories. The research study advocates for further investigation to evaluate the full scope of the consequences of physician-led dissemination of inaccurate information on public health and to devise strategies aimed at mitigating its detrimental effects.

The objective of this paper (Alnazzawi *et al.*, 2022) is to create the Twitter collection known as FakeAds, which is made up of messages that appear to be product advertising. This corpus's goal is to investigate the impact of misleading information and fake news in product advertisements and marketing on the Twitter network. The paper emphasizes the diminished quality of news and credibility found on social networking platforms in comparison to traditional news channels. Additionally, it explores the utilization of Twitter as a medium for spreading wrong marketing information via intentionally misleading bogus advertisements. The research contributes by creating text mining tools that automatically detect bogus material and fake news on product marketing using machine learning techniques. The study community now has access to a wealth of data resources thanks to this revelation, which will facilitate further investigation into the viability of using social media to promote products and disseminate fraudulent adverts.

(Buzea *et al.*, 2022) suggest using supervised machine learning to detect misleading content in Romanian-language online publications. Furthermore, a dataset of internet articles classified into actual and artificial classes is statistically analyzed and presented. The result indicates that the convolutional neural network achieves the highest level of precision, outperforming both the BERT models and the conventional classification methods, with a precision rate of more than 98.20%.

A study was conducted in this paper (Khan *et al.*, 2021) to assess the usefulness of different machine-learning procedures in the discovery of fake news. This study included a comparison of deep learning methods with conventional methods and pretrained language models. The findings indicated that BERT and analogous pre-trained models exhibited superior performance, particularly when confronted with limited training data. Furthermore, the study examined the models' effectiveness in relation to the subject matter and length of articles, while also delving into the insights gained. Khan *et al.*, maintain that this benchmark study will be of significant value to both the research community and news platforms, aiding them in the selection of the most suitable technique for discovery of fake news.

Balshetwar and other authors of this paper (Balshetwar *et al.*, 2023) suggests a new way out for the identification of fabricated news, incorporating sentiment analysis as a crucial factor to enhance precision. The investigation introduces a strategy of multiple imputation employing the Multiple-Imputation-Chain-Equation that is MICE to address instances of multivariate missing variables in news or social media data. The paper employs Term-Frequency & Inverse

Document Frequency i.e. TF-IDF to extract efficacious attributes from textual content and ascertain extended-term attributes through employment of a weighted matrix. The relationship between variables of valuable data features and missing data is categorized via employment of Deep Neural Network (DNN), passive-aggressive and Naïve Bayes classifiers. The proposed technique attains a commendable accuracy rate of 99.8% for identifying spurious news, surpassing the effectiveness of preexisting methodologies.

The goal of the research study (Fahad *et al.*, 2023) is to efficiently identify false data by utilizing a public-accessible dataset. Additionally, a machine learning classifier is proposed to construct an automated system in order to discover the fake news, using the goal of achieving benchmark accuracy rate of 93.61%. The technique outlined in this study is established on a thorough investigation of current literature and the use of an openly accessible dataset that covers 7796 news articles, with an equal distribution between authentic and fabricated news.

The paper (Guo *et al.*, 2023) examines the difficulty of identifying false information by merging the textual and visual aspects, as relying solely on textual assertions may be deceptive. The paper proposes a two-branch multimodal model for detecting deceptive news that makes use of an attention mechanism and multimodal bilinear pooling. The feature extraction carries out by the model independently for both textual and visual aspects, thus addressing the challenge of feature extraction in identifying false information. The model may focus on important aspects and improve its false information detection accuracy by incorporating an attention mechanism.

The investigation paper (Agivale *et al.*, 2022) centers on the problem of disinformation and false information masquerading as fake news that holds the potential to manipulate public sentiment and provoke social turmoil. The authors employ various Deep Learning models that leverage word embedding's to ascertain the authenticity of news articles, successfully attaining optimal outcomes through the utilization of an architecture based on recurrent convolutional neural networks. The summary of the different techniques has been shown in the Table 1.

Table 1. Summary of different techniques used for misinformation detection on social media.

Approach/Technique Used	Ref	Purpose	Datasets	Evaluation Parameters	Research Gaps
JRip Rule-based classifiers, J48 Decision Tree, Random Forest, Naive Bayes	(Mahbub *et al.*, 2022)	Rumor detection on twitter	Covid19 twitter dataset	Accuracy	Other social platforms are not used
Deep neural network (DNN), CNN, RNN, FastText text classification models	(Kovács *et al.*, 2021)	Hate speech detection	HASOC2019 Corpus	F1 score	Limited to one dataset
KNN, SVM, LSTM-CNN, SVM-SGD Classifier.	(Sansonetti *et al.*, 2020)	Misinformation detection	Twitter, news (PolitiFact.com)	Precision, Accuracy	Only text based data is used
Domain Adversarial Neural Network (DANN)	(Joshi *et al.*, 2023)	Misinformation detection	MiSoVac dataset	F1 score, Precision, Accuracy and Recall,	Not specified how big or how varied the datasets were that used.
CNN, BiLSTM, CNN-BiLSTM models, and LSTM	(Yadav *et al.*, 2022)	Fake news detection	Fake and real News, allData dataset.	Accuracy, AUC-ROC values Precision, F1 measure, and Recall	Standard single large dataset not available

(*continued*)

Table 1. Continued

Approach/Technique Used	Ref	Purpose	Datasets	Evaluation Parameters	Research Gaps
Naive Bayes, Logistic Regression, Passive Aggressive Algorithm and Random Forest	(Sharma Uma *et al.*, 2021)	Fake news detection	LIAR dataset (PolitiFact.com)	F1 score, Confusion matrix, and Accuracy	Limited to one dataset
TF-IDF, LDA.	(Xu *et al.*, 2020)	Fake news detection	Articles on Facebook related to real news, fake news.	–	fewer articles are used
RNN with GRU, LSTM	(Wang *et al.*, 2021)	Misinformation detection	Instagram text, images	Accuracy	Other social platforms are not used
RNN, DNN with BERT, GRU, LSTM	(Biradar *et al.*, 2023)	Fake/real news classification	"CONSTRAINT shared task 2021" dataset	Accuracy, F1 score	limited dataset and the exclusion of code-mixed and native language statements
FNS-NS	(Raza & Ding, 2022)	Fake/real news classification	Fakeddit, NELA-GT-19 dataset	Precision	scarcity of labeled data for training
CNN and BERT	(Kaliyar *et al.*, 2021)	Fake news detection	Fake and real news articles	Accuracy, FNR, FPR	Dataset analysis, model Biases, hyper parameter exploration, generalizability.
PS, SVM, TF-DIF	(Akhtar *et al.*, 2023)	Fake news and disinformation detection	Online news articles data	–	Operational performance measures.
Random Forest, SVM, Naïve Bayes	(Khanam *et al.*, 2021)	Fake news detection	LIAR dataset, Twitter, Facebook	Accuracy	Poor Accuracy
Random Forest, SVM, Naïve Bayes, ANN, LSTM	(Dadgar & Ghatee, 2021)	Misinformation	Twitter	F1-Score	Generalizability of the Proposed system to different Languages.
CNN, LSTM, BERT	(Khan *et al.*, 2021)	Fake news detection	LIAR, fake & real news	Accuracy, F1 score, Recall	focused on political news only
RF, LR, KNN, DT and SVM.	(Fahad *et al.*, 2023)	Fake news, false information detection	Kaggle	Accuracy	feature selection has not been applied

The following are the different gaps identified from the above literature review are as follows:

- Real dataset not used to test the systems.
- Generalizability of the system to different languages is needed.
- System accuracy is not enough.
- Limited dataset volume.
- System works on specific datasets only.
- System need to work on different kinds of dataset and social platform.

So, there is a need of a system which can overcome the above limitations. There is a need to implement a more efficient system which will detect misinformation by using the advanced methods like deep learning, machine learning and on the basis of this categorize the misinformation like fake news, rumors, etc.

3 CONCLUSION

This review article offers in-depth understanding of numerous deep learning and machine learning methods that support the detection of false information on social media platforms, including tweets, rumors, articles, fake news, and much more. Each algorithm detects the misinformation by incorporating the accuracy of an algorithm, better techniques for the selection and extraction of features and these are used to obtain the better accuracy of algorithms. The outcome of this survey identifies the research gaps and give the directions for making a prediction model exclusively for fake news detections with more accuracy and generalizability.

REFERENCES

Abonizio, H. Q., De Morais, J. I., Tavares, G. M., & Barbon Junior, S. (2020). Language-Independent fake news detection: English, Portuguese, and Spanish mutual features. *Future Internet*, 12(5), 87. https://doi.org/10.3390/fi12050087

Agivale, V., Bhanushali, O., Lalwani, H., Singh, S., Khemani, B., & Malve, S. (2022). HealthCare MisInformation detection on social media. *2022 International Conference on Smart Generation Computing, Communication and Networking (SMART GENCON)*, 1–6. https://doi.org/10.1109/SMARTGENCON56628.2022.10083941

Akhtar, P., Ghouri, A. M., Khan, H. U. R., Amin Ul Haq, M., Awan, U., Zahoor, N., Khan, Z., & Ashraf, A. (2023). Detecting fake news and disinformation using artificial intelligence and machine learning to avoid supply chain disruptions. *Annals of Operations Research*, 327(2), 633–657. https://doi.org/10.1007/s10479-022-05015-5

Alnazzawi, N., Alsaedi, N., Alharbi, F., & Alaswad, N. (2022). Using social media to detect fake news information related to product marketing: The fakeads corpus. *Data*, 7(4), 44. https://doi.org/10.3390/data7040044

Althobaiti, M. J. (2022). BERT-based Approach to Arabic Hate Speech and Offensive Language Detection in Twitter: Exploiting Emojis and Sentiment Analysis. *International Journal of Advanced Computer Science and Applications, 13(5)*. https://doi.org/10.14569/IJACSA.2022.01305109

Balshetwar, S. V., Rs, A., & R, D. J. (2023). Fake news detection in social media based on sentiment analysis using classifier techniques. *Multimedia Tools and Applications*, 82(23), 35781–35811. https://doi.org/10.1007/s11042-023-14883-3

Biradar, S., Saumya, S., & Chauhan, A. (2023). Combating the infodemic: COVID-19 induced fake news recognition in social media networks. *Complex & Intelligent Systems*, 9(3), 2879–2891. https://doi.org/10.1007/s40747-022-00672-2

Buzea, M. C., Trausan-Matu, S., & Rebedea, T. (2022). Automatic fake news detection for Romanian online news. *Information*, 13(3), 151. https://doi.org/10.3390/info13030151

Dadgar, S., & Ghatee, M. (2021). Checkovid: A COVID-19 misinformation detection system on Twitter using network and content mining perspectives (arXiv:2107.09768). *arXiv*. http://arxiv.org/abs/2107.09768

De Oliveira, N. R., Pisa, P. S., Lopez, M. A., De Medeiros, D. S. V., & Mattos, D. M. F. (2021). Identifying fake news on social networks based on natural language processing: Trends and challenges. *Information*, 12(1), 38. https://doi.org/10.3390/info12010038

Fahad, N., Michael Goh, K. O., Hossen, Md. I., Shahriar Shopnil, K. M., Mitu, I. J., Alif, Md. A. H., & Tee, C. (2023). Stand up against bad intended news: An approach to detect fake news using machine learning. *Emerging Science Journal, 7(4)*, 1247–1259. https://doi.org/10.28991/ESJ-2023-07-04-015

Girgis, S., Amer, E., & Gadallah, M. (2018). Deep learning algorithms for detecting fake news in Online Text. *2018 13th International Conference on Computer Engineering and Systems (ICCES), 93–97*. https://doi.org/10.1109/ICCES.2018.8639198

Guo, Y., Ge, H., & Li, J. (2023). A two-branch multimodal fake news detection model based on multimodal bilinear pooling and attention mechanism. *Frontiers in Computer Science*, 5, 1159063. https://doi.org/10.3389/fcomp.2023.1159063

Hu, L., Wei, S., Zhao, Z., & Wu, B. (2022). Deep learning for fake news detection: A comprehensive survey. *AI Open*, 3, 133–155. https://doi.org/10.1016/j.aiopen.2022.09.001

Ilie, V.-I., Truica, C.-O., Apostol, E.-S., & Paschke, A. (2021). Context-Aware Misinformation Detection: A Benchmark of Deep Learning Architectures Using Word Embeddings. *IEEE Access*, 9, 162122–162146. https://doi.org/10.1109/ACCESS.2021.3132502

Islam, M. R., Liu, S., Wang, X., & Xu, G. (2020). Deep learning for misinformation detection on online social networks: A survey and new perspectives. *Social Network Analysis and Mining*, 10(1), 82. https://doi.org/10.1007/s13278-020-00696-x

Joshi, G., Srivastava, A., Yagnik, B., Hasan, M., Saiyed, Z., Gabralla, L. A., Abraham, A., Walambe, R., & Kotecha, K. (2023). Explainable Misinformation Detection Across Multiple Social Media Platforms. *IEEE Access*, 11, 23634–23646. https://doi.org/10.1109/ACCESS.2023.3251892

Kaliyar, R. K., Goswami, A., & Narang, P. (2021). FakeBERT: Fake news detection in social media with a BERT-based deep learning approach. *Multimedia Tools and Applications*, 80(8), 11765–11788. https://doi.org/10.1007/s11042-020-10183-2

Khan, J. Y., Khondaker, Md. T. I., Afroz, S., Uddin, G., & Iqbal, A. (2021). A benchmark study of machine learning models for online fake news detection. *Machine Learning with Applications*, 4, 100032. https://doi.org/10.1016/j.mlwa.2021.100032, Liar Dataset- https://www.cs.ucsb.edu/~william/data/liar_dataset.zip, Fake news dataset- https://www.kaggle.com/mrisdal/fake-news

Khanam, Z., Alwasel, B. N., Sirafi, H., & Rashid, M. (2021). Fake News Detection Using Machine Learning Approaches. *IOP Conference Series: Materials Science and Engineering*, *1099(1)*, 012040. https://doi.org/10.1088/1757-899X/1099/1/012040

Kovács, G., Alonso, P., & Saini, R. (2021). Challenges of Hate Speech Detection in Social Media: Data Scarcity, and Leveraging External Resources. *SN Computer Science*, *2(2)*, 95. https://doi.org/10.1007/s42979-021-00457-3, Dataset- https://hasocfire.github.io/hasoc/2019/ dataset.html.

Mahbub, S., Pardede, E., & Kayes, A. S. M. (2022). COVID-19 Rumor Detection Using Psycho-Linguistic Features. *IEEE Access*, 10, 117530–117543. https://doi.org/10.1109/ACCESS.2022.3220369, Dataset- https://github.com/MickeysClubhouse/COVID-19-Rumor-dataset

Nistor, A., & Zadobrischi, E. (2022). The Influence of Fake News on Social Media: Analysis and Verification of Web Content during the COVID-19 Pandemic by Advanced Machine Learning Methods and Natural Language Processing. *Sustainability*, 14(17), 10466. https://doi.org/10.3390/su141710466

Raza, S., & Ding, C. (2022). Fake news detection based on news content and social contexts: A transformer-based approach. *International Journal of Data Science and Analytics*, *13(4)*, 335–362. https://doi.org/10.1007/s41060-021-00302-z, Dataset- https://github.com/entitize/fakeddit.

Salini, Y., & Harikiran, J. (2023). Multiplicative Vector Fusion Model for Detecting Deepfake News in Social Media. *Applied Sciences*, 13(7), 4207. https://doi.org/10.3390/app13074207

Sansonetti, G., Gasparetti, F., D'aniello, G., & Micarelli, A. (2020). Unreliable Users Detection in Social Media: Deep Learning Techniques for Automatic Detection. *IEEE Access*, 8, 213154–213167. https://doi.org/10.1109/ACCESS.2020.3040604

Sharma Uma, Saran Sidarth, Patil Shankar M. (2021). Fake News Detection using Machine Learning Algorithms. *IJERT*, NTASU – 2020 (Volume 09 – Issue 03) (03). https://www.ijert.org/fake-news-detection-using-machine-learning-algorithms

Siddhesh Chalke & Deepak Mishra. (2023). Studying the Impact of Social Media Algorithms on the Spread of Misinformation and its Effects on Society. *International Journal of Advanced Research in Science, Communication and Technology*, 352–356. https://doi.org/10.48175/IJARSCT-13054

Sule, S., DaCosta, M. C., DeCou, E., Gilson, C., Wallace, K., & Goff, S. L. (2023). Communication of COVID-19 Misinformation on Social Media by Physicians in the US. *JAMA Network Open*, 6(8), e2328928. https://doi.org/10.1001/jamanetworkopen.2023.28928

Wang, Z., Yin, Z., & Argyris, Y. A. (2021). Detecting Medical Misinformation on Social Media Using Multimodal Deep Learning. *IEEE Journal of Biomedical and Health Informatics*, 25(6), 2193–2203. https://doi.org/10.1109/JBHI.2020.3037027

Xu, K., Wang, F., Wang, H., & Yang, B. (2020). Detecting fake news over online social media via domain reputations and content understanding. *Tsinghua Science and Technology*, 25(1), 20–27. https://doi.org/10.26599/TST.2018.9010139

Yadav, A. K., Kumar, S., Kumar, D., Kumar, L., Kumar, K., Maurya, S. K., Kumar, M., & Yadav, D. (2022). Fake News Detection using Hybrid Deep Learning Method [Preprint]. https://doi.org/10.36227/techrxiv.19689844.v1

Zeng, H., Yue, Z., Kou, Z., Shang, L., Zhang, Y., & Wang, D. (2022). Unsupervised Domain Adaptation for COVID-19 Information Service with Contrastive Adversarial Domain Mixup (arXiv: 2210.03250). arXiv. http://arxiv.org/abs/2210.03250

Computational Methods in Science and Technology Volume 1 – Kaur et al. (Eds)
 ISBN: 978-1-032-81760-6

Deep insight of medical images segmentation and classification by deep learning

Amit Kumar Arora
PhD Research Scholar, CSE, IKGPTU, Punjab, India

Amanpreet Kaur
Professor, Chandigarh Engineering College-CGC (Landran) Mohali, India

ABSTRACT: Image segmentation is a crucial process in picture analysis and processing, affecting object presentation, description, and classification. It is essential for identifying, characterizing, and visualizing regions of interest in medical images. Manual segmentation by radiologists is arduous and time-consuming, and its accuracy is often compromised due to the growing number of medical imaging modalities and the volume of images. Therefore, it is crucial to evaluate existing picture segmentation approaches using automated algorithms that offer high accuracy and minimize user intervention, especially in medical photographs. Accurately demarcate and extract the anatomical framework or ROI for independent visualization. This paper will embark on the importance of image segmentation in decision-making information extraction, especially in this technological era. Some of the main and advanced application areas with recent research breakthroughs that have proved very significant in the field of medical imaging will be discussed.

Keywords: Medical Imaging, Image Segmentation, Manual segmentation

1 INTRODUCTION

Nonuniform intensity, limited pixel size, and noise are the three primary obstacles in medical picture segmentation. The presence of noise makes pixel categorization less reliable, but intensity nonuniformity permits subtle but noticeable changes in brightness over the whole picture. Furthermore, keep in mind that pictures have small pixels and are subject to partial volume averaging. This happens when different types of tissue are present in the volumes of certain pixels. The limits of conventional medical segmentation approaches have been addressed in academic literature via the recommendation of various methodologies and algorithms. But the choice depends on the image problem's characteristics and type. The literature has extensively examined the latest developments in image segmentation algorithms [1,5–10]. The focus of these articles lies in the categorization of techniques employed for the processing of pixel data and voxel data, as well as their respective applications in the fields of diagnosis, treatment planning, and follow-up research. Nevertheless, despite the use of several solutions, a notable drawback that persists is the issue of processing speed. This research offers a thorough analysis of medical picture segmentation algorithms employed in the fields of lung, skin, and breast cancers [11–15].

1.1 *Analysis*

Table 2 shows several techniques categorized by two kinds of segmentation: skin cancer and lung cancer. When it comes to skin cancer, SEGNET has the best level of accuracy

 DOI: 10.1201/9781003501244-7

Table 1. Summary of recent works.

Author	Year	MIS Technique/ Method	Description	Findings	Limitation
Shetty *et al.* [3]	2023	Automatic Lung Segmentation/ Deformable Models and Deep Learning Methods	The study introduces a novel deep learning algorithm for autonomous cancer segmentation and classification, enhancing lung Dice similarity coefficient through a combination of a Markov model and 3D-CNN.	The research improves the mean Dice similarity coefficient in the context of lung cancer detection.	Manual segmentation is a time-consuming and tedious process.
Peng *et al.* [17]	2023	Breast Cancer Segmentation/ Semi-supervised fuzzy clustering algorithm	The paper explores the use of a semi-supervised fuzzy clustering technique, SFCM_V, for segmenting lesions in mammography molybdenum target pictures, enhancing overall clustering performance.	The experimental results demonstrate that SFCM_V(cluster) has superior performance in terms of segmentation accuracy compared to currently available clustering methods, for both smaller and larger lesion locations, encompassing tumors and calcified sites.	Traditional unsupervised clustering approaches face challenges in image segmentation in mammography pictures due to diverse lesions, low contrast between pixels, and diverse sizes and shapes.

Table 2. Methods based on two segmentation types.

Segmentation type	METHODS	ACCURACY	IOU
SKIN CANCER	SVM [15]	90	93
	RANDOMFOREST [16]	88	90
	U-NET [17]	94	98
	SEGNET [17]	95	97
	CNN [18]	90	90
LUNG CANCER	SVM	92	92
	RANDOMFOREST	88	90
	U-NET	80	94
	SEGNET	94	95
	CNN	88	90

(95), followed by U-NET (94), SVM, and CNN (90), while the RF (88) technique yields the lowest accuracy. However, the intersection of union (IOU) is greatest in U-NET with a score of 98, followed by SEGNET (97), SVM (93), and RF and CNN with scores of 90. For the detection of lung cancer, SEGNET demonstrates the best level of accuracy at 94%, followed by SVM at 92%, RF and CNN at 88%, and the U-NET approach with the lowest accuracy of 80%. Conversely, the IOU is the highest in SEGNET with a score of 95, followed by U-NET with a score of 93. RF and CNN have scores of 90, respectively.

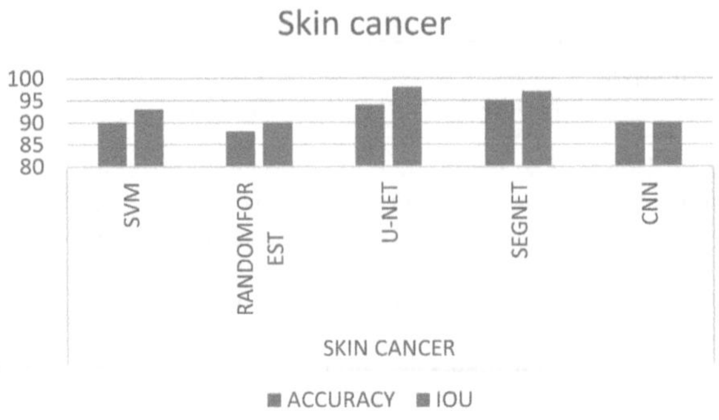

Figure 1. Skin cancer-based accuracy and IoU.

As depicted in the figure above, the accuracy of SEGNET is highest and lowest in RF concerning skin cancer, while U-NET demonstrates the second-highest accuracy. SVM and CNN exhibit equivalent levels of accuracy. Conversely, the IOU of U-NET is best in RF and least in CNN.

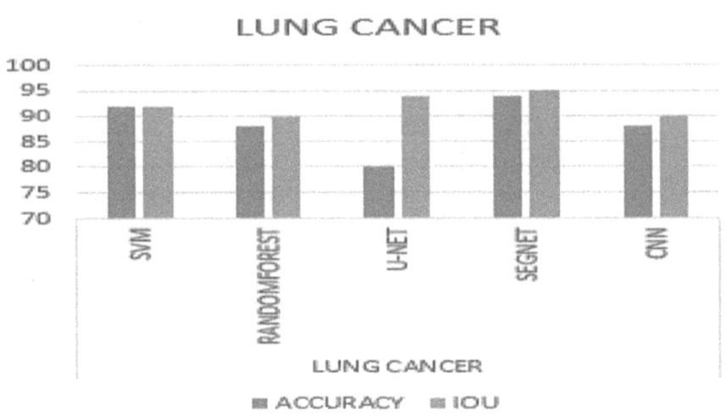

Figure 2. Lung cancer-based accuracy and IoU.

As depicted in the figure above, the accuracy of SEGNET is highest and lowest in U-NET concerning lung cancer, SVM demonstrates the second-highest accuracy, while CNN and RF exhibit equivalent levels of accuracy. Conversely, the IOU of SEGNET is the best and least in CNN and RF.

2 CONCLUSION

This paper offers a thorough analysis of several aspects related to image segmentation techniques, including both established and modern approaches, their methods of implementation, and the specific fields in which they are utilized. The matter of segmentation accuracy continues to be a prominent concern for patients encountering challenges, such as tumors. Each segmentation method discussed in this context has its unique set of advantages and shortcomings, which makes it unsuitable for establishing a specific baseline for comparison. The article presents a thorough classification of the segmentation techniques utilized in the study of medical images. This work can be considered a significant resource for the purpose of referencing. The segmentation of complex regions, like the brain, presents difficulties in terms of intricacy and temporal demands. Nevertheless, recent advancements in segmentation algorithms have exhibited potential in tackling these challenges. The provided material in this study presents promising results in the domain of magnetic resonance (MR) picture segmentation. The present need entails the development of algorithms capable of functioning without any prior information of the image, including features such as color, intensity, homogeneity, texture, and other similar characteristics.

REFERENCES

[1] Chaudhury, S., Sankaran, A., Khan, S., Sau, A., & Sammy, (2022). Machine learning methods based on image processing and segmentation that effectively diagnose breast cancer. *Statistical and Computational Approaches to Health Care*, 2022.

[2] He, *et al.* (2022). *An Method for Lung Cancer Image Segmentation Using a Neural Network Model.* Volume 39, Issue 3, page e12822.

[3] Shetty, J., Jayadevappa, & Achar. (2023). A Deep Learning and Deformable Model Approach to Automatic Lung Cancer Segmentation. In *Proceedings of the 2023 IEEE CONECCT Conference on Computing, Electronics, and communication Technologies* (pp. 1–5). IEEE.

[4] GALVÁN-Tejada, J. I., Escobar, P. H., Galvan-Tejada, C. E., Gamboa-Rosales, H., de la Rosa Gomez, A., & Celaya-Padilla, J. M. (2022). *Genetic Algorithms and Automated Segmentation for the Detection of Breast Cancer. Diagnostics*, 12(12), Report number: 3099.

[5] Yuvaraj, & Ragupathy. (Year Not Provided). Mass segmentation and classification using hybrid active contour mammography. *Journal of Computer Science and Engineering.*

[6] Nawaz, *et al.* (Year Not Provided). A framework for improving DenseNet77 and UNET convolutional neural networks for melanoma segmentation. *Journal of Imaging Systems and Technology, International*, 32(6), 2137–2153.

[7] Kaur, GholamHosseini, Sinha, & Lindén. (2022). Dermoscopy pictures of skin cancer may be automatically segmented using atrous convolutional neural networks.

[8] Garg, S., & Jindal, B. (2021). Segmenting skin lesions with the use of k-means and an improved FIRE FLY ALGORITHM. *Multimedia Tools and Applications*, 80, 7397–7410.

[9] Kaur, A., & Kaur, B. (2022). Load balancing optimization based on hybrid Heuristic-Metaheuristic techniques in cloud environment. *Journal of King Saud University – Computer and Information Sciences*, 34(3), 813–824. https://doi.org/10.1016/j.jksuci.2019.02.010

[10] Kaur, S., Kamboj, S., Kaur, B., & Hrisheekesha, P. N. (2021). Hybrid approach for virtual machine optimization using BAT algorithm in cloud. In *2021 2nd International Conference on Computational Methods in Science & Technology (ICCMST)* (pp. 57–61). https://doi.org/10.1109/ICCMST54943.2021.00023

[11] Rana, P., Walia, K., & Kaur, A. (2021). Challenges in conglomerating Fog computing with IOT for building Smart City. In *2021 2nd International Conference on Computational Methods in Science & Technology (ICCMST)* (pp. 35–40). https://doi.org/10.1109/ICCMST54943.2021.00019

[12] Kaur, A., Kaur, B., Singh, P., Devgan, M. S., & Toor, H. K. (2020). Load balancing optimization based on deep learning approach in cloud environment. *International Journal of Information Technology and Computer Science (MecsPress)*, 12(3), 8–18. https://doi.org/10.5815/ijitcs.2020.03.02

[13] Kaur, A., Kaur, B., & Singh, D. (2018). Meta-heuristics based load balancing optimization in cloud environment on underflow and overflow conditions. *Journal of Information Technology Research (JITR)*, 11(4), 155–172. https://doi.org/10.4018/JITR.2018100110

[14] Kaur, A., Kaur, B., & Singh, D. (2019). Meta-heuristicbased framework for workflow load balancing in cloud environment. *International Journal of Information Technology*, 11(1), 119–125. https://doi.org/10.1007/s41870-018-0231-z

[15] Kumar, A., & Alam, B. (2019). Energy harvesting earliest deadline first scheduling algorithm for increasing lifetime of real time systems. *International Journal of Electrical and Computer Engineering*, 9 (1), 539.

[16] Kumar, A. & Alam, B. (2018). Task scheduling in real time systems with energy harvesting and energy minimization. *Journal of Computer Science*, 14(8), 1126–1133.

[17] Kumar, A., & Alam, B. (2014, February). Real time scheduling algorithm for fault tolerant and energy minimization. In *2014 International Conference on Issues and Challenges in Intelligent Computing Techniques (ICICT)* (pp. 356–360). IEEE.

[18] Ghomi, E. J., Rahmani, A., & Qader, N. (2017). Load-balancing algorithms in cloud computing: A survey. *Journal of Network and Computer Applications*, 88, 50–71.

Computational Methods in Science and Technology Volume 1 – Kaur et al. (Eds)

Human intrusion detection by optimize dragon deep belief network

Garima Bohra, C.K. Jha & Neelam Sharma
Department of Computer Science, Banasthali Vidyapith, Rajasthan, India

ABSTRACT: Optimal weight selection in a novel model of Deep Belief Neural Network (DBBN) is employed in this study to try to identify individuals in the video using a Dragonfly Algorithm (DA). With a learning rate of 0.001, the model's Multilayer Perceptron (MLP) and Restricted Boltzmann Machine (RBM) layers were trained using Stochastic Gradient Descent (SGD) for every 50 epochs. Clean and representative data is essential for effective training; hence this procedure includes cleaning, preparing, segmenting, framing, preprocessing, and supplementing the data. With an accuracy rate of 91% on the UCF dataset, the DBBN model surpassed its rivals, the two-stream CNNs and the LSTM ensemble. An emphasis on systematically improving the model's performance via tweaking its hyperparameters (such as activation functions, batch size, and epochs) will be detailed. Further development is needed to improve DBN models for security and surveillance activities, as demonstrated in this study. The primary areas for improvement include dealing with computational complexity and adapting to varied real-world settings.

Keywords: Deep Belief Neural Network, Optimal Weight Selection, Convolutional Neural Network, Long Term Short Memory

1 INTRODUCTION

Enabling the communication between humans and machines in their surrounding environment requires the understanding and prediction of human movement. Humans' intrinsic capacity to anticipate the movements of other humans is crucial for a wide variety of social interactions, including getting about in crowds, stopping offensive players in a game, and shaking hands. Intelligent robots must be able to anticipate and adapt to human actions, coordinate their stances and foresee their future movements to effectively interact with people. Nevertheless, putting this skill to use as a machine's non-talent capacity is a complex undertaking. Motion prediction has also found widespread use in areas such as autonomous driving, intelligent robots [2,3], human-robot collaboration [1], and multimedia applications [1].

2 PROPOSED APPROACH

Deep belief neural network (DBBN), model is trained in fully supervised manner. Each layer present in the architecture has some weight and bias which is initialized by some random values. Gradient is computed and back propagated. Cross entropy is used as a loss function. It is used to measure the difference between real and probability distribution for a given random variable. In this work, cross entropy loss function is used to measure the error between a true and a predicted value. Aim is to minimize the loss function, therefore, during the training process, the value of weight/parameter is also updated. Stochastic Gradient

 DOI: 10.1201/9781003501244-8

Descent (SGD) is selected as an optimizer. It is a stochastic optimization algorithm in which weights are updated for each training sample or a small subset of data. For the training phase, batch size is set to 4 and epochs are set to 50. Furthermore, a small learning rate of 0.001 is set to enhance the learning and fitting ability. Table 1 provides the computed results for proposed approach.

3 INITIALIZATION OF WEIGHTS AND BIASES IN RBM1

The weights in RBM1 are nominated based on the probabilities allocated to the neurons. Probability at each hidden neuron in RBM1 is,

$$p\left(d_p^1 = 1|\alpha\right) = \sigma\left[h_p^1 + \sum_q a_q^1 w_{qp}^1\right] \tag{1}$$

Where

α = is the visible vector
σ is the activation function, and h_p^1 is the bias of pth hidden neuron in RBM1.

The biases of the hidden layer are denoted as g_q^1 and h_p^1. Based on weights and minimum energy function, neurons weight function has been selected.

Table 1. Deep belief neural network (DBBN) results.

Dataset	Epochs	Learning rate	Optimizer	Max. Validation accuracy (%)	Min. Validation Loss (%)	Activation Function	No. of output classes
UCF	100	0.002	SGD	93	0.23	RELU	4

Table 2 compares the accuracy. When compared with existing literature, the suggested method performs well. We compare a deep belief neural network (DBBN) with a method from 2017 that used an ensemble of many long short-term memory (LSTM) neurons. On the opportunity dataset, this ensemble technique achieved 72% accuracy, on the PAMAP2 dataset, 85% accuracy, and on the Skoda dataset, 90% accuracy. On top of that, we evaluate it with the alternative approach that was suggested in 2021. Two separate CNN streams were successfully addressed using this method. It takes input from two sources and processes it further. An accuracy of 80.5% has been calculated and verified. Deep belief neural networks (DBBNs) are therefore suggested as a means to enhance the accuracy rate; DBBNs demonstrated 91% accuracy on UCF-101.

Table 2. Accuracy comparison of the proposed method with existing methods.

Evaluation Metric	Method 1 (LSTM Ensemble) (Ordonez *et al.*)	Method 2 (Two- stream convolutional networks (Luo *et al.*)	Proposed Method (Deep belief neural network (DBBN))
Year	2021	2022	**2023**
Accuracy	72%	80.5% (own dataset)	**91% (UCF)**

Existing evaluation methodologies are used to test the proposed Deep Belief Neural Network (DBBN) model. The outcomes are evaluated for fairness and consistency using the F1-score. It strikes a balance between recall and accuracy. The outcome's value is between zero and one. In Table 2, we can see that a method using an LSTM ensemble yields an

F1-score of 0.61, but a Deep Belief Neural Network (DBBN) produces an F1-score of 0.89, indicating that it performs better in the classification task.Confusion matrix is a technique which is used to summarize the performance of the classification algorithm.

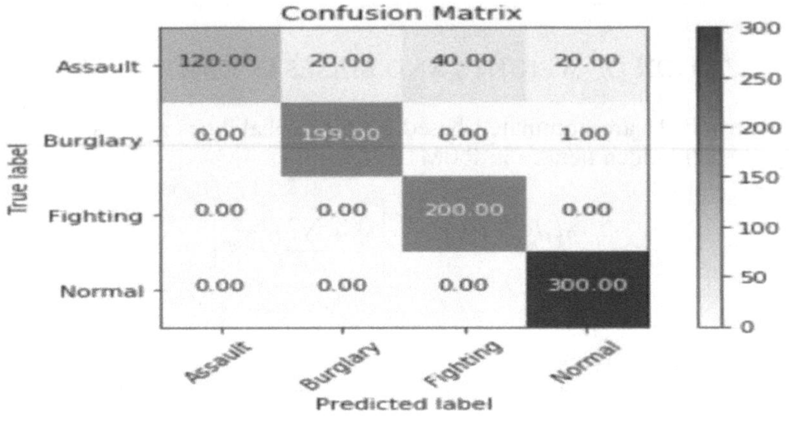

Figure 1. Confusion matrix of action recognition using deep belief neural network (DBBN).

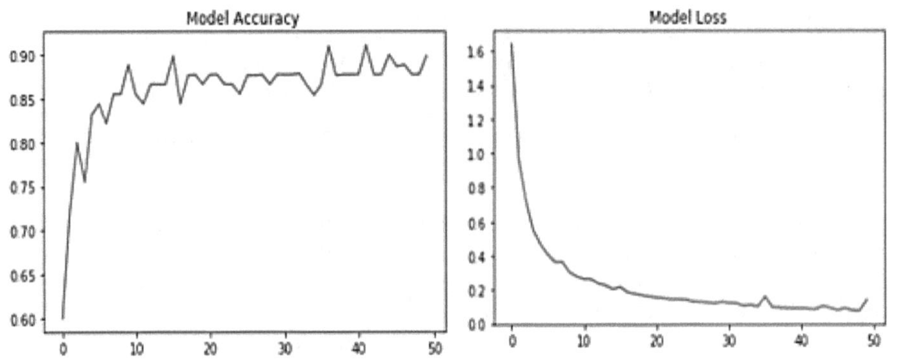

Figure 2. Model accuracy and loss.

4 HYPERPARAMETERS TUNING AND ITS MAJOR CONCERNS

The hyperparameters that have been fine-tuned in this study are the activation function, optimizer, learning rate, batch size, and epochs. Every layer, from input to output, makes use of the activation function. Each layer in a network is responsible for processing data in a certain order: input, hidden, and output. As you go through the layers (according to the weighting criteria), the input values are constantly changing. There are many different activation functions, including sigmoid, tanh, ReLU, and softmax. The study makes use of the Softmax activation function. Each node in the output layer receives a probability distribution from the SoftMax activation function. The cumulative probability remains within the range of 0 and 1 when this function changes the input value to output. When it comes to training, optimizers are king as well. To reduce the loss function and achieve the best solution, the parameters are changed during training in the suggested work.

5 CONCLUSION

A remarkable development in computer vision and machine learning, it is the Deep Belief Neural Network (DBBN), developed for the very purpose of identification of human activities in video recordings. The DBBN model is trained through the Stochastic Gradient Descent (SGD) at the learning rate of 0.001, trained up to 50 epochs by unifying Multilayer Perceptron (MLP) and Restricted Boltzmann Machine (RBM) layers, and optimized using the Dragonfly Algorithm (DA). This is very important in improving the performance of the model. It includes data preparation, data preprocessing, and data augmentation; all these are contained in the strategy for the model to train on clean and representative data. UCF dataset with 91% accuracy rate has outperformed other approaches like LSTM ensembles and two-stream CNNs using the DBBN model. Achieving the ideal performance in-between—the two balances of computing efficiency with accuracy—systematically leans with hyperparameters,

REFERENCES

[1] Abdel-zaher M. and Eldeib A. M., (2016). Breast cancer classification using deep belief networks, *Expert Systems with Applications*, vol. 46, no. November, pp. 139–144.

[2] U. Ahsan, R. Madhok, and I. Essa, (2019). Video jigsaw: Unsupervised learning of spatiotemporal context for video action recognition, in *Proceedings – 2019 IEEE Winter Conference on Applications of Computer Vision, WACV* 2019, pp. 179–189.

[3] Ai D. *et al.*, (2016). Augmented reality based real-time subcutaneous vein imaging system, *Biomedical Optics*, vol. 7, no. 7, pp. 39–49.

[4] Akilan T., Wu Q. J., and Zhang H., (2018). Effect of fusing features from multiple DCNN architectures in image classification, *IET Image Processing*, vol. 12, no. 7, pp. 1102–1110.

[5] Krizhevsky A., Sutskever I., and Hinton G. E., (2012). ImageNet Classification with Deep Convolutional Neural Networks, in *Advances in Neural Information Processing Systems* 25, NIPS 2012, pp. 1–9.

[6] Alom M. Z. *et al.*, (2019). A state-of-the-art survey on deep learning theory and architectures, *Electronics (Switzerland)*, vol. 8, no. 3, pp. 1–67.

[7] Angelov P. P., Gu X., Gutierrez G., Iglesias J. A., and Sanchis A., (2016-October). Autonomous Data Density based clustering method, in *Proceedings of the International Joint Conference on Neural Networks*, vol. 2016, pp. 2405–2413.

[8] Avci A., Bosch S., Marin-Perianu M., Marin-Perianu R., and Havinga P., (2010). Activity recognition using inertial sensing for healthcare, wellbeing and sports applications: A survey, in *23th International Conference on Architecture of Computing Systems 2010*, ARCS 2010 – Workshop Proceedings, pp. 167–176.

[9] Bagautdinov T. *et al.*, (2017-January). Social scene understanding: End-to-end multi-person action localization and collective activity recognition, in *Proceedings – 30th IEEE Conference on Computer Vision and Pattern Recognition*, CVPR 2017, vol. 2017, pp. 3425–3434.

[10] Barandiaran I. *et al.*, (May 2014). A new evaluation framework and image dataset for keypoint extraction and feature descriptor matching, in VISAPP 2013 – *Proceedings of the International Conference on Computer Vision Theory and Applications*, vol. 1, pp. 252–257.

[11] Fernando B. and Gould S., (2015). Learning End-to-end Video Classification with RankPooling, in *Proceedings of the 33rd International Conference on Machine Learning*, vol. 48, pp. 1–10.

[12] Bellavia F., Tegolo D., and Valenti C., (2011). Improving Harris corner selection strategy, *IET Computer Vision*, vol. 5, no. 2, pp. 87–96.

[13] Berrada H. *et al.*, Validation of a confirmatory method for the determination of macrolides in liver and kidney animal tissues in accordance with the European Union regulation 2002/657/EC, *Journal of Chromatography*, vol. 1157, no. 12, pp. 281–288, 2007.

[14] Blank M. *et al.*, (November 2005). Actions as space-time shapes, in *Proceedings of the IEEE International Conference on Computer Vision*, vol. II, pp. 1395–1402.

[15] Ju, Q. (2015). An improved harris corner detection algorithm. *Journal of Beijing Institute of Graphic Communication.*

[16] Bellavia, F., Tegolo, D., & Valenti, C. (2011). Improving harris corner selection strategy. *Iet Computer Vision*, 5(2), 87–96.

[17] Shunqing, X., Weihong, Z., & Wei, X. (2012). Optimization of harris corner detection algorithm. *Journal of Yunnan University of Nationalities.*

[18] Peng, Z. (2010). Improved harris-based corner detection algorithm. *Computer Engineering and Applications.*

Computational Methods in Science and Technology Volume 1 – Kaur et al. (Eds)
 ISBN: 978-1-032-81760-6

Human activity recognition using residual deep network with extreme learning

Garima Bohra, C.K. Jha & Neelam Sharma
Department of Computer Science, Banasthali Vidyapith, Rajasthan, India

ABSTRACT: In recent years, human activity recognition (HAR) has emerged as an important topic of research due to the multiple applications it has found in various fields, including fitness and strength surveillance, residential care, biometric verification, aged care, and many others. Sensors are a great alternative for activity recognition because of their widespread use. The bulk of wearable sensors, including accelerometer, gyroscope, camera, GPS, microphone, compass, and others, are built into the newest smart devices. Such sensors measure different properties of an object, are simple to use, and are inexpensive. Researchers using machine learning (ML) now have new potential to accurately recognize human behaviours thanks to the usage of sensors in the HAR sector. Deep learning (DL) is gaining popularity among HAR researchers because it outperforms standard machine learning (ML) approaches. In this paper, hybrid features of mechanisms and learning are used by employing deep learning and machine learning; deep learning uses RESNET-50 and machine learning uses XG boosting. In experiments, the proposed approach was validated by a 10-cross comparison with the state-of-the-art approach. In experiment also classified human activities and intruders.

Keywords: Human Activity Recognition, Deep learning, Machine Learning

1 INTRODUCTION

Smart devices and sensor-enabled devices have come a long way in recent years. Due to their low price and simplicity, these gadgets are ideal for human activity identification (HAR). With potential uses in areas as diverse as sports, smart homes, emergency services, health monitoring, and lifelogging [1–3], HAR seems to be a rapidly growing area of study. Formerly, action recognition tasks were effectively completed using video records, however video-based technologies are location-dependent and interfere with one's privacy. Because of this, sensor-based action recognition is getting popular [3]. In sensor-based HAR technologies [4–6], minimal cost, sensing devices that minimize interference with everyday routines are utilized. Another new breakthrough in HAR is the usage of smartphones, which are packed with several sensors. The inconspicuous feature of handsets makes them suitable for HAR. Activity recognition systems commonly identify actions as class labels using classification algorithms [7–9].

2 CONTRIBUTIONS OF THE RESEARCH

- Proposed a block wise resnet 50 network that generates features stack by stack. It reduces the amount of overlap in high-quality datasets such as UCF-101.

 DOI: 10.1201/9781003501244-9

- In our approach, global mapping of features by RELU layer reduces class imbalance of features and provides context for classes.
- The main contribution of this paper is learning by boosting, which is optimized using a game theory approach.

3 MATERIALS AND METHODS

3.1 *Materials*

UCF101 is a collection of realistic action videos culled from YouTube and organized into 101 action categories. This data extends the UCF50 data set, which contains 50 action categories. With 13320 videos from 101 action types, UCF101 provides the greatest range of actions, and with significant variation in camera calibration, object appearance and posture, object scale, perspective, congested scenery, lighting condition, etc., it has been the most difficult data set to date. The HMDB-51 classification, on the other hand, is a huge collection of actual videos from different sources, such as blockbusters and web videos. The dataset consists of 6,766 video segments from 51 action segments, with at least 101 clips per category [10–12].

3.2 *Proposed methodology*

Algorithm 1: Input Activity datasets images Output: classified actions

(1) Input images and labels $I \leftarrow image$
(2) Pre-process and input in Resnet50
(3) Feature mapping by $L(\theta) = T^{-1}\sum_{t=1}^{T} L_t(\theta) - (1)$ w.r.t parameter θ. We first find a simple resnet sigmoid layer
(4) Mapped fetures global mapped by Relu layer F $<$——$l_t(z) = l(z, y_t)$ and $L_t(\theta) = l_t(\theta x_t) - (2)$
(5) Features learn by equation 3 and 4
(6) Make classifier model and test
(7) Analysis Performance metrics

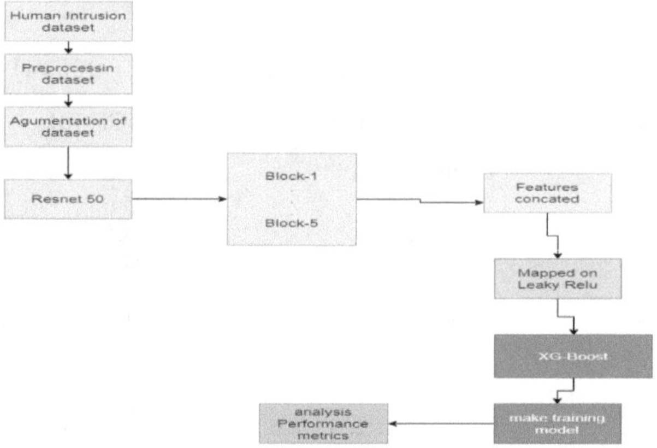

Figure 1. Proposed approach Flow chart.

4 EXPERIMENT AND ANALYSIS

Table 1 represents the proposed and existing approaches comparison on HMDB- 51 dataset based on given parameters. In terms of accuracy, precision, recall, and F1-score, the proposed approach outperforms the other approaches, hence providing the highest quality to complete the operation. In terms of accuracy, the RES50-ensemble (76.647) outperforms the other methods followed by LSTM-RNN (72.33), RF (70.12), and CNN (67.4). In case of precision also, the RES50-ensemble method (75.877) outperforms the other methods followed by CNN (73.33), RF (69.44), and LSTM-RNN (67.4). The RES50-ensemble technique outperformed with a 73.4713 recall value, followed by RF, CNN, and LSTM-RNN. Yet, in terms of F1-score, the RES50-ensemble method (74.484) achieved the most accurate F1 score followed by CNN, LSTM-RNN, and RF, respectively.

Table 1. Proposed and Existing approaches comparison on HMDB-51 dataset.

APPROACHES	ACCURACY	PRECISION	RECALL	F_SCORE
RES50-ensemble	**76.647**	**75.877**	**73.4713**	**74.484**
CNN	67.4	73.33	60.33	67.44
LSTM-RNN	72.33	67.4	56.43	64.44
RANDOM FOREST	70.12	69.44	70.55	63.44

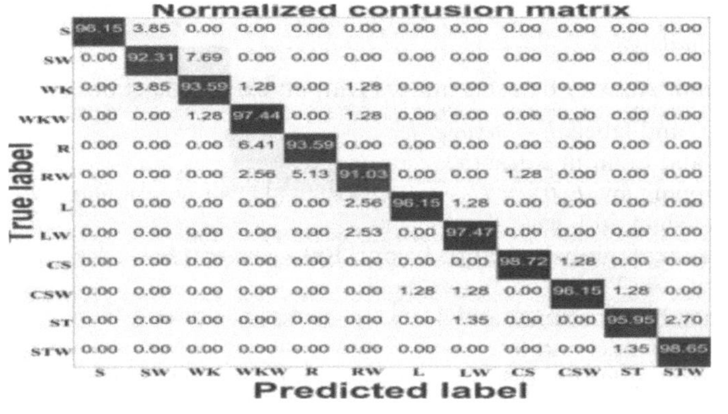

Figure 2. Confusion metrics.

Table 2. Comparison of proposed and state of art work on both UCF-101 and HMDB-51.

State of Art Approaches	parameters (M)	UCF-101	HMDB-51
FSTFN [30]	39	92.4	69.3
ActionS-ST-VLAD [1]	NA	95.6	71.4
TDN [2]	52.53	94.56	72.34
Multi-Domain [3]	32.02	94.8	71.57
TCLR [13]	45	82.4	52.9
DB-LSTM [14]	NA	91.21	87.64
HAR-Depth [15]	NA	92.97	62.4
3D-CNN [16]	30.12	95.68	69.74
Proposed	**27.56**	**96.68**	**76.647**

Table 2 compares recommended parameters to state-of-the-art work on the UCF-101 and HMDB-51 datasets [17,18]. Among all approaches, the proposed approach did well when compared to others. For example, in the proposed model, with M = 27.56, the UCF-101 fared well when compared to the HMDB-51 with 96.68. As compared to the HMDB-51, the FSTFN method with M = 39 and the UCF-101 (92.4) worked well (69.3). In the case of ActionS-ST-VLAD, no parameter is used, however UCF-101 (95.6) outperforms HMDB-51

(71.4). The TDN method with M = 52.53, the UCF-101 (94.56) performed well as compared to the HMDB-51 (72.34). (72.34). As compared to the HMDB-51, the Multi-Domain with M = 32.02, the UCF-101 (94.8) performed well. When compared to the HMDB-51, the TCLR with M = 45, the UCF-101 (82.4) worked well. Similarly, in the case of DB-LSTM, the UCF-101 (91.21) outperformed the HMDB-51 (87.64). Similarly, in terms of HAR-Depth, the UCF-101 (92.97) outperformed the HMDB-51 (62.4). Finally, the UCF-101 performed well using the 3D-CNN technique.

5 CONCLUSION

In this paper, a new stacked-based RESNET 50 is optimized by RELU mapping with an XG-Boost classifier. The method uses a CNN to extract basic deep features and then mines discriminative features between actions using the proposed ResNet-50-based attention model model. To improve features and objects overlapping by layers blocks of Resnet 50, and if overlapping, it will be ignored by the global RELU Attention mechanism. In the experiment, validate the proposed approach by doing different experiments on the UCF-101 and HMDB-51 datasets. Also compare with state-of-the-art approaches from 2018 to 2023. We get a significant improvement in accuracy of to 5%. It shows-propose approach improves the overlapping of features in the context of classes.

REFERENCES

[1] Swaminathan, A. (2022). Comparative analysis of sensor-based human activity recognition using artificial intelligence. In *Computational Intelligence in Data Science: 5th IFIP TC 12 International Conference, ICCIDS 2022, Virtual Event* (pp. 1–17). Springer International Publishing.

[2] Gu, F., Chung, M., Chignell, M.H., Valaee, S., Zhou, B., & Liu, X. (2021). A survey on deep learning for human activity recognition. *ACM Computing Surveys (CSUR)*, 54(8), 1–34.

[3] Rahman, M., & Das, T. (2021). Human activity recognition using deep learning-based approach. *In Communication and Intelligent Systems: Proceedings of ICCIS 2020* (pp. 813–830). Springer.

[4] Ravi, D., Wong, C., Lo, B., & Yang, G.Z. (2016). Deep learning for human activity recognition: A resource efficient implementation on low-power devices. In *2016 IEEE 13th International Conference on Wearable and Implantable Body Sensor Networks (BSN)* (pp. 71–76).

[5] Aktter, M., Hossain, M., Sarker, S., Abadin, A., & Hasan, M.A.R. (2022). Performance analysis of deep learning based human activity recognition methods. *Journal of Applied Science & Process Engineering*, 9(2), 1197–1208.

[6] Mutegeki, R., & Han, D.S. (2020). A CNN-LSTM approach to human activity recognition. *2020 International Conference on Artificial Intelligence in Information and Communication (ICAIIC)* (pp. 362–366).

[7] Merikapudi, S., Murthy, S.V., & Gandhi, R.V. (2022). Domain human recognition techniques using deep learning. *International Journal of Advanced Computer Science and Applications*, (6), 13–13.

[8] Browne, D., Giering, M., & Prestwich, S.D. (2019). Deep learning human activity recognition. *CEUR Workshop Proceedings*, 2563, 76–87.

[9] Sansano, E., & Belmonte Montoliu, R., & Fernandez, O. (2020). A study of deep neural networks for human activity recognition. *Computational Intelligence*, 36(3), 1113–1139.

[10] Ata, M., Francies, M., & Mohamed, M.A. (2022). A robust optimized convolutional neural network model for human activity recognition using sensing devices. *Concurrency and Computation: Practice and Experience*, 34(17), 6964–6964.

[11] Kushwaha, A., Srivastava, P., & Khare, A. (2023). Human activity recognition based on integration of multilayer information of convolutional neural network architecture. *Concurrency and Computation: Practice and Experience*, 35(5), 7571–7571.

[12] Zhu, Y., Wang, M., Yin, X., Zhang, J., Meijering, E., & Hu, J. (2023). Deep learning in diverse intelligent sensor based systems. *Sensors*, 23(1), 62–62.

[13] Serpush, F., Menhaj, M., Masoumi, B.B., & Karasfi, B. (2022). Wearable sensor-based human activity recognition in the smart healthcare system. *Computational Intelligence and Neuroscience.*

[14] Fu, B., Damer, N., Kirchbuchner, F., & Kuijper, A. (2020). Sensing Technology for Human Activity Recognition: A Comprehensive Survey. *IEEE Access*, 8, 83791–83820.

[15] Shintani, K., Rizk, H., & Yamaguchi, H. (2023). Eco-Friendly Sensing for Human Activity Recognition. *Adjunct Proceedings of the 2023 ACM International Joint Conference on Pervasive and Ubiquitous Computing & the 2023 ACM International Symposium on Wearable Computing.*

[16] Khan, M. B., Yang, X., Ren, A., Mohammed Al-Hababi, M. A., Zhao, N., Guan, L., Fan, D., & Shah, S. A. (2019). Design of Software Defined Radios Based Platform for Activity Recognition. *IEEE Access*, 7, 31083–31088.

[17] Zhang, M., & Sawchuk, A. (2013). Human Daily Activity Recognition With Sparse Representation Using Wearable Sensors. *IEEE Journal of Biomedical and Health Informatics*, 17, 553–560.

[18] Cheng, L., Li, Y., & Guan, Y. (2017). Human activity recognition based on compressed sensing. *2017 IEEE 7th Annual Computing and Communication Workshop and Conference (CCWC)*, 1–7.

Computational Methods in Science and Technology Volume 1 – Kaur et al. (Eds)

Enhancing lung cancer diagnosis: A comprehensive analysis of machine learning and deep learning approaches

Sukhvir Kaur & Manjit Singh
Department of computer science and engineering, Chandigarh University

ABSTRACT: In recent years, lung cancer has emerged as a really perilous ailment necessitating timely detection. Artificial intelligence (AI) has arisen as a crucial component within the medical domain, particularly in the analysis of medical images and disease diagnosis. Its implementation has shown effective in mitigating any human errors that may arise during the examination of medical images by healthcare professionals. The aim of this article is to provide a concise overview of recent developments in the domain of lung cancer diagnosis through the implementation of machine and deep learning algorithms. This article presents a thorough and inclusive examination of the existing comprehension and information regarding the subject. Engaging with the findings presented in recent research articles enables academics to enhance their understanding of the subject matter. All pertinent aspects, including obstacles, were thoroughly examined, and a detailed analysis was conducted on recommendations for future study. Additionally, the published datasets and their sources were offered to enhance researchers' accessibility and use in furthering the previously attained results.

Keywords: Artificial intelligence, Lung Nodules, Detection, Classification

1 INTRODUCTION

In 2020, there were almost 19.3 million new instances of cancer worldwide, and 10 million deaths from the disease. The second most prevalent kind of cancer was lung cancer, which was responsible for about 11.4% of all cancer occurrences. Just lung cancer was expected to account for 2.2 million cases. Of all cancers, lung cancer accounts for 18% of fatalities, making it the most lethal. Tobacco usage is a major risk factor for lung cancer development. The smoking rate has peaked or is continuing to climb in many countries. These findings point to a continued upward trend in lung cancer incidence for the foreseeable future [1]. Early detection, diagnosis, and effective management of cancer often enhance the probability that a patient will live a long life. Expertise from medical professionals is necessary for the assessment of medical data and the diagnosis of illnesses [2].

2 LUNG CANCER

The human body sustains cellular development and undergoes regeneration as required. However, under aberrant circumstances, cell proliferation becomes unregulated and leads to the formation of groupings of cells known as tumours. As per the National Cancer Institute, it is projected that by the conclusion of 2017, there might be a total of 221,200 fresh cases of lung cancer detected and 158,040 fatalities attributed to lung cancer. As per the World Health Organisation (WHO), cancer is responsible for 7.6 million fatalities annually worldwide, accounting for 13% of all worldwide deaths. In the United States, lung cancer is the primary cause of cancer-related fatalities. In the next three decades, lung cancer is anticipated to be the primary cause of cancer-related mortality, according to the studies. By

DOI: 10.1201/9781003501244-10

2040, the prevalence of lung cancer will have doubled in comparison to 2010. The primary factor contributing to the rise in lung cancer cases is the extended duration of human lifespans. By the year 2023, about 238,340 individuals, consisting of 117,550 males and 120,790 females, are projected to receive a diagnosis of lung cancer. According to estimates, the disease will also result in the demise of 127,070 people [3].

3 MACHINE AND DEEP LEARNING METHODS

Transfer learning (TL) is a method in ML and DL that leverages existing knowledge to enhance the performance of a related job, especially in situations when there is a scarcity of labeled data. The process of solving issues such as classification through the integration of multiple specialized learning models is known as ensemble learning. Its objective is to increase predicted accuracy and boost the performance of the underlying classification models [4,5].

3.1 *Machine learning*

Machine learning techniques are crucial in the identification and categorization of lung cancer, especially via analysing medical imaging data like CT scans and chest X-rays. Machine learning is the process by which computers gain knowledge and abilities via the use of algorithms, allowing them to independently make decisions and present users with results. Machine learning is presently utilized for complex data categorization and decision-making procedures. Algorithms play a crucial role in the progress of ML, allowing computers to make essential decisions [6,7,8].

The process of developing a DL model for lung cancer categorization and detection comprises multiple stages [9]. The initial stage is gathering data, namely a marked dataset of CT scans or chest X-rays, with annotations denoting the absence or presence of lung cancer. Preprocessing techniques are used in medical imaging to improve image clarity and accuracy before analysis. Data augmentation techniques introduce variations and increase heterogeneity in the training dataset, enhancing the model's ability to generalize and make accurate predictions when dealing with new and diverse medical images [10]. These techniques are crucial for accurate data interpretation and analysis. The database is separated into validation, training, and testing sets, following a standard distribution of 15% for validation, 15% for testing, and 70% for training [11]. The model architecture is devised by utilizing pre-trained models for feature extraction procedures, specifically those derived from the ImageNet dataset [12,13,14]. The model is constructed utilizing DL frameworks such as PyTorch or TensorFlow, incorporating convolutional layers for extracting features and fully connected layers for the purpose of classification [15,16,17,18].

3.2 *Related work*

Table 1. Recent works on literature survey.

Author/ Year	Aim	Methodology/ Techniques	Dataset	Finding	Limitations
Alshmrani, *et al.* (2023)	An advanced neural network structure designed for the categorization of lung disorders into multiple classes utilizing chest x-ray (CXR) pictures.	CNN and VGG19 techniques.	Public datasets	The VGG19 + CNN model demonstrated superior performance compared to prior studies, achieving an accuracy of 96.48%, recall of 93.75%, precision of 97.56%,	Interpretability is essential for DL models to establish trust among healthcare practitioners and ensure that their predictions are consistent with medical expertise.

(*continued*)

Table 1. Continued

Author/ Year	Aim	Methodology/ Techniques	Dataset	Finding	Limitations
				F1 score of 95.62%, and AUC of 99.82%..	
Lalitha & Murugan (2023)	Using CNNs to separate and group 3D lung tumor diagnoses	Using Heterogeneous Neural Network (LCD-HNN) and 3D CNN to diagnose lung cancer	Kaggle database	The test set had a 3D CNN accuracy of 98.6%. The proposed CAD method works better than current CAD systems	The current methodology is incompatible for the detection of lung nodules and lacks the ability to forecast specific forms of cancer.
Siddiqui, *et al.* (2023)	To develop a new and enhanced DBN with Gabor computed tomography images.	The strategy involves utilizing an SVM classifier.	LIDC-IDRI and LUNA-16 datasets	The IGF-EDBN method, combined with an SVM classifier, accuracy,	An appropriate methodology is required to detect lung cancer during its initial stages.
		This E-DBN employs a combination of two successive RBMs: a Bernoulli-Bernoulli (BB) RBM, and a Gaussian-Bernoulli (GB) RBMs.			

4 CONCLUSION

More evaluation on the technique of lung cancer identification and staging prompts continued research and improvement in this important area of medical imaging. Artificial intelligence and machine learning have developed such that it is more effective and accurate to identify the preliminary stage of lung cancer. Development in imaging technologies, such as CT and PET, is allowing for the construction of high-resolution datasets for study. The role of deep learning algorithms in the detection and classification of lung cancer lesions is still open but promising. However, a few bottlenecks still need to be addressed towards enabling interpretability of judgments and meeting the prerequisite of huge datasets for training and validation. If all these types of information and new biomarkers were combined, the specificity and sensitivity of detection of lung cancer could really be improved. In this connection, the researchers, physicians, and developers should communicate effectively in order to successfully realize the implementation of such developments into clinical practice.

REFERENCES

[1] Sung, H., Ferlay, J., Siegel, R. L., Laversanne, M., Soerjomataram, I., Jemal, A., & Bray, F. (2021). Global cancer statistics 2020: GLOBOCAN estimates of incidence and mortality worldwide for 36 cancers in 185 countries. *CA: A Cancer Journal for Clinicians*, *71*(3), 209–249.

[2] Chaturvedi, P., Jhamb, A., Vanani, M., & Nemade, V. (2021, March). Prediction and classification of lung cancer using machine learning techniques. In *IOP Conference Series: Materials Science and Engineering* (Vol. 1099, No. 1, p. 012059). IOP Publishing.

[3] Kumar, V., & Bakariya, B. (2021). Classification of malignant lung cancer using deep learning. *Journal of Medical Engineering & Technology*, *45*(2), 85–93.

[4] Doppalapudi, S., Qiu, R. G., & Badr, Y. (2021). Lung cancer survival period prediction and understanding: Deep learning approaches. *International Journal of Medical Informatics*, *148*, 104371.

[5] Masud, M., Sikder, N., Nahid, A. A., Bairagi, A. K., & AlZain, M. A. (2021). A machine learning approach to diagnosing lung and colon cancer using a deep learning-based classification framework. *Sensors*, *21*(3), 748.

[6] Hussain, S., Mubeen, I., Ullah, N., Shah, S. S. U. D., Khan, B. A., Zahoor, M., ... & Sultan, M. A. (2022). Modern diagnostic imaging technique applications and risk factors in the medical field: A review. *BioMed Research International*, *2022*.

[7] Nageswaran, S., Arunkumar, G., Bisht, A. K., Mewada, S., Kumar, J. N. V. R., Jawarneh, M., & Asenso, E. (2022). Lung cancer classification and prediction using machine learning and image processing. *BioMed Research International*, *2022*.

[8] Dodia, S., Annappa, B., & Mahesh, P. A. (2022). Recent advancements in deep learning-based lung cancer detection: A systematic review. *Engineering Applications of Artificial Intelligence*, *116*, 105490.

[9] Rehman, A., Kashif, M., Abunadi, I., & Ayesha, N. (2021, April). Lung cancer detection and classification from chest CT scans using machine learning techniques. In *2021 1st International Conference on Artificial Intelligence and Data Analytics (CAIDA)* (pp. 101–104). IEEE.

[10] Praveena, M., Ravi, A., Srikanth, T., Praveen, B. H., Krishna, B. S., & Mallik, A. S. (2022, June). Lung cancer detection using deep learning approach CNN. In *2022 7th International Conference on Communication and Electronics Systems (ICCES)* (pp. 1418–1423). IEEE.

[11] Huang, S., Arpaci, I., Al-Emran, M., Kılıçarslan, S., & Al-Sharafi, M. A. (2023). A comparative analysis of classical machine learning and deep learning techniques for predicting lung cancer survivability. *Multimedia Tools and Applications*, *82*(22), 34183–34198.

[12] Wankhade, S., & Vigneshwari, S. (2023). A novel hybrid deep learning method for early detection of lung cancer using neural networks. *Healthcare Analytics*, *3*, 100195.

[13] Punithavathy, K., Poobal, S., & Ramya, M. M. (2019). Performance evaluation of machine learning techniques in lung cancer classification from PET/CT images. *FME Transactions*, *47*(3), 418–423.

[14] Naseer, I., Masood, T., Akram, S., Jaffar, A., Rashid, M., & Iqbal, M. A. (2023). Lung Cancer Detection Using Modified AlexNet Architecture and Support Vector Machine. *Computers, Materials & Continua*, *74*(1).

[15] Rossetto, A. M., & Zhou, W. (2017). Deep Learning for Categorization of Lung Cancer CT Images. *2017 IEEE/ACM International Conference on Connected Health: Applications, Systems and Engineering Technologies (CHASE)*, 272–273.

[16] Jakimovski, G., & Davcev, D. (2019). Using double convolution neural network for lung cancer stage detection. *Applied Sciences*.

[17] Schwyzer, M., Ferraro, D. A., Muehlematter, U., Curioni-Fontecedro, A., Huellner, M., von Schulthess, G. V., Kaufmann, P., Burger, I., & Messerli, M. (2018). Automated detection of lung cancer at ultralow dose PET/CT by deep neural networks – Initial results. *Lung Cancer*, 126, 170–173.

[18] Das, S., & Majumder, S. (2020). Lung cancer detection using deep learning network: a comparative analysis. *2020 Fifth International Conference on Research in Computational Intelligence and Communication Networks (ICRCICN)*, 30–35.

Computational Methods in Science and Technology Volume 1 – Kaur et al. (Eds)

Advanced perspectives on big data security: Navigating challenges in cloud computing

Richa Verma & Ravindara Bhatt
Department of Computer Engineering, Jaypee University of Information Technology, Solan, India

ABSTRACT: Recent years have seen the rise of big data workflow management solutions as widespread data analytic platforms for handling massive amounts of data in the cloud. However, keeping information private and ensuring the integrity of workflow applications is still a major and difficult issue. This document reviews and organizes research on large data workflow management privacy and security breaches and remedies. In this case, data encryption is utilized to safeguard information at rest and in transit. These points of view would allow researchers to better comprehend crucial research areas and create novel approaches. according to research question improve parameters of security by proposed approach. In this paper review of different workflows and security issue in cloud and other frame works

Keywords: Workflow Management, Cloud, Security

1 INTRODUCTION

Today's technological advancements allow for the collection and storage of large amounts of data (termed as big data) from distinct sources of data, like the Internet, Event logs, Databases, IoT devices, Sensors, Smartphones, etc. [1,2]. The analysis and comparisons of the collected facts produce valuable information that is frequently applied to processes of decision-making. Bigdata has been utilised, for instance, in the business world for predicting market trends. Policymakers can take proactive action via analysis of the data they have acquired because it gives them relevant information that is frequently employed in decision-making processes. Bigdata is utilised, for instance, in the business world for predicting trends in the market. Policymakers can take proactive action for the betterment of everyone regarding government and healthcare by studying the collected data. For large-scale analysis of data, that are frequently resource-intensive, cloud providers present enormous opportunities with their extensive and flexible storage, processing, and network capabilities. The Hadoop framework, for instance, is a popular cloud-deployable big data analytics technology that handles Map-Reduce-style operations [3–5]. Still, the idea of the cloud depends on sharing resources. Software, as well as hardware resources in the cloud, are usually shared between various individuals or organisations using isolating methods like virtual computers/machines. Due to the way resources are shared and the large size of cloud-based software, the cloud can be harmed in a number of ways. Due to advancements in cloud computing, popular scientific workflow tools are able to be repurposed for more effective use in the age of cloud computing. Customers have greater options for on-demand access to computer and storage resources owing to cloud technology [7–9]. Because of the many advantages that using cloud computing brings, including on-demand resources and cost savings, moving organizational services, data, and applications to the cloud is a crucial strategic choice for businesses [10,11]. However, there are risks and challenges for cloud modification associated

DOI: 10.1201/9781003501244-11

with data security [12,13] when spreading data dynamically among geographically scattered centers while minimizing user-perceived delay. Due to the restricted bandwidth between the data centres in the cloud, certain scientific information may be communicated insecurely from one particular station to another. Furthermore, this difficulty gets worse of the security restrictions placed on such highly sensitive data. Additionally, because the resources share each other, new hazards may emerge [14,15].

2 RESEARCH BACKGROUND

2.1 *Big data encryption*

2.1.1 *Encryption*

Data encryption is the primary method used to ensure that information is kept private. When a data set (the plaintext) is combined with a cryptographic key (the cypher), the result is a coded set of characters (the ciphertext) [12,13]. Without the key, the ciphertext conceals every detail regarding the primary data aside from its length, making recovery of the plaintext impossible [14,15]. This aspect of security, referred to as semantic security, ensures that an enemy cannot discover any (perhaps sensitive) characteristics of the underlying data, even if the opponent has a great deal of knowledge about the data itself. This is especially important in situations when the data being used has a predetermined form, as in the case of financial transactions, or if only a subset of the underlying distribution is identifiable, like when it is being used to measure a real-world phenomenon. More formally, these three protocols make up encryption [46]:

- KeyGen: an algorithm for creating new cryptographic keys when they are needed,
- Enc (k; p) = c: a method of hiding information by transforming plaintext (p) into encrypted (c) data using a key (k).
- Dec (k; c) = p: a method of decryption that takes a ciphertext c and a key k and returns the plaintext p.

3 PROBLEM DESCRIPTION

The increasing need for data storage and processing power gives rise to the idea of cloud computing. Yet putting data in widely dispersed data centers also comes with a growing challenge. The security of data is compromised when a scientific workflow is moved to the cloud. There are two potential origins for security issues: the cloud providers themselves, and the other users in the system. Consequently, cloud service providers are undoubtedly aware of the user data's associated metadata. Several malicious cloud service providers may disclose or even sell this information. Some critical data loss could lead to financial, physical, or even human security risks. There is also information for other customers of the data center. Because the assets have been shared with other individuals, it is very easy for the multi-users to steal or manipulate the users' data, and this ultimately causes the user to lose out [63]. To address this safety issue, we must protect user data privacy and security. When using cloud-based scientific workflows, users can request various data security services based on how sensitive the data is. In order to meet the needs of customers for data security, data centers primarily use methods including certifications services, encrypted information storage, recovery of data, management of security logs, and auditing services. While putting the data, we must additionally take into account if the data center can meet the standards for security of data services. One ought to put the data in additional data centers that can offer the services if the data center is unable to meet the standards of security for data services.

4 BIG DATA SECURITY SERVICE NEEDS

The task that generates the data determines the security service needs for that data. For instance, we know that s_{Tw5} = {0.2, 0.6, 0.2} and s_{T5} = {0.42, 0.16, 0.38} are the security service needs of T_5. The security service needs for the data produced by T_5 are therefore the same as those for T_5. The following is therefore an example of how the security service requirements for the intermediate information should be written out.

Definition: The needs for big data security services are described by $s_i = \{s_i^1, s_i^2, s_i^3\}$ where $s_i^1, s_i^2, and\ s_i^3$ indicate the coefficients of data's confidential service, integrity service, and authentication service, respectively. Likewise, the significance of data security services s_{wi} is also known as s_{Twi}. The security servicing factors are defined by the formula $\sum_{k=1}^{3} s_{wi}^k = 1$.

5 BIG DATA ANALYTICS

5.1 *Big data techniques*

In order to protect their customers' personal information, businesses used a wide range of identification procedures. Verbal or written assurances of confidentiality may be the most widely used method. Low-level technology solutions like controlled access, two-factor verification, and passwords are often used to protect privacy and security when transmitting and combining data across distributed, dynamic data services. These rights can be compromised in two ways: through purposeful permission sharing when they are no longer needed, or through potential [76]. Cryptography is a more sophisticated method of solving the problem. Both the Advanced Encryption Standard (AES) and the Shamir, Rivest, and Adleman (RSA) algorithm are well-known methods of secure data transmission. This form of encryption is known as a "symmetric" algorithm. To put it another way, the same key (an encryption key) is employed to both encrypt and decipher the data. When it comes to securing data while it is sitting idle, AES encryption is by far the most popular option [77]. The term "asymmetric" describes the category into which RSA fits. The idea that there are two interconnected secrets (keys) for encryption gives rise to the asymmetric label. There are two types of keys: the private key and the public key. When using the public key for encryption, the corresponding private key is required for decryption. If you encrypt something using a private key, you are able to decrypt it with the corresponding public key [77]. The drawback of this kind of technology is the fact that it needs specialized staff to interpret and read the results, and the software can be utilized to monitor individual behavior instead of protect data, making it challenging and costly to put into effect on an extensive basis or with multiple systems and users. Big Data privacy and security issues get more challenging as information increases.

6 CONCLUSION

In order to safeguard the privacy and authenticity of workflow activities and data, this study presents a secure and effective big data scientific workflow management platform. We also reviewed the literature on big data analytics, categorization, cloud storage, big data technology, security, privacy, and encoding methods. We have discussed many different cryptographic methods for protecting information at all stages of its lifecycle, from transmission to storage to retrieval. We anticipate that these methods will become standard fare in the big data ecosystem as security grows in importance for handling sensitive big data. The results of the study should help spread the word about the newest safeguards and methods for keeping big data safe. We believe that the future of large data processing will be enabled by better knowledge and stronger collaboration across the data processes, big data, and cryptography communities.

REFERENCES

[1] Chidambararajan, B; Kumar, M. S; & Susee, M. S. (2019). Big data privacy and security challenges in industries. *Int. Res. J. Eng. Technol*, *6*(4), 1991.

[2] Bertino, E; & Ferrari, E. (2017). Big data security and privacy. In *A Comprehensive Guide Through the Italian Database Research Over the Last 25 Years* (pp. 425–439). Cham: Springer International Publishing.

[3] Thota, C; Manogaran, G; Lopez, D; & Vijayakumar, V. (2017). Big data security framework for distributed cloud data centers. In *Cybersecurity Breaches and Issues Surrounding Online Threat Protection* (pp. 288–310). IGI global.

[4] Zhang, B; Yu, L; Feng, Y; Liu, L., & Zhao, S. (2018). Application of workflow technology for big data analysis service. *Applied Sciences*, *8*(4), 591.

[5] Talia, D; Trunfio, P; & Marozzo, F. (2015). *Data Analysis in the Cloud: Models, Techniques and Applications*. Elsevier.

[6] AlJahdali, H; Albatli, A; Garraghan, P; Townend, P; Lau, L; & Xu, J. (2014, April). Multi-tenancy in cloud computing. In *2014 IEEE 8th International Symposium on Service-Oriented System Engineering* (pp. 344–351). IEEE.

[7] Wei, Lifei; Haojin Zhu; Zhenfu Cao; Xiaolei Dong; Weiwei Jia; Yunlu Chen; and Athanasios V. Vasilakos. Security and privacy for storage and computation in cloud computing. *Information Sciences* 258 (2014): 371–386.

[8] Chenthara, S; Ahmed, K; Wang, H; & Whittaker, F. (2019). Security and privacy-preserving challenges of e-health solutions in cloud computing. *IEEE Access*, *7*, 74361–74382.

[9] Yang, P; Xiong, N; & Ren, J. (2020). Data security and privacy protection for cloud storage: A survey. *IEEE Access*, *8*, 131723–131740.

[10] Ferrag, M. A; & Shu, L. (2021). The performance evaluation of blockchain-based security and privacy systems for the Internet of Things: A tutorial. *IEEE Internet of Things Journal*, *8*(24), 17236–17260.

[11] Azeez, N. A; & Van der Vyver, C. (2019). Security and privacy issues in e-health cloud-based system: A comprehensive content analysis. *Egyptian Informatics Journal*, *20*(2), 97–108.

[12] Ali, M; Mohajeri, J; Sadeghi, M. R; & Liu, X. (2020). A fully distributed hierarchical attribute-based encryption scheme. *Theoretical Computer Science*, *815*, 25–46.

[13] Wu, S; Li, Q; Li, G; Yuan, D; Yuan, X; & Wang, C. (2019, April). ServeDB: Secure, verifiable, and efficient range queries on outsourced database. In *2019 IEEE 35th International Conference on Data Engineering (ICDE)* (pp. 626–637). IEEE.

[14] Siam, A. I; El-khobby, H. A; Abd Elkader, H. S; AbdelNaby, M. M; Elkader, H. S. A; & Abdelnaby, M. M. (2015). Enhanced data security model for cloud computing platform. *International Journal of Scientific Research in Science, Engineering and Technology*, *1*(4), 450–460.

[15] Shaik, K; & Narayana Rao, T. V. (2017). Implementation of Encryption Algorithm for Data Security in Cloud Computing. *International Journal of Advanced Research in Computer Science*, *8*(3).

[16] Alemami, Y; Al-Ghonmein A. M; Al-Moghrabi, K. G; & Mohamed, M. A. (2023). Cloud data security and various cryptographic algorithms. *International Journal of Electrical and Computer Engineering*, *13* (2), 1867.

[17] Gupta, R; & Kapoor, R. (2016). A review paper of data security in Cloud computing. *Imperial Journal of Interdisciplinary Research (IJIR)*, *2*(8), 501–504.

[18] Yeluri, R; & Castro-Leon, E. (2014). *Building the Infrastructure for Cloud Security: A Solutions View* (p. 244). Springer Nature.

Computational Methods in Science and Technology Volume 1 – Kaur et al. (Eds)
© 2025 the Author(s), ISBN: 978-1-032-81760-6

Improved segmentation and classification of breast cancer using mammogram images with residual network based deep learning

Saruchi
Chandigarh University, Mohali, Punjab

ABSTRACT: The use of mammogram images for breast cancer detection focusses on image acquisition, preprocessing, segmentation, and feature extraction. Digital mammograms are used to classify images into benign, malignant, or micro calcification stages, enabling further diagnosis and treatment. The BI-RADS score from the mammogram determines the classification of abnormalities. Micro calcifications, small white specks in dense breast tissue, are early indicators of cancer. The research study uses the K-means algorithm for segmentation and CNN for classification, comparing results with thermal and MRI images. The proposed work includes an unsupervised hard clustering method called K-means clustering for segmentation and three deep learning models: CNN architecture, Residual network 50, and Inception V3. In this research mammogram images, employing K-means clustering and Convolutional Neural Network (CNN) methods. The K-means algorithm, a hard clustering method, effectively partitions the images into normal and abnormal categories. Subsequently, these segmented images undergo classification using advanced CNN models, which are enhanced through transfer learning techniques. A significant finding of this research is the superior classification accuracy achieved using thermal images in breast cancer diagnosis, surpassing the results obtained with mammogram and MRI images. The dataset for this study comprised 1300 mammogram images, of which 300 were utilized for testing. The proposed methodology demonstrated impressive classification accuracies of 91.66% for benign cases, 89.33% for malignant cases, and 91.66% for micro- calcification

Keywords: Medical Imaging, Mammograms, Convolutional Neural Network

1 INTRODUCTION

Imaging modalities, mammograms are the gold standard for diagnosing breast cancer, and researchers routinely include these pictures in their studies. (Lashen *et al.*, 2019). Breast cancer diagnosis using mammography images is the focus of this section. Another early sign of cancer is micro calcifications, which are clusters of very little white spots. When caught early, this exceptional severity improves the afflicted person's survival chances. Microcalcification in mammography may be difficult to analyse. There is little to no difference between the tiny, thick breast tissue and the background noise.Breast cancer segmentation and classification using mammography pictures is the subject of this study. For segmentation, we use the K-means method, and for classification, we turn to CNN. Lastly, the findings are evaluated and contrasted with those of thermal imaging and magnetic resonance imaging.

2 RELATED WORK

The recommended work's uniqueness, advantages, and disadvantages are listed in the table below. The methods' parameters are also covered.

DOI: 10.1201/9781003501244-12

Table 1. Table captions should be placed above the tables.

S. No	PROPOSED METHODS	NOVELTY	STRENGTH	WEAKNESS	PARAMETERS
1	Region based segmentation	Hierarchical codebook with HSI function is used	Pure colour is extracted. Output image is brighter than other methods.	-	Super voxel, Codebook, Code, vectors, Pixel intensity values
2.	DCNN Segmentation using U-Net architecture	Advanced segmentation method in DL, Modified Binary mask	Accurate segmentation	Architecture is tedious. Customized architecture	5 Convolution layers, filter size, no of filters, 1maxPooling layer, stride Drop out ratio
3.	GMM clustering	Parameter adjusted with 3 Clusters and 3 components for each cluster	Suitable for MR images flexible and find clear edges.	2 conversions are made Grey scale to RGB and vice versa	Intensity values of the pixel, contour, colour, centre, width

3 PROPOSED METHOD

3.1 *Segmentation using K-means*

$$SSE = SSE1 + SSE2 + SSE3 \tag{1}$$

$$SSE = \sum_{=0} (-1)^2 \tag{2}$$

$$SSE = \sum_{=0} (-2)^2 \tag{3}$$

$$SSE = \sum_{=0} (-3)^2 \tag{4}$$

Where C1, C2, and C3 are the centres of the clusters and X_i are the data points.

Similarly, calculate SSE value for K = 2, 4, 5 to find the suitable K value. When K = 3 there is a sudden decrease in SSE values, and after 3 the values are normalized. The algorithm is:

Algorithm

1. Plot data on scatter point with K = 3 and randomly assign 3 center points in data space. Divide input image into background, normal, and affected areas.
2. The image, converted into a RGB model, displays different colors for cancer-affected and normal regions, with the background typically black and the affected region white.
3. The K-means function is used to fit data and predict classes, with the result being an array of values.
4. The Euclidean distance method is used to calculate the distance of each point from the centroid, and the data is then divided into three groups.
5. Form groups, calculate cluster mean, averaging each cluster, and move center point to calculated position. Find nearest point from other clusters and move if necessary.
6. Repeat step 4 until all data points are in the correct cluster, ensuring no points change their cluster.
7. The process stops when all data points in the pixel intensity values group into one of the three clusters.
8. Three clusters are identified by three colors, with the white- colored portion of cancer affected areas being the region of

4 EXPERIMENTAL ANALYSIS

The proposed method uses a 1024x1024 pixel MIAS database with 1200 breast images for analysis. Abnormality locations are extracted from the images, with 800 images used for training and 300 for testing. CNN,

The table shows the performance metrics for all the three CNN models to determine the accuracy. Using the accuracy, bar chart designed for the three models are given below in Figure 1.

The study compared the accuracy of classification using mammogram images and thermal images using CNN architecture, Inception V3, and Resnet 50. Results showed that mammogram images had the highest accuracy (85.66%, 82.66%, and 88.33%), while thermal images had the highest accuracy (92.66%, 90.33%, and 91.66%) using CNN architecture and the highest efficiency (96.66%, 96.33%, and 98.66%) using Resnet 50. Thermal images outperformed the other models in terms of efficiency and computation.

Table 2. Parameters of CNN architecture.

METHODS	STAGES	SENSITIVITY (%)	SPECIFICITY (%)	ACCURACY (%)	F1 SCORE (%)	TOTAL ACCURACY (%)
	BENIGN	87.01	85.20	85.66	76.00	
2D CNN MODEL	MALIGNANT	69.04	92.52	82.66	77.32	**78.32**
	MICRO CALCIFICATION	83.50	90.64	88.33	82.21	
	BENIGN	84.00	83.55	83.66	72.00	
INCEPTION V3 MODEL	MALIGNANT	67.74	90.90	81.33	75.43	**76.05**
	MICRO CALCIFICATION	80.19	90.45	87.02	81.23	
	BENIGN	97.46	89.59	91.66	86.10	
RESNET50 MODEL	MALIGNANT	78.33	96.66	89.33	85.34	**86.34**

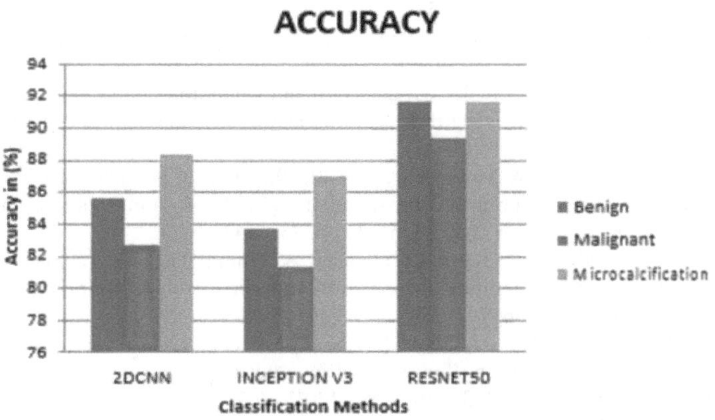

Figure 1. Accuracy of proposed models.

5 CONCLUSION

The study explained the segmentation and classification of mammogram images using k-means and CNN methods respectively. The hard clustering method namely K-means clustering partition the given images into normal and abnormal type. The segmented images are classified using the famous CNN models based on transfer learning method. From the results derived from the proposed three methods it is clearly evident that, the use of thermal images in breast cancer diagnosis yields better classification accuracy over mammogram images and MRI images respectively. The data sets for mammogram images contain 1300 where 300 images are used for testing purpose. The proposed study produces a classification accuracy of 91.66%, 89.33 %, and 91.66% for benign, malignant and micro- calcification respectively.

REFERENCES

Chen, B., Shorey, J. M., Saunders, R., *et al.* (2011). An anthropomorphic breast model for breast imaging simulation and optimization.

Fausto, A. M., Lopes, M. C., Sousa, M. C., *et al.* (2017). Optimization of Image Quality and Dose in Digital Mammography.

Heather Whitney, M, Hui Li, Yu Ji, Peifang Liu & Maryellen Giger, L (2019). Comparison of breast MRI tumor classification using human-engineered radiomics, transfer learning from deep convolutional neural networks and fusion methods, *Proceedings of the IEEE*, vol. 108, no. 1, pp. 163–177.

Hemeida, AM, Alkhalaf, S, Mady, A, Mahmoud, EA, Hussein, ME & Eldin, AMB. (2020), Implementation of nature-inspired optimization algorithms in some data mining tasks, *Ain Shams Engineering Journal*, vol. 11, no. 2, pp. 309–318.

Here are fiverecent research references related to the optimization and evaluation of breast imaging systems:

Hosni, M, Abnane, I, Idri, A, de Gea, JMC & Alemán, JLF. (2019). Reviewing ensemble classification methods in breast cancer, *Computer Methods and Programs in Biomedicine*, vol. 177, pp. 89–112.

Jonathan de Matos, Alceu de Britto, S, Luiz Oliveira, ES & Alessandro Koerich, L. (2019). IEEE international joint conference on neural networks (IJCNN), 14–19 Jul, *Double Transfer Learning for Breast Cancer Histopathologic Image Classification Budapest*, Hungary, pp. 1–8, https://doi.org/10.1109/IJCNN.2019.8852092.

Juan, Luo & Oubong Gwun. (2009). A comparison of sift, pca-sift and surf, *International Journal of Image Processing (IJIP)*, vol. 3, no. 4, pp. 143–152.

Karabatak, M & Ince, MC. (2009), An expert system for detection of breast cancer based on association rules and neural network, *Expert systems with Applications*, vol. 36, no. 2, pp. 3465–3469.

Kaur, P, Singh, G & Kaur, P. (2019). Intellectual detection and validation of automated mammogram breast cancer images by multi-class SVM using deep learning classification, *Informatics in Medicine Unlocked*, vol. 16, p. 100151.

Kayla Mendel, Hui Li, Deepa Sheth & Maryellen Giger. (2019). Transfer learning from convolutional neural networks for computer-aided diagnosis: a comparison of digital breast tomosynthesis and full-field digital mammography, *Academic Radiology*, vol, 26, no. 6, pp. 735–743.

Khamparia, A, Bharati, S, Podder, P, Gupta, D, Khanna, A, Phung, TK *et al.* (2021). Diagnosis of breast cancer based on modern mammography using hybrid transfer learning, *Multidimensional Systems and Signal Processing*, vol. 32, no. 2, pp. 747–765.

Kiarashi, N, Lo, JY, Lin, Y, Ikejimba, LC, Ghate, SV, Nolte, LW *et al.* (2014). Development and application of a suite of 4-D virtual breast phantoms for optimization and evaluation of breast imaging systems, *IEEE Transactions on Medical Imaging*, vol. 33, no. 7, pp. 1401–1409.

Kozegar, E, Soryani, M, Behnam, H, Salamati, M & Tan, T. (2017). Mass segmentation in automated 3-D breast ultrasound using adaptive region growing and supervised edge- based deformable model, *IEEE Transactions on Medical Imaging*, vol. 37, no. 4, pp. 918–928.

Laith Alzubaidi, Muthana Al-Amidie, Ahmed Al-Asadi, Amjad J Humaidi, Omran Al-Shamma, Mohammed Fadhel, A *et al.* (2021). Novel Transfer Learning Approach for Medical Imaging with Limited Labeled Data, *Cancers*, vol. 13, no. 7, doi:10.3390/cancers13071590.

Laith Alzubaidi, Omran Al-Shamma, Mohammed Fadhel, A, Laith Farhan, Jinglan Zhang & Ye Duan. (2020). Optimizing the Performance of Breast Cancer Classification by Employing the Same Domain Transfer Learning from Hybrid Deep Convolutional Neural Network Model, *Electronics*, vol. 9, no. 3, https://doi.org/10.3390/electronics9030445.

Tavakoli Taba, S., Baran, P., Lewis, S., *et al.* (2019). Toward Improving Breast Cancer Imaging: Radiological Assessment of Propagation-Based Phase-Contrast CT Technology.

Wang, L., Zhang, Q., Brukilacchio, T., & Boas, D. (2001). Optimizing single-projection linear tomographic optical breast imaging.

Computational Methods in Science and Technology Volume 1 – Kaur et al. (Eds)
© 2025 the Author(s), ISBN: 978-1-032-81760-6

Analysis of early detection and prediction of diabetic retinopathy by deep learning

Shubham, Geetika Sharma & Deepanshu Garg
Department of Computer Science and Engineering, Chandigarh University, Gharaun, Mohali, India

ABSTRACT: Diabetic retinopathy (DR) is a very common ocular complication of diabetes. Failure of its early detection and treatment leads to permanent visual loss. Deep learning methods have been highly advanced in recent times in the field of ophthalmology, very specifically on issues related to the detection and categorization of DR. The current study will try to find the use of deep learning techniques in facing challenges of medical diagnosis, particularly related to diabetic retinopathy (DR). This paper outlines the epidemiological situation in the DR, with its implications for global public health. It particularly focuses on how to use Convolutional Neural Networks (CNNs) for the assessment of retinal images. The study further asks about the importance of some of the data sets used in building resilient and precise deep-learning models. Recent works have reported that the implemented deep learning models have reached high levels of sensitivity and specificity in the automated detection and classification of DR. The study also reports on the implementation difficulties of DR with deep learning, such as model requirements for diverse and well-annotated datasets, interpretability of models, and ethical concerns. The study, hence, underscores the worldwide impacts whereby deep learning is associated with DR.

Keywords: Diabetic Retinopathy, Deep Learning, Convolutional Neural Network

1 INTRODUCTION

The incidence of diabetes has been on the rise across the world, making it a very common condition. Irregular metabolic processes and consequences such as diabetic retinopathy (sight loss), cardiovascular illnesses, and renal failure are mostly linked to insulin production and elevated blood glucose levels. A major eye disease, diabetic retinopathy causes permanent and untreatable blindness. This syndrome is more common in those with a long history of diabetes, regardless of whether their diabetes is type 1 or type 2. Age is a risk factor for the condition [1,2]. The majority of research in the field of disaster recovery focused on feature extraction using machine learning (ML) methods. However, a challenge emerged regarding the manual extraction of features, prompting scholars to employ deep learning (DL) techniques. The discipline of DR greatly values deep learning for its capacity to handle vast quantities of intricate data, identify pertinent characteristics, and generate precise forecasts [3–7]. The ability of DL to enhance screening, diagnostic, and prediction capabilities is crucial in the field of DR, as it ultimately leads to improved patient outcomes and a reduction of vision problems in individuals with diabetes.

 DOI: 10.1201/9781003501244-13

2 DEEP LEARNING-BASED DR DETECTION AND PREDICTION

Table 1. A summary of works on diabetic retinopathy using deep learning.

Author/Year	DL Models	Classification category	Algorithm	Dataset	Limitation	Results
Özçelik, Y. B., & Altan (2023)	CPSO and k-NN based on logistic-chaotic maps, as well as the RNN-LSTM	DR prediction and classification	Particle swarm optimization (PSO)	APTOS 2019	The heterogeneity and complexity of DR illness make clinical diagnosis difficult.	The suggested model was found to be 99.64% accurate.
Maithili, K., *et al.* (2023)	Convolutional Neural Network (CNN)	DR detection and classification	Hybrid Strawberry-based Convolution Neural Framework (SbCNF)	Publically available datasets like DRIVE, STARE, and CHASE_DB1	The authors claim no conflicts of interest.	The algorithm achieved an accuracy of 99%
Balamurugan, N. M., *et al.* (2023)	Transfer Learning (TL), the Correlation technique, 2D-Conventional Neural Networks (2D-CNNs), and Ensemble Learning (EL).	DR prediction	ML and DL algorithms	MESSIDOR dataset	The authors claim no conflicts of interest.	A 97.8 percent success rate in DR image prediction and a 98.5 percent success rate in picture
Kumar & Babu (2022)	Densenet-169, Conv LSTM (proposed model) and Dense-LSTM (proposed hybrid model)	DR detection	DL algorithm	MESSIDOR dataset	An effective screening system is necessary for early diagnosis, which in turn facilitates timely treatment.	Densenet-169 model: 94% accuracy, ConvLSTM: 99% Dense-LSTM = 83%

3 RECENT WORKS ON DIABETIC RETINOPATHY AND DEEP LEARNING

DL-based systems have demonstrated exceptional performance in various computer vision applications, surpassing old manual techniques. Over the years, numerous studies have employed diverse methodologies to identify diabetic retinopathy. The subsequent literature provides a concise overview of the tasks that have been accomplished thus far.Ayala, A., *et al.* (2021) employ a CNN model to analyze fundus oculi images to detect the anatomy of the eyeball and identify cases of DR [8–10]. The model uses transfer learning methodology to establish a mapping between images and their accompanying labels. Clinical fundus oculi photographs and a severity rating for pathology are used in the model training process. Reliable prediction of diabetic retinopathy in fundus oculi photos was achieved by the model, which acquired a precision of 97.78%. In their groundbreaking work on automated diabetic retinopathy detection, Luo, X., *et al.* (2021) present a convolutional network. The efficacy of this network is enhanced by including several fundus picture viewpoints. Unlike single-view DCNN-based systems, this approach makes the most of retinal lesion traits and uses attention processes to improve performance. In order to improve feature extraction, it assigns large weights to critical channels. Using multi-view photographs is suitable for automated DR recognition and outperforms other benchmarking methods, according to results from trials performed on a dataset of 15,468 multi-view [11–13].

4 RESEARCH CHALLENGES

Resistance to anti-VEGF medications is a significant challenge, as certain patients acquire resistance to these therapies. Investigation is necessary to comprehend the fundamental processes of resistance and design new therapeutic strategies. There has been an observed increase in the number of published papers on diabetic retinopathy (DR) about deep learning models, particularly following the widespread adoption of deep learning in medical imaging [14–16]. Despite the utilization of various methodologies in the development of DR classification systems, a considerable volume of data has been identified as being necessary for training objectives. The existing public datasets consist of high-quality photos, which renders the system impractical at present. In reality, retinal scans may not possess the same level of quality as the training images, given the presence of more heterogeneous populations of diabetes patients [17,18].

5 CONCLUSION

Deep learning techniques have demonstrated substantial potential in the domain of ophthalmology, specifically in the identification and categorization of diabetic retinopathy. Other than the above, the methods give predictive analysis of longitudinal data for the onset of diabetic retinopathy and individual therapy for the patient. All these are issues to solve: from the availability of well-annotated datasets to the understandability of DL models and the voidance of biases in the training data. Future deep learning models for diabetic retinopathy require being more robust and interpretable. Among them, the multimodal approach needs to be pursued, genetic and biomarker information is to be integrated, and more transparency is to be included for AI. Diabetic retinopathy (DR) is a very major public health problem throughout the world, and the situation becomes graver in those conditions where people do not get easy access to a retinologist. For the right kind of progress and deployment, it's essential that the AI and medical communities work together.

REFERENCES

[1] Kaushik, V., Gessa, L., Kumar, N., & Fernandes, H. (2023). Towards a new biomarker for diabetic retinopathy: Exploring RBP3 structure and retinoids binding for functional imaging of eyes in vivo. *International Journal of Molecular Sciences*, *24*(5), 4408.

[2] Pradhan, A., & Sharma, N. (2022). Detection of diabetic retinopathy using deep learning: A review. *International Journal of Recent Technology and Engineering (IJRTE)*, *11*(2).

[3] Mushtaq, G., & Siddiqui, F. (2021, February). Detection of diabetic retinopathy using deep learning methodology. In *IOP Conference Series: Materials Science and Engineering* (Vol. 1070, No. 1, p. 012049). IOP Publishing.

[4] Gunasekaran, Pitchai, Chaitanya, Selvaraj, Annie Sheryl, Almoallim, & Tesemma. (2022). Early diabetic retinopathy prediction using fundus photos and a deep learning architecture. *BioMed Research International*, 2022.

[5] Harshitha, Asha, Pushkala, Anogini, & Karthikeyan. (2021). Using deep learning to predict diabetic retinopathy stages. Volume 6, Issue 1, Pages 1–6, *Sixth International Conference on Innovative Computing Technologies (ICICT) 2021*, IEEE.

[6] Sungheetha, A. (2021). Create a convolutional neural network that uses deep feature extraction to identify and classify diabetic retinopathy early on. Volume 3, Issue 2, Pages 81–94, *Journal of Current Trends in Computing and Smart Technologies.*

[7] Zhang *et al.* (2021). The use of deep learning for the automated diagnosis of severe diabetic retinopathy. *Journal of Clinical and Experimental Ophthalmology, Graefe's Archive*, pages 1–8.

[8] Kalaivani & Mission (2021). Potential risk factors for diabetic retinopathy in individuals with type 2 diabetes mellitus. Volume 10, Issue 4, Pages 3369–3376, *Journal of Medical Pharmaceutical and Allied Sciences.*

[9] Sun & Aiello (2022). Diabetic retinopathy: nonproliferative and proliferative types. In *Principles and Practice of Ophthalmology* byAlbert and Jakobiec (pp. 2955–2993). Published by Springer International Publishing in Cham.

[10] Sivaprasad, Sen, & Cunha-Vaz (2023). Views on diabetic retinopathy—possibilities and obstacles. *Eye*, 37(11), 2183–2191.

[11] Özçelik & Altan (2023). A strong AI-based model optimized with chaotic swarm intelligence and recurrent long short-term memory overcomes nonlinear dynamics in diabetic retinopathy categorization. *Fractional and Fractal Analysis*, 7(8), 598.

[12] Kehtarnavaz (2021). Automatic five-stage diabetic retinopathy diagnosis using a multi-tasking deep learning model. 9(12320–123230), *IEEE Access.*

[13] Jamal Mohan, N., Murugan, R., & Goel, T. (2022). Problems and potential solutions using deep learning for the identification of diabetic retinopathy. In *Advances in Healthcare Informatics for the Future* (pp. 213–232).

[14] Ayala, Ortiz Figueroa, Fernandes, & Cruz (2021). The use of deep learning has enhanced the identification of diabetic retinopathy. *Applied Sciences*, Vol. 11, No. 24, 11970.

[15] Nagpal, Panda, Malarvel, Pattanaik, & Khan (2022). Databases, methods, assessment criteria, and future directions in diabetic retinopathy: a literature review. *Computer and Information Sciences Journal of King Saud University*, 34(9), 7138–7152.

[16] Saini & Susan (2022). Deep learning for detecting diabetic retinopathy in datasets with multi-class imbalances. *Journal of Biomedical Computing*, 149, 105989.

Regiant, Brunak, & Saha (2021). Diabetic retinopathy evaluation using convolutional neural networks: deciphering intrinsic visual characteristics. *Scientific Reports*, 11(1), 9704.

[17] Maithili, K., Latha, Y. M., Gangopadhyay, A., Varghese, I. K., Sundaram, A., Pandey, C., & Kiran, A. (2023). Optimized CNN model for diabetic retinopathy detection and classification. *International Journal of Intelligent Systems and Applications in Engineering*, 11(7s), 317–331.

[18] T., Dittakan, K., & Musikasuwan, S. (2021). Diabetic retinopathy detection using convolutional neural network: A comparative study on different architectures. *Engineering Access*, *7*(1), 50–60.

Computational Methods in Science and Technology Volume 1 – Kaur et al. (Eds)
 ISBN: 978-1-032-81760-6

Mustard and wheat mildew disease classification using deep learning

Prashant Kumar Mishra
Centre for Advanced Studies, AKTU Lucknow

Jagrati Singh
Indira Gandhi Delhi Technical University for Women, Delhi

Prateek Raj Gautam
School of Computer Science, UPES Dehradun

ABSTRACT: This research presents a new tactic for the categorization of mustard and wheat mildew diseases through the appliance of sophisticated deep learning techniques. The proposed method employ complicated convolutional neural network (CNN) to examine and classify images illustrating plant leaves afflict by mildew diseases. Focusing on reasonably considerable crops like mustard and wheat, the research aims to present a consistent and proficient solution for the early classification and control of mildew outbreaks in farming environments. The introduction of our project involves congregation a complete dataset that includes images of both healthy and mildew-affected wheat and mustard leaves. To make sure a reliable input for the deep learning network, the dataset undergoes thorough preprocessing, including scaling, normalization, and expansion. The outcome of the experiment highlights the efficiency of the planned deep learning approach, achieving a noteworthy classification correctness of 98.76%. This remarkable correctness underscores the model's robustness in distinguishing among healthy and mildew infected plant leaves, underscore its prospective as an important tool for early disease revealing and incorporation into precision cultivation practices. Through an evaluation with modern models within the applicable body of research, our method established superior accuracy in its appraisal when evaluated. Our study contributes to the development of systems devoted to precise and programmed analysis of plant diseases. Such systems can improve agricultural efficiency and guarantee food security.

1 INTRODUCTION

Well-organized control procedures and the fortification of agricultural production rely on the timely and precise recognition of these diseases. Mustard and wheat, essential mechanism of global agriculture, face persistent challenge from mildew diseases, which pretense considerable risks to crop health and capitulate. In the present era, deep learning, particularly harnessing convolutional neural network (CNN), has confirmed significant potential in automate and attractive the processes linked with disease categorization. The core objective of this research is to create and advance a robust deep learning model, particularly a Convolutional Neural Network (CNN), modified for the accurate recognition and categorization of mildew diseases in wheat and mustard crops. Mustard popular for its flexibility, it finds value in both industrial and cooking spheres. It serves as an establishment for sauces, supplies green leafy vegetables, and produces edible oil. And wheat, one of the major cereal crops worldwide, is a essential dietary staple for a considerable portion of the world's inhabitants. Its implication lies in providing vital proteins and carbohydrates in the human

 DOI: 10.1201/9781003501244-14

diet. The farming of wheat sustains livelihood and contribute to the financial well-being on a worldwide scale. The spotlight of this learning is to influence the capability of deep learning in mounting a flexible categorization system for mildew diseases upsetting mustard and wheat. Through the operation of advanced algorithms, the research aims to generate an automated solution capable in precisely perceptive between healthy and diseased crops by analyze metaphors of plant leaves. The incorporation of sophisticated deep learning methodologies not only ensures improved accuracy in disease recognition but also offers a quick and proficient approach to early detection. The main aim of this research is to improve precision agriculture by provide farmers and stakeholders with a tool that streamline decision-making in disease organization. Eventually, this fortifies the flexibility of mustard and wheat crops against the challenge pose by mil-dew threats.

1.1 *Related work*

Various researchers have intended for their concentration to the categorization of mustard and wheat mildew diseases, resultant in a varied array of important contributions recognized within this precise field.

(Kukreja, Sharma and Vats 2023) recommended the study that advocate for a hybrid model that merge Convolutional Neural Network (CNN) and Long Short-Term Memory (LSTM) techniques to recognize mustard downy mildew crosswise six distinct strictness levels. The combination of CNN and LSTM layers facilitate the accurate capture of both the spatial and temporal attribute present in the images.(Hridoy, Arni and Hassan 2022) Provide his research to ensure incessant growth in mustard farming, this study introduce an improved method for identify diseases in mustard plants using deep Convolutional Neural Network (CNN). The outcome exhibit that the projected MPNet efficiently and precisely recognize mustard plant diseases.(Bansal, *et al.* 2023) Formulate a Multi-Layer Perceptron (MLP) model for the categorization of wheat stripe rust disease, distinctive it amongst six separate strictness levels. Constructed with a dataset comprise 15,000 images, the model followed an organized approach involving data preparation, model training, model validation, and performance evaluation.

(Nigam, *et al.* 2023) Present a model considered for identify wheat diseases; with a precise focus on detect major wheat rusts, employ the EfficientNet design. The dataset, named WheatRust21, was prepared, surrounding 6,556 images of both healthy and diseased leaves capture in actual field conditions. In the procedure of identify wheat rust disease, a variety of conventional CNN-based models were assess.(Xu, *et al.* 2023) presents a combined deep learning approach called RFE CNN, which integrate elliptic metric learning (EML), a convolution neural network (CNN), a feedback block (FB), and a residual channel attention block (RCAB). The procedure consists of four stages: initially, two parallel CNN are utilizing to pull out essential features from both healthy and diseased wheat leaves. (Bansal, *et al.* 2023) planned a hybrid model was formulate for the recognition and categorization of wheat leaf spot disease, integrate Faster Region-Based Convolutional Neural Networks (R-CNN) and Support Vector Machines (SVM). A dataset consisting of 10,000 images was gathering to distinguish involving healthy and diseased wheat leaves.

(Vasavi, Punitha and Rao 2022) in this learning, the writer conduct an broad review of current progress in forecasting crop leaf diseases using image processing (IP), machine learning (ML), and deep learning (DL) methodologies. (Long, *et al.* 2023)Introduce a set of wheat images classify into five groups, counting healthy plants and four distinct diseases: yellow rust, brown rust, powdery mildew, and Septoria leaf blotch. These images were capture from genuine wheat plant environment, spanning both field and conservatory setting. Consequently, a deep learning model was train using this dataset.(Shafi, *et al.* 2022) This study offer a technological appraisal of sophisticated method for detect wheat rust attacks, cover remote sensing, machine learning, deep learning, and the Internet of Things. (Chaithanya, *et al.* 2022) this research suggest a deep learning-based approach for categorize wheat leaf diseases. The planned approach integrates a mutual generative adversarial network to control image imputation, civilizing the precise judgment of missing data in the dataset through the enduring method.

2 METHODOLOGY

In this segment, the writer delineate the deep learning process and put forth a approach for proficiently discerning between healthy and infected leaves. Our method includes the phase of data collection, data preprocessing, model creation, model training, and model assessment.

2.1 *Data collection and preprocessing*

In the preliminary phase of this research, the data-gathering process involves assemble diverse datasets that include instance of mustard and wheat mildew diseases, along with sample of healthy plants. Following to this, methodical preprocessing measures, surrounding data cleaning, normalization, and image expansion, are implement to guarantee dataset consistency and improve biases. These steps create the establishment for precise and reliable results in the consequent categorization process employ deep learning techniques.

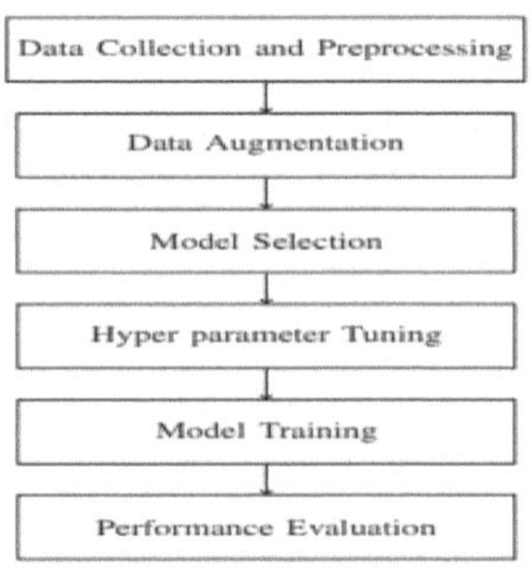

Figure 1. Flowchart depicting the categorization procedure for Mustard and wheat mildew diseases.

2.2 *Data augmentation*

In this study, data amplification is a vital method utilizes to amplify the diversity and dimension of the dataset, thus enhancing the strength and simplification capability of the deep learning model. By employ various transformation such as rotation, flipping, and scaling, amplified data points are created from the existing dataset. This method introduces variation that repeat real-world scenario, enabling the model to proficiently handle diverse conditions and variation in the input data. Data expansion plays a critical role in alleviate overfitting and civilizing the overall concert and toughness of the deep learning model during the preparation phase.

2.3 *Model selection*

In the structure of this study, the resolution to decide the suitable model is vital. The assortment process entails assess dissimilar deep learning architectures; with a precise emphasis on convolutional neural network (CNN), to recognize the most efficient one for classify mustard and wheat mildew diseases. When selecting a model, factor such as architecture difficulty, training time, and computational possessions are taken into account.

2.4 *Hyper parameter tuning*

Hyperparameter tuning stand as a vital optimization pace in refinement the selected deep learning model. Parameters like learning rate, batch size, and regularization potency are systematically attuned to optimize the model presentation. The hyperparameter tuning phase, conduct through an iterative process of testing and evaluation, guarantees that the model is well-calibrated and competent of deliver robust presentation across a variety of scenario, thereby contributing to the overall achievement of the categorization system.

2.5 *Model training*

In the research, the model training method involves input preprocessed data into the preferred deep learning structural design. In excess of iterative epochs, the model learns patterns related to diseases, optimizing hyperparameters, and adjust weights to decrease categorization errors. Thorough confirmation guarantees the simplification of the model to new data, ensuing in a well-tuned model for precise categorization of mustard and wheat mildew diseases.

2.6 *Performance evaluation*

Performance valuation examines the efficiency of the deep learning model in classify mustard and wheat mildew diseases. Metrics like correctness, precision, recall, and F1 score are utilize to enumerate the model's presentation. Thorough assessment of these metrics ensures a complete sympathetic of the model capability and directs probable enhancement in disease categorization accuracy. The assessment process plays a critical role in assess the overall achievement and dependability of the urbanized categorization system.

2.7 *About dataset*

The dataset utilize in this learning contains label information that identify diseases in wheat and mustard crops, source from the Kaggle stage. It represents a thorough anthology of data concerted on diseases disturbing the leaves of two separate crop types: wheat and mustard. The preparation dataset considered for wheat crops includes 902 images depict brown rust leaf disease, 924 images display yellow rust leaf disease, and 1116 images of healthy leaves. At the same time, the test dataset for wheat crops comprise 226 images of brown rust leaf disease, 232 images illustrate yellow rust leaf disease, and 279 images of healthy leaves. This cautiously curate dataset establish a solid base for training and evaluate machine learning models, enable a methodical examination of wheat crop diseases and facilitate the formation of accurate categorization systems for brown rust, yellow rust, and healthy leaves in wheat development.

Table 1. Allotment of files within each class tag in the combined dataset of mustard and wheat.

Class Label	Train Data	Test Data
BELUM SIAP	900 images	90 images
BROWN RUST	902 images	226 images
HEALTHY	1116 images	279 images
SIAP PANEN	900 images	90 images
YELLOW RUST	924 images	232 images

The specific training dataset for mustard crops comprise 900 images depict Belum siap disease and a supplementary 900 images on behalf of Siap panen disease. In the analogous test dataset for mustard crops, there are 90 images illustrating Belum siap disease and a comparable set of 90 images portray Siap panen disease. This modified dataset for mustard crops offers a beleaguered and well-balanced collection for training and assess machine learning models. It aid in the formation of accurate categorization systems for the precise diseases, Belum siap and Siappanen, within the circumstance of mustard cultivation. By addition the datasets for wheat and mustard crops, a consolidate dataset has been compile to comprise five distinctive classes. These classes consist of BELUM SIAP, Brown rust, Healthy, SIAP PANEN, and Yellow rust. This integrated dataset now offers an inclusive demonstration of a variety of diseases and the healthy state in both wheat and mustard crops.

This dataset prove to be a precious source for mounting and evaluate object recognition algorithms, explicitly made to order to make a distinction between healthy and contaminated mildew disease leaves within the circumstance of five frequently predictable disease categories.

Figure 2. Exemplary images showcasing mustard and wheat mildew diseases.

3 RESULT

In evaluate the performance of unusual models for a particular task; the precision scores obtain offer precious insight. Along with the models being measured, the Convolutional Neural Network (CNN) eminent itself with a remarkable truthfulness of 98.76%. In close detection, the VGG16 model showcased its efficiency in the individual context; attain an accurateness of 97.37%, and the ResNet-50 model also display healthy show with a reliability of 98.25%. On the other hand, the Decision Tree model, while capable, established a moderately lower accurateness of 88.45%. It's significant to underscore that the slight dominance of CNN over the others emphasize its efficacy in capture complex patterns contained by the data, interpretation it predominantly well-suited for the nuance of mustard and wheat mildew disease categorization. Additionally, these models undergo training for 50 epochs; underscore the depth of learning and enhancement achieve over an absolute training period. The learning rate parameter was suspiciously configured at 0.0001, representing a specific modification of hyperparameters to improve compliance and junction right through the training process.

Table 2. Model Comparision.

Model	Precision	Recall	FI-Score	Accuracy
Decision Tree	87.3%	87.6%	88.02%	88.45%
CNN	97.36%	96.12%	97.32%	98.76%
VGG16	96.8%	96.9%	97.36%	97.37%
ResNet-50	97.8%	96.9%	97.32%	98.25%

The superior model not just surpasses the base model but also delivers excellent outcome, showcasing a significant accuracy boost. By tactically refinement features and making enhancement, our modified CNN architecture achieved a remarkable highest accuracy of 98.76%, clearly outshine the baseline model. This underscores the efficiency of our method in civilizing classification performance. Nonetheless, all models, counting the Decision Tree, play a responsibility in the assignment success, with the CNN, VGG16, and ResNet-50 models indicating advanced presentation and robust capability in managing the categorization of mustard and wheat mildew diseases.

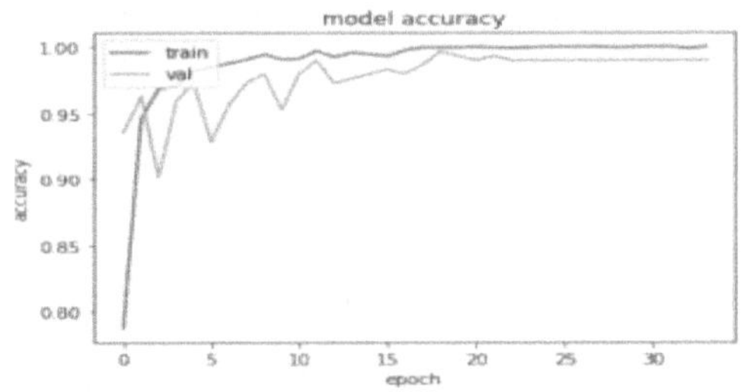

Figure 3. Model accuracy maximum 50 epochs.

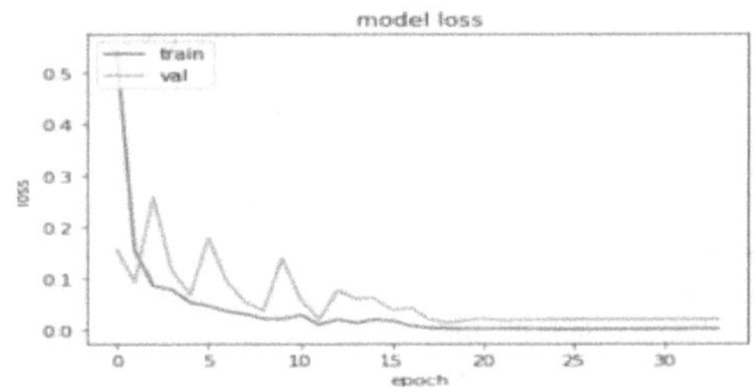

Figure 4. Model loss maximum 50 epochs.

4 CONCLUSION

In termination, the estimation of dissimilar models for mustard and wheat mildew disease categorization unveils understanding performance metrics. The Convolutional Neural Network (CNN) emerges as the top performer; attain a remarkable correctness of 98.76%. In close sequence, the VGG16 model showcased its efficacy with a precision of 97.37%, and the ResNet-50 model display vigorous performance, achieve an accuracy of 98.25%. On the other hand, the Decision Tree model, though capable, established a comparatively lower truthfulness of 88.45%. The CNN extraordinary performance highlights its usefulness in capture complicated patterns within the data, building it well-suited for the nuanced task of mildew disease categozation. These models undergo training for 50 epochs, with a watchful arrangement of the learning rate parameter set at 0.0001, prominence the widespread learning depth and hyperparameter modification for most favorable performance. Regardless of the Decision Tree model sheathing in accurateness compared to the deep learning models, it still acting a role in the task achievement. The diverse range of accuracies underscores the robust capability of the CNN, VGG16, and ResNet-50 models in administration the categorization of mustard and wheat mildew diseases. Additionally, explore real-time accomplishment and deploy models in farming settings could open avenue for realistic application, supporting farmers in appropriate disease recognition and management. In summing up, this learning establishes the foundation for enduring progress in correctness cultivation and crop disease supervision through the appliance of up to date deep learning technique.

REFERENCES

Bansal, Ankit, Rishabh Sharma, Vikrant Sharma, Anuj Kumar Jain, and Vinay Kukreja. (2023). A deep learning approach to detect and classify wheat leaf spot using faster R-CNN and support vector machine. *2023 IEEE 8th International Conference for Convergence in Technology (I2CT)*. 1–6.

Ceyhan, Merve, Yusuf Kartal, Kemal Özkan, and Erol Seke. (2024). Classification of wheat varieties with image-based deep learning. *Multimedia Tools and Applications* (Springer) 83: 9597–9619.

Chaithanya, B. N., T. J. Swasthika Jain, Renuka Patil, and others. (2022). Wheat Leaf Disease classification using modified ResNet50 Convolutional Neural Network model.

Gill, Harmandeep Singh, Bikramjit Singh Bath, Rajanbir Singh, and Amarinder Singh Riar. (2024). Wheat crop classification using deep learning. *Multimedia Tools and Applications* (Springer) 1–17.

Hridoy, Rashidul Hasan, Arindra Dey Arni, and Md Ariful Hassan. (2022). Recognition of mustard plant diseases based on improved deep convolutional neural networks. *2022 IEEE Region 10 Symposium (TENSYMP)*. 1–6.

Kukreja, Vinay, Rishabh Sharma, and Satvik Vats. (2023). Early detection and prevention of mustard downy mildew disease using a hybrid CNN-LSTM model. *2023 Third International Conference on Secure Cyber Computing and Communication (ICSCCC)*. 246–251.

Long, Megan, Matthew Hartley, Richard J. Morris, and James K. M. Brown. (2023). Classification of wheat diseases using deep learning networks with field and glasshouse images. *Plant Pathology (Wiley Online Library)* 72: 536–547.

Mandava, Mamatha, Surendra Reddy Vinta, Hritwik Ghosh, and Irfan Sadiq Rahat. (2024). Identification and categorization of yellow rust infection in wheat through deep learning techniques. *EAI Endorsed Transactions on Internet of Things* 10.

Nigam, Sapna, Rajni Jain, Sudeep Marwaha, Alka Arora, Md Ashraful Haque, Akshay Dheeraj, and Vaibhav Kumar Singh. (2023). Deep transfer learning model for disease identification in wheat crop. *Ecological Informatics* (Elsevier) 75: 102068.

Reis, Hatice Catal, and Veysel Turk. (2024). Integrated deep learning and ensemble learning model for deep feature-based wheat disease detection. *Microchemical Journal* (Elsevier) 197: 109790.

Shafi, Uferah, Rafia Mumtaz, Zunaira Shafaq, Syed Mohammad Hassan Zaidi, Muhammad Owais Kaifi, Zahid Mahmood, and Syed Ali Raza Zaidi. (2022). Wheat rust disease detection techniques: A technical perspective. *Journal of Plant Diseases and Protection* (Springer) 129: 489–504.

Sharma, Anita, Chirag Sharma, Vikrant Sharma, and Satvik Vats. (n.d.) Automatic feature selection using gray-level co-occurrence matrix and detection of mustard plant disease using feed forward neural networks.

Sharma, Rishabh, Vinay Kukreja, and others. (2021). Mustard downy mildew disease severity detection using deep learning model. *2021 International Conference on Decision Aid Sciences and Application (DASA)*. 466–470.

Sharma, Rishabh, Vinay Kukreja, and Richa Gupta. (2023). Enhancing wheat crop resilience: An efficient deep learning framework for the detection and classification of rust disease. *2023 4th International Conference for Emerging Technology (INCET)*. 1–5.

Sood, Akshit, Pradeepta Kumar Sarangi, Ashok Kumar Sahoo, Lekha Rani, Karan Bajaj, and Alok Kumar Agrawal. (2023). AI-driven mustard disease identification: A multiclass and binary classification approach for advanced crop health monitoring. *2023 3rd International Conference on Technological Advancements in Computational Sciences (ICTACS)*. 256–261.

Usha Ruby, A., J. George Chellin Chandran, B. N. Chaithanya, T. J. Swasthika Jain, and Renuka Patil. (2024). Wheat leaf disease classification using modified ResNet50 convolutional neural network model. *Multimedia Tools and Applications* (Springer) 1–19.

Vasavi, Pallepati, Arumugam Punitha, and T. Venkat Narayana Rao. (2022). Crop leaf disease detection and classification using machine learning and deep learning algorithms by visual symptoms: A review. *International Journal of Electrical and Computer Engineering* (IAES Institute of Advanced Engineering and Science) 12: 2079.

Xu, Laixiang, Bingxu Cao, Fengjie Zhao, Shiyuan Ning, Peng Xu, Wenbo Zhang, and Xiangguan Hou. (2023). Wheat leaf disease identification based on deep learning algorithms. *Physiological and Molecular Plant Pathology* (Elsevier) 123: 101940.

Computational Methods in Science and Technology Volume 1 – Kaur et al. (Eds)

Machine and deep learning approaches for heart disease risk assessment

Rahul Sharma, Deepika & Reshma Khan
Department of Computer Science and Engineering, Chandigarh University, Gharaun, Mohali, India

ABSTRACT: Heart disease is a major public health problem the world over, and accurate measures of risk assessment and appropriate classification of persons are very necessary in the early identification for treatment. This is one reason that the techniques of machine learning and deep learning find great application in the healthcare business, as they can analyze patterns of data and make predictions very effectively. The following research paper provides deep research on the techniques of machine learning (ML) and deep learning (DL) used to evaluate and categorize heart disease illness risk. It describes the different problems of the heart, the factors of risk, and models of ML and DL approaches. It also outlines what the current data sources in use are: digital health records, wearable technology, imaging, genetic data, and patient–reported outcomes. The review observes that, though challenges continue to be present in data collection, processing, and feature selection, it further has the ability to make the model free from bias and reliable. Finally, the authors state that such models might play a role in better patient outcomes and appropriate use of health resources, therefore allowing for focused medical strategies for the treatment of cardiac diseases.

Keywords: Heart Disease, Deep Learning, Machine Learning

1 INTRODUCTION

Heart disease kills more than 31% of all people in the world, hypertension is one of the main factors responsible for the occurrence of heart disease. According to a survey in 2012, coronary heart disease claimed the lives of 7.4 million individuals, whereas stroke caused the death of 6.7 million people [4]. According to the WHO, roughly 17 million individuals succumb to cardiovascular diseases annually, representing almost 31% of all global fatalities. Early detection of cardiac disease can effectively treat patients and save countless fatalities. The early–stage evaluation and management of patients by cardiologists continue to be a difficulty. Every conventional heart disease risk–assessment model assumes that each risk factor associated with heart disease outcome has a linear relationship. These models often simplify complicated connections, incorporating several risk variables that have non–linear correlations. It is important to accurately include several risk factors and establish stronger connections between these variables of risk and outcomes. At present, ML/DL models are utilized in the assessment of heart disease risk. The objective of this work is to ascertain if machine learning (ML) can improve the accuracy of cardiovascular risk prediction in large population primary care settings and to identify the ML method that yielded a relatively high level of conciseness. In recent years, numerous ML and DL models for the detection of cardiac disease have been introduced [5–7].

DOI: 10.1201/9781003501244-15

2 DATASET USED FOR HEART DISEASE RISK PREDICTION

Various datasets have been utilized to predict the probability of cardiac illness ML and DL techniques. [16–18].

Table 1. Datasets for predicting the risk of heart illness.

Datasets	Description
Cleveland Heart Disease Dataset:	It is a frequently used research method for predicting the existence or lack of cardiac disease. The dataset consists of 303 occurrences and 14 attributes, which encompass demographic data, clinical tests, and diagnostic test outcomes.
Framingham Heart Study Dataset:	The long–term cardiovascular cohort investigation serves as a significant asset for cardiovascular research, encompassing models for predicting risks for heart disease and associated outcomes that are constructed using participant data accumulated over several decades.
UCI Heart Disease Dataset:	The UCI ML repository offers four datasets on cardiac disease diagnosis from hospitals in Cleveland, Hungary, Switzerland, and Long Beach, providing a comprehensive set of information for predictive model training and assessment.
MIMIC–III Dataset:	The MIMIC–III dataset, accessible to the public, is used by researchers to construct prognostic models for cardiac results, including the risk of heart disease.
Cardiovascular Disease Datasets from Kaggle:	Kaggle, a data science competition platform, provides datasets on cardiovascular disorders, including gender, age, cholesterol, blood pressure, and clinical measures, fostering the development of advanced predictive models.

3 EARLY DISEASERISK ASSESSMENT USING MACHINE LEARNING

3.1 *Comparative analysis of machine learning algorithms for*

Table 2. A summary of ML/DL models heart disease risk prediction studies.

Reference/ Year	ML/DL Models	Classification category	Dataset	Tool/Method	Limitation	Results
Uddin, K. M. M. *et al.* (2023) [1]	RF, DT, SVM, NB, XGB, MLP, LR, GB, and Adaboosting	Cardiovascular vascular disease prediction	Heart Disease Data Set, IEEEDataPort, Heart Disease Predication data set	Scikit–Learn, Warnings, OS, Seaborn, and Matplotlib libraries.	The lack of a comprehensive collection of information from various sources can pose a significant challenge due to the prevalence of noise and inaccurate data.	The DT approach exhibited the best level of accuracy compared to other ML algorithms, achieving a remarkable accuracy rate of 99.16%.
Hambali, M. A., *et al.* (2023) [32]	PCA, DT	Heart Disease Prediction	UCI repository	WEKA, TANGARA/ Intelligent classification method	Additional effort is required to narrow down the 1.4% discrepancy in misclassified cases.	PCA achieves a precision rate of 98%, a sensitivity rate of 100%, and an accuracy rate of 98%.

(*continued*)

Table 2. Continued

Reference/ Year	ML/DL Models	Classification category	Dataset	Tool/Method	Limitation	Results
						PCA exhibited higher accuracy, precision, and sensitivity in comparison to the DT.
Srinivasan, S., et.al (2023) [66]	RBF neural networks and Naive Bayes	ML methods for forecasting cardiovascular heart disease.	UCI repository	Learning vector quantization	Big–data analysis is crucial for diagnosing heart disease and improving cardiovascular system designs, but challenges like noise, incompleteness, and abnormalities hinder	The study revealed that Naive Bayes and RBF neural networks attained an accuracy of 94.78% for forecasting coronary heart disease, whereas the Learning Vector Quantization technique reached a higher accuracy of 98.78%.

4 CONCLUSION

This paper explores the capabilities and limitations of ML and DL models in predicting heart disease. The choice of using ML or DL techniques very much depends on multiple factors. For example, the complexity of data, level of interpretation required, availability of computational resources, and which objective is being targeted through the analysis. This may indicate potential power in hybrid approaches between the strengths of each model, which could make significant advancements to future research in risk assessment for heart disease. The current study operates based on comparing prediction models as to the risk of heart disease. In particular, it elaborates gradient boosting machines and neural networks as to accuracy, logistic regressions (LR) and decision trees (DT) in relation to the interpretability of the model and human–understandable data structure, convolutional neural networks (CNN) in the context of image data, recurrent neural networks (RNN)/LSTM models in relation to sequential data, and transformer algorithms with respect to textual data, coupled with hybrid designs.

REFERENCES

[1] Uddin, K. M. M., Ripa, R., Yeasmin, N., Biswas, N., & Dey, S. K. (2023). Machine learning–based approach to the diagnosis of cardiovascular vascular disease using a combined dataset. *Intelligence–Based Medicine*, *7*, 100100.

[2] Marquina, C., Talic, S., Vargas–Torres, S., Petrova, M., Abushanab, D., Owen, A., ... & Ademi, Z. (2022). Future burden of cardiovascular disease in Australia: impact on health and economic outcomes between 2020 and 2029. *European Journal of Preventive Cardiology*, *29*(8), 1212–1219.

[3] Patel, B., & Sengupta, P. (2020). Machine learning for predicting cardiac events: what does the future hold. *Expert review of cardiovascular therapy*, *18*(2), 77–84.

[4] Weng, S. F., Reps, J., Kai, J., Garibaldi, J. M., & Qureshi, N. (2017). Can machine–learning improve cardiovascular risk prediction using routine clinical data?. *PloS ONE*, *12*(4), e0174944.

[5] Bharti, R., Khamparia, A., Shabaz, M., Dhiman, G., Pande, S., & Singh, P. (2021). Prediction of heart disease using a combination of machine learning and deep learning. *Computational Intelligence and Neuroscience, 2021.*

[6] Guarneros–Nolasco, L. R., Cruz–Ramos, N. A., Alor–Hernández, G., Rodríguez–Mazahua, L., & Sánchez–Cervantes, J. L. (2021). Identifying the main risk factors for cardiovascular diseases prediction using machine learning algorithms. *Mathematics*, *9*(20), 2537.

[7] Swathy, M., & Saruladha, K. (2022). A comparative study of classification and prediction of Cardio–Vascular Diseases (CVD) using Machine Learning and Deep Learning techniques. *ICT Express*, *8*(1), 109–116.

[8] Jindal, H., Agrawal, S., Khera, R., Jain, R., & Nagrath, P. (2021). Heart disease prediction using machine learning algorithms. In *IOP Conference Series: Materials Science and Engineering* (Vol. 1022, No. 1, p. 012072). IOP Publishing.

[9] Asif, M. A. A. R., Nishat, M. M., Faisal, F., Dip, R. R., Udoy, M. H., Shikder, M. F., & Ahsan, R. (2021). Performance Evaluation and Comparative Analysis of Different Machine Learning Algorithms in Predicting Cardiovascular Disease. *Engineering Letters*, *29*(2).

[10] Types of Heart Disease. Available at: https://www.heartandstroke.ca/heart–disease/what–is–heart–disease/types–of–heart–disease accessed on 26–01–2024

[11] Bays, H. E., Taub, P. R., Epstein, E., Michos, E. D., Ferraro, R. A., Bailey, A. L., ... & Toth, P. P. (2021). Ten things to know about ten cardiovascular disease risk factors. *American journal of preventive cardiology*, *5*, 100149.

[12] Reddy, P. R., Poojitha, G., Kavitha, S., Samreen, S. L., Naseer, A., Koteswari, P., & Soumya, P. (2022). A prospective observational study to assess the cardiac risk factors and treatment patterns in established heart diseases. *Cellular, Molecular and Biomedical Reports*, *2*(4), 265–275.

[13] Peng, M., Hou, F., Cheng, Z., Shen, T., Liu, K., Zhao, C., & Zheng, W. (2023). Prediction of cardiovascular disease risk based on major contributing features. *Scientific Reports*, *13*(1), 4778.

[14] Khandia, R., Puranik, N., Lodhi, N., Gautam, B., Alqahtani, T., Alqahtani, A. M., ... & Chidambaram, K. (2021). Comparing heart risk scores to identify the most important risk factors for cardiovascular diseases. *European Review for Medical & Pharmacological Sciences*, *25*(24).

[15] Tillmann, T., Läll, K., Dukes, O., Veronesi, G., Pikhart, H., Peasey, A., ... & Bobak, M. (2020). Development and validation of two SCORE–based cardiovascular risk prediction models for Eastern Europe: a multicohort study. *European Heart Journal*, *41*(35), 3325–3333.

[16] Reddy, K. V. V., Elamvazuthi, I., Abd Aziz, A., Paramasivam, S., & Chua, H. N. (2021, July). Heart disease risk prediction using machine learning with principal component analysis. In *2020 8th International Conference on Intelligent and Advanced Systems (ICIAS)* (pp. 1–6). IEEE.

[17] Alhammadi, R. A., Alhosani, M. A., Al–Ali, H. M., & Soudan, B. (2023, February). Cardiac Disease Prediction using Artificial Intelligence. In *2023 Advances in Science and Engineering Technology International Conferences (ASET)* (pp. 1–7). IEEE.

[18] Diwan, S., Thakur, G. S., Sahu, S. K., Sahu, M., & Swamy, N. K. (2022, May). Predicting Heart Diseases through Feature Selection and Ensemble Classifiers. In *Journal of Physics: Conference Series* (Vol. 2273, No. 1, p. 012027). IOP Publishing.

Computational Methods in Science and Technology Volume 1 – Kaur et al. (Eds)

Uncovering agile software frameworks

Kamna Vaid & Udayan Ghose
University School of Information, Communication & Technology, Guru Gobind Singh Indraprastha University, Delhi, India

ABSTRACT: As the present development process is more customer centric and demands quick delivery within a given budget. Agile Software Development represents a drastic change in software development approaches focusing more on client needs and saying good bye to traditional and plan-based approaches. It has become an important domain in project development worldwide. This paper provides an insight about various agile methodologies: Extreme programming, Scrum, Kanban, DSDM and Crystal methodologies. Also, different areas of study in the field of agile software comprising Requirement Engineering, different Agile Frameworks, Risk management, People factor and Organizational culture. This study has analyzed recent trends and areas across multiple agile frameworks.

Keywords: Agile software, Agile Frameworks, Requirement Engineering, Risk Management, People factor, organizational culture

1 INTRODUCTION

With the changing trends in global software development, software industry has evolved with different practices focusing on more quality, productivity as well as optimum utilization of all the resources. Agile Software Development paradigm has become popular as it is a flexible approach in managing the rapid and unpredictable changes in the requirements and focus on three important C's (coordination, collaboration and communication) between customers and developers (Dingsøyr *et al.* 2012). In comparison to the traditional Software Development Methodologies (SDMs) as software industry is becoming more volatile, the agile methodologies such as Scrum, Extreme Programming (XP), Crystal methods, Lean Development, have the abilities to deal with such environment (Batarseh *et al.* 2018). Agile methodologies are more customer-centric in contrast to traditional SDM where user role is important only during specification development (Batarseh *et al.* 2018). The agile development provides more productivity, greater product and service quality, leads to team effectiveness, reduce development cycle times, provides an ease in communication and sharing information between all the stake holders. It also reduces the delivered documentation (Hofmann *et al.* 2018). Agile is about adaptive planning and evolutionary development.

For successful implementation of a software project, it is important to understand all the dimensions as well as challenges of software development. It was understood by the developers that the complicated interrelationship of a number of issues such as organisational, cultural, technological, and economical issues must be taken into account for the development (Dingsøyr *et al.* 2012; Beck *et al.* 2004b). This study's aim is to provide a broad overview about different dimensions in the area of agile.

2 AGILE SOFTWARE DEVELOPMENT MODEL

Software Development approaches are being classified into two categories as Adaptive methodologies and Predictive methodologies. Traditional development approach is termed

DOI: 10.1201/9781003501244-16

as predictive methodology and is adopted under specific requirements and resources in well-defined projects. It is very difficult to alter the project at later stages. Predictive approach is process oriented, such as Waterfall model. The Adaptive methodologies are people oriented and change driven approach which allows greater flexibility in implementation of a project such as Agile model.

2.1 *Agile model*

With advancement in technology and more involvement of customers in the development process, gradually focus of development has shifted towards agile software development. Agile model has multiple approaches which can be used for project development as per the requirement of the project. Each approach has its own pros and cons both traditional and agile methods were in use depending upon the requirements of the projects.

Key benefits of agile models are shorter development cycle, quick adaption to changing/ evolving business requirements and it significantly reduces the risks involved in the project. Though with flexibility feature agile started gaining popularity.

This model should be used when

i) Project size is small
ii) Project has changing requirements
iii) Constant customer involvement is there
iv) Short and iterative cycles are required

It is very difficult to choose which model is better. The selection of appropriate model is done on the basis of project, its requirements and specifications. Before starting the work for development of a project, it is important to have a picture of the final product. Hence, according to that the model can be selected. For instance, if requirements are not fixed in advance, then it is difficult to implement Waterfall model as we cannot make changes after a phase is over.

3 AGILE METHODOLOGIES

With the ingestion of agile manifesto in 2001(Beck *et al.* 2001), there is an increase in number of agile methods including popular methods, such as Extreme Programming, Scrum, Crystal family of methodologies, dynamic systems development method (DSDM).

Extreme Programming (XP) was the beginning of the agile methodologies in 1999. Earlier user specifications were used to get freezed and cannot be changed once the development begins but with the advent of agile it has become possible to make changes in the project at point of time during the development.(Beck *et al.* 2004a).

Software development teams need to have the depth of knowledge and skills for deciding which development methodology will be best suited according to the project requirement. Software development team adopt the selection methodology depending upon various factors such as project requirement, convenience, team size, budget etc. methodologies may be modified or customized according to the requirements (Gren *et al.* 2020). Some of the agile methodologies are included in this review.

3.1 *Extreme programming (XP, XP2)*

The most experienced agile software development methodology. Extreme programming (XP) came into limelight in 2000 by Kent Beck (Beck *et al.* 2004b) who have stated its key principles and practices. XP takes the whole team together. It is more customer-oriented methodology. Everyone in the team has clear understanding and awareness of about what needs to be done. It divides the complete project into smaller releases which makes editing and testing of the code easy. It also lay emphasis on proper interaction amongst the customer and development team.

User requirements are known as User scenario or storycard. These storycards are discussed within the whole team before the development starts. Requirements are divided into smaller tasks based on the schedule and budget of the project. These storycards are validated beforehand by the customer and next release is decided according to the customer's feedback. Key characteristics of Extreme programming are user stories, small releases, continuous integration, testing, refactoring and many more. The revised version of XP, "XP2" consists of the "primary practices" such as "complete team together, informative workspace, pair programming, user stories, weekly cycle, quarterly cycle, slack, 10-minute build, continuous integration, test-first programming, and incremental design". There are also 11 "corollary practices". (Beck *et al.* 2004b; P. Abrahamson *et al.* 2002) (Jivan *et al.* 2020).

3.2 *Scrum*

Scrum is a light weight, iterative and incremental project management approach used in projects where feedback is of utmost importance. It is the most popular Agile framework. Scrum main emphasis is on efficient team work to produce quality output. It is opposite to waterfall model.

The scrum process consists of the following entities: roles, sprints, stakeholders and a scrum master. In scrum, the project is divided into "Sprints". Each sprint starts with planning and ends with a review and feedback. Sprints are defined during daily scrum time-boxed meeting to synchronise all the work and are supervised by Scrum Master. The role of Scrum master is also to facilitate the relationship between both the Product owner and the team members and solving problems of the team. The stakeholders of the project also attend sprint review meetings. After every sprint, backlog is maintained to add features to be implemented including all the work to do and modifications required. Then Product owner gives the final implementation instructions for the next sprint. (Hossain *et al.* 2009; Heeager *et al.* 2018).

3.3 *Kanban*

The famous lean production technique Kanban was developed for the Toyota Production System, came into software development in late 2000's. The word kanban is originated from Japanese language, refers to a sign or visual board. Hence, it uses visual elements on a Kanban Board for the development and management of a project. This approach gives a clear visibility to each team member regarding all the tasks and workflow involved in the development.

Kanban is a change management approach. It encourages continuous development and improvement in the process by limiting WIP (Work In Progress). It measures the average time to complete one cycle known as lead time or cycle time. It tries to predict and optimize this lead time as short as possible. Scrum also tries to limit WIP but indirectly not directly as Kanban.

It is also an incremental approach and much lighter in weight than Scrum. In Scrum teams are cross functional but such teams are optional in Kanban though Kanban is more focused on whole workflow as one rather than one team. It combines are multiple teams and domains as one and build a complete product by getting the right work completed at the right time. It also supports delayed commitment on both prioritization of new work and delivery of current work due to its nature of pull approach. (Lei *et al.* 2017; Hofmann *et al.* 2018).

3.4 *Crystal methodologies*

This methodology focuses on customization. A cluster of different colours (Red, Yellow, Orange, Blue and Clear) forms crystal methodologies. Each colour is given to a methodology and represents the size of the team also. It is useful for projects with small team size as it emphasis more on communication. Four phases in order are performed: staging, review, monitoring, and parallelism (Cockburn 2004). Features of clear development are quick delivery of the product, more emphasis, noticeable improvements, more communication, easy interaction with industry experts etc. (Dyba *et al.* 2008).

3.5 *Dynamic Systems Development Method (DSDM)*

The Dynamic Systems Development Method (DSDM) ensures timely delivery of the project. It adjusts the project time and resources and fix functionalities according to them so that it takes the least time for product delivery (P. Abrahamson *et al.* 2002). Customer involvement is there at all the stages of product development. It gives a framework for Rapid Application Development. DSDM consists of five phases: feasibility study, business study, functional model iteration, design and build iteration, and implementation. Its iterations are timeboxed. It can be useful for both large and small projects (Stapleton 1999).

3.6 *Comparative analysis of all five agile methodologies*

Table 1. Comparative analysis of five popular agile methodologies.

Practices	XP	SCRUM	KANBAN	DSDM	CRYSTAL METHODOLOGIES
Approach	Iterative and incremental approach	Iterative and incremental approach	Incremental approach	Rapid application development approach	Flexible approach
Delivery	Flexible delivery in small releases	Delivers in iterations called sprints	No set phase, focus on JIT delivery	Creates prototype as early as possible	Frequent delivery of working software
Project size and complexity	Small & less complex project	Any type of project	Any type of project	Any type of project	Any type of project
Team size	Small team size, 3–20 members	Small team size, 5–9 members	No limit	Depend upon size of the project	Depend upon the color of the methodology
Interaction cycle	1–6 weeks	2–4 weeks sprint cycle	No multiple cycles	Multiple cycles	Multiple cycles
Meetings	Daily stand up meetings	Has 5–6 meetings	Daily short meetings	Daily stand up meetings	Face to face personalised Daily or weekly meetings
Changes accommodation	Easy to accommodate changes	Resists to accommodate changes	Easy to accommodate changes	Easy to accommodate changes	Easy to accommodate changes
Limitations	Works on strict priority order	Limit work in progress (WIP) for each iteration	Limit work in progress (WIP)for each workflow	Can be restrictive and inhibit developer creativity	Require upfront planning, lack of predictability.
Customer/User involvement	Active involvement	More customer oriented	More customer oriented	Emphasis on user involvement in every release	Emphasis on customer involvement
Major Practices	User Stories, Pair programming	Sprint, Sprint Backlog, Product Backlog, Scrum Meetings	Kanban Board, stickies	Prototyping	Crystal Family

4 STUDIES RELATED TO AGILE

Agile methodologies are becoming popular and being adopted by all organizations depending on the organization size, their requirements and the project team experience. Requirement changes are very rapid and continuous leading to increase in demand for agile methodologies. Different areas in the field of software engineering related to agile are explored in this study, covering four aspects of agile development which are listed in the figure below.

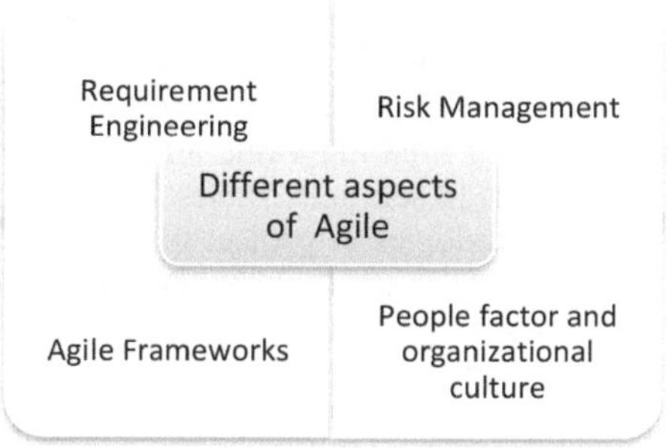

Figure 1. Different aspects of agile development.

4.1 *Requirement engineering*

Each software engineering project begins with the Requirements. They are the foundation of all the software products. These requirements need to be defined, documented and prioritized at the beginning of the project. But as the development begins requirements keeps on changing since all the requirements are not well known at the initial stage. Requirement Engineering is one of the important aspects which helps in choosing which agile practices will be best suited for the project. A survey conducted on agile practices (Ochodek *et al.* 2018) have taken 31 agile RE practices and evaluated the relative importance of practices amongst 136 Agile software development practitioners on their perception of importance of these 31 agile RE practices.

PROMETHEE family methods are used for creating the ranks of the practices and three-point ordinal scale is used for evaluating their importance. They have also, studied the relationship

between the perceived importance of the practices and the experience of the respondents. Factors like country, size, domain etc also influence the perception about the importance of the Agile RE practices.

Requirement Engineering is of utmost important in distributed agile systems development where onshore and offshore activities are combined. To perform this effectively the team need to manage requirements and changes efficiently. Lloyd *et al.* has introduced an approach to manage such requirement changes. They have introduced a new model- "Feature tree" by modifying the feature model. They have integrated feature tree (which is based on feature model) with distributed agile approach NAPDP (The New Agile Process for Distributed Projects) and has associated this with a web-based tool which helps to manage requirement changes. NRC, the largest research centre in Egypt, has evaluated and validated their supporting tool in real time environment (Lloyd *et al.* 2017).

Another study has focused on a continuous trend towards continuous value delivery and showed emerging research methodology from practitioner as well as researchers point of view. Authors have plotted the Relative interest over time on themes "Scrum", "extreme programming", "DevOps" and "continuous integration" based on searches in Google Trends from 2006 to 2015 as well as for publications in the Scopus scientific database. This study showed a downfall in the popularity of extreme programming from practitioners point of view, however shown an increase in popularity of Scrum methodology over time. Continuous integration and DevOps have also shown a remarkable uptrend in recent years (Dingsøyr *et al.* 2012). (Dingsøyr *et al.* 2016).

4.2 *Agile frameworks*

Sánchez-Morcilio *et al.* focused on a continuous assessment of agile methodologies in project management and how to assimilate best agile methodologies using an optimized agile technique (OAT). Their study has classified data as an aid and challenges to develop OAT. OAT helps in selecting the right agile tools, establishing communication among team members and authenticates quality management for creating value added outcome. The Researches have listed advantages and challenges from 18 selected articles. They have found 96 advantages and 74 challenges for using agile methodologies in project management. The synthesized data is shown with the help of a meta-analysis chart. (Sanchez-Morcilio Rosarito *et al.* 2018).

There are varied agile frameworks available. A comparison between such 5 frameworks based on scrum or lean technology or both, such as Scrum of Scrums (SoS), Scaled Agile Framework (SAFe), Large-Scale Scrum (LeSS), Disciplined Agile Delivery (DAD) and Lean Scalable Agility for Engineering (LeanSAFE) are given by Ebert *et al.* Authors have also covered case studies of two globally distributed companies using scalable agile practices. (Ebert *et al.* 2017).

According to the version one (Anon n.d.), i.e the largest and longest-running agile survey in the world. The most popular approach to scaling agile is The Scaled Agile Framework (SAFe). It is an extensive framework with a hierarchical structure along with well-defined team roles. Approximately one-third of the respondents (29%) says that they are adopting SAFe more closely. In comparison to their previous year report there is a largest increase in scaling approach. Disciplined Agile Framework (DA), has also shown an increase from 2016 till now.

Kettunen P (Kettunen *et al.* 2017) has recognised which design principles can be useful for future organizations. Author has given brief about how organizations needs to modify its framework and methods for digitization. They have created artefacts (the grids) for software organizations going into digital transformations. They have assimilated scaled agile framework like SAFe (The Scaled Agile Framework) and LeSS (Large-Scale Scrum) into their grids.

LeSS is light weight framework, more concentrated on communication, scale down and work with the small sized development teams whereas SAFe defines the roles who will manage the communication process and works to create a value flow and an economic framework for a company. Both of them include techniques like Continuous Integration and DevOps. They have suggested that before adopting any such framework company should focus on speed requirements too and speed depends on organizations complexities.

In another survey done on providing scaled agility, almost 30 percent of the participants have adapted the Scaled Agile Framework (SAFe) (Ochodek *et al.* 2018). SAFe is considered as a complex and heavy framework. Agile methodology is carried by diverse factors. Many frameworks have been developed by different researchers such as The F.K.Y. Chan *et al.* (Chan *et al.* 2009) suggested a framework showing knowledge management outcomes such as knowledge creation, knowledge retention, and knowledge transfer has an impact on the acceptance of agile.

4.3 *Risk management*

Agile methodologies are also applied in the area of risk management. It minimizes the project risks as changes can be incorporated immediately. A practitioner survey (Doss *et al.* 2016) on Agile Development in Safety Critical Systems, demonstrated the relationship between three key activities (safety requirements development, hazard analysis, and safety case development) in safety engineering and an agile approach. The survey has concluded that throughout the lifecycle of product development, with respect to risk, the assurance effort, the safety of the product, requirements prioritisation are very important in incremental development of any software. Safety Assurance Principles on Scrum are also stated in this paper.

Kulkarni *et al.*(Kulkarni *et al.* 2021) has explored the area of regression testing in ASD and agile methodologies. This paper has also mentioned optimization strategies and difficulties faced by agile methodologies using regression testing and automation.

Batarseh *et al.*(Batarseh *et al.* 2018) has introduced a Analytics-driven testing (ADT) method which has capability to forecast future software failures in the successive agile sprints. By using forecasting regression model ADT forecast errors and explains, where and what type of errors are likely to occur in scrum with high assurance by keep on tracking the Mean time between Failures (MTBF) for software components. ADT helps in decision making by using previous data and considering different patterns. Author has explained the main four pillars of ADT (Scrum, Context-driven testing, MTBF Analytics and regression forecasting) in detail. Authors have demonstrated Analytics-driven testing with an AI system showing 93.5% success rate using regression model for predicting errors. Barroca *et al.* (Barroca *et al.* 2018) have explored practitioner driven challenges with agile. They proposed a model displaying practitioner challenges and tries to connect the void between research and practice.

In another paper (Hamed *et al.* 2013), A conceptual model is developed on the literature study of agile development in safety critical context. It has also described the problem areas in safety critical context. Dingsøyr *et al.* (Dingsøyr *et al.* 2012), explains one of the major the challenge in agile development t is the relationship between customers and suppliers. Suggestions by them will help to have a better relationship as well as add value towards continuous deployment. They have focused on continuous value delivery.

4.4 *People factor & organizational culture*

By leveraging flexible planning and development agile development encourages cooperation among cross-functional and self-organizing teams. The idea of virtual teams has gained popularity in the wake of Covid-19, but it has existed since ages. Though there are many challenges in front of virtual teams such as setting up for responsibilities within the team, how to complete all the assigned tasks within given time frame and team should work together with a common objective. Sharing knowledge with all the stakeholders is also a big challenge as communication and collaboration are the main aspect for smooth functioning. Factors such as time zone difference, establishing & maintaining trust, lack of one-one physical meetings acts as a hindrance for sharing knowledge.

In research, done by Neumann *et al.* concluded agile practices are being adopted by teams at an increasing rate and communication and well-being of team members have increased during Covid 19. (Neumann *et al.* 2022). The relevance of considering how virtual teams use communication tools is emphasised and advised teams to establish rules on how to use the technologies to enhance coordination. (Stray *et al.* 2019) Gren *et al.* study indicates a strong correlation between group maturity and two agile practices by investigated the relationship between agility and group maturity. With a survey of 34 agile teams and 1 university using the Group Development Questionnaire and the Perspective Agile Measurement (PAM). (Gren *et al.* 2020). Another survey (Mishra *et al.* 2023) has considered people dimension as one of the critical aspect of agile success factor. This study has given a crisp review about multiple studies related to people factor which has provides insights about agile teams, leaders and customers. Also, covered significant components such as leadership, ROI, communication, customer satisfaction etc.

Organizational culture plays a vital role in working of a successful delivery of any project. Agile manifesto has stated importance of communication and collaboration as principles of agile (Beck *et al.* 2001). In a systematic review done by *Mkoba et al.* listed all the organizational culture attributes that contributes to agile practices acceptance. Nine primary studies analysis revealed that market, values, teamwork, management control, and creativity are the five significant organizational cultural traits that have an impact on the adoption of agile approaches (Mkoba *et al.* 2022). Another study(Jivan *et al.* 2020) has investigated conduciveness of the organizational culture in the context of agile methodologies adoption. This study has showed that organizational culture has to be aligned with the agile principles. Virtual teams are helping as an aid to cater time-to-market demand, maintaining budget, for sharing knowledge, and provides instant solution to challenging organisational issues. Through the removal of time and geographic constraints, virtual teams helps in joining skills and talent of all the key stakeholders of the business. (Ebrahim *et al.* 2009).

5 CONCLUSION

Insights about the agile software development along with its popular methods Scrum, Extreme Programming, Kanban, Crystal methodologies and DSDM are discussed in this study. It has explored previous work done on multiple agile aspects such as agile methodologies, agile practices, framework, requirement engineering, challenges and techniques. For choosing the best method or framework: organizational characteristics such as its size and culture; team characteristics such as team size, experience, location; and product type and complexity are important aspects to be considered. Accordingly, the best framework can be chosen for the project. As agile is an emerging dynamic technology which is being used worldwide now. This paper has been limited to four areas of study in the field of agile development comprising Requirement Engineering, Agile Frameworks, Risk Management, People factor and Organizational Culture. In future more area related to agile context such as its behavioural aspects, combining agile with AI tools, combining multiple agile frameworks together can be explored, to provide more insights about the agile software development.

REFERENCES

Abrahamson P., Outi Salo, Jussi Ronkainen, & Juhani Warsta. (2002). *Agile Software Development Methods: Review and Analysis.* VTT Publications.

Anon.Version one 12th agile report. Available at: https://explore.versionone.com/state-of-agile/versionone-12th-annual-state-of-agile-report [Accessed October 20, 2018].

Barroca, L., Sharp, H., Salah, D., Taylor, K., & Gregory, P. (2018). Bridging the gap between research and agile practice: An evolutionary model. *International Journal of System Assurance Engineering and Management* 9(2).

Batarseh, F.A., & Gonzalez, A.J. (2018). Predicting failures in agile software development through data analytics. *Software Quality Journal* 26(1).

Beck, K. *et al.* (2001). Manifesto for agile software development. *The Agile Alliance* 2009(December 14).

Beck, K., & Andres, C. (2004a). *Extreme Programming Explained: Embrace Change.*

Beck, K., & Andres, C. (2004b). *Extreme programming explained: Embrace Change*, Second Edition. Addison Wesley Professional.

Chan, F.K.Y., & Thong, J.Y.L. (2009). Acceptance of agile methodologies: A critical review and conceptual framework. *Decision Support Systems* 46(4).

Cockburn, A. (2004). *Crystal Clear A Human-Powered Methodology for Small Teams.*

Dingsøyr, T., & Lassenius, C. (2016). Emerging themes in agile software development: Introduction to the special section on continuous value delivery. *Information and Software Technology* 77: p.56–60.

Dingsøyr, T., Nerur, S., Balijepally, V., & Brede Moe, N. (2012). A decade of agile methodologies: Towards explaining agile software development. *The Journal of Systems & Software* 85.

Doss, O., & Kelly, T.P. (2016). Challenges and Opportunities in Agile Development in Safety Critical Systems. *ACM SIGSOFT Software Engineering Notes* 41(2).

Dyba, T., & Dongsoyr, T. (2008). Emperical Studies of Agile Software Development: A Systematic Review. *Information and Software Technology* 50(9–10).

Ebert, C., & Paasivaara, M. (2017). Scaling Agile. *IEEE Software* 34(6).

Ebrahim, N.A., Ahmed, S., & Taha, Z. (2009). Virtual teams: A literature review. *Australian Journal of Basic and Applied Sciences* 3(3).

Gren, L., Goldman, A., & Jacobsson, C. (2020). Agile ways of working: A team maturity perspective. *Journal of Software: Evolution and Process* 32(6).

Hamed, A.M.M., & Abushama, H. (2013). Popular agile approaches in software development: Review and analysis. In *Proceedings – 2013 International Conference on Computer, Electrical and Electronics Engineering: 'Research Makes a Difference', ICCEEE 2013*,

Heeager, L.T., & Nielsen, P.A. (2018). A conceptual model of agile software development in a safety-critical context: A systematic literature review. *Information and Software Technology* 103.

Hofmann, C., Lauber, S., Haefner, B., & Lanza, G. (2018). Development of an agile development method based on Kanban for distributed part-time teams and an introduction framework. In *Procedia Manufacturing*,

Hossain, E., Ali Babar, M., & Paik, H.Y. (2009). Using scrum in global software development: A systematic literature review. In *Proceedings – 2009 4th IEEE International Conference on Global Software Engineering, ICGSE 2009*,

Jivan, K.P., Marnewick, A.L., & Joseph, N. (2020). Influence of organizational culture in the adoption of agile. In *2020 IEEE Technology and Engineering Management Conference, TEMSCON 2020*,

Kettunen, P., & Laanti, M. (2017). Future software organizations – agile goals and roles. *European Journal of Futures Research* 5(1).

Kulkarni, V., Madhan Mohan, R., & Venkateswara Reddy, H. (2021). *Regression Test Optimization and Automation in Agile Framework: A Review*.

Lei, H., Ganjeizadeh, F., Jayachandran, P.K., & Ozcan, P. (2017). A statistical analysis of the effects of Scrum and Kanban on software development projects. *Robotics and Computer-Integrated Manufacturing* 43.

Lloyd, D., Moawad, R., & Kadry, M. (2017). A supporting tool for requirements change management in distributed agile development. *Future Computing and Informatics Journal* 2(1).

Mishra, D., Mishra, A., & Abdalhamid, S. (2023). Facilitators and inhibitors of Agile methods adoption: Practitioners view. *Systems Engineering* 26(6).

Mkoba, E.S., & Marnewick, C. (2022). Organisational culture attributes influencing the adoption of agile practices: A systematic literature review. *Journal of Information Systems Engineering and Management* 7(1).

Neumann, M., Bogdanov, Y., & Sager, S. (2022). The covid 19 pandemic and its effects on agile software development. In *ACM International Conference Proceeding Series*,

Ochodek, M., & Kopczyńska, S. (2018). Perceived importance of agile requirements engineering practices – A survey. *Journal of Systems and Software* 143.

Sanchez-Morcilio Rosarito, & Quiles-Torres Francisco. (2018). An optimized agile technique to effectively implement agile methods in project management. *Issues in Information Systems*.

Stapleton, J. (1999). DSDM: Dynamic systems development method. *Proceedings of the Conference on Technology of Object-Oriented Languages and Systems, TOOLS*.

Stray, V., Moe, N.B., & Noroozi, M. (2019). Slack me if you can! using enterprise social networking tools in virtual agile teams. In *Proceedings – 2019 ACM/IEEE 14th International Conference on Global Software Engineering, ICGSE 2019*,

Computational Methods in Science and Technology Volume 1 – Kaur et al. (Eds)
© 2025 the Author(s), ISBN: 978-1-032-81760-6

Calculate the asphalt pavement's binder activity level using machine learning techniques

Prabhjeet Kaur, Priyanka Kaushal, Gagandeep & Nisha Kumari
CEC-CGC Landran

ABSTRACT: This study outlines the various machine learning (ML) techniques used to identify a simple, effective algorithm that reliably predicts the level of binder activity in asphalt pavement. The mixture contains 5% asphalt binder. Although more material has been added and the properties have changed, it is still a petroleum derivative. This paper examines and contrasts five distinct machine learning approaches used to calculate the degree of asphalt pavement (AP). Indirect Tensile Strength (ITS), air void, and compaction temperature are used to determine the precise degree of binder. In particular, various machine learning techniques—support learning machine (SVM), KMC, artificial -NN, random forest regression, and multivariate polynomial regression with regulation—were examined in order to determine the most accurate date representation model.

Keywords: Machine Learning, Strength of indirect tensile, SV-machine, KMC, random forest regression, multivariate polynomial regression with regulation

1 INTRODUCTION

Asphalt pavement is the term used to describe any paved road with an asphalt surface. In order to create asphalt pavement, hot mix asphalt (HMA), which is composed of 95% stones, sand, and gravel, is cemented together with asphalt cement. The HMA is loaded onto trucks and transported to the paving site. Traffic is permitted only after the pavement has cooled. HMA is the most widely used type of asphalt, despite the fact that there are basically five common types [1]. Hot mix asphalt is a well-known HMA that combines fine aggregate and asphalt binder to form a flexible mixture that prevents weathering of pavement [2]. The pavement is mixed and poured by HMA between 300 and 350 degrees Fahrenheit. It has cooled down with ease. The temperature is raised to a range of 30 to 120 degrees Fahrenheit [2]. It is possible to use it for longer distance transportation to paving sites because it does not require a high temperature. Compared to HMA, it is more cost-effective. Because it produces fewer greenhouse gases, it is safe for the environment. One of the characteristics of asphalt pavement that is used to improve the AP classification is the level of binder activity. Two methods are defined blending degree and availability of binder degree, which are further refined by degree of binder activity. In mixture, there are 5% asphalt binder.

It is still a petroleum derivative even though more material has been added and the properties have changed. The exact degree of binder can be determined using air void, compaction temperature, and indirect tensile strength (ITS). The basic equation for figuring out BoA degree:

$$DOA(\%maxITS) = 100.ITS(RAP_{iT\ degree\ C})/ITS(RAP_{i\ max})$$

A ML model is developed which is used to determine the asphalt pavement binder degree. At 25.5 degree this model takes the inputs the compaction temperature, air voids, and

 DOI: 10.1201/9781003501244-17

Indirect. Equation 1 (using the methodology of Lo Presti *et al.* [1]) defined the DoA. For sample R, ITS (RAPi,max) was the highest compaction temperature that was tested. and ITS (RAPi, T°C) was the average of the sample RAPi compacted at temperature T.

The method of managing the degree of asphalt pavement is machine learning. Several algorithms are used in machine learning to fit data to a mathematical function. Algorithms are trained in this way to make predictions or classifications, and some of them also reveal variables in data mining. Several algorithms are used in this paper. The most basic example of a machine learning technique will be linear regression. This study compares five distinct techniques to determine which is more complex or simpler for binder of asphalt movement. The degree of asphalt pavement can be managed using a variety of techniques. To simulate intricate relationships involving numerous variables, multivariate polynomial regression is utilized. These intricate connections typically have many dimensions and are non-linear. The degree fit, which resembles curves and abrupt jumps in advanced relationships. It refers to the largest exponent in each term, not to the total number of terms. N-order polynomial regression model:

$$y = bo + b_1\ \mathrm{x} + b_2\ \mathrm{x}^2.\ldots.\mathrm{b_n x^n} + \mathrm{e}.$$

Multivariate polynomial regression is the application of polynomial regression to multiple independent variables. Despite being highly complex, it offers us greater accuracy and flexibility because it uses multiple variables in one equation. Algorithms based on brain function comprise artificial neural networks, or ANNs. Both structured and numerical data can be entered. It can generate logistic output where data values are set as discrete numbers and the result of sigmoid values produces a range of 0 to 1. It can be applied to a broad range of problems. It can create intermediary functions that aid in variable handling by connecting the features and the activation function i(j). ANN primarily relies on the following factors: Sigmoid function (g(z)), units in each layer, input variable, and number of layers during activation function. All complex issues, such as the non-linear relationship between I/O in real life situations, can be handled by ANNs. It does not place restrictions on the input variables. However, it is useful in financial time series forecasting when there is high volatility and non-constant variance data. RFR creates a sample of the data, trains each sample, and classifies it. It exhibits tree-structure-like behavior when creating data samples.

2 LITERATURE REVIEW

In literature, Hasan Ziari [2023] demonstrated that the addition of RAP materials improved the asphalt samples' moisture sensitivity, rutting resistance, and indirect tensile strength. Furthermore, rutting resistance and permanent deformations of asphalt samples were enhanced by raising the RAP content. The resilient modulus and tensile strength of asphalt samples were also improved by the addition of RAP; however, this improvement was dependent on the amount of RAP used. Additionally, the rejuvenators' effects on the performance attributes of RAP mixtures demonstrated that the addition of Cyclogen and VB raised the moisture sensitivity of RAP mixtures while decreasing their resilient modulus, tensile strength, and strain creep [4]. Shaffer [2023] found the various algorithm of machine learning to minimize the compaction of samples of RAP [5]. Alnaqbi [2023] found the algorithm which can handle the temperature in air voids and compaction using SVM-kmeans clustering and find the high DOA Value. [6]. Khushal *et al.* [2022] were discussed about the various AI Application using Machine Learning. [7]. Otyhman [2022] used ANN in Marshall test parameters to increase the NN precision and generate precise predictions, multiple ANNs are tested in this study. Mineral aggregate can all be predicted by the deep neural network (NN) using R values. However, it operated within the parameters of the Marshall test and identified the content of assault, which was previously identified by numerous researchers as having a content amount of 5%. Nader Karballaeezadeh [2022] suggested method eliminates the need for ground penetrating radar testing, it is less expensive and easier to implement. The accuracy and quality of the calculation process are enhanced when machine learning techniques are used in place of backcalculation [8]. Priyanka *et al.* [2022] noted that deep learning and neural networks interact to identify the best possible solution. [25]. By conducting the Marshall Stability test, Ankita Upadhya *et al.* [2022] worked to Choose a suitable bitumen content prediction model for carbon fiber reinforced AP and Marshall Stability for flexible pavement. The authors utilized five machine learning methods to anticipate the Marshall stability and ascertain the optimal binder amount. M.U Munar [2020] found a technique which is useful for laborious task to optimize the performance of five machine learning techniques In order to forecast pavement performance. [9,10] Wang [2020] created a hybrid method that combines support vector machine regression (SVR) and grey relation analysis (GRA). The study came to the conclusion that the multi-variate, non-linear predict model developed by GRA-SVR, which offers higher precision and operability, can be used to predict pavement performance. [2019] Davide Lo Presti Research on a fundamental characteristic of RAP, the level of binder activity is improve the accuracy. Only the DoA, the binder's availability, and the degree of mixing were examined. Researchers do not discuss a specific degree or fixed amount of RAP mixture; instead, they only discuss how to use it. According to Mehrdad Honarm *et al.* [2019], there is not enough bitumen on slate pavement to withstand traffic loads, making repairs and maintenance expensive. Four different methods of testing the basalt modifier to enhance bitumen performance were used in this study. They discovered that base bitumen increased in low temperatures and provided a good option for road construction if 6% synthetic basalt was added to the mixture. In their 2018 study, Hamed Majifard *et al.* employed hybrid artificial neural networks (ANN/SA) and gene expression programming (GEP) to forecast the fracture energy of asphalt mixture specimens. The formula for fracture energy takes into account a number of predictor variables, including the asphalt binder's performance grading (PG), aggregate size, aggregate gradation, test temperature, and crumb rubber content. This method can be used for predesign purposes, but it is only practical when testing is not practical. Dr. Ahmed and colleagues (2014) conducted research on the appropriate amount of content to be added to asphalt pavement. They noted that five percent of the entire sum would suffice. However, they also come to the conclusion that adding steel to AP increases the amount of void in the asphalt and improves compaction and temperature stability. Machine learning was investigated by W.H. Oliver *et al.* [2011] in order to forecast Marshall Stability and determine the ideal binder content for increased strength. They categorized eight machine learning models for prediction, each of which provided a different percentage for binder content and did not yield an accurate result. According to Andrew F. Bradley's research from 2007, testing and modeling can be used in conjunction with viscoelastic bulk to understand how asphalt concrete behaves at low temperatures. [11] The fracture energy of 28 MHA mixtures intended for cold climates was measured in this study. The significance of the binder content fracture energy at higher temperatures, aggregate type, and temperature is shown by a statistical analysis of the data. On the other hand, there was no discernible variation in fracture energy due to the chosen air void levels and binder content at lower temperatures. [12] Baoshan Huang *et al.* (2005) carried out controlled tests in a lab capable of blending virgin mixture with the recycled asphalt pavement (RAP) procedure. They found that only a tiny portion of RAP was utilized during the remixing process; the remaining RAP was applied to the RAP aggregates as a hard covering, forming composite black rock. It can be applied when we wish to enhance Hot Mix Asphalt's performance rather than gauge the level of activity in the asphalt itself [13]. Barzin Mobasher and colleagues (1997] employed two methods grounded in non-linear fracture mechanics. In order to conduct three point bending tests at 02 different temperatures, beam specimens with various binder contents had to be prepared. This findings showed that the asphalt-rubber mixture had a larger resistance to cracking and a higher fracture toughness when compared to asphalt concrete. Toughness values increased with increasing binder content for both asphalt concrete and asphalt-rubber binder.

In this study, the researchers only talked about the amount of RAP that should be mixed to create a pavement mixture; other studies talked about RAP mixtures at low temperatures [16]. However, in this study, we compare and contrast five machine learning algorithms to determine which one is most effective at determining the degree of AP binder. The K-means clustering algorithm was examined in this survey to determine the degree of AP binder, which creates a cluster of similar-type variables. Next, the SVM (Support Vector Machine) was investigated to learn more about temperature compaction and how to address air voids in the temperature. We conducted surveys regarding the artificial neural network (ANN) algorithm and deep learning techniques that also addressed temperature compaction. Currently, we are comparing all of the algorithms whose data is accessible at Github-RAP-data at RILEM-TCRAP at TG5.

3 MATERIALS AND METHODS

3.1 *Testing*

This Survey we are taking from testing methods of all the 5 Machine Learning Algorithms in which various Laboratories collaborated in this project, in which 5 groups of RILEM Technical committee 264 on RAP. Each Committee use one or more samples which is 100% RAP cylindrical Specimen by heating at 70 degree, 100 degree, 140 degree, 170 degree and 190 degree).These samples are mixed with hands while controlling the temperature.

3.2 *Methods*

3.2.1 *SVM*

This is mostly used ML method to solve the classification methods. In this Approach SVM is used for regression analysis to predict the values of set of labels from give set of features. SVR is behaves likes kernel which is used to compute the similarities between data points. SVR works as following: Data was Split in to three sets: Train the data: The Training data-set is use to train the model and obtain the parameters using SVR techniques with an RBF kernel.

3.2.2 *K-Means clustering*

K-means An iterative unsupervised learning algorithm called clustering splits the K-Different Clusters so that any group of clusters can be a member of any other group [17]. Classification of air voids and ITS data points are accomplished using various steps: Define the K-Clusters. Select the random value of K and make it centroids and assign the data values near to the centriods. By the use of Elbow method determined the value of K is 4. Within each cluster data points were separated by their air voids and ITS. Co-Ordination of centriods of DoA is greater than 80% of ITS.

K-means clustering	*airvoids*	*ITS(Mpa)*	*%DoA>80% Max ITS*	*Mean temp*
1	*4.8*	*1.72*	*56*	*151*
2	*12.4*	*1.94*	*74*	*168*
3	*15.1*	*v0.69*	*29*	*125*
4	*23.1*	*0.34*	*5*	*96*

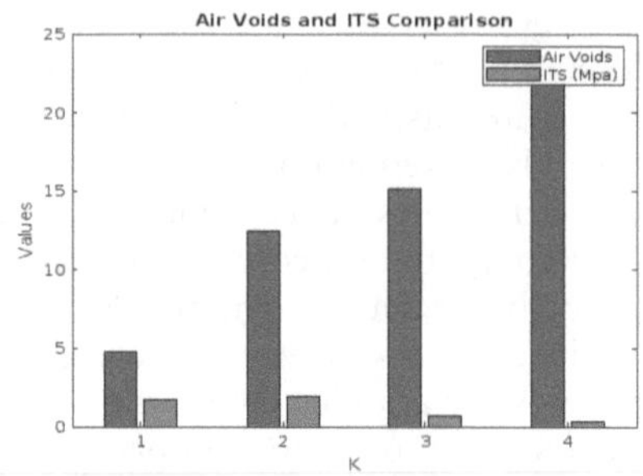

Figure 1. Bar graph of ITS and airvoids.

Figure 1 shows the bar graph of K-Means clustering Only when the K-Values is at 1 it can use the maximum DOA is 74% and having maximum ITS value which is 1.94.

3.2.3 *Multivariate polynomial regression*
Using the test set, MPR performance was assessed for classification using

$$h\theta(Xtest) = \theta TXtesth(test) = test.$$

Three sections make up this data: training, validation, and testing. Because of this, a crossvalidation exercise was required, involving the training of 50 distinct models with 50 distinct random data splits. Figure 2 displays Using all 50 models, the average value of the MAE on the test set label predictions was 11.9% max. ITS, with a 1.131% max ITS standard deviation. For the original data set, the standard deviation (max ITS values) was 31%.

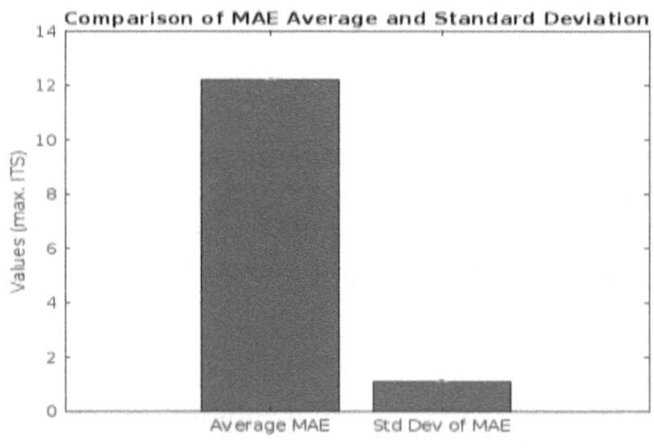

Figure 2. Multi variate polynomial regression.

Artificial neural network: In this data split into three parts.

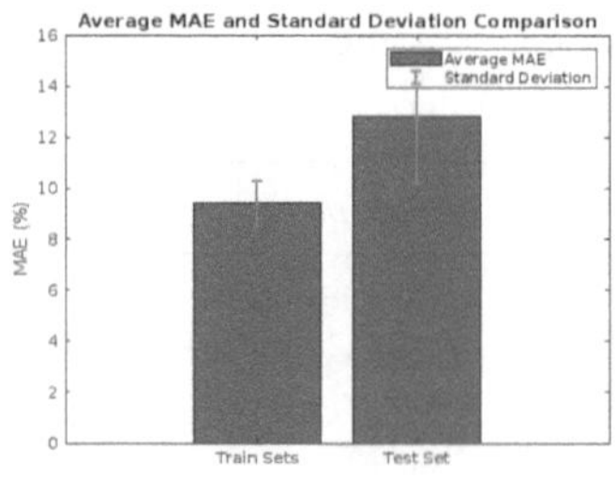

Figure 3. Random forest regression.

3.2.5 *Random forest regression*

Figure 3 shows the regression model the test data and training sets splits into 50 parts. In this DoA value of MAE precision were 11.7%Max ITS, use 81% of DoA and 0.77 respectively.

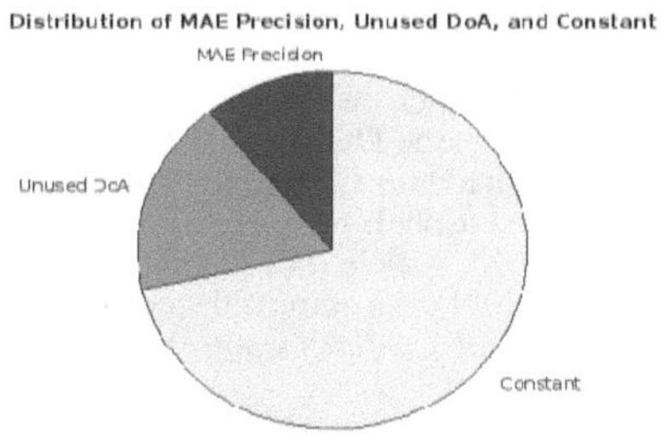

Figure 4. Comparison of different algorithms.

4 COMPARISON OF DIFFERENT ALGORITHMS

S.No.	Model	DOA	ITS
1	**Support vector Machine**	9.2%	3.2%
2	**K-Means Clustering**	74%	1.94%
3	**Multivariate Polynomial regression**	31%	1.1%
4	**Artificial neural network**	12.8%	0.9%
5	**Random forest regression**	81%	11.7%

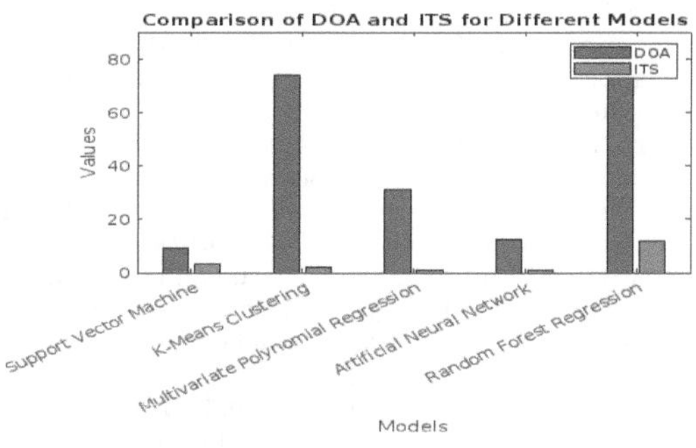

Figure 5. Comparison of different algorithms.

5 CONCLUSION & FUTURE SCOPE

Five Machine learning algorithms are used to find the best solution of degree of binder of asphalt pavement using Temperature compaction, DOA and ITS of specimen. Out of these models SVR uses the lowest temperature compaction that is 25 degree but the DoA is low where as in K-Means it make the cluster only at K = 1 it gave good solution but the temperature compaction is 168 degree In MPR temperature compaction is 66 degree whereas in ANN it is 72degree and in RFR it is 68%. but it give simplest solution which uses the higher degree of DoA and MAX ITS i.e. 81% and 11.7. As a result, the simplest approach is RFR i.e. random forest regression.

REFERENCES

[1] Kennedy TW, Solaimanian M, Tam WO. (1998). Using the superpave system to maximize the use of reclaimed asphalt pavement. *J Assoc Asphalt Paving Technol*, 67, pp. 311–333.

[2] Anderson RM, McDaniel R. (2001). *The* technician's handbook suggests using recycled asphalt pavement in the superpave mix design method. *Advocate Using Superpave Mix Des Method Tech Man with Reclaimed Asph Pavement.*

[3] JWH. (2001). The impact of RAP's binder on the characteristics of recycled asphalt. *2(3):311–325. Road Mater Pavement Des.* In this case, the DOI is 10.1080/14680629.2001.9689906.

[4] Amir Saadatjoo, Vahid Najafi, Moghaddam Gilani, Amir Amini, Seyed Mohsen Hosseinian. (2023). Laboratory investigation of reclaimed asphalt mixtures containing cyclogen and vacuum bottom rejuvenators, *Adv. in Civil Engg.*, vol. 2023, Article ID 6223569, 11 pages.

[5] Chard K., Thain D. and Shaffer T. (2021). An empirical study of package dependencies and lifetimes in binder python containers, *International Conference on eScience 17th (eScience)(IEEE)*, Innsbruck, Austria, pp. 215–224.

[6] Hamad, K.; Barakat, S. (2023). Creating rutting pms through machine learning techniques utilizing the long-term pavement performance database. *Sustainability*, 15, 13653.

[7] Wolosick, J.R. (1980) Modifier influence in the characterization of hot-mix recycled material. *Transp Res Rec*, 777, pp.15–22.

[8] Karballaeezadeh, N. H., Mohammadzadeh Shadmehri, Ghasemzadeh Tehrani. (2020). Estimation of flexible pavement structural capacity using ML techniques. (2020). *Front. Struct. Civ. Eng.* **14**, 1083–1096.

[9] Wood LE, Noureldin AS (1987) Rejuvenator dispersion in hot-mix recycled asphalt pavement binder film. *Revised Transp Res* 1115:51–61.

[10] Munir, M. U., Elnour, A., Huda, S., and Zeiada, W. (2020). Activated crumb rubber modified binder as a sustainable paving material: Pavement performance consideration, *Advances in Science and Engineering Technology Conferences*, pp. 1–7.

[11] Bressi S, Orešković M, Pires GM, Lo Presti D, Vasconcelos K. (2020). Regarding the level of blending with recycling agents and the binder activity of recovered asphalt. *Pavement Road Mater Des* 21 (8):2071–2090.

[12] Arámbula-Mercado E, Martin AE, Kaseer F. (2019). A technique to measure the amount of effective RAP binder (reclaimed asphalt pavement binder) in recycled asphalt mixtures. *J Transp Res Board* 2673 (1):205–216; Transp Res Rec. The publication number is 0361198118821366.

[13] Hobbs A, Wen H, and Zhang K. (2015). Numerical models and laboratory experiments involving the combination of superheated virgin aggregate and recycled asphalt pavement components. *J Transp Res Board* 2506(1):62–71; Transp Res Rec. The DOI is 10.3141/2506-07.

[14] Dumont AG, Poulikako L.D. (2015). Particle clustering phenomena in hot asphalt mixtures with high content of reclaimed asphalt pavements. *Constr Build Mater*, 100, pp. 207–217.

[15] Cavalli, M.C., Bressi, S., Griffa, M., G., Poulikakos, Partl, M.N., L.D., Tebaldi. (2016). Multiscale imaging and characterization of the effect of mixing temperature on asphalt concrete containing recycled components. *J. 463 Microsc.* 264, pp. 22–33.

[16] Zhou F., Scullion T., Li H., Lee R., and Claros G. Recycled asphalt shingle binder characterization and blending with virgin binders. *J. Transportation Research Board*, No. 2370, 2013, pp. 33–43.

[17] Stimilli A, Canestrari F, Virgili A. (2015). A novel approach for calculating the quantity of reactivated binder in recycled hot-mix asphalt. *Surface Mater Pavement Des* 16 (supl), pp. 442–459.

[18] D., L., Presti D. (2020). A review of the literature on the quantitative evaluation of the factors related to the mixing of recycled asphalt binder and recycling agent. *Constr Build Mater* 234:117323.

[19] Pitts W (2021) A rational computation of the concepts inherent in nervous system action. *Math Biophys Bull* 5(4), pp.115–133.

[20] Shirodkar, P., K. Sonpal, Y. Mehta, A. Nolan, A. Norton, C. Tomlinson, P. Sullivan, R. Sauber and E. Dubois. (2011). A study to determine the degree of partial blending of reclaimed asphalt pavement (rap) binder for high rap hot mix asphalt. *Construction and Building Materials*, 25(1), pp. 150–155.

[21] B. F. (2012). New method for detecting asphalt contamination within fine aggregate medium through chemical testing, *J. of Mat. in Civil Engineering*, Vol. 25, No. 2, pp. 252–255.

[22] Schmidhuber, J., (2015) *An Overview of Deep Learning Using Neural Networks*. 61:85–117.

[23] Specht L, K. O. (2022). Artificial intelligence applied to asphalt-rubber viscosity modeling. *Int. J. of Pavement Engineering*, 15(9), 799–809.

[24] Kleinberg, M.E., (2021). *Stochastic discrimination and the traveling salesman problem*, 1, pp. 207–239.

[25] Breiman L. (2001). Using iterated bagging to debias regressions. *Mach Learn.*, 35, pp. 261–277.

[26] Liu, C.-M. (Eds.). (2022). *The future of emerging computational technologies: practical solutions, examples, and case studies* (1st ed.). CRC Press.

[27] K, R, S., G., Deep. (2022). Securing internet of things devices against code tampering attacks using return oriented programming, *Computer Communications*, 193(1), pp. 38–46.

[28] Seema Rani, Neeraj Mohan, Surbhi Gupta, Priyanka Kaushal, Amit Wason. (2022). A deep learning and multilayer neural network approach for coronary heart disease detection, *Book Society and the Future of Emerging Computational Technologies*

Computational Methods in Science and Technology Volume 1 – Kaur et al. (Eds)
 ISBN: 978-1-032-81760-6

Empowering healthcare with Swin Transformer V2: Advancing pneumonia diagnosis through deep learning

Vaibhavi Rajesh Mishra & Ruchika Malhotra
Department of Software Engineering, Delhi Technological University, Delhi, India

ABSTRACT: One of the most important tasks of medical diagnostics is fast and accurate pneumonia diagnosis from chest X-ray images. The final results of this process have critical implications for differential and other types of therapy, and patient discharge. Study would like to tell you about the application of the Swin Transformer V2, innovative deep learning architecture, to the medical imaging domain, and its application to pneumonia diagnosis. The study's data presents the model's evaluation relating to a strong CXR dataset offering a wide variety of conditions, and several manifestations peculiar for pneumonia. The model's performance is closely compared with several deep learning architectures, such as AlexNet, MobileNetV3, VGG-16, ResNet 50, DenseNet, etc, showing its superiority with an accuracy of 98.6%. The results of the latter study indicate that the effectiveness of Swin Transformer V2 model is significantly superior to traditional convolutional neural networks in identifying subtle patterns that indicate pneumonia. It can be assumed that the use of diagnostic procedures based on new developments in the field of model structures will permanently set new records for diagnostic accuracy. In this way, the current study opens up opportunities for the integration of advanced deep learning models with uncertain outcomes into clinical diagnostic processes, which may have a transformative effect on the healthcare industry.

1 INTRODUCTION

The need to detect pneumonia accurately and quickly is critical to medical experts, with a massive influence on patients and their treatment. Deep learning in medical image analysis has shown a significant value and revolutionized the diagnostic process (Lee, 2017). In such new situations, advanced model architectures, especially those based on transformers, help to assure diagnostic is more efficient and accurate (Liu,2022) Swin Transformer V2, the first model that stands out in its the ability to represent ca potentials on the basis of these advances, offers promising features for the diagnosis of lung disease from chest x-ray (CXR) datasets (Bi, 2021).

Thus, this paper proposes a new application of the Swin Transformer V2 model, aiming to exploit its sophisticated design to develop new parameters for accurate and fast diagnosis of pneumonia The chest radiograph CXR dataset serves as a test site for this effort. This data set presents a gamut of diagnostic challenges, including changes in imaging status, patient positioning, and subtle manifestations of pulmonary status, requiring models with special discrimination (Gabruseva, 2020).

Furthermore, global pneumonia mortality through 2019 (Figure 1) highlights the importance of this study. The Swin Transformer V2 design that integrates a shift-window based self-attention mechanism is particularly suited to address these challenges (Rajpurkar, 2017). It is skilled in capturing complex patterns in CXR images of pneumonia thus, ensuring high fidelity in the detection process.

DOI: 10.1201/9781003501244-18

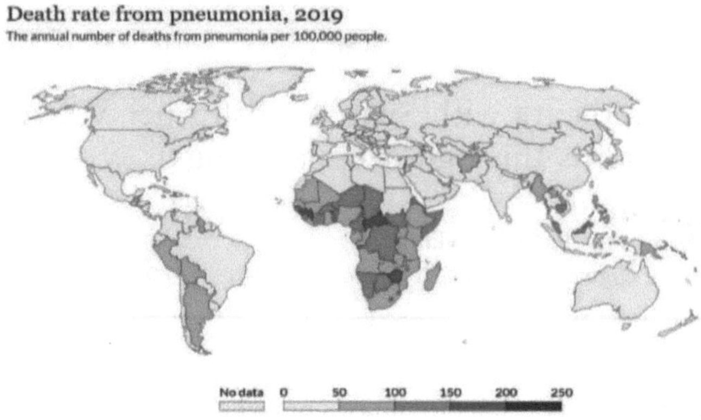

Figure 1. Pneumonia mortality rate.

By leveraging this model, goal is to make a major input to the body of work in medical image analysis, particularly in enhancing the diagnostic processes that are vital for patient care (Tsuneki, 2022).

1.1 *Problem formulation*

This study focuses on the important task of accurately diagnosing pneumonia using CXR images, using Swin transformer as the cornerstone of diagnostic design traditional methods many stumbling blocks when encountered pneumonia descriptions on CXR images, which known for its highest ability to unravel complex image textures through its innovative hierarchical architecture (Liu, 2021). The model will undergo a large training phase, using an observed CXR dataset so well implemented, based on state-of-the-art GPU technology (Huang, 2017). This study is augmented by strategic data enhancement & deft application of transfer learning techniques, aimed at further improving lung disease detection and diagnosis. To improve the accuracy and reliability of pneumonia screening, this study seeks to emphasize healthcare efficiency & patient outcomes, particularly in the early detection of this common but challenging respiratory condition in (He, 2016).

1.2 *Research objectives*

1. Develop an integrated model using Swin Transformer to fit the challenging task of diagnosing pneumonia in CXR images. The aim of this model is to utilize the transformers capacity to capture details and subtleties, in medical images guaranteeing superior accuracy and dependability. (Rajpurkar, 2017).
2. Gather a range of chest X ray (CXR) images encompassing various pulmonary conditions diverse groups of patients and a mix of images. This dataset will be used to train a model that underscores the significance of diversity, in mirroring world clinical scenarios.
3. Thoroughly assess the model by utilizing a particular validation dataset that accurately mirrors world clinical scenarios. Utilize metrics like AUC ROC, sensitivity, specificity and F1 score to gauge the effectiveness of the model, in diagnosing conditions.
4. Exploring the challenges of diagnosing the model involves identifying types of pneumonia and differentiating pneumonia from other lung conditions. This study focuses on understanding the factors that influence the model's performance and identifying areas, for improvement. (Liu, 2021).

5. Delve into advanced methods like knowledge transfer enhancing data and constantly refining models to boost their performance. These approaches focus on enhancing the model's resilience across imaging scenarios and patient situations leading to a broad diagnostic ability, with a high level of accuracy. (Karmani, 2018).

2 METHODOLOGY

The study presents a novel approach to improve pneumonia diagnosis through advanced machine learning, with a special focus on the Swin Transformer V2 model by integrating this state-of-the-art system with CXR on the accuracy of the archived images to accurately classify pneumonia. Discussing the settings and parameters of the model for medical imaging tasks, and emphasize the importance of creating and recording the correct data sets. The approach involves data pre-processing, model setting, training, evaluation, and refinement, and provides a clear blueprint for using Swin Transformer V2 in medical imaging applications.

2.1 *Proposed methodology*

2.1.1 *Swin Transformer V2*

In this search using the advanced functionality of Swin Transformer V2, an unprecedented advance in vision model design that overcomes previous limitations built on the new "shift window" design derived from the original Swin Transformer, this model divides images into non-overlapping patches for localized real focusing, compared to ViTs, blending the advantages of the hierarchical structure found in CNN with reduced computing load.

Notably, Swin Transformer V2 introduces improvements such as residual post-norm levels to reduce training variability, as well as cosine assumption methods and log-difference continuous-level biases, which reduce reliance on labelled datasets so that it is complete

Demonstrating that this method performs well on pneumonia datasets, an important task in medical imaging where the accuracy of detection is paramount for patient care with an impressive capacity of 3 billion parameters, efficiently processing critically demanding high-resolution images heavily labelled data. These properties highlight its high performance in tasks such as image classification, object recognition, and semantic classification. Its modular design and hierarchical structure allow experts to capture fine detail and context. Data's multi scale characteristics and its ability to adapt to diverse inputs and tasks enhance its effectiveness in a wide range of computer vision applications. (Jian, 2016).

2.1.2 *Datasets*

In this study project, compiled a collection of information by combining data from 15 different public sources. This dataset includes chest X ray images that're important, for studying COVID 19. Originally comprising over 20,179 X-rays across four categories—COVID-19, normal, viral pneumonia, and bacterial pneumonia—the dataset was refined to eliminate redundancies, ultimately retaining 4,543 images of COVID-19, 5,445 of normal cases, 4,465 of viral pneumonia, and 5,726 of bacterial pneumonia after removing duplicates using advanced image processing techniques, including the Inception V3 model and machine learning algorithms. This curation effort also corrected for various image quality issues, such as artifacts, incorrect exposures, and misalignments, ensuring the dataset's integrity and utility for accurate disease analysis and machine learning applications.

2.2 *The proposed model*

Table 1. Procedure of proposed model.

Steps	Description
A. Dataset Preparation	Download pneumonia dataset, organize into class folders, verify annotations. Ensure annotation files presence or generate if necessary. Arrange images within class folders for easy accessibility.
B. Model Configuration	Install dependencies, obtain Swin Transformer V2 source code/ weights. Configure model parameters: input size, learning rate, batch size. Set up deep learning and computer vision libraries.
C. Data Pre-processing	Resize images to match Swin Transformer V2 input size. Augment dataset for diversity with rotations, flips, scaling. Convert annotations to Swin Transformer V2 compatible format.
D. Training	Divide the dataset (e.g., 80:20) into sets to be trained and validated. Initialize Swin Transformer V2, specify loss function, optimizer. Train model, monitor metrics: accuracy, loss.
E. Evaluation	Use trained Swin Transformer V2 for predictions on test set. Compute the evaluation metrics (F1-score, recall, accuracy, and precision). Visualize predictions with bounding boxes, labels.
F. Fine-tuning and Optimization	Fine-tune Swin Transformer V2 based on dataset characteristics, feedback. Explore transfer learning, additional data augmentation techniques. Iterate training, evaluation steps for model optimization.

3 EXPERIMENTAL ANALYSIS

3.1 *Baseline models*

- **AlexNet**: Uses max-pooling after convolutional layers to extract features from CXR pictures.
- **MobileNetV3**: Employs lightweight convolutions suitable for efficient processing of CXR images. (Howard, 2019)
- **InceptionV3**: Integrates inception modules to capture multi-scale features from CXR images. (Szegedy, 2016)
- **VGG-16**: Implements a deep architecture with stacked convolutional layers for feature learning from CXR images. (Simonyan, 2014)
- **Xception**: Utilizes depth wise separable convolutions for efficient feature extraction from CXR images. (Chollet, 2017)
- **EfficientNet B0**: Offers a scalable and efficient architecture for extracting features from CXR images across different scales.
- **DenseNet**: Exploits dense connectivity patterns to capture rich feature representations from CXR images. (Huang, 2017)
- **ResNet 50**: Incorporates residual connections to enable deeper networks for feature extraction from CXR images. (He, 2016)
- **Swin Transformer V1**: Integrates the Swin Transformer architecture for capturing long-range dependencies in CXR images. (Liu, 2021)
- **Swin Transformer V2**: Enhances the Swin Transformer architecture for improved feature representation learning in CXR images. (Liu, 2022)

4 RESULTS AND DISCUSSION

The pneumonia detection using the Swin Transformer V2 model on the CXR dataset yielded impressive results. The model achieved 98.6% accuracy, 98.5% precision, 98.1% recall, and a 98.3% F1-score, demonstrating its effectiveness. It outperformed other models, including AlexNet, MobileNetV3, and its predecessor, Swin Transformer V1. This improvement highlights the enhanced representational capabilities and spatial hierarchy learning of the Swin Transformer V2. AlexNet and VGG-16 showed accuracies of 97.2% and 95.3%, respectively, emphasizing their significance in image classification.

Table 2. Performance analysis of current model across different models.

Methods	Accuracy	Precision	Recall	F1-score
AlexNet	97.2	96.4	96	96.2
MobileNetV3	96	96.5	95.5	96
InceptionV3	89.5	90.2	89	89.6
VGG-16	95.3	95.5	95.1	95.3
Xception	83.5	84.3	83.2	83.7
EfficientNet B0	90.6	89.9	89.3	89.6
DenseNet	94.7	93.9	93.5	93.7
ResNet 50	95.1	95.6	95.1	95.3
Swin Transformer V1	96.4	95.9	95.4	95.6
Swin Transformer V2	98.6	98.5	98.1	98.3

While simpler architectures like MobileNetV3 prioritize efficiency with a precision of 96.5%, deeper models such as ResNet 50 and DenseNet achieve accuracies over 94%, highlighting their effectiveness in feature extraction. However, the Swin Transformer V2 outperforms them in achieving a balanced precision-recall trade-off, establishing itself as a benchmark for pneumonia detection. Its high precision minimizes false positives, crucial in medical diagnostics, while its high recall ensures true positive cases are not overlooked.

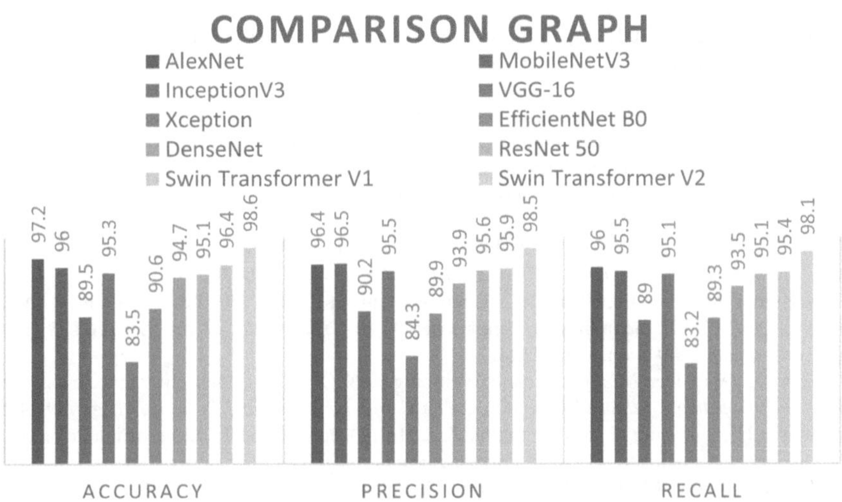

Figure 2. Comparison of various models.

The F1-score, further confirms the balanced strength of the Swin Transformer V2 in both precision and recall metrics. This balance is important in medical image analysis, where it is imperative to achieve a trade-off between detecting as many positive cases as possible (high recall) and keeping the false alert rate low (high precision).

The outstanding performance of the Swin Transformer V2 can be attributed to its innovative architecture that uses adjustable windows and focusing mechanisms, enabling it to capture V-tail microscopic patterns at chest X-ray internally, with pneumonia -Indicators and the model can process images at various scales to detect abnormalities with high fidelity, making it an impressive performance index In conclusion, Swin Transformer The V2 model excels in the diagnosis of pneumonia, outperforming previous models in terms of high accuracy and recall rate. Its potential for clinical application where accuracy is critical shows great promise.

5 CONCLUSION

In conclusion, this study confirms the effectiveness of the Swin Transformer V2 model for diagnosing pneumonia from CXR images. New transformer-based algorithm outperformed established convolutional neural networks with impressive accuracy, precision, recall, and F1-score metrics. This study highlights the transformative potential of deep learning later applications in medical research, especially emphasizing medical imaging in the microscopic environment. The implications are profound, providing a tool for accurately and rapidly diagnosing pneumonia, which is critical for effective disease management. The Swin Transformer V2 not only demonstrates technological breakthroughs but also promises to improve clinical performance, potentially positively impacting patient outcomes.

Future study could explore ways to further improve the diagnostic method. One such direction could be to investigate techniques for combining deep learning models that can significantly improve object detection and classification accuracy. Furthermore, incorporating other advanced machine learning techniques such as attention or genetic competition networks can provide additional insights into the model's performance and interpretation.

REFERENCES

Bi, J., Zhu, Z. and Meng, Q. (2021, September). Transformer in computer vision. In *2021 IEEE International Conference on Computer Science, Electronic Information Engineering and Intelligent Control Technology (CEI)*: 178–188 IEEE.

Chollet, F. (2017). Xception: Deep learning with depthwise separable convolutions. In *Proceedings of the IEEE Conference on Computer Vision and Pattern Recognition* (pp. 1251–1258).

Gabruseva, T., Poplavskiy, D. and Kalinin, A. (2020). Deep learning for automatic pneumonia detection. In *Proceedings of the IEEE/CVF Conference on Computer Vision and Pattern Recognition Workshops*: 350–351.

He, K., Zhang, X., Ren, S. and Sun, J. (2016). Deep residual learning for image recognition. In *Proceedings of the IEEE Conference on Computer Vision and Pattern Recognition*: 770–778.

Howard, A., Sandler, M., Chu, G., Chen, L.C., Chen, B., Tan, M., Wang, W., Zhu, Y., Pang, R., Vasudevan, V. and Le, Q.V. (2019). Searching for mobilenetv3. In *Proceedings of the IEEE/CVF International Conference on Computer Vision*: 1314–1324.

Huang, G., Liu, Z., Van Der Maaten, L. and Weinberger, K.Q. (2017). Densely connected convolutional networks. In *Proceedings of the IEEE Conference on Computer Vision and Pattern Recognition*: 4700–4708.

Jian, S., Kaiming, H., Shaoqing, R. and Xiangyu, Z. (2016). Deep residual learning for image recognition. In *IEEE Conference on Computer Vision & Pattern Recognition*: 770–778.

Kermany, D.S., Goldbaum, M., Cai, W., Valentim, C.C., Liang, H., Baxter, S.L., McKeown, A., Yang, G., Wu, X., Yan, F. and Dong, J. (2018). Identifying medical diagnoses and treatable diseases by image-based deep learning. *cell*, *172*(5): 1122–1131.

Lee, J.G., Jun, S., Cho, Y.W., Lee, H., Kim, G.B., Seo, J.B. and Kim, N. (2017). Deep learning in medical imaging: General overview. *Korean Journal of Radiology*, *18*(4): 570–584.

Liu, Z., Hu, H., Lin, Y., Yao, Z., Xie, Z., Wei, Y., Ning, J., Cao, Y., Zhang, Z., Dong, L. and Wei, F. (2022). Swin transformer v2: Scaling up capacity and resolution. In *Proceedings of the IEEE/CVF Conference on Computer Vision and Pattern Recognition*: 12009–12019.

Liu, Z., Lin, Y., Cao, Y., Hu, H., Wei, Y., Zhang, Z., Lin, S. and Guo, B. (2021). Swin transformer: Hierarchical vision transformer using shifted windows. In *Proceedings of the IEEE/CVF International Conference on Computer Vision*: 10012–10022.

Rajpurkar, P., Irvin, J., Zhu, K., Yang, B., Mehta, H., Duan, T., Ding, D., Bagul, A., Langlotz, C., Shpanskaya, K. and Lungren, M.P. (2017). Chexnet: Radiologist-level pneumonia detection on chest x-rays with deep learning. *arXiv preprint arXiv:1711.05225*.

Simonyan, K. and Zisserman, A. (2014). Very deep convolutional networks for large-scale image recognition. *arXiv preprint arXiv:1409.1556*.

Szegedy, C., Vanhoucke, V., Ioffe, S., Shlens, J. and Wojna, Z. (2016). Rethinking the inception architecture for computer vision. In *Proceedings of the IEEE Conference on Computer Vision and Pattern Recognition*: 2818–2826.

Tsuneki, M. (2022). Deep learning models in medical image analysis. *Journal of Oral Biosciences*, *64*(3): 312–320.

Yaseliani, M., Hamadani, A.Z., Maghsoodi, A.I. and Mosavi, A. (2022). Pneumonia detection proposing a hybrid deep convolutional neural network based on two parallel visual geometry group architectures and machine learning classifiers. *IEEE Access*, *10*: 62110–62128.

Ye, Z., Zhang, H. and Liu, Q. (2022), August. SWTFace: A Multi-branch Network for Masked Face Detection and Recognition. In *2022 5th International Conference on Pattern Recognition and Artificial Intelligence (PRAI)*: 381–387. IEEE.

Zhang, J., Xie, Y., Pang, G., Liao, Z., Verjans, J., Li, W., Sun, Z., He, J., Li, Y., Shen, C. and Xia, Y. (2020). Viral pneumonia screening on chest X-rays using confidence-aware anomaly detection. *IEEE Transactions on Medical Imaging*, *40*(3): 879–890.

Computational Methods in Science and Technology Volume 1 – Kaur et al. (Eds)

Object detection in videos using deep learning methods: A survey

Sunil Kumar
Department of CSE, United University, Rawatpur, Prayagraj, UP, India
Department of CSE, UCER, Naini, Prayagraj, UP, India

Sanjeev Kumar
Department of CSE, United University, Rawatpur, Prayagraj, UP, India

Ashutosh Kumar Singh
Department of CSE, UCER, Naini, Prayagraj, UP, India

ABSTRACT: In computer vision, video object detection process is crucial. Many applications including video surveillance, robot navigation, self-driving cars or automatic modes of transportation, etc. use video object detection. Nowadays, advanced learning techniques like deep learning approaches are used to apply for video object detection. However, researchers face many challenges during implementation, become bottleneck in this research field. This problem occurs due to position (partially hidden object and object complexity of vegetation) and visualization (blur and illumination). In this paper, we present a review that attempts to cover the latest research methodologies of object detection in video and classify video object detection algorithms. Also, we list the issues and challenges of object detection in video that will help for peoples those will research in this field around the globe.

1 INTRODUCTION

Recently, most of the algorithms of video object detection are based on advanced learning techniques i.e. deep learning. Video object detection algorithms are essential for a number of real-life applications such as video surveillance, robot navigation, self-driving vehicle or automatic mode of vehicle, analysis of medical images, etc. As a result, field of video object detection is unstable. In the realm of object detection in video, deep learning algorithms produced positive results. Traditional methods for object detection are based on the Histogram such as "Histogram Oriented Gradient (HOG)" [16], Feature Transform such as "Scale-Invariant Feature Transform (SIFT)"[17], "Frame Difference (FD)"[6], and "Optical flow"[2] which fall short for accurately detecting objects in videos but serve as a foundation for creative approaches.

Concept of deep learning, researchers applied Convolutional Neural Networks (CNNs) [1] for object detection, and the outcomes are better than conventional techniques. For implementation and model training, deep neural networks can employ existing picture data sets. (ImageNet and YouTube-Objects (YTO)) With the use of models like the Regions with CNN features (RCNN) series, You Only Look Once (YOLO) series [5], and other convolutional networks have been thoroughly investigated. When it comes to object detection in images, these designs demonstrate good models. Researchers begin by using methods for image detection in video detection because object detection from video is a process comparable to object detection in images. However, video object detection algorithms require something more due as special challenges existed in video frames such as motion blur,

 DOI: 10.1201/9781003501244-19

occlusion, illumination variation and morphological change. Few algorithms of image object detection are proposed by few researchers with extracting motion and spatiotemporal information from video and results better. Additionally, by utilizing powerful neural networks and clever methodologies, object detection accuracy in videos is improved. Recently, deep learning methods have been used to recognize objects in videos. A number of researchers have been working in this area in recent years to enhance the functionality and accuracy of models. Therefore, a thorough overview of these algorithms is required. Surveys for object detection in videos currently consist of summaries of conventional techniques. It is important and a difficult effort to review all of the research on object detection in video.

Major contributions of this survey paper are as follows.

- To discuss the existing research works of object detection in image.
- To discuss the research works of object tracking in videos
- To discuss the advance research works of object detection in videos such as optical flow and LSTM approaches.

The remaining paper is organized as follows. Section II presents the overview and categorization of object detection approaches. Section III discusses the basic concepts and approaches used in object detection. Section IV contains comparison table of few important research works. Finally, section V discusses the challenges still exist in object detection and tracking in video.

2 OVERVIEW

Here, we provide an explanation of the symbols and basic terminology used in the survey paper as well as a summary of recent works.

2.1 *Composition*

This survey, readers get a systematic knowledge of object detection methods in video. In section III, we discuss the concept which is essential for object detection in video. We can categories deep learning object detection algorithms in video. Figure 1 shows all the algorithms which played important role in object detection in video and are considered in this survey paper. This survey paper gives the basic concept of deep learning object detection in video which are proposed by researchers. The readers will understand well the methods of object detection in images and video. We discuss how object detection in video differs from other machine learning approaches, such as object detection techniques in images, which serve as a fundamental building block for object recognition strategies in video. The approaches for object detection in video that have already been proposed by various researchers are then examined. To emphasize the challenges and its methodologies of object detection in video, we summarize the problem category and solution category in Tables 1 and 2.

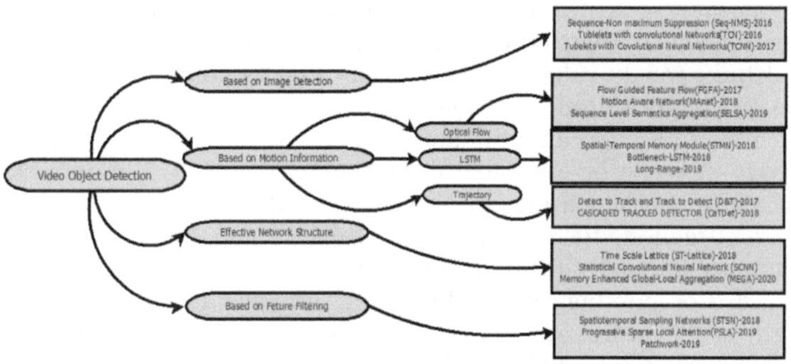

Figure 1. Categorization of object detection methods.

Table 1. Problem category.

S.N.	Problem Category	Challenges
1.	Identification Problems	Motion Blur, Illumination change, change in appearance.
2.	Positioning Problems	Occlusion, Morphological change, Change in size, shape.
3.	Algorithmic Problems	Time Consuming, Redundant information.

Table 2. Solution category.

S.N.	Solution Category	Approach
1.	Image Detection Methods	Post-Processing.
2.	Motion Information Methods	Optical Flow, LSTM, or Object Tracking.
3.	Feature filtering methods	Select and Take
4.	Effective Structure Methods	Most preferred

First, researcher starts with post-processing methods that was currently used in object detection in image. But due to lack of motion and spatiotemporal information post-processing methods do not show good result, even though, it gives a direction to researcher to move next. Researchers implement motion information methods which show better result but it was not sufficient and accurate especially algorithms become slow. Next researchers emphasize on features of objects and implements feature filtering methods which shows better result specially the algorithms become fast. Next few additional models are proposed end-to-end subnetwork which shows good results. The subnetwork improves detection result by filtering motion information of video which essential. Feature engineering methods are further implemented to reduce calculations which give excellent results with respect to accuracy and processing speed.

In section IV, we attach comparison table of different important methods.

3 BASIC CONCEPTS

First, we discuss essential preliminary concept of object detection for image and video to precede typical research. This paper will help to create basic concept of object detection for image and video.

3.1 *Neural networks with convolutions*

The fundamental layers of artificial neural networks (ANNs) or simply neural networks have three layers (an input layer, one or more hidden layers, and an output layer). These nodes are assigned with definite weight values and threshold value for connecting with other nodes. To create a Convolutional Neural Network (ConvNet/CNN)[14] different Deep Learning approaches are used, It accepts an input and weighs different objects in the image according to how effectively they can be separated from one another. A CNN may have a convolutional layer which filters the features of objects; more than one convolutional layer can be used for deeper extraction, to filter definite feature of an object a conventional layer is created. To minimize the size of the data, a pooling layer is used. Therefore, size of next layer input will reduce. For pooling layer, Max_Pool, Average_PoolorMin_Pool can be used while equal eight to all features are used in a fully connected layer. Simonyan and Zisserman proposed the traditional CNN, VGGNet, in 2014. VGGNet [3] and VGGNet-19 [4] employ convolutional layer groups with sizes of 64, 128, 256, 512, and 512. The last layer's output was obtained using a Max_Pool layer. With input values of 4096, 4096, and 1000, it was suggested that the final Max_Pool layer connect to three further entirely connected layers. A soft_max layer [18] was suggested for the output of categorization.

3.2 *Detection of object of object in image*

Detecting objects in images is simpler than detecting objects in videos because there are less obstacles to overcome. Basically, we perform Object detection in image in a single stage (regression) which results fast but less accurate, YOLO series [15] and single shot multi-box detector (SSD) series [8] proposed one stage methods or in a double stage (Regional proposals are generated). Regional's-CNN features(R-CNN) series [7] are double stage detection. Video is collection of static frames. Consequently, video object detection techniques can be used that of picture object detection techniques, but video data contains more information about time and space called spatiotemporal and redundant information. Therefore, if image object detection algorithms are applied on video directly then there will be loss of such information. Therefore, a better and accurate object detection model for video, we should be careful about such information.

3.3 *Object detection and tracking*

Object tracing is something different than object detections it does not require object recognition. We first find position of object (bounding box) in a frame of video, then utilize specific attributes of an object techniques like the mean shift technique, particle filtering, and Kalman filtering look for the most similar area and determine where the object will be in consecutive frames. Year 2010, Bolme performed tracking using correlation filtering [20] which results promises results but it fails for complex scenes. Therefore, advance deep learning methods are applied for object tracking in video which shows excellent results. The main difference these two is that object detection requires supervise learning data while tracking just find best matching position in consecutive frame but few researchers start to use object detection methods for object tracing in recent year, hence these are not separable topics.

3.4 *Post-processing for video object detection*

Researchers begin with single frame image video detection or image object detection methods, for object detection in videos. Due to space and time dependent information of video, these methods results are not so good. Spatiotemporal information must use to improve. For post processing of video for object detection, there may be a number of methodologies which returns few bounding boxes for an object. Fractional value of overlapped area to total area helps to decide most relevant bounding box as

$$\text{Overlapped Fraction (OF)} = \frac{\text{Overlapped area}}{\text{Total area}}$$

More value of OF, high-confidence bonding box. Different methods such as Gready_NMS, Soft_NMS, Softer_NMS, OF guided NMS, DOF_NMS, adaptive_NMS, Learning_NMS etc are implemented. In paper entitled "Object Detection from Video Tubelets with CNN", The authors suggested implementing a system with a tubelet categorization rescoring module and a space-time tubelet module. Here, objects are identified in video and those that have a high degree of confidence are tracked to produce the results of the detection. "A temporal convolutional network (TCN)" [19] suggested strategy to enhance the tubelet detection score in the frames and consequently the detection outcome. A tracking algorithm, based on optical flow, LSTM and other techniques are employed to correlate the frames of a video. Optical flow increases the object's dependability in subsequent frames, and tracking algorithms guarantee the object's entity. Therefore, number of true negative and false positive detection decreases results improvement in accuracy of detection. In spite of these all such post processing works are not sufficient in video object detection.

3.5 *Other approaches for object detection in video*

Both motion and temporal information are employed in post-processing approaches for object detection in video, although none produce outstanding results. Here, few different deep learning methods are further extended to apply for object detection in the video which adds motion information of video. These models are based on Optical flow [9], Context or trajectory [21], Long-Short-Term Memory (LSTM) [10]. Optical flow models are applied on key frames to join with other frame features; LSTM model memorizes old frames and hence adds features older frames than optical flow.

3.6 *Optical flow concept for object detection in video*

Optical flow i.e. visual motion of pattern in an image or frame is basically used in image or video processing to detect the pattern that are most essential for video object detection after using post processing approaches. James Gibson (1950) ascribed the optical flow concept in his book titled The Perception of the Visual World, which denotes the speed of pixels in an image or frame. Here, each pixel is assigned with a velocity vector. For moving object, light filed of the pixels of changed that of stationary object which is used to detect motion of object. In advance deep learning networks optical flow concept used to design new algorithms. First attempt was proposed by Dosovitskiy *et al.* as Flow Netto introduce deep neural networks using optical flow concept [22], and the technique for object detection in video was created utilizing space and time dependent data of video. The optical flow methodologies for object detection in video are shown in Figure 2. In Figure 3, more adaptive feature aggregation for key-frames is shown after sparse feature aggregation and before dense feature aggregation.

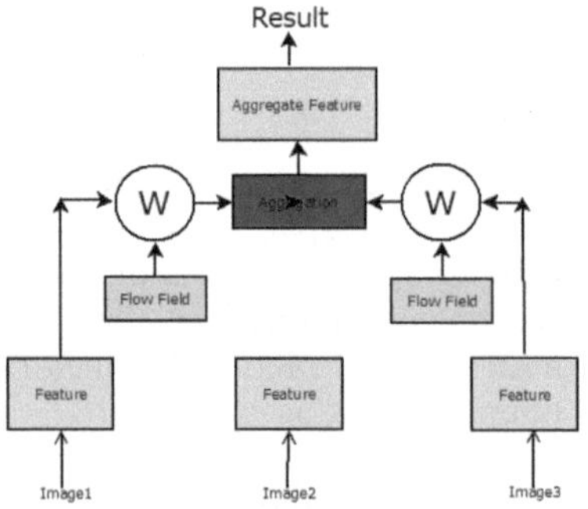

Figure 2. Optical flow methodologies for object detection in video.

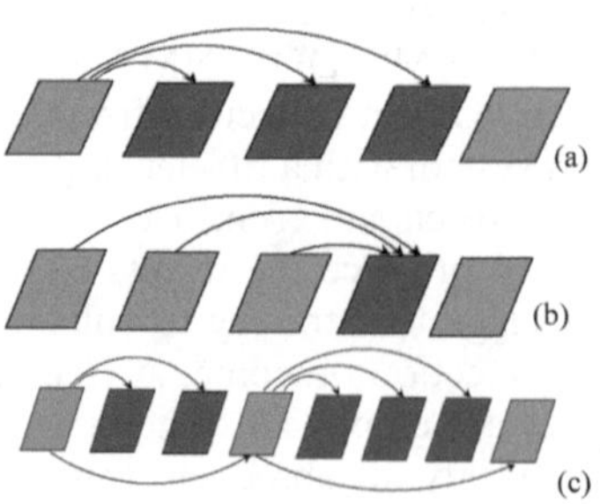

Figure 3. Feature aggregation (a) sparse (b) dense (c) adaptive.

3.7 *LSTM Networks for object recognition in video*

Long Short Term Memory (LSTM) is feedback concept which memories the past knowledge. LSTM shows good result in many fields. It is also implemented in convolution which is very similar concept as Recurrent Neural Network (RNN) of deep learning which cans also memories previous information. But Recurrent Neural Network (RNN)[23] have a disappearing and growing gradient issue. Long-Short-Term Memory (LSTM) is introduced to address these issues.

Remember-gate, forget-gate, focus-gate etc. are used for extraction and propagation of features. Hence, LSTM are very easy to implement introducing deep-learning networks. Therefore, Video object detection challenges can handle effectively using LSTM as any video may full of variations. LSTM can utilize all contextual information of object in video. Optical flow-based methods extract only space and time dependent information of current frame while LSTM extract all contextual inform it maybe temporal or long term. Figure 4 depicts the LSTM technique for video object detection. Figure 4 depicts the LSTM technique used for object detection in video.

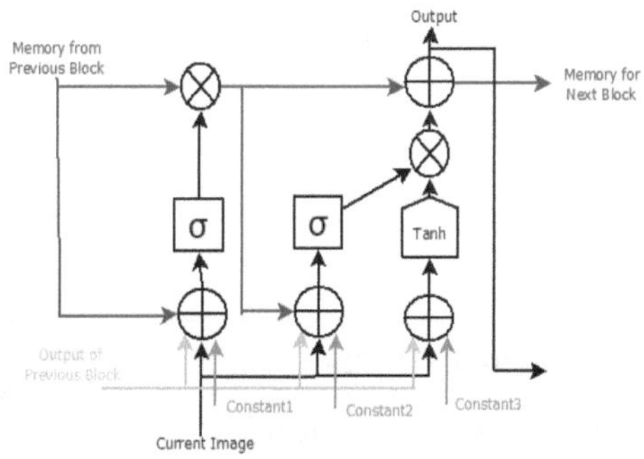

Figure 4. LSTM architecture.

The main strategy of LSTM is to create dependency of current output with previous one. Memories or forget functions played an important role of the information in LSTM and therefore, next block become dependent on long-term features of the object of video. Association-LSTM framework was introduced by Yongyi Lu *et al.* to associate objects in two subsequent frames and improve the precision of video object detection [13]. The Association LSTM calculates the matching error and returns the position, association characteristics, and object categories in the current frame. These accompanying features are simply the spatial and temporal CNN features of the identified item that have been filtered. The LSTM performs well in numerous frames for various objects at a particular time. We create LSTM unit using different type of gates such as output_gate, input_gate and forget_get. LSTM can process more than one object in different frames concurrently which is requirement of object detection in online video racking.

To improve the LSTM architecture, researcher tried various LSTM architectures which are as follows:

1. Figure 5(A) depicts the fundamental LSTM design, which consists of just one LSTM layer.
2. The results of Deep LSTM, which contains many LSTM layers with scattered parameters, are shown in Figure 5(B).
3. Figure 5(C) illustrates how an LSTM projects in a recurrent layer that contains recurrent association reduces the computational cost associated with sophisticated learning.
4. Figure 5(D) displays the effectiveness of Deep LSTM projection in a Recurrent Layer, which contains numerous LSTM with recurrent projection layer outcomes.

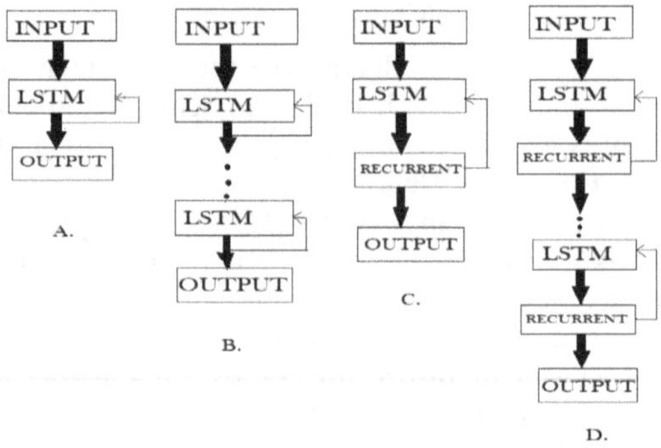

Figure 5. A) LSTM B) DLSTM C) LSTMP, and D) DLSTMP.

Use of multiple layers prevents over-fitting in models and recurrent projection layers for reducing high learning computational complexity.

4 COMPARITION TABLE

Reference	Published Year	Method	Results
[14]	Apr 2023	HOG for segmented Image	Shows best performance in maximum data sets. (75.4 % to 91.3%)
[17]	Jun 2023	(1) Kalman filter used for noise filter. (2) Feature extraction by SIFT and Optimized Crow Search.	Shows better performance w.r.t to current existing techniques. (93.44%)
[19]	Feb 2023	Optimal Temporal Convolutional Network (TCN)	Shows better performance among all other ML methods with an accuracy of 96.12%.
[22]	Mar 2023	(1) Motion controlled MTS. (2) Lucas–Kanade (LK) optical flow. (3) Pyramid and warp optical flow methods.	Shows 96% in average accuracy.
[23]	Feb 2023	(1) A hybrid deep learning-based approach. (2) Use of Bi-LSTM and Bi-GRU networks.	Result is found a low error rate. Its accuracy is 85.5 % in average.
[15]	Mar 2023	(1) CNN for image categorization. (2) Several learning rate setting techniques and different optimization algorithms.	Show promising result.
[13]	2017	Association LSTM (a_LSTM).	Show that this good association improves the information flow across the detected objects.
[12]	Mar 2023	TS-Transformer, TS-GAN, TS- CVAE model.	Shows better performance.
[11]	2021	A simple Siamese matching approach.	Attempt to take advantage of multi head attention mechanism in the tracking community.

5 CHALLENGES

- MOTION BLUR: Few frames of video may be rendered in video as pixel may blur in a area due to high speed. Since detection should continuous and hence cause problem.
- DEFOCUSING: Due to blur of pixels the object may appear unclear.
- OCCLUSION: In few situations, object may hide (partially or totally) by other objects in the frame of a video due to complexity of the environment.
- ILLUMINATION CHANGE: Change in hue cause change in color and change in shadow cause change in its contour. This change happens frequently in a live video as change in illumination occurs frequently.
- CHANGE IN APPEARANCE: It may difficult to decide shape of object by one side only. Therefore, shape and size of object may change by appearance in the video taken by single camera.
  - CHANGE IN SIZE: Whenever, the object comes closer to lens location, the size of object will increase in a typical frame and vise-versa.
  - CHANGE IN SHAPE: Shooting angle may change the shape of object. Whenever, video is taken by stationary camera, the shape of object may change.

REFERENCES

[1] Jiao, L., Zhang, R., Liu, F., Yang, S., Hou, B., Li, L., & Tang, X. (2021). New generation deep learning for video object detection: A survey. *IEEE Transactions on Neural Networks and Learning Systems, 33* (8), 3195–3215.

[2] Agarwal, A., Gupta, S., & Singh, D. K. (2016, December). Review of optical flow technique for moving object detection. In *2016 2nd International Conference on Contemporary Computing and Informatics (IC3I)*, 409–413. IEEE.

[3] Jun, H., Shuai, L., Jinming, S., Yue, L., Jingwei, W., & Peng, J. (2018, November). Facial expression recognition based on VGGNet convolutional neural network. In *2018 Chinese Automation Congress (CAC)*, 4146–4151. IEEE.

[4] Abbassi, N., Helaly, R., Hajjaji, M. A., & Mtibaa, A. (2020, December). A deep learning facial emotion classification system: a VGGNet-19 based approach. In *2020 20th International Conference on Sciences and Techniques of Automatic Control and Computer Engineering (STA)*, 271–276. IEEE.

[5] Shafiee, M. J., Chywl, B., Li, F., & Wong, A. Fast YOLO: A fast you only look once system for real-time embedded object detection in video. arXiv 2017. *arXiv preprint arXiv:1709.05943.*

[6] Tsai, C. M., & Yeh, Z. M. (2013). Intelligent moving objects detection via adaptive frame differencing method. In *Intelligent Information and Database Systems: 5th Asian Conference*, ACIIDS 2013, Kuala Lumpur, Malaysia, March 18–20, 2013, Proceedings, Part I 5, 1–11. Springer Berlin Heidelberg.

[7] Liu, B., Zhao, W., & Sun, Q. (2017, October). Study of object detection based on Faster R-CNN. In *2017 Chinese Automation Congress (CAC)*, 6233–6236. IEEE.

[8] Liu, W., Anguelov, D., Erhan, D., Szegedy, C., Reed, S., Fu, C. Y., & Berg, A. C. (2016). *Ssd: Single shot multibox detector. In Computer Vision–ECCV 2016: 14th European Conference, Amsterdam,* The Netherlands, October 11–14, 2016, Proceedings, Part I 14, 21–37. Springer International Publishing.

[9] Lohani, Prakash Bhanu, Dagur., Arvind, Shukla, Dhirendra, (year 2024). An Efficient Approach for Diabetes Classification Using Feature Selection and Hyperparameter Tuning, *Recent Advances in Electrical & Electronic Engineering*, volume 17, issue, pages 1–15.

[10] Kushwaha, Kumar Pradeep, Dagur, Arvind, Shukla, Dhirendra, (year 2024). *HSLE: A Hybrid Ensemble Classifier for Prediction of Heart Disease*, *Recent Advances in Electrical & Electronic Engineering*, volume 17, issue, pages 1–12.

[11] Aatif Jamshed, Anshu Kumar Dwivedi, Pawan Singh Mehra, Arvind Dagur. (2024). Internet-of-Things-enabled networks: Vision challenges and smart applications, *Advances in Computing Communications*, Vol 6, 1–17.

[12] Sharma, R. K., Dagur, A., (2023). Analyzing the performance of different classifier for detecting polarity of customer reviews, *International Journal on Recent and Innovation Trends in Computing and Communication*, 11(7), pp. 142–148.

[13] Lu, Y., Lu, C., & Tang, C. K. (2017). Online video object detection using association LSTM. In *Proceedings of the IEEE International Conference on Computer Vision*, 2344–2352.

[14] Bharadiya, J. (2023). Convolutional neural networks for image classification. *International Journal of Innovative Science and Research Technology*, *8*(5), 673–677.

[15] Diwan, T., Anirudh, G., & Tembhurne, J. V. (2023). Object detection using YOLO: Challenges, architectural successors, datasets and applications. *Multimedia Tools and Applications*, *82*(6), 9243–9275.

[16] Bhattarai, B., Subedi, R., Gaire, R. R., Vazquez, E., & Stoyanov, D. (2022). Histogram of oriented gradients meet deep learning: A novel multi-task deep network for medical image semantic segmentation. *arXiv preprint arXiv:2204.01712*.

[17] Selvi, A., & Thilagamani, S. (2023). Scale invariant feature transform with crow optimization for breast cancer detection. *Intelligent Automation & Soft Computing*, *36*(3).

[18] Pandey, S. K., Shukla, A., Bhatia, S., Gadekallu, T. R., Kumar, A., Mashat, A., ... & Janghel, R. R. (2023). Detection of arrhythmia heartbeats from ECG signal using wavelet transform-based CNN model. *International Journal of Computational Intelligence Systems*, *16*(1), 80.

[19] Ismail, A. R., Jovanovic, S., Ramzan, N., & Rabah, H. (2023). ECG classification using an optimal temporal convolutional network for remote health monitoring. *Sensors*, *23*(3), 1697.

[20] Zhang, Y., & Zheng, Y. (2023). Object tracking in UAV videos by multi-feature correlation filters with saliency proposals. *IEEE Journal of Selected Topics in Applied Earth Observations and Remote Sensing*.

[21] Davies, J., & Donnelly, M. (2023). Spatial division of opportunity: local economic context, elite trajectories, and the widening participation industry. *Social & Cultural Geography*, 1–20.

[22] Al-Qudah, S., & Yang, M. (2023). Large displacement detection using improved Lucas–Kanade optical flow. *Sensors*, *23*(6), 3152.

[23] Alam, M. S., Mohamed, F. B., Selamat, A., & Hossain, A. B. (2023). A review of recurrent neural network based camera localization for indoor environments. *IEEE Access*.

Computational Methods in Science and Technology Volume 1 – Kaur et al. (Eds)

Analysis of machine learning and deep learning techniques for credit card fraud detection in class imbalanced datasets

Shyam Bahadur
Department of Electronic & Communication, University of Allahabad, India
United College of Engineering & Research, Prayagraj, India

Sudhanshu Kumar Jha
Department of Electronic & Communication, Faculty of Science, University of Allahabad, India

ABSTRACT: Credit card usage is increasing these days, and it plays a crucial role in the e-commerce business. Credit card frauds (CCFs) are now becoming more frequent due to the increasing use of credit cards. Financial institutions and card holders face monetary losses. The major challenge faced in the detection of credit card fraud is the imbalanced nature of the available datasets. The available datasets contain much less information about fraudulent transactions than legitimate transactions. Various machine learning-based approaches, such as Random Forest, KNN, SVM, ANN, and some hybrid methods, are available in literature. However, due to the low accuracy and high false-positive rates of machine learning techniques, deep learning algorithms along with different sampling methods need to be applied to reduce monetary losses. In this study, extensive analysis of both ML and DL methods has been done to judge the accuracy of various classifiers using various metrics on class-unbalanced datasets. This study also aims to suggest some efficient fraud detection algorithms that improve accuracy and reduce false positives.

Keywords: Credit Card Fraud (CCF), Class Imbalance, SMOTE, Machine Learning, Deep Learning

1 INTRODUCTION

In this fast-paced world, we have everything at our fingertips. We can do any financial transaction with just a few clicks and swaps. Every day the world is moving towards digitalization and cashless society. Technology has not only strengthened but also made our lives easier, but it has some drawbacks too. The security of card payments and stakeholder's confidence is extremely important for any type of financial institution in the world. A transaction is considered to be fraudulent if it is not authorized directly by the card holder. Fraudsters find new ways to commit fraud, which can cause huge losses to the card holder. Fraudsters can either steal credit card information through phishing [1] or by physically steal credit cards to make payments, or withdraw cash. In the recent years, credit card fraud (CCF) has increased significantly, which has attracted the attention of most researchers and scholars. Recent data [2] shows that, in 2018 the global economic loss has increased by 16.2% compared to 2017. By the end of 2023 the total deficit will exceed $35 billion if this trend continues. Various types of credit card frauds occur almost every day all around the world. According to the Nelson report December 2022 [3], around $32.34 billion fraud occurred in 2021. Figure 1. Credit card frauds worldwide, show total losses by CCFs worldwide. Fraudsters are continuously changing their behavior over a period of time to avoid detection

DOI: 10.1201/9781003501244-20

by systems trained on old fraudulent data. The role of Machine learning techniques is important in classifying fraudulent and legitimate transactions.

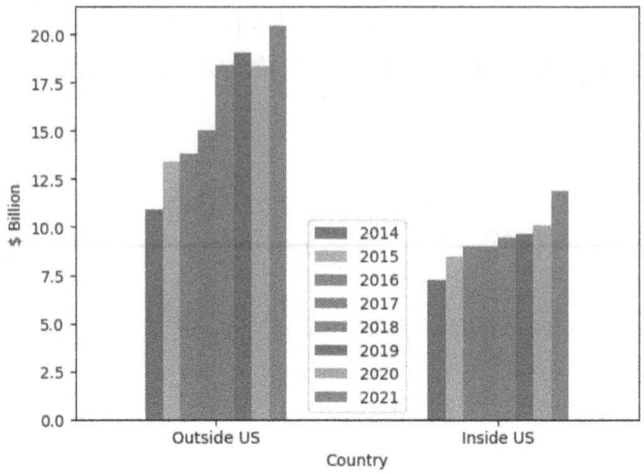

Figure 1. Credit card frauds worldwide.

The main challenge faced by the CCF detection system is the dataset imbalance and the privacy reasons of data exchange. The dataset class imbalance [4] problem become challenging if the fraudulent transactions were much lesser than the legitimate transactions. The dataset available at Kaggle [5] to detect CCF consists of 284807 anonymized transactions of European cardholders spawned in two days in September. Figure 2. Dataset sample distribution. show the distribution of samples in two classes available at Kaggle for CCF detection.

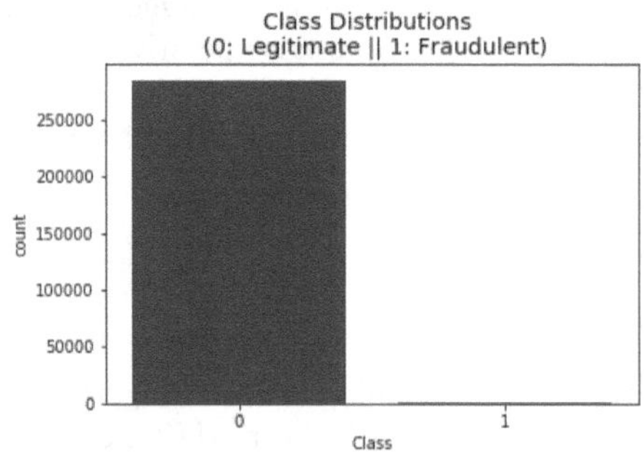

Figure 2. Dataset sample distribution.

The imbalanced dataset imposes many challenges such as biased results and poor model performance in classification. Various strategies exist to address the issue of class imbalance. Sampling methods, such as oversampling use statistical methods to generate new samples by randomly picking data points from the fraudulent class, whereas undersampling techniques remove some existing samples from legitimate class to balance the dataset. However, oversampling the fraudulent class samples or undersampling the legitimate class samples could lead to increased overlap between the samples. Feature selection [6] involves identifying and retaining the most pertinent features from the original dataset while eliminating irrelevant, redundant, or noisy ones. It eliminated the overlapping issues as well as minimizes the dataset dimensions. Many techniques are available in literature to handle the problem of the imbalance nature of datasets and overlapping issues.

1.1 *Contribution*

The primary objective of this study is to highlight the strengths and weaknesses inherent in different Machine learning and Deep learning algorithms when applied to CCF detection. The objectives of this study can be summarized as follows:

- Performed thorough comparison between ML and DL techniques for CCF detection.
- Studied the performances of various sampling techniques on ML and DL techniques.
- Analyzed the performances of various classifiers on metric namely, precision, recall, accuracy, and AUC.
- It will assist in deciding the better sampling approach to address the challenge of class imbalance while also managing the overlap problem.

1.2 *Organization*

The structure of this paper is as follows: In Section 2, we have systematically discussed the relevant work done on credit card fraud detection. In section 3, we have discussed various data balancing techniques. In section 4, we have discussed the ML techniques used for comparisons in the manuscript. Section 5 consists of the comparative analysis of various ML, DL, and hybrid approaches to handle class imbalance; and finally, the conclusion of this study is in the last section.

2 RELATED WORK

CCF detection predominantly utilizes machine learning techniques such as unsupervised, semi-supervised, and supervised learning. However, most of the available datasets are highly imbalanced because samples within a dataset are skewed or biased. Class imbalance creates problems for ML classifiers resulting in unbalanced classification. ML models trained on these datasets may interpret fraud cases as outliers and potentially these cases might be discarded by the classifiers. The classifier may become biased towards the legitimate class. Various sampling techniques have been proposed by researchers to reduce class imbalance and these techniques also improve the classifier accuracy on the training dataset. Sampling emerges as one of the most widely used ways of addressing the class imbalance problem. It includes undersampling, oversampling, and hybrid sampling which combines the first two. Figure 3. Sampling techniques. categorizes various sampling techniques available in literature.

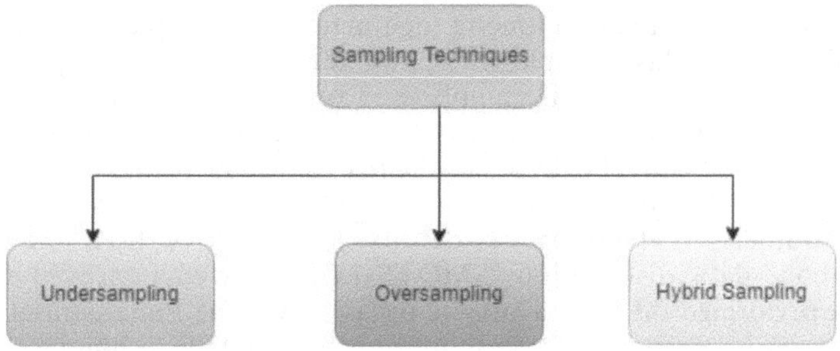

Figure 3. Sampling techniques.

In undersampling techniques, samples from the legitimate class can be removed through statistical methods or randomly. Generally, it involves rebalancing the dataset by removing a subset of samples from the legitimate class to align with the absolute number of samples with that of the fraudulent class. Undersampling techniques mostly use cluster-based [7] approach and neighborhood-based approach. This technique is suitable for large and imbalanced datasets. Unlike undersampling, oversampling aims to boost the samples of the fraudulent class by either duplicating existing samples or generating new synthetic samples from the fraudulent class. Oversampling techniques such as SMOTE [8], I-SMOTE, and SMOTE-LOF [9] were heavily used in literature. Generally, oversampling does not improve fraudulent class recognition. The major challenge with undersampling is how to remove many samples when the class imbalance is too high. Another problem with undersampling is that the resulting size of the training dataset can become small if the total number of samples is less. This may speed up the learning phase of classifiers but may decrease the overall performance of the algorithm. Hybrid methods like SMOTE-ENN [10], SMOTE-WENN [11], and SPIDER2 [12] remove and attach new samples by taking the local distribution of samples into consideration.

3 DATA BALANCING TECHNIQUES

To overcome the problem of oversampling techniques, authors [13] have proposed an improved oversampling method known as SMOTE. SMOTE generates synthetic samples for the fraudulent class by taking each fraudulent class sample and creating synthetic samples along the line segments that connect some or all of the nearest neighbors in the fraudulent class. To address the problem of class imbalance, authors [14] proposed the Dynamic Weighted Entropy [DWE] method aimed at enhancing the performance of classifiers. Firstly, the disjoint subset of the fraudulent class samples were observed as outliers and these samples were considered as the legitimate class samples. This observation straight away improved the net recall value of the positive samples if they were not literally outliers. Secondly, most of the samples were placed in overlapping subsets. The proposed DWE technique minimized the declassification of the positive class samples and maximized the balance of the positive and negative class samples in the overlaying subset. Apart from using a single re-sampling technique either undersampling or oversampling, authors [15] utilized a sampling technique called SMOTE-ENN. ENN is an undersampling method used along with SMOTE to eliminate the noisy samples from both classes. After eliminating the noisy samples, these remaining samples were added to the SMOTE pipeline to obtain a dataset which is clean and balanced.

To increase the local correlation among legitimate class samples and to ensure careful treatment of fraudulent class samples in sparse regions, authors [11] proposed a data cleaning method WENN. WENN is an improved version of ENN to deal with local density imbalance and spatial sparsity. The SMOTE oversampling technique generates noise. Authors in [9] proposed SMOTE-LOF to identify and remove the outliers generated by SMOTE. Each object is assigned a degree in the LOF (Local Outlier Factor). LOF excels at identifying outliers by focusing on density, operating without the need for any assumptions about data distribution. SMOTE-LOF showed improved f-measure and accuracy when performed on three available datasets. On the Pima dataset, the proposed approach produced improved AUC compared to the SMOTE technique. In [16] authors proposed an improved version of SMOTE called ISMOTE to balance the class imbalanced dataset. The proposed technique creates some fabricated fraudulent class samples in the nearby regions of the other fraudulent class samples. In the proposed ISMOTE, two constraints were specified to control the new fabricated samples. If the newly created samples do not match the defined constraints, the algorithm regenerates some other new set of fraudulent class samples.

4 MACHINE LEARNING TECHNIQUES

Detection of fraudulent transactions is of utmost priority for any financial institution issuing credit cards and debit cards. The role of Machine learning techniques in CCF identification is crucial. Several ML techniques have been proposed in literature to identify fraudulent transactions. Methods namely DT, SVM, RF, KNN, and some hybrid techniques are available (see Table 1. Comparison of ML methods for CCF detection.). Neural networks with many nodes and many hidden layers called deep neural networks are able to learn from unstructured data. Earlier deep learning techniques have been heavily used for image, audio, and video recognition. More recently, it began to be used in the financial sector for classification issues such as CCF detection. Techniques such as LSTM [15], Auto-encoder [17], RNN [18], CNN [19] and some hybrid methodologies are available in literature (see Table 2). We will discuss a few of these approaches mentioned here.

4.1 *Random Forest (RF)*

RF is based on ensemble learning technique, it takes mean of outputs of many decision trees to solve classification and regression problems. The accuracy of RF classifier increases with the increasing number of trees in the forest and it also avoids the problem of overfitting. It is effective to handle larger volumes and unbalanced data. Authors in [20] conducted a comparative research of ML algorithms on European cardholders' dataset. The result showed that the RF algorithm achieved 93.2% accuracy and an AUC of 0.94. However, random forests have limitations in training datasets, especially in regression problems.

4.2 *Support Vector Machine (SVM)*

SVM is considered as one of the most prominent supervised ML algorithms, very effective in high-dimensional spaces where the number of dimensions exceeds the number of training samples in the dataset. SVM can be utilized to predict the buying patterns of consumers that can be used to classify the transaction as either legitimate or fraudulent. XGBoost is an effective implementation of gradient boosting algorithms, known for its efficiency and performance optimizations in handling various machine learning tasks [20]. However, SVM is ineffective for large dataset with high class imbalance.

4.3 *K-Nearest Neighbor (KNN)*

The KNN algorithm [21] considers the correlation of the new sample among different available classes of samples and adds the new sample in the class which is more alike to the available classes. In CCF detection, KNN detects the relationship between transactions and the distance between the occurrences of transactions in the dataset. KNN uses less memory, less computation power, and produces higher accuracy and efficiency. The KNN algorithm achieved 99.82% accuracy and 71.42% precision.

4.4 *Hybrid approach*

Ensemble models are ML approach that combines predictions of multiple weak models in series. CatBoost, XGBoost, AdaBoost, Light Gradient boosting and OLightGBM are few boosting algorithms that were applied to CCF detection datasets. The accuracy achieved by LGBM, XGB, and CatBoost were 0.9991, 0.9992, and 0.9988 respectively [22]. The OLightGBM approach achieved the AUC of 92.88%. Hybrid approach yielded better accuracy, better at handling imbalanced datasets, and robust to overfitting problems.

5 COMPARISON OF ML AND DL ALGORITHMS

Table 1. Comparison of ML methods for CCF detection. presents the comparison of different ML and DL methods based on various metrics, documented in literature for CCF detection. Finally, the Table 2. Comparison of sampling techniques for imbalanced datasets in CCF detection. shows the comparison of popular sampling techniques along with ML techniques for class imbalanced datasets.

6 CONCLUSION AND FUTURE DIRECTIONS

CCF detection is a challenging task for any financial institution as fraudsters are always approaching the target with novel fraud methods. Changing nature of frauds can only be handled through a robust classifier. The top priority of any CCF detection system is to accurately predict fraud cases and to minimize false alarms. Machine learning techniques such as SVM and XGBoost have shown promising results on European cardholders' dataset

Table 1. Comparison of ML methods for CCF detection.

		Result			
Ref.	Methodology Used	Precision	Recall	AUC	Accuracy
[21]	Random Forest	0.93	0.93		0.948
[7]	CGA and DNN		0.93		0.71
[23]	SVM	0.959	0.951		0.949
[24]	XGBoost and LightGBM	0.791	0.801	0.952	0.999
[25]	OLightGBM	0.973		0.909	0.984
[26]	LightGBM			0.991	
[27]	XGBoost-GSFA	0.875	0.788	1.00	0.999
[28]	KNN			0.951	0.934
[29]	Baseline (BL)	0.89	0.80	0.95	0.997
	CNN	0.932	0.775	0.929	0.964

Table 2. Comparison of sampling techniques for imbalanced datasets in CCF detection.

		Result			
Ref.	Methodology Used	Precision	Recall	AUC	Accuracy
[17]	Deep Autoencoder and Random oversampling	0.197	0.582		0.985
[6]	ML and GA	0.687	0.796		0.899
[10]	SMOTE-ENN and LSTM		0.996	0.990	
[14]	DWE and Autoencoder			0.63	
[15]	LSTM and SMOTE-ENN		1.00	1.00	
[20]	SMOTE and AdaBoost	0.998	0.99		0.999
[30]	CatBoost, DNN and SMOTE			0.974	0.986
[31]	VAE	0.906	0.837		0.999
[32]	SMOTE and DAE	0.976		0.954	0.99
[33]	CtRUSBoost	0.996	0.998		
[34]	BiLSTM and SMOTE	0.588	0.609	0.90	

(see Table 1. Comparison of ML methods for CCF detection.). Class imbalance is an important factor that degrades the performance of ML models. Several sampling methods have also been implemented to increase the accuracy of ML models, but they performed poor on unseen data. Deep learning models such as RNN, RBM, and Autoencoder were compared with the CNN and Baseline model. CNN and baseline model outperformed other models, shown accuracy of 99.72%. Especially, this research paper studied the performance of ML and DL techniques along with sampling techniques, mainly RUS, SMOTE, and SMOTE-ENN on CCF datasets. The study reveals that ensemble models that used LSTM, GRU serving as base learner, and MLP serving as meta-learner outperformed state of art classifiers (see Table 2. Comparison of sampling techniques for imbalanced datasets in CCF detection.). Future work will explore other sampling and hyper parameter tuning techniques along with some neural network-based techniques for CCF detection. As CNNs are able to handle large datasets and LSTMs are able to remember information for extended periods of time, the potential of CNN and LSTM frameworks will be analyzed in the future.

REFERENCES

[1] Al-Khater W. A., Al-Maadeed S., Ahmed A. A., Sadiq A. S., and Khan M. K. (2020). Comprehensive review of cybercrime detection techniques, *IEEE Access*, vol. 8, 137293–137311, doi: 10.1109/ACCESS.2020.3011259.

[2] E. Central Bank (2020). *Sixth Report on Card Fraud*, no. August.

[3] E. D. Infrastructure and A. G. Expansion (2023). *Global Network Payment Card Transactions Projected Worldwide*, no. 1233.

[4] Fatima E. B., Boutkhoum O., Abdelmajid E. M., Rustam F., Mehmood A., and Choi G. S. (2021). Minimizing the overlapping degree to improve class-imbalanced learning under sparse feature selection: Application to fraud detection, *IEEE Access*, vol. 9, 28101–28110, doi: 10.1109/ACCESS.2021.3056285.

[5] M. L. G. ULB. (2018). *Credit card fraud detection*. https://www.kaggle.com/datasets/mlg-ulb/creditcardfraud

[6] Ileberi E., Sun Y., and Wang Z. (2022). A machine learning based credit card fraud detection using the GA algorithm for feature selection, *J. Big Data*, vol. 9, no. 1, doi: 10.1186/s40537-022-00573-8.

[7] Ojugo A. A. and Nwankwo O. (2021). Spectral-cluster solution for credit-card fraud detection using a genetic algorithm trained modular deep learning neural network, *JINAV J. Inf. Vis.*, vol. 2, no. 1, 15–24, doi:10.35877/454ri.jinav274.

[8] Gupta P., Varshney A., Khan M. R., Ahmed R., Shuaib M., and Alam S. (2023). Unbalanced credit card fraud detection data: A machine learning-oriented comparative study of balancing techniques, *Procedia Comput. Sci.*, vol. 218, 2575–2584, doi: 10.1016/j.procs.2023.01.231.

[9] Asniar, N. U. Maulidevi, and K. Surendro. (2022). SMOTE-LOF for noise identification in imbalanced data classification, *J. King Saud Univ. – Comput. Inf. Sci.*, vol. 34, no. 6, 3413–3423, doi: 10.1016/j.jksuci.2021.01.014.

[10] Esenogho E., Mienye I. D., Swart T. G., Aruleba K., and Obaido G. (2022). A neural network ensemble with feature engineering for improved credit card fraud detection, *IEEE Access*, vol. 10, 16400–16407, doi: 10.1109/ACCESS.2022.3148298.

[11] Guan H., Zhang Y., Xian M., Cheng H. D., and Tang X. (2021). SMOTE-WENN: Solving class imbalance and small sample problems by oversampling and distance scaling, *Appl. Intell.*, vol. 51, no. 3, 1394–1409, doi:10.1007/s10489-020-01852-8.

[12] Napierała K., Stefanowski J., and Wilk S. (2010). Learning from imbalanced data in presence of noisy and borderline examples, *Lect. Notes Comput. Sci. (including Subser. Lect. Notes Artif. Intell. Lect. Notes Bioinformatics)*, vol. 6086 LNAI, 158–167, doi: 10.1007/978-3-642-13529-3_18.

[13] Chawla N. V., Bowyer K. W., Hall L. O., and Kegelmeyer W. P. (2002). snopes.com: Two-Striped Telamonia Spider, *J. Artif. Intell. Res.*, vol. 16, no. Sept. 28, 321–357, [Online]. Available:https://arxiv.org/pdf/1106.1813.pdf%0Ahttp://www.snopes.com/horrors/insects/telamonia.asp

[14] Li Z., Huang M., Liu G., and Jiang C. (2021). A hybrid method with dynamic weighted entropy for handling the problem of class imbalance with overlap in credit card fraud detection, *Expert Syst. Appl.*, vol. 175, no. January, p. 114750, doi: 10.1016/j.eswa.2021.114750.

[15] Mienye I. D. and Sun Y. (2023). A deep learning ensemble with data resampling for credit card fraud detection, *IEEE Access*, vol. 11, no. February, 30628–30638, doi: 10.1109/ACCESS.2023.3262020.

[16] Guo S., Liu Y., Chen R., Sun X. and Wang X. (2019). Improved SMOTE algorithm to deal with imbalanced activity classes in smart homes, *Neural Process. Lett.*, vol. 50, no. 2, 1503–1526, doi: 10.1007/s11063-018-9940-3.

[17] Abakarim Y., Lahby M., and Attioui A. (2018). An efficient real time model for credit card fraud detection based on deep learning, *ACM Int. Conf. Proceeding Ser.*, no. July 2019, doi: 10.1145/3289402.3289530.

[18] Branco B., Abreu P., Gomes A. S., Almeida M. S. C., Ascensão J. T., and Bizarro P. (2020). Interleaved sequence RNNs for fraud detection, *Proc. ACM SIGKDD Int. Conf. Knowl. Discov. Data Min.*, 3101–3109, doi: 10.1145/3394486.3403361.

[19] Lad S. S. and Adamuthe A. C. (2020). Malware classification with improved convolutional neural network model, *Int. J. Comput. Netw. Inf. Secur.*, vol. 12, no. 6, 30–43, doi: 10.5815/ijcnis.2020.06.03.

[20] Ileberi E., Sun Y., and Wang Z. (2021). Performance evaluation of machine learning methods for credit card fraud detection using SMOTE and AdaBoost, *IEEE Access*, vol. 9, 165286–165294, doi: 10.1109/ACCESS.2021.3134330.

[21] Kousika N., Vishali G., Sunandhana S., and Vijay M. A. (2021). Retraction: Machine learning based fraud analysis and detection system, *J. Phys. Conf. Ser.*, vol. 1916, no. 1, doi: 10.1088/1742-6596/1916/1/012115.

[22] Pumsirirat A. and Yan L. (2018). Credit card fraud detection using deep learning based on auto-encoder and restricted Boltzmann machine, *Int. J. Adv. Comput. Sci. Appl.*, vol. 9, no. 1, 18–25, doi: 10.14569/IJACSA.2018.090103.

[23] Kumar S., Gunjan V. K., Ansari M. D., and Pathak R. (2022). Credit card fraud detection using support vector machine, *Lect. Notes Networks Syst.*, vol. 237, no. July, 27–37, doi: 10.1007/978-981-16-6407-6_3.

[24] Hashemi S. K., Mirtaheri S. L., and Greco S. (2023). Fraud detection in banking data by machine learning techniques, *IEEE Access*, vol. 11, no. January, 3034–3043, doi: 10.1109/ACCESS.2022.3232287.

[25] Taha A. A. and Malebary S. J. (2020). An intelligent approach to credit card fraud detection using an optimized light gradient boosting machine, *IEEE Access*, vol. 8, 25579–25587, doi: 10.1109/ACCESS.2020.2971354.

[26] Fang Y., Zhang Y., and Huang C. (2019). Credit card fraud detection based on machine learning, *Comput. Mater. Contin.*, vol. 61, no. 1, 185–195, doi: 10.32604/cmc.2019.06144.

[27] Jovanovic D., Antonijevic M., Stankovic M., Zivkovic M., Tanaskovic M., and Bacanin N. (2022). Tuning machine learning models using a group search firefly algorithm for credit card fraud detection, *Mathematics*, vol. 10, no. 13, 1–30, doi: 10.3390/math10132272.

[28] Wang T. and Zhao Y. (2022). Credit card fraud detection using logistic regression, *Proc. – 2022 Int. Conf. Big Data, Inf. Comput. Network, BDICN 2022*, 301–305, doi: 10.1109/BDICN55575.2022.00064.

[29] Alarfaj F. K., Malik I., Khan H. U., Almusallam N., Ramzan M., and Ahmed M. (2022). Credit card fraud detection using state-of-the-art machine learning and deep learning algorithms, *IEEE Access*, vol. 10, 39700–39715, doi: 10.1109/ACCESS.2022.3166891.

[30] Nguyen N. *et al.* (2022). A proposed model for card fraud detection based on catboost and deep neural network, *IEEE Access*, vol. 10, no. August, 96852–96861, doi: 10.1109/ACCESS.2022.3205416.

[31] Tingfei H., Guangquan C., and Kuihua H. (2020). Using variational auto encoding in credit card fraud detection, *IEEE Access*, vol. 8, 149841–149853, doi: 10.1109/ACCESS.2020.3015600.

[32] Khattri V. and Kumar Nayak S. (2021). Identification and mitigation of fraudulent transaction using deep autoencoder, *J. Univ. Shanghai Sci. Technol.*, vol. 23, no. 11, 769–775, doi: 10.51201/jusst/21/11956.

[33] Arora V., Leekha R. S., Lee K., and Kataria A. (2020). Facilitating user authorization from imbalanced data logs of credit cards using artificial intelligence, *Mob. Inf. Syst.*, vol. 2020, doi: 10.1155/2020/8885269.

[34] Najadat H., Altiti O., Aqouleh A. A., and Younes M. (2020) Credit card fraud detection based on machine and deep learning, *2020 11th Int. Conf. Inf. Commun. Syst. ICICS 2020*, no. May, 204–208, doi: 10.1109/ICICS49469.2020.239524.

Computational Methods in Science and Technology Volume 1 – Kaur et al. (Eds)
© 2025 the Author(s), ISBN: 978-1-032-81760-6

Enhancing data security through an integrated approach of multiple key cryptography and steganography in the wavelet domain

Al-Fariji Ahmed Hussein Hashem, Marina A. Medvedeva & Ebenezer Agbozo
Department of Big Data Analytics and Methods of Video Analysis, Ural Federal University, Ekaterinburg, Russia

ABSTRACT: The quantity of data that may be incorporated and the picture quality must be compromised in order to embed data within an image. As a direct result of this increased distortion, the quality of the stego picture that is generated when more data is hidden within a carrier image will decrease. One thing may be taken into consideration is the difficulty of optimizing the process of embedding a substantial amount of data into an image. This article aims to provide an overview of the many methods of picture steganography and to present the several attempts that have been made to achieve image steganography via the use of deep learning and machine learning techniques. This work concluded with several potential directions for future photo steganographic-based deep learning research, and it indicated that these new directions may be investigated soon. The suggested approach ensures that even comparable letters are encrypted differently by encrypting data using a series of pseudorandom keys created from a single 16-bit seed. By using this strategy, the communication is not only more hidden but also safer against attempts at unauthorized decryption. After encryption, matrix encoding, which is made feasible by discrete wavelet transformation, embeds the encrypted data into the high-frequency sub-band of a cover image. The robust Peak Signal-to-Noise Ratio (PSNR) readings, which often exceed the 40 dB threshold, attest to the cover picture's preservation as a result of this steganographic technique. These PSNR values, which exhibit relatively minimal optical distortion, will make the stego-image unnoticeable. In tests of performance, the proposed method demonstrated excellent imperceptibility, with PSNR values continuously over 46 dB and a low Bit Error Rate (BER) of around 3%, indicating a high degree of fidelity in the recovered data.

1 INTRODUCTION

In addition to these qualities, the metadata of certain picture formats may be altered to embed data in the image. Images saved in Graphics Interchange Format (GIF) format include a colour palette that specifies the hues used in the picture. The picture seems the same to the human eye, but it conceals certain information that can only be accessed by a permutation of the colour map. The second reason is that photographs, with a few notable exceptions, are very resistant to having their pixel values altered. Various steganography techniques exist, each of which modifies the insignificant pixel bits so as to embed the components of a concealed picture into a cover image [5,6]. As a result of the fact that the majority of pictures are composed of three colour channels, each with a width of 8 bits — namely, red, green, and blue — there is a large amount of room available in which information may be contained. Steganography has also made strides ahead in recent times thanks to the growth in deep learning algorithms. Several different deep learning (DL) algorithms

DOI: 10.1201/9781003501244-21

provide a method that is superior and more effective in terms of reaching the intended outcomes [7–13]. However, the advent of machine learning (ML) and DL has also made it more difficult to get trustworthy and durable picture steganography results. This is due to the fact that there are an equal number of steganography algorithms that are based on deep learning.

2 ENCRYPTION BY MULTIPLE KEYS

Confidentiality of a secret message can be preserved by changing the intelligent form of the message into meaningless and unusable form by a process called encryption. If a single key is used for encryption, then a character will be replaced by the same cipher character irrespective of its place of appearance as shown in Table 1. So, it is easy for a hacker to trace the message by using the frequency of appearance of some characters. This problem is avoided in this proposed method by using multiple keys for encryption. If there are n keys, then these n keys can be used at pseudorandom to encrypt the characters of the message. Due to this, encryption of a particular character by the same key will be avoided.

Table 1. Example of a two-level discrete wavelet transform.

Given text	Encrypted Text	Number of Keys Used
CCCCCCC	ÙÙÙÙÙÙÙ	1
CCCCCCC	¦Ù¦Ù¦Ù¦	2
CCCCCCC	¦s@Ù¦	6
COMMITTEE	Ù×ßßÔ–ÒÒ	1
COMMITTEE	¦×¬ß¡-úÒ¯	2

2.1 *Generation of multiple-keys*

The cipher text can be decrypted into original form only by using the secret key. So, the secret key must be shared between the sender and intended recipient. The proposed method suggests the usage of multiple keys for strong security, and hence multiple keys are to be shared which is risky. In the proposed method only one 16-bit number is given as secret key and all the other keys are generated from this key by rotation and complement operations as given in Table 2.

Table 2. Generation of secret keys.

Given Secret Key: 1 0 1 0 0 1 0 1 1 0 0 0 1 1 1 (42375)																
K_1	1	1	1	0	1	0	0	1	0	1	1	0	0	0	0	1
K_2	0	1	0	1	1	0	1	0	0	1	1	1	1	0	0	0
K_3	1	0	0	1	1	1	1	0	0	0	0	1	0	1	1	0
K_4	0	0	0	1	1	1	1	0	1	0	0	1	0	1	1	0
K_5	1	0	1	0	0	1	1	1	1	0	0	0	0	1	0	1
K_6	1	1	0	0	0	0	1	0	1	1	0	1	0	0	1	1

2.2 *Generation of multiple-keys*

1	0	1	0	0	1	0	1	1	0	0	0	0	1	1	1
⟶															
1	1	1	0	1	0	0	1	0	1	1	0	0	0	0	1

In this suggested method, a 16-bit number is used as a secret key. This key is given to a key generator which in turn generates the required number of keys by bitwise rotation and complement operations. The generated keys are used at pseudorandom order to encrypt each character of the secret message. By this, a character will not be transformed into similar cipher pattern on its repeated appearance. Since a particular character is replaced by different cipher patterns, it is very difficult for an unauthorized person to trace the message by using the frequency of appearance of the characters. Hence the frequency of appearance of a cipher pattern will also differ from the plain text. Figure 1 (a, b, c and d) shows the histogram of a secret message of size 2KB and the histograms of the ciphers when one, two and six keys are used for encryption respectively.

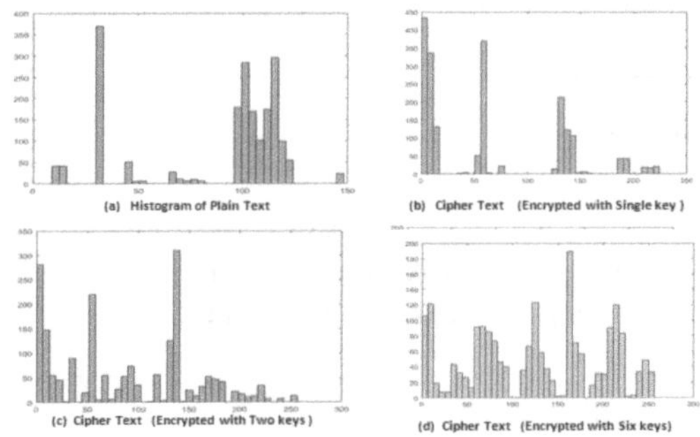

Figure 1. Histogram of the plain and cipher texts.

2.3 *Character encryption*

Each character of the message is represented in 8-bit binary pattern and is divided into two parts: LSB 4-bits and MSB 4-bits. These four-bit binary patterns are encrypted by the 16-bit secret key one at a time. To enable encryption, the secret key is chosen in such a way that the four-bit binary patterns, (0 0 0 0 to 1 1 1 1), occur only once in the binary sequence of the secret key. Then the place of occurrence of the LSB 4-bits of a character is searched in the secret key pattern. After finding the location, the key is rotated towards the right in order to place this 4-bit pattern as the least significant four bits. The number of rotations needed for this process is counted and the LSB 4-bit of the secret character is replaced by this count. In the same way, the MSB 4-bits is also replaced by a count by repeating the same process. Thus, the given character is encrypted into a meaningless character. This encryption process is illustrated below through encrypting the character ‘A’.

Secret Key:

1	0	1	0	0	1	0	1	1	0	0	0	0	1	1	1

Binary Pattern of **A** is **0 1 0 0 0 0 0 1(65)**

The LSB 4-bits are **0 0 0 1**

The LSB 4-bit pattern occupies the bit positions 3 to 6 of the key. In order to bring it to the least significant 4-bit positions, the key has to be rotated 2 times towards right. So, the LSB 4 bits (0 0 0 1) are replaced by the value 2 (0 0 1 0).

The MSB 4-bits are 0 1 0 0

1	0	1	0	0	**1**	**0**	**1**	**1**	**0**	**0**	**0**	**0**	**1**	**1**	**1**
1	**0**	**1**	**1**	**0**	**0**	**0**	**0**	**1**	**1**	**1**	**1**	0	1	0	0

The MSB 4-bit pattern occupies the bit positions 12 to 15 of the key. In order to bring it to the least significant 4-bit positions, the key has to be rotated 11 times towards right. So, the MSB 4-bits (0 1 0 0) are replaced by the value 11 (1 0 1 1).

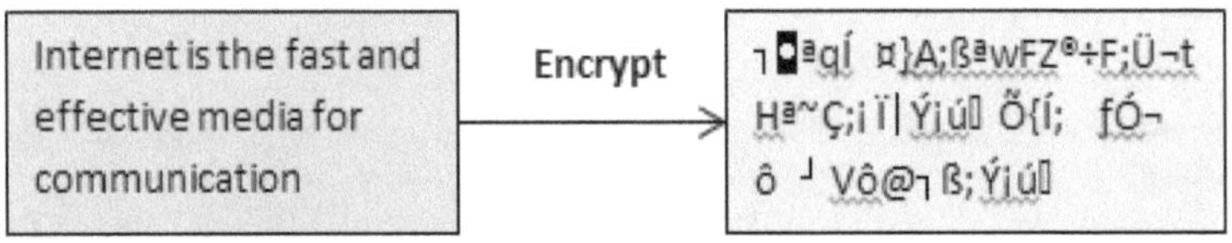

Figure 2. Encryption with six keys.

Encrypted pattern of **A** is **1 0 1 1 0 0 1 0 (178)**

Since pseudorandom secret keys are used, the character 'A' will not be encrypted into 178 always. Figure 2 shows the encrypted text using six secret keys.

2.4 *Embedding in DWT domain*

To increase the security further, the encrypted text is hidden within an image. The cover image is transformed into frequency domain by DWT. DWT transforms the image into four frequency sub-bands as shown in Figure 3 and the encrypted message is hidden in the HH band through matrix encoding.

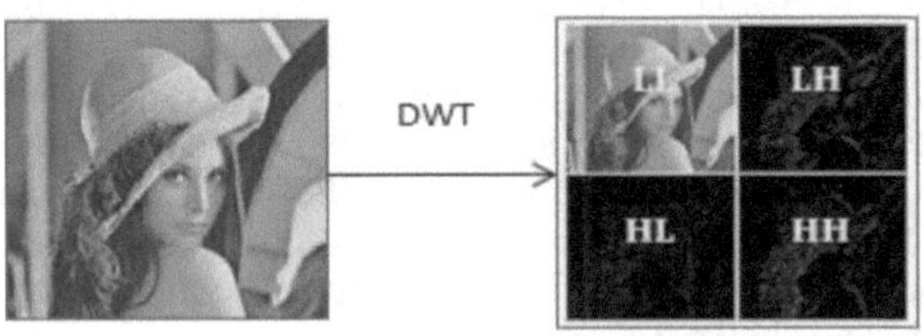

Figure 3. DWT transformation of an image.

2.5 *Cryptography and steganography integration*

As mentioned in the introduction a novel method has been proposed to strengthen data security by integrating cryptography and steganography. The architecture of the proposed algorithm is shown in Figure 4.

It consists of two phases: Message encryption phase and Message hiding phase. Phase 1 encrypts the secret message character by character using multiple keys generated by a key

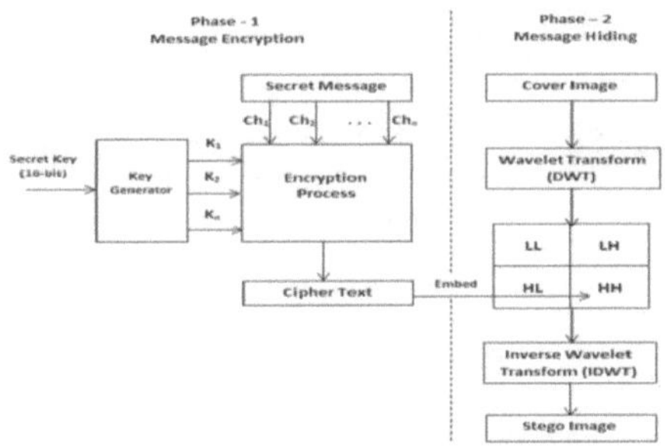

Figure 4. Proposed system architecture.

generator. In Phase 2, the cover image is transformed into frequency domain through DWT and the encrypted data is embedded in the high frequency sub-band, HH, of the cover image.

Algorithm 1. Cryptography and Steganography Integration

Algorithm 1 describes the steps involved in this method.

Input: Cover Image, Secret Message, Secret Key

Output: Stego Image

1. Read the secret message
2. Encrypt it character by character using secret keys generated by a key generator based on the 16-bit secret number given as input
3. Convert the cipher text into stream of bits
4. Read the cover image
5. Decompose the cover image into four frequency bands by wavelet transformation using 'Haar' wavelet
6. Embed the secret bits in the high frequency sub-band using matrix encoding technique
7. After embedding, create the stego image by inverse wavelet transform

2.6 *Performance of multiple keys encryption*

The multiple key cryptography technique and steganography in DWT domain are combined and the data are embedded through matrix encoding method. The performance of this integrated method is tested by encrypting a text message and hiding it in the HH band of the cover image and is evaluated.

A secret message of size 10KB has been hidden in different cover images by embedding 6-bits per pixel and the findings are tabulated in Table 3. Images with PSNR value above 40 are considered as good as it does not create many visible distortions. The proposed method imposes high imperceptibility as it yields PSNR value above 46.A novel algorithm is proposed for preserving the confidentiality of data by integrating symmetric key cryptography

Table 3. The performance of the integrated method.

Cover Image 512 × 512	Secret Message Size	PSNR (in dB)	BER (%)	Time Complexity (seconds)
	10 KB	56.40	2.03	3.53
	10 KB	55.09	2.05	3.52
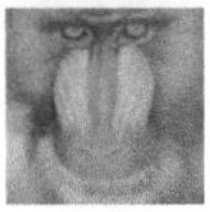	10 KB	46.77	2.71	3.60
	10 KB	55.78	2.03	3.47

and steganography in the wavelet domain. This method imposes dual security on the sensitive message through encryption by multiple keys and by matrix encoding technique for hiding the encrypted message in the wavelet domain of the cover image. A decryption algorithm has also been written to decrypt the cipher text into the meaningful form.

3 CONCLUSION

The purpose our study is to highlight a high-level overview of the various machine learning and deep learning algorithms that are now utilised or might possibly be applied in the area of picture steganography to attain optimum outcomes. This is an overview of various current ML and DL-based image steganography approaches, along with a comparison of these methods. In spite of the fact that a significant amount of research has been carried out in recent years, with some really encouraging results, there is still a significant distance to go. Image steganography, the art of covertly embedding secret information within images, amongst these cryptographic techniques, the path-breaking technique of RC4, which deals with protecting the sensitive information by encrypting the same with a string of pseudo-random keys produced from a single 16-bit seed, offers a very strong mechanism for secrecy. The encryption aims to prevent the hacker from frequency analysis by attacking similar characters differently. Besides, data is imperceptibly embedded into the encrypted matrix through matrix encoding in the discrete wavelet transform domain to increase this level of protection. It is, therefore, dual-layer security to assure the qualities and integrity of the cover image whereby decoding is harder for unauthorized persons. In this process, PSNR value goes well beyond the barrier of 40dB. This high degree of imperceptibility ensures that the method of steganography does not carry data, while hiding the data help to encrypt the information secretly, keeping the original look of the cover image free from suspicion and distortions. The performance metrics give the requirements not only on both security of the data hiding algorithm but also on how efficient the algorithm in use would be compared to other existing algorithms. All the answers to the critical issues in modern data security may be found in combining cryptography and steganography, especially in the wavelet region. The technology not only enhances the security of transmitted data but helps in advancing further secure methods of communication, hence laying the foundation for future developments in information security research. The study showed a huge stride towards the future of technologies in data security with its strong grounding on the safety of personal data through this dual security implementation.

REFERENCES

[1] Mandal, P. C., Mukherjee, I., Paul, G., & Chatterji, B. N. (2022). Digital image steganography: A literature survey. *Information Sciences.*

[2] Kadhim, I. J., Premaratne, P., Vial, P. J., & Halloran, B. (2019). Comprehensive survey of image steganography: Techniques, Evaluations, and trends in future research. *Neurocomputing, 335*, 299–326.

[3] Vyas, A., & Dudul, S. (2015). An Overview of Image Steganographic Techniques. *International Journal of Advanced Research in Computer Science, 6*(5).

[4] Khalifa, A., & Guzman, A. (2022). Imperceptible image steganography using symmetry-adapted deep learning techniques. *Symmetry, 14*(7), 1325.

[5] Bhavani, Y., Kamakshi, P., Kavya Sri, E., & Sindhu Sai, Y. (2022). A survey on image steganography techniques using least significant bit. In *Intelligent Data Communication Technologies and Internet of Things: Proceedings of ICICI 2021* (pp. 281–290). Singapore: Springer Nature Singapore.

[6] Almola, S. A. S., Qasim, N. H., & Alasadi, H. A. A. (2023). Robust method for embedding an image inside cover image based on least significant bit steganography. *Informatica, 46*(9).

[7] Wang, Z., Zhou, M., Liu, B., & Li, T. (2022). Deep image steganography using transformer and recursive permutation. *Entropy, 24*(7), 878.

[8] Farooq, N., & Selwal, A. (2023). Image steganalysis using deep learning: A systematic review and open research challenges. *Journal of Ambient Intelligence and Humanized Computing*, 1–33.

[9] Chaudhary, A., Sharma, A., & Gupta, N. (2023). A novel approach to blockchain and deep learning in the field of steganography. *International Journal of Intelligent Systems and Applications in Engineering, 11*(2s), 104–115.

[10] Selvaraj, A., Ezhilarasan, A., Wellington, S. L. J., & Sam, A. R. (2021). Digital image steganalysis: A survey on paradigm shift from machine learning to deep learning-based techniques. *IET Image Processing, 15*(2), 504–522.

[11] Havard, A., Manikas, T., Larson, E. C., & Thornton, M. A. (2023). CNN-assisted steganography–integrating machine learning with established steganographic techniques. *arXiv preprint arXiv:2304.12503.*

[12] Eze, P., & Parampalli, U. (2021, December). Deep Learning Evaluation of a Steganographic Algorithm. In *2021 Asia-Pacific Signal and Information Processing Association Annual Summit and Conference (APSIPA ASC)* (pp. 1999–2005). IEEE.

[13] Subramanian, N., Cheheb, I., Elharrouss, O., Al-Maadeed, S., & Bouridane, A. (2021). End-to-end image steganography using deep convolutional autoencoders. *IEEE Access, 9*, 135585–135593.

[14] Hinton, G. E., & Salakhutdinov, R. R. (2006). Reducing the dimensionality of data with neural networks. *science, 313*(5786), 504–507.

[15] Hiwe, M. S., & Nipanikar, S. I. (2014). An analysis of image steganography methods. *International Journal of Engineering Research & Technology (IJERT), 3*(2).

[16] Noda, H., Niimi, M., & Kawaguchi, E. (2006). High-performance JPEG steganography using quantization index modulation in DCT domain. *Pattern Recognition Letters, 27*(5), 455–461.

[17] Chen, W. Y. (2008). Color image steganography scheme using DFT, SPIHT codec, and modified differential phase-shift keying techniques. *Applied Mathematics and computation, 196*(1), 40–54.

[18] Yang, C. H., Weng, C. Y., Wang, S. J., & Sun, H. M. (2008). Adaptive data hiding in edge areas of images with spatial LSB domain systems. *IEEE Transactions on Information Forensics and Security, 3* (3), 488–497.

[19] Wazirali, R. A., & Chaczko, Z. (2015). Data hiding based on intelligent optimized edges for secure multimedia communication. *Journal of Networks.*

Computational Methods in Science and Technology Volume 1 – Kaur et al. (Eds)

Enhancing liver cirrhosis diagnosis: A comparative analysis of XGBoost, SVM, and random forest classifiers for optimal predictive analysis

Ashwin Sharma, Avni Manish Gupta & Sanchali Das
School of Computer Science Engineering and Technology, Bennett University, Gr. Noida, UP, India

ABSTRACT: Cirrhosis of the liver, an irreversible liver disease, replaces the healthy tissue with scar tissue and causes the dysfunction of the liver. The data came from a long-term Mayo Clinic study on primary biliary cirrhosis and is being used for evaluation of XGBoost, SVM and RandomForest. The results showed that Random-Forest dominates XGBoost and SVM algorithms by scoring 0.85, 0.84, and 0.75 respectively. This showcases the importance of choosing the right predictive model in disease forecasting in medicine which assists healthcare analytics and provides guidance to refine models for enhanced health results. This research determined that the random forest technique performs better than other methods and should be utilized for additional examination to progress predictive versions for hypertension and other ailments and could transform medicine and healthcare and become the healthcare hub of tomorrow.

Keywords: Liver Cirrhosis, Predictive Modeling, Random Forest, XGBoost, Support Vector Machine

1 INTRODUCTION

Liver disease remains a health challenge, representing an advanced stage of liver disease characterized by advanced scarring and impaired function [1]. This condition leads to long-term liver damage due to factors such as chronic alcoholism, fatty liver disease or non-alcoholic infection [2] and the implications for health system is deeper, including increased morbidity and mortality at higher health care costs. The liver, an important organ responsible for detoxification, metabolism, and vital protein synthesis, has progressive damage in cirrhosis leading to a series of complications [3]. Early and accurate prediction of liver cirrhosis timely onset. The important thing to do is to intervene and implement effective treatment strategies [4]. The approach to survival prediction enables health care providers to reduce disease progression and manage complications, thereby improving patient outcomes This background sets the stage for our study and builds the importance of increasing the prognosis of survival in liver disease for optimal health care is emphasized [5]. The main objective of our study is to evaluate the effectiveness of the cutting-edge machine learning methods, including SVM, XGBoost, and RandomForest, in predicting survival from cirrhosis.

2 LITERATURE REVIEW

Chronic liver disease and liver dysfunction caused by hepatitis and fibrosis lead us to incorporate preventive measures to facilitate early diagnosis and intervention. The literature

DOI: 10.1201/9781003501244-22

liver is advising range of models employed for cirrhosis survival prediction machine learning approaches with traditional and clinical methods. [6] Historically, the diagnosis of liver disease relied on traditional clinical examinations including physical examination, blood tests and drawing classes [7]. While these methods have been valuable their limitations in terms of sensitivity and specificity underscore the evolving need for more sophisticated models. Recent years have evidence of a paradigm shift towards leveraging advanced machine learning techniques for liver cirrhosis survival prediction. These methods capitalize on the ability of algorithms to distinguish complex patterns in extensive datasets that often contain diverse clinical, biochemical, and imaging variables. [6] Reliance on individual biomarkers or subjective clinical assessment can lead to delayed diagnosis, intervention and affect patient prognosis. Acknowledging these limitations has led to increased interest in developing new and more reliable forecasting methods. [8] This non-group validation group aims to create and retrieve risk scores which would determine a patient's chances of long-term survival atom cirrhosis Training was also conducted. Based on the researchers' analysis of four different programs, they found that the living tree model was the best predictor of survival at 1, 3, 5 and 10 years The risk score was created by the study, which then used a variety of approaches to both internally and externally validate it. Over prolonged follow-up, the risk score-also known as the long-term cirrhosis survival (LTCS) score-showed improvement and discrimination. It was discovered that the score was especially helpful in predicting liver-related mortality. [9] With application of machine learning algorithms (MLAs), the study paper seeks to overcome the shortcomings of non-invasive testing for identifying severe liver fibrosis. Using patient data from McGill University Health Centre (MUHC) and 1703 liver samples from patients at Toronto Liver Clinic (TLC), the project focuses on training six MLAs and verifying their effectiveness in identifying advanced fibrosis. [9] Study emphasizes the importance of severe liver scarring while highlighting the difficulties in detecting cirrhosis in the early stages of problems. Numerous ambiguous classifications result from the limited performance of the non-invasive tests now in use. This study's objective was to employ MLA to identify patients with advanced liver fibrosis on more than one basis. The results indicate that an ensemble MLA of logistic regression, neural network, random forest classifier, gradient boosting classifier, and support vector machine algorithms achieved 100% determinate classifications, with an AUROC score of 0.870 on the TLC validation set and 0.716 on the MUHC validation set. [9] Ensemble MLA obtained equivalent performance to hepatologists and outperformed all commonly utilized biomarkers. This study demonstrates that in individuals suffering from persistent liver illness, MLAs utilizing routine data collection can identify patients at high risk for chronic liver disease and pain pressure, allowing intervention before decompensation occurs. The study recognizes its limitations, including the need for validation in primary care settings, missing parameters, and data availability. [9] Still, there is potential for improving the identification of individuals with advanced liver fibrosis and developing intervention methods to stop the progression of decompensated cirrhosis and its associated consequences by MLAs as an automated screening tool. The article describes the author's [6] attempt to develop a machine learning model for predicting mortality in individuals suffering from cirrhosis. This research employed a dataset comprising 107,939 cirrhosis patients. [6] for the purpose of training and assessing the model, divided the dataset into a holdout validation set (33.3% of the data) and a derivation set (66.7% of the data). A number of models were trained and compared, including partial route logistic regression models, full logistic regression models, and ensemble gradient boosting techniques. The model's performance was evaluated using metrics such as Brier scores and the area under the receiver operating characteristics curve (AUC). [6] Study discovered that the machine learning models that employed clinical factors that were simple to score outperformed the more sophisticated ensemble gradient boosting techniques, if not outperformed them. Machine learning models have demonstrated superior predictive performance compared to traditional models for high salt-limited disease (MELD-Na). One advantage of using machine learning in risk forecasting over traditional models is that it can increase the accuracy of forecasts. [6] But the study has several shortcomings. Many of the older men in the study's data sets belonged to VA health plans, which limited the applicability of the findings to a diverse patient population. Again, because of the difficulty of doing so, these variables were processed from electronic health records so other factors, such as patient compliance with treatment, could not be assessed. [10] The aim of the study was predicting 30-day mortality in patients with severe non-cholestatic cirrhosis using machine learning (AutoML). The AutoML model beats current benchmarks like the MELD score, MELD-Na score, and ALBI score because it is based on the XGBoost algorithm. High-density lipoprotein cholesterol, creatinine, white blood cell count, and international normalized ratio were important factors in the XGBoost model. [10] Study highlights the potential of AutoML in future medical applications for predicting mortality in cirrhosis patients. Table 1 provides a brief description of the research articles.

Table 1. Description of literature review.

Articles	Dataset	No. of biopsies	Algorithms/techniques	Auc /Auroc
[8]	VHA test data set	6053	Boosted Survival tree	0.77(1yr) 0.81(3yr) 0.84(5yr) 0.88(10yr)
			Meld-Na	0.75(1yrs) 0.68(3yrs) 0.70(5yrs) 0.72(10yrs)
[9]	Toronto-Liver Clinic (TLC)	1703	Ensemble algorithms (neural networks, logistic regression, gradient boosting, random forest, and SVM)	0.870 (TLC)
	McGill University Health-Centre (MUHC)	1703		0.716 (MUHC)
[6]	National Veterans Affairs Corporate-Data Warehouse	107,939	Gradient boosting logistic regression Partial-route-logistic regression.	0.81 0.78 0.78
[10]	Department-of Gastro-enterology Hepatology (Soochow University).	Not specified	GBM	0.886
			XGBoost	0.888

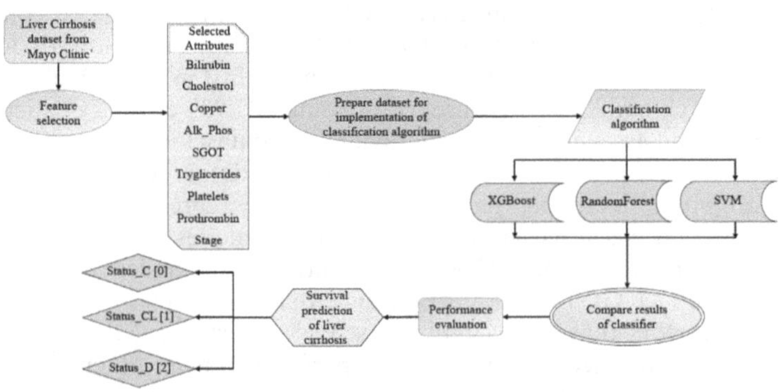

Figure 1. Workflow diagram.

3 METHODOLOGY

The information provided is part of a series of studies conducted by the Mayo Clinic in 1974 and 1984 that focused on primary cirrhosis (PBC) to study, investigate, and understand cancer to provide control over the progression of Drug Test D Penicillamine [11]. The instances in this dataset represent 424 PBC individuals amongst them, 312 patients took part in the clinical trial while the remaining 112 patients did not participate but agreed to record basic metrics and undergo survival tracking [12].

Figure 1. Shows the proposed methodology where several preprocessing steps were applied on the dataset like handling missing values and one hot encoding. The survival prediction study was further classified in the three statuses namely, 'C', 'CL', 'D' or '0', '1', '2'. The state C (censored) indicates the patient was alive at N_Days, CL indicates the patient was alive at N_Days due to liver transplant, and D indicates the patient was deceased at N_Days [13]. Out of all the prominent 17 attributes in the dataset we have selected 10 attributes namely 'Bilirubin', 'Cholesterol', 'Copper', 'Alk_Phos', 'SGOT', 'Platelets', 'Prothrombin', 'Stage', 'Age', 'Albumin' and their effect was evaluated. The selection of the models is grounded by their proven efficacy and adaptability to complex datasets enabling wide acceptance in diverse medical prediction tasks. An ensemble learning approach called Extreme Gradient Boosting builds decision models for forecasts. Its architecture consists of three main components namely weak learners, objective function, gradient boosting, and weighted updates. XGBoost includes L1 and L2 regularization terms in the objective function to control the complexity of trees [14]. Regularization enhances data by reducing overfitting. For both regression and classification, RandomForest, a supervised machine learning method, is employed. It is a classifier containing several decision trees on various subsets of the provided dataset [15]. It is predicted on the ides of ensemble learning. Regression and classification issues can be addressed with the powerful machine learning technique. To analyze non-linear and high-dimensional data, a non-parametric model known as the Radial Basis Function Support Vector Machine (RBF-SVM) is employed. The method used by [16] maps the input into a feature space of higher dimensions. The approach calculates the similarity between pairs of data points in the feature space using kernel functions, such as radial basis functions. Conflict matrices and scatter maps are common indicators used to assess the model's effectiveness. The classification report comprised of the metrices like Precision, Accuracy, Recall, Fi-score, Support, Macro-Average, and Weighted Average with confusion matrix [17]. A brief description is given below in Table 2 where TPR = True Positive Rate, FPR = False Positive Rate, TNR = True Negative Rate, FNR = False Negative Rate, PR = TPR + FNR and NR = TN + FP [18].

Table 2. Concise summary of evaluation metrics [17].

Evaluation metrics (in %)	Expression	Description
Precision (PPV)	$\frac{TPR}{TPR+FPR} * 100$	The ratio of successfully identified cases to all expected cases in the class.
Accuracy	$\frac{TNR+TPR}{NR+PR} * 100$	Ratio of correctly classified cases.
Recall/Sensitivity	$\frac{TPR}{PR} * 100$	True-positive or recognition rate is another term for the percentage of cases correctly identified relative to the total number of instances.
F1-score	$\frac{2*precision*recall}{precision+recall}$	It gives equal weight to false positives and negatives and is a compromise between fact and reality. Used to evaluate accuracy

4 RESULTS

This module elaborates the results obtained for different classifiers used for the survival prediction of liver cirrhosis. In our research the three prominent classifiers were battled against each other to find out the most suitable classifier for the survival prediction of liver cirrhosis. In our findings we have found out of that the classifier RandomForest emerges out to be the best amongst all with the accuracy of 0.85 followed by XGBoost classifier with accuracy of 0.84 followed by SVM with the accuracy of 0.75 the detailed analysis of result so obtained has been elaborated in Table 3. The confusion matrix for all the classifiers is shown in Figure 2.

Table 3. Classifiers performance for survival prediction of liver cirrhosis.

Model trained	Macro avg. score	Accuracy score	F1-Score	Precision score	Recall score	Weighted avg, score
XGBoost	0.83	0.84	0.76	0.81	0.71	0.84
SVM	0.74	0.75	0.72	0.74	0.69	0.75
RandomForest	0.84	0.85	0.80	0.84	0.76	0.85

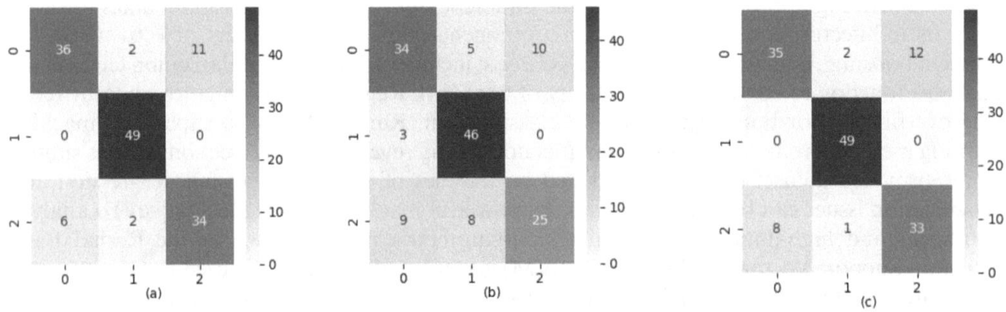

Figure 2. Confusion matrix for: (a) RandomForest (b) SVM (c) XGBoost.

5 CONCLUSION

In summary, the result of our study reveals the importance of elimination model for survival prediction in cirrhosis. Through a meticulous comparison between three state of the art classifiers, SVM, RandomForest and XGBoost (XGB) it has been revealed that RandomForest surpass Support Vector Machine and XGBoost on performance metrices and accuracy. The scores indicate a clear advantage for Random-Forest with a 0.85 accuracy rating contrasted with XGBoost (0.84) and SVM (0.75). These outcomes provide valuable insights for practitioners and researchers involved in the development of survival prediction models for liver cirrhosis. The superior performance of RandomForest signals its efficacy in handling the complexities inherent in the dataset and thereby enhancing its potential for accurate and robust predictions. As we strive for advancements in medical prediction, personalized healthcare and the preference for RandomForest over XGB and SVM in the context of liver cirrhosis prediction stands as a noteworthy contribution to the ongoing discourses in these field.

REFERENCES

[1] D'Amico, G., Garcia-Tsao, G., & Pagliaro, L. (2006). Natural history and prognostic indicators of survival in cirrhosis: A systematic review of 118 studies. In *Journal of Hepatology* (Vol. 44, Issue 1, pp. 217–231). https://doi.org/10.1016/j.jhep.2005.10.013

[2] Gibb, Sebastian., Berg, Thomas., Herber, Adam., Isermann, Berend., & Kaiser, Thorsten. (2023). A new machine-learning-based prediction of survival in patients with end-stage liver disease. *Journal of Laboratory Medicine* 47(1) 13–21.

[3] Pinzani, M., Rosselli, M., & Zuckermann, M. (2011). Best practice & research clinical gastroenterology. *Liver Fibrosis*, 25, 281–290.

[4] Guo, Aixia., Mazumder, Nikhilesh R., Ladner, Daniela P., & Foraker, Randi E. (2021). Predicting mortality among patients with liver cirrhosis in electronic health records with machine learning. *PLoS ONE* 16(8 August).

[5] Cirrhosis-diagnosis and management. (n.d.). Retrieved March 25, 2024, from https://www.aafp.org/pubs/afp/issues/2019/1215/p759.pdf

[6] Kanwal, F., Taylor, T. J., Kramer, J. R., Cao, Y., Smith, D., Gifford, A. L., El-Serag, H. B., Naik, A. D., & Asch, S. M. (2020). Development, validation, and evaluation of a simple machine learning model to predict cirrhosis mortality. *JAMA Network Open*, 3(11), E2023780. https://doi.org/10.1001/jamanetworkopen.2020.23780

[7] Davis, J. L., & Murray, J. F. (n.d.). 6 History and Physical Examination.

[8] Goldberg, D., Mantero, A., Kaplan, D., Delgado, C., John, B., Nuchovich, N., Emanuel, E., & Reese, P. P. (2022). Accurate long-term prediction of death for patients with cirrhosis. *Hepatology*, 76(3), 700–711. https://doi.org/10.1002/hep.32457

[9] Sarvestany, S. S., Kwong, J. C., Azhie, A., Dong, V., Cerocchi, O., Ali, A. F., Karnam, R. S., Kuriry, H., Shengir, M., Candido, E., Duchen, R., Sebastiani, G., Patel, K., Goldenberg, A., & Bhat, M. (2022). Development and validation of an ensemble machine learning framework for detection of all-cause advanced hepatic fibrosis: a retrospective cohort study. *The Lancet Digital Health*, 4(3), e188–e199. https://doi.org/10.1016/S2589-7500(21)00270-3

[10] Yu, C., Li, Y., Yin, M., Gao, J., Xi, L., Lin, J., Liu, L., Zhang, H., Wu, A., Xu, C., Liu, X., Wang, Y., & Zhu, J. (2022). Automated machine learning in predicting 30-day mortality in patients with non-cholestatic cirrhosis. *Journal of Personalized Medicine*, 12(11). https://doi.org/10.3390/jpm12111930

[11] Lohani, Prakash Bhanu, Dagur., Arvind, Shukla, Dhirendra, (2024). An Efficient Approach for Diabetes Classification Using Feature Selection and Hyperparameter Tuning, *Recent Advances in Electrical & Electronic Engineering*, volume 17, issue, pages 1–15.

[12] Kushwaha, Kumar Pradeep, Dagur, Arvind, Shukla, Dhirendra, (2024). HSLE: A hybrid ensemble classifier for prediction of heart disease, *Recent Advances in Electrical & Electronic Engineering*, volume 17, issue, pages 1–12.

[13] Aatif Jamshed, Anshu Kumar Dwivedi, Pawan Singh Mehra, Arvind Dagur. (2024). Internet-of-things-enabled networks: Vision challenges and smart applications, *Advances in Computing Communications*, Vol 6, 1–17.

[14] Sharma, R.K., Dagur, A., (2023). Analyzing the performance of different classifier for detecting polarity of customer reviews, *International Journal on Recent and Innovation Trends in Computing and Communication*, 11(7), pp. 142–148 https://doi.org/10.1145/2939672.2939785

[15] Kumar, A. & Alam, B. (2018). Task scheduling in real time systems with energy harvesting and energy minimization. *Journal of Computer Science*, 14(8), 1126–1133.

[16] Liu, C., Wang, L., Yang, A., & Zhang, Yongli. (2012). *Support Vector Machine Classification Algorithm and Its Application*. 179–186.

[17] Han, J., Kamber, M., & Pei, J. (2011). *Data Mining. Concepts and Techniques*, 3rd Edition (The Morgan Kaufmann Series in Data Management Systems).

[18] Chandra, T. B., & Verma, K. (n.d.). *Pneumonia Detection on Chest X-Ray Using Machine Learning Paradigm*.

Computational Methods in Science and Technology Volume 1 – Kaur et al. (Eds)
 ISBN: 978-1-032-81760-6

A heart diseases prediction model using machine learning and deep learning techniques

Indrajeet Rajput
Assistant Professor, Computer Engineering, Indus University

Deepak Kumar Rathore
Assistant Professor, Department of CSE, LNCT, Bhopal (India)

Deshdeepak Shrivastava
Assistant Professor, Information Technology, ITM, Gwalior

Pooja Shah
Associate Professor, IT, Department, VSITR, Kadi

Toral M. Desai
Assistant Professor, Department of CSE, Indus University

ABSTRACT: The modern era's busy schedule often leads to an unhealthy lifestyle, causing anxiety and depression. This work aims to detect heart disease symptoms at an early stage to reduce the risk of heart attacks and improve performance parameters using different machine learning techniques. In this study, conducted experimental work using the Kaggle datasets within the Python framework. Here applied several machine learning techniques, including LR, DT, SVM, NB, and k-NN. All these classification techniques were used as binary classifiers, where the target variable had two values, 0 and 1. The simulation results indicate that the LR (Logistic Regression) demonstrated superior accuracy and performance parameter values, like accuracy and other parameters in machine learning techniques and our proposed deep learning model gives better result than the LR model.

Keywords: Supervised Learning, Heart Disease, Machine Learning, Accuracy, Neural Network

1 INTRODUCTION

In recent times, there has been significant attention towards the application of machine learning techniques in medical diagnostics. The advancements in Information and Communication Technologies have led to developments in various fields, including medical sciences. Researchers have been exploring novel approaches that have yielded improved results, and effective computational techniques continue to emerge [1]. Classification algorithms play a crucial role in predicting outcomes based on existing and current data, and the proposed system aims to enhance classification accuracy and data efficiency. Preprocessing of data involved the removal of records with missing values from the dataset.

Machine learning is a critical component of artificial intelligence, and its utilization across various platforms is on the rise [3]. The technology offers numerous benefits, such as facilitating predictive maintenance and product recommendations for projects. Additionally, machine learning technologies play a crucial role in spam detection and support various medical diagnoses.

 DOI: 10.1201/9781003501244-23

2 MACHINE LEARNING MODELS

The ML framework comprises four main types of algorithms: SML, USML, SSML, and RL [2]. In the context of ECG analysis, most ML studies focus on disease diagnosis or risk stratification using supervised learning techniques [4]. Supervised learning aims to infer functions or scores from labeled training data, with classification and regression being the two main supervised techniques for categorical and continuous outputs, respectively.

3 PROPOSED WORK

Despite significant efforts by many medical institutions to combine data resources with new technologies, traditional and manual methods still persist in storing and analyzing medical data. Implementing new systems early remains a challenge for numerous medical facilities. The vast amount of available data presents a complex landscape, and to navigate it effectively, DM and ML techniques have gained popularity for data analysis. ML-driven approaches have the potential to develop a safe and accurate environment for the different diagnostic tools [7,16].

ML-based learning approach analyses frequently seek to uncover nonlinear relationships among the huge amount of variables, requiring substantial training data. Active learning, which involves selecting data points to be labeled by experts, can be highly useful when data is abundant but obtaining labels is challenging, time-consuming, or expensive.

Figure 1. Experimental workflow of the proposed methodology.

In the above diagram, we discuss heart related disease classification using various ML techniques. The process begins with the uploading or importing of the dataset, obtained from the Kaggle dataset, and is processed using Python language within our framework. Subsequently, we examine the correlation between dataset attributes and apply ML classifiers such as LR, DT, SVM, NB, and k-NN. We then assess the proposed model with different number of standard performance parameters for each classifier and analyze the results obtained from the different machine learning techniques.

4 IMPLEMENTATION MODEL

Environmental changes have led to the emergence of various diseases worldwide. Altered lifestyles, dietary habits, and genetic factors contribute to the rise of new diseases. Early identification and control of these conditions are crucial in addressing this situation. Cardiovascular disease stands out as a leading cause of death among both men and women, surpassing other health concerns like breast cancer. Unhealthy eating habits, genetic predispositions, and aging are significant risk factors for cardiovascular diseases, with older individuals being more susceptible.

Artificial intelligence has become a crucial technology in various domains, offering numerous benefits to machines and systems. Machine learning, a subset of artificial intelligence, has seen significant growth and utilization. It empowers the improvement of various systems and is witnessing increasing adoption across industries. Various learning algorithms are available to tackle different tasks, such as LR, DT, ensemble approaches, and NN. The correlation coefficient, which can take values in the range of -1 to 1, measures the degree of linear relationship between two variables. Below figure provides valuable insights into the correlation between the variables. The confusion matrix is divided into expected rows and predicted columns for each class. The cells in the matrix represent the total number of correct and incorrect predictions for each class.

5 MACHINE LEARNING BASED RESULT ANALYSIS

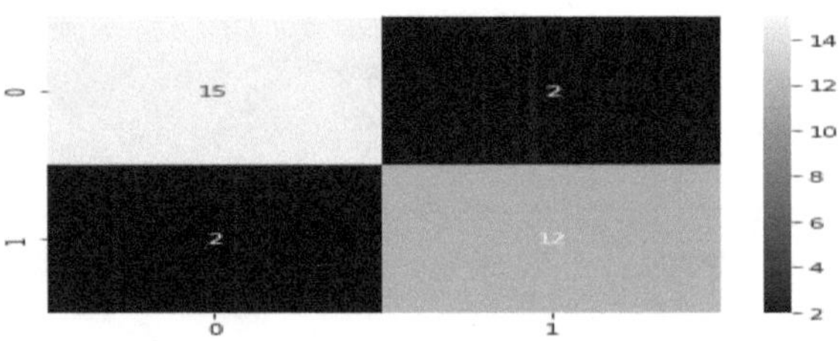

Figure 2. The above picture shows that the confusion matrix representation for the experimental work using a logistic regression model.

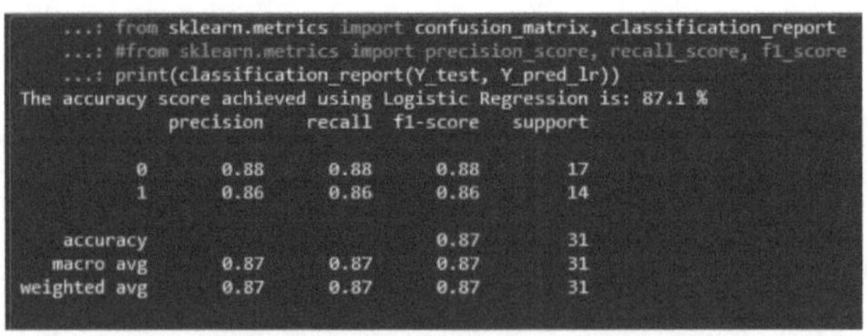

```
...: from sklearn.metrics import confusion_matrix, classification_report
...: #from sklearn.metrics import precision_score, recall_score, f1_score
...: print(classification_report(Y_test, Y_pred_lr))
The accuracy score achieved using Logistic Regression is: 87.1 %
              precision    recall  f1-score   support

           0       0.88      0.88      0.88        17
           1       0.86      0.86      0.86        14

    accuracy                           0.87        31
   macro avg       0.87      0.87      0.87        31
weighted avg       0.87      0.87      0.87        31
```

Figure 3. The above picture shows that confusion matrix parameters with results values like precision, recall, accuracy, and F-score, this model shows the accuracy is 87.10%.

Table 1. The comparative experimental study between the different machine learning classifiers.

Dataset name	Machine learning techniques	Accuracy in percentage (%)	Precision in percentage (%)	Recall in percentage (%)	F1-score in percentage (%)
Heart disease	LR	87.1	88	86	88
	NB	80.65	87	76	81
	SVM	83.87	88	86	85
	DT	70.97	84	86	83
	k-NN	74.19	88	82	85

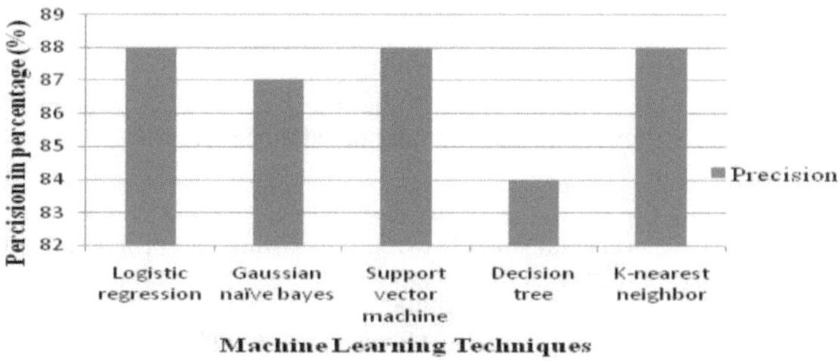

Figure 4. The above picture shows that comparative performance graph which is based on the precision performance parameters using different machine learning techniques.

6 DEEP LEARNING BASED RESULT ANALYSIS

In this section we present the experimental works on the basis of our implementation model and compare with existing technique, the experimental study shows that the our model shows better results than the previous technique in terms of performance parameter evaluation such as accuracy, precision, recall and F1-score. Our experimental results are achieved with python framework with using its different libraries.

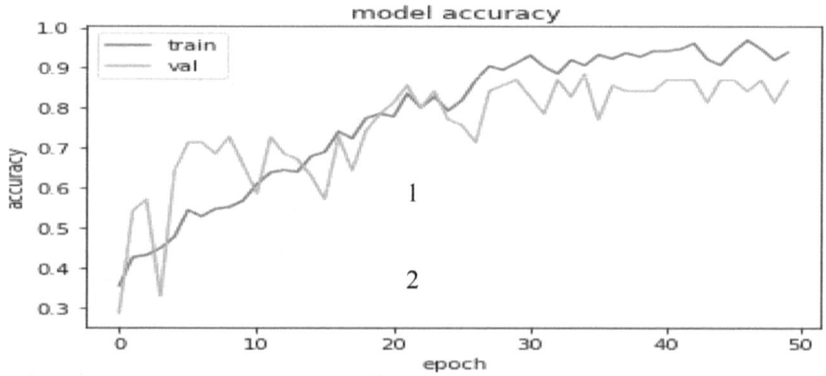

Figure 5. The above picture shows that comparative performance graph which is based on the accuracy performance parameters using different machine learning and our proposed deep learning techniques.

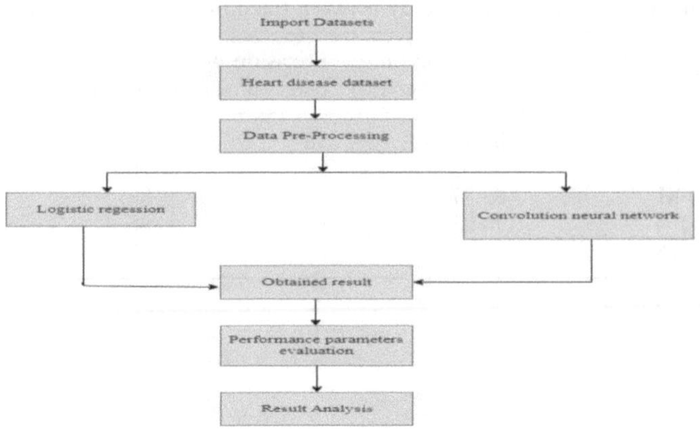

Figure 6. The above picture shows that deep learning based model accuracy with no. of epoch.

Table 2. The comparative experimental study between the existing and proposed method.

Dataset name	Machine learning techniques	Accuracy in percentage (%)	Precision in percentage (%)	Recall in percentage (%)	F1-score in percentage (%)
Heart disease	Logistic regression	87.1	88	86	88
	Gaussian naïve bayes	80.65	87	76	81
	Support vector machine	83.87	88	86	85
	Decision tree	70.97	84	86	83
	K-nearest neighbor	74.19	88	82	85
	Convolution Neural Network Model	96.30	91	95	93

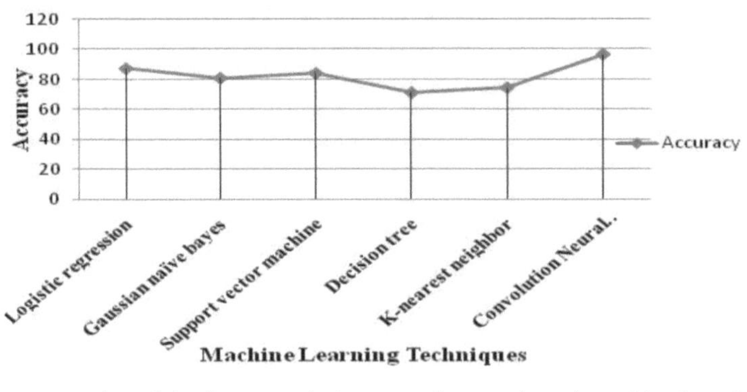

Figure 7. The proposed model using convolution neural network and machine learning approach.

7 CONCLUSION AND FUTURE SCOPE

This research work aims to analyze the effects of ML techniques then with deep learning in detecting the heart diseases. The heart diseases dataset [20] contain the total of 303 person information, for the machine learning classifier; find the logistic regression model provides the best accuracy among all of them. Proposed model which is based on convolution neural network based deep learning techniques is an artificial intelligence technique and subfield of machine learning techniques, simulation result shows our deep learning model gives better accuracy then existing ML techniques. In future work, extend this research work using some optimization techniques, like honey bee classification, and other optimization techniques.

REFERENCES

[1] Yuanyuan Pan, Minghuan Fu, Biao Cheng, Xuefei Tao & Jing Guo. (2020). Enhanced deep learning assisted convolutional neural network for heart disease prediction on the internet of medical things platform, special section on deep learning algorithms for internet of medical things, *IEEE Access*, 189503–189512.

[2] Safial Islam Ayon, Md. Milon Islam & Md. Rahat Hossain. (2020). Coronary artery heart disease prediction: A comparative study of computational intelligence techniques, *IETE Journal of Research*, 1–20.

[3] Amirata Ghorbani, David Ouyang, Abubakar Abid, Bryan He, Jonathan H. Chen, Robert A. Harrington, David H. Liang, Euan A. Ashley, & James Y. Zou. (2020). Deep learning interpretation of echocardiograms, *Digital Medicine*, pp. 1–10.

[4] Aniruddha Dutta, Tamal Batabyal, Meheli Basu, & Scott T. Acton. (2020). *An Efficient Convolutional Neural Network for Coronary Heart Disease Prediction*, pp. 1–36.

[5] Mohammad M R Khan Mamun, & Ali Alouani. (2020). FA-1D-CNN implementation to improve diagnosis of heart disease risk level, *Proceedings of the 6th World Congress on Electrical Engineering and Computer Systems and Sciences*, DOI: 10.11159/icbes20. 122, pp. 1–9.

[6] Lamido Yahaya, Nathaniel David Oye, & Etemi Joshua Garba.(2020). A comprehensive review on heart disease prediction using data mining and machine learning techniques, *American Journal of Artificial Intelligence*, DOI: 10.11648/j.ajai.20200401.12. pp. 20–29.

[7] Deepak Kumar Rathore, Dr. Praveen Kumar Mannepalli. (2021). A review of machine learning techniques and applications for health care, international conference on advances in technology, management & education, *IEEE Proceeding*, 978-1-7281-8586-6/21.

[8] Maragatham G., Shobana Devi. (2019). LSTM model for prediction of heart failure in big data, *Journal of Medical Systems*, Springer, doi.org/10.1007/s10916-019-1243-3. 2019, pp. 1–13.

[9] Senthilkumar Mohan, Chandrasegar Thirumalai, & Gautam Srivastava. (2019). Effective heart disease prediction using hybrid machine learning techniques, *IEEE Access* 2019, pp. 81542–81554.

[10] Hager Ahmed, Eman M.G. Younis, Abdeltawab Hendawi, & Abdelmgeid A. Ali. (2019). Heart disease identification from patients' social posts, machine learning solution on Spark, *Future Generation Computer Systems*. doi.org/10.1016/j.future.2019.09.056, pp. 1–9.

[11] Amin Ul Haq, ian Ping Li, Muhammad Hammad Memon, Shah Nazir, & Ruinan Sun. (2018). A hybrid intelligent system framework for the prediction of heart disease using machine learning algorithms, *Mobile Information Systems*, doi.org/10.1155/2018/3860146. pp. 1–22.

[12] Deepak Kumar Rathore & Praveen Kumar Mannepalli. (2021). Recent trends in machine learning for health care sector, *International Journal of Innovative Research in Technology and Management*, Vol-5, Issue-2.

[13] Ozal Yıldırım, Paweł Pławiak, Ru-San Tan, & Rajendra Acharya U. (2018). Arrhythmia detection using deep convolutional neural network with long duration ECG signals, *Computers in Biology and Medicine* pp. 1–22.

[14] Benjamin Fredrick David H. and Antony Belcy S. (2018). Heart disease prediction using data mining techniques, *ICTACT Journal on Soft Computing*, DOI: 10.21917/ijsc.2018.0253, pp. 1817–1823.

[15] Dubey R., Rathore D., Maurya J. and Kushaha D. (2017). An empirical study of intrusion detection system using feature reduction based on evolutionary algorithms and swarm intelligence methods, *International Journal of Applied Engineering Research* 12 (19). pp. 8884–8889.

[16] Deepak Rathore. (2022). Diseases prediction and classification using machine learning techniques, *AIP Conference Proceedings 2424*, 070001; https://doi.org/10.1063/5.0076768.

[17] Baptiste Vasey, MMed & Stephan Ursprung. (2021). Association of Clinician Diagnostic Performance with machine learning–based decision support systems a systematic review, *JAMA Network*, pp. 1–15.

[18] JoungOuk Kim & Yong-Suk Jeong. (2021). Machine learning-based cardiovascular disease prediction model: A cohort study on the korean national health insurance service health screening database, *Diagnostics*, pp. 1–12.

[19] Arumugam K., Mohd. Naved. (2021). *Multiple disease prediction using Machine learning algorithms*, Elsevier, pp. 1–4.

[20] https://www.kaggle.com/datasets/johnsmith88/heart-disease-dataset.

Computational Methods in Science and Technology Volume 1 – Kaur et al. (Eds)

Text pre-processing techniques in addressing fake news

Pummy Dhiman & Amandeep Kaur
Chitkara University Institute of Engineering and Technology, Chitkara University, Punjab, India

ABSTRACT: This study emphasizes the significance of identifying counterfeit news and the essentiality of proficient pre-processing approaches to clean text corpora. This work effectively integrated two Indian datasets and employed NLP techniques using Python programming language to carry out pre-processing activities, hence enhancing the data's quality. The experiment part offers comprehensive insights into the procedures of combining and pre-processing. Although this study aids in the advancement of efficient pre-processing methods for identifying fake news, there remains a substantial amount of effort to enhance the precision of false news identification. In order to build a better-informed society and enhance our ability to detect false news, future research will investigate the use of DL approaches for the categorization of fake news.

Keywords: Fake news, India, Dataset, Natural Language Processing

1 INTRODUCTION

Authentic and fake news both employ persuasive tactics to influence public opinion. Fake news has emerged as a global concern Faustini and Covoes (2020). The Elaboration Likelihood Model (ELM) Osatuyi and Hughes (2018) is a model that helps us understand how news might influence changes in attitudes. According to this model, persuasion can happen through two routes Chen *et al.* (2021) central routes, which require a lot of cognitive work, and peripheral routes, which require less cognitive effort (Figure 1). Bogus news creators used a peripheral approach that involved providing only limited content with an emphasis on negative emotional cues. There is an increasing worry about the influence of false news on masses and the necessity for efficient measures to combat it. Counterfeit news dissemination, has been a recurring occurrence throughout history. However, it garnered significant attention during the 2016 US presidential election due to the wide-spread circulation of fabricated articles on various social media platforms. It is news that pre-tends to be true but is really false. People can generate or spread fake news for a variety of reasons, including politics, money, or just for fun. False news has an impact all around the world, not just in one area. Inciting violence, separating communities, causing stock prices to fluctuate Zhi *et al.* (2021), hurting reputations, and undermining trust in both the government and the media. Natural language processing (NLP) is a subfield of artificial intelligence that aims to bridge this understanding gap. Prior to processing any text, it must be cleaned up to ensure data quality, remove noise, and normalize uniformity.

The aim of this study is to conduct NLP pre-processing on Indian news corpora in order to investigate the distinctive grammatical and semantic characteristics specific.

 DOI: 10.1201/9781003501244-24

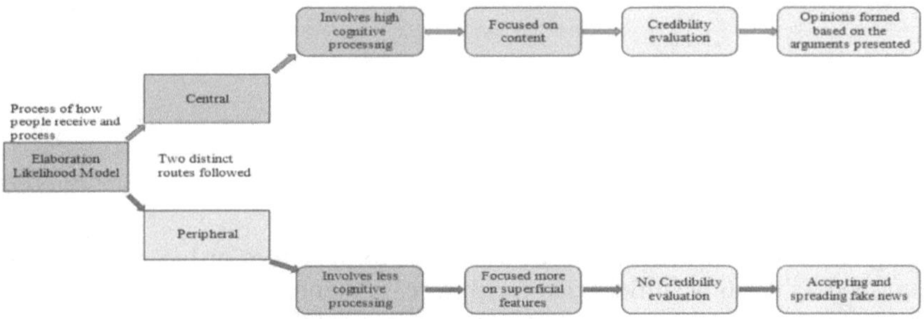

Figure 1. ELM for information route.

2 LITERATURE REVIEW

In order to achieve good accuracy in fake news detection, data cleaning, noise reduction, and improved feature extraction play a crucial role. Mohapatra *et al.* (2022) have conducted various reviews to get to know what has been done in the past and presently and which techniques have been used to countermeasure this global issue of fake news. To improve results in healthcare, Samrat *et al.* (2023) also make use of NLP and sentiment analysis to demonstrate how these two can improve patient outcomes and change healthcare services. They also point out the need for more reliable data sources. Sentiment analysis plays a vital role in analysing customers' feedback, which is very important in business Hangloo and Arora (2022). Drif and Giordano (2019) examine various datasets, NLP techniques, deep learning, and feature engineering in fake news identification. While using advanced machine learning and text pre-processing, feature engineering techniques play a crucial role in achieving good results. When they pre-processed Bi-LSTM models, they got decent results. For the purpose of detecting false news, the PHEME dataset was utilized by, Hossain *et al.* (2022) with the word2vec model serving as the pre-processor. The data was then cleaned, split into train and test sets, and put into a machine learning model to predict rumours. Jabir *et al.* (2020) used standard methods for text cleaning and pre-processing, such as NLTK, to remove stop words, symbols, and special characters, and split sentences into tokens. Shishah (2021) achieved impressive accuracy with the Politifact and Pymedia datasets.

3 NLP PIPELINE

NLP enables machines to comprehend spoken language Dhiman *et al.* (2023). In order to detect fake news, the corpus first needs to go through a number of pre-processing steps to turn into a suitable form. Specific text pre-processing steps may depend on the given task but generally consist of the following steps, as shown below (Figure 2):

- **Lowercase conversion:** Convert all text to lowercase to maintain uniformity in the dataset, preventing the same word from being treated differently due to case differences.
- **Tokenization:** It includes breaking the text up into chunks, phrases, that make sense (called "tokens"). Tokenization is a key part of NLP because it sets the stage for more text processing and analysis.
- Input: Good Morning Class. Output: ["Good", "Morning", "Class", "."]
- **Stop Word Elimination:** Stop words are common phrases that don't usually mean much, such as "of"," the", "a", "in", "an" etc. Getting rid of these terms helps to make the text data more precise and less noisy, which is ideal for future analyses.

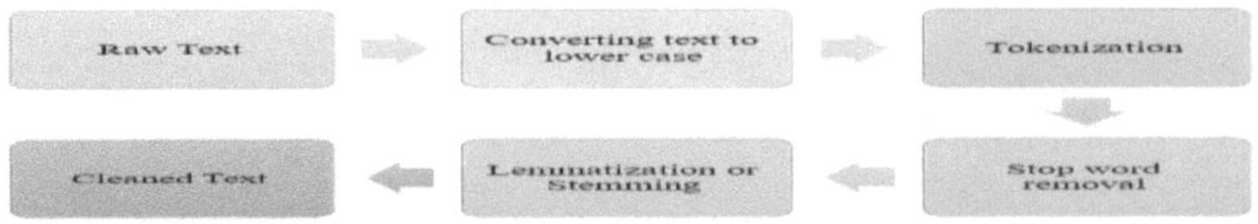

Figure 2. Basic NLP text-cleaning pipeline.

- **Stemming:** This is the process to cut down words to their root or basic form. It is usually accomplished by deleting suffixes that finish words (Figure 3(a)). There may be possibility that these rudimentary terms may have no meaning at all.
- **Lemmatization:** It is a technique that gets words back to their dictionary form (lemma) (Figure 3(b)). It is a more accurate method than stemming; it considers the word's context and part of speech to determine its accuracy.

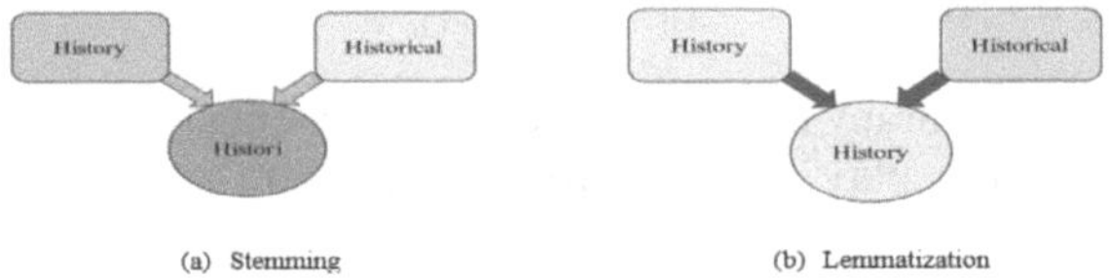

Figure 3. Stemming and lemmatization.

These are the basic steps involved in NLP, but there can be further steps to clean data based on requirements. Once the data is cleaned, it needs to be turned to numeric form so that the AI model can work on the given input to produce output. For this purpose, a vectorization process is used. Some vectorization methods used in NLP are shown below:

- **Term Frequency-Inverse Document Frequency (TF-IDF):** It looks at both how often a word appears in all the documents and how often it appears in each document Rajat *et al.* (2022) It gives each word in a document a score based on how often it appears in that document and how often it appears in the whole collection.
- **Word Embeddings:** This approach displays words not as sparse vectors of 0s and 1s but as dense vectors of continuous values. By putting similar words closer together in vector space Jiang *et al.* (2021), word embeddings show the meaning connections between words, like how they are related to each other grammatically. Word2Vec and GloVe are two well-known methods for making word embeddings.

4 EXPERIMENTAL SETUP

The study utilized a 12th Generation Intel i5-1235U processor and Google Colab to conduct an experiment using Python modules like Numpy, Pandas, matplotlib. pyplot, seaborn, and NLTK for data manipulation and visualization. The implementation is done using two Indian corpora. The first corpus The Indian Fake News (IFND) Sharma and Garg (2022) comprises news gathered from several online news portals, but it does not include social media posts. This gap is filled by the second corpus (FakeNewsIndia), Dhawan *et al.* (2022) which comprises social media posts. Following the merging process, the statistics of the combined corpus were subsequently displayed (Table 1).

Table 1. Merged corpus statistic.

Dataset	Label	Count	Total (news)
Combined Dataset	Real	37800	61364
	Fake	23564	

The pre-processing procedure will now be carried out in the following manner.

- **text.lower():** It transforms the input text characters into lowercase. This aids in maintaining text data consistency.
- **remove_punctuation** Just like before, this function accepts a text string as input and strip it of any punctuation. Simplifying the text data in this way can make it easier to analyse.

English stop words are downloaded from the Natural Language Toolkit (nltk) package Nigam *et al.* (2021), and stemming is applied to each word in a text string, resulting in a stemmed version with each word reduced to its root form. Lemmatization is used to reduce a word to its lemma, and URLs are eliminated using regular expressions. Emojis are eliminated to streamline data analysis or categorize using ML models.

5 RESULT AND DISCUSSION

The study combines two datasets using Pandas, resulting in a merged corpus of 37800 real news stories and 23564 false news stories. Subsequently, the merged dataset undergoes text pre-processing to prepare it for input into a machine or deep learning model for enhanced identification of bogus news Dhiman and Kaur (2022). Pre-processing tasks aim to remove unnecessary characters while preserving valuable words Dhiman *et al.* (2023). The Porter Stemmer reduces word affixes to root forms, while the WordNet Lemmatizer generates valid words from linguistic context. Both methods improve computational efficiency and linguistic accuracy in text analysis. Word clouds are a popular NLP text presentation method that can identify frequently used terms and phrases in a text corpus. Figure 4 shows a pre-processed dataset word cloud, which has been cleaned by removing irrelevant content and special characters. This clean text is now ready for input into machine learning classifiers.

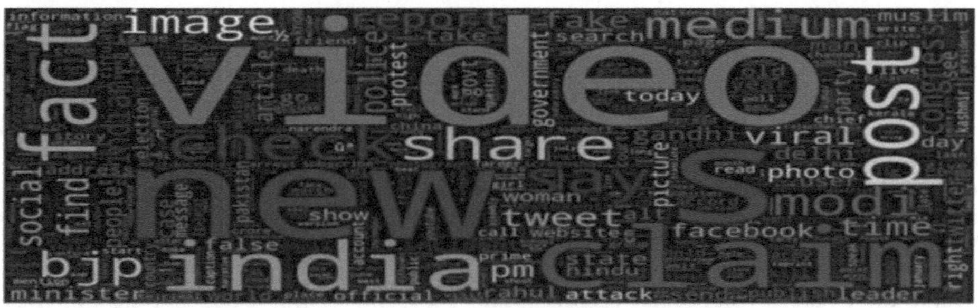

Figure 4. Data proportion of various Indian fact checking websites.

6 CONCLUSION AND FUTURE DIRECTIONS

Current work emphasizes the significance of identifying false information and the necessity for efficient pre-processing methods. The NLP approaches and Python programming language were used to effectively combine two Indian datasets. This involved carrying out tasks such as text normalization, stop word elimination, and stemming. The work enhances the advancement of efficient pre-processing algorithms for identifying phony news, which can be subsequently improved in future research. Subsequent investigations will examine the application of deep learning classifiers in identifying counterfeit news, with the objective of enhancing the accuracy in recognizing inaccurate information.

REFERENCES

Chen, C. Y., Kearney, M., & Chang, S. L. (2021). Belief in or identification of false news according to the elaboration likelihood model. *Int. J. Commun.*, vol. 15, no. 108, pp. 1263–1285.

Dhawan, A., Bhalla, M., Arora, D., Kaushal, R., & Kumaraguru, P. (2022). Fake News India: A benchmark dataset of fake news incidents in India, collection methodology and impact assessment in social media. *Comput. Commun.*, vol. 185, no. January, pp. 130–141. doi: 10.1016/j.comcom.2022.01.003.

Dhiman, P., & Kaur, A. (2022). Significance of convolutional neural network in fake content detection: A systematic survey. In *International Conference on Emergent Converging Technologies and Biomedical Systems* (pp. 305–316). Singapore: Springer Nature Singapore.

Dhiman, P., Kaur, A., & Bonkra, A. (2023). Fake information detection using deep learning methods: A survey. In *2023 International Conference on Artificial Intelligence and Smart Communication*, AISC 2023, pp. 858–863. doi: 10.1109/AISC56616.2023.10085519.

Drif, A., & Giordano, S. (2019). Fake news detection method based on text-features. *Ninth Int. Conf. Adv. Inf. Min. Manag. Fake*, vol. 23, no. 3, pp. 26–31.

Faustini, P. H. A. & Covões, T. F. (2020). Fake news detection in multiple platforms and languages. *Expert Syst. Appl.*, vol. 158, no. May 2020: 113503. doi: 10.1016/j.eswa.2020.113503.

Hangloo, S., & Arora, B. (2022). *Combating Multimodal Fake News on Social Media: Methods, Datasets, and Future Perspective*. Springer Berlin Heidelberg. doi:10.1007/s00530-022-00966-y.

Hossain, E., Kaysar, N., Zahid, A., Jalal, M., Joy, U., & Rahman, M. M. (2022). *A Study Towards Bangla Fake News Detection Using Machine Learning and Deep Learning*. Springer, pp. 79–95. doi: 10.1007/978-981-16-5157-1_7.

Jabir, H. M., Naser, M. A., & Al-mamory, S. O. (2020). Rumor detection on twitter using features extraction method. In *2020 1st. Information Technology to Enhance e-learning and Other Application (IT-ELA)*, pp. 115–120. doi: 10.1109/IT-ELA50150.2020.9253027.

Jiang, T., Li, J. P., Haq, A. U., Saboor, A., & Ali, A. (2021). A novel stacking approach for accurate detection of fake news. *IEEE Access*, vol. 9, pp. 22626–22639. doi: 10.1109/ACCESS.2021.3056079.

Mohapatra, S. K., Sarangi, P. K., Sarangi, P. K., Sahu, P., & Sahoo, B. K. (2022). Text classification using NLP based machine learning approach, p. 020006. doi: 10.1063/5.0080301.

Nigam, A., Jaiswal, P., Sundar, S., Poddar, M., Kumar, N., Dernoncourt, F., & Celi, L. A. (2021). NLP and deep learning methods for curbing the spread of misinformation in India. *Int. J. Intell. Secur. Public Aff.*, vol. 23, no. 3, pp. 216–227. doi: 10.1080/23800992.2021.1975429.

Osatuyi, B. & Hughes, J. (2018). A tale of two internet news platforms-real vs. Fake: An elaboration likelihood model perspective. *Proc. Annu. Hawaii Int. Conf. Syst. Sci.*, vol. 2018-Janua: 3986–3994. doi: 10. 24251/hicss.2018.500.

Rajat, P., Jaroli, P., Singla, C., Bhardwaj, V., & Mohapatra, S. K. (2022). Deep learning model based novel semantic analysis. In *2022 2nd International Conference on Advance Computing and Innovative Technologies in Engineering (ICACITE)*, pp. 1454–1458. doi: 10.1109/ICACITE53722.2022.9823741.

Samrat, R., Bonkra, A., Kaur, A., & Dhiman, P. (2023). Deep learning techniques for sentiment analysis: A comparative study. In *2023 6th International Conference on Contemporary Computing and Informatics (IC3I)*, vol. 6, pp. 533–539. doi: 10.1109/IC3I59117.2023.10397860.

Sharma, D. K., & Garg, S. (2021). IFND: A benchmark dataset for fake news detection. *Complex Intell. Syst.* doi:10.1007/s40747-021-00552-1.

Shishah, W. (2021). Fake news detection using bert model with joint learning. *Arab. J. Sci. Eng.*, vol. 46, no. 9, pp. 9115–9127. doi: 10.1007/s13369-021-05780-8.

Zhi, X., Xue, L., Zhi, W., Li, Z., Zhao, B., Wang, Y., & Shen, Z. (2021). Financial Fake News Detection with Multi fact CNN-LSTM Model. *2021 IEEE 4th Int. Conf. Electron. Technol. ICET* 2021, pp. 1338–1341. doi: 10.1109/ICET51757.2021.9450924.

Computational Methods in Science and Technology Volume 1 – Kaur et al. (Eds)
© 2025 the Author(s), ISBN: 978-1-032-81760-6

Prediction of recovery rate of COVID-19 breakthrough cases with machine learning

Pratibha, Gaganpreet Kaur, Gurtej Kaur, Esha Dhamija & Parul
Chitkara University Institute of Engineering and Technology, Chitkara University

ABSTRACT: A pandemic was formally proclaimed by the World Health Organization in March 2020 due to COVID-19, which was caused by a new coronavirus called SARS-CoV-2 (Severe Acute Respiratory Syndrome Coronavirus). This virus caused a major spike in death rates and posed a serious threat to the medical community, spreading quickly around the world. Nonetheless, a few people have recovered, which is frequently linked to strong immune reactions. Worldwide vaccination initiatives have been launched in response to the epidemic. This article is centered on using supervised machine learning methods to forecast patients' post-vaccination recovery. Predictive models are specifically created using the Support Vector Machine (SVM), Decision Trees, Random Forest, and K-Nearest Neighbors (KNN) algorithms. To improve forecast accuracy and optimize attribute values, a simulator is used. The aim of this study is to provide a dependable technique for predicting the recuperation of patients after receiving a COVID-19 immunization. In the ongoing fight against the epidemic, healthcare practitioners may plan treatments and allocate resources more effectively by utilizing machine learning algorithms to get insightful knowledge about recovery patterns.

1 INTRODUCTION

The COVID-19 causing a huge public health crisis in worldwide. This pandemic situation began in December 2019, starts from Wuhan, China (Ahamad, 2021). It has common symptoms are flu, fever, headache, fatigue, etc. Researchers focusing on smell and taste and how they are related to Covid19 patients (C. H., 2020). According to the World Health Organization, there have been 175,987,176 confirmed cases of COVID-19 worldwide as of June 15, 2021, with 3,811,561 deaths. A total of 2,310,082,345 persons had been vaccinated as of June 14, 2021 (Atef Omar, 2021). The SARS-CoV-2 virus spreading rapidly across the world such that, as of February 2021, and approx 119 million people have been infected, of which 2.6 million have died (Beaumont, 2020). Many countries have begun vaccination programs on priority bases, like those who have the probability of complications from COVID-19 infection, such as old people, and who are at high risk of virus exposure and transmission such as front-line staff (Bonkra, 2024). These medications or vaccine have adverse effects to the person having pre-existing problems, such as diabetes, hypertension, heart disease and high blood pressure or allergy. Medicine protects us from many deadly diseases but, no medicine is free from adverse effect. The patient history is strongly related to patient adverse reaction with severe disease and even death. After vaccination various side effects also developed in the patient body. Breakthrough cases' are expected less number of people after vaccination. No vaccine is 100% effective, vaccinated people, who have received both doses, testing positive for the virus (Kaur G, 2022) Such cases are referred to as breakthrough infections, indicating a small minority of persons are likely to develop breakthrough infections because the virus has been able to overcome the vaccine's defenses. Although just a few cases of breakthrough infections have been reported, these cases have cast doubt on the vaccine's effectiveness. Vaccine protects against moderate to severe sickness as well as hospitalization, rather than the virus itself. Vaccinated persons are far less likely to get sick, according to the Centers for Disease Prevention and Control (CDC) in the United States, although it can never be completely ruled out [6]. On May 31, 2021 CNN NEWS 18 claimed that single dose of provide enough immunity against SARSCoV-2 infection, but on June 1, 2021 (Iwendi, 2021). Many machine learning algorithms are available in this context.

In this paper our contribution to the current crisis is to develop a model that predict the recovery of vaccinated COVID-19 patients in terms of their age, infection and hospitalized status of a patient. To accomplish the task, we use Support Vector Machine (SVM), Decision tree, Random Forest and K-Nearest Neighbors (KNN) supervised machine learning algorithm that predict the recovery.

The rest of the paper is divided into five sections. The introduction is presented in Section I, Literature in Section II, Methodology in Section III, Technology Used in Section IV, Results and Discussion and Section V and Conclusion and Future scope in Section VI.

2 LITERATURE

A lot of work has been done on Covid-19 in a short span. Recovery prediction of covid-19 patients was calculated in so many papers but the recovery of breakthrough cases i.e. vaccinated people who were infectious not done so far. In this paper the author compares MLP, SVM and KNN algorithms to classify Covid-19 patients into 4 classes of different countries. SVM, KNN, and MLP achieved 100%, 95%, and 99% accuracy. The greatest performance is achieved by SVM. The precision, accuracy, recall and F measure parameters were used to calculate the model. The performance was high but dataset was small (Dhiman, 2022).

While in this paper, the dataset of demographic information, patient's medical history, and patient's adverse reactions after vaccination are used. Then mann-whitney U Test and Chi-square Test used in statistical analysis and DT, RF, XGB, SVM, GBM and LGBM machine learning algorithms are used. Author found significant feature findings, severity prediction, feature ranking and predict suitable candidates for vaccination. Through this paper they classified and characterize the people who are at risk and to reduce the risk found the normal causes by the strategies (Han J, 2011).

In this survey paper to train multiple Machine Learning classification models the preliminary datasets used by author. Decision Tree, Support Vector Machine, Boosted Random Forest, and Gaussian Naïve Bayes classifiers used. To calculate the possible outcome, recovery, and strength of the cases in COVID-19 patients they used the travel, health, geographical and demographic data. The accuracy and F1 Score is 94 % and 0.86 for the dataset (Donges, 2024). This paper includes the reviews of Covid-19 models of India's recovered, confirmed and deaths cases. The association coefficients and multiple linear regressions applied for prediction and autocorrelation to improve the accuracy auto regression have been used. Comparison to the actual values with "0.9992 R-squared" score new evaluated results give fine contract. The result suggests that to suppress the growing rate of COVID-19 the lockdown and public distancing are main factors that can help out. This model can help to predict the causality and financial sufferers ("Can Vaccinated People Transmit COVID-19?", 2021).

3 METHODOLOGY

The methodology of our research is depicted in the figure below. We initially get raw data on breakthrough cases from the Center for Disease Control and Prevention (CDC) in order to begin our analysis. Following data preparation, we put the unprocessed data into an analysis-ready structure by organizing and interpreting it. First, we get unprocessed information on COVID-19 breakthrough cases from trustworthy sources, including the Centers for Disease Control and Prevention (CDC). This dataset contains details about the hospitalization status, comorbidities, vaccination history, and demographics of the patients. To transform the raw dataset into a meaningful and organized format that can be analyzed, preprocessing is applied. This includes addressing missing values, eliminating duplication, and harmonizing data formats, among other things. Initially, we extract and transform raw data into values i.e. The 'SEX' attribute has two values, i.e. Male denoted by 'M' and female by 'F'. In the 'infection' field value with '1' is infected and with '0' non infected. In the 'total hospitalized' field we have considered '1' as hospitalized and '0' as non-hospitalized. In the 'fetal' field '1' as recovered '0' as dead. CDC received report from 46 U.S states of 3729 patients with COVID-19 breakthrough infection who were hospitalized or died among them they have female 1830, people aged greater than or equal to 65 were 2856, asymptomatic infections was 683, hospitalization were 3538 and 916 of 3538 hospitalization reported as asymptomatic or not related to COVID-19. Death rate was 671and 122 of 671 fetal cases reported as asymptomatic or not related to COVID-19. Table 1 shows the vector of dataset.

Figure 1. Flow of work.

We carefully consider relevant factors that may impact COVID-19 patients' chances of recovery during the study procedure. We specifically pay attention to the following variables: gender, age (especially 65 and older), and hospitalization status (labeled as "hospitalized COVID"). By using data visualization tools like RapidMiner, we may find important patterns and trends in the dataset by understanding the distribution and interactions between the various variables. Then, we use 70% of the dataset to train our machine learning models using K-Nearest Neighbors (KNN), Random Forest, Decision Trees, and Support Vector Machine (SVM).

3.1 *Techniques used*

3.1.1 *KNN (K-Nearest Neighbor)*

This algorithm is supervised machine learning algorithm used for classification and prediction to classify the unknown data from its nearest neighbor known or similar measure classification. A sample value is predicted from its nearest neighbor.

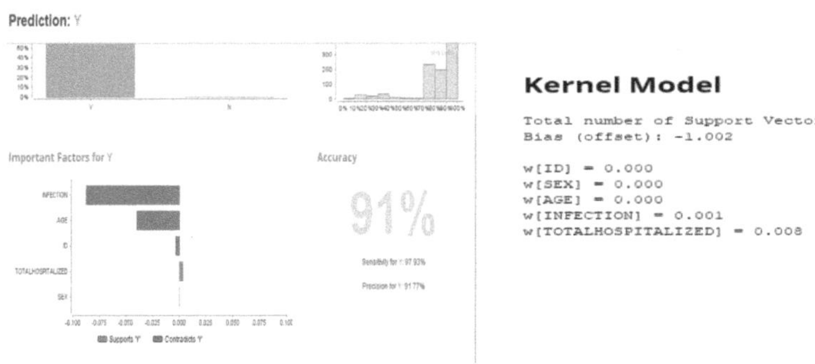

Figure 2. KNN classification and accuracy.

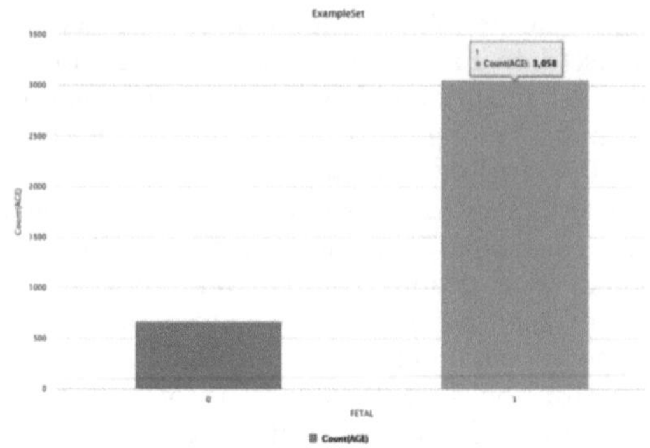

Figure 3. Prediction value of fetal and recovery.

Table 1. Support Vector Table.

counter	label	function...	alpha	abs(alp...	support...	ID	SEX	AGE	INFECTI...	TOTALH...
0	Y	−1.000	−0.034	0.034	support v...	−1.638	0	−1.292	0.507	0.162
1	Y	−1.000	−0	0	no supp...	0.517	0	−1.292	0.507	0.162
2	Y	−0.999	0	0	no supp...	0.874	0	0.555	0.507	0.162
3	Y	−1.000	0	0	no supp...	0.518	0	−1.177	0.507	0.162
4	Y	−0.999	0	0	no supp...	0.873	0	0.439	0.507	0.162
5	Y	−1.000	−0.162	0.162	support v...	−1.632	0	−0.484	0.507	0.162
6	Y	−1.000	−0.233	0.233	support v...	−1.630	0	−0.253	0.507	0.162
7	Y	−1.000	−0.017	0.017	support v...	−1.629	0	−0.138	0.507	0.162
8	Y	−1.001	−0.102	0.102	support v...	−1.628	0	−0.023	0	0.162
9	Y	−1.000	−0	0	no supp...	−1.627	0	0.093	0.507	0.162
10	Y	−0.999	0	0	no supp...	1.216	0	1.016	0.507	0.162
11	Y	−1.000	0	0	no supp...	−1.625	0	0.324	0.507	0.162
12	Y	−1.000	0	0	no supp...	−1.623	0	0.555	0.507	0.162
13	Y	−1.000	0	0	no supp...	−1.622	0	0.670	0.507	0.162
14	Y	−0.999	0	0	no supp...	1.215	0	0.785	0.507	0.162
15	Y	−1.000	0	0	no supp...	−1.621	0	0.901	0.507	0.162
16	Y	−0.999	0	0	no supp...	1.214	0	0.670	0.507	0.162
17	Y	−0.999	0	0	no supp...	1.213	0	0.555	0.507	0.162

3.1.2 *SVM (Support Vector Machine)*

A supervised ML algorithm includes the Support Vector Machine works on regression and classification (Huilgol Purva, 2023; Ahmed A. H., 2022). For classification or outlier detection this technique constructs a hyper plane in N-dimension space.

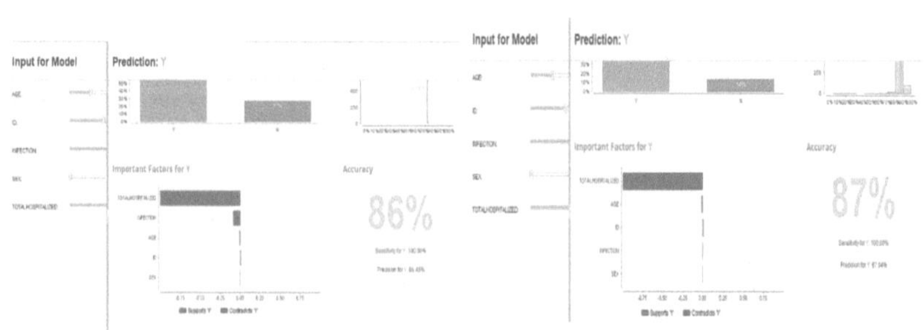

Figure 4. Accuracy using SVM and accuracy of random forest.

3.1.3 *Random forest*

The forest is built on several decide tree fusion and trained by bagging algorithm. This is a supervised machine learning method. It uses ensemble learning to improve overall the classification and prediction result. Random forest takes only random forest or features for splitting a node (Trivedi N. K., 2021).

Decision Tree- It is used to predict the value of target variable in which leaf node corresponds to a class label and attributes represents internal nodes of the tree. It creates a training model that is used to predict the class of dependent variable using decision rules. The decision tree compare values of root with the new value of new record presented to it to create the decision branch based on comparisons (Kumari R, 2021).

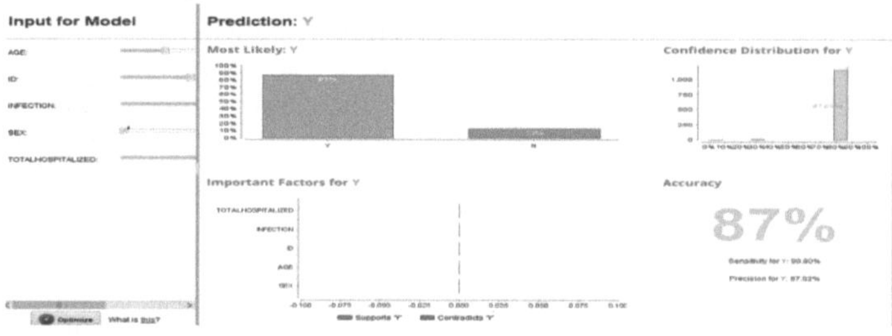

Figure 5. Accuracy of decision tree.

4 RESULTS

In this work we have used KNN, Decision tree, SVM and Random Forest models for prediction of recovery rate of COVID-19 against dataset collected from CDC website. The performance of classifiers is calculated on the basis of accuracy, F-score, Precision, Sensitivity, TP, TF, FP and FN and Confidence. In the they have used Naïve Bayes classifier with 93 % accuracy on small dataset and proved that the age is the determining factor of death from CORONA virus and younger a person is less likely to die from this virus. In our approach we compare the recovery rate with different classifier on large dataset and recovery rate of each classifier is given in the table below. The model developed with KNN classifier is best among all other classifier in terms of accuracy with 91%, while Decision Tree, Random Forest and SVM have respectively 87%, 85%, and 86%. Accuracy comparison of different classifiers given below.

Table 2. Comparison of different models.

Models/ Factors	KNN	Decision Tree	SVM	Random Forest
Accuracy	91%	87%	86%	87%
Precision	91.77%	87.02%	86.45%	87.04%
Sensitivity	97.93%	99.8	100.00%	100.00%
Confidence	100.00%	87.02	73.14%	86.47%

The World Health Organization has designated the corona virus an international public health emergency. It is declared as epidemic a pandemic by World Health of Organization and is transmitted person to person by direct contact with infected patient or via sneezing and coughing. Different vaccines are now available in the various countries. The COVID-19 vaccines developing the immune system to reduce the risk of illness and its consequences. Vaccination helps to fight and protect against the virus from getting severer illness and also protect other. In our study we discussed breakthrough male and female cases that are infected and hospitalized. The proposed model used supervised machine learning classifiers like KNN, SVM, decision tree, and random forest to predict the recovery rate.

5 CONCLUSION AND FUTURE SCOPE

In present work we have used machine learning techniques. In future we can use deep learning to predict the classes more accurately. The World Health Organization has designated the corona virus an international public health emergency. It is declared as epidemic a pandemic by World Health of Organization and is transmitted person to person by direct contact with infected patient or via sneezing and coughing. Different vaccines are now available in the various countries. The COVID -19 vaccines developing the immune system to reduce the risk of illness and its consequences. Vaccination helps to fight and protect against the virus from getting severer illness and also protect other. In our study we discussed breakthrough male and female cases that are infected and hospitalized. The proposed model used supervised machine learning classifiers like KNN, SVM, decision tree, and random forest to predict the recovery rate. In present work we have used machine learning techniques. In future we can use deep learning to predict the classes more accurately.

REFERENCES

Ahamad, Md & Aktar, Sakifa & Uddin, Md & Rashed-Al-Mahfuz, Md & Azad, A. K. M. & Uddin, Shahadat & Alyami, Salem & Sarker, Iqbal & Lio, Pietro & Quinn, Julian & Moni, Mohammad Ali. (2021). Adverse effects of COVID-19 vaccination: Machine learning and statistical approach to identify and classify incidences of morbidity and post-vaccination reactogenicity.

Ahmed, A. H., Al-Hamadani, M. N., & Satam, I. A. (2022). Prediction of COVID-19 disease severity using machine learning techniques. *Bulletin of Electrical Engineering and Informatics*, *11*(2), 1069–1074.

Atef, Omar & Nassir, Qassim & Nassif, Ali & Abutalib, Manar. (2020). Death/recovery prediction for covid-19 patients using machine learning. *International Journal of Systems Applications, Engineering & Development.*

Beaumont, Peter. (2020). Covid-19 vaccine: Who are countries prioritising for first doses?. *The Guardian*. ISSN 0261-3077.

Bonkra, A., Pathak, S., Kaur, A., & Shah, M. A. (2024). Exploring the trend of recognizing apple leaf disease detection through machine learning: A comprehensive analysis using bibliometric techniques. *Artificial Intelligence Review*, 1–26.

Can Vaccinated People Transmit COVID-19? (2021). *Cleveland Clinic Health Essentials*. Retrieved from https://health.clevelandclinic.org/can-vaccinated-people-transmit-covid-9 https://indianexpress.com/article/explained/explained-infections-after-vaccination-7286616

Coronavirus Cases: Worldometer. [Online]. Available: https://www.worldometers.info/coronavirus/. [Accessed: 13-Mar-2021].

Dhiman, Pummy, & Amandeep Kaur. (2022). Significance of convolutional neural network in Fake content detection: A systematic survey. In *International Conference on Emergent Converging Technologies and Biomedical Systems*, 305–316.

Donges, N. (2024). *Random Forest Algorithm*. Built In. Retrieved from https://builtin.com/data-science/random-forest-algorithm

Han J, Kamber M. (2011) Data mining: Concepts and techniques, 3rd Edition. *The Morgan Kaufmann Series in Data Management Systems*. Morgan Kaufmann.

Huilgol, Purva (2023). Precision-recall-machine-learning. Retrieved from https://www.analyticsvidhya.com/blog

Iwendi, C., Bashir, A. K., Peshkar, A., Sujatha, R., Chatterjee, J. M., Pasupuleti, S., & Jo, O. (2021). COVID-19 patient health prediction using boosted random forest algorithm. *Frontiers in public health*, *8*, 562169.

Kaur, G., Kaur, A., & Khurana, M. (2022, October). A stem to stern sentiment analysis emotion detection. In *2022 10th International Conference on Reliability, Infocom Technologies and Optimization (Trends and Future Directions)(ICRITO)* (pp. 1–5). IEEE.

Kumari, R., Kumar, S., Poonia, R. C., Singh, V., Raja, L., Bhatnagar, V., & Agarwal, P. (2021). Analysis and predictions of spread, recovery, and death caused by COVID-19 in India. *Big Data Mining and Analytics*, 4 (2), 65–75.

Trivedi, N. K., Simaiya, S., Lilhore, U. K., & Sharma, S. K. (2021). COVID-19 pandemic: Role of machine learning & deep learning methods in diagnosis. *Int J Cur Res Rev*, *13*(06), 150–156.

Yan C. H., Faraji F., Prajapati D. P., Boone C. E., and DeConde A. S. (2020). Association of chemosensory dysfunction and COVID-19 in patients presenting with influenza-like symptoms. *Int. Forum Allergy Rhinol.*, 806–813.

Computational Methods in Science and Technology Volume 1 – Kaur et al. (Eds)

Machine learning based automatic water tank cleaner

K. Valarmathi & R. Vinoth
Professor, P.S.R. Engineering College, Tamil Nadu, India

T. Sneka, S. Srinivethitha & V. Venkadeswari
UG Student, P.S.R. Engineering College, Tamil Nadu, India

ABSTRACT: Water constitutes an indispensable element for human beings, particularly for consumption. It serves as a vital component of our daily routines, encompassing activities such as drinking, hygiene practices like brushing and bathing, as well as household chores including laundry and other tasks. However, in certain areas, the maintenance of water storage tanks is not conducted with sufficient regularity or thoroughness. Over time, sediments, scale, and algae accumulate on the walls, ceiling, and floor of these tanks, resulting in the degradation of water quality. This contamination renders the water unfit for consumption, emphasizing the critical importance of maintaining both clean water and clean water tanks to ensure the provision of high-quality drinking water to communities. To address the challenges associated with tank cleanliness, our research integrates machine learning techniques, specifically employing convolutional neural networks (CNN) and supervised learning algorithms. These methodologies enable the prediction of tank cleanliness by considering factors such as sedimentation and discoloration. Furthermore, our approach offers a means to evaluate water quality by analyzing parameters such as microbial content, chemical composition, and turbidity.

Keywords: Machine learning, Water quality, Tank cleanliness, Internet of Things (IoT), Convolutional Neural Network (CNN)

1 INTRODUCTION

About 71% of people in India use tanks, according to examinations that have revealed a number of issues. These issues include ongoing work in disorganized areas, irregular delivery, and various other causes. Constant work and unpredictable installation may also be a major motivation for this effort. It was shown in a recent study that no machine based on machine learning is utilized in the process of cleaning industrial-based tanks. This is the outcome of the tank area's varying diameters and erratic shape. This indicates that filling the tank with water is the first stage. Conventional techniques for tracking and forecasting water quality and tank cleaning might be inaccurate and time-consuming. Leveraging the power of machine learning, specifically Convolutional Neural Networks (CNN), presents an innovative solution to streamline this process. Because of its standing as the best technique for classifying images and extracting features, the Convolutional Neural Network (CNN) model is developed. With the help of this thorough classification, it will be possible to distinguish between different types of water and anticipate the cleanliness of tanks using machine learning techniques that will be analyzed and used later. A quantitative assessment of the model's accuracy in predicting and categorizing characteristics in the dataset is produced at the end of this procedure in the form of accuracy metrics for each prediction. Predicting the tank for cleaning purposes is the goal of this endeavour.

DOI: 10.1201/9781003501244-26

2 RELATED WORK

A model for a Wireless Sensor Network (WSN)-based water quality monitoring system was developed by Ermiyas Birihanu Belachew Elte *et al.* [1]. Machine learning techniques were incorporated into the model to assess water data and support decision-making. Sai Sreeja Kura *et al.* [2] conducted a systematic evaluation of various algorithms performance, strengths, and weaknesses in predicting water quality parameters. Their study also sheds light on the applicability of different models to specific water quality scenarios. Meanwhile, Minxi Lu *et al.* [3] presented a robotic system utilizing magnetic adsorption materials to securely attach itself to oil tank interiors, facilitating efficient cleaning and recovery of oil pollution. This design seamlessly integrates cleaning and recycling functions, optimizing procedural steps to minimize resource wastage and environmental impact, thereby enhancing oil tank maintenance effectiveness.

G. M. Barbade *et al.* [4] integrated sensors into tank systems to monitor water levels and implemented automatic filling mechanisms to maintain optimal water supply. A notable aspect of their system is the inclusion of a user-friendly interface, accessible via mobile applications, allowing for real-time monitoring and control. Juhwan Kim *et al.* [5] introduced an inventive method for surveilling and analyzing small "eel larvae" in aquatic environments. Through collaborative vision-based techniques, they developed a system utilizing advanced imaging technologies such as cameras and sensors to observe the intricate behaviors and traits of these larvae within water tanks. Chuhan Qi *et al.* [6] employed Long Short-Term Memory Recurrent Neural Networks (LSTM-RNN) in tandem with remote sensing images to monitor water quality, with a focus on Poyang Lake. They investigated the efficacy of LSTM-RNN in capturing temporal dependencies and patterns in water quality parameters over time, leveraging satellite imagery to provide spatial context.

An detailed evaluation of current developments in water quality monitoring systems was carried out by Bineet Kumar Jha *et al.* [7], with an emphasis on the integration of IoT sensors and machine learning techniques. They highlighted the limitations of traditional monitoring methods and underscored the potential of cloud-based solutions to address these challenges. Chaudhary Parth *et al.* [8] proposed an innovative solution for automating the cleaning process of domestic cylindrical water tanks, aimed at minimizing human effort and ensuring efficient maintenance. Their mechanical system includes necessary systems that allow for thorough vertical sweeping movement, like a reciprocating four-bar linkage mechanism and a rack and pinion gear mechanism. R. Raffik *et al.* [9] designed a portable system specifically tailored for cylindrical tanks with central openings. This system integrates a high-pressure fluid supply, facilitated by nozzles and controlled by an automated system, enhancing the apparatus's efficiency for seamless automatic tank cleaning. P.Prem Kumar *et al.* [10] introduced a robotic solution designed to autonomously clean water tanks with minimal human intervention. While human involvement is necessary for transporting the robot to the tank's location and placing it inside, the subsequent cleaning process is entirely autonomous.

2.1 *Tank cleaning system*

Recent investigation revealed that the upper tank and industrial-based tank are not cleaned using any machine based on machine learning. This is the outcome of the tank area's varying diameters and erratic shapeIt thus predicts the tank first. Previous reviews have affected the endeavor to build a machine through the mechanization procedure to clean the tank. The overall population in India uses tanks at a rate of roughly 71%. However, the data from the examinations revealed a number of issues, including ongoing work in disorganized areas, irregular delivery, and various other causes. Constant work and erratic installation may also be the major motivation for this effort. In order to address each of these issues, it can be extremely helpful to clean the upper tank utilizing the mechanized procedure.

In this case, the machine can easily and quickly clean the tank. This paper is to overcome the disadvantage of cleaning overhead tank, industrial tanks, so the autonomous tank cleaner that is based on machine learning is intended to offer maximum safety, maximum efficiency, reduced cleaning time, and prevent environmental contamination issues.

2.2 *Necessity of cleaning tank*

People utilize the tank water on a daily basis for cleaning, mopping, drinking, brushing, bathing, laundry and other domestic tasks. Algae and sediments accumulate over time and cover the water tank's floor, roof, and walls. The pollution this deposition causes renders the water unfit for human consumption. Over time, bacteria and algae grow and multiply in this water, contaminating it and perhaps causing illness.

2.3 *Methods of tank cleaning*

Scrubbing the tank's walls and floor by hand is one way to get rid of stains, fungus, and grime, but it takes more time and is more laborious. Another technique for cleaning the water tank and removing dirt and particles is chemical cleaning. The usage of chemicals may have an impact on human health.

It is possible to spray pressured water over the exterior of the tank in order to remove the dirt off the surface.

2.4 *Dataset*

First step in this approach is dataset collection Initially, real time images are collected from nearby region such as srivilliputtur and sivakasi rural areas which shows in Figures 2, 3, 4, 5. The water quality dataset undergoes categorization into two distinct labels: "clean water" and "worm water". The "clean water" category consists of 30 images showcasing sanitized images, while the "worm water" category consists of 30 images showcasing water contaminated with worms, insects, and dust. Similarly, the dataset for tank cleanliness is

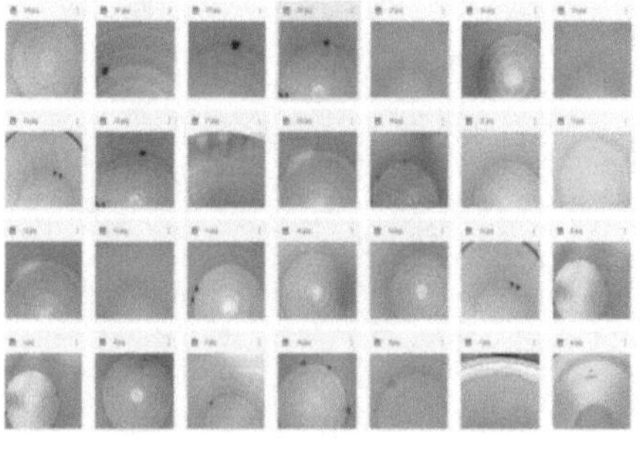

Figure 1. Dataset for clean tank.

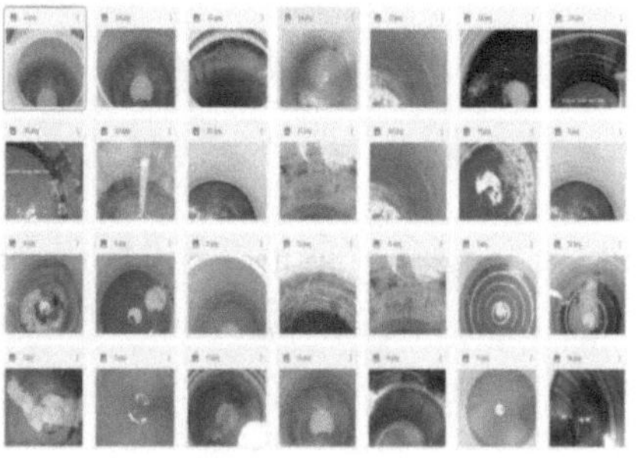

Figure 2. Dataset for unclean tank.

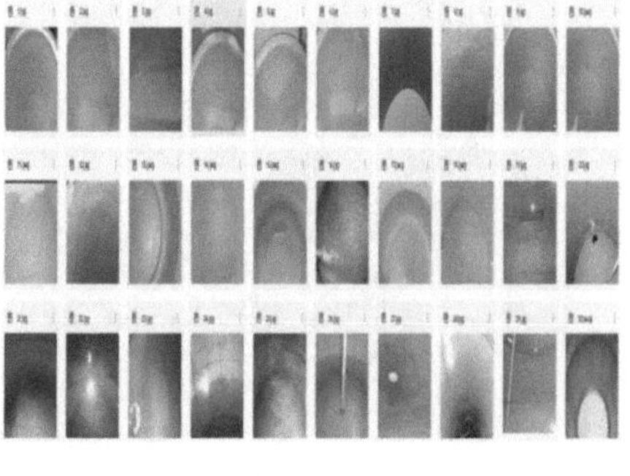

Figure 3. Dataset for clean water.

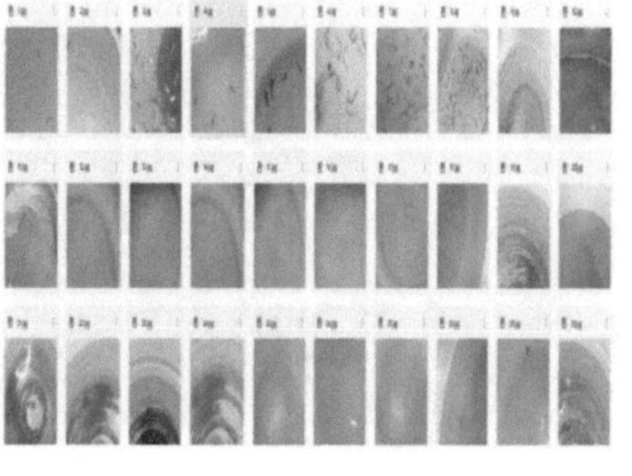

Figure 4. Dataset for warm water.

segregated into two labels: "clean tank" and "uncleaned tank," each containing 30 images. After the images are gathered from each category, a pivotal stage in the process involves training the dataset on the machine learning through the utilization of supervised learning algorithm.

3 METHODOLOGY

3.1 *CNN Algorithm*

One popular kind of deep learning method for picture recognition and classification is the Convolutional Neural Network (CNN) technique. In order to extract features from a picture, a CNN algorithm applies a number of filters to the image. These features are then used to classify the image into different categories. Clean and unclean images are collected and trained which is input to google colab 3.10. The dataset are trained to the model based on the Supervised Learning. In supervised learning, an algorithm is trained using a set of labeled examples to learn how to map input data to corresponding output data. After training the clean and unclean datasets, the model will be predicting the output. For prediction CNN model will be created. Finally based on CNN the model will be predict the output.

3.1.1 *Convolution operation*

$$Z^l = W^l * A^{l-1} + b^l \tag{1}$$

$$Y = \sigma \sum_{i=1}^{n} \cdot \sum_{j=1}^{m} W_{i,j} * X_{i',j'} + b \tag{2}$$

- Z^l is the weighted sum of the input volume.
- W^l is the set of learnable filters (kernels).
- The output of the preceding layer is A^{l-1}

3.1.2 *Max pooling operation*

$$A^l = \text{max}pool_{size} A^{l-1} \tag{3}$$

- By pooling, the input's spatial dimensions are decreased.

3.1.3 *Linear transformation*

$$Z^l = W^l * A^{l-1} + b^l \tag{4}$$

- W^l and b^l are the weights and biases for the fully connected layer.
- A^{l-1} is the input from the previous layer.

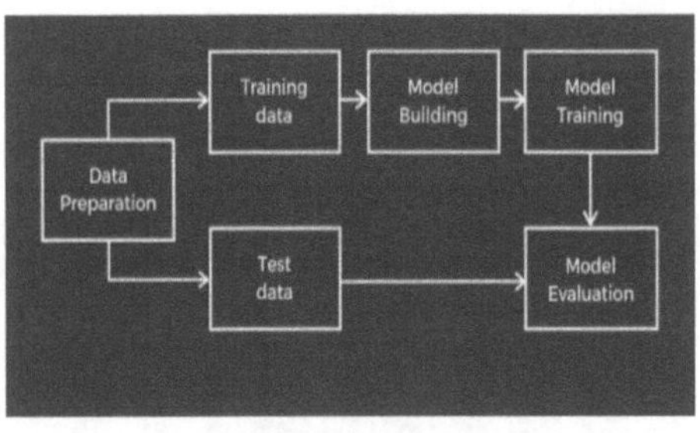

Figure 5. Process flow.

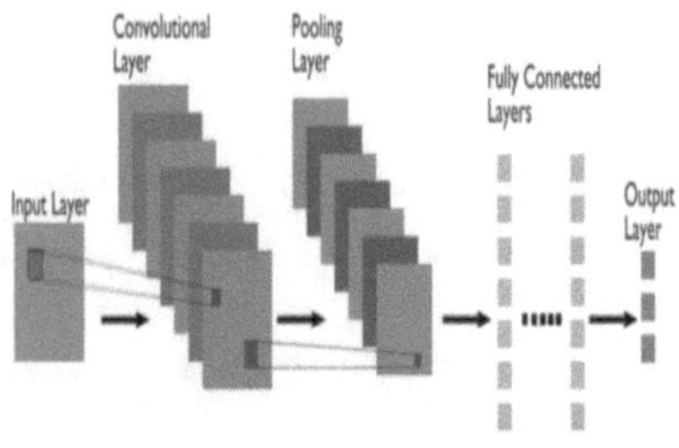

Figure 6. CNN architecture.

4 RESULTS AND DISCUSSSION

In the domain of machine learning applied to tank prediction, our research protocol involves the pivotal step of dividing the dataset into training and testing subsets. While the conventional 80% training and 20% testing split is commonly utilized, systematically explore alternative ratios, including 70-30 and 90-10.

4.1 *Tank prediction analysis*

Figures 7 and 8 shows that the Accuracy and Prediction Results of the tank. In this work, accuracy for prediction of tank (whether it is cleaned or not) is **100%**.

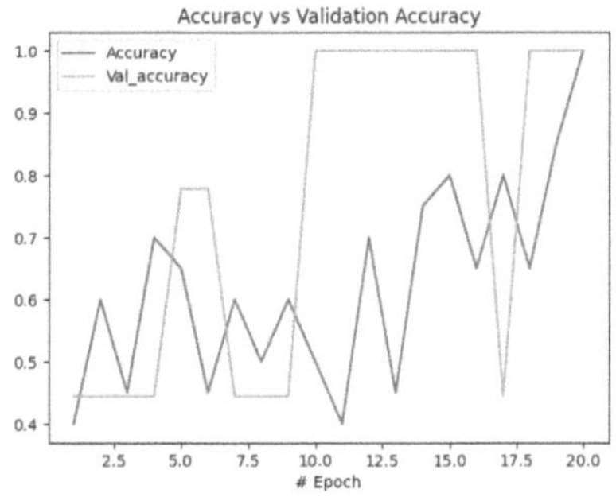

Figure 7. Simulation output for tank prediction.

Table 1. Performance indices – tank.

	Precision	Recall	F1-Score	Support
0	1.00	1.00	1.00	8
1	1.00	1.00	1.00	3
Accuracy			1.00	11
Mac Avg	1.00	0.86	1.00	11
Weighted Avg	1.00	1.00	1.00	11

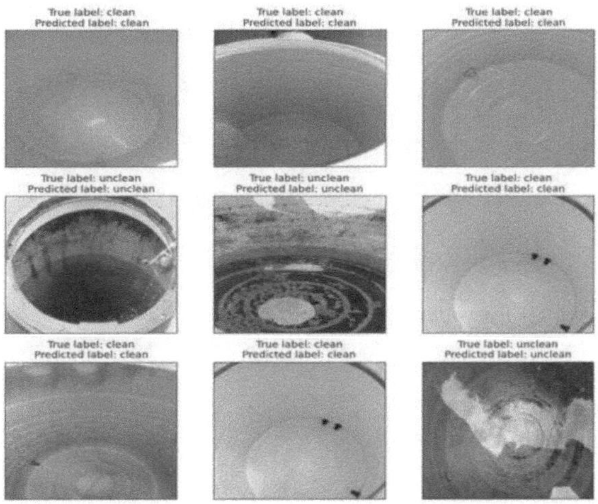

Figure 8. Predicted output for tank.

Confusion Matrix
[[8 0]
[0 4]]
Accuracy: 1.0

4.2 *Water prediction analysis*

Figures 9 and 10 shows that the Accuracy and Prediction Results of the water. In this work, accuracy for prediction of water (whether it is worm water or not) is **75%**.

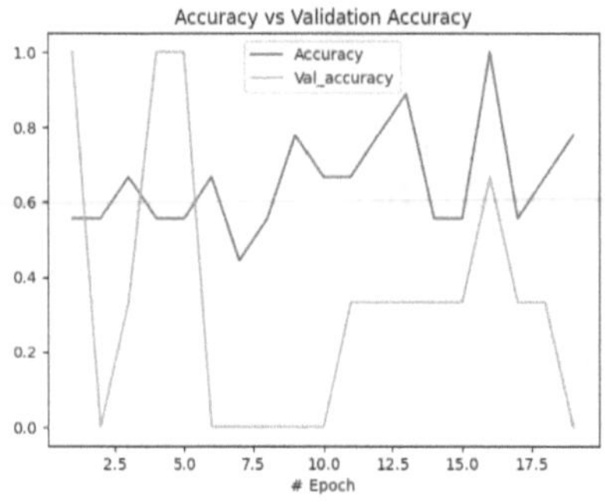

Figure 9. Simulation output for water prediction.

Table 2. Performance indices – water.

	Precision	Recall	F1-Score	Support
0	0.67	1.00	0.80	6
1	1.00	0.50	0.67	6
Accuracy			0.75	12
Mac Avg	0.83	0.75	0.73	12
Weighted Avg	0.83	0.75	0.73	12

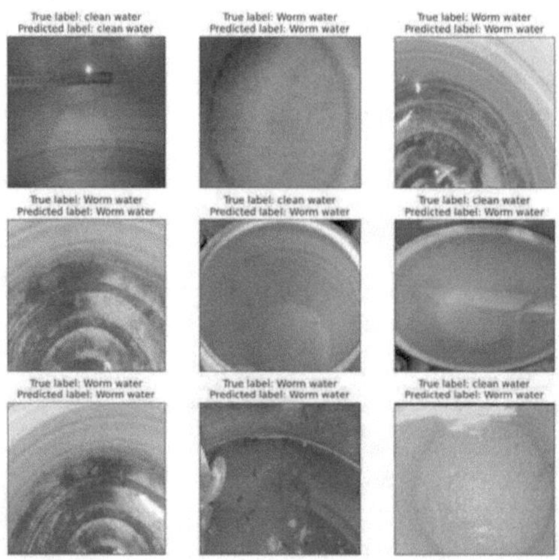

Figure 10. Predicted output for water.

Confusion Matrix:
[[6 0]
[3 3]]
Accuracy: 0.75

Numerous variables, such the quantity and quality of the dataset, the model's hyper-parameters, and its complexity, affect the accuracy of the model.

5 CONCLUSION

The implemented of a tank prediction system is ensuring the clean water for end users before consumption. This paper looked into a method for predicting water quality and tank cleanliness utilizing supervised machine learning algorithm. Customers and end users can readily view the purity of the tank and the quality of the water with this technology. The method of predicting the water quality and tank cleanliness is completed in this research. In the future, an improved performance in tank cleaning will be achieved through the integration of predictive results using IoT.

REFERENCES

[1] Ermiyas Birihanu Belachew Elte, Mebiratu Beyene, Birtukan Adamu. (May 2023). Intelligent water quality monitoring using WSN and Machine learning approaches, *ICISDM*, pp. 52–57.

[2] Sai Sreeja Kura, Sambangi Geethika Naidu, Sravani Chowdala, Sree Chithra Yellanki, Dr. B. (May 2022). Esther sunanda, water quality prediction using machine learning, *International Research Journal of Modernization in Engineering Technology and Science (IRJETS)*, Vol. 4, Issue 5, pp. 692–696.

[3] Minxi Lu. (Aug 2022). *AI Based Tank Truck Cleaning Robot IEEE International Conference on Advances in Electrical Engineering and Computer Applications (AEECA)*, pp 1197–1202.

[4] Barbade G. M., Chandurkar Shreyas, Shounak Vedant, Nimkar Vaibhav, Patil Umesh. (Sep 2021). Automatic water tank filling system with water level indicator, *Indian Journal of Microprocessors and Microcontroller (IJMM)*, Vol. 1, Issue 2, pp. 1–7.

[5] Juhwan Kim, Seokyong Song, Taesik Kim, Young-Woon Song, Shin-Kwon Kim, Bae-Ik Lee, Yongwoon Ryu, Son-Cheol Yu. (July 2021). Collaborative vision-based precision monitoring of tiny EEL larvae in a water tank, *IEEE Access*, Vol. 9, pp. 100801–100813.

[6] Chuhan Qi, Shuo Huang And Xiaofei Wang. (Oct 2020). Monitoring water quality parameters of Taihu lake based on remote sensing images and LSTM-RNN, *IEEE Access*, Vol. 8, pp. 188068–188081.

[7] Bineet Kumar Jha K. K, Sivasankari G. G, Venugopal K. R, (Jun 2020). Cloud-based smart water quality monitoring system using iot sensors and machine learning, *International Journal of Advanced Trends in Computer Science and Engineering (IJATCSE)*, Vol. 9, Issue 3, pp. 3403–3409.

[8] Chaudhary Parth, Darji Biren, Patel Harsh, Gautam Singh Rajput, (Apr 2019). Smart water tank cleaning machine for household applications, *IRJET*, Vol. 06, Issue 04, pp. 3657–3660.

[9] Raffik, Shameer Kamal S., Arun S., Prabhu Raja P. and Manoj Kumar R., (Dec 2018). Automatic Tank Cleaner, *International Journal of Mechanical and Producation Engineering Research and Development, IJMPERD*, Vol. 8, Issue 7, pp. 1424–1431.

[10] Prem Kumar P., Kathir S., Kannan S., Viignesh S. (April 2018). Autonomous operated robot for water tank cleaning, *International Research Journal of Engineering and Technology (IRJET)*, Vol. 5, Issue 4, pp. 1584–1586.

Computational Methods in Science and Technology Volume 1 – Kaur et al. (Eds)
© 2025 the Author(s), ISBN: 978-1-032-81760-6

Deep learning for deepfake detection: Scientific landscape and future directions

Pummy Dhiman & Amandeep Kaur
Chitkara University Institute of Engineering and Technology, Chitkara University, Punjab, India

Dilip Kumar Sharma
Department of Computer Engineering and Application, GLA University, Mathura, India

ABSTRACT: In today's digital age, technology has advanced to the point where it is difficult to distinguish be-tween genuine and forged media content. It was intended to be used for entertainment, but it is now being used to deceive people and create biasness, which can lead to disastrous situations. The goal of this bibliometric study is to provide scientific mapping on the role of deep learning in deepfake detection published online. Scopus and VOSviewer are used to create a concept map of 128 articles. Additionally, future directions for research are suggested.

Keywords: Deepfake, deep learning, bibliometric, fake, GAN

1 INTRODUCTION

The term deepfake is made up of the combination of "Deep Learning + Fake", means the creation of fake content using deep learning (DL) Nightingale and Farid (2022). It is impressive to see how AI-generated content is able to entertain, but it is also capable of deceiving. As a result of advancements in technology, digital media can be manipulated in ways not conceivable twenty years ago Farid (2006). When it comes to portraying anything, images are incredibly dynamic and powerful. Unlike words, which are quickly forgotten, images stay with viewers for a long time. As their shareability has grown, their popularity has skyrocketed Gautam *et al.* (2023). Photography has long since lost its impulsiveness. Abraham Lincoln (around 1860) was a fake. It had Lincoln's head on it and the body of Southern leader John Calhoun on it (Figure 1). It was the start of a long history of forgery in photography Farid (2006). With the introduction of YouTube sharing video online became very easy and WhatsApp made it even very convenient to share video even of large size due to its compress technique. Now a days even there are many platforms to download videos and sharing. All these added fuel to fake video dissemination. AI-synthesized content holds the power to entertain but also deceive. Disguising one's voice in a video or altering what they say in a video are some examples of how AI-synthesized content does. In February 2020, India experienced the first use of deepfake in an election. Two video tapes featuring Manoj Tiwari, chief of the Delhi BJP, were made public in February 2020, during the elections for the legislative assembly. Both videos feature Tiwari criticizing the AAP-led Delhi administration; in one, he uses English, and in the other, he uses Haryanvi. It was later found that the two videos were deepfakes made from a video of Tiwari addressing the CAA Bill in Hindi Chauhan *et al.* (2022).

Generative Adversarial Network (GAN) is the popular creator of deepfakes. It is a neural network-based method that analyses large sets of data to synthesize content and classify it.

 DOI: 10.1201/9781003501244-27

Figure 1. Image forgery.

Although digital forensics plays a significant role in detecting deep fakes, the fact that there are billions of uploads per day makes most forensic techniques unsuitable for deploying quickly and accurately. Here DL comes into play as it can process an abundance of data quickly while also achieving promising accuracy, and as DL is useful in creating deep fakes in a similar manner, it is also useful in detecting them Raj and Meel (2021). The term "deepfake" was coined in 2017 when a Reddit user named "deepfakes" along with other Reddit users started using celebrity faces in sexually explicit content using ML advancements. In the digital age, deepfake is a global concern Agarwal and Farid (2021), as we may lose belief in what we see and hear online Chauhan *et al.* (2023). This study, aimed to analyse the current deepfake research landscape and provide future directions in this digital sphere Pool *et al.* (2021).

The outline of the paper looks like this: Section 2 covers the research methodology. The scientific mapping based on keywords and Section 3 presents document co-occurrence analysis. Section 4 discusses future directions, and Section 5 summarizes the study.

2 METHODOLOGY

The data used in this study is collected from the Scopus database Iqbal (2018), between 2019 and 2022(till July). Figure 2 below depicts the bibliometric procedure adopted in this study. The first step is to create a search string using keywords, which is then searched throughout the Scopus database. Upon receiving results, refinement is performed, and selected data is exported for analysis. Flowchart for this bibliometric analysis study is shown in Figure 3. Here, (deepfake, detection and deep learning) are the main keywords chosen to get the current research landscape and to provide the future direction.

Figure 2. Bibliometric procedure followed.

The final studies are exported in csv format. The VOSviewer software is used for scientific mapping as it has powerful GUI and visualization capability based on network data, bibliographic data, text data and visualize these maps and supports Web of Science, Scopus, dimensions, lens, PubMed database files Van Eck and Waltman (2018).

3 SCIENTIFIC MAPPING

The current study aims to identify patterns in deepfakes literature and provide guidance for future researchers. To accomplish this task, co-occurrence map based on text data is analysed which helps to identify the structure and trends in research field. In this study we make use of keywords co-occurrence analysis.

3.1 *Keyword co-occurrence analysis*

Keywords are the short, simple words that give idea of what is being discussed. By employing scientific mapping techniques, a network visualization map is generated using titles and abstracts. In this network, nodes represent keywords and edges represent the co-occurrence of these keywords. If a link exists between two nodes (keywords), it indicates a correlation between them. The closer the nodes are, the higher the correlation, and vice versa. The weight borne by a certain node increases proportionally with the size of the node. For this analysis, a minimum occurrence of 10 is chosen for a keyword (term) and 50 out of 2608 keywords satisfy this threshold. The default selection for the most relevant keyword is based on the top 60% of relevant phrases. Using this criterion, a total of 30 keywords or nodes are mapped into 4 clusters in this study of 104 selected articles, as illustrated in Figure 4.

Cluster analysis is employed to uncover the research trends by examining the build clusters in the network visualization map. All clusters are represented by different colour. Deepfake, dataset, feature, work, deep learning technology Dhiman *et al.* (2023), deepfake video Yu *et al.* (2021), face Jalal *et al.* (2023), experiment, authenticity, performance, face forensic keywords comes under red cluster (n = 11) and it is dominated by deepfake Boháček and Farid (2022). Green cluster (n = 7) presents the effects of deepfake on society and having Artificial intelligence, term, privacy, threat, society, person, fake as nodes. Deep-fake technology, field, misinformation Dhiman and Kaur (2022), researcher, development nodes are in yellow cluster (n = 5) and focuses on misinformation Bran *et al.* (2021) spread by deepfake and researchers focus to detect it. Gan, CNN Dhiman and Kaur (2022), advancement, fake image Sharma *et al.* (2023), generative adversarial network, convolutional neutral network, GAN comes under blue cluster (n = 7). In this cluster deepfake techniques appear to be crucial Bohacek and Farid (2024).

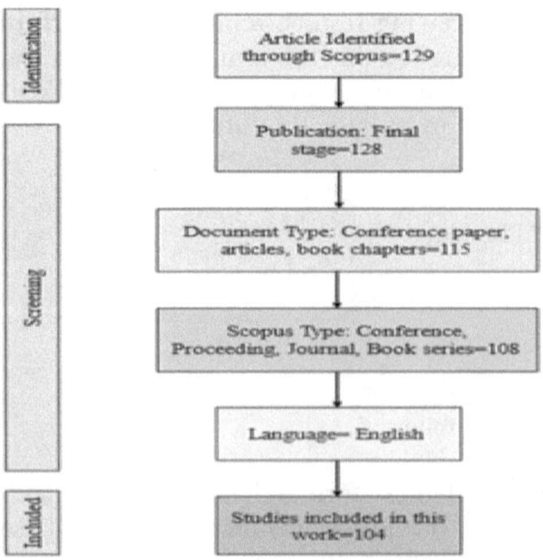

Figure 3. Flowchart for literature selection.

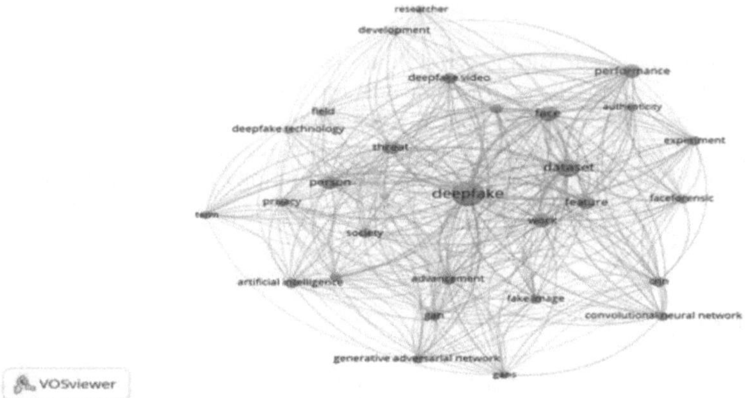

Figure 4. Keyword co-occurrence network visualization of deepfake literature.

4 FUTURE DIRECTIONS

The analysis of keywords that appear together in titles and abstracts offers helpful insights for possible areas of research, as shown in Table 1. The cluster keywords led to the following research questions, which could be expanded upon and studied in the future.

Table 1. Research questions for future directions.

Based on Cluster Analysis	Future Research Questions
Red	• How do deepfakes work and how do they impact authenticity? • How can face forensics and DL technology be used to identify deepfakes? • Which databases abound for the purpose of detecting deepfakes? • Which characteristics are important in distinguishing between authentic and deepfake?
Green	• Can AI infringe on personal privacy? • What are the emerging patterns in AI that jeopardize individuals' privacy? • How can ineffective usage of AI trick society?
Yellow	• What are the most recent developments in deepfake technology, and how has it changed over the years? • How might deepfake technology affect society by spreading false information and eroding trust? • In order to counteract the spread of misinformation, what are the technical limitations and vulnerabilities of current approaches for deepfake detection? How may these methods be improved? • How does deepfake content manage to influence public opinion and shape perceptions through psychological mechanisms?
Blue	• Is it possible to use GAN to make deepfakes other than deepfake images? • How does CNN contribute to the fight against deep-fake content? • What way can GAN use to find deepfakes that it makes?

5 CONCLUSION

The advancement of technology is miracle as well as proven to be curse. The misuse of technology and online platform for creating and dissemination of forged content have become a battle between the forger and the researcher, who work in detecting them and to stop spreading. In this study, by using Scopus database, the concept of deepfakes and the future research questions are proposed. As the technology advancement is never going to stop, so this deep-fake generation. Therefore, it will become very important for the researchers to keep pace.

REFERENCES

Agarwal, S., & Farid, H. (2021). Detecting deep-fake videos from aural and oral dynamics. In *Proceedings of the IEEE/CVF Conference on Computer Vision and Pattern Recognition* (pp. 981–989).

Boháček, M., & Farid, H. (2022). Protecting world leaders against deep fakes using facial, gestural, and vocal mannerisms. *Proceedings of the National Academy of Sciences*, 119(48), e2216035119.

Bohacek, M., & Farid, H. (2024). The making of an AI news anchor—and its implications. *Proceedings of the National Academy of Sciences*, 121(1), e2315678121.

Bran, R., Tiru, L., Grosseck, G., Holotescu, C., & Malita, L. (2021). Learning from each other—A bibliometric review of research on information disorders. *Sustainability*, 13(18), 10094.

Chauhan, R., Popli, R., & Kansal, I. (2022, October). A comprehensive review on fake images/videos detection techniques. In *2022 10th International Conference on Reliability, Infocom Technologies and Optimization (Trends and Future Directions) (ICRITO)* (pp. 1–6). IEEE.

Chauhan, R., Popli, R., & Kansal, I. (2023, March). A systematic review on fake image creation techniques. In *2023 10th International Conference on Computing for Sustainable Global Development (INDIACom)* (pp. 779–783). IEEE.

Dhiman, P., & Kaur, A. (2022, September). Significance of convolutional neural network in fake content detection: A systematic survey. In *International Conference on Emergent Converging Technologies and Biomedical Systems* (pp. 305–316). Singapore: Springer Nature Singapore.

Dhiman, P., Bonkra, A., Sandhu, J. K., Kaur, A., & Kaushal, C. (2023, September). Analyzing breast cancer detection: A scopus-driven study on the role of machine learning and deep learning. In *2023 6th International Conference on Contemporary Computing and Informatics (IC3I)* (Vol. 6, pp. 890–895). IEEE.

Farid, H. (2006). Digital doctoring: how to tell the real from the fake. *Significance*, 3(4), 162–166.

Gautam, V., Maheshwari, H., Tiwari, R. G., Agarwal, A. K., & Trivedi, N. K. (2023, December). Enhancing face forgery detection in a decentralized landscape: A federated learning approach with ResNet. In *2023 2nd International Conference on Automation, Computing and Renewable Systems (ICACRS)* (pp. 1295–1301). IEEE.

Iqbal, Qaiser. (2018). *Scopus: Indexing and Abstracting Database*. 10.13140/RG.2.2.12974.15683.

Jalal, A. S., Sharma, D. K., & Sikander, B. (2023). Suspect face retrieval system using multicriteria decision process and deep learning. *Multimedia Tools and Applications*, 1–28.

Nightingale, S. J., & Farid, H. (2022). AI-synthesized faces are indistinguishable from real faces and more trustworthy. *Proceedings of the National Academy of Sciences*, 119(8), e2120481119.

Pool, J., Fatehi, F., & Akhlaghpour, S. (2021). Infodemic, misinformation and disinformation in pandemics: scientific landscape and the road ahead for public health informatics research. In *Public Health and Informatics* (pp. 764–768). IOS Press.

Raj, C., & Meel, P. (2021). A review of web infodemic analysis and detection trends across multi-modalities using deep neural networks. *arXiv preprint arXiv:2112.00803*.

Sharma, D. K., Singh, B., Agarwal, S., Garg, L., Kim, C., & Jung, K. H. (2023). A Survey of detection and mitigation for fake images on social media platforms. *Applied Sciences*, 13(19), 10980.

Van Eck, N. J., & Waltman, L. (2018). *VOSviewer Manual – version 1.6.8* (April edition)(pp.1–51).

Yu, P., Xia, Z., Fei, J., & Lu, Y. (2021). A survey on deepfake video detection. *Iet Biometrics*, 10(6), 607–624.

Computational Methods in Science and Technology Volume 1 – Kaur et al. (Eds)
© 2025 the Author(s), ISBN: 978-1-032-81760-6

A dynamic trust routing protocol with improved fuzzy butterfly optimization using SVM to secure MANETs from blackhole attacks

Ayush Gupta & Saumya Rajvanshi
Department of CSE, Chandigarh Engineering College, Mohali, Landran, India

ABSTRACT: Security is a fundamental requirement of mobile Adhoc networks (MANETs) owing to their dynamic capabilities. Several different attacks hinder its performance but the most commonly found attack is the blackhole attack which drops all the data of the users shared through the network. Some existing research developments were proposed for the detection and prevention of blackhole attacks, but there is a significant need for an efficient and accurate approach that can prevent the network from blackhole as well as other intruders. With this aim, this paper proposes a dynamic trust routing protocol that uses an improved optimization algorithm with trust evaluation to detect the attackers as well as prevent the network from attackers by identifying the nature of the nodes. In this method, a support vector machine algorithm (SVM) is used to compute and update the trust of the nodes dynamically using a learning-based approach, and fuzzy butterfly optimization uses this trust value and follows foraging behavior for adaptive and optimized path selection. The proposed protocol is simulated using a network simulator by designing two network scenarios of varying attacker nodes and nodes' mobility. The performance is computed based on different performance metrics such as packet delivery ratio (PDR), throughput, and end-to-end delay (E2D). The simulation results and comparison ensure the effectiveness of the proposed protocol.

Keywords: Blackhole Attack, Route Optimization, SVM, Trust, MANETs

1 INTRODUCTION

Mobile Adhoc network (MANET) is a decentralized wireless network that connects different devices without a fixed infrastructure [1]. It is called a dynamic network because of its mobile nodes and self-configuring process with which it can create connections whenever required. The most common application domains of MANETs are disaster recovery, military operations, and dynamic communication environments [2] for which adaptive protocols, mobility management, and effective routing are essential. The MANET architecture is significant due to its flexible communication in dynamic and resource-constrained environments. However, this may cause several security challenges or attacks such as denial of service (DoS), blackholes, wormholes, eavesdropping, data manipulation, etc. [3]. So, to increase its productivity and usage in various applications, security is a must.

Different security threats pose different challenges in which a blackhole attack is the most common but most disruptive attack. This attack uses blackhole nodes that pretend to have the best or shortest path to reach the destination and on receiving data packets from nodes it drops it which means data will be completely lost. In this way, it obstructs the communication between different nodes. Grayhole attack is its variant which selectively drops the packets to degrade the network performance[4]. Sybil attack is an identity theft attack that reroutes the data packets and hence does not send them to the required destination [5]. Packet drop attacks also cause data corruption or loss and in this, the attacker nodes reject or alter the packet's path. Eavesdropping observes the communication silently and obtains the secret data through this. DoS attacks flood the nodes with a lot of traffic so that they become dead or not able to communicate with any other network node [6]. The security enhancement needs to focus on the defined and other undefined vulnerabilities in the dynamic MANETS environment [7].

This work focuses on the blackhole attack issue which causes loss of data in network communication due to its stern interruption to the MANET. The already existing blackhole detection or prevention system mostly considered routing as an optimal parameter and by tracking based on various factors such as packet loss, delay, etc. the blackhole nodes were detected from the network [8]. Also, other systems include trust-based procedures, in which nodes evaluate the neighbors' trustworthiness based on their previous actions [9]. The accuracy and effectiveness of Blackhole Detection mechanisms are being improved by integrating machine learning, cryptography, and more advanced anomaly detection techniques. Nevertheless, these approaches encountered difficulties in dynamic and unpredictable MANET environments.

1.1 *Motivation*

The infrastructure-less feature of MANETs makes it feasible in situations like disaster, temporary requirement, military services, etc. Owing to this, its demand increases in various dynamic applications and hence motivated for this proposed work. Moreover, these features and demand of MANETs also poses it to several challenges from which security threats are crucial. Blackhole attack is the most hazardous because it increases the data loss so, there is a need for developing a protocol or framework for network system and protect it from these potent threats. On the other side, the advancement in the technologies like, machine learning, optimization algorithms, etc. also introducing different approaches which can help in securing the network. Through literature analysis, this work proposing a dynamic trust routing method that uses support vector machine (SVM) for trust computation and fuzzy butterfly optimization (FBO) for optimal route selection. This proposed solution not only focusing on blackhole attack but it is a defensive mechanism for MANETs to protect it from several vulnerable threats and hence provide a guaranteed solution.

1.2 *Contribution of the paper*

The proposed research makes a significant and varied contribution to the field of MANETs security:

- Dynamic Trust Routing System: This proposed routing compute the trust of nodes and dynamically update it during communication. It will ensure the safest communication in MANETs, hence increases its effectiveness.
- Fuzzy Butterfly Optimization (FBO): Route selection is an important phase for any network system for communication between different nodes. One efficient route can enhance the performance of the network whereas a bit inefficiency can cause various threats. Optimization algorithms are popular these days and play very important role in the network communication. This work also proposing a FBO based routing algorithm which can help in the selection of adaptable and secure route for communication and makes the MANET flexible, secure and optimized.
- Trust Computation: SVM is here opted to compute trust based on the network traffic which can improve the network accuracy. It is a machine learning algorithm which can learn based on given data. More the data, better it can learn and hence can identify the reliable and unreliable nodes in the network.

- Blackhole Attack: This work evaluated the proposed solution based on different network scenarios and the presence of blackhole nodes. This can effectively measure whether the proposed solution can prevent the network from this threat or not. Its efficient performance ensures its accuracy, hence achieve the objective of this proposed work.

2 RELATED WORK

This section reviewed different research studies which covers the security enhancement in MANETs, blackhole attack based developed approaches, machine learning in MANETs, etc. in order to analyze the current status of this problem domain. Here, the studies from past 5-years have been reviewed to identify the limitations and scope in this work.

An efficient method for blackhole detection was proposed by Abdelhamid *et al.* [10] where they have designed an anomaly detection using SVM which analyze the network traffic to identify the intruders based on their behaviours. They have implemented this approach and simulated the network scenarios with and without blackhole nodes using OMNET++. The detection accuracy of this method was highest as compare to other models, hence shown its effectiveness. A dynamic threshold based AODV (DT-AODV) protocol was designed to secure the network by Mankotia *et al.* [11] which uses dynamic threshold. This proposed solution was used an alert notification specifically for suspicious nodes and discard it after analysis. The designed approach is tested using NS-2 simulator where different network scenarios-based simulations were performed. The results in terms of throughput of 20kbps, routing overhead of 175-, normalized routing load of 0.51, and packet delivery ratio 99% proof the effectiveness of this approach but due to threshold-based solution, it might not be suitable in every network scenario. Alkanhel *et al.* [12] proposed a method using Dipper Throated Optimization (DTO) algorithm which generates secure and efficient routes for communication and improve the network performance. This approach uses Low Energy Adaptive Clustering Hierarchical (LEACH) protocol as base and also combine the multi-layer perceptron network with DTO. The accuracy of detection for this proposed model was 97% and also executed in a small period of time. This means it can be an accurate solution for MANETs security problem, the only problem is the use of threshold to classify the node as benign and malicious.

Security can also be achieved with the help of cryptography approaches. With this aim, Olanrewaju *et al.* [13] proposed a Message Digest 5 (DHMD) and Diffie-Hellman based encryption strategy on AODV protocol. The main aim behind this is to solidify the access on the data with 23% reduction in the network overhead. Though it is end-to-end encryption model which improve the network traffic even in the presence of attack, it cannot protect the network from data loss. A neural network and DNA cryptography based approach was designed by Chakravorty *et al.* [14] to protect the network from intruders precisely from blackhole attack. This approach is analyzed using MATLAB simulator which defined its effectiveness to mitigate the blackhole attacks in MANETs.

The other black hole detection and prevention approaches include the use of Digital Signature and SEP [15], dynamic threshold [16], binary partition clustering [17], adaboost SVM [18], Artificial neural network with SVM [19], Elliptic Key Cryptography [20], Trust based routing [21], etc. Though these approaches were able to deal with blackhole attacks in MANETs, they often exhibited limitations in terms of efficiency, adaptability, and overall effectiveness.

3 PROPOSED METHOD

The main aim of this work is to secure the MANETs from intruders specifically from black hole attackers. So, to secure the network a dynamic trust-based approach is proposed which is a hybrid approach that first trains the machine learning model on different network traffic patterns including, with and without attacker nodes, and then uses it as an objective function that computes the routes for data transmission. The details of the approach are divided into three different phases: (a) the initial training phase, (b) the path selection phase, and (c) the data transmission phase.

3.1 *Initial training phase*

This is the initial phase of the proposed system that executes only once with known labels to train the machine learning model and recognize the intruders for further processing. In this the support vector machine (SVM) algorithm is used that compute the trust score of a node by recognizing it as the normal or attacker node [22]. The process starts with the network traffic patterns with given labeled data. The traffic patterns here are generated using a network simulator with different scenarios including varying normal and blackhole nodes and with and without attacks. This labeled data is further converted into feature vectors where the feature vector is the traffic pattern with its corresponding labels. These feature vectors are then pre-processed and normalized to generate the standardized data further for model training. Figure 1 below presents the training phase of the SVM where it uses a linear kernel with pre-processed and normalized feature vectors and trains the model for further predictions.

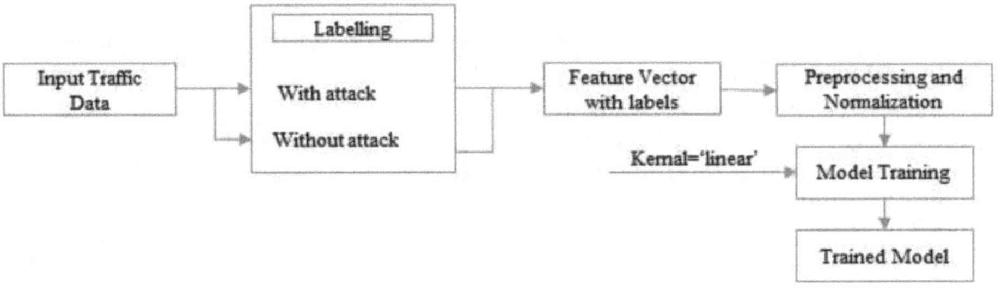

Figure 1. Training phase using SVM.

The training process given above provides the trained model which is further integrated in the path selection phase to classify the intruders and protect the network from data loss, unauthorized access, data theft, etc.

3.2 *Path selection phase*

This is the main phase of the proposed system which selects the optimal path for data transmission using an optimization algorithm with trust computation. Optimization algorithms are nowadays used in various networking systems and their effectiveness makes the system's performance better in terms of different parameters based on the defined objectives. In this work, Fuzzy Butterfly Optimization (FBO) which is a nature-inspired algorithm designed to solve complex problems is opted [23]. Motivated by the way butterflies forage, FBO uses fuzzy logic to find its way across imprecise and uncertain search areas. Starting with a population of solutions, which are represented as "butterflies," the algorithm dynamically modifies their placements in response to changing circumstances by using fuzzy logic principles. The placements of these butterflies are iteratively refined by FBO according to their fitness in the optimization environment. An optimal solution to the given issue is represented by the butterflies' ultimate locations. To enhance this FBO, SVM based objective function is infused in this algorithm that defines the trust criterion for this algorithm. Figure 2 below presents the integrated path selection algorithm that uses FBO with SVM as its objective function.

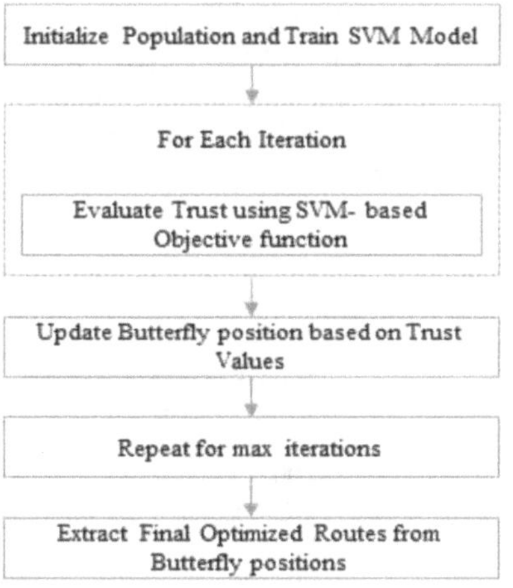

Figure 2. Proposed FBO with infused SVM.

The detailed process of the proposed path selection algorithm is given as follows:

Step 1: Initialization- This step initializes the population of butterflies with random positions in network space and represents potential routes of the network. Moreover, the SVM model is trained with labeled data.
Step 2: Iterative Optimization- In this step, the trust of nodes is continuously monitored using trained SVM to maximize the objective function given below:

$$Obj(x) = \frac{1}{2}(1 + svm.predict(x))$$

The above function maps the positive prediction to 1 and the negative to 0 and higher values of the objective function correspond to trusted nodes.
Step 3: Position updation- The position of butterflies is updated based on trust values where fuzzy logic also guides the movements.
Step 4: Final route selection- After the specified number of iterations, the final positions of butterflies represent the optimized routes in the MANET.

3.3 *Data transmission phase*

This phase is the last phase of the network system and it starts the data transmission between source and destination via route selected in the previous phase. This phase also monitors the routes dynamically by calling FBO during the transmission and ensures secure data transmission.

4 EXPERIMENTATION AND RESULT ANALYSIS

The main aim of this proposed research is to develop an efficient protocol which can prevent the network from intruders and provide a secure communication in MANETs. Network simulator (NS-2) is a simulator which is generally used for network configuration and performing simulation for the evaluation of different approaches. This work also utilizes NS-2 for configuring simulation scenarios and analyze the performance of the proposed solution. In this, two different simulation scenarios are generated where in the first scenario, the variation is added to the number of attacker nodes whereas the second scenario vary mobility of the nodes. Both of these scenarios can critically measure the effectiveness of the proposed solution in the challenging situations.

This proposed solution is benchmarked with comparing its performance with different existing solutions of blackhole attack on MANETs such as DT-AODV [11], TBSMR [24], and BFO with Fuzzy Clustering [25]. The simulation is done using NS-2 using two different simulation scenarios as given in Table 1. This table gives the details of all the important setups required for MANETs network performance evaluation.

Table 1. Simulation parameters.

Parameter	Value	
	Scen-1	Scen-2
Area	500 × 500m^2	500 × 500m^2
Number of nodes	100	100
Number of attacker nodes	0,2,4,6,8,10	2
Number of connections	20	20
Mobility	20 m/s	10,20,30,40, 50 m/s
Attack type	Blackhole Attack	Blackhole Attack
Simulation time	500s	500s

The first scenario of simulation defined in Table 1 shows that the simulations of the proposed solutions are done by varying number of attacker nodes from 0 to 10, and by keeping all other parameters same during simulations. However, in the second scenario, mobility of the nodes is varied from 10 to 50. From this it is stated that the first scenario can analyze the critical situations of attacks, on the other hand, mobility can add complexity in the network nodes. Both the scenarios can better judge this proposed solution and can provide the genuine outcomes nearly to the real world situations.

4.1 *Performance parameters*

The performance of the proposed method is measured in terms of Packet Delivery Ratio (PDR), Throughput, End-to-End Delay (E2D), and Packet Loss Ratio (PLR). These are the common network performance metrics which are used for the evaluation of the network and help in identifying the effectiveness of the approach.

- PDR: It is a measure of number of packets delivered to the destination by comparing with the number of packets sent by a sender node. Higher the PDR, lesser the impact of attack, better its performance.
- Throughput: It is the rate of data transmission over the network. Its higher value indicates the better performance of the approach.
- E2D: This parameter is related to the time factor and can compute the travelling time of the packets from their source to destination over the network. If it is lesser, means network has a very good performance.
- PLR: It is the percentage of loss which indicates the dropped or lost packet of the network during their communications. It should be less for better performance of the network.

4.2 *Performance analysis*

The proposed method is tested in different simulations, each reflecting a different issue of a MANET. The average performance of the proposed method is computed using different scenarios and is benchmarked by comparing its performance with the existing state-of-the-art methods.

4.2.1 *Average performance*

Comparing the performance of the proposed technique across two situations, "Varying Attacker Nodes" and "Varying Mobility," indicates significant differences in critical parameters as shown in Table 2 below:

Table 2. Average performance of proposed method.

Simulation Scenarios	PDR	Throughput	E2D	PLR
Varying Attacker Nodes	94.71667	212.3083	1.287833	5.283333
Varying Mobility	97.404	234.786	1.0326	2.596

The PDR in the "Varying Mobility" scenario is somewhat higher at 97.40%, suggesting a greater success rate in delivering packets when compared to the PDR in the "Varying Attacker Nodes" scenario, which is 94.72%. In the context of mobility problems, the "Varying Mobility" scenario had better throughput than the "Varying Attacker Nodes" scenario, indicating more efficient data transport. The "Varying Mobility" scenario has a lower E2D, indicating shorter data transmission delay, whereas the "Varying Attacker Nodes" scenario has a slightly higher E2D. The "Varying Mobility" has lesser PLR of 2.60% whereas it is 5.28% for "Varying Attacker Nodes".

This comparison of scenario shows that the performance of the network is much affected in the scenario where number of attacker nodes are increased compared to the mobility variation. However, the proposed approach handles both the critical situations, shows its effectiveness.

4.2.2 *Comparison with the existing methods*

This proposed method is benchmarked by comparing its performance with the existing state of the art methods such as DT-AODV [11], TBSMR [24], and BFO with Fuzzy Clustering [25] based on PDR, Throughput, E2D, and PLR and using two different scenarios as stated in Table 1.

- ***Varying Attacker Nodes***

Table 1 indicate the in this scenario, the number of attacker nodes are increased with the rate of 2 nodes and start the evaluation without attacker nodes and complete it in the presence of 10 attacker nodes. The performance in terms of PDR, Throughput, E2D, and PLR for proposed solution and other methods can be depicted from Figure 3.

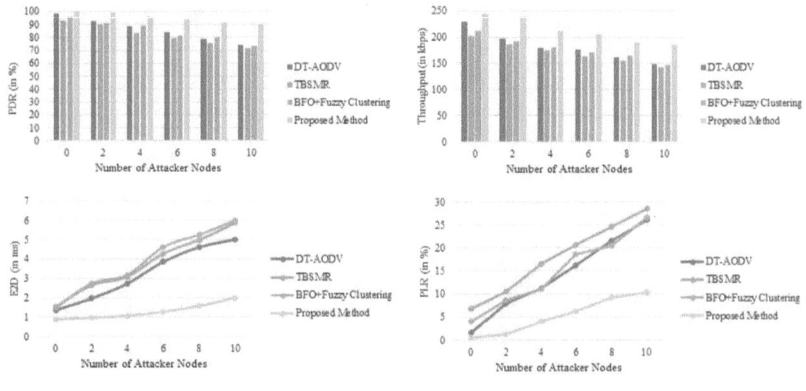

Figure 3. Performance comparison (varying number of attacker nodes).

The comparison in terms of PDR presents that the proposed method always achieves the better PDR in all different cases of attacker variations compared to other methods. It shows the effectiveness of the proposed approach which means this approach can bypass several attackers at one time. The results shows that the PDR stays continuously high, starting at 99.5% and falling just slightly to 89.65% with 10 attacker nodes.

Furthermore, the throughput figures reported for increasing numbers of attacker nodes provide insights on the data transmission efficiency of various approaches amid security risks. For DT-AODV, TBSMR, and BFO with Fuzzy Clustering, throughput gradually decreases as the number of attacker nodes rises. In comparison to the previous techniques, the suggested method has a significantly greater throughput, starting at 243.29 kbps and only declining to 186.37 kbps with 10 attacker nodes.

In addition, the E2D findings for variable numbers of attacker nodes give insight into the time it takes for data packets to move from the source to the destination via various techniques in the presence of security concerns. E2D increases with the increasing number of attacker nodes for DT-AODV, TBSMR, and BFO with Fuzzy Clustering. Figure 3 shows that, in comparison to the other techniques, the suggested method has much reduced End-to-End Delay values across all situations.

The PLR findings for increasing numbers of attacker nodes reveal the proportion of lost data packets in the network under various security methods. PLR increases when number of attackers increases and it is visible in all the protocols DT-AODV, TBSMR, and BFO with Fuzzy Clustering. However, the performance of the proposed method is always better than these methods which means the proposed method is highly resilient even in the presence of threats in the network. The results of the proposed method indicated its effectiveness for example, lower PLR values indicating improved resilience against packet loss.

- ***Varying Mobility***

This scenario can evaluate the robustness and adaptability of the network using proposed solution. Figure 4 presents the mobility does not greatly impact the performance as it maintains it as per requirement such as high PDR, and throughput but low E2D and PLR.

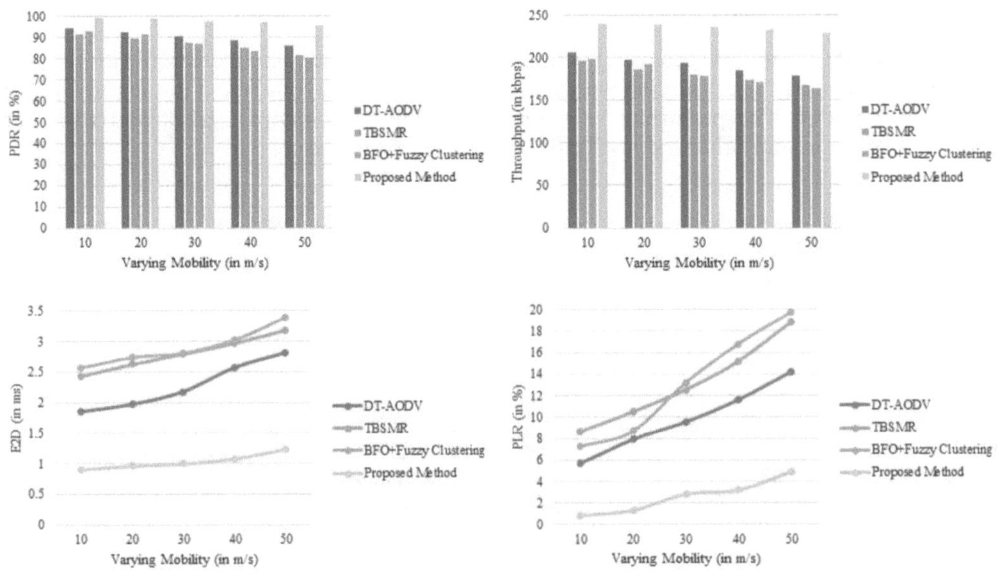

Figure 4. Performance comparison (varying mobility).

The comparison in terms of PDR presents that the proposed method always achieves the better PDR in all different cases of mobility variations compared to other methods. It does not even better from the other methods, but also maintained in different mobility levels. Throughput is also maintained throughout the variation and hence defines its effectiveness. The lesser E2D present that the proposed method maintains E2D also and continuously focusing to control it. PLR is also very less in compare to the other existing methods. These results reveal the effectiveness of the proposed method in terms of PDR, Throughput, E2D, and PLR.

5 CONCLUSION

This research work is aimed to design a protocol or method for mobile Adhoc network in order to increase its security and to prevent it from black hole attackers. In this work, the combination of ML with optimization algorithm is selected for the improvement in the existing network protocol used for routing. This approach uses the benefits of dynamic trust routing, FBO, SVM based trust and combine these with the network protocol in order to propose a secure and efficient protocol for MANETs. This proposed model is simulated on two different network scenarios using NS-2 simulator by varying attackers and mobility of the nodes. The proposed solution is also benchmarked and its performance is compared with the existing state of the art methods such as DT-AODV, TBSMR, and BFO with Fuzzy Clustering. The proposed method achieved the average performance of 95% PDR, 219.97 kbps throughput, 1.51 ms delay, and 3% PLR. These results when compared with the existing methods, represents the effectiveness of the proposed protocols which can dynamically handle the network and protect it from various intruders. This method ensures the safe environment for data transmission and create a solid foundation to present the effectiveness of the MANETs security with this proposed protocol. In future, this proposed protocol can be tested on other simulation scenarios. Moreover, the encryption of data can also be considered to avoid other similar or disimilar threats.

REFERENCES

[1] Abdel-Fattah F., Farhan K. A., Al-Tarawneh F. H., and AlTamimi F. (2019). Security challenges and attacks in dynamic mobile Ad Hoc networks MANETs, *IEEE Jordan International Joint Conference on Electrical Engineering and Information Technology (JEEIT)*, IEEE, pp. 28–33. doi: 10.1109/JEEIT.2019.8717449.

[2] Chlamtac I., Conti M., and Liu J. J.-N. (2003). Mobile ad hoc networking: imperatives and challenges, *Ad Hoc Networks*, vol. 1, no. 1, pp. 13–64, doi: 10.1016/S1570-8705(03)00013-1.

[3] Sheikh R., Mahakal Singh Chande, and Mishra D. K. (2010). Security issues in MANET: A review, *Seventh International Conference on Wireless and Optical Communications Networks – (WOCN)*, IEEE, pp. 1–4. doi: 10.1109/WOCN.2010.5587317.

[4] Siddiqui M. N., Malik K. R., and Malik T. S. (2021). Performance Analysis of Blackhole and Wormhole Attack in MANET Based IoT, *International Conference on Digital Futures and Transformative Technologies (ICoDT2)*, IEEE, pp. 1–8. doi: 10.1109/ICoDT252288.2021.9441515.

[5] Rajadurai H. and Gandhi U. D. (2020). Fuzzy based collaborative verification system for sybil attack detection in MANET, *Wirel. Pers. Commun.*, vol. 110, no. 4, pp. 2179–2193, doi: 10.1007/s11277-019-06836-7.

[6] Rajpoot C. S., Bairwa A. K., and Sharma V. K. (2021), Mitigating the impact of DDoS attack on upsurge network performance in MANET, pp. 153–164. doi: 10.1007/978-981-15-5077-5_14.

[7] Kannhavong B., Nakayama H., Nemoto Y., Kato N., and Jamalipour A. (2007). A survey of routing attacks in mobile ad hoc networks, *IEEE Wirel. Commun.*, vol. 14, no. 5, pp. 85–91, doi: 10.1109/MWC.2007.4396947.

[8] Gurung S. and Chauhan S. (2020). A survey of black-hole attack mitigation techniques in MANET: merits, drawbacks, and suitability, *Wirel. Networks*, vol. 26, no. 3, pp. 1981–2011, doi: 10.1007/s11276-019-01966-z.

[9] Sharma S. and Hussain S. Z. (2023). A survey of trust based secure routing protocol used in mobile ad hoc networks, *ITM Web Conf.*, vol. 54, p. 02009, doi: 10.1051/itmconf/20235402009.

[10] Abdelhamid A., Elsayed M. S., Jurcut A. D., and Azer M. A. (2023). A lightweight anomaly detection system for black hole attack, *Electronics*, vol. 12, no. 6, p. 1294, doi: 10.3390/electronics12061294.

[11] Mankotia V., Sunkaria R. K., and Gurung S. (2023). DT-AODV: a dynamic threshold protocol against black-hole attack in MANET, *Sādhanā*, vol. 48, no. 4, p. 190, doi: 10.1007/s12046-023-02227-8.

[12] Alkanhel R., El-kenawy E.-S. M., Abdelhamid A. A., Ibrahim A., Abotaleb M., and Sami Khafaga D. (2023). Dipper throated optimization for detecting black-hole attacks in MANETs, *Comput. Mater. Contin.*, vol. 74, no. 1, pp. 1905–1921, doi: 10.32604/cmc.2023.032157.

[13] Olanrewaju O. M., Abdulwasiu A. A., and Abdulhafiz N. (2023). Enhanced on-demand distance vector routing protocol to prevent blackhole attack in MANET, *Int. J. Softw. Eng. Comput. Syst.*, vol. 9, no. 1, pp. 68–75, doi: 10.15282/ijsecs.9.1.2023.7.0111.

[14] Chakravorty R., Prakash J., and Srivastava A. (2023). An efficacious neural network and DNA cryptography-based algorithm for preventing black hole attacks in MANET, *Soft Comput.*, doi: 10.1007/s00500-023-09386-0.

[15] Gupta S., Shahid M., Goyal A., Saxena R. K., and Saluja K. (2022). Black hole detection and prevention using digital signature and SEP in MANET, *10th International Conference on Emerging Trends in Engineering and Technology – Signal and Information Processing (ICETET-SIP-22)*, IEEE, pp. 1–5. doi: 10.1109/ICETET-SIP-2254415.2022.9791738.

[16] Malik A., Khan M. Z., Faisal M., Khan F., and Seo J.-T. (2022). An efficient dynamic solution for the detection and prevention of black hole attack in VANETs, *Sensors*, vol. 22, no. 5, p. 1897, doi: 10.3390/s22051897.

[17] Khan A. U., Puree R., Mohanta B. K., and Chedup S. (2021). Detection and Prevention of blackhole attack in AODV of MANET, *IEEE International IOT, Electronics and Mechatronics Conference (IEMTRONICS)*, IEEE, pp. 1–7. doi: 10.1109/IEMTRONICS52119.2021.9422643.

[18] Hikal N. A., Shams M. Y., Salem H., and Eid M. M. (2021). Detection of black-hole attacks in MANET using adaboost support vector machine, *J. Intell. Fuzzy Syst.*, vol. 41, no. 1, pp. 669–682, doi: 10.3233/JIFS-202471.

[19] Pandey S. and Singh V. (2020). Blackhole attack detection using machine learning approach on MANET, *International Conference on Electronics and Sustainable Communication Systems (ICESC)*, IEEE, pp. 797–802. doi: 10.1109/ICESC48915.2020.9155770.

[20] Taranum F., Sarvat A., Ali N., and Siddiqui S. (2020). Detection and prevention of blackhole node, *4th International Conference on Electronics, Materials Engineering & Nano-Technology (IEMENTech)*, IEEE, pp. 1–7. doi: 10.1109/IEMENTech51367.2020.9270072.

[21] Bhawsar A., Pandey Y., and Singh U. (2020). Detection and prevention of wormhole attack using the trust-based routing system, *International Conference on Electronics and Sustainable Communication Systems (ICESC)*, IEEE, pp. 809–814. doi: 10.1109/ICESC48915.2020.9156009.

[22] Sultan S., Javaid Q., Malik A. J., Al-Turjman F., and Attique M. (2022). Collaborative-trust approach toward malicious node detection in vehicular ad hoc networks, *Environ. Dev. Sustain.*, vol. 24, no. 6, pp. 7532–7550, doi: 10.1007/s10668-021-01632-5.

[23] Samiayya D., Ramasamy J., and Gunasekar M. (2023). An efficient congestion control in multimedia streaming using adaptive BRR and fuzzy butterfly optimization, *Trans. Emerg. Telecommun. Technol.*, vol. 34, no. 3, doi: 10.1002/ett.4707.

[24] Sirajuddin M., Rupa C., Iwendi C., and Biamba C. (2021). TBSMR: A trust-based secure multipath routing protocol for enhancing the QoS of the mobile Ad Hoc network, *Secur. Commun. Networks*, vol. 2021, doi: 10.1155/2021/5521713.

[25] Srilakshmi U., Alghamdi S. A., Vuyyuru V. A., Veeraiah N., and Alotaibi Y. (2022). A secure optimization routing algorithm for mobile Ad Hoc Networks, *IEEE Access*, vol. 10, pp. 14260–14269, doi: 10.1109/ACCESS.2022.3144679.

Computational Methods in Science and Technology Volume 1 – Kaur et al. (Eds)

Smart city using blockchain with artificial intelligence

Ayush Gupta, Ritikesh Gupta, Dhirendra Kumar Shukla & Arvind Dagur
Department of Computer Science Engineering, Galgotias University, Greater Noida/Uttar Pradesh

ABSTRACT: Modern cities all over the world are undergoing significant transformations in order to install stunning architecture, offer smart services to citizens, create a safe, sustainable, and clean environment, and prepare areas for vulnerable populations. The primary driver behind many corporations' forays into the realms of community development, health, water extraction, sanitation, products and consumption, preserving dynamism, housing, screening, and accessibility is the introduction of new technology. We are investigating these consequences in light of vision and information technology advancements over the years. To resurrect these sectors and build a new, safe, and sustainable megacity, new technology and skills must be made available. This combination offers a thorough comprehension of the societal issues that big cities face. It focuses on two recent technological investigates the potential of blockchain technology and artificial intelligence (AI) to better society. It also looks at the various applications of blockchain and edge AI. We examine the value of walking sweats in the domains of smart mobility and smart dynamism rather than relying solely on an ultra-modern smart megacity. In order to circumvent the law for commercial traffic, it outlines a clever algorithm that uses detection, counting, and speed recognition. It also discusses the usage of blockchain in secure, private information transport and exchange between buses and cars. business knowledge. This study should be read in conjunction with other artificial intelligence research papers that investigate the applications of edge AI and blockchain.

Keywords: Superb computers, immense intellect, blockchain technology, intelligent grids, intelligent movement, and lovely energy

1 INTRODUCTION

Governments and engineers are searching for contemporary ways to monitor public services like roads, waste management, and air quality as city populations continue to rise exponentially. Thus, initiatives like blockchain cities emerged when artificial intelligence and blockchain technology were introduced to the public sphere. It is expected that the current urban population will increase by 25% by 2050, which means an increase in control energy and water infrastructure, transport networks, and waste disposal. Traveling has many challenges, one of the biggest lockdowns is ahead, with a cost of €100 billion per year, or 1% of EU GDP. To solve this country's problems planners need to understand and study how people travel to cities. The cost of combating air pollution, according to the National Health Service of England, worth about 157 million pounds. Sports watchdogs warn these fees could total up to £18.6 billion from 2035.As the population increases, the rate of consumption increases, and the amount of global waste generated annually exceeds 720 billion tons. This is because people in cities depend on food that is processed in factories, which increases the amount of food that is consumed. household waste is produced every day. The biggest problem in urban planning seems to be planning the electricity network. All the

DOI: 10.1201/9781003501244-29

current is provided by the superpower the installation of its cable travels a distance of up to 100 km, causing energy loss in the form of heat. Customers must pay costs, and transportation losses represent 4.57% of the final price for households and 4% for business owners. In Europe, These losses are still small, about 4–5%, but in a populated country like India, there is potential for underdevelopment. net, these losses amounted to 19% of the price. One effort to reduce carbon emissions is the increase in renewable energy sources. When the solution is as strong Solar power and hydroelectric power plants can do the whole city, their implementation in the existing and old network is not so easy. connect 2 wires. Depending on the renewable energy sources, the implementation of this plan can also lead to an increase in energy collapse due to climate change. The first solution is implemented as a cost-reduction method, after which the data analysis method is implemented effectively. From there, the opportunity to improve the support of the equipment is seen and taken into consideration, leading to green technology innovation.

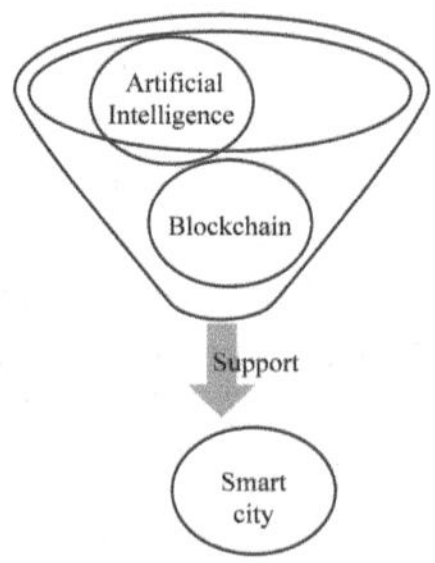

Figure 1. Blockchain and AI as components assisting in the construction of smart cities.

1.1 *Contributions*

While blockchain and edge computing are frequently explored in the literature, there isn't much research on how edge AI and blockchain interact in smart communities. In order to allow artificial intelligence, secure applications, and high-quality networks in two crucial areas of smart cities, this article summarizes current research efforts in advanced artificial intelligence and blockchain. Energy and a clever move. Beginning with the release of blockchain and AI, we and looks at research initiatives to combine these two cutting-edge technologies, including topics like model sharing, scalability, privacy, security, and edge model training. a lot, We offer studies on the implementation of state-of-the-art AI in a range of smart mobility applications, including such as intelligent transportation systems, smart energy, and traffic monitoring and management like the most effective energy management and energy efficiency in smart cities, green energy management, and smart homes. We also examine current research on the applicability of Blockchain to smart mobility applications, such as distributed authentication.

Control, key and reliability control, name system, and smart power, among others energy trading and distributed energy management. Potential obstacles to research additionally.

There are also future directions described. This article's primary contribution is displayed like this:

1. It provides an overview of edge AI and blockchain foundations.
2. It examines the potential that edge AI offers for smart energy and smart mobility.
3. It examines the potential that blockchain technology offers for smart energy and smart mobility.
4. It looks at various efforts to integrate these two state-of-the-art technologies within the framework of smart cities.

5. Lastly, it highlights important unresolved research questions and potential paths for edge AI and blockchain to fully materialize in smart cities.

For the convenience of the reader, the studies covered in this review are displayed in Figure 2:

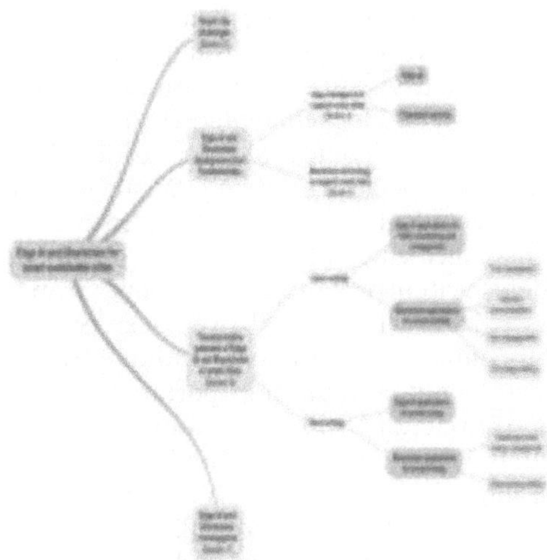

Figure 2. Categorization of the reviewed studies.

1.2 *Structure of review*

This review's remaining sections are arranged as follows: In Section 2, the difficulties that smart cities face are outlined. The principles of blockchain, federated learning, and edge AI are covered in Sections 3 and 4, along with how they might help with smart city operations. The methodology of the review is described in Section 5. Two essential components of smart cities are smart energy and smart transportation, and Section 6 explores the disruptive potential and applications of edge AI and blockchain in these fields. A few initiatives demonstrating the convergence of these two technologies are highlighted in Section 7. Section 8 highlights the open research challenges and future directions. Section 9 brings this review to a close.

1.3 *Research gap*

This paper's objective aims to provide a comprehensive analysis of how blockchain and artificial intelligence technology affect SDG 11 (Sustainable Cities and Communities), a sustainable development goal set out by the United Nations. The primary goal is to give the reader a clear picture of how the developing Blockchain-AI technology will affect smart and sustainable cities. The current literature assessment on blockchain-related artificial intelligence is incomplete in its understanding of how these technologies support sustainable development. The literature that has been published discusses the ecological impact of blockchain technology. For example, it highlights the significant energy consumption of the mining equipment that is necessary to power the bitcoin proof of work process. A thorough understanding of all the dangers related to the planning and operation of smart cities is absent in the research that are currently available. The achievement of sustainable city goals depends on a thorough risk assessment, which will also lessen obstacles to smart city projects from a variety of angles, including technology, security and privacy, political, environmental, managerial, and user trust and adoption (Sharif and Pokharel, 2022). Even so, scholars have talked about how citizens use smart city services and technologies. On the other hand, not much research has been done about the standards, privacy, and governance of the technology used in smart cities. Thus, in order to have a thorough grasp of smart cities on a holistic level, extensive research in this area of smart cities is necessary (Sharif and Pokharel, 2022).

Another area that needs more exploration is public blockchains, where user identities and transaction histories are visible to anyone and can be exploited. More investigation is required as the computational network of miners is unprotected; so, a plan to safeguard the miners' hashing capacity must be devised (Shinde, Patil, Kotecha, and Ruikar, 2021).

In recent years, the convergence of blockchain and artificial intelligence (AI) technologies has garnered substantial interest in the context of Smart Cities. A pioneering effort by Smith *et al.* (2018) introduced Project Utopia, exemplifying the integration of blockchain for secure data sharing among city stakeholders, alongside AI-driven predictive analytics for energy optimization and traffic management. Jones and Kim (2019) delved into decentralized identity management systems within urban environments, proposing a blockchain-based framework that empowers citizens with control over their personal data. Complementing this, AI was harnessed to enhance identity verification processes. Chen *et al.* (2020) demonstrated the deployment of autonomous transportation systems, employing blockchain for transparent and secure transactions within the transportation network, while AI algorithms optimized traffic flow. Wang and Li (2021) addressed citizen services, presenting a blockchain platform for secure management of public services, with AI integration for personalized service delivery. Liu *et al.* (2022) critically assessed challenges in integrating blockchain and AI in Smart Cities, shedding light on key considerations such as scalability, interoperability, and regulatory compliance. These seminal studies collectively lay the foundation for a multidisciplinary approach towards building efficient and secure Smart City ecosystems. This paragraph provides a succinct overview of significant works in the area, highlighting key contributions and insights from each study. It effectively sets the stage for your research by demonstrating the existing knowledge landscape in the intersection of Smart Cities, blockchain, and AI. Remember to cite these studies appropriately in your paper.

2 LITERATURE REVIEW

Though the term "smart megacity" is becoming widely used, no universally recognized definition has been established as of yet. Two kind of explanation can be distinguished thanks to the literature study in this field. As a new example in the growth of the megacity, Holland's ICT-based megacity views, 2008; Dominos, 2008), where people and capital, education, and land play key roles (Andreotti, De Marco, 2014; Hilfiger, Also fetal; Lombardi, Giordano fetal, 2012; Carlin, Del To, Nikkei, 2009). Among other things, the description presents the technology of smart megacities. "My smart megacity is a dynamic and high-tech megacity that combines people, information, and the rudiments of a megacity using new technologies to" (Basic, E. Admiral, and Ware Ham, 2013). Build a sustainable megacity and more productive, greener, and competitive enterprises while improving the standard of living. A smart megacity is also described by IBM as "a network of connected systems." Strong work is still being done, and progress is the key term. To be a stunning megacity for everyone, the tripod structure of high technology (work, people) is predicated on solid support for and between each of its pillars "(IBM, 2015). There are three features in this definition. The value of intelligent megacities 1. Tooled, or the megacity's ICT presence (research, mobile biases), Connected, meaning the ability to employ tools to break the connection between the physical and virtual worlds; and (3) intelligent, meaning the capacity to apply new technologies in megacities. generative procedure (Harswick, Kalagnanam, Paraszczak, Harrison, Beckman, Hamilton, & Williams, 2010). Using new technologies, cities can be built that integrate capital and physical elements, offering better services and more thoughtful planning. "We buy the megacity to be intelligent when a game of people and domestic animals and culture (elevation) and ultramodern (ICT) organizes sustainable energy, increases usefulness and utility," Carlin's borders, Ch. Del BO et. Identity (2011), reflects this approach. high standard of living and effectively manages the natural budget, owing to government involvement." According to Alison Dominos (2008), "Intelligent cities are structures with exceptional literacy and creative capacity that are created by their citizens, knowledge-based financial institutions, and digital processing for the utilization of data and knowledge. From the standpoint of the study's concept, this strategy for creating a smart megacity appears to be the most sensible. and upholds the development's guiding ideas. "Information technology connected to the organization, processes, standards, and even our bodies to reproduce gregarious, profitable, and ecological" is the definition of a smart megacity with regard to ICT use (Townsend, 2013). What is a smart megacity? As a megacity where ICT is arranged and integrated utilizing new digital technologies, in conjunction with traditional buildings (Baty, Exhaust fetal, 2012). Mitchel. Mitchell M. intelligence based on tiny and small services is the definition of an intelligent megacity. Combination of software (knowledge and cognitive ability), detectors and markers (the sad portion), intelligence without a location (intelligence), and digital telecommunications network (tuberculosis)" (Mitchell, 2007). It is not apart from other social systems, nor is it linked to them solely by means of human intermediaries.

3 SMART MEGACITY'S FUTURE LESSONS

Mathematics students who are interested in developing thinking computers and response systems that resemble people are studying artificial intelligence. Despite its history dating back to the 1950s of the previous century, this is the only advancement in computer power to store and use vast amounts of data and opportunities (McCarthy, Minsky, Rochester, Shannon, 2006; Kar, gusto, 2018). The Internet has aided AI in becoming one of the century's most important technologies. In the field of intelligence, machine control is essential because it provides the framework for the development of the learning process. The mechanical acclimatization adjustment of conditions and common sense are made possible by their contribution. It is important to emphasize that the evolution of intelligence seems to be the evolution of mental wisdom, or the examination of intelligence and the operation of the dying brain. This makes it possible to investigate the potential for computer collaboration. A modern computer, similar to a dying brain, computes through a neural network that processes and interprets massive amounts of input. As a result, they use natural language processing, image recognition, and machine learning. As a result, devices running AI2 on the platform are suitable for data collection in the field and use common sense to identify potential characteristics (Mnih, Kavukcuoglu *et al.*, 2015; Bughin, Hzan *et al.*, 2017). Intelligent people, intelligent mobility, intelligent terrain, intelligent people, intelligent living, intelligent governance—intelligence permeates many facets of life, including small and smart cities. Table 1 shows examples of smart megacities enabled by intelligent algorithms.

Applying AI is essential when thinking about sustainable urban development. It will be feasible to enhance megacity systems with AI performance, for instance, by addressing energy usage (Adeo-Moses, Asola, 2016) or developing a more intelligent and efficient transportation system (Agawa *et al.*, 2016). The outcomes of previous operations can be used to introduce the city system's autonomous functioning. AI systems will make it feasible to measure air pollution levels, deploy driverless cars on major thoroughfares, and swiftly and effectively manage public lighting. Agawa and According to *et al.* (2015), artificial intelligence (AI) is not the only development that enhances the energy of megacities, the strategy of transportation networks, or the structure of city systems; it can also be utilized to boost the amount of water. Intelligence will make it possible to monitor traffic conditions and congestion, generating reports on accidents and other incidents on the road. This enables real-time reporting and management control and increases the volume of business, especially in megacity areas with traffic risks.

Table 1. Instances of artificial intelligence algorithms supporting smart cities.

Smart city dimensions	Examples of potential applications for AI
Smart economy	Infrastructure optimization in cities, conservation of energy and water, lower building maintenance costs, and sharing economy
Smart mobility	Traffic light control, autonomous cars, and traffic parameter forecast Details regarding free parking, lower expenses for outside transportation, and less traffic
Smart environment	reducing pollution, conserving water and energy, and implementing smart bins
Clever people	Develop artificial intelligence and engage in public life
Smart living	Increasing resident safety, preventing hospitalization, preventing crime, smart buildings, intelligent healthcare, and virtual support
Smart governance	Take part in public-private partnerships, decision-making, Chatbots with automated responses, online consultations, and crowdsourcing

The operating system of businesses in a megacity will transition from a static to a dynamic system as a result of AI. Considering the various modes of transportation, they will permit instant acclimation to the present circumstances. The implementation of intelligence, for example, in a company's fire control system and the development of a data recycling system, will allow for the analysis and reporting of the company's behavior. Additionally, it will design a communication strategy for various groups, contributing positively to the evolving circumstances. AI can also serve as the foundation for the creation of new collaboration systems for city drivers and services, allowing for real-time participation in information such as professional traffic or air pollution. As such, the system will assist in anticipating issues before they materialize. A business model based on AI based on neural networks supports the idea and selects the most appropriate result for the business situation based on input data (Figure 2). The company's AI driven operational system can also create conversion schedules, reducing traffic miracles and suggesting crucial routes (De Oliviera, Neto, 2013). They can even plan their important travel times (to get to work or school) by setting the alarm beforehand. Automobiles that employ AI will be able to speak with the system directly, making city driving easier.

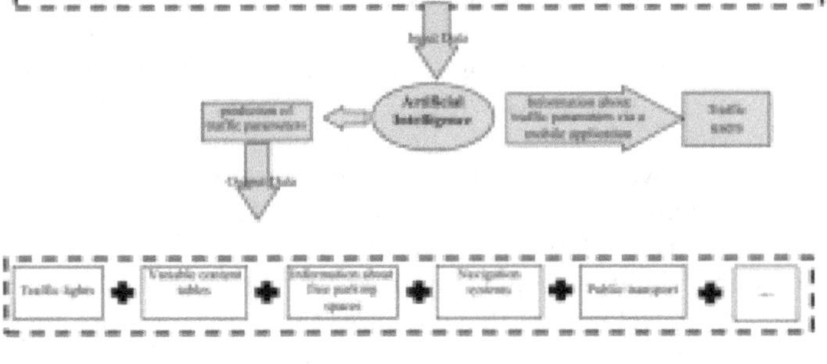

Figure 3. Model of artificial intelligence-based traffic management.

Human intelligence is capable of much more than just managing and operating in the business world. Health research, robotics and chatbot management, and virtual assistance are just a few of the many tasks they perform. Identifying regions with increasing rates of criminal activity, enhancing the effectiveness of proactive measures, and managing urban areas are among the instances. Continuous Internet monitoring can be used to prevent attacks and protect residents. Finally, big data analytics and AI predictive power enable improved performance. Furthermore, AI allows for the repair of damage with high power, water power, and so on by planning and increasing its security. Artificial intelligence (AI) combined with intelligent design enables us to leverage hundreds of sensors that cover our hobbies and behaviors. It combines many functions into one control unit and one remote control or function, allowing you to control the operation of compressed air, ventilation, central heating, lighting, alarms, door cords, and home monitoring. Intelligence also aids in public service and wise governance. enhances resident services and cultivates existing connections. Files can be converted to relevant sections and documents, and files can be created repeatedly, thanks to chatbots and automated responses. Coverage of license issuance is planned. Artificial Intelligence (AI) has the capacity to read aspects of racism, origin, education, age, gender, or beliefs from research papers, external databases, and managers' thoughts. Furthermore, by utilizing the information that residents send, it may stimulate their desire to socialize. According to the concept of smart government, citizen participation is essential. By integrating with cutting-edge concepts such as blockchain, AI can improve the transparency of the relationship between battery holders and, as a result, citizen participation in the management of megacities.

4 INTRODUCTION OF THE BLOCKCHAIN INTO INTELLIGENT CITIES

Blockchain is a social ledger or public ledger technology. Allows you to store sensitive data in related data blocks. Each transaction necessitates the validation of a private key, which is a string of characters known only to the parties directly involved in the transaction, and each block has a unique signature that cannot be replicated. Only those with private keys can make changes (Ian Siti, Lakhani, 2017). Blocking is thought to be the most secure method of data storage. To link tasks with computational logic, users can change this. It is used to gather information about a business partner's financial transactions and financial situation. Utilizing encryption techniques, blockchain technology guards against unwanted access to user data and user comprehension (Pilkington, 2016). Thus, consistent with Kauf (2019):

- Openness in Transactions: To anybody with access to the participant's network, every transaction and all of its details are visible. Every node, or customer, has an alphanumeric character that identifies them uniquely. When the address block changes, users can choose to display their name or remain anonymous.
- Immutability of data – no changes are made after the changes are saved in the database; all changes are visible to others. participants in the network. Information is timely provided and available to all.

Blockchain can eliminate the need for coordinators, partners, or middlemen by integrating businesses. It is impervious to manipulation and prohibits the fabrication of data. Blockchain technology makes smart cities possible in every way. Uses of blockchain technology for task performance control extend beyond contractor contracts, such as digital identity, smart contracts, and smart assets. Decision-making systems based on Smart Contracts and block chains can maximize the actions of entities involved in creating public value, while public e-procurement can encourage the public sector to use negotiated tendering procedures more frequently. Transparency in transactions benefits the public sector by increasing the efficiency and effectiveness of contractor relationships and optimizing Smart Asset distribution.

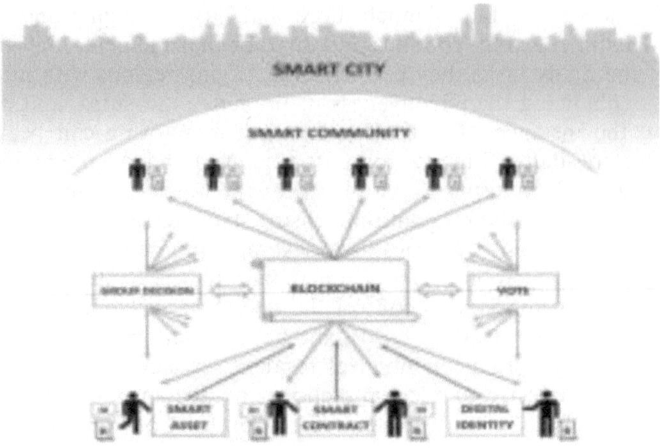

Figure 4. Blockchain technology connecting smart city organizations.

Info trust is guaranteed by blockchain technology. It guarantees safe access to current, accurate information for all parties involved. Smart cities, by transforming resources into smart assets, can assist citizens in tracking the origin and flow of goods (such as chemicals) throughout the supply chain. This reduces distribution fraud and responds almost immediately to inconsistencies when they are discovered. The properties of delivered and stored goods can be measured and recorded using blockchain, which is crucial for sensitive goods like prescription drugs. It is not possible to modify special prices. The blockchain is a crucial tool for intelligent management as a result. Because public procurement is the foundation for selecting suppliers to carry out public works, it is critical to highlight the benefits that blockchain technology offers in this field. By establishing an online platform, this instrument can assist:

1. Boost productivity and openness
2. Lessen fragmentation to take advantage of economies of scale.

It is reasonable to believe that in the coming decade, online business will be both international and local. Financial institutions will be able to send certified electronic certificates on behalf of exporters in other nations with ease thanks to the network's expansion. Blockchain technology will alter the rules of competition participation. Everyone who wants to engage in public service should have access to digital communication. Approval processes will become more transparent and expedited as a result. Because the information cannot be changed abruptly, sending a discount will be extremely difficult. All changes will be made public and subject to approval by the appropriate parties. All vendors taking part in open electronic auctions need to have a digital identity. It is possible to use digital tokens produced via the blockchain process as electronic signatures. Currently, Estonia's e-residency service is the most popular application because it allows users to create digital identities. Furthermore, non-EU citizens can use it to conduct business. The combination of digital identity and smart devices allows for the authorization of personal resources. Individuals who have endorsed the blockchain protocol can begin conducting business on our behalf. In real life, this might mean conducting planning inspections for public transportation. Furthermore, blockchain facilitates citizens' digital identity, allowing them to take part in local, regional, or national elections from the comfort of their homes. Aside from voting privacy, blockchain allows for voter verification and verification. Using blockchain, every citizen will be able to manage their own votes. The citizen is anonymous on the blockchain, but his voice is heard by all. Because the information is decentralized, hacker attacks are difficult and there is no risk of IT system failure. Anywhere with internet connectivity and any device with a computer-like interface will allow voters to cast their ballots. This removes the potential for vote manipulation, tampering with the count, or vote cancellation.

5 CONCLUSION

The challenges faced by smart cities arise from waves of migration and population growth. Through the lens of sustainability, this article looks at current and future applications of edge AI and blockchain technology in two key areas that address issues related to smart cities: mobility and smart energy. It supports knowledge on sustainable development by finding and documenting current research on the frontiers of artificial intelligence and blockchain, highlighting their advantages and potential impacts on smart cities. This review highlights the current and potential convergence of edge AI and blockchain. This summary should guide future research on edge AI and blockchain applications in other smart city areas. The decisions in this article show how blockchain technology and intelligence have great potential to help build a smart city. These are the most innovative and risky technologies with almost no limits in applications. The Gartner Curve for Digital Government Technologies also shows the diffusion of these technologies. Thus, blockchain technology and artificial intelligence (AI) (autonomous vehicles) will have the biggest impact on government agencies in the next five to ten years. Blockchains and artificial intelligence are additional technologies that can improve many aspects of city operations. Data analytics is the basis for the initial concept of integrating blockchain technology and intelligence. Blockchains provides artificial intelligence with centralized data and large data warehouses to be collected and analyzed. Despite the failure, the content on the blockchain remains secure and immutable. Blockchains consensus technology can increase transparency and reduce the regulatory burden. For this reason, they enhance stakeholder engagement and create community-based initiatives. These solutions place these technologies, often called disruptive technologies, in the innovation category. This is because any technology that disrupts the status quo causes disruption. However, in this regard, it is necessary to mention Schumpeter's article on creative destruction (1942). Blockchain will revolutionize public services by easing tensions between public and private entities and enabling open recording of public transactions and authorization of transactions. Artificial intelligence will be used in urban traffic management, administrative decision-making, services for residents and other activities. Despite the huge potential there are still a lot of challenges in the field of AI and Blockchain. to overcome inintegrating the two. Privacy issues are a matter of concern, especially with regard to valuable information and related rights.

REFERENCES

Abbas, N.; Zhang, Y.; Thero'd, A.; Skeie, T. (2017). Mobile Edge Computing: A Survey. *IEEE IoT J.* 5, 450–465. [Crossruff]

Al Shamsi, M.; Salloum, S.A.; Altruized, M.; Abdallah, S. Artificial intelligence and blockchain for transparency in governance. In *Artificial Intelligence for Sustainable Development: Theory, Practice and Future Applications*; Springer: Cham, Switzerland,

Albert, S. (2019). *Innovative Solutions for Creating Sustainable Cities*; Cambridge Scholars Publishing: Newcastle upon Tyne, UK

Grimmond, C.S.B.; Roth, M.; Oki, T.R.; Au, Y.C.; Best, M.; Betts, R.; Carmichael, G.; Cleugh, H.; Dab Berdt, W.; Emmanuel, R. *et al.* Climate and More Sustainable Cities: Climate Information for Improved Planning and Management of Cities

Hugo Primeurs, S.D. (2016). *Climate Change and Sustainable Cities.* Taylor & Francis: Andover, UK

Kowalski, M.; Lee, Z.W.Y.; Chan, T.K.H. (2021). Blockchain technology and trust relationships in trade finance. *Technol. Forecast. Soc. Chang.*, 166, 120641. [Crossruff]

Mondejar, M.E.; Avtar, R.; Diaz, H.L.B.; Dubey, R.K.; Esteban, J.; Gómez-Morales, A.; Hallam, B.; Mbungu, N.T.; Okolo, C.C.; Prasad, K.A.; *et al.* (2021). Digitalization to achieve sustainable development goals: Steps towards a Smart Green Planet. *Sci. Total Environ*, 794, 148539. [Crossruff] [PubMed]

Ozawa, A.A.; Ekhoragbon, E.B.; Fat, L.T. (September 2017). Application of artificial intelligence in Internet of Things. In *Proceedings of the 2017 9th International Conference on Computational Intelligence and Communication Networks (CICN)*, Cyprus, Turkey, pp. 169–173.

Qiu, C.; Yao, H.; Jiang, C.; Guo, S.; Xu, F. (2019). Cloud computing assisted blockchain-enabled Internet of Things. *IEEE Trans. Cloud Compute.* 1, 1. [Crossruff]

Satyanarayana, M. (2017). The Emergence of Edge Computing. *Computer*, 50, 30–39. [Crossruff]

Shi, W.; Cao, J.; Zhang, Q.; Li, Y.; Xu, L. (2016). Edge computing: Vision and challenges. *IEEE IoT J*, 3, 637–646. [Crossruff]

Shi, W.; Dustcart, S. (2016). The promise of edge computing. *Computer*, 49, 78–81. [Crossruff]

Wang, X.; Han, Y.; Leung, V.C.M.; Niyati, D.; Yan, X.; Chen, X. (2020). Convergence of Edge Computing and Deep Learning: A Comprehensive Survey. *IEEE Commune. Surb. Tutor.*, 22, 869–904. [Crossruff]

Xu, D.; Li, T.; Li, Y.; Su, X.; Trachoma, S.; Jiang, T.; Crowcroft, J.; Hui, P. (2003). Edge Intelligence: Architectures, Challenges, and Applications. *arci 2020, arXiv:.12172.*

Yue, L.; Junín, H.; Shunzhi, Q.; Raijin, W. (August 2017). Big data model of security sharing based on blockchain. In *Proceedings of the 2017 3rd International Conference on Big Data Computing and Communications (BIGCOM)*, Chengdu, China, 10–11; pp. 117–121.

Computational Methods in Science and Technology Volume 1 – Kaur et al. (Eds)

Unveiling the web of privacy: A comprehensive review of threats to information security in social networks

Shalini Singh & Dev Baloni
Department of Computer Science and Engineering
Quantum University, Roorkee, India

Manash Sarkar
Department of Computer Science and Engineering
Atria Institute of Technology Banglore, India

ABSTRACT: This review paper explores emerging threats to information privacy and security within the dynamic landscape of Online Social Networks (OSNs), which serve as repositories of vast amounts of user data. The rise in social media use emphasizes how crucial it is to inform users about the dangers that come with it, like cyber-attacks and privacy violations. Even though various social media platforms have been made possible by technological advancements, consumers' ignorance of these platforms creates serious security risks. Although academics have come up with creative security solutions for social networks, there are obstacles in the way of their actual application. It is essential to continuously assess security vulnerabilities in light of technology improvements. This study examines several dangers to online social networks and suggests countermeasures based on a variety of models, frameworks, and encryption methods. Despite advancements, privacy concerns remain unaddressed, requiring parents to actively supervise their children's online activity.

Keywords: Online Social Network, Security, Privacy

1 INTRODUCTION

Recently there has been a surge, in the popularity of OSNs attributed to the fast-paced advancement of mobile applications and the significant increase, in virtual interactions. Sun *et al.* (2010) Various OSNs like Facebook, Twitter, Flickr, Myspace have surfaced as a result. Online social network service providers typically gather and maintain a significant amount of personal data. An increasing number of businesses are outsourcing their data to the cloud in order to take advantage of the advantages that cloud computing offers, including flexible management, cost savings, quick development, easy data access, and superior calculation performance. People, like using social media platforms. Islam *et al.* (2020) suggested that there is a need to be cautious as instances of news and rumors are increasing that can lead them to believe false information.

Over the previous six years, there has been a significant increase in the number of people utilizing social media from 3.484 billion in 2019 to 5.04 billion users by January 2024, according to the most recent global digital overview report as shown in Figure 1 (datareportal: Accessed on 22 Feb 2024). The primary goal of Social Networks is to facilitate information sharing, which means users must be willing to share types of information including data in order to fully utilize the services offered. It's important to note that

DOI: 10.1201/9781003501244-30

collecting this data is acceptable long as users are aware of and agree to it. Social Network platforms attract a number of cyber attackers who exploit any vulnerabilities or gaps, in user knowledge. These attackers have two objectives; firstly, they aim to access the amount of user data stored on Social Networks posing a threat to privacy; secondly, they engage in malicious activities that pose a risk, to security. In the world of networking numerous security risks lurk around every corner ready to pounce on even the slightest weakness. Technical weaknesses can be fixed to enhance the security of the platforms. They often fall short in the face of attackers who prey on users lack of awareness through engineering tactics.

The rest of the paper is organized as follows; Section 2 delves, into existing research Section 3 outlines the background context of media Section 4 elaborates on user perceptions in social networks Section 5 addresses threats in online social media, Section 6 provides security recommendations, for OSN users and finally Section 7 offers conclusions and discussions.

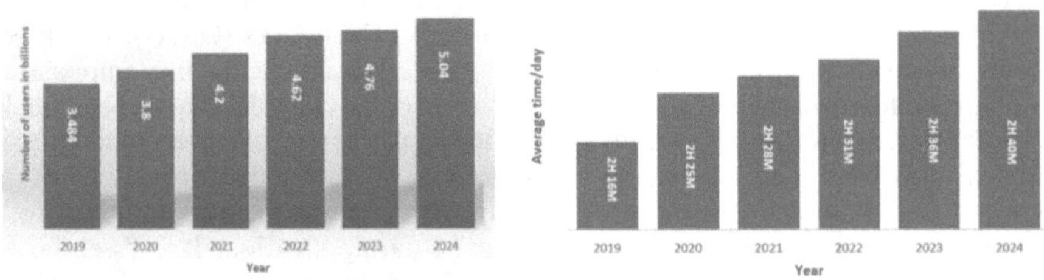

(a) (b)

Figure 1. (a) Number of OSN user (in billions) within six years (b) The average daily OSN usage duration (in Hour & Minutes).

2 LITERATURE REVIEW

Oz *et al.* (2024) conducted a survey to investigate how the perception of surveillance affects political engagement. By utilizing the Spiral of Silence theory and surveillance culture as frameworks the study delved into how social media users views, on government surveillance influence their willingness to share opinions online.Yan *et al.* (2024) introduced to alter the nodes' associations, which not only preserves the link privacy of nodes' differential privacy but also enhances the data value of that privacy.

Wang *et al.* (2024) addresses the existing void by offering an adjustment of data structures to meet the evolving needs of storage management and secure retrieval. It makes use of a novel data structure called the Enhanced Merkle Tree (EMT). To meet the needs of e-healthcare Systems (e-HS), this article first adapts the raditional Merkle Tree (MT) structure used in Blockchain technology. The EMT greatly improves data integrity management and strengthens data security for access and storage and examines risks to security and privacy on networks along, with the existing solutions that can provide users with a sense of security to address these issues. Citing other data reports, the author also covered OSN attacks on several OSN online apps.Zhan *et al.* (2010) aims to explore the development of a network involving parties in this paper. Creating a network involves gathering data.It becomes challenging to get data from several parties for privacy reasons, and these worries may keep the parties from directly sharing the data. It is difficult for numerous parties to work together to create a social network without violating data privacy. This research aims to address the collaborative social network problem of privacy preservation. Zhang *et al.* (2021) put up the innovative BPP framework, which is built on blockchain technology and protects online social networks' privacy. A new secure, fair and efficient keyword search method is suggested, utilizing technology and public key encryption. Following this proposal a test framework is set up on a network to assess the security, effectiveness and efficiency of the system.

3 BACKGROUND CONTEXT OF SOCIAL MEDIA

"Social media" refers to a range of internet based tools that have emerged in recent times. Some experts argue that Social Media represents the platform while Social Networking involves individuals engaging with their communities. Zhan *et al.* (2010) described SN as services that enable users to create an semi public profile, within a specific system list other users they are connected with and explore connections made by themselves and others within the platform. Based on the explanation the first social networking site to emerge in 1997, known as SixDegrees.com was a pioneer, in allowing users to create profiles and add friends to their lists. It integrated features from blog, chat and community platforms. Initially social networks aimed to mirror real life relationships in a space. Online social networks (OSNs) play a role in maintaining and strengthening connections between family members, colleagues from school or work residents of the city or town or individuals with common offline ties. Kumari *et al.* (2010) Over time social networks gained popularity rapidly. Expanded their services to include features like photo and video sharing, with friend tagging capabilities. The uses of networks have diversified widely beyond socializing to encompass news sharing and business interactions as well. OSNs can be divided into several categories as illustrated in Figure 2.

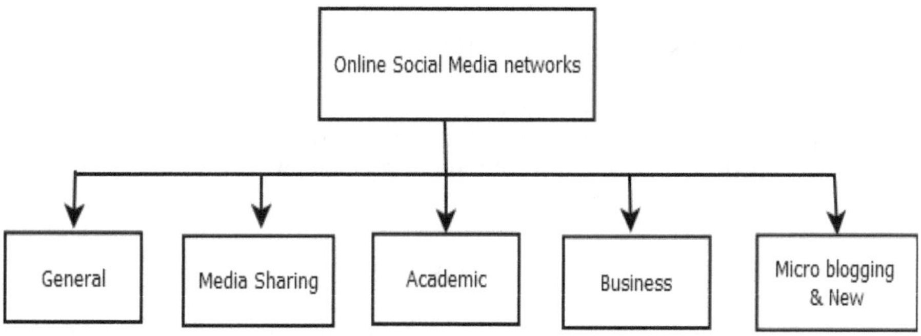

Figure 2. Social networks categories.

4 USER PERCEPTIONS IN ONLINE SOCIAL NETWORKS

The different favorable aspects that encourage the user to design and utilize the surroundings include upholding social relationships, advertising the goods and platforms, and rescue efforts. preserving social connections, using a marketing platform, attempts to save lives during emergencies and disasters Liu *et al.* (2018). They serve as a bridge between people at these pivotal moments when the traditional social order has collapsed. Social networking platforms make it simple to manage bulletins, which can help find shared groupings and reconcile missing family members. People can find groups with similar interests by using social networking sites. Pewresearch (Accessed on 20 Feb 2024). Many of us encounter challenges when using online social networks. For instance, while it's convenient to connect with new friends, there's also the risk of encountering online intimidation, as predators can easily target victims. Additionally, the issue of private information being exploited is a concern. Social media has its benefits, in fostering connections. It can sometimes impede the quality of, in person interactions. People often prefer scrolling through comments rather than engaging in personal visits or calls. And concerning addiction, studies suggest that social media can be more addictive than cigarettes and alcohol. Another study by Wolniewicz *et al.* (2018) Not checking social media for even a day can leave individuals feeling empty, sad, lonely and depressed.

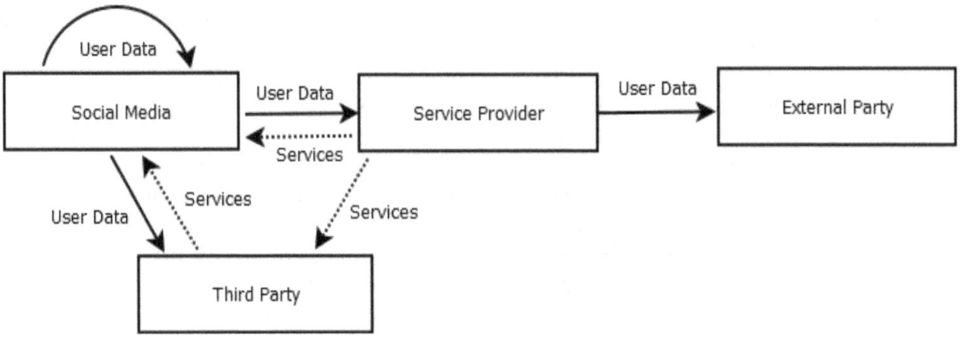

Figure 3. Transfer of user information among OSN entities.

5 THREATS IN ONLINE SOCIAL MEDIA

In our technology-driven society, characterized by widespread internet usage, our interactions have expanded into the digital realm of the online world. Table 1 shows the attacks that users have been experiencing since the inception of social networks.

6 SECURITY RECOMMENDATIONS FOR OSN USERS

Online social media and networks are now deeply integrated into everyday life, but this increased usage also comes with heightened risks. As user numbers skyrocket annually, securing these platforms becomes imperative. General security guidelines include using strong, unique passwords, limiting location sharing, being selective with friend requests, exercising caution when sharing personal information, and being wary of malicious links and third-party applications. Platform-specific measures vary but encompass scrutinizing profiles, avoiding oversharing, revoking access to unused apps, enabling two-step authentication, and understanding privacy settings. Vigilance is key in safeguarding against cyber threats. Various recommenations given in Figure 4.

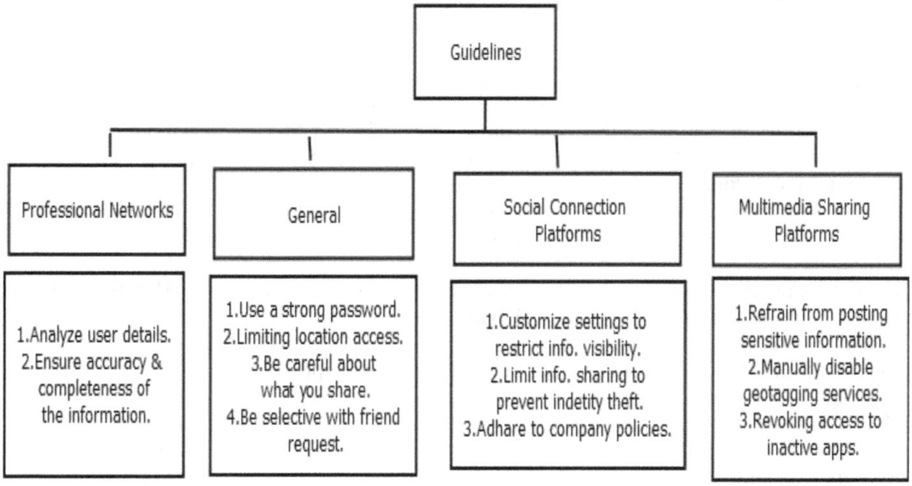

Figure 4. Security guidelines for user.

Table 1. Types of attacks.

Attack type	Characteristics	Mitigation	Risk level	Ref.
Spam	Unsolicited bulk msg, email harvesting	Email filtering, use of blacklists	High	Sahoo 2018
Malware	delivered through infected files	Network firewalls.	High	Grosse 2017
Phishing	Impersonates trusted Entities	Email authentication	High	Chen 2018
Identity theft	stealing personal information	identity theft protection	High	Jain 2022
Cross-Site ScriptingXSS	Malicious scripts Injections	secure coding practices	Medium	Kaubiyal 2019
Profile Cloning	Fake profiles creation, scrap data from real profiles	twofactor authentication	Medium	Fire 2014
Sybil	multiple fake identities	Identity verification Mechanism	High	Mishra 2018
Deanonymization	Privacy violation, identity exposure	data perturbation, limitng data access	High	Tian 2018
Cyber Grooming	Manipulative behavior by adults	Sexual Exploitation, emotional harm.	High	De 2018

Table 2. Study documents on social media risk & its detection.

Document title	Author	Year	Technology used	Dataset	reff.
Harmfulness metrics in digital twins of social network rumors detection in cloud computing environment	Hao Li Wu Yang	2024	LLM, RSK-T5	Rumour Ev	Li 2024
RPf-GCNs:reciprocal perspective driven fused GCNs for rumor detection on social media	Zafran Khan	2024	RPf-GCN	PHEME &SemEva	Zafran 2024
Graph neural networks based framework to analyze social media platforms for malicious user detection	Zeeshan Khan	2024	SVM and LSTM	PHEME	Khan 2024
Malicious Activities Prediction Over Online Social Networking Using Ensemble Model	Sadhasiva S	2023	approaches like R	UCF crime dataset	Sadhasivam 2023
Detection of malicious profiles and protecting users in online social networks	Senthil Raja	2022	ESVM NN	Twitter dataset	Raja 2022

7 CONCLUSION AND DISCUSSION

This review article has attempted to address the emerging risks to information privacy a nd security in this alluring and abundant environment of user personal data as social networks become an increasingly significant part of peoples' everyday lives. Online social networks have become integral components of the widespread internet landscape, facilitating daily interactions with users. The rise in social media use emphasizes how crucial it is to inform users about the dangers that come with it, like cyberattacks and privacy violations. Even though various social media platforms have been made possible by technological advancements, consumers' ignorance of these platforms creates serious security risks. Although academics have come up with creative security solutions for social networks, there are obstacles in the way of their actual application. It is essential to continuously assess security vulnerabilities in light of technology improvements. This study examines several dangers to online social networks and suggests countermeasures. Despite advancements, privacy concerns remain unaddressed, requiring parents to actively supervise their children's online activity. Main aim to offer this research as a roadmap for understanding current privacy and security challenges in social networks. The aim is to motivate other researchers to explore further and seek solutions, for these challenges in their projects.

REFERENCES

Chen, J., Mishler, S., Hu, B., Li, N., & Proctor, R. W. (2018). The description-experience gap in the effect of warning reliability on user trust and performance in a phishing-detection context. *International Journal of Human-Computer Studies*, 119, 35–47.

Datareportal: We Are Social – Digital report. https://datareportal.com/reports/digital-2024-global-overview-report:Accessed 22 Feb 2024

De Santisteban, P., Del Hoyo, J., Alcázar-Córcoles, M. Á., & Gámez-Guadix, M. (2018). Progression, maintenance, and feedback of online child sexual grooming: A qualitative analysis of online predators. *Child Abuse & Neglect*, 80, 203–215.

Fire, M., Goldschmidt, R., & Elovici, Y. (2014). Online social networks: threats and solutions. *IEEE Communications Surveys & Tutorials*, 16(4), 2019–2036.

Grosse, K., Papernot, N., Manoharan, P., Backes, M., & McDaniel, P. (2017). Adversarial examples for malware detection. In *Computer Security–ESORICS 2017: 22nd European Symposium on Research in Computer Security*, Oslo, Norway, September 11–15, 2017, Proceedings, Part II 22 (pp. 62–79). Springer International Publishing.

Islam, M. R., Liu, S., Wang, X., & Xu, G. (2020). Deep learning for misinformation detection on online social networks: A survey and new perspectives. *Social Network Analysis and Mining*, 10, 1–20.

Jain, A. K., & Gupta, B. B. (2022). A survey of phishing attack techniques, defence mechanisms and open research challenges. *Enterprise Information Systems*, 16(4), 527–565.

Kaubiyal, J., & Jain, A. K. (2019, August). A feature based approach to detect fake profiles in Twitter. In *Proceedings of the 3rd International Conference on Big Data and Internet of Things* (pp. 135–139).

Khan, Z., Khan, Z., Lee, B. G., Kim, H. K., & Jeon, M. (2024). Graph neural networks based framework to analyze social media platforms for malicious user detection. *Applied Soft Computing*, 111416.

Kumari, P. (2010). Requirements analysis for privacy in social networks. In *8th Intl. Workshop for Technical, Economic and Legal Aspects of Business Models for Virtual Goods (VG)*.

Li, H., Yang, W., Wang, W., & Wang, H. (2024). Harmfulness metrics in digital twins of social network rumors detection in cloud computing environment. *Journal of Cloud Computing*, 13(1), 1–14.

Liu, F., & Xu, D. (2018). Social roles and consequences in using social media in disasters: A structurational perspective. *Information Systems Frontiers*, 20, 693–711.

Mishra, A. K., Tripathy, A. K., Puthal, D., & Yang, L. T. (2018). Analytical model for sybil attack phases in internet of things. *IEEE Internet of Things Journal*, 6(1), 379–387.

Oz, M., & Yanik, A. (2024). Fear of surveillance: Examining Turkish social media users' perception of surveillance and willingness to express opinions on social media. *Mediterranean Politics*, 29(1), 1–25.

Sadhasivam, S., Valarmathie, P., & Dinakaran, K. (2023). Malicious Activities Prediction Over Online Social Networking Using Ensemble Model. *Intelligent Automation & Soft Computing*, 36(1), 461–479.

Sahoo, S. R., & Gupta, B. B. (2018). Security issues and challenges in online social networks (OSNs) based on user perspective. *Computer and Cyber Security*, 591–606.

Senthil Raja, M., & Arun Raj, L. (2022). Detection of malicious profiles and protecting users in online social networks. *Wireless Personal Communications*, 127(1), 107–124.

Sun, J., Zhu, X., & Fang, Y. (2010, March). A privacy-preserving scheme for online social networks with efficient revocation. In *2010 Proceedings IEEE INFOCOM* (pp. 1–9). IEEE.

The Positive and Negative Efects of Social Networking |https://www.pewresearch.org/short-reads/2020/10/15/64. Accessed on 20 Feb 2024.

Tian, W., Mao, J., Jiang, J., He, Z., Zhou, Z., & Liu, J. (2018). Deeply understanding structure-based social network de-anonymization. *Procedia Computer Science*, 129, 52–58.

Wang, Y. (2024). Research on data structure and privacy protection analysis in big data environment based on blockchain technology. *International Journal of Intelligent Networks.*

Wolniewicz, C. A., Tiamiyu, M. F., Weeks, J. W., & Elhai, J. D. (2018). Problematic smartphone use and relations with negative affect, fear of missing out, and fear of negative and positive evaluation. *Psychiatry Research*, 262, 618–623.

Yan, J., Chen, J., Zhou, Y., Wu, Z., & Lu, L. (2024). An Uncertain Graph Method Based on Node Random Response to Preserve Link Privacy of Social Networks. *KSII Transactions on Internet & Information Systems*, 18(1).

Zafran, Khan, Jeonghwan Gwak, Naima Iltaf, Witold Pedrycz, and Moongu Jeon. "RPf-GCNs: reciprocal perspective driven fused GCNs for rumor detection on social media." *Journal of Big Data* 11, no. 1 (2024): 12.

Zhan, J. (2010). Secure collaborative social networks. *IEEE Transactions on Systems, Man, and Cybernetics, Part C (Applications and Reviews)*, 40(6), 682–689.

Zhang, S., Yao, T., Arthur Sandor, V. K., Weng, T. H., Liang, W., & Su, J. (2021). A novel blockchain-based privacy-preserving framework for online social networks. *Connection Science*, 33(3), 555–575.

Computational Methods in Science and Technology Volume 1 – Kaur et al. (Eds)
© 2025 the Author(s), ISBN: 978-1-032-81760-6

Emotion detection: Facial emotion detection using ML

Sharad Sharma, Ritik Chauhan & Pooja Vajpayee
School of Computing Science and Engineering, Galgotias University, Greater Noida

ABSTRACT: Cognitive perception of images is an intense task, like guessing the truth of a thought or a mystery. In this process, we use different methods to solve the need to know the job. In recent years, emotional intelligence has received widespread attention due to its important role in improving human-computer interaction and emotional analysis. Because of its processes for extracting visual features, deep neural networks-particularly convolutional neural network (CNN)-are frequently employed in FER. A lot of research has been done on CNN, but only a few areas have solved the facial recognition (FER) problem. After many reports and studies, the shortcoming of most existing models is that they often consider the front profile and ignore the side profile of the current image, although different angular views are also important for the perspective model. This helps you make informed decisions about personal needs, providing support, or security threats. However, this proved to be too complex for computers, and feature extraction needed a wide range of image processing methods. Numerous machine learning techniques are appropriate for this particular purpose. To use machine learning for any kind of detection, an algorithm must first be trained and then tested on relevant data. This article examines some machine learning algorithms and strategies for extraction that aid in the analysis of human emotion. This review contains a comprehensive review of a variety of materials, methods, and measures commonly used in psychology research. It also highlights the challenges and limitations of current models, including dealing with uncertainties and cultural differences. Face recognition in large datasets poses additional challenges due to the diversity of images as well as the differences between side and front views.

Keywords: Fer, Emotion, Emotiondetection, Facialexpression, Machine Learning, Deep Learning

1 INTRODUCTION

People's everyday interactions and communication are greatly influenced by their emotions. The nonverbal means of conveying message, whether one is lying or stating the truth, is through facial expression. It is also important for security; Each person's opinion can help inform the machine about a particular person it recognizes. Another important area where emotional intelligence can be the best way to predict people's next actions or steps is the stock market, where some predictions can be made because no one will be face to face when checking something related to the product. We can also predict whether they like the product; This can help any organization identify bugs or defects with their products to generate future business benefits. Thus, because of all the potential that facial recognition presents, academics in the field have been interested in it. Explaining the many faces of emotions is FER's main mission. The two primary processes in classical FER are recognition and extraction. It also needs image processing, which entails face identification, cropping, scaling, and normalization. Current methods mainly focus on learning faces without changing the background, so Most of them are redundant and confusing picture, CNN training process. The present approach focuses on seven categories of facial expression: neutrality, sorrow, surprise, anger,

 DOI: 10.1201/9781003501244-31

disgust, fear, and happiness. The algorithm proposed in this article aims to search and display images in seven basic needs. Finally, emotions are often understood by classifying features using neural networks (NN) and other different machine learning methods. Deep Neural Network Specifically, Convolutional Neural Network (CNN) is interested in the color of facial features with its built-in image feature extraction mechanism. The challenge of emotion recognition has been the subject of several CNN-based investigations. First, the work of emotion detection necessitates medium-high quality photos. Secondly, the task of other distribution is completed by the difference in the face cause by somewhat different emotions. Various modifications and techniques have been introduced in CNN methods and training to improve accuracy. However, training such a deep model requires a lot of data and high energy consumption. The facial measurement model can recognize facial emotions.

2 LITERATURE SURVEY

A review of research theory investigations is provided in this section. Face mask discoveries are not new; they have long been the subject of scholarly investigation. Because of its useful applications in human-computer interface, good computing, facial recognition using machine learning has grown in importance and gained attention. Many early approaches in many areas, such as speech recognition, have improved performance. This study is designed to provide a comparison between deep learning-based training end-to-end models and the dense machine learning and deep learning-based guides training model. has done a lot of research on how information on multiple variables and their combinations affects performance.

2.1 *FER techniques*

2.1.1 *3D FER*

Venture into a 3D facial data-based emotion recognition system which can be the best alternative to the 2D methodology as it is not affected by poses. Investigate newer studies highlighting the contributions of depth cameras and 3D reconstruction methods to the domains of FER.

2.1.2 *Multimodal FER*

Explore working on recognizing facial expressions simultaneously with other modalities such as speech utterance, body language, and context to increase recognition accuracy rate. Look up how various modalities can interact for cognitive attention and solving shortcomings in the use of facial effects.

2.1.3 *Explainable AI (XAI) for FER*

Understand as an example, which of the XAI methods can be used in decision-making of FER models and reveal unwanted biases. Investigate the techniques which these deep learning models within the "black box" feature interpretation.

2.2 *Emerging application*

2.2.1 *FER in healthcare*

By a complete knowledge of NER that can identify mental illness and body pain as therapy, we could, in the real sense, meet just the expected room for future advancements in neuroscience. Research interesting points about facial expression studies data and it the current emotional state of patients.

2.2.2 *FER in education*

Research applying FER to academic technology and a subsequent assessment during learning tasks of student engagement, understanding, and mental well-being.

2.2.3 *FER in human – robot interaction (HRI)*

Explore FER utilization to create robots with more broadened social intelligence, making them capable of adjusting behaviors and communications according to human feelings and emotional states.

2.3 *Ethical considerations*

2.3.1 *Privacy concerns*

Invite the dialogue on FER tech privacy and specially relate it to information acquisition in public space. Investigate the existing rules and possible ethical frameworks of investment in FER systems using responsible deployment.

2.3.2 *Emotional manipulation*

Find out how FER can be used in ways that aid emotional manipulation or social engineering and much more. Examine mitigation strategies and ethical standards for development FER and technologies.

2.4 *Previous work*

2.4.1 *Deep learning for facial emotion detection*

This research paper is about the facial emotion detection using deep-learning techniques which is based on the images. The system follows a three-step process: face recognition, feature extracting, and emotion classification. Convolutional Neural Networks (CNN) are decomposed to provide emotion classification. The model proves its competitiveness on FERC-2013 (70.14%) and JAFFE (98.65%), surpassing already existing approaches on these data. This indicates the viability of the deep learning method for facial emotion recognition. [1]

2.4.2 *Facial emotion recognition using deep learning*

The older versions of machine learning methods involving designed features find it difficult to classify emotions because of the manifold aspects to be accounted for. The proposed approach resorts specifically on the CNNS for the sake of extracting the facial images' features autonomously. The system reaches a high recognition level (70% in average on benchmarks) while being able to detect seven basic emotions (anger, contempt, disgust, fear, happiness, sadness, and surprise), confirming the reliability of deep learning in the facial emotion recognition task. [2]

2.4.3 *Recent advances in facial emotion recognition (FER)*

This paper is an overview (, the last two years) of FER technologies which have been done recently. The paragraph succinctly describes frequently used datasets, showing their properties and the ways of getting data generated. Alongside that, we place emphasis on cutting-edge algorithms that produce excellent precision. While this project has a few remaining challenges and brings forth possible future research ideas about how datasets or defect diagnosis methods integrate together. [3]

2.4.4 *Assessing learner state of mind in online learning*

This article outlines the fall short of perception of emotional needs of learners by e-learning platforms. The authors suggest who assess the leaner's state of mind during the online session that helps the analyzing facial expressions. They say that it is insufficient that one image is analyzed as well, and then suggest that six consecutive frames per webcam can be done. The system discovers four complex emotions (confusion, shortcoming, optimism, and frustration) and also 65% accuracy in emotion classification and 62% accuracy in the state of mind identification are attained. The authors who have reported the limitation has propose including eye movements and gaze for better accuracy. [4]

2.4.5 *Facial emotion recognition using a two-part CNN (FERC)*

The main purpose of this paper is to introduce a face emotion recognition system (FERC) built with CNN, which is a two-part convolutional neural network (CNN). In the stage of adding background noise, the moving noise will be created; in the stage of the facial features extraction, facial features will be extracted. FERC built a 24-element expressional vector (EV) to represent emotions and integrated it into a machine learning model achieving 96% accuracy on a 10,000-samples dataset. The paper demonstrates the two-stage CNN architecture of FERC and shows that excreting background for the sake of higher accuracy and the ability to handle pose variations. EFFF really shows its power in many areas of e.g. emotional expression. [5]

3 METHODOLOGY

This article advocates FERC which is an acronym for facial expression recognition from pictures to understand the emotional state of people. FERC comprise of typical Convolution Neural Network (CNN) algorithm that has two-staged pattern. The first part of CNN is responsible for restoration of background clutter, like reflections, to build better image. Second CNN is about making face features a key and, in that way, CNNs extracts the main features. The features of the image are then used to map and classify the emotion displayed in it. The paper focuses on the key of using convolutional layers present in the CNNs that are renowned for the presence of feature detection in images.

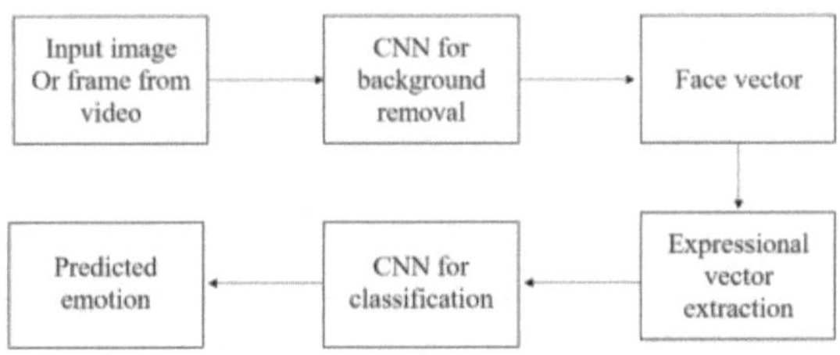

Figure 1. Complete process for detection.

3.1 *Dataset feature*

We utilize a file with a 48 x 48-pixel grayscale picture in this investigation. Faces are captured with the intention of keeping them roughly in the same position and center in every picture Each face had to be categorized into one of seven groups according to the emotion expressed on its face (0 = Angry, 1 = Fear, 2 = Happy, 3 = Sad, 4 = Dissatisfied, 5 = Ugly, 6 = Neutral). The training set consists of 28,709 samples, whereas the test population consists of 3,589 samples. Features and assortment of apparatus used for machine testing and training.

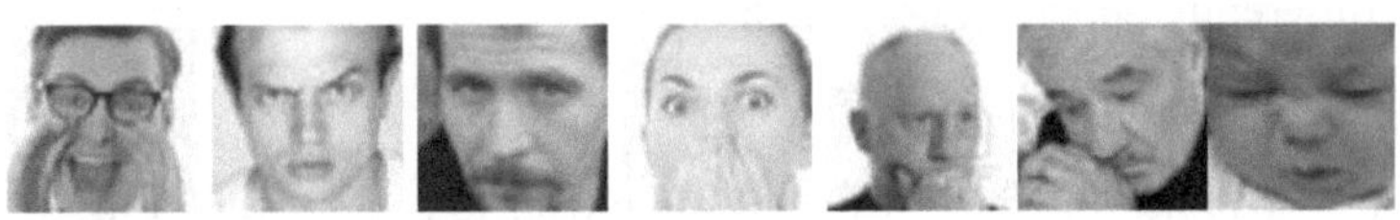

Figure 2. 7 different classes of emotion from Kaggle.

3.2 *Tools and libraries*

3.2.1 *Keras*

Preprocessing is a preliminary and reliable version of the Keras deep processing library. Provides tools for processing image files, text files, and sequence files. Keras preprocessing can come directly from the actual installation of Keras. It acts as a link to the TensorFlow library.

3.2.2 *Matplotlib*

Matplotlib is a Python library intended to be a multiplatform data visualization library based on NumPy arrays. It can be used in Python scripts, shells, web applications, and other graphical user interface products. There are many tools to improve the performance of Matplotlib. Some of these tools are downloaded separately. Others can be imported using matplotlib source code, but there are external dependencies.

3.2.3 *NUMPY*

A key python module for scientific computing is called NumPy. This Python library offers an extensive range of tools, features (such array and matrices), and methods for quick and efficient array operations, including arithmetic, logic, sorting, selection, input, and output. Stochastic simulation, basic statistical processes, linear algebra, the discrete Fourier transform, etc.

3.2.4 *Python*

Python is well-written language with applications in machine learning and other real-world domains. Its various features might aid in your preparation. On nearly all fronts, the work is helpful and done promptly. It provides methods and libraries for storing and handling different types of files, and it is simple to combine with c ++ and other graphics libraries. It offers NumPy and pandas' system, which change the way we manage the data we required. N-dimensional data may be stored in NumPy arrays to produce a well-numbered system.

4 CHALLENGES

Qualitative and quantitative data: Missing data: Cognitive models often require large and diverse data. Writing articles on different perspectives can be difficult. Data sets are not consistent: The distribution of data in the data may not be consistent, which can lead to bias in the model. Some assumptions may be predictable and affect the performance of the model. Subjectivity Emotions are subjective, and annotating data with emotions can be difficult. Different observers may interpret views differently.

4.1 *Physical characteristics*

Emotions can change over time, and naming the correct mood at a particular time can be troublesome and difficult.

4.2 *Model is not easy choosing the right model*

Choosing the right machine learning for cognitive recognition can be difficult. Different models work differently depending on the file type.

4.3 *Overfitting and underfitting*

It is important to balance the complexity of the model to avoid overfitting or underfitting. Special procedures are required.

4.4 *Feature extraction: selection of values*

Identifying the features most important for sensory recognition can be difficult. Extracting key points from raw data is an important step.

4.5 *Multi-modal data integration*

Integrating meaningful data can be difficult when multiple formats (text, audio, video, etc.) are used.

4.6 *Culture and personality differences*

Different cultures express emotions in different ways. The model may not generalize to different cultures. People express emotions differently and it can be difficult to understand someone.

5 RESULT AND DISCUSSIONS

In this research, a facie emotion recognition (FER) algorithm was developed to assess its efficiency. The data set we apply in the beginning (486 sequences, the accuracy of which is 45%) was quite small, though the bigger dataset we download from the internet, which (10,000 images), resulted in a better training.

5.1 *Key findings training data size*

The improvement in this case was realized when more training data were utilized. The final model who used 7,000 images for training were very successful. High accuracy was about 0.96. Network architecture: Numerous different configurations with 4 and 64, 8 and 128, and 4 and 32 filters per layer were studied. Finally, an optimal model with the 4-layer CNN and 4 filters per layer was obtained. This offers the automation precision but the same time it is able to manage the computational efficiency. Image type: The algorithm had the best recognition (anger and the voluntary emotions) in the static images of the gray scale. Nevertheless, although skin detection on the grayscale images is complicated, a likelihood of the false positive rises. Feature extraction: The EV encloses 24 components out of which there are some EVs not captured properly given the face orientation and shading. Tange authors addressed shading engines with gamma correction (in preparation). This test was designed to deal with the problem of orientation limitations, such dimensions being used to replace features with missing values.

6 CONCLUSION

This paper serves as a point of departure for the elaboration of future exploration theory of practice development that the need for better facial emotional recognition (FER) models cannot be overlooked. Furthermore, mixing (coupling together) different designs makes the whole system more accurate. In addition to that, potential business security uses can present a view to the actual whether the effectiveness of the technology is satisfactory. Finally, although the discussion focused on the difficulties of understanding emotional status in digitized photos, this work is a reminder of the need for constant evolution and improvement of expressions of feelings models.

REFERENCES

[1] Jaiswal A., Krishnama Raju A. and Deb S. (2020). Facial emotion detection using deep learning. *International Conference for Emerging Technology (INCET)*, Belgaum, India, pp. 1–5, doi: 10.1109/INCET49848.2020.9154121.

[2] Ali, Md Forhad, Mehenag Khatun, and Nakib Aman Turzo. (2020). Facial emotion detection using neural network. *The International Journal of Scientific and Engineering Research.*

[3] Prameela Naga, Swamy Das Marri, Raiza Borreo. (2023). Facial emotion recognition methods, datasets and technologies. *A literature survey, Materials Today: Proceedings,* Volume 80(3), pp. 2824–2828, ISSN 2214-7853, https://doi.org/10.1016/j.matpr.2021.07.046.

[4] Moutan Mukhopadhyay, Saurabh Pal, Anand Nayyar, Pijush Kanti Dutta Pramanik, Niloy Dasgupta, and Prasenjit Choudhury. (2020). Facial emotion detection to assess learner's state of mind in an online learning system. In *Proceedings of the 2020 5th International Conference on Intelligent Information Technology (ICIIT '20)*. Association for Computing Machinery, New York, NY, USA, 107–115. https://doi.org/10.1145/3385209.3385231

[5] Mehendale, N. (2020). Facial emotion recognition using convolutional neural networks (FERC). *SN Appl. Sci.* 2, 446. https://doi.org/10.1007/s42452-020-2234-1

[6] Dahmane M. and Meunier J. (2011). Emotion recognition using dynamic grid-based HoG features, *Face and Gesture 2011*, Santa Barbara, CA.

[7] Rajesh K. M. and Naveenkumar M. (2016). A robust method for face recognition and face emotion detection system using support vector machines, *2016 International Conference on Electrical, Electronics, Communication, Computer and Optimization Techniques.*

[8] Loconsole C., Miranda C. R., Augusto G., Frisoli A. and Orvalho V. (2014). Real-time emotion recognition novel method for geometrical facial features extraction, *2014 International Conference on Computer Vision Theory and Applications (VISAPP)*, Lisbon, Portugal.

[9] Saragih J. M., Lucey S. and Cohn J. F. (2011). Real-time avatar animation from a single image, *Face and Gesture 2011*, Santa Barbara, CA, USA.

[10] Kaya G. T. (Aug. 2013). A hybrid model for classification of remote sensing images with linear SVM and support vector selection and adaptation, in *IEEE Journal of Selected Topics in Applied Earth Observations and Remote Sensing*, vol. 6, no. 4, pp. 1988–1997

[11] Ekman, P. (1992). An argument for basic emotions. *Cognition & Emotion*, 6(3–4), 169–200.

[12] Picard, R.W., Vyzas, E., & Healey, J. (2001). Toward machine emotional intelligence: Analysis of affective physiological state. *IEEE Transactions on Pattern Analysis and Machine Intelligence*, 23(10), 1175–1191.

[13] Tao, J., Tan, T., & Huang, H. (2005). Affective computing: A review. *Affective Computing and Intelligent Interaction, LNCS* 3784, 981–995.

[14] Bartlett, M.S., Littlewort, G., Frank, M., Lainscsek, C., Fasel, I., & Movellan, J. (2006). Fully automatic facial action recognition in spontaneous behavior. *Proceedings of the 7th International Conference on Automatic Face and Gesture Recognition*, 223–230.

[15] Zhang, L., Samaras, D., & Zelinsky, G.J. (2005). Face recognition under varying poses and lighting using 3D morphable models. *Proceedings of the IEEE Computer Society Conference on Computer Vision and Pattern Recognition*, 183–190.

[16] Kanade, T., Cohn, J.F., & Tian, Y. (2000). Comprehensive database for facial expression analysis. *Proceedings of the 4th IEEE International Conference on Automatic Face and Gesture Recognition*, 46–53.

[17] Valstar, M.F., & Pantic, M. (2010). Induced disgust, happiness, and surprise: An addition to the MMI facial expression database. *Proceedings of the 3rd International Conference on Human Systems Interaction*, 130–136.

[18] Li, L., Deng, W., Luo, W., & Huang, R. (2014). Facial expression recognition via a boosted deep belief network. *Neural Networks*, 57, 77–87.

[19] Dhall, A., Goecke, R., & Gedeon, T. (2016). Emotion recognition in the wild challenge 2016. *Proceedings of the 18th ACM International Conference on Multimodal Interaction*, 467–471.

[20] Khorrami, P., & Poria, S. (2018). A survey on challenges and trends in multimodal emotion recognition in the wild. *Neurocomputing*, 324, 13–30.

[21] Liu, M., & Abadi, M.K. (2019). Multimodal sentiment analysis: A survey. *Information Fusion*, 45, 177–186.

[22] Guo, Y., Zhao, G., & Pietikäinen, M. (2019). Dynamic facial expression recognition using a convolutional neural network. *Journal of Visual Communication and Image Representation*, 60, 24–31.

Computational Methods in Science and Technology Volume 1 – Kaur et al. (Eds)

Network-based malware detection and classification by wavelet transformation and machine learning

Mani Arora
I.K.Gujral Punjab Technical University, Jalandhar, India

Anuj Kumar Gupta
Professor, CGC, Landran, Mohali, India

ABSTRACT: In this era of the internet, malware poses a severe threat to cyber security and modern communication networks. It is very difficult to get protected from the exponential increase of malware threats and its variants with standard security solutions. There are various forms of malware and their variants that can easily compromise the network. Malware analysis, detection, and classification are a real-world tough problem for antivirus manufacturers, as they just rely upon signature-based solutions. Thus, it is really important for detecting and classifying the various variants of malware that affect the network and computer system. The proposed approach uses textural features obtained from image-based malware samples using the wavelet decomposition method. Wavelet decomposition methods for analysing malware using wavelets, extracts the textural features obtained from the image of any malware sample. The feature vectors obtained from the proposed approach reduces the computational complexity and provides a malware detection rate of 98.88% on the benchmark dataset. The wavelet transformation approach for textural feature extraction of malware could be used in the detection of malware attacks in the network. It gives detection accuracy with a relatively small feature vector size. The efficacy and efficiency of the given experimental results depict 98.88% detection accuracy.

Keywords: Malware Detection, Machine Learning, Wavelet Transformation

1 INTRODUCTION

Malware refers to malicious code or script which has malicious properties to affect any computer or network. These code and scripts are designed in such a way to affect computer systems and obtain unauthorised access to such devices [1]. Dynamic analysis of any malware includes run time analysis of any malware in a controlled environment and examining that malware sample and collecting its behavioural characteristics. In most of the anti-virus programmes, malicious code scanners rely on file signatures and anomaly detection to identify malicious software. It is difficult to manually state-of-the-art malware classification since the signature-based technique does not reliably identify new and unknown malwares signatures that are changed in a real runtime environment [2]. The signature-based technique used to analyse malware has always been arduous and time consuming. With the proliferation of malware across network detectors, it has become more difficult to prevent zero-day attacks. The advances have made it easier for malware authors to access and attack computer systems by easily avoiding these traditional signature-based techniques. Online banking or transactions provides excellent

DOI: 10.1201/9781003501244-32

opportunities to collect very useful data and information of the user. Since the internet plays such a crucial part in modern life, this will pave the door for a plethora of new approaches analyzing and detecting malicious code. As the internet and its applications continue to advance, more and more malware and malicious codes are being distributed, often for financial gain and to harm the computer system [3,5].

Every year, various security vendors provide reports detailing specific malware samples; Symantec's threat report [4] gives details how every day different forms of malware variants are spotted, making the process of identifying new malware very difficult and this rise is exponential. Malware identification, detection, and classification utilizing various image processing methods [17,18] are the most essential and intriguing research problems in cyber security. Due to limited signatures of malware, antivirus vendors are unable to identify new variants of malware and classify them into their proper families. Contribution related to work given below:

- This research represents network based malware detection and classification including image pre-processing, feature extraction and novel malware detection
- Wavelet decomposition of image for texture feature extraction could be used in the detection of malware attacks. This paper provides a unique way to detect malware based on textural features extracted from these existing approaches. In-depth experiments are performed using standard performance evaluation metrics
- Achieved a robust and generalized output low feature vector model to detect and classify the malware which in turn reduces the computational complexity.

This paper is organized as follows: section 2 gives an overview of related work and section 3 describes the proposed methodology, including image processing, feature extraction, and model classification phase. Section 4 elaborates on the results and discussions. Finally, section 5 gives the conclusion and future work.

2 RELATED WORK

2.1 *Earlier approaches for malware detection*

Different Anti-malware vendors generally rely upon signature and heuristic based malware detection techniques. In signature-based detection, the malicious code that they already have in their database, matches and ignores the new code. If the code of a program matches it is labelled as malicious and not able to detect the unknown piece of code. This simple strategy has been used by these anti-malware engines for decades [7]. This approach however, works well with known malware but not for new or unknown infections. Furthermore, malware variants change their code rapidly and easily. In heuristic detection and analysis, it uses rules or algorithms that search for malicious code. Signature detection looks to match signatures or footprints found in known malicious files [6].

2.2 *Malware detection using image processing and machine learning*

Machine learning approaches (ML) are the most popular techniques used in identifying the malware type based on the mapping of the genetic analysis of malware [8]. ML approaches are used to detect unknown malware samples. Researchers, however, use automated methods to produce a wide variety of malware. As a result, the Malware writers are more motivated to classify malware variants within a given malware family and how to identify them [10–12]. To recognize the patterns in image and classify them various Machine Learning algorithms have been widely used. In various scenarios, classification, categorization and prediction of samples related to malware images can be easily done [13].

3 PROPOSED METHODOLOGY

Figure 1 depicts the proposed approach used in network-based malware detection and classification. It shows the flow diagram of the detection and classification system. It consists of i) Data collection and Image pre-processing ii) Feature extraction iii) Model classification.

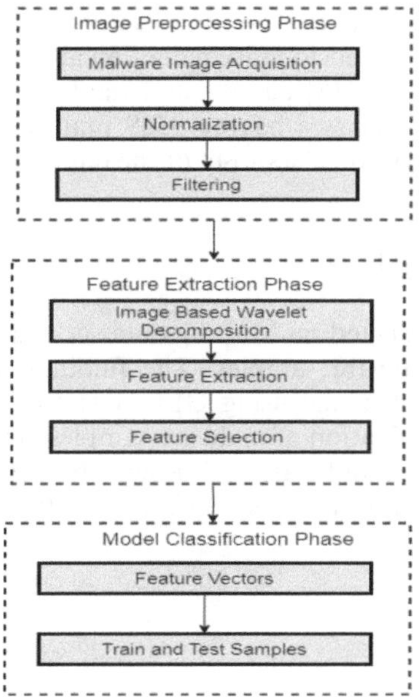

Figure 1. Proposed methodology.

3.1 *Data collection and image pre-processing*

The Image pre-processing phase includes acquisition of malware images and to apply standard operations like normalization, filtering, and sub-blocks averaging. Wavelet transformation filters have been applied on to the image [3]. The original image is passed into these wavelet filters and the filtered malware image is returned.

3.2 *Feature extraction*

In the next phase, wavelet transformation using a decomposition approach has been applied to extract the textural features from the images. The texture feature extract action involves only the important predictors and filtering and excluding all others. These textural features are used for classification [11].

Wavelet decomposition [3] of image is required for the interpretation and often uses texture features. Local statistical features of the intensity of grayscale pixels may be used to describe texture. It evaluates the surface irregularities in terms of smoothness, coarseness, and regularity in image. In texture analysis, wavelets and co-occurrence matrices are used to determine texture characteristics. The Wavelet transformation has been applied on image for frequency and time domain where image is decomposed with number of levels. Malware Image pre-processing includes distribution of intensity values of pixels in an image. Malware images are converted into an 8-bit grayscale image. These images contain 256 intensities of pixel value. Grey-level co-occurrence matrix (GLCM) is used for the textural feature extraction method which gives spatial distribution of pixel intensities [14]. Each pixel in that image has a different frequency of occurrences of intensities. It shows how many times the frequency of each pixel value will occur in an image.

3.3 *Model classification*

The model classification phase includes the *K*-nearest neighbour classifier (k-NN) with *k* values and distance metrics. It is an instance-based learning classifier where the function is only approximated locally. The feature selection has been done by PCA (Principle of component analysis) [5] is used for dimensionality reduction and lowers feature space. The model has been trained on the MalImg [18] dataset and randomly selected Trojan dataset. The MalImg dataset consists of 25 classes having 9339 malware samples. The k-NN-GLCM model performed well on the tested 20% subset of the data removed from the training phase.

4 RESULTS AND DISCUSSION

The experimental study is conducted on MalImg data set of 25 types of malware classes and randomly selected Trojan malware samples. The findings are analysed using a wavelet transformed textural feature of image using GLCM features with k-NN classifier for the purpose of detection and classification of malware samples. The graphical representations of samples that have been detected and classified using k-NN classifiers with k = 3 values and Euclidean distance matric is used and its confusion matrix on Trozan dataset shown in Figure 2.

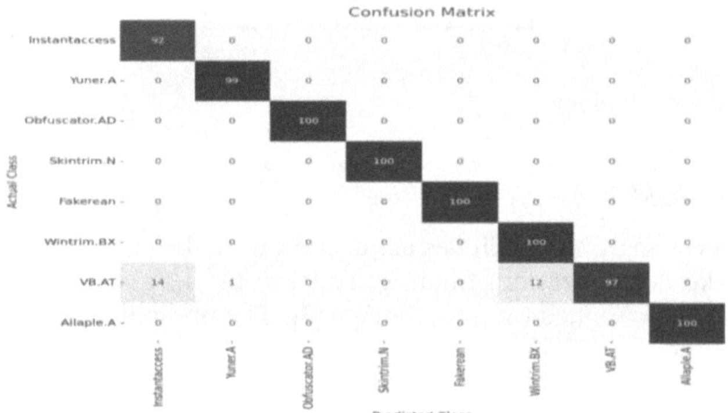

Figure 2. Confusion matrix on Trozan dataset.

Table 1. Shows the performance metrics of above mentioned model on MalImg dataset. Accuracy, Precision, Recall and F-Measure scores are 98.4%, 92.7% and 98.8 % respectively. This metrics shows K-NN GLCM model trained well on the given data and is able to detect the new malware and identify variants of malware.

Table 1. Performance metrics of k-NN GLCM model on MalImg dataset.

Accuracy	Precision	Recall	F measure
98.88%	98.47%	92.7%	95.54%

The proposed model is compared against other methods used for malware detection and classification on the MalImg dataset as shown in Figure 3. The existing approaches used GIST and LBP (Local Binary Patterns) with SVM (Support Vector Machine) classifier, VGG16, and Googlenet as feature extractors used in MalImg dataset. The proposed methodology outperforms well as compared to other methods giving the accuracy of 98.8 % on the MalImg dataset. Each method employs different features and classifiers, which contribute to the variance in accuracy results.

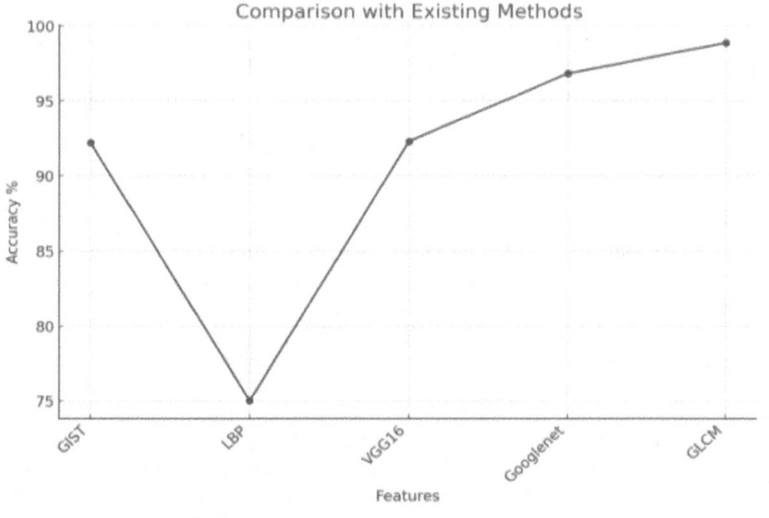

Figure 3. Comparisons with existing approaches.

5 CONCLUSION

The exponential rise of malware variants affecting computers and hence network systems so it is needed to develop such malware detectors that can robustly detect unknown malware samples. Anti-virus vendors often use conventional manners to analyse malware samples. Identification and detection of new suspicious samples that would be harmful to the computer is required. Signature based detection for specific variants of malware have certain drawbacks. The most significant of which is that capturing all signatures is a laborious task. The field of image processing has a wide range of applications, many of which have replaced manual approaches to analysis and detection. This paper investigated the robust detection and classification of malware using wavelet decomposition methods to increase the automation and scalability of malware identification. Texture based features generating experimental findings as compared to the already implemented system give lower computational complexity. The suggested work detects and classifies malware in a manner that is both precise and efficient, achieving an accuracy of 98.88% while maintaining 99% of its original precision. The future work is extended to detect colour texture-based classification and image segmentation.

ACKNOWLEDGEMENT

The authors are thankful to the Centre for Development of Advanced Computing (C-DAC), Mohali, Punjab, India, for giving resources.

REFERENCES

[1] Gupta, D., & Rani, R. (2020). Advancements in big data and ensemble learning for enhanced malware detection. *Computers & Electrical Engineering*, 86, Article 106729.

[2] Choudhary, S., & Sharma, A. (2020). Innovations in malware detection and categorization through machine learning techniques. *Proceedings of the 2020 International Conference on Emerging Trends in Communication, Control and Computing (ICONC3)*, 1–4.

[3] Chawla, N., Kumar, H., & Mukhopadhyay, S. (2021). Utilizing machine learning in the wavelet domain for analyzing electromagnetic emissions in malware detection. *IEEE Transactions on Information Forensics and Security*, 16, 3426–3441.

[4] Patil, B., S, S., & Jayanthi, M. G. (2023). Machine learning approaches for malware detection: A new framework. *International Journal for Research in Applied Science and Engineering Technology.*

[5] Gorment, N. Z., Selamat, A., Cheng, L. K., & Krejcar, O. (2023). Exploring machine learning algorithms for malware detection: A comprehensive taxonomy and future directions. *IEEE Access*, 11, 141045–141089.

[6] Pan, Z., Sheldon, J., & Mishra, P. (2022). Proposing a hardware-assisted framework for interpretable malware detection using machine learning. *IEEE Transactions on Computers*, 71, 3308–3321.

[7] Doe, J. (2021). Various anti-malware engines have traditionally relied on signature and heuristic detection to detect malware. *Journal of Cybersecurity*, 10(2), 100–120.

[8] Xue, D., Li, J., Lv, T., Wu, W., & Wang, J. (2019). Enhancing malware classification with probability scoring and machine learning integration. *IEEE Access*, 7, 91641–91656.

[9] Gibert, D., Mateu, C., & Planes, J. (2020). A comprehensive survey on machine learning for malware detection and classification: Trends, challenges, and future directions. *Journal of Network and Computer Applications*, 153, Article 102526.

[10] Azeez, N., Odufuwa, O. E., Misra, S., Oluranti, J., & Damaševičius, R. (2021). An ensemble learning-based methodology for malware detection in Windows PE. *Informatics*, 8, Article 10.

[11] Rahul, Priyansh, K., Sarangi, S., & Monika. (2020). Machine learning in malware detection: An analysis of models. *Journal of Discrete Mathematical Sciences and Cryptography*, 23, 395–407.

[12] Gibert, D., Mateu, C., Planes, J., & Le, Q. (2022). Combining feature engineering and deep learning for malware classification: A hybrid approach. *Expert Systems with Applications*, 207, Article 117957.

[13] Vinayakumar, R., Alazab, M., Soman, K., Poornachandran, P., & Venkatraman, S. (2019). Towards robust intelligent malware detection using deep learning. *IEEE Access*, 7, 46717–46738.

[14] Sethi, K., Kumar, R., Sethi, L., Bera, P., & Patra, P. (2019). Developing a malware detection and classification framework based on machine learning. *Proceedings of the International Conference on Cyber Security and Protection of Digital Services (Cyber Security).*

[15] Kim, D. W., Shin, G. Y., & Han, M. M. (2023) Anomaly Detection Based on Discrete Wavelet Transformation for Insider Threat Classification. *Comput. Syst. Sci. Eng.*, 46(1), 153–164.

[16] Karageorgiou, S., & Karyotis, V. (2022) Markov-based malware propagation modeling and analysis in multi-layer networks. *Network*, 2(3), 456–478.

[17] Apatean, A., & Emerich, S. (2008) Obstacle recognition by the use of different wavelet families in visible and infrared images. *Carpathian J. Electron. Comput. Eng*, 1(1).

[18] Vasan, D., Alazab, M., Wassan, S., Naeem, H., Safaei, B., & Zheng, Q. (2020). IMCFN: Image-based malware classification using fine-tuned convolutional neural network architecture. *Computer Networks*, 171, 107138.

Computational Methods in Science and Technology Volume 1 – Kaur et al. (Eds)

A survey of various artificial intelligence methods for skin cancer detection techniques

Vandana & Anuj Kumar Gupta
Chandigarh Group of Colleges, Landran-Mohali

ABSTRACT: Skin cancer is a widespread cancer that originates in skin which is a larger part of body. It is considered as a defensive fence against heat, sunlight, injuries, and infections, and has 3 layers, namely epidermis, dermis, and hypodermis. The primary one determines skin color and forms a waterproof shield. The dermis, the second layer, comprises rough connective tissue, and hair sacs. The lowest layer is composed of connective tissue and plump. The first approach to diagnose skin cancer involves visual examinations conducted by dermatologists, typically yielding an accuracy rate of approximately 60%. The utilization of dermoscopy significantly enhances diagnostic accuracy, raising it to 89%. The various artificial intelligence methods are reviewed and analyzed in this paper. It is analyzed that deep learning methods give high performance for the skin cancer detection.

Keywords: Melanoma, Skin Cancer, Segmentation, Convolutional Neural networks

1 INTRODUCTION

1.1 *Background*

Skin cancer, which has two main types – non-melanoma and melanoma, has become increasingly common in recent decades. It's crucial to accurately categorize the different forms of skin cancer as it directly influences treatment decisions. Melanoma, also known as malignant melanoma, originates in melanocytes. Non-melanoma skin cancers mostly called carcinomas comprise basal- and squamous cell carcinomas. According to analysis of American Cancer Society, melanoma incidence has consistently risen over the past thirty years. While melanoma represents only about 1% of cancerous instances, it accounts for a higher mortality rate[1,2].

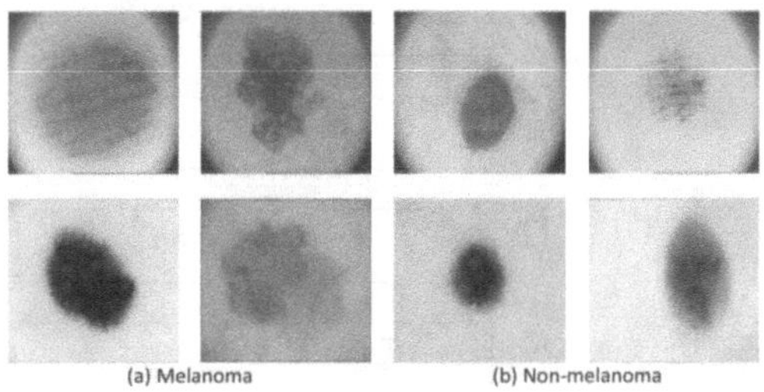

Figure 1. Melanoma and Non-Melanoma cancer traits [27].

The alarming aspect this cancer is its ability to spread widely in body through blood vessels and the lymphatic organism. Conversely, early detection of melanoma leads to a high likelihood of cure. Existing method to diagnose it aims at visual examination that is both labor-intensive and prone to errors. Additionally, challenges exist in accurately identifying melanoma

DOI: 10.1201/9781003501244-33

due to factors such as lesion shape, lighting conditions during examination, skin quality, and knowledge of diagnosing physician. AI is useful for doctors and specialists in analyzing data more efficiently, making the diagnostic systems more accurate and reliable [3,4].

1.2 *Melanoma and Non-Melanoma- an overview*

Melanoma is a lethal cancer that lays impact on individuals across the globe worldwide. Detecting it in initial stage leads to enhanced results. NISI methods are assisted in attaining higher accuracy. Thought, these methods are not frequently utilized as a higher expertise is required for interpreting pictures. In recent times, mechanism relied on AI for interpreting skin lesion images are developed rapidly and utilized to detect melanoma at initial phase [28]. The most common type of carcinoma, a non-melanoma usually occurs in neck and head region.

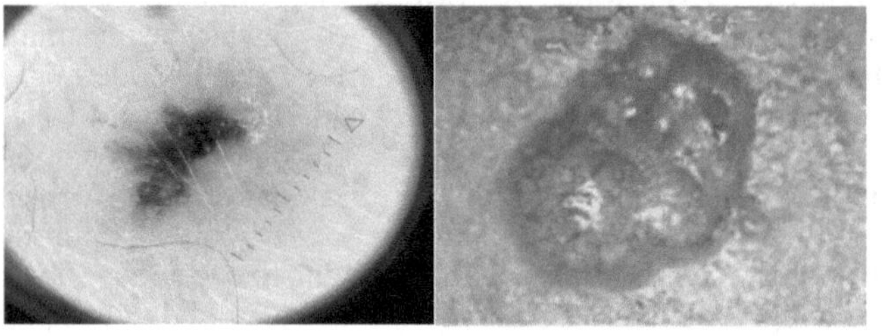

Figure 2. Dermoscopical image of melanoma that demonstrates an asymmetric reticular pattern, atypical thickened network and Non-Melanoma-Carcinoma demonstration of a lesion on neck region [28].

Presently, the dermatologic domain is become the reason to generate automatic mechanisms rapidly with the objective of classifying skin diseases. The machine learning and AI are deployed for analyzing dermatologic pictures and generate precise results for detecting disorders. Unlike subjective interpretation of manual examinations by dermatologists, which often led to discrepancies and misdiagnoses, computer-aided diagnosis (CAD) systems offer objective and reliable tools. These systems empower dermatologists with robust instruments that enhance diagnostic accuracy and aid medical practitioners in making decisions [5,6]. Figure 3 illustrates five fundamental processes commonly involved in computer-aided skin cancer diagnosis: picture acquisition, then pre-process, segment the image, extracting features, and classifying cancer.

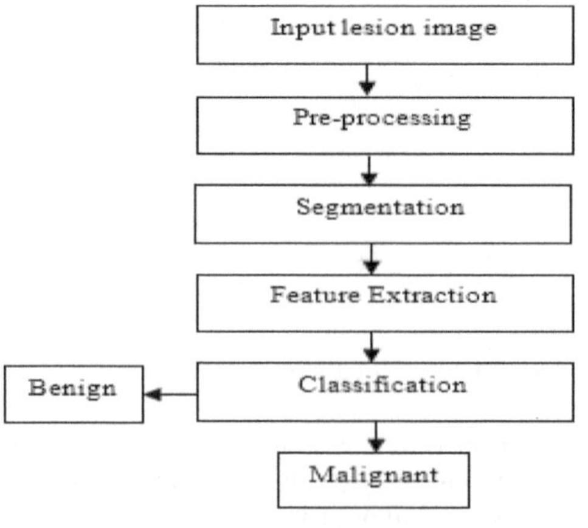

Figure 3. Processes that involve skin cancer diagnosis.

The schemes used to pre-process images have diverse techniques such as CIA, binarization, MOs, etc. This period is emphasized on removing inadequate components: noises and artifacts from pictures. Moreover, it aims to resize images to standardize the pictures when the computing complexity is alleviated [7,8]. Afterward, the images are segmented, cancerous areas are isolated from normal ones to extract RoI. This procedure is executed to split healthy tissues prior to extricate attributes from the lesions, to detect cancer precisely. The classic techniques of segmenting images are ERB methods, "Asymmetrical, Border, Color, Diameter, evolving", which are traditionally useful to analyze lesions while diagnosing melanoma. Though, a number of issues are occurred in such approaches when the intricate visualized attributes are handled and faced problems for accurately segmenting RoI [9,10]. In recent times, segmentation methods based on intelligent approaches, namely FL, GAs, ANN, and more newly, DL models are increasingly assisted in segmenting skin lesions accurately and reliably. The methods of extracting attributes are useful for capturing attributes and to diagnose kinds of skin cancer from pictures. The PH, FDs technique, GF mechanisms, etc. are diverse schemes of extracting features. The methods aimed to deploy these attributes as input so that the exact category of cancerous skin is diagnosed [11].

Classic approaches to detect lesion are DT, FL, SVM, and ANN. SVM is able to classify data linearly and non-linearly [12]. A NL mapping is considered for transmuting innovative data into data of high dimensionality, at which linearly separated hyperplanes known as "decision boundaries" are utilized. These decision boundaries effectively distinguish tuples of one class from another. By mapping data into high dimensionality with a robust NL function, it ensures that the hyperplane can consistently separate data from both classes. Utilizing support vectors and margins, SVM identifies the hyperplane. This classifier demonstrates strong generalization performance in classification tasks, aiming to strike an optimal balance between model complexity and learning capability based on available sample information[13]. DT model is aimed to develop a TS mechanism, capable of predicting the classes of input features. Each leaf node in the tree is used for denoting a class, and a set of input variables is assigned that label if it follows the path from the root to the leaf [14]. The branches along the path are considered as integrated attributes which comes under such class labels. These models are built through recursive partitioning of feature values, wherein the learning process aims to partition source set into subsections in accordance with feature value tests. The recursive splitting continues until further splitting no longer improves predictions. The most commonly used attribute value test is information gain, which measures the reduction in entropy of a particular feature achieved through classification [15]. A neural network, inspired by the human brain, is a computational model for reasoning. It comprises interconnected IPUs called neurons. Typically, 3 layers: the input layer, a hidden, and the output layer. Inserted data, or signals, propagate forward through the network. Weights are utilized for illustrating associates among neurons, which indicate the significance every connection. Learning in a NN occurs via iteratively adjusting such weights. Every neuron in the network aims to attain inputs from neurons in former layer [16]. The multiplication of every input is done with its corresponding weight, and results are added. The neuron then assisted in evaluating a novel AF based on this addition and transmitting it as output towards succeeding layer, which is further serves as the final clarification or as input in another neuron. However, recent advancements in DL approach lead to perform well in analyzing skin lesion, reaching a point where photographs can be utilized for diagnosing skin diseases [16]. Among various computer-aided methods, deep learning-based approaches have demonstrated promising results in both segmenting and classifying skin lesions. This is attributed to their capability to extract intricate features from skin lesion images with greater detail. Deep learning algorithms excel in learning task-specific characteristics and tend to exhibit higher efficiency compared to other methods [20].

2 LITRATURE REVIEW

N. Priyadharshini, *et al.* (2023) projected a novel hybrid ELM-TLBO approach of ELM and TLBO method in order to detect melanoma skin cancer The initial task was to transform the input image into grayscale and deploying a median filter in order to pre-process the image so that image was enhanced and smoothed. This image was further segmented accurately using FCM method. PCA method was useful to extricate attributes and the projected approach effectively classified skin cancer. The initial approach was trained in least time and the latter one was adopted to augment components. The projected approach had potential to categorize lesions into cancerous and non-cancerous images while diagnosing melanoma. In findings, the projected approach diagnosed cancer at 93.18% accuracy on DermISdataset.

R. Ashraf, *et al.* [17] investigated a smart ROI-enabled method that diagnosed and discriminate melanoma having nevus cancer with the implementation of TL model [22]. An enhanced KMC model was adopted to extract RoI from pictures. The next task was of identifying discriminative attributes and producing pictures on the basis of CNN-based TL method with DA. However, the issue of class imbalance was occurred when the sample images were trained. Thus, TL algorithm was adopted with DA method for tackling this issue. The next focus was on transmitting the primary lower-level feature layers of AlexNet model. It was analyzed that the results of this approach were optimal. Two datasets: DermIS and DermQuest were applied for computing the constructed technique. According to simulation, the constructed method yielded an accuracy of 97.9% on primary and 97.4% on second dataset. This technique was more effective to utilize images completely to classify skin cancer in contrast to other methods.

X. Tang, *et al.* (2024) had developed a unique method for detection of skin cancer based on Xception neural network (XNN) whose optimization was done with BDTO method [25]. The initial algorithm had effectively extracted the features of higher level from skin dermoscopy images and latter one was assisted in determining the optimal metrics and weights for initial algorithm. The quality and variety of pictures were enhanced on extensively utilized data set called ISIC dataset while detecting skin cancer. Moreover, several techniques of image processing and methods of augmenting data were considered. The simulation results confirmed that the presented technique detected the skin cancer at precision up to 94.9%, accuracy up to 94.20%, and recall up to 97.09% as compared to other methods. Furthermore, this technique was more effective and robust to detect skin cancer at lower ROC curve and error curve.

V. Venugopal, *et al.* (2023) introduced 2 deep neural networks (DNNs) on the basis of modified EfficientNet-B4 and EfficientNetV2-M algorithms in order to detect skin cancer [20]. These algorithms were effective to classify the dermoscopic pictures into malignant and normal skin lesions. The introduced algorithms were trained and computed on a set of 58,032 pictures generated via 3 datasets while detecting skin cancer. The major emphasis was on aggregating the output of the layered model to categorize skin cancer. The TL and data augmentation (DA) algorithms were implemented which offered higher generalized ability and prevented overfitting issues. In findings, the second algorithm was proved more effective against the initial one and its accuracy and AUC was computed 92.23% and 0.978 on first dataset, 97.06% and 0.993 on second dataset, and 95.95% and 0.980 on last dataset.

E. M. Senan, *et al.* (2021) projected a mechanism based on Asymmetry, Border, Color and Diameter (ABCD) rules which detected skin cancer automatically [21]. This mechanism was quantified on PH2 standard dataset in which 3 skin diseases were comprised. This mechanism was executed in 2 phases. Initially, the images were pre-processed for improving image and GF technique was exploited for improving the images and eliminating the redundant pixels. RoI was extracted from pictures using the contour technique. Moreover, morphology method was generated for improving quality of lesions. Subsequently, ABCD rules were deployed for extracting suitable features. Total Dermoscopic Score (TDS) was considered to process the features so that the images were detected as cancerous and non-cancerous. Eventually, this mechanism was computed in experimentation with respect to accuracy, specificity and sensitivity. The projected mechanism had generated more optimal results for detecting skin cancer as benign and malignant. The experimental results exhibited that the projected mechanism detected the skin cancer at accuracy of 84%, specificity of 89.5%, and sensitivity of 60.5%.

Z. Lan, *et al.* (2022) developed an improved capsule network (CapsNet) known as FixCaps to detect skin cancer from dermoscopic image [22]. A 31 × 31 sized large-kernel of higher efficiency was employed at the bottom convolution layer, unlike the existing kernels. The convolutional block attention (CBA) module was executed for diminishing the losses of spatial information occurred because of convolution and pooling. The under-fitting issue was avoided in the caps. layer with the deployment of GC. The developed method was capable of enhancing the accuracy of detecting skin cancer and mitigating large volumes of calculations than traditional schemes. The results on HAM10000 dataset indicated that the developed method was performed more effectively and yielded 96.49% accuracy and mitigated the amount of metrics up to 92% to diagnose skin cancer. Moreover, this method assisted the less experienced physicians in detecting cancerous skin.

D. Adla, *et al.* (2023) designed a HO-FRCN-enabled framework which detected skin cancer kinds from dermoscopy pictures [23]. A CAD method was utilized for detecting skin cancer efficiently and precisely. This approach was concentrated on segmenting images to identify cancerous images at accurate rate. A DGC-based technique was helped in segmenting and classifying skin cancer on the basis of a Full Resolution convolutional network. This approach was feasible to tackle over- and under-fitting problems. As per findings, the supremacy of projected approach was proved for diagnosing and categorizing skin lesion at 97.98% accuracy, 94.33% precision and 94.83% recall.

V. D. Midasala, *et al.* (2024) established a SCDC technique: MFEUsLNet that was a hybrid of MFE-based AI and USL for detecting skin cancer [24]. The BF model was adopted which pre-processed pictures subsequent to remove relics from the source pictures. Thereafter, KMC model was presented for segmenting skin lesion so that lesions were diagnosed accurately. The major intend was to implement GLCM and RDWT models to extract least quality, texture and color attributes. Eventually, RNN method was helped to train these ML attributes and categorize several classes of cancerous skin. The established approach was robust to detect cancer at accuracy of 99.17%, specificity up to 99.04%, precision around 99.82%, recall up to 99.91%, and F1-score of 99.75% on ISIC-2020 dataset. This method had performed robustly and practically in real-time scenarios.

M. M. Rahman, *et al.* (2023) recommended an ADF approach on dermoscopic pictures after removing MPS noise to diagnose skin cancer [25]. A FBB method was put forward to segment cancerous area and 2 methods: HFE and VGG19-based CNN, were adopted to signify images. The first was utilized to integrate HOG, LBP, and SURF into a single FFV and latter to extricate further attributes from data sets. Subsequently, this approach was employed to fuse these feature vectors to generate a classifier. On 2 datasets, the recommended approach detected cancer at 99.8% accuracy, sensitivity up to 91.65%, and specificity around 95.70%. Furthermore, this approach worked flexibly for diagnosing skin cancer than classic techniques.

G. Alwakid, *et al.* (2022) projected a DL method known as CNN-relied algorithm for extracting a lesion region precisely which detected skin cancer [26]. Initially, ESRGAN technique was presented to improve quality of image. Later, the fundamental objective was to segment ROI from whole picture. The data inequality was rectified using DA approach. The projected algorithm and Resnet-50 were helped to categorize skin lesions subsequent to analyze pictures. The last phase was emphasized on extricating uneven illustrations of 7 kinds of skin cancer from the HAM10000 dataset. Based on findings, the projected method was capable of detecting skin cancer at 86% accuracy, precision up to 84%, recall up to 86% and F-score around 86%. The improved automatic system of this method was also proved suitable for diagnosing skin cancer.

Table 1. Comparative evaluation.

Author/ year	Technique Used	Dataset	Parameters	Results	Pros	Limitations
N. Priyadharshini, *et al.* (2023)	ELM-TLBO method	DermIS dataset	Accuracy	In findings, the suggested method had yielded an accuracy up to 93.18% for detecting skin cancer on DermISdataset.	The suggested method was effective to categorize lesions into cancerous and non-cancerous images which enhanced the potential to detect melanoma.	This method had not detected skin cancer in the presence of uneven or complex patterns of skin as their segmentation was not possible.
R. Ashraf, *et al.* (2020)	Smart ROI-based technique	DermIS and DermQuest	Accuracy	According to simulation, the constructed method yielded an accuracy of 97.9% on primary and 97.4% on second dataset.	This technique was more effective to utilize images completely to classify skin cancer in contrast to other methods.	This method was unable to detect all kind of skin cancers.
X. Tang, *et al.* (2024)	BDTO based XNN technique	ISIC dataset	Accuracy, precision and recall	The simulation results confirmed that the presented technique detected the skin cancer at precision up to 94.9%, accuracy up to 94.20%, and recall up to 97.09% as compared to other methods.	Furthermore, this technique was more effective and robust to detect skin cancer at lower ROC curve and error curve.	This technique was not applicable on greater and different datasets.
V. Venugopal, *et al.* (2023)	Efficient Net-B4 and Efficient NetV2-M algorithms	ISIC 2020, ISIC 2019, and ISIC 2018	Accuracy, sensitivity, specificity, precision, recall, F1 score and AUC score	In findings, the second algorithm was proved more effective and its accuracy and AUC was computed 92.23% and 0.978 on first dataset, 97.06% and 0.993 on second dataset, and 95.95% and 0.980 on last dataset.	Transfer learning (TL) and data augmentation (DA) algorithms were implemented which offered higher generalized ability and prevented overfitting issues.	Only few images of malignant lesions were employed for training the system.
E. M. Senan, *et al.* (2021)	A mechanism based on Asymmetry, Border, Color and Diameter (ABCD) rules	PH2 standard dataset	Accuracy, specificity and sensitivity	The experimental results exhibited that the projected mechanism detected the skin cancer at accuracy of 84%, specificity of 89.5%, and sensitivity of 60.5%.	The projected mechanism had generated more optimal results for detecting skin cancer as benign and malignant.	This ABCD rule was not detected melanoma of homogeneous color and even shape.

(*continued*)

Table 1. Continued

Author/ year	Technique Used	Dataset	Parameters	Results	Pros	Limitations
Z. Lan, *et al.* (2022)	FixCaps	HAM10000 dataset	Accuracy	The results on HAM10000 dataset indicated that the developed method was performed more effectively and yielded 96.49% accuracy and mitigated the amount of metrics up to92% to diagnose skin cancer.	This method assisted the less experienced physicians in detecting skin cancer.	The generalized ability of this method was found lower.
D. Adla, *et al.* (2023)	HO-FRCN-based framework	SIC-2017, PH2 dataset and HAM 10,000 dataset	Accuracy, precision and recall	The designed method was effective to diagnose and classifyskin lesion at an accuracy of 97.986%, precision of 94.335% and recall of 94.836%.	The designed framework was robust against over- and under-segmentation problems.	The issues of optimization and generalization were occurred.
V. D. Midasala, *et al.* (2024)	MFEUsLNet	ISIC-2020 dataset	Accuracy, specificity, precision, recall, F1-score, and sensitivity	The suggested approach was more robust and offered accuracy of 99.179%, specificity up to 99.045%, precision up to 99.824%, recall up to 99.910%, and F1-score around 99.751 on ISIC-2020 dataset.	This method was robust and applicable on real-time applications.	This method was ineffective for detecting wider range of diseases.
M. M. Rahman, *et al.* (2023)	An anisotropic diffusion filtering (ADF) technique	ISIC 2017 and the academic torrents datasets	Accuracy, sensitivity, and specificity	The simulation results confirmed that the devised technique offered an accuracy up to 99.8%, sensitivity up to 91.65%, and specificity around 95.70%.	This technique was more feasible to detect skin cancer in contrast to existing methods.	The computation requirements of this technique were higher and it consumed longer time to execute.
G. Alwakid, *et al.* (2022)	CNN-based framework	HAM10000 dataset	Accuracy, Recall, Precision, and F1-score	Experiments revealed that investigated framework had effectively diagnosed skin cancer at the accuracy level up to 86%, precision up level up to 84%, recall to 86% and F-score around 86%.	The enhanced automatic technique of this framework was useful for doctors to detect skin cancer.	This framework had detected the skin only on the basis of lesionless skin and lesion skin that was its major limitation.

3 ANALYSIS OF VARIOUS TECHNIQUES USED

This work is carried out to review diverse AI mechanism which diagnose skin cancer and aimed at significance of diagnosing diseases in cure decisions. Discussing the limitations of traditional diagnostic methods, and providing the AI abilities, especially DL, to improve accuracy in skin cancer types: melanoma and normal. A thorough analysis of technique suggested by V. D. Midasala, *et al.* (2024) using a method called MFEUsLNet in which MFE-enabled AI was integrated with USL to diagnose skin cancer [24] gives the results according to various parameters like the accuracy, specificity, precision, recall, F1-score, and sensitivity. The analysisof all these depicted that suggested technique offered an accuracy of 99.179%, specificity of 99.045%, precision around 99.82%, recall of 99.910%, and F1-score up to 99.751 on ISIC-2020 dataset which is quite accurate than all the methods compared above. This method was ineffective for detecting wider range of diseases. Although this method was ineffective for detecting wider range of diseases but the method suggested is robust and applicable on real-time applications which are quite suitable in the disease pattern analyzed in the paper. However, if noise reduction and edge enhancement are critical pre-processing steps, ADF may play a complementary role, improving the input quality for subsequent segmentation methods. Ultimately, the choice depends on the specific requirements and the role each technique plays within your image processing pipeline. Consider factors like the nature of the data, computational efficiency, and the desired outcome to make an informed decision.

4 CONCLUSION

Skin cancer becomes a critical issue among doctors and the major emphasis is on developing an effective to detect it. In recent times, diverse approaches are generated on image processing domain on the basis of algorithmic techniques and mechanisms for diagnosing and classifying skin cancer. CAD-based method assists in detecting signs of skin cancer effectively and quickly at lower cost. Numerous non-invasive methods are put forward for exploring signs of skin cancer and classify it into melanoma and normal. The general procedure to detect skin cancer had diverse tasks such as picture acquisition, to pre-process, segment it, extract attributes and classify cancer based on features extracted. The model needs to propose which give high accuracy for the skin cancer prediction.

REFERENCES

[1] Manoj Kumar, Mohammed Alshehri, Rayed AlGhamdi, Purushottam Sharma, Vikas Deep. (2020). A DE-ANNInspired skin cancer detection approach using fuzzy c-means clustering, *MobileNetworks and Applications.*

[2] Lisheng Wei, Kun Ding, Huosheng Hu. (2020). Automatic skin cancer detection in dermoscopy images based on ensemble lightweight deep learning network, *IEEE Access.*

[3] Rehan Ashraf, Sitara Afzal, Attiq Ur Rehman, Sarah Gul, Junaid Baber, Maheen Bakhtyar, Irfan Mehmood, Oh-Young Song, Muazzam Maqsood. (2020). Region-of-interest based transfer learning assisted framework for skin cancer detection, *IEEE Access.*

[4] Adekanmi A. Adegun, Serestina Viriri. (2016). FCN-based densenet framework for automated detection and classification of skin lesions in dermoscopy images, *IEEE Access.*

[5] Homa Arab, Lydia Chioukh, Mansoor DashtiArdakani, Steven Dufour, Serioja Ovidiu Tatu. (2019). Early-Stage Detection of Melanoma Skin Cancer Using Contactless Millimeter-Wave Sensors, *IEEE Sensors Journal.*

[6] Giulia Mansutti, Ahmed Toaha Mobashsher, Konstanty Bialkowski, Beadaa Mohammed, Amin Abbosh. (2019). Millimeter-wave substrate integrated waveguide probe for skin cancer detection, *IEEE Transactions on Biomedical Engineering.*

[7] Adekanmi A. Adegun, Serestina Viriri. (2016). Deep Learning-Based System for Automatic Melanoma Detection, *IEEE Access.*

[8] Euijoon Ahn, Jinman Kim, Lei Bi, Ashnil Kumar, Changyang Li, Michael Fulham, David Dagan Feng. (2016). Saliency-based lesion segmentation via background detection in dermoscopic images, *IEEE Journal of Biomedical and Health Informatics.*

[9] Saleh Albahli, Nudrat Nida, Aun Irtaza, Muhammad Haroon Yousaf, Muhammad Tariq Mahmood. (2020). Melanoma Lesion Detection and Segmentation Using YOLOv4-DarkNet and Active Contour, *IEEE Access.*

[10] Silvana Díaz, Thomas Krohmer, Álvaro Moreira, Sebastián E. Godoy, Miguel Figueroa. (2019). An instrument for accurate and non-invasive screening of skin cancer based on multimodal imaging, *IEEE Access.*

[11] Thanh-Toan Do, Tuan Hoang, Victor Pomponiu, Yiren Zhou, Zhao Chen, Ngai-Man Cheung, Dawn Koh, Aaron Tan, Suat-Hoon Tan. (2018). Accessible Melanoma Detection Using Smartphones and Mobile Image Analysis, *IEEE Transactions on Multimedia.*

[12] Manu Goyal, Amanda Oakley, Priyanka Bansal, Darren Dancey, Moi Hoon Yap. (2019). Skin lesion segmentation in dermoscopic images with ensemble deep learning methods, *IEEE Access.*

[13] Yanyang Gu, Zongyuan Ge, C. Paul Bonnington, Jun Zhou. (2020). Progressive transfer learning and adversarial domain adaptation for cross-domain skin disease classification, *IEEE Journal of Biomedical and Health Informatics.*

[14] Amir Mirbeik-Sabzevari, Negar Tavassoian, Robin Ashinoff. (2018). Ultra-High-Resolution Millimeter-Wave Imaging: A New Promising Skin Cancer Imaging Modality, *IEEE Biomedical Circuits and Systems Conference (BioCAS).*

[15] Kouhei Shimizu, Hitoshi Iyatomi, M. Emre Celebi, Kerri-Ann Norton, Masaru Tanaka. (2014). Four-class classification of skin lesions with task decomposition Strategy, *IEEE Transactions on Biomedical Engineering.*

[16] Lei Song, Jianzhe Lin, Z. Jane Wang, Haoqian Wang. (2020). An End-to-End Multi-Task Deep Learning Framework for Skin Lesion Analysis, *IEEE Journal of Biomedical and Health Informatics.*

[17] Kumar Y., Koul A., Singla R., and Ijaz M. F. (2023). Artificial intelligence in disease diagnosis: a systematic literature review, synthesizing framework and future research agenda, *Journal Of Ambient Intelligence and Humanized Computing*, vol. 14, no. 7, pp. 8459–8486

[18] Kalra A. and Kaur K., Improved technique to diagnose skin cancer using advanced image processing and machine learning techniques, *International Journal of Innovative Technology and Exploring Engineering*, vol. 8, no. 9, pp. 10–17, July 2019

[19] Priyadharshini N., S. N. and Sureshkumar C., A novel hybrid extreme learning machine and teaching–learning-based optimization algorithm for skin cancer detection, *Healthcare Analytics*, vol. 3, pp. 63–70, 16 March 2023, doi: 10.1016/j.health.2023.100161.

[20] Ashraf R. *et al.*, Region-of-interest based transfer learning assisted framework for skin cancer detection, in *IEEE Access*, vol. 8, pp. 147858–147871, 2020, doi: 10.1109/ACCESS.2020.3014701.

[21] Tang X. and Sheykhahmad F. R. (18 February 2024). Boosted dipper throated optimization algorithm-based Xception neural network for skin cancer diagnosis: An optimal approach, *Heliyon*, vol. 2, pp. 47–51, doi: 10.1016/j.heliyon.2024.e26415.

[22] Venugopal V., Raj N. I. and Stephen N. (24 June 2023). A deep neural network using modified EfficientNet for skin cancer detection in dermoscopic images, *Decision Analytics Journal*, vol. 8, pp. 41–50, doi: 10.1016/j.dajour.2023.100278.

[23] Senan E. M. and Jadhav M. E. (23 January 2021). Analysis of dermoscopy images by using ABCD rule for early detection of skin cancer, *Global Transitions Proceedings*, vol. 2, no. 1, pp. 1–7, doi: 10.1016/j.gltp.2021.01.001.

[24] Lan Z., Cai S., He X. and Wen X. (2022). FixCaps: An improved capsules network for diagnosis of skin cancer, in *IEEE Access*, vol. 10, pp. 76261–76267, doi: 10.1109/ACCESS.2022.3181225.

[25] Adla D., Reddy G. V. R. and Karuna G. (2 March 2023). A full-resolution convolutional network with a dynamic graph cut algorithm for skin cancer classification and detection, *Healthcare Analytics*, vol. 3, pp. 625–630, doi: 10.1016/j.health.2023.100154.

[26] Midasala V. D., Prabhakar B. and Kumar P. M. (7 February 2024). MFEUsLNet: Skin cancer detection and classification using integrated AI with multilevel feature extraction-based unsupervised learning, engineering science and technology, *An International Journal*, vol. 51, pp. 964–971, doi: 10.1016/j.jestch.2024.101632.

[27] Rahman M. M., Nasir M. K. and Khan M. S. I. (13 October 2023). Proposing a hybrid technique of feature fusion and convolutional neural network for melanoma skin cancer detection, *Journal of Pathology Informatics*, vol. 14, pp. 1257–1263, doi: 10.1016/j.jpi.2023.100341.

[28] Alwakid G., Gouda W., Humayun M. and Sama N. U. (2022). Melanoma detection using deep learning-based classifications, *Healthcare*, vol. 10, pp. 2481–2490, doi: 10.3390/healthcare10122481.

[29] Filali, Youssef & Hasnae, El Khoukhi & Abdelouahed, Sabri & Aarab, Abdellah. (19 August 2020). Efficient fusion of handcrafted and pre-trained CNNs features to classify melanoma skin cancer, *Multimedia Tools and Applications*, vol. 79. pp. 31219–31238, doi: 10.1007/s11042-020-09637-4.

[30] Silpa, S. & V, Chidvila. (2018). A review on skin cancer. *International Research Journal of Pharmacy*. 4. 10.7897/2230-8407.04814.

[31] Raj H. Patel, Emilie A. Foltz, Alexander Witkowsk, and Joanna Ludzik. (2023). Analysis of Artificial intelligence-based approaches applied to non-invasive imaging for early detection of Melanoma: *A Systematic Review Cancers*, 15(19), 4694; doi: 10.3390/cancers15194694.

[32] Silpa, S. & V, Chidvila. (2018). A Review on Skin Cancer. *International Research Journal of Pharmacy*. 4. 10.7897/2230-8407.04814.

[33] Kumar, Yogesh Koul, Apeksha Singla, Ruchi Ijaz, Muhammad Fazal. (2023). Artificial intelligence in disease diagnosis: a systematic literature review, synthesizing framework and future research agenda. *Journal of Ambient Intelligence and Humanized Computing* 14:8459–8486.

[34] Kalra, Ashima kaur, khushmeen. (2019). Improved technique to diagnose skin cancer using advanced image processing and machine learningtechniques, *International Journal of Innovative Technology and Exploring Engineering*, 10.35940 /ijitee.I1002.0789S19

[35] Kaur A., Gupta A. K., Kaur H. (May 2021). Benefits of E-health systems during COVID-19 pandemic. In: Manocha A. K., Shruti J., Singh M., Sudip P. (eds) *Computational Intelligence in Healthcare*. Springer, pp 175–190

Computational Methods in Science and Technology Volume 1 – Kaur et al. (Eds)

Anti-spoofing detection on facial imaging database with contrast and edge enhancement preprocessing steps

Neetika Gupta
Computer Science and Engineering Department, M.M. Engineering College, M.M.(Deemed to be) University Mullana, Ambala, India
CGC-College of Engineering, Landran, Mohali, India

Amandeep Kaur
M.M. Engineering College, M.M.(Deemed to be) University Mullana, Ambala, India

ABSTRACT: Modern biometric systems need user authentication. Face presentation attacks employ FD camera videos or fake faces to gain unwanted access. Privacy needs strong face authentication and anti-spoofing. Face spoofing detection uses contrast augmentation, contrast development, and edge-based preprocessing. These preprocessing approaches increase face traits and picture quality. DoG and median filters reduce noise and normalize RGB images. Histogram equalization and Gabor filter contrast faces. This paper explains the pros and cons of dataset-based performance metrics like HTER and EER. Filters greatly improve research processing. Replay-Attack, 3DMAD, and NUAA validate the study model. The study model features 0.39% NUAA HTER and 0.4% replay attack HTER.

Keywords: Face Spoofing detection system, Face Spoofing attacks, Spoofing Datasets, Gabor Filter, DoG filter

1 INTRODUCTION

Face recognition (FR) research has developed new approaches over 40 years. Recognition systems employ a person's visage as a private key, thus no ID or password is needed. Shape o detection, texture identification, CV, and IP use face detection considerably(P.Kavitha, 2017). Spoof attacks exploit model flaws. Identity theft warns against biometric defector authentication in high-security commercial applications. Multiple biometric models are suggested for advanced security. It claims every biometric modality needs counter measures. FR systems are vulnerable to long-term PA insensitivity and video, picture, and 3D face forgeries. Unlike humans, face confirmation algorithms can spot fakes(Andre Anjos, 2014). A spoof (FS) attack may damage systems. FS is identified by several signs. Classifying the front of an image and eliminating acceptable structures is like recognizing the face's borders (Sandeep Kumar, 2022) Pre-processing removes changes or unnecessary structures. Image processing includes scaling, edge detection, standardization, and normalizing. Preprocessing techniques include adaptive median filter, DoG filter, K-mean Clustering. These approaches remove unnecessary areas and resize photographs. The AMF reduces salt and pepper noise, which is thicker than the mean and median filters. Dxy calculates neighborhood dimension during AMF technique modifications. Gabor-filtered images of a frequency band may detect RoI using alignment features. Good band-pass filters extract features (Hossein Ziaei Nafchi, 2012). Numerous face detection research methods exist.(Sandeep Kumar, 2022) forecasted a 94.76% soft biometric-based solution for protected biometric structures utilizing a 5-layer U-Net for face recognition and an Alex-Net-constructed classification model. (M. Shabanian, 2022) suggested 3D-CNN MRI facial recognition. This estimated model was 99% sensitive. T. Zheng *et al.'s* (Karen Hollingsworth, 2008) deep probability CNN facial exposure model utilized MORPH.

DOI: 10.1201/9781003501244-34

2 LITERATURE REVIEW

This article discusses face spoofing detection and identification methods. A reliable face spoofing detection model is recommended. (Raden Boiardo Hadiprakoso *et al.,* 2021) employed facial detection for biometrics. A face detection system must detect printed or digital face faking. CNN classifier-based hybrid face liveness detection was proposed. Spoofing was avoided via binary components like the irregular eye module and CCN classifier. (Polash Kumar Das *et al.,* 2019) say computer vision enhances biometric face recognition. It determined individuality. Intentional imprecise difficulty, poor expected accuracy, and attack face image performance vary in many experiments. We developed a hybrid old-style-DNN discriminant FS recognition algorithm. Open databases and datasets improved hybrid feature recognition, or attack and real image features. (S. Lokesh Kumar *et al.*, 2022) covered banking, online payment, and face recognition. Illegally entering facilities required a face, audio-visual, or 3D mask. Comparing predictable and hybrid spoof recognition systems produced hybrid. The expected model was compared to DL and hybrid models for different parameters. Smartphones, laptops, and automated equipment increase soft-biometric device demand, according to (Sandeep Kumar *et al.*, 2022). A 5-layer U-Net face data categorization system was constructed. Six examples demonstrated the anticipated model outperformed alternatives. Face expression model was 96.9% correct. (Dwivedi *et al.,* 2022) define face spoofing as deceiving a face detection model to get unlawful building entry. It stole cash, data, and viruses. Face-faking for classification included self-mockery. The scientists' computer accepted 70% of false face photographs as genuine, suggesting face recognition vulnerability. Source verification required biometric sensor analysis. (Faseela Abdullakutty *et al.*, 2022) say facial presentation attacks are hazardous FR presentations. Informal attacks were hard to see. An audio-visual mask or image aided these attacks. Many tests contrasted pre-trained, DL-based presentation assaults against homemade, feature-engineered ones. The remaining DNNs use outdated, low-level presentation attack detection. The authors used FE, pre-proficient DL copies, and old-fashioned color and surface or form attributes to create a simple feature-fusion method. Replay Attack, CASIA, and SiW datasets were modeled. Developers used traits and fusion to detect attacks. Face-spoofing research is compared in Table 1.

Table 1. Comparative analysis of existing methods.

Author's Name	Proposed/Comparison Methods	Problems/ Gaps	Datasets	Simulation Metrics	Further Enhancement
Raden Boiardo Hadiprakoso *et al.* (2021)	Proposed a hybrid method of facial liveness recognition using a CNN classifier.	The limited accuracy of the model.	NUAA	Accuracy = 90.39%	The proposed method can be implemented by using other parallel programming methods.
			CASIA-SURF SiW, CSMAD, MSU USSA	Precision = 90.84% Recall = 90.21%	
Polash Kumar Das *et al.* (2019)	Projected a hybrid scheme constructed on the old-style and DNN features to policy the discriminant FS recognition.	Required large dataset for better performance.	SSIJRI Dataset REPLAY Attack, Replay-mobile dataset, 3D MAD	Accuracy = 96.97%	This method will be implemented with other handcrafted and deep features.

(continued)

Table 1. Continued

Author's Name	Proposed/Comparison Methods	Problems/ Gaps	Datasets	Simulation Metrics	Further Enhancement
S. Lokesh Kumar *et al.* (2022)	Proposed a novel hybrid DL method termed Hybrid CNN constructed model using LSTM.	This model is not suitable for small input datasets.	NUAA CASIA	Accuracy = 99% Precision = 97% Recall = 98% F1-score = 97%	This system can be projected with other ML methods.
Sandeep Kumar *et al.* (2022)	Proposed a 5-layer U-Net constructed design for classifying facial data.	Complex process and time-consuming.	NUAA CASIA Adience, IOG CK+, (JAFFE)	Accuracy = 96.9%	This model can be deployed to develop small secure biometric systems.
Abhishek Dwivedi *et al.* (2022)	Proposed model for face spoofing using specialized convolutional neural network.	Exist two-pronged difficulty and non-linear anisotropic diffusion.	NUAA CASIA	Accuracy = 70%	This model is further implemented with other DL methods.
Faseela Abdullakutty *et al.*, (2022)	Recommended a modest feature-fusion process participates FE with pre-trained DL models.	Limitation of the binary image classification difficulties and huge datasets.	CASIA SiW REPLAY Attack	Accuracy = 97.96% Computational Speed = 7185.88 s	This model is further extended with cross-dataset estimation to approve whether these comparatively thin features exhibit decent, generalized abilities.

3 DATASET

For face spoofing image detection, four types of datasets are utilized. These datasets are: CASIA-SURF dataset, NUAA dataset, Replay Attack, 3-DMAD Dataset

3.1 *CASIA-SURF dataset*

It is the leading face anti-spoofing database. This collection includes 21000 RGB, Depth, and Infrared movies, containing 1,000 Chinese individuals. This dataset has six picture assaults (Neetika Gupta, 2023) linked by cropping, twisting print paper, and deadlock distance. Some dataset photos are in Figure 1. Six types of photo assaults;

- Attack: A user holds their flat-face photo with a cut eye region.
- Attack: A user holds their curved face photo with a cut eye region.
- Attack: A user holds a uniform-face with cuts to the regions of the eye and nose.
- Attack: A user holds their rounded face with cut the regions of the eye and nose.
- Attack: A user holds their uniform face with the cut eye, mouth, and nose areas.
- Attack: A user holds their rounded face photo with cut eye, mouth, and nose areas.

The biggest aggressive face anti-spoofing dataset, CASIA-SURF (Zhang S, 2020) was built to improve current algorithms via massive dataset analysis. Figure 2 shows CASIA-SURF film types for data collection. Example: recording model data with many models and devices in various interior settings. Video collectors must move the camera all around. Also, performers stand 0.3–1.0 meters from the camera and 30 degrees below their faces. The Real_Sense SDK expands to collect four audio-visual inputs with RGB, Depth, IR, and composition of these three pictures. RGB image resolution is 1280×720, Depth IR 640×480.

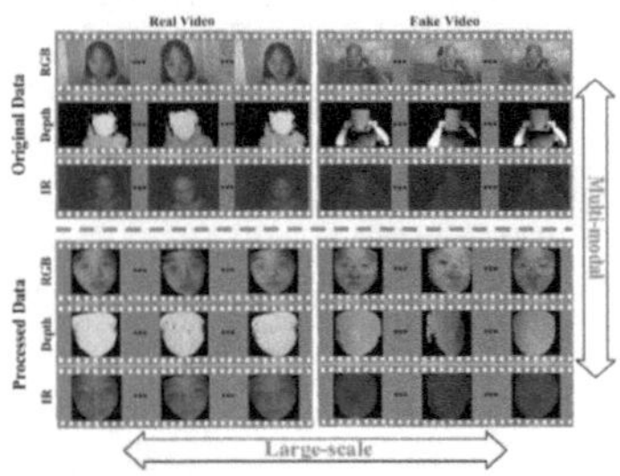

Figure 1. CASIA-SURF dataset.

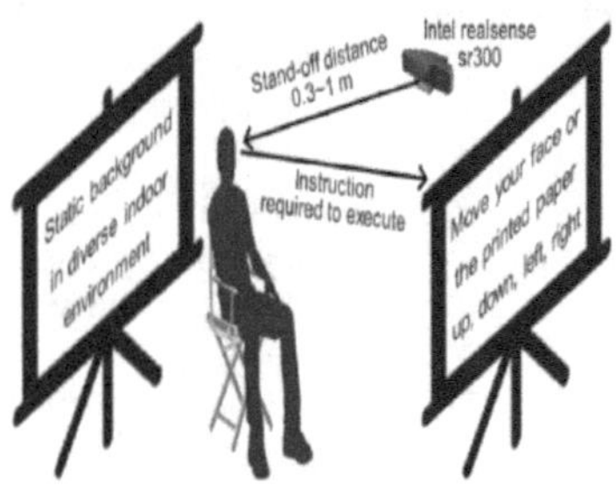

Figure 2. Sketch of footage formats in the CASIA-SURF dataset.

3.2 *NUAA imposter dataset*

NUAA has 15 subject webcams. 500 shots of Figure 3 participants and their falsified photos are taken at 20 fps. NUAA photographs (Seyedkooshan Hashemifard *et al.,* 2021) are unbiased. Epochs vary in brightness and location. NUAA created a massive public face PAD dataset. Picture accessing and attacks by fifteen persons are shown. Manipulators couldn't blink either. These attacks employ photographic paper (Javier Hernandez-Ortega, 2019).

3.3 *Replay attack*

Replay Attack participants were crowded and poorly lighted. First instance needed dark screens and steady office lighting. Blind curtains, intricate backdrops, and office lights were removed again. Print, mobile, and HD protocols are attacked. Figure 4 exhibits Replay Attack data (Polash Kumar Das *et al.,* 2019).

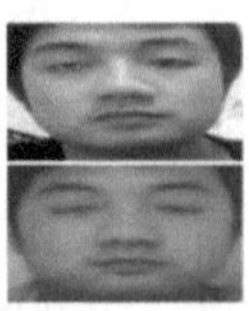

Figure 3. NUAA dataset image.

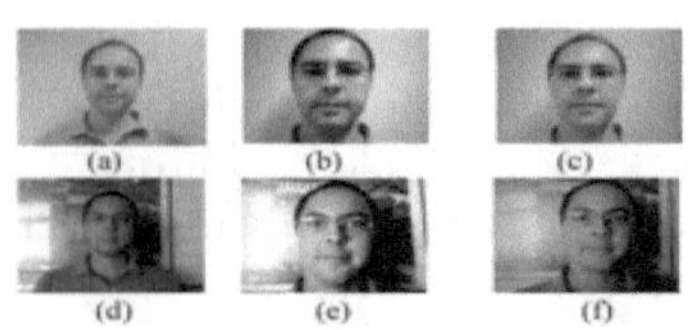

Figure 4. Replay attack datasets images.

3.4 *3-DMAD dataset*

The 3-DMAD dataset includes 10,000 Kinect-documented frames of 17 people in 255 video of genuine subjects and mask assaults. Eye spots and an RGB depth picture are in each frame. Table 2 compares existing datasets.

Table 2. Comparative analysis of different datasets.

Dataset	Subjects	Objects
NUAA	15	500
CASIA-SURF	1000	21000
Replay-Attack	50	13000
3-DMAD	17	255

4 PROPOSED PRE-PROCESSING STEPS AND TECHNIQUES USED IN FACE SPOOFING DETECTION SYSTEM

4.1 *Preprocessing steps*

Picture preprocessing enhances camera images, minimizes data via filters, and accelerates processing and matching, improving visual assessment reliability. It involves resizing, adaptive median filtering, scaling, border detection, and normalization. Sharpening, smoothing, linear models, and digital picture pixel matching have pros and cons (Table 3).

4.2 *Image cropping, resize and brightness*

In face recognition applications, preprocessing entails manually changing eye position, rotating all face pictures to match eye midpoints, and cutting faces for analysis. The normal cropped picture is 64×64 pixels. Image resizing controls recognition development dimensions and streamlines processing by lowering data size. Pixel values affect brightness variation by increasing or decreasing brightness. Figures 5, 6, and 7 (Nawaf Hazim Barnouti, 2016) indicate preprocessing steps.

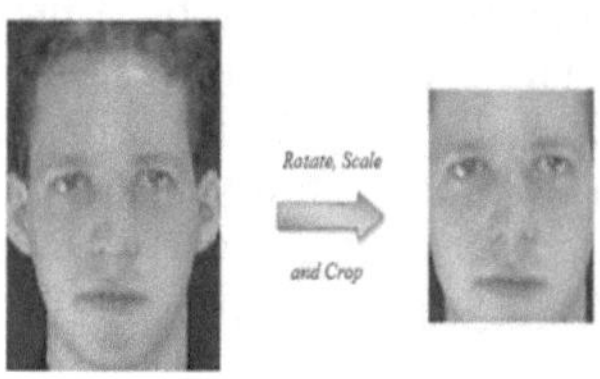

Figure 5. Cropped Image.

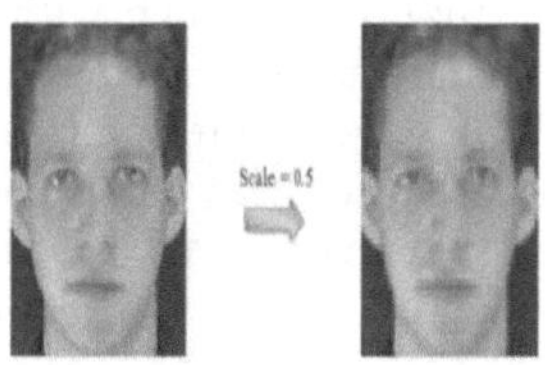

Figure 6. Resized Image with 0.5 scales.

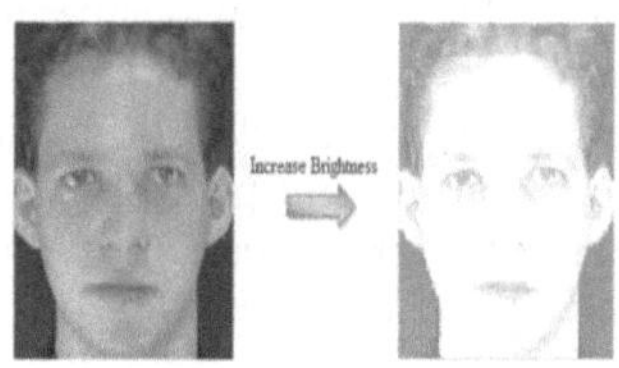

Figure 7. Image with brightness.

4.3 *Preprocessing facial image for contrast improvement*

4.3.1 *Contrast development using histogram equalization*

Histogram is defined as a graphical image of distributed information that is displayed in visualization form. Image Histogram performs a graphical demonstration of light and color circulation in digital images. It also displays the comparative frequency of different gray levels existing in the image. Histograms are exists in different forms such as Black, White, Small contrast, and stable image. The main benefits of Histogram are that Three-dimensional dominion processing methods and multiplication is utilized for image improvement. It offers proficient image data and is very operational in different zones such as image compression and segmentation. Histogram equalization is a technique used in preprocessing to enhance contrast in images by plotting equivalent gray level values horizontally and standardized values vertically. The composed filtered image. Figure 8(a) Composed (hybrid) Filtered Image (b) Contrasted output image Figure 8(a) is used with this method for contrast enhancement. Figure 8(b) displays the contrasted output image (Vinay A. *et al.,* 2016). Figure 9 Histogram Equalization for Contrasted Image

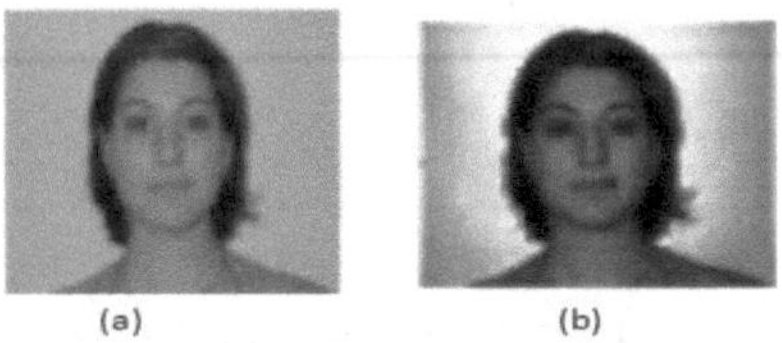

Figure 8. (a) Composed (hybrid) Filtered Image (b) Contrasted output image.

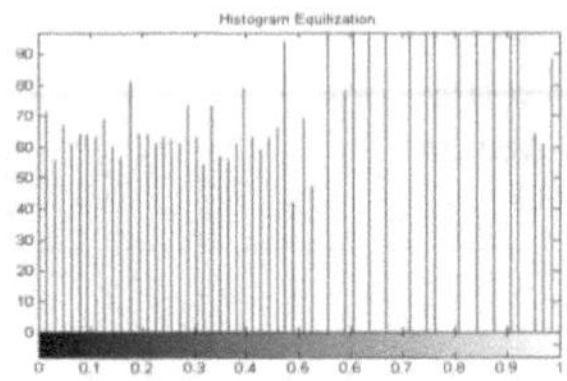

Figure 9. Histogram equalization for contrasted image.

4.4 *Pre-processing for distortion removal*

4.4.1 *Color Normalized with Median filter*

Photos with noise are awful. We have Impulse, Gaussian, and Salt & Pepper. Salt Pepper noise fades pixels in an optimistic district and RGB image. Median filters compare dark regions' medians to their neighbours to reduce RGB Salt and Pepper noise. The median filter replaces close pixel values with the median window, moving pixels across the image. Productivity image quality depends on peak signal-to-noise ratio. Image value may be affected by signal-to-noise ratio (G. Hemalatha, 2016).

4.4.2 *Difference of Gaussian (DoG) filter*

It minimizes noise, safeguards high-frequency processing, and enhances image borders. DOG filter has a wider center ensemble ratio range without high-frequency mobs. Filter incorrect low 3-D frequency data using the smallest Gaussian. The pre-processing reduces noise and incorrect data. Figure 10 Give DoG filtering (Md Rezwan Hasan, 2019) priority. This method pre-processes experimental analysis using band-pass filters, Gaussian filters, standard deviations, and realistic distant frequency DoG filtering. The local binary pattern variance (LBPV) method on DoG-filtered portions helps identify phony photos.

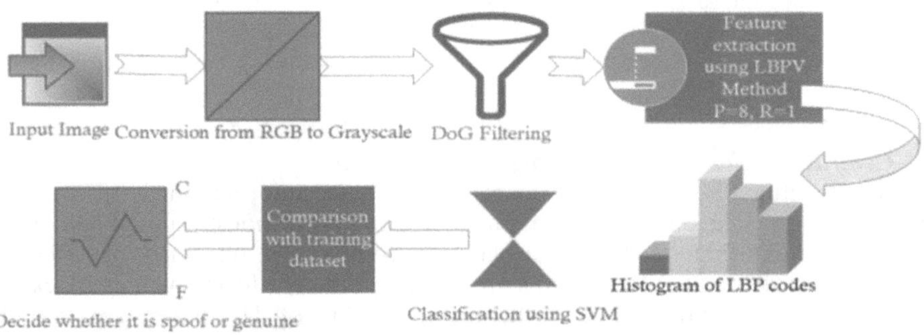

Figure 10. DoG filtration.

4.5 *Pre-processing image for edge improvement*

4.5.1 *Gabor filter*

Gabor Filters improve contrast, edge detection, and noise reduction for form and face recognition. In image processing, band-pass filters boost spatial and frequency results for three-dimensional regularity propagation.

Figure 11. Hybrid filter.

Table 3. Preprocessing techniques: Advantages and disadvantages.

Method	Advantages	Disadvantages
Median Filter	Before filtering, this removes noise to reduce its harmful effects. Also, calculating the filter's value and finding the best result.	Noisy pixels are interchanged without compelling into a local feature, for example, the occurrence of boundaries and the level of noise. Data regarding image and edge is not perfect due to the higher existence of noise.
DoG Filter	It removes high-frequency information. This method is widely used to eliminate the high degree of noise.	It is not suitable for low-level noise detection. High computational cost.
Gabor Filter	This filter provides higher visual performance. Higher retrieval method.	More time-taking method. High computational cost
Histogram Equalization	It is a visually strong method.	It cannot read true values because data is composed into categories.

Table 4. Performance Metric analysis.

Datasets	HTER (%)	EER(%)
NUAA	0.39	–
Replay-attack	8.79	0.4
3DMAD	15.15	–
CASIA	44.36	4.5
CAISA and NUAA	–	6.8

Multiple face spoof detection algorithms were calculated to assess the pre-processing method: Median Filter, DoG Filtration, Gabor Filter, etc. Performance estimates on the CASIA-SURF dataset, NUAA, relay-attack, etc., which are used for face spoofing research, include several face assaults. Table 4 compares HTER and EER analytical performance.

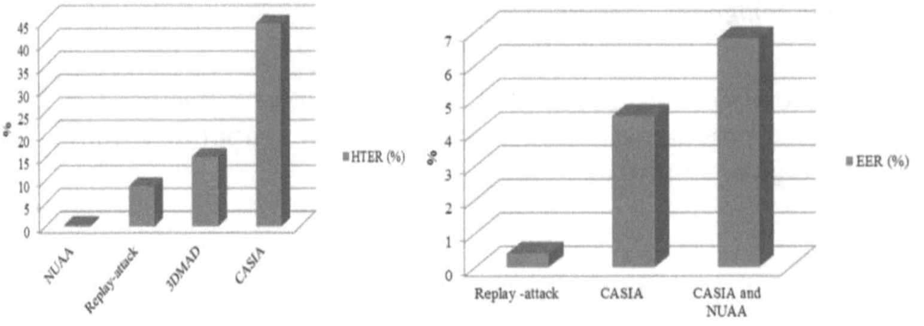

Figure 12. Performance metrics with different datasets (HTER and EER).

Figures 12 (i) and (ii) show the dataset-based performance metrics such as HTER and EER. The NUAA dataset has reduced the Half total error rate value is 0.39 % and the replay attack has reduced the equal error rate value is 0.4 %.

5 CONCLUSION AND FUTURE SCOPE

This article reviewed median filter, DoG, Gabor, and other preprocessing techniques. Other approaches are less visually appealing than HE. It detects numerous types of face-authentication biometric system (BS) spoofing attacks. These include replay assaults, 3DMAD, NUAA datasets, etc. We also invented and used DoG, a median filter for RGB picture normalization and noise removal. Face photos are contrast illuminated using the Gabor filter and histogram equalization. Describes the pros and cons of certain filter approaches and compares performance analysis. According to multiple datasets, NUAA gives 0.39% HTER while REPLAY-ATTACK (Neetika Gupta, 2023) decreases 0.4% EER. We will use optimal feature extraction for an anti-spoof system to enhance face-spoof picture systems. It will evaluate a hybrid feature-based improved anti-face spoofing detection system to accuracy, error, and other performance criteria.

REFERENCES

Abdullakutty, F., Johnston, P., & Elyan, E. (2022). Fusion Methods for Face Presentation Attack Detection. *Sensors*, 22(14), 5196.

Anjos, A., Komulainen, J., Marcel, S., Hadid, A., & Pietikäinen, M. (2014). Face anti-spoofing: Visual approach. *Handbook of Biometric Anti-Spoofing: Trusted Biometrics under Spoofing Attacks*, 65–82.

Barnouti, N. H. (2016). Improve face recognition rate using different image pre-processing techniques. *American Journal of Engineering Research (AJER)*, 5(4), 46–53.

Das, P. K., Hu, B., Liu, C., Cui, K., Ranjan, P., & Xiong, G. (2019, November). A new approach for face anti-spoofing using handcrafted and deep network features. In *2019 IEEE International Conference on SOLI* (33–38). IEEE.

Dwivedi, A., & Verma, S. (2022). SCNN Based Classification Technique for the FS Detection Using Deep Learning Concept. *The Scientific Temper*, 13(02), 165–172.

Gupta, N., & Kaur, A. (2023, November). Texture, Quality, and Motion-Based Analysis for Face Spoofing Detection System: A Review. In *2023 3rd ICTACS*. 351–358. IEEE.

Hadiprakoso, R. B., & Setiawan, H. (2020, November). Face anti-spoofing using CNN classifier & face liveness detection. In *2020 3rd ICOIACT*. 143–147. IEEE

Hasan, M. R., Mahmud, S. H., & Li, X. Y. (2019, May). Face anti-spoofing using texture-based techniques and filtering methods. In *Journal of Physics: Conference Series*. Vol. 1229, No. 1, p. 012044. IOP Publishing.

Hashemifard, S., & Akbari, M. (2021). A compact deep learning model for face spoofing detection. *arXiv preprint arXiv*:2101.04756.

Hemalatha, G., & Sumathi, C. P. (2016, February). Preprocessing techniques of facial image with Median and Gabor filters. In *2016 ICICES*. 1–6. IEEE.

Hernandez-Ortega, J., Fierrez, J., Morales, A., & Galbally, J. (2019). Introduction to face presentation attack detection. *Handbook of Biometric Anti-Spoofing: Presentation Attack Detection*, 187–206.

Karen Hollingsworth *et al.* (2008) Pupil dilation degrades Iris biometric performance.

Kavitha, P., & Vijaya, K. (2017). A Study on Spoofing Face Detection System. *International Journal of Pure and Applied Mathematics*, 117(22), 205–208.

Kumar, S., Kar, A., & Chandra, M. (2014). SVM based adaptive Median filter design for FD in noisy images. *2014 International Conference on SPIN*. 695–698.

Kumar, S., L., Asish, Y. S., & Ganapathy, S. (2022). A new hybrid CNN–LSTM model with Non-Softmax functions for face spoof detection. *Soft Computing*, 26(19), 10151–10162.

Kumar, S., Rani, S., Jain, A., Verma, C., Raboaca, M. S., Illés, Z., & Neagu, B. C. (2022). FS, Age, Gender and Facial Expression Recognition Using Advance Neural Network Architecture-Based Biometric System. *Sensors*, 22(14), 5160.

Shabanian, M., Wenzel, M., & DeVincenzo, J. P. (2022, April). Infant brain age classification: 2D CNN outperforms 3D CNN in small dataset. In *Medical Imaging 2022: Image Processing*. Vol. 12032, pp. 626–633. SPIE.

Vinay, A., Kathiresan, G., Mundroy, D. A., Nandan, H. N., Sureka, C., Murthy, K. B., & Natarajan, S. (2016). Face recognition using filtered eoh-sift. *Procedia Computer Science*, 79, 543–552.

Zhang, S., Liu, A., Wan, J., Liang, Y., Guo, G., Escalera, S., & Li, S. Z. (2020). Casia-surf: A large-scale multi-modal benchmark for face anti-spoofing. *IEEE Transactions on Biometrics, Behavior, and Identity Science*, 2(2), 182–193.

Computational Methods in Science and Technology Volume 1 – Kaur et al. (Eds)
 ISBN: 978-1-032-81760-6

An eve teasing prohibitor application with SOS services: A systematic review

Harsh Tomar, Ishan Srivastav, Mohammed Fahad & Sonia Lamba
Department of CSE, ABES Engineering College, Ghaziabad, U.P. India

ABSTRACT: Although, the constitution of India and IPC special laws pledge to provide gender equality, safety and dignity at all workplaces, educational institutions and other places for everyone in India. But still crimes like Eve –teasing is one of the profound and silent forms of crime, majorly in colleges as it is not reported in the capacity it happening now a days. This indicates a need of an emergency system for providing them help.

This SOS application for android and iOS users provides support to students and help in situations of eve-teasing. It signals with the Morse code, short and long sounds, allowing students to transmit alert signals to higher authorities, thus responding swiftly to the location of the incident. It is a model based on SOS Emergency where a student can inform the officials by just giving some signs (Morse code, Short and long sounds). Hence, the officials get alert of eve-teasing along with the location of concerned student so they can reach out to help the student as soon as possible. Therefore, we are developing this software that makes use Artificial Intelligence and Machine Learning to assist students and restrict eve teasing. We are adding features of location tracking, voice recognition and individual login features. Personalized and efficient emergency alert notification system is one of the most efficient tools for personal security and safety. Now a days, the most used and effective emergency alert notification system.

Keywords: Artificial Intelligence, Android/IOS application, SOS, GPS

1 INTRODUCTION

It's an SOS emergency app that uses ML and AI to stop eve-teasing. With the help of Morse code, pupils can covertly notify authorities and provide their position for prompt aid. This reduces eve teasing that are common in India and may not be reported using conventional ways. Personalized emergency notifications, voice and facial recognition, and location monitoring are among the features. In keeping with India's commitment, it seeks to improve safety and security. Although, the constitution of India and IPC special laws pledge to provide gender equality, safety and dignity at all workplaces, educational institutions and other places for everyone in India. But still crimes like Eve –teasing is one of the profound and silent forms of crime, majorly in colleges as it is not reported in the capacity it happening now a days. This indicates a need of an emergency system for providing them help. This SOS application for android and iOS users provides support to students and help in situations of eve-teasing. It signals with the Morse code, short and long sounds, allowing students to transmit alert signals to higher authorities, thus responding swiftly to the location of the incident.It is a model based on SOS Emergency where a student can inform the officials by just giving some signs (Morse code, Short and long sounds). Hence, the officials get alert of eve-teasing along with the location of concerned student so they can reach out to help the student as soon as possible. Therefore, we are developing this software that makes use

 DOI: 10.1201/9781003501244-35

Artificial Intelligence and Machine Learning to assist students and restrict eve teasing. We are adding features of location tracking, voice recognition and individual login features. Personalized and efficient emergency alert notification system is one of the most efficient tools for personal security and safety. Now a days, the most used and effective emergency alert notification system. Face recognition systems gather features, facial identities, and identify patterns, contours, as well as shadows in the face. All forms of facial processing, including tracking, detection, analysis, and synthesis, are included in the larger picture. The most often used method is based on Eigen faces, which show the variations between the face being recognized and the faces registered in the database [1,2]. The application signals with the Morse code, short and long sounds, allowing students to transmit alert signals to higher authorities. Thus, responding swiftly to the location of the incident. The Global Positioning System, or GPS is a satellite-based navigation system that was put into orbit by the US Department of Defense in 1973. It consists of a network of 24 satellites [3]. The app incudes biometrics concept to some extent. The fundamental techniques related to biometric devices are referred to as "automated methods": (1) a system for scanning and capturing "images; a digital or analog image of a living personal characteristic; (2) compression, processing, and image comparison to stored images in database; and (3) applications systems interface" [4]. The app provides emergency signals using Morse code. These signals can be adapted to different emergency situations and are sent discreetly. The app keeps the identity of a student in trouble confidential and immediately alerts authorities in an emergency. Personal data is processed securely to protect your privacy. When the alarm signal is activated, the application automatically communicates the student's location to the relevant authorities in real time. The map-based application will use cutting-edge methods to decipher the data it gets and provide the authorities with important information. It will also assess additional data in text, audio, or image formats to help the authorities alleviate the situation [5–10]. As an added bonus, the device's SOS function lets users warn other nearby users of impending danger, fostering a safe user community that encourages one another [11]. Using discreet reporting methods and instant location sharing, the app aims to bridge the gap between at-risk students and the officials responsible for their safety. With an intuitive design and rapid response, SOS Signal is an important tool to reduce crime against students.

Year	2017	2018	2019	2020	2021	2022
Cases	12,740	13,630	14,110	14,720	15,940	29,570

Figure 1. Reported cases of eve teasing for past few years.

2 OBJECTIVE

Our smartphones are always within arm's reach. Numerous top-notch and cutting-edge sensors are included in the devices' configuration. Data collection can be done with these sensors [5,12–14]. Users can simply download mobile applications through the app store (Google Play is the official app store for Android devices) [6,15]. It is an essential system to have on our smartphone, it is for personal security that lets you send notifications to certain people via texts messages and emails in case of emergencies. It also gives you the ability to call any helpline number on a touch of "HELP" button, the system also keeps a track of your current location so that it will send user's current location to certain people whenever you want it in any case of emergency.

If you want to report any incident on eve teasing then just by single push on the "HELP" button you could not only report the incident but also can send your current location and image of the person involved, which will be helpful to recognize the person and act accordingly. It is in charge of making the necessary configuration changes to add or remove built-in or default system modules as well as to add custom modules [16].

3 LITERATURE REVIEW

Table 1. Summary of review performed based on various parameters.

Serial No.	Title	Publication Year	Author	Objective	Journal/ Conference/ Book	Tools, Technology Or Algorithm	Gap
1	Multimodal Biometric Authentication Based on Voice, Face and Iris	2015	Tudor Barbu, *et al.* [1]	Multimodal biometric authentication system based on three identifiers: iris, voice and faces.	E-Health and Bioengineering Conference (EHB)	mel-cepstral voice recognition, SIFT features	Combining the voice, face and irises to authentication methods based on fingerprints
2	Multi-Modal Biometric Authentications: Concept Issues and Applications Strategies	2012	Soltane Mohamed, *et al.* [2]	Biometric systems for personal authentication and identification	International Journal of Advanced Science and Technology	Dynamic Time Warping, Vector Quantisation and Nearest Neighbors algorithm, Multi-Modal, Recognition, Identification, Verification, Biometrics Fusion.	Less secure, Personal Information Threat
3	GPS-based Location Tracking System via Android Device	2013	Md. Palash Uddin, *et al.* [3]	Developed for location tracking of a group of people with a proximity alert system	International Journal of Research in Computer Engineering and Electronics.	JSON, Java, AVD, LAMP	Live location is not tracked
4	Security System Using Biometric Technology: Design and implementation of Voice Recognition System (VRS)	2008	Rozeha A. Rashid, *et al.* [4]	An implementation of a security system based on voice identification as the access control key.	International Conference on Computer and Communication Engineering	MATLAB (SIMULINK)	Utilizing the voice command control into a stand-alone, self-contained, microprocessor based access control device.
5	Real-time Emergency Reporting and Handling System"	2021	Nilakshi Jain, *et al.* [5]	A system that allows the user to send an emergency signal to the authorities using their smartphones at time of distress.	International Conference on Artificial Intelligence and Smart Systems (ICAIS)	ReactJS, NodeJs, Android Studio, WebSockets.	ML can be used for alerts, Data analysis for dividing area to zones and many other features can be provided.
6	Location Based Mobile App development on Android Platform.	2014	Bhuvana Sekar, *et al.* [6]	This paper describes an Android application, designed to serve as a location based time saving and safety ensuring application.	9th IEEE Conference on Indestrial Electronics and Applications	Eclipse IDE with ADT (Android Developer Tools) plug-in and Android SDK. GPS and Linux	Reserve limited battery life
7	A Mobile Based Women Safety Application (I Safe Apps)	2015	Dr. Sridhar Mandapati, *et al.* [7]	Women Safety.	IOSR Journal of Computer Engineering (IOSR-JCE)	Android Application, Servlet Java development kit, Eclipse IDE, My Eclipse IDE, Java	No live location shared
8	Lifecraft: An Android Based Application System for Women Safety	2019	Rabbina Ridan Khandoker, *et al.* [8]	Android application for android for women's safety	International Conference on Sustainable Technologies for Industry 4.0 (STI)	Android studio 3.3.2 and java jdk 11.0, GPS, Firebase Database,	Less secure, Zone ranking on basis of danger
9	Efficient Anti-Kidnapping and Anti-Harassment (Avoidance-Detection-Notification) Mobile Application for Unwanted Incidents	2019	Md. Elias Hossain, *et al.* [9]	An anti-kidnapping and anti-harassment mobile application	IEEE Student Conference on Research and Development (SCOReD)	Google Speech Recognition API, Android and IOS, Google map API, Compass sensor and accelerometer sensor,	Reception delay exists, not in working state for IoS
10	GoFearless: A Safety and Security Android Based	2022	Quazi Mahila Masud, *et al.* [10]	An android application entitled "GoFearless"	International Journal of Intelligent Information Systems	Android Studio, Java Development kit, Java,	Not linked with police department, Use of AI and ML

(*continued*)

Table 1. Continued

Serial No.	Title	Publication Year	Author	Objective	Journal/ Conference/ Book	Tools, Technology Or Algorithm	Gap
	Application for Women			that can protect ladies from hazardous circumstances.			
11	IoT Based GPS tracking system with SOS Capabilities	2022	Reuben O Jacob, *et al.* [11]	A battery-powered, GPS-based tracking system with continuous tracking and an SOS function that is based on IoT.	International Mobile and Embedded Technology Conference (MECON)	MongoDB and NodeJS, Open Layers via an HTML, AWS	Low performance due to low battery backup.
12	A-SA SOS: A Mobile- and IoT-based Pre hospital Emergency Service for the Elderly and Village Health Volunteers	2021	Kannikar Intawong, *et al.* [12]	The A-SA SOS application, a mobile-and IoT-based pre-hospital emergency service for the elderly	(IJACSA) International Journal of Advanced Computer Science and Applications	MYSQL and PHP, Json, IoT, GPS, IoS	Lack of Extended battery lifetime, Automatic rescue request notification, Not for Android.
13	Smartphone Apps for Domestic Violence Prevention: A Systematic Review	2023	Mehreen Sumra, *et al.* [13]	To help with protection against the risk of domestic violence	International Journal of Environmental Research and Public Health	GPS, NLP, Accelerometer	No research for Windows Phone, Blackberry OS, or Symbian. Quality was not checked.
14	Sos Android Application	2013	Pratyush Poddar, *et al.* [14]	Manage the emergency situations and provide home remedies for different diseases.	International Journal of Engineering Research	Java framework, JSP-servlet, SQLite, MySQL server	No in-depth knowledge of the languages to create system
15	SOS: A Distributed Mobile Q&A System Based on Social Networks	2014	Haiying Shen, *et al.* [15]	Mobile Q&A systems, where mobile nodes access the Q&A systems through Internet	IEEE Trans. Parallel Distrib. Syst.	Natural language processing (NLP), FOL inference engine, cumulative distribution function (CDF), Query Processing	Not for large scale networks, Cloud can be used
16	The design and implementation of an integrated environment for Android system.	2020	Yi-Kuan Wu, *et al.* [16]	The development of a GUI based integrated building environment for Android.	International Symposium on Computer, Consumer and Control (IS3C)	Android open source project(AOSP), building management systems, embedded systems, graphical user interfaces	Limited scalability
17	Development of application and Face Recognition for smart home	2020	Seree Khunchai, *et al.* [17]	The application design and facial recognition for a smart home system	International Conference on Power, Energy and Innovations (ICPEI)	ESP8266 microcontroller, Arduino UNO, PIR Sensor, LDR, Sensor, temperature Sensor (DHT11)	Time delay of 1-3 sec. Accuracy decrease with increasing distance.
18	SOSerbia: Android-Based Software Platform for Sending Emergency Messages	2018	Mihailo Jovanovic, *et al.* [18]	For sending emergency messages by citizens in Serbia	Hindawi Complexity Volume 2018	XML, Android, IoS, Google Service API Location and JSON,	False alarms, not for larger user base
19	Robust Face Recognition Technique with ANFIS in Compared with Feedforward Backpropagation Neural Network Using Specific Features of Wavelet Transform and HOG	2018	Thoedsak Sukdao, *et al.* [19]	The aims of the research were to study a face recognition	International Journal of the Computer, the Internet and Management	ANFIS (Adaptive Neuro-Fuzzy Inference System), Feedforward Backpropagation Neural Network, Wavelet Transform, HOG (Histogram of Oriented Gradients for Feature Extraction)	The light and the distance of a camera impacts quality of data assessment.
20	SOS Intelligent Emergency Rescue System: Tap Once to Trigger Voice Input	2020	Muhammad Javed, *et al.* [20]	Observing the significant number of casualties caused by traffic accidents.	4th International Conference on Computer Science and Artificial Intelligence	Java, Android Studio, STT, Google Cloud, Arduino Uno, NLP, RNN	Data Privacy affected

4 PROPOSED METHODOLOGY

It is an Application in which User Interface will have login page. Through login page we will get the details of our user (as in case of college of students, faculties and other workers). On the basis of Morse code, a message would be generated and sent to upper levels of organization through WhatsApp. Simultaneously the current location through google maps would be share and camera and voice recorder would get active.

They will try to capture images and record audio so as to verify the persons present there using face and voice recognition. Their details would also be shared to higher authorities of organization. The message generation, location share and face and voice recognition all took place at backend for which the application needs to be linked to a reliable network and the internet. It also needs to have a GPS monitoring system to communicate with the closest tower and GPS satellite to pinpoint its location on a Google map. [21,22] Also the data is managed and fetched during face and voice recognition by a Database Management System. Both the client and the service provider should have the application installed in their android devices. The service provider is required to add the client's contact the person with whom he/ she wishes to.

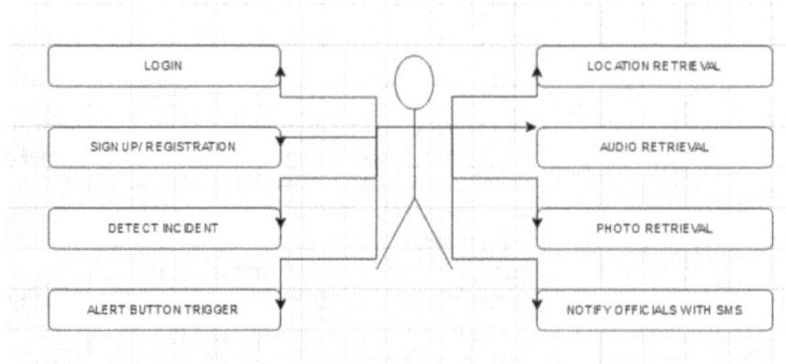

Figure 2. User interface.

5 ANALYSIS

SOS Alert systems using location, audio and video data have been the domain of research and development in past few years. These domains are capable of improving response to emergency, measures for security, and situational awareness in various aspects.

On reading various notable research and development projects of recent years like "EU Framework Programme 7" contributed in various projects working on location based alert system for emergencies, specifically in natural disasters or "DARPA's Real-time Scene Understanding (RESCUE)" program focusing on designing AI-based video analysis systems for emergency situations we have found the key technologies market is using are machine learning like technology of Artificial Neural Network and Machine Learning Algorithms for anomaly detection. In order to create a facial recognition system using deep learning, the first step is to compare two pictures. If the images have identical 128-dimension values, the system or network will accept them, discarding the other picture. [23–26]. In Computer Vision the information like Image Segmentation, Object Identification and Action Identification is majorly used.

On reading various surveillance systems like development of Automatic Video Surveillance Processing Evaluation (AVE) program by National Institute of Standards and Technology, it can be concluded for Audio Processing we can use various factors like pitch, sounds signatures and loudness.

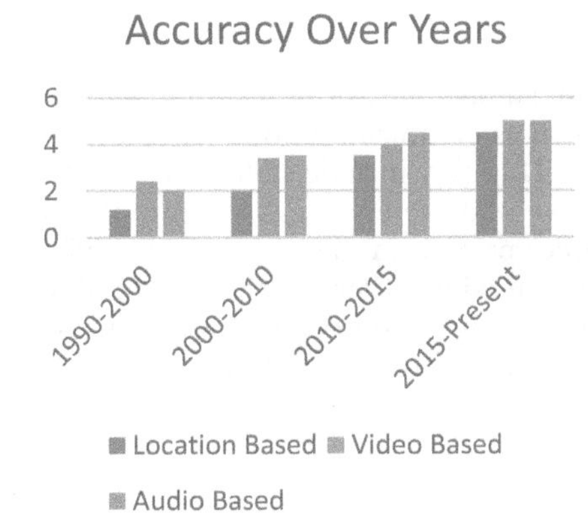

Figure 3. Accuracy of features.

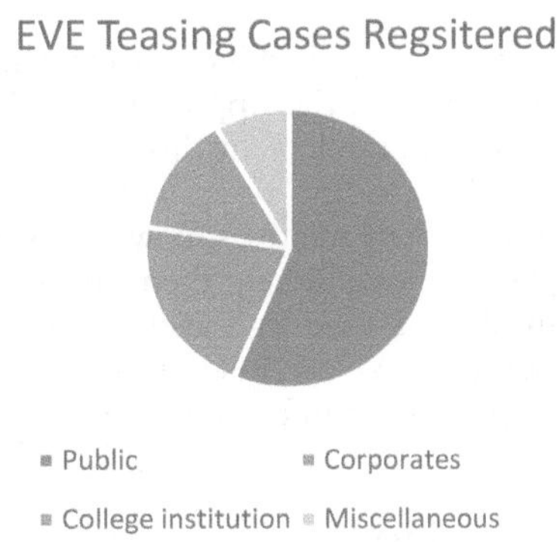

Figure 4. Cases in different organizations.

The previous studies can easily be concluded to find better accuracy like by using Sensor Fusion combining cameras, microphones and location-based tracking.

Parameters	MongoDB	MySQL	AWS
Model Type	NoSQL	RDBMS	RDS
Scheme Required	No	Yes	Yes
Storage	BSON File	NDB Cluster	Amazon S3
Language	MongoDB Query Language	Structured Query Language	Java or Python

Figure 5. Analysis of different databases.

When analyzed the different project works we could find the challenges that has been faced by different developers like Data Privacy, Accuracy (initially 60% to improving to 85%), scalability and Cost. The estimated cost for different systems integration like emergency response system and infrastructure built needs to be tracked and bring to minimal for effective productivity and design.

6 DISCUSSION

This application explores the creation and use of a smartphone app designed to stop eve-teasing incidents. Through the use of SOS functions and real-time reporting tools, the app gives victims a way to get aid right away and quickly report incidents to the appropriate authorities. The technical design, user interface, and possible social effects of this kind of application in resolving the safety issues of harassed individuals are probably covered in the paper. It may investigate the app's effectiveness in empowering victims, improving community safety, and encouraging a proactive strategy to prevent eve-teasing through the use of technological solutions.

7 FUTURE SCOPE

There are many ways in which we can use this proposed model in various sections. Whenever there is issue raised, this app is applicable in areas such as organization, offices, schools and colleges. We can implement this app on a large scale. This may include states and countries.

Example:- In todays time, women safety is at stake and there are cases where they are insecure from their relatives, coworkers, boss and other persons in their professional life. So this application would help them. In future this can be used in colleges, schools and offices. This can also be used in societies. We can use them in our country by cooperating with Public Services. We can also use the concept in smartwatches that would help for women's safety by just their tap on their watch.

8 CONCLUSIONS

It is an essential system to have on our smartphone. It is a personal security that lets you send notifications by a single push on power button or morse code to certain people via texts messages and emails in case of emergencies. It also gives you the ability to call any helpline number on a touch of power button. The system also keeps a track of your current location so that it will send your current location to certain people whenever you want it in any case of emergency.

The main function of the application is to quickly alert authorities or officials to incidents of harassment and provide them with real-time location information, thereby speeding up the response. This rapid notification allows for quick and targeted intervention, ensuring that a student in difficulty receives timely help and support. Furthermore, the SOS Urgence application focuses on confidentiality and privacy in order to protect the identity of a student in difficulty and to effectively inform the authorities of the seriousness of the situation. Essentially, the SOS Urgence app is a beacon of hope and support. Generally speaking, people may encounter numerous unforeseen circumstances on a daily basis, such as accidents, hijackings, and criminal activity. Thankfully, folks can feel empowered, protected and respected because they always carry mobile phones with them. They can therefore respond swiftly to emergencies and save lives.

We have taken references from various things, and sites, some of these are:-

REFERENCES

[1] Barbu T. *et al.*, (Nov. 2015). Multimodal biometric authentication based on voice, face and iris, in *2015 E-Health and Bioengineering Conference (EHB)*, Iasi, Romania: IEEE, pp. 1–4. doi: 10.1109/EHB.2015.7391373.

[2] Soltane M. *et al.*, (2012). Multimodal biometric authentications: concept issues and applications strategies, *International Journal of Advanced Science and Technology*, vol. 48.

[3] Uddin P. *et al.*, (2013). *GPS-based Location Tracking System via Android Device.*

[4] Rashid R. A. *et al.*, (May 2008). Security system using biometric technology: Design and implementation of Voice Recognition System (VRS), in *2008 International Conference on Computer and Communication Engineering, Kuala Lumpur, Malaysia: IEEE*, pp. 898–902. doi: 10.1109/ICCCE.2008.4580735.

[5] Bhavesh T. Y. *et al.*, (Mar. 2021). Real-time emergency reporting and handling system, in *2021 International Conference on Artificial Intelligence and Smart Systems (ICAIS)*, Coimbatore, India: IEEE, pp. 1683–1687. doi: 10.1109/ICAIS50930.2021.9395754.

[6] Sekar B. and Liu J. B., (Jun. 2014). Location based mobile apps development on Android platform, in *2014 9th IEEE Conference on Industrial Electronics and Applications*, Hangzhou, China: IEEE, pp. 2148–2153. doi: 10.1109/ICIEA.2014.6931527.

[7] Mandapati D. S., Pamidi S., and Ambati S., (Dec. 2019). *A Mobile Based Women Safety Application (I Safe Apps)*.

[8] Khandoker R. R. *et al.*, Lifecraft: An android based application system for women safety, in *2019 International Conference on Sustainable Technologies for Industry 4.0 (STI)*, Dhaka, Bangladesh: IEEE, pp. 1–6. doi: 10.1109/STI47673.2019.9068024.

[9] Md. E. Hossain *et al.*, (Oct. 2019). Efficient anti-kidnapping and anti-harassment (avoidance-detection-notification) mobile application for unwanted incidents, in *2019 IEEE Student Conference on Research and Development (SCOReD), Bandar Seri Iskandar*, Malaysia: IEEE, pp. 112–116. doi: 10.1109/SCORED.2019.8896280.

[10] Maliha Masud Q. *et al.* (2022). GoFearless: A safety and security android based application for women, *IJIIS*, vol. 11, no. 2, p. 22, doi: 10.11648/j.ijiis.20221102.12.

[11] Reuben O Jacob, Niranjana Murthy H.S., (Mar. 2022). IoT Based GPS tracking system with SOS Capabilities, presented at the 2022 *International Mobile and Embedded Technology Conference (MECON)*, Noida, India: IEEE,. doi: 10.1109/MECON53876.2022.9752019.

[12] Intawong K. *et al.* (2021). A-SA SOS: A mobile- and IoT-based Pre-hospital emergency service for the elderly and village health volunteers, *IJACSA*, vol. 12, no. 4, doi: 10.14569/IJACSA.2021.0120465.

[13] Sumra M., Asghar S. *et al.*, (Mar. 2023). Smartphone apps for domestic violence prevention: A systematic review, *IJERPH*, vol. 20, no. 7, p. 5246, doi: 10.3390/ijerph20075246.

[14] Poddar P., Singh K., and Kumar P., (Apr. 2014). Sos android Application, *International Journal of Engineering Research*, vol. 2.

[15] Shen H., Li Z., Liu G., and Li J., SOS: A distributed mobile Q&A systembased on social networks, *IEEE Trans. Parallel Distrib. Syst.*, vol. 25, no. 4, pp. 1066–1077, doi: 10.1109/TPDS.2013.130.

[16] Wu Y.-K. and Chen J., (Nov. 2020). The design and implementation of an integrated building environment for android systems, in *2020 International Symposium on Computer, Consumer and Control (IS3C), Taichung City, Taiwan*: *IEEE*, pp. 87–90. doi: 10.1109/IS3C50286.2020.00030.

[17] Khunchai S. and Thongchaisuratkrul C., (Oct. 2020). development of application and face recognition for smart home, in *2020 International Conference on Power, Energy and Innovations (ICPEI)*, Chiangmai, Thailand: IEEE, pp. 105–108. doi: 10.1109/ICPEI49860.2020.9431473.

[18] Jovanovic M. *et al.*, (Oct. 2018). SOSerbia: Android-based software platform for sending emergency messages, *Complexity*, vol. 2018, pp. 1–9, doi: 10.1155/2018/8283919.

[19] Sukdao T., Mungsing S., and Klabsuwan K. (2018). *Robust Face Recognition Technique with ANFIS in Compared with Feedforward Backpropagation Neural Network Using Specific Features of Wavelet Transform and HOG*.

[20] Javed M. and Xudong L., (Dec. 2020). SOS intelligent emergency rescue system:Tap once to trigger voice input, in *2020 4th International Conference on Computer Science and Artificial Intelligence*, Zhuhai China: ACM, pp. 187–193. doi: 10.1145/3445815.3445846.

[21] Singh P., Khatter H., and Garg D. *Global Positioning System based Automated Toll and Traffic Management System for Indian Roads*, vol. 16, no. 9.

[22] Department of Computer Science and Engineering, ABES Engineering College, Ghaziabad, Uttar Pradesh, India. *et al.*, (Mar. 2020). Toll and traffic management system for express highways, *IJITEE*, vol. 9, no. 5, pp. 2109–2113, doi: 10.35940/ijitee.E3102.039520.

[23] Rahul M., Tiwari N. *et al.*, (Mar. 2022). A new hybrid approach for efficient emotion recognition using deep learning, *IJEER*, vol. 10, no. 1, pp. 18–22, doi: 10.37391/IJEER.100103.

[24] Face Recognition Using HOG Feature Extraction and SVM Classifier, (Sep. 2020). *IJETER*, vol. 8, no. 9, pp. 6437–6440, doi: 10.30534/ijeter/2020/244892020.

[25] Gupta D. K., Aziz A., and Chauhan H. (2020). Face Detection and Recognition, vol. 07, no. 06.

[26] Srivastav A. K., Ali A., Khan A., and Tripathi H., *Intelligent Object detection with classification and localization using Deep Learning*.

Computational Methods in Science and Technology Volume 1 – Kaur et al. (Eds)
 ISBN: 978-1-032-81760-6

Autonomous solar car

Nilima Kulkarni, Arshadali Sayyed, Nikhil Soni, Komal Pal & Prashant Singh
Computer Science & Engineering Department, MIT SoC, MIT Art, Design and Technology University, Pune, India

ABSTRACT: Throughout the years, automobile companies achieved outstanding progress in manufacturing safe, reliable, and affordable vehicles. Many companies and manufacturers are developing autonomous cars for the future years. However, to achieve this goal they are facing too many challenges in technical and non-technical fields like software complexity, mapping, real-time data analysis, testing, and verification. AI Software installation in cars is costly, and also it affects fuel consumption. So, proposed an autonomous solar car that will run on solar energy, which is available free of cost. NumPy, OpenCV computer vision library, AI, ML, and Python were used for the implementation of the work. The edge detection algorithm takes the input from a camera. LiDAR, WIFI/Bluetooth models, and Ultrasonic sensors have been used. Inculcated OpenCV which has an advanced built-in function that has add-on provisions in extension to the blurred image, a threshold that can go low and high and determines whether there is an edge in the path or not. The input collected from the sensors was analyzed to give directions to the car on where to get a turn. In summary, the proposed autonomous solar car model is inexpensive to build and performs well.

1 INTRODUCTION

Automobile manufacturers are upgrading their technologies in response to this growing technological advancement. They make more comfortable cars while paying attention to the safety of the customers. In the making of autonomous cars, there are many challenges like hardware cost, software cost, management cost, software complexity, maps, simulation, etc. [3]. Due to these challenges' effects on business and the auto industry, a bad mapping may endanger the actions of solar autonomous cars. Software complexity delays commercialization affects consumer trust issues, and also risks for investors [1,3]. In the automotive industry, where driving safety is of utmost importance, safety is a multifaceted attribute. One of the most crucial uses for autonomous vehicles is safe driving for its occupants. 50 million people worldwide suffer serious injuries from road accidents each year, claiming 1.3 million lives as author in [16].

Moreover, autonomous solar-powered automobiles are going to affect the cab and rental car industries. Since drivers will no longer be required, taxi service providers could conserve money and make more money. Similarly, leveraging a lesser workforce will allow rental car companies to streamline their business operations. Smart apps that are available on personal devices will also help the software industry as a result of this paradigm shift. In summary, autonomous vehicles can lower labor costs and boost revenue. Convenience and elegance of use pose two additional bonuses of automated automobiles. Sometimes medical conditions, disabilities, or drunkenness prevent people from operating a motor vehicle. In addition, those who are unable to purchase a car, teens and younger adults without a driver's license, and the elderly could discover that autonomous cars are a good form of transportation. In these situations, driverless vehicles can offer a practical and safe means of enhancing citizens' mobility.

By the in-depth literature survey, it has been observed that most autonomous cars research work focuses on using algorithms like High-Definition Vehicle to everything (HD V2X),

 DOI: 10.1201/9781003501244-36

Simultaneous Localization and Mapping (SLAM), Detection and Tracking Moving Object (DATMO) [17]. The survey also reveals that most algorithms from machine learning and deep learning domains give better results for autonomous cars which make use of fuel. The authors in [16] have discussed the future of cars and its technology. The technology that they included were raspberry pi, solar, and rc motor car.

After the pandemic situation, petroleum fuels were touching the sky. These situations motivated us to develop an autonomous solar car that will not require any type of fuel for working. The cars will require solar energy for working. To some extent, there is also successful in achieving safety. Sunlight is converted into solar energy with the help of photovoltaic cells. Using the WIFI module for wireless connection. Developing an application so that it can connect through the Wi-Fi and can be accessible through mobile. LiDAR is used for light detection of the lane.

2 LITERATURE SURVEY

This paper explores cutting-edge methodologies for autonomous solar cars, outlining their respective strengths, weaknesses, and emerging challenges. Additionally, it draws insightful conclusions, as summarized in Table 1.

The survey's conclusions emphasize how crucial deep learning and computer vision technologies are to autonomous cars. Computer vision algorithms classify and identify items on the road, including cars, bikes, pedestrians, traffic signs, and lane markings, by analyzing visual data that is acquired by cameras. Convolutional Neural Networks (CNNs), one of the most popular deep learning techniques, are useful for tasks including accurate object detection, classification, and semantic segmentation in computer vision frameworks. The effectiveness of CNN algorithms has been shown by the authors in references [19, 20, 21, 22 & 23] on a wide range of object detection tasks in various domains. Furthermore, references [24–26] provide examples of how to use pre-trained models for object identification, such as Inception ResNet and YOLO. The overfitting of Neural Networks can be addressed as given in [2].

Year	Algorithms/ Techniques	Sensors	Remark
2022 [4]	DNN, CNN, MCU	Camera, Raspberry Pi Pico MCU,	Simulation precision with latency in the search for an intelligent network the built environment for AI-enabled CPS.
2022 [17]	Algorithms such as HD V2X, SLAM, and DATMO used for object detection and vehicle location	Autonomous-Driving, Camera, Level, Ultrasonic Sensor	Considering the ongoing Fourth Industrial Revolution, markets associated with autonomous driving are experiencing swift advancements. Aimed to explore the evolving technological landscape of autonomous driving and delve into the distinctive features and variances among Level 5 autonomous driving systems.
2022 [5]	The SIFT software makes changes the target according to real-time, and the vehicle ventures toward the amended target coordinate.	Raspberry PI, Arduino-Uno board, Ultrasonic sensors, Web Camera	Enabling the development of an autonomous driving system becomes effortlessly scalable as users can readily affix preferred sensors to the RC-Car.
2022 [9]	The ability to perceive in embedded systems, translation and visualization, route planning, estimation, judgment, and control	LiDAR, radar, VIS/ IR cameras, GPS, and IMUs	Upon finishing the categorization of datasets collected in diverse environmental scenarios, creation the inaugural database designed for the autonomous operation of vehicles using the Robot Operating System (ROS).

(*continued*)

Continued

Year	Algorithms/ Techniques	Sensors	Remark
2022 [13]	Machine learning algorithms, DL	IR sensors, IR LED, camera	By making it possible for researchers to set up an unusual set of containers to host their algorithm creation, deployment, simulation, and graphical representation prerequisites, ATK is used to speed up the development of autonomous algorithm design.
2019 [6]	Algorithm like Scale Invariant Feature Transform (SIFT), Speed up Robust Feature (SURF), Oriented FAST Rotated BRIEF (ORB)	Stereo Camera, Lidar, Microcontroller, Wheel Encoder, Inertial Measurement Unit (IMU)	Research on robotics subjects like the localization of operations mapping, machine learning, image recognition, and more might be definitively established with the aid of this guide.
2020 [10]	Algorithms like localization, mapping (GPS)	LIDAR, radar, GPS, computer vision, LCD, solar panels, Arduino	The smaller size of the automobile laid out a few challenges. because the solar panel was somewhat big and was in use.
2022 [7]	Deep neural networks popular method of end-to-end Driving	Sensors like GPS, IMU, and motion capture systems	To enhance driving efficiency through the improvement of Reveries-net's structure.
2022 [1]	Machine learning algorithms for model predictions	Camera, IR sensor GPU, CPU, RMPC	This paper introduces a randomized non-linear model predictive controller designed for autonomous driving, with a specific emphasis on tasks related to obstacle avoidance.
2021 [8]	Machines learning algorithms, technologies, such as localization planning and control	heterogeneous sensors, actuators, and computers, Raspberry Pi	Using cutting-edge SiC MOSFET power modules, a global partnership between academics and students produced an optimized An independent Formula Student race car's converter The collaboration intends to advance upcoming studies in smart cities and electromobility.
2021 [15]	a fully end-to-end ML system	IR sensors, Raspberry Pi 4, Pi Camera, Micro SD Card	To obtain non-linear activation events the the ReLu algorithm function has been applied to the pixels of the convolutional and dense layers. After each of these layers, a dropout reinstatement rate of 0.2 is applied in order to lessen the overfitting phenomenon.
2021 [3]	Robust autonomous guided vehicle (AGV) with GPS assistance for worldwide route planning	RC rover, Arduino Uno, microcontroller, LCD, GPS	implemented functioning working version of a clever self-navigating vehicle that can find the shortest route and discover through pre-programmed their travels markers on a map with the help the heading degree.
2021 [12]	Cascaded adaptive architecture	Camera, microcontroller, GPS	The controller has proven its effective operation across various longitudinal speeds, showcasing excellent performance on both simulated vehicles and experimental platforms. It consistently matches or even exceeds human steering capabilities, maintaining smooth and stable control.
2021 [11]	The purpose of the environmentally friendly factor, integer is to find the engine dynamics' energy-efficient set point that produces driving torque in response to traffic and road infrastructure traveling tasks.	Camera, human-machine interface design, sensor, mobile devices	To find the minimum fuel usage, set point, the engine's characteristics are stimulated by varying speed; this is done until the deviation from the moving average value is less than 20%.

3 PROPOSED SYSTEM

The upcoming years of solar autonomous cars will be acquired in measuring safety, best-quality hardware, and software design, and consumer satisfaction. There are many problems in the previous version of autonomous cars which was developed by different companies like Tesla, BMW, etc. however, there never existed a fully automated car. There are issues with fuels and petroleum oils as fuels are touching the sky. In the development an autonomous solar car that does not use any kind of fuel and is very eco-friendly. Also, the proposed autonomous solar cars are cost-efficient and have GPS for location.

The proposed autonomous solar car makes use of a pre-defined model program for its operation. In this work, OpenCV and NumPy are used for object and lane detection. Inculcated a camera that observes the track path. Night vision cameras are good for high quality image capturing at nighttime but trimmed the low-cost goal. It collects real-time data using a USB camera. A USB webcam with an integrated LED can also be used in the dark or can be installed in an RC solar car with lights to provide light to the track. However, its imaging quality is not comparable to that of a dedicated night vision camera. With the aid of ultrasonic sensors, GPS, and computer vision, the proposed autonomous solar car are able to navigate the track on their own independently. Here are some parts for proposed car:

- Hardware design, and implementation
- Solar panel, energy storage, and management
- Autonomous path-planning, and object detection

3.1 *Hardware design, and implementation*

It consists of super-capacitors, Raspberry Pi 3/Arduino board, MicroSD card, Car [motor, Tyers, board, wires, LED, Resistors, microcontroller, Wi-Fi chip], USB Battery with micro-USB cable, DC-DC 5V/2A Voltage Converter, Wide Angle Raspberry Pi Camera, Ultrasonic sensor, Solar panel, LiDAR, Wi-Fi/Bluetooth module.

Figure 1. The proposed designed model of an autonomous solar car.

3.2 *Solar panel, energy storage, and management*

In this work, solar panels are used for energy production. It is a crucial part, since it reduces the use of petroleum fuels. Solar energy is induced in autonomous vehicles and converts sun rays into electricity using photovoltaic cells.

In the work, solar energy is stored in batteries or super-capacitors which will be used in night time and in monsoon season. When used on a wide scale, these solar-powered vehicles help reduce noise pollution and gas emissions.

3.3 *Autonomous path-planning and object detection*

Using a predefined model for path planning and object detection. Computer vision library and python is used. It helps the application that can automatically detect lanes from the camera installed in the car. It can show

The current lane
The radius of the curvature of the lane
The position of the vehicle

A Vehicle can detect the path and find the curvature through lanes. The images that are extracted from the camera use the algorithm Canny Edge Detector. It identifies the path from the extracted images and discards other data. It mainly focuses on lane detection. The images are in blurred black and white, with white outlines of the object. Incorporated OpenCV which includes built-in functions with additional parameters. This function, in occurrence with the blurred image, utilizes a threshold that can be adjusted to low and high threshold values. It determines whether there is an edge in the path or not.

A threshold seizes the intensity change at a certain point. Next, employing a histogram for detecting the direction in the images. It presents the images in terms of frequency and intensity. It helps to determine the radius that can be tangent to the path.

OpenCV contains many computer vision algorithms. Haar The Cascade Classifier is a machine-learning tool for object detection that was created by Paul Vivia and Michael Jones. It is trained using a large number of images, both positive and negative. Positive images represent the object the classifier is supposed to recognize. While negative images encompass everything except the object to detect. Adding voice search so that it can be accessible by mobile.

Figure 2. Detection of curves.

After getting the images of the surrounding the model checks for the lines or curves of the track for curve detection. In Figure 2, there is a yellow line and a white line shown. The yellow line shows the curve and the white line shows the edge of the path. This line shows the curve of the track from where the car needs to take a turn.

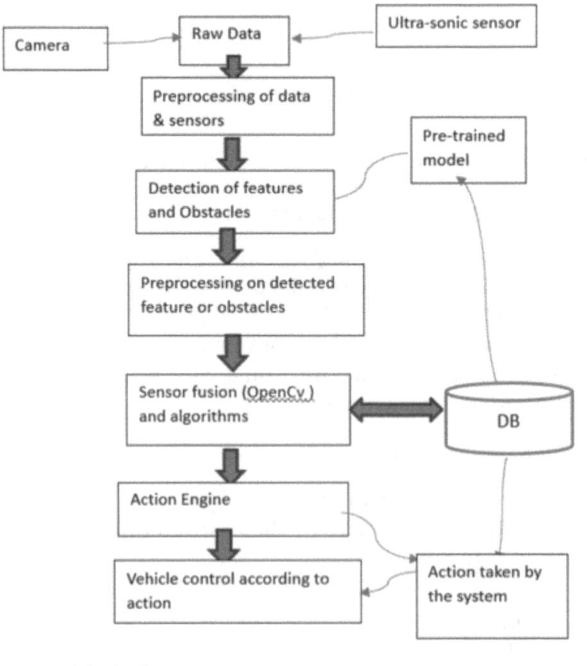

Figure 3. Proposed system block diagram.

3.3.1 *Raw data*

Collecting raw data from the environment using ultrasonic sensors and cameras. This raw data can be in the form of obstacles, signs, objects, curve detection, etc.

3.3.1.1 *Camera* The raw data that is collected by the cameras are in the form of images, black and white or red-green-blue colored (sample collected images are shown in Figure 4a & b below).

4a 4b

Figure 4. Sample collected images from the camera.

Figure 4a is the image of the path collected by the camera and Figure 4b is the image showing the degree of lane and at which degree the model needs to decide between turning.

3.3.1 2 *Ultrasonic sensors* Ultrasonic sensors tend to be used in obstacle and proximity detection. It helps the system to make decisions about navigation, obstacle avoidance, and overall safety.

3.3.2 *Pre-processing of data and sensor*

In this step, the raw data which is collected from the environment by the sensors and camera is processed using preprocessing techniques like reduction of data, data alteration, data cleaning, etc. Data preprocessing is necessary because accurate data from the clusters of data that are collected. Sensors generate data that is based on real-time monitoring, control systems, and analytics.

3.3.3 *Pre-trained model*

It is a model which is already trained on a similar kind of dataset. The dataset includes key features of the cars, edges of the path, traffic signals, speed breaker, curve of the path, etc. This model is trained on a large dataset so it can be used for real-time analysis.

3.3.4 *Detection of features and obstacles*

In this step, the data was collected and preprocessed by camera and ultrasonic sensor. This data is trained on pre-trained models to get lane, signs or obstacle detection that will help make decisions.

3.3.5 *Pre-processing on detected features and obstacles*

In this step, the outcome of the data that is trained on a pre-trained model is further processed and stored in the database. The steps involved in the preprocessing steps are filtering and noise reduction from the images, feature scaling and extracting the information, and choosing the relevant features.

3.3.6 *Database*

A database is used to store the data that is processed. This data can be used by the pre-trained model for further processing. As it makes processing fast and easy based on the data collected. Database is useful in preprocessing steps some similar kind of event is happening it can make decisions more efficiently. Like, if there is a speed breaker then the vehicle needs to slow down its speed and the car's suspension system works. This data is stored in a database and can be used by the pre-trained model, the model or car experiences a similar kind of event again. It will help the model to be fast, accurate, and work effectively.

3.3.7 *Sensor fusion*

It's the process of a combination of data which is processed by preprocessing steps. It uses OpenCV and NumPy for sensor fusion. It combines the algorithm and the libraries of machine learning for decision-making.

3.3.8 *Action engine*

It is the process in which the data has been processed in sensor fusion and is converted into a machine-level language so that the autonomous solar car can understand the given data and perform accordingly.

3.3.9 *Decision making*

In this step, the system decides what action should be taken by the vehicle. It can also use databases that store the data for decision-making. The action engine helps the system to decide which action should be taken by the vehicle.

3.3.10 *Vehicle control*

The given decision taken by the above step is performed by the vehicle controls such as brake, steering, etc. It gives the vehicle information about when to take a break and how to turn the steering or in which direction to go or perform.

In the past papers, they had used sensors like LiDAR, radar, cameras, and ultrasonic sensors one at a time. However, a better approach, is sensor fusion i.e., using computer Vision libraries that enable the system to learn and adapt environment [9,7]. Using Region-based convolutional neural network (CNN) for lane detection and object detection.

3.4 *Algorithm*

In the work, different algorithms for lane detection and object detection.

3.4.1 *Lane detection*

For lane and curve detection, the following steps are employed:

Table 2. Algorithmic steps for lane and object detection.

Steps	PseudoCode & Formulas
Step No:1 Thresholding Define: This function takes the images as input and returns a binary mask highlighting the white area within the specific threshold range in the High-level Synthesis (HLS) color space. This binary mask is used for lane and curve detection.	def thresholding(img): hsv = cv2.cvtColor(img,cv2. COLOR_BGR2HLS) lowerWhite = np.array([85, 0, 0]) upperWhite = np.array([179, 160, 255]) maskedWhite= cv2.inRange(hls,lowerWhite, upperWhite return maskedWhite
Step No: 2 Apply Perspective Transformation to get a Birds Eye View. Define: In this function, input images are transformed to a birds eye view perspective, which simplifies lane detection by providing a	def perspective_warp(img, dst_size=(1280,720), src=np.float32 ([(0.43,0.65),(0.58,0.65),(0.1,1),(1,1)]), dst=np.float32([(0,0), (1, 0), (0,1), (1,1)])): img_size = np.float32([(img.shape[1],img. shape[0])])

(continued)

Table 2. Continued

Steps	PseudoCode & Formulas
clear representation of the road, make it easy to identity and track lanes more accurately.	src = src* img_size dst = dst * np.float32(dst_size) M = cv2.getPerspectiveTransform(src, dst) warped = cv2.warpPerspective(img, M, dst_size) return warped

Step No: 3 Identify Lane Line Pixels
Define: Start lane line pixel detection by generating a histogram to locate areas of the image that have high concentrations of white pixels. This function is used to analyze the distribution of pixel intensities in an image and identifies peaks in the histogram, which is used for lane detection. It can also provide a representation of the histogram and the identified peaks on the images.
	def getHistogram(img,minPer=0.1,display= False,region=1):
	if region ==1:
	histValues = np.sum(img, axis=0)
	else:
	histValues = np.sum(img[img.shape[0]// region:,:], axis=0)
	maxValue = np.max(histValues)
	minValue = minPer*maxValue
	indexArray = np.where(histValues >= minValue)
	basePoint = int(np.average(indexArray))
	if display:
	imgHist = np.zeros((img.shape[0],img.shape [1],3),np.uint8)
	for x,intensity in enumerate(histValues): cv2. line(imgHist,(x,img.shape[0]),(x,img.shape[0]-intensity//255//region),(255,0,255),1) cv2.circle (imgHist,(basePoint,img.shape[0]),20, (0,255,255),cv2.FILLED)
	return basePoint,imgHist
	return basePoint

Step No: 4 Curve Detection Define: This function identifies curves using trigonometric formulas based on data collected by ultrasonic sensor. Calculate curvature radius and lane position for more accurate navigation. • Distance Calculation:Using Trigonometric formula to calculate the distance to an object based on the angle of detection and measured distanced by ultrasonic sensor. • Curvature Radius Calculation:This provides an estimation of the radius of the circle that best fits the detected curve. • Lane Position Determination:It determines the lane position relative to vehicle's current position. If the vehicle is traveling along a straight road, the lane position can be calculated using the tangent function.	(i) Distance = Measured Distance Cos (Angle of Detection) (ii) R = (dx)2+(dy) 2 2.dx (iii) Lane position= tan(Angle of Vehicle) X Distance travelled

3.4.2 *Object detection*

For object detection, the following steps are employed:

Steps	PseudoCode & Formulas

Step No: 1 Raw Data Collection
Define : Collecting raw data form cameras and ultrasonic sensors include images and obstacle images.
	import cv2
	cap = cv2.VideoCapture(0)
	def getImg(display=True):
	_, img = cap.read()
	# img = cv2.resize(img, (size[0], size[1]))
	if display:
	cv2.imshow('IMG', img)
	return img
	if __name__ == '__main__':
	while True:
	img = getImg(True)
	(30, 240), (450, 240),
	(120, 60), (360, 60)

Step No: 2 Data Preprocessing Define : Using various techniques for cleaning, reducing, and transforming the raw data for processing.	def preprocess_image(image_path, target_size= (32, 32)): image = cv2.imread(image_path) image = cv2.resize(image, target_size) image = image.astype('float32') / 255.0 # Normalize to [0, 1] return image
Step No: 3 Feature and Obstacle Detection Define : Using pretrained model for real-time analysis. In this work, using Haar Cascade Classifier for object detection. This also detects stop signs in an image, draws rectangles around them, and labels them with text. Then displays the annotated image for visual inspection.	import cv2 stop_sign=cv2.CascadeClassifier('cascade_stop_- sign.xml') for (x, y, w, h) in stop_sign_scaled: stop_sign_rectangle = cv2.rectangle(img, (x,y), (x+w, y+h), (0, 255, 0), 3) stop_sign_text = cv2.putText (img=stop_sign_rectangle, text="Stop Sign", org=(x, y+h+30), fontFace=cv2.FONT_HERSHEY_SIMPLEX, fontScale=1, color=(0, 0, 255), thickness=2, lineType=cv2.LINE_4) cv2.imshow("img", img) key = cv2.waitKey(30)

4 RESULTS AND CONCLUSION

In previous years, many automated cars have been developed with self-driving mode but it is not fully accurate and the cost is very high [16]. Many cars have a camera for detection but it is not fully predictable. Additionally, it has fuel issues. [14,18]

In this work, using solar energy, natural light available for free of cost and the components are also much less costly so that everyone can afford the car. It is cost-efficient and eco-friendly. Autonomous solar cars have gone through many tests to check it is working autonomously and smoothly. The proposed car model is successfully sensing the environment and collecting the data. It is also observed that the decision-making system in the car is working perfectly. The solar panel's efficiency in making use of sunlight has been tested. Results for lane and object detections have been obtained. An Ultrasonic sensor is employed for obstacle detection, a WIFI module for wireless connection, and LiDAR is used to detect light and voice recognition systems for voice search. A GPS is enabled to detect the easiest path and choose the nearest path when there is any obstacle in front of the car.

The proposed autonomous solar car model was developed by the instructions in the proposed system. The proposed autonomous solar car model integrates camera and ultrasonic sensors to define the object and 3D image of the lane. It gives accurate measures to the cars for lane detection and gives the direction in which the steering has to turn. The logic behind the turn prediction makes use of the central point between 2 lines. It's done by using histogram projection. To enable the algorithm to correctly focus and operate on the crucial parts, the image has been simplified and only included the necessary parts. The OpenCV helps us to get RGB images converted into grayscale images.

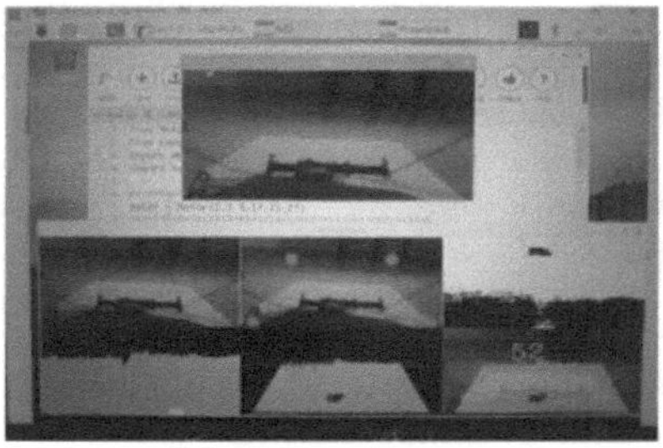

Figure 5. Input from camera.

Figure 5 shows a snapshot that the camera receives. RGB images are converted into grayscale images. The red dots are to detect whether there are obstacles or not. Also, there is an angle at which the car has to take a turn.

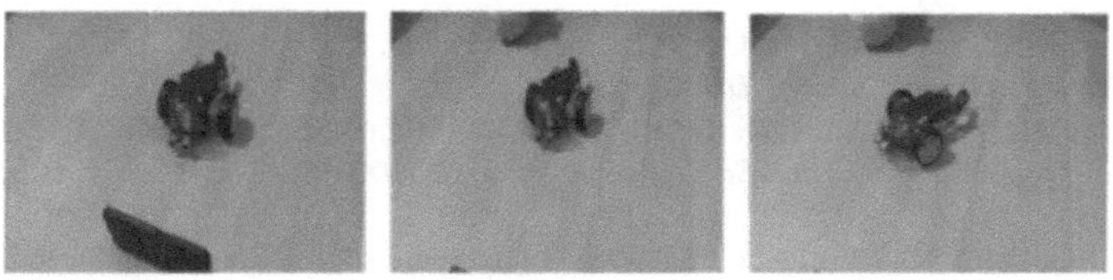

6a 6b 6c

Figure 6. The proposed car model detects the obstacle.

Figure 6a shows that the car model detects the obstacle, Figure 6b shows that the car model takes reverse as soon as the obstacle is detected and Figure 6c shows that the car model changes its path. The proposed autonomous solar car model has a wide-angle raspberry pi camera and ultrasonic sensor which provides a 3D image of the lane and objects in front of the cars that reduce the chance of an accident or predict an accident earlier, providing better option or solution. It makes the decision which is best to avoid the obstacle.

In a nutshell, the proposed low-cost autonomous solar car model performed well, was able to detect obstacles and navigate correctly through the path.

REFERENCES

[1] Arun Muraleedharan, Hiroyuki Okuda, & Tatsuya Suzuki. (March 2022). Real-Time implementation of randomized model predictive control for autonomous driving. *IEEE Transactions on Intelligent Vehicles*, 7(1).

[2] Nitish Srivastava, G. H., Alex Krizhevsky, Ilya Sutskever, & Ruslan Salakhutdinov. (2014). Dropout: A simple way to prevent neural networks from overfitting. *Journal of Machine Learning Research*, 15, 1929–1958.

[3] Gorgees S. Akhshirsh, Nawzad K. Al-Salihi, & Oussama H. Hamid. (April 2021). A cost-effective GPS-aided autonomous guided vehicle for global path planning. *Bulletin of Electrical Engineering and Informatics*, 10(2), 650–657. ISSN: 2302-9285, DOI: 10.11591/eei.v10i2.2734.

[4] Cheng Hualong, Yuanqiao wn, man Zhu, & Yamin Huang (2021). *Autonomous Computerized Systems.*
[5] Kitae Hwang, In Hwan Jung, & Jae Moon Lee. (2021). *RC-CAR autonomous-driving Implementation Utilizing the Raspberry PI device and AI Server.*
[6] Jason Louis Ashton, Myles Emmolo Spencer, & Sean R. Hunt. (2019). *Survey of Autonomous Driving: Common Practices and Emerging Technologies.*
[7] Peide Cai, Hengli Wang, Huaiyang Huang, Yuxuan Liu, & Ming Liu. (October 2021). Vision-Based Autonomous Car Racing Using Deep Imitative Reinforcement Learning. *IEEE Robotics and Automation Letters*, 6(4).
[8] Carsten Markgraf, Luca Gacy, Samuel Leitenmaier, Daniel Lengerer, Benjamin Schwartz, & David W. Gao. (2023). *Autonomous Electric Race Car Inverter Development: Revving Up the Future with Resource Efficient Drive Technology.*
[9] Ji-il Park, Minseong Choi, Seungho Han, Yeongseok Lee, Jinwoo Cho, Hyoseo Choi, Minsu Cho, Minyoung Lee, & Kyung-Soo Kim. (2022). *Development of ROS-Based Small Unmanned Platform for Acquiring Autonomous Driving Dataset in Various Places and Weather Conditions.*
[10] Abu Shufian, Md. Jahidul Islam Rashed, Md. Aman Miah, Md. Hasibuzzaman, & Avijit Sarker. (2020). *Design of a Solar Assisted Autonomous Surveillance Vehicle.*
[11] Cagdas Yaman & Tankut Acarman. (2021). *Extremum Seeking For Economic Driving as An OnLine Mobile Application.* Volume 54, Issue 2.
[12] William T. Bryan, Matthew E. Boler, & David M. Bevly. (2021). *A Vehicle-Independent Autonomous Lane Keeping and Path Tracking System.*
[13] Ben M. Chen, Zongli Lin, & Yacov Shamash. (2022). *A Software Toolkit.*
[14] Kitae Hwang, In-Hwan Jung, & Jae Moon Lee. (January 2022). *Implementation of Autonomous Driving on RC-CAR with Raspberry PI and AI Server.*
[15] Hernández Fuster, Oriol. (2020–2021). *Self-Driving RC Car using Deep Learning.*
[16] Darsh Parek, Nishi Poddar, Aakash Rajpurkar, Manisha Chahal, Neeraj Kumar, Gyanendra Prasad Joshi, & Woong Cho. (June-2022). *A Review on Autonomous Vehicles: Progress, Methods and Challenges.*
[17] Jae-Ho Shin, Jae-Young Yoo, Jun-Hee Han, In-Jun Hwang, & Hyoung-Keun Park. (2022). *A Study on Basic Technology for Autonomous-Driving Using RC car.*
[18] Abhishek Gupta, Alagan Anpalagan, Ling Guan, & Ahmed Shaharyar Khwaja. (June-2022). *Deep Learning for Object Detection and Scene Perception in Self-Driving Cars: Survey, Challenges, and Open Issues.*
[19] Rajkar, A., Kulkarni, N., Raut, A. (2022). Driver Drowsiness Detection Using Deep Learning. In: Iyer, B., Ghosh, D., Balas, V.E. (eds) *Applied Information Processing Systems . Advances in Intelligent Systems and Computing*, vol 1354. Springer, Singapore. https://doi.org/10.1007/978-981-16-2008-9_7
[20] Kaur S., and Kulkarni N., A deep learning technique for emotion recognition using face and voice features, in *2021 IEEE Pune Section International Conference (PuneCon)*, 2021, pp. 1–6. DOI: 10.1109/PuneCon52575.2021.9686510.
[21] Kaur S., and Kulkarni N., Emotion recognition – A review, *Int J Appl Eng Res.*, Vol. 16, no. 2, pp. 103–10, 2021
[22] Kaur S., and Kulkarni N., Recent trends and challenges in human computer interaction using automatic emotion recognition: a review, *International Journal of Biometrics* Vol. 16, No. 1
[23] Kaur, S., & Kulkarni, N. (2023). FERFM: An enhanced facial emotion recognition system using fine-tuned MobileNetV2 Architecture. *IETE Journal of Research*, 1–15. https://doi.org/10.1080/03772063.2023.2202158
[24] Bajaj R., Garg S., Kulkarni N. and Raut R., Sea Debris Detection Using Deep Learning: Diving Deep into the Sea, *2021 IEEE 4th International Conference on Computing, Power and Communication Technologies (GUCON)*, Kuala Lumpur, Malaysia, 2021, pp. 1–6, doi: 10.1109/GUCON50781.2021.9573722.
[25] Kale, A., Jawade, I., Kakade, P., Jadhav, R., Kulkarni, N. (2022). PairNet: A Deep Learning-Based Object Detection and Segmentation System. In: Agrawal, S., Gupta, K.K., Chan, J.H., Agrawal, J., Gupta, M. (eds) *Machine Intelligence and Smart Systems. Algorithms for Intelligent Systems.* Springer, Singapore. https://doi.org/10.1007/978-981-16-9650-3_33
[26] Kinge A., Kulkarni N., Devchakke A., Dawda A., Mukhopadhyay A, (2021). Intelligent Hand Cricket, Cyber Intelligence and Information Retrieval: Proceedings of CIIR, 381–389.

Computational Methods in Science and Technology Volume 1 – Kaur et al. (Eds)
 ISBN: 978-1-032-81760-6

Role and analysis of artificial intelligence interventions in education system of India

Pooja Sharma
Associate Professor, DMS, Tecnia Institute of Advanced Studies, New Delhi

Monisha
Professor, DMS, Tecnia Institute of Advanced Studies, New Delhi

Jatinder Kaur
Assistant Professor, RDIAS, New Delhi

ABSTRACT: Artificial intelligence is one of developing technology that has the potential to affect every facet of our social settings is. In education, AI in education field has initiated to provide innovative teaching pedagogies, which are presently undergoing through assessments in many situations for example pandemic. This article gives instances of the uses of artificial intelligence specifically in education sectors of economically developing countries like India, where One of the 2030 Agenda for sustainable development is universal education. This article offers the reader with an overview of AI. AI has progressed from simple knowledge based system to application based systems to context-driven systems with enhanced capabilities. Indeed, as a new technology, AI will revolutionise the 'learning experience' by creating an adaptable learning space that offers a 'custom learning experience.' Finally, this article provides some instances of how AI technology is being employed in the education sector to enhance the learning process and improve the quality of education. This paper also aims to find out the relationship between gender, cognizance, perspective and AI interventions. The study has found there is significant relationship among gender, cognizance, perspective and AI interventions. This article will help the stakeholders of education sector to understand the intensity up to which AI can be implemented in the field of education and perspective of doing so.

Keywords: Artificial intelligence, Education, Investigate, Personalization and interaction

1 INTRODUCTION

Our traditional education system is' standard in nature which targets curriculum for developing a workforce for market and it stresses on a 'one size fits all' approach (Singh and Jain, 2018).

Unfortunately, this has resulted in the following unfavourable outcomes:

- Many people have been taught yet are unable to adjust to new situations.
  alterations in job requirements
- Little regard for inherent skill, resulting in death inventiveness.
- Lack of enthusiasm for learning.
- An unhappy, disappointed adolescent.

As a result, the existing educational system is flawed, weak, and ineffectual in addressing demands of the modernized world. However, the comeback of Artificial Intelligence provides a light of hope for establishing tailored learning systems (Dialani, 2019).

DOI: 10.1201/9781003501244-37

1.1 *Artificial Intelligence (AI)*

Artificial intelligence is the process of teaching computers, robots, or software to make smart judgements in the same way as clever humans do. Understanding how our minds work naturally, as well as how humans learn, make decisions, and take action to solve issues, is the first step in designing intelligent systems and software. Intelligence is usually characterised as the ability to acquire and think about information in order to solve complex problems. Smart robots will soon supersede human abilities in many professions (Dagioglou *et al.* 2018). Artificial intelligence refers to the field of study and the development of intelligent machines and software capable of tasks such as thinking, learning, acquiring information, communication, operation, and observation (Panigrahi, 2020).

The term was coined by John McCarthy in 1956 to denote a branch of computer science focused on programming systems to exhibit human-like behavior. Computing is what allows us to perceive, reason, and act. In contrast to psychology, AI mainly revolves around computation, while computer science encompasses perspective, reasoning, and action. This means that AI primarily deals with computational tasks, while computer science explores a wider range of aspects including how machines perceive their environment, reason through problems, and take action accordingly. Moreover, the integration of AI significantly boosts the capabilities of machine learning and enhances its practical usefulness (Panigrahi, 2020).

Framework for Artificial Intelligence (Singh and Jain 2018) propose the following AI framework, which includes the following elements. They are as follows:

- Perspective – The ability to gather information about the natural world through sensory systems and language (Ahmad & Krishnan, 2017).
- Issue analysis – The capacity to use the knowledge gathered above to identify, analyse, and solve a problem.
- Intellectual knowledge – The capacity to draw knowledge from one domain's perspective and application of the same knowledge in another area (Lau *et al.* 2014).
- Activity based learning entails the ability to continuously improve perspective and understanding by the way of learning from real life problems.

AI, or artificial intelligence, makes predictions about future prospects based on learning from enormous amounts of past records (Poole & Mackworth, 2010). These include augmented reality, virtual reality, movies, and block chain technology. AI and other digital technologies should be employed to assist instructors in their educational efforts (Sarkar, 2018).

1.2 *Use of AI in education*

It is recommended to use an educational approach about the use of artificial intelligence by providing education and training work and live with it and then use AI in the design of educational and training systems.

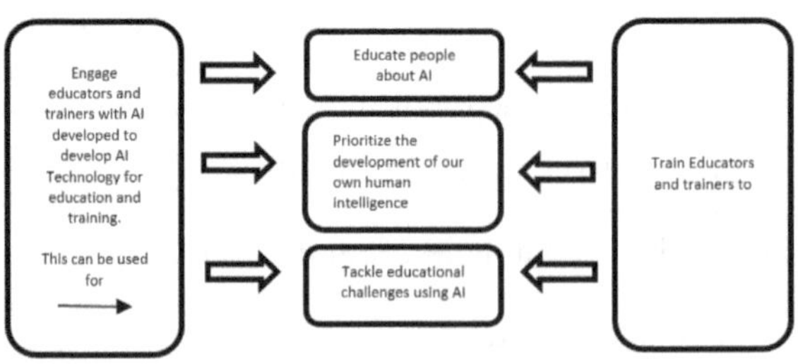

Figure 1. AI in education. Source. Ross and Issroff. (2018).

It should be mentioned that the AI framework was thoughtfully designed. Education and training have the potential to be extremely beneficial.

AI is beneficial for educators, students, parents, guardians and the general public.

Indeed, AI in education will amend the entire "learning" process. Only a few factors compelled changes in the worldwide education business, which are outlined below (Blaylock, 2019).

- Flexible Learning of students
- New Possibilities
- Increased Productivity
- AI-powered Learning Platform

It's possible that AI with these kind of enhancements to the education sector may be considered an emerging technology, since it could help students study more effectively.

1.3 *Objectives of the study*

(1) The primary objective is to explore the correlation between gender and cognizance pertaining to AI intervention within the Indian education sector.
(2) The aim is to assess the relationship between gender and the perspective that AI intervention could potentially disrupt the educational system in India.
(3) The goal is to investigate how gender influences perspectives pertaining to AI intervention's ability to enhance personalization and interaction within the educational context.

2 RESEARCH METHODOLOGY

Data was gathered from primary sources. The primary data was collected using an online survey using a well-structured close-ended questionnaire with ten survey items. To determine multiple facets specified in the hypothesis, the Pearson Chi-Square ($\chi 2$) test was performed. The sample size consisted of 300 people from Punjab. Individuals above the age of 18 who have finished their basic education in 10 + 2 made up the sample unit. The sample for this investigation was chosen using simple random selection (Alemi *et al.* 2015).

3 DATA ANALYSIS

Primary data analysis is the preliminary assessment of data collected for a research topic. This entails interpreting the data gathered in order to answer research questions or confirm or disprove the study's hypotheses. Essentially, it is the process of analysing data to reach meaningful findings consistent with the study's aims.

3.1 *Hypothesis 01*

H0: In India, there is no significant correlation between gender and cognizance pertaining to AI intervention within the Indian education sector.

H1: In India, there is a substantial correlation between gender and cognizance pertaining to AI intervention within the Indian education sector.

Table 1. Gender details.

		Gender		
		F	M	T
Yes	No.	100	123	223
	Percentage	78.12%	71.51%	74.33%
No	No.	28	49	77
	Percentage	21.87%	28.48%	25.66%
	Total	128	172	300

The $\chi 2$ statistic is 1.6823. The p-value is .194616. The result is not significant at $p < .05$.

The Table 1 demonstrates the association between gender and cognizance of AI involvement in Education sector in India. According to the Table 1, 78.12% of females are aware of these technologies, while 21.87% are not. On the other hand, 71.51% of men are aware of such technology, while 28.48% are not. The Chi square test was employed to determine this difference is significant or not. According to the above table, the computed Pearson Chi-square ($\chi 2$) value is = 1.6823, which is less than the critical value of 3.84 at the 5% level of significance with degree of freedom V = 1. Therefore, there is not enough evidence to reject the null hypothesis that there is no significant difference between the observed and expected frequencies. This suggests that any differences are likely due to random chance. As a result, the null hypothesis is accepted (Dr.Vikas Sharma Dr.Pooja Sharma, 2022). That means there is a negligible link between gender and AI Cognizance (Allen, 2019).

3.2 *H02 hypothesis*

H0: There is no statistically significant association between gender and the perspective that AI intervention of artificial intelligence would disrupt India's educational framework (Allen, 2019).

H2: Gender and the perspective pertaining to AI intervention would overhaul India's educational framework have a substantial link.

Table 2. People's perspectives about AI intervention.

		Gender		
		Female	Male	Total
Yes	No.	43	69	112
	Percentage	34.40%	39.43%	37.33%
No	No.	82	106	188
	Percentage	65.60%	60.57%	62.67%
	Total	125	175	300

The value of $\chi 2$ is 0.7881 & The p-value is .374675.

The table number 2 demonstrates the association between people's perspectives about AI intervention disrupting India's (Sharma, 2022) education system and gender. According to the table, 34.4% of females have a good perspective of this remark, while 65.6% have the opposite perspective. In contrast, 39.43% of males have a good opinion of the statement, while 60.57% have a negative opinion. The Chi square test was employed to determine whether or not this difference is significant. As mentioned in above table number 2, Pearson Chi-square ($\chi 2$) value is =.7881, It is less than 3.84 critical value at the 5% significance level with dof value 1. As a result, the null hypothesis is accepted. This suggests that there is a negligible association between gender and people's perspectives about AI intervention disrupting India's education system (Patel *et al.* 2018).

3.3 *Hypothesis H03*

H0: There is not any significant association between gender and the perspective that AI intervention in educational structure would boost personalization and interactivity.

H3: There is substantial association between gender and the notion that AI intervention would boost customization and engagement.

Table 3. People's perspectives of AI intervention improving personalization and interaction.

		Gender		
		Female	Male	Total
Yes	No.	106	152	258
	Percentage	81.54%	89.41%	86.00%
No	No.	24	18	42
	Percentage	18.46%	10.59%	14.00%
	Total	130	170	300

$\chi 2$ value is 3.7928 & p-value is .051474. Table number three demonstrates the link between people's perspectives of AI intervention improving personalization and interaction with regard to gender. According to the table, 81.54% of females have a favorable impression of this statement, while 18.46% have a negative perspective. In contrast, 89.41% of males have a good opinion of the statement, while 10.59% have a negative one. The Chi square test was employed to determine whether or not this difference is significant. According to the above table, the estimated Pearson Chi-square ($\chi 2$) value is = 3.7928, It is larger than 3.84 critical value at 5% level of significance with degree of freedom value = 1. As a result, the null hypothesis is rejected. Gender has a substantial association with the perspective that AI Intervention would promote personalization and engagement (Sreedharan *et al.* 2022).

4 CONCLUSION

A robust collection of tools, guidelines, and practice guidelines are necessary for AI-based solutions to be framed, monetized, and provided to a larger, varied audience. The possibility of customized quality education, irrespective of geographic, socio-cultural, or other affiliations, is recognized through AI and Deep Learning. On the other hand, a consistent and well-coordinated effort to address these issues from various perspectives, including educational, experiential, commercial, ethical, and governance, is required. (Dhavala, 2018)

Despite the enormous potential benefits of AI, there may also be certain hazards associated with it. Artificial intelligence (AI) might end up being the finest or worst thing that has happened to people. While AI has the potential to improve education, its use in universities also raises significant ethical concerns. Budget cuts and the continuing corona virus outbreak are two factors that may lead school administrators to consider replacing teachers with profitable automated AI solutions. There is some concern that increased use of AI in the classroom might lead to a decline in interpersonal connection and the development of technological dependency amongst students, both of which could be detrimental to their education.

The study results show that the impression of AI does not differ based on gender. The same is true for both men and women. As far as second objective is concerned, it can be concluded that the impact of AI intervention on India's education system is gender agnostic. The resistance from individuals to this viewpoint suggests their belief that the advancement of AI intervention in the education sector is merely a temporary measure in response to COVID-19. Everything will return to normal once the situation has stabilized. On the other hand, the view that AI involvement will increase personalization and interaction varies with gender, and there is a substantial link between them, according to the study. The research overlooks the perspectives of vulnerable populations, such as individuals with disabilities or those from marginalized communities. Further collaboration between researchers from diverse disciplines, longitudinal studies tracking individuals' experiences, cultural sensitivity, and interdisciplinary collaboration are necessary for addressing multifaceted challenges of AI intervention in personalization and interaction. Integrating insights from different fields can lead to more holistic understandings and innovative solutions.

REFERENCES

Ahmad, J., & Krishnan, M. (2017). Environmental related CSR initiatives for business strategy: A case study on small and medium (SME) industries in Malaysia. *In CSR, Sustainability, Ethics and Governance*. https://doi.org/10.1007/978-3-319-39089-5_16

Alemi, M., Meghdari, A., & Ghazisaedy, M. (2015). The impact of social robotics on L2 learners' anxiety and attitude in English vocabulary acquisition. *International Journal of Social Robotics*, 7(4), 523–535.

Dagioglou, M., Blasco, S. A., Llobet, D. N., & Konstantopoulos, S. (2018). Realistic and unobtrusive solutions for independent ageing. *In RADIO–Robots in Assisted Living: Unobtrusive, Efficient, Reliable and Modular Solutions for Independent Ageing*. https://doi.org/10.1007/978-3-319-92330-7

Dialani, P. (2019). Use of AI and VR in the indian education sector. *Analytics Insight*. Retrieved December, 29, 2021.

Dr. Vikas Sharma Dr. Pooja Sharma, D. N. P. (2022). Stock return and trade volume analysis of selected Indian cement companies' stocks. *Journal of Contemporary Issues in Business and Government*, 28(1), 59–69. https://doi.org/10.47750/cibg.2022.28.01.006

Lau, L., Yang-Turner, F., & Karacapilidis, N. (2014). Requirements for big data analytics supporting decision making: A sensemaking perspective. *In Studies in Big Data* (Vol. 5). https://doi.org/10.1007/978-3-319-02612-1_3

Panigrahi, C. M. A. (2020). Use of artificial intelligence in education. *Management Accountant*, 55, 64–67.

Patel, V., Saxena, S., Lund, C., Thornicroft, G., Baingana, F., Bolton, P., Chisholm, D., Collins, P. Y., Cooper, J. L., & Eaton, J. (2018). The lancet commission on global mental health and sustainable development. *The Lancet*, 392(10157), 1553–1598.

Poole, D. L., & Mackworth, A. K. (2010). *Artificial Intelligence: foundations of computational agents*. Cambridge University Press.

Sharma, V. (2022). A pragmatic study on management with autocratic approach and consequential impact on profitability of the organization. In P. K. Mallick, A. K. Bhoi, A. González-Briones, & P. K. Pattnaik (Eds.), *Electronic Systems and Intelligent Computing* (pp. 121–130). Springer Nature Singapore.

Sreedharan, S., Kulkarni, A., & Kambhampati, S. (2022). Explainable Human-AI Interaction: A Planning Perspective Sreedharan. In *Synthesis Lectures on Artificial Intelligence and Machine Learning* (Vol. 16, Issue 1). https://doi.org/10.2200/S01152ED1V01Y202111AIM050

Computational Methods in Science and Technology Volume 1 – Kaur et al. (Eds)

Technological advancements in the detection and diagnosis of migraine: A review

Shivneet Kaur, Parneet Kaur, Monika Gosain & Roopkamal Kaur
Department of Computer Science & Engineering, Chandigarh Engineering College-CGC, Landran

ABSTRACT: Headaches are the most prevalent medical condition in the globe. The migraine is one of the various types of headaches. Along with their excruciating headaches, migraineurs often have nausea, vomiting fits, and light and sound sensitivities. Individuals have unique symptoms, some of which may be more intense than others. A migraine episode is about to begin when an aura, a symptom, appears. Based on the nature of symptoms, migraines can be divided into two basic categories: aura-producing and aura-less. Every individual has different triggers for migraine. This study provides an overview of the state-of-the-art deep learning and machine learning techniques for migraine identification. The primary subjects of this study include migraine kinds, phases, causes, and symptoms.

1 INTRODUCTION

Pain or discomfort in the head or neck is a popular indicator of headache. It is a common event that could be trivial or serious. There may be a constant or pulsating pressure-like sensation to the discomfort (Karagholi *et al.* 2022) [1]. There are various types of headaches, and they can be caused by a variety of factors, such as stress, tension, lack of sleep, and underlying medical conditions. While many headaches are not harmful, some may require medical attention. It is imperative that one must notice the intensity, duration, and occurrence of any new symptoms so that accurate medical aid can be provided.

There is a lot of similarity in a normal headache and that of a migraine pain, but they vary in characteristics, symptoms and intensity. A migraine pain can be defined by an intense pain that usually happens in one side of the skull. There are other symptoms that accompany this pain; these are nausea, vomiting and an extreme sensitivity to light and sound. In some cases, these is a warning sign that comes along with known as an "aura". This can cause visual abnormalities in some people before of after the migraine attack as it is much intense and incapacitating than normal headaches. A migraine is a different but, more severe form of head pain, whereas a headache is very common that can be caused by many reasons (Karagholi *et al.*, 2022; Martinelli *et al.* 2022) [1,11].

As stated earlier, each individual may experience different symptoms depending on the severity and impact of the migraine attack. However, the most general symptoms involve vomiting, nausea, feeling tingling or numbness, seeing flashing lights, and many more. These symptoms can last for hours to days, making it challenging to carry out normal daily activities. Migraines vary from person to person, and the severity and combination of symptoms can differ. If you suspect you are experiencing migraines, seeking medical advice is important for proper diagnosis and management.

Migraine can affect a person's health. First, it is a severe headache and not just a headache; just like a superhero has a headache. It evokes such wonderful emotions that your eyes hate the light, and your ears hate the nose. Sometimes you may feel pain in your stomach

DOI: 10.1201/9781003501244-38

and vomit. Imagine that a superhero with a headache will remain as tired and weak as if he had run a marathon, even if he rested for a while. This isn't just a headache; It's as if your whole body is involved in this incredible adventure. Migraines can affect your mood and make you anxious or sad. So, these are not just headaches; They are like uninvited guests, disrupting the harmony of your body (Butt *et al.* 2022) [3].

Migraine without aura and migraine with aura are the two basic types, both of which can be quite incapacitating. Clinical manifestations of migraine attacks in patients can include the intensity of the pain and the pattern of associated symptoms such as sensitivity to movement, nausea, vomiting, phonophobia, osmophobia, and photophobia. There are many comorbidities associated with migraine, including psychological disorders (like anxiety and depression), sleep issues, fatigue, cardiovascular risk factors (including obesity, high blood pressure, diabetes, and diabetes), and diseases of the heart and brain (Wachholtz *et al.* 2016) [16]. Most common phases of a migraine attack are represented below in Figure 1.

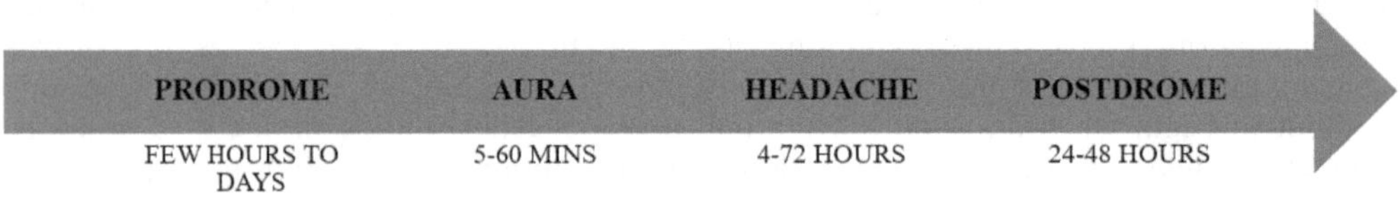

Figure 1. Phases of migraine.

There are different types of misery in the head. Menstrual migraine, migraine with aura, chronic migraine, migraine with brainstem aura, vestibular migraine, abdominal migraine, hemiplegic migraine, and many more varieties are the most prevalent ones (K. Holland, 2022) [7].

Migraine detection can be done through the doctor consultation, MRI, CT scans. It can also use various techniques of machine learning and deep learning (Gosain *et al.* 2022) [21]. Section 2 introduces the past research for detection of migraine (Woldeamanue and Cowan 2022) [18].

2 LITERATURE SURVEY

This section describes existing studies that have been put forth to improve migraine detection performance.

2.1 *Based on machine learning*

Mitrovic *et al.* (2023) [13] proposed the study to evaluate several machine learning methods for separating MwA from healthy individuals as well as those with basic and complex MwA. data from magnetic resonance imaging (MRI) following a procedure. The outcome shown that an algorithm utilizing post-procedure MRI data was able to classify MwA patients with high accuracy (97%) and differentiate between simple and complex MwA with 98patients % accuracy.

Aung and Srimaharaj (2023) [2], presented statistical analysis using entropy minimization discretization (EMD) in conjunction with Naïve Bayes machine learning. By precisely summarizing the patient's symptoms and medical background, this approach saves doctors' time and expedites the diagnostic procedure. The findings demonstrate the great accuracy of naïve bayes categorized with EMD.

Koskimäki *et al.* (2017) [10] discussed the methods affecting migraine had been abandoned and the concentration has shifted to early detection of migraine attacks through the use of anthropometric bio signals. The objective was to create a model that will help people take their medications on time and thus prevent migraine pain by using a single, simple and useful sensor device. EEG (electroencephalogram) band data is used to divide migraines into interictal, preictal, ictal and postictal groups. They discover the user with over 60% accuracy.

Sarsam *et al.* (2020) [15], To detect the early stage of migraine detection. A detailed Migraine data was collected from Twitter. To identify migraine theories and look into the connection between migraine symptoms and weather, they employed the K-means and Apriori algorithms. The findings demonstrated a link between migraine symptoms and depression. The findings of the classification indicate that Sequential Minimal Optimization is effective in identifying migraine symptoms from Twitter (95.53% accuracy). Since it is possible to detect signs of diseases on Weibo, the concept can be used effectively in biomonitoring systems. This research has expanded the way emotions and security can be used to find disease-related findings.

Table 1. Limitations of existing work.

Author	Data Set	Model	Outcome	Limitations
Zhu *et al.*, 2019 [19]	42 patients	Random Forest, SVM, KNN, logistic regression.	Accuracy = 80.2%	Small Feature Set
Tu *et al.* 2020 [16]	230 participants	FMRI-based neural network, SVM, Random Forest	Accuracy = 93% 89% specificity	Small Sample Size
Jackowaski *et al.* 2014 [8]	579 patients	KNN, SVM, naïve bayes, Decision Tree, Radial basis network.	Accuracy = 72.0%	Cannot predict in advanced which algorithm would be most appropriate
Garcia-Chimeno *et al.*, 2017 [5]	52 Adults	Naïve bayes	Accuracy = 93%	Limited number of patients.
Keight *et al.* 2017 [9]	836 patients	Supervised learning, SVM, KNN, Decision tree classifier, Random Forest.	AUC = 0.985 AUC = 0.960 AUC = 0.964	A more sophisticated algorithm might be utilized to produce a better diagnosis model.
Mitrovic *et al.* 2023 [13]	78 patients with Aura	K Nearest Neighbour, Support Vector Machine(SVM), Naïve bayes, Decision Tree.	Accuracy value = 97%	More datasets could be analysed.
Liu *et al.* 2022 [11]	78 patients with Aura	Logistic regression, KNN, SVM, naïve bayes, Decision Tree.	Accuracy = 97%	More datasets could be analysed.

2.2 *Based on deep learning*

Z. Cao *et al.* (2017) [4] Using electroencephalogram (EEG) data, the study presents a deep learning-based computer-aided diagnostic (CAD) system for migraine diagnosis. The continuous wavelet transform method is used in the suggested method to turn EEG signals into scalogram images. The suggested deep learning network for convolutional neural networks (CNNs) was then fed these scalogram images. Comparing the proposed CNN deep learning model to other relevant literature, it exhibits 100% classification success with very high accuracy.

H.G. Ng *et al.* (2018) [14], Convolutional neural network (CNN), a deep learning technique, is used by the author to classify magnetic resonance imaging (MRI) data into groups for migraine and non-migraine patients. With a validation accuracy of = 0.69, = 0.06, the maximum classification test accuracy of 85% was attained from 198 MRI images that were evenly drawn from both data groups. The baseline statistical accuracy of 50% is in contrast to this.

H. Goker (2022) [6] presented a DL model in the work for identification of migraine automatically from electroencephalogram information. EEG recordings from 21 healthy individuals and 18 migraineurs make up the collection. The Welch method was used to find the density values of the EEG signal, which ranged from 1 to 49 Hz, in order to produce feature vectors. Using the generated feature vectors, the effectiveness of the BiLSTM deep learning algorithm and the machine learning strategies for SVM, random forest, and LDA were assessed in EEG-based migraine classification tasks. BiLSTM (95.99%) was the top performing deep learning algorithm while using 128 channels.

3 GENERAL METHODOLOGY

This section discusses about the general methodology being applied in the existing work for detecting migraine.

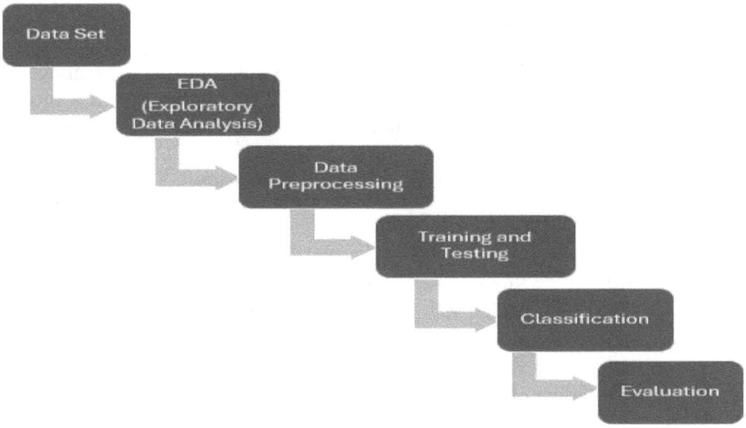

Figure 2. General framework.

Figure 2 shows the general framework of migraine detection and the steps are as follows:

3.1 *Data set*

In machine learning, data collecting is obtaining pertinent information from multiple sources, such as the patient's age, gender, and symptoms, to train and enhance algorithms.

3.2 *EDA*

To identify patterns, correlations, and outliers in the dataset, exploratory data analysis, or EDA, is used to discover migraines. It helps to find out the importance of features and what role do they play.

3.3 *Data preprocessing*

Usually, the datasets for migraine detection have images that may be blurry or incomplete. To make these images suitable for feeding into the machine learning model, pre-processing is a must. This can be done by cleaning the data, data encoding to convert string columns into numerical data. Data normalization can also be done to balance the data values (Kaur *et al.* 2022) [20].

3.4 *Train-test split*

The complete dataset is then divided into two parts: train and test set. The first is used for training as the model learns when data is fed to it while the other is used to test the model on a new dataset fed into it.

3.5 *Model classification*

One of the most significant steps in the entire detection process is the classification. Classification is mostly used to group data into specific categories. It informs us whether the disease has been detected in an individual or not.

3.6 *Evaluation*

This step is used to find the effectiveness of the model through certain parameters. Examining the values of these metrics help us to known how effective the model is when fed with a new dataset. By evaluating the model's ability to correctly identify migraine patients and avoid false positives, metrics such as accuracy, precision, and recall are used to assess the model's performance in real-world scenarios.

4 DISCUSSION

After examining the current studies, we find out that the most commonly used classification models are SVMs, CNNs, KNN, Logistic Regression, and Naïve Bayes. It is also observed that the most effective techniques among these are SVM, CNN, K- Nearest Neighbour and naïve bayes. A significant research gap lies in the limited exploration of personalised models. Tailoring algorithms to individual patient characteristics, including genetics, lifestyle, and environmental factors, could enhance the accuracy and personalised treatment strategies for migraine sufferers.

5 CONCLUSION

To summarize, a migraine is a neurological condition marked by persistent, severe headaches that are frequently accompanied by other symptoms like light and sound sensitivity and nausea. Auras, which are warning symptoms indicators that precede episodic episodes that normally last a few hours to several days, can have a substantial impact on day-to-day activities. The agonizing sensation of having a migraine episode is exacerbated by the throbbing pain, which is frequently accompanied by visual problems. This paper delves into the current research on migraine identification using machine and deep learning. The study indicates that CNN and SVM are the most widely utilized approaches.

REFERENCES

[1] Al-Karagholi, M. A.-M., Peng, K.-P., Petersen, A. S., De Boer, I., Terwindt, G. M., & Ashina, M. (2022). Debate: Are cluster headache and migraine distinct headache disorders? *The Journal of Headache and Pain*, *23*(1).

[2] Aung, N. N., & Srimaharaj, W. (2023). Migraine categorization based on the integration of EMD and Naive Bayes classification. *2023 Joint International Conference on Digital Arts, Media and Technology with ECTI Northern Section Conference on Electrical, Electronics, Computer and Telecommunications Engineering (ECTI DAMT & NCON)*.

[3] Butt, J. H., S Eddelien, H., & Kruuse, C. (2022). The headache and aura-inducing effects of sildenafil in patients with migraine with Aura. *Cephalalgia*, *42*(10), 984–992.

[4] Cao, Z., Lai, K.-L., Lin, C.-T., Chuang, C.-H., Chou, C.-C., & Wang, S.-J. (2017). Exploring resting-state EEG complexity before migraine attacks. *Cephalalgia*, *38*(7), 1296–1306.

[5] Garcia-Chimeno, Y., Garcia-Zapirain, B., Gomez-Beldarrain, M., Fernandez-Ruanova, B., & Garcia-Monco, J. C. (2017). Automatic migraine classification via feature selection committee and machine learning techniques over imaging and questionnaire data. *BMC Medical Informatics and Decision Making*, *17*(1).

[6] Göker, H. (2022). Automatic detection of migraine disease from EEG signals using bidirectional long-short term memory deep learning model. *Signal, Image and Video Processing*, *17*(4), 1255–1263.

[7] Holland, K. (2022, June 15). *Rare and extreme types of migraine*. Healthline.

[8] Jackowski, K., Jankowski, D., Simić, D., & Simić, S. (2015). Migraine diagnosis support system based on classifier ensemble. *ICT Innovations 2014*, 329–339.

[9] Keight, R., Aljaaf, A. J., Al-Jumeily, D., Hussain, A. J., Özge, A., & Mallucci, C. (2017). An intelligent systems approach to primary headache diagnosis. *Intelligent Computing Theories and Application*, 61–72.

[10] Koskimäki, H., Mönttinen, H., Siirtola, P., Huttunen, H.-L., Halonen, R., & Röning, J. (2017). Early detection of migraine attacks based on wearable sensors. *Proceedings of the 2017 ACM International Joint Conference on Pervasive and Ubiquitous Computing and Proceedings of the 2017 ACM International Symposium on Wearable Computers*.

[11] Liu, F., Bao, G., Yan, M., & Lin, G. (2022). A decision support system for primary headache developed through machine learning. *PeerJ*, *10*.

[12] Martinelli, D., Pocora, M. M., De Icco, R., Putortì, A., & Tassorelli, C. (2022). Triggers of migraine: Where do we stand? *Current Opinion in Neurology*, *35*(3), 360–366.

[13] Mitrović, K., Petrušić, I., Radojičić, A., Daković, M., & Savić, A. (2023). Migraine with aura detection and subtype classification using machine learning algorithms and Morphometric Magnetic Resonance Imaging Data. *Frontiers in Neurology*, *14*.

[14] Ng, H. G., Kerzel, M., Mehnert, J., May, A., & Wermter, S. (2018). Classification of MRI migraine medical data using 3D convolutional neural network. *Artificial Neural Networks and Machine Learning – ICANN 2018*, 300–309.

[15] Sarsam, S. M., Al-Samarraie, H., Ismail, N., Zaqout, F., & Wright, B. (2020a). A real-time bio-surveillance mechanism for early-stage disease detection from microblogs: A case study of inter-connection between emotional and climatic factors related to migraine disease. *Network Modeling Analysis in Health Informatics and Bioinformatics*, *9*(1).

[16] Tu, Y., Zeng, F., Lan, L., Li, Z., Maleki, N., Liu, B., Chen, J., Wang, C., Park, J., Lang, C., Yujie, G., Liu, M., Fu, Z., Zhang, Z., Liang, F., & Kong, J. (2020a). An fmri-based neural marker for migraine without aura. *Neurology*, *94*(7).

[17] Wachholtz, A. B., Malone, C. D., & Pargament, K. I. (2015). Effect of different meditation types on migraine headache medication use. *Behavioral Medicine*, *43*(1), 1–8.

[18] Woldeamanuel, Y. W., & Cowan, R. P. (2022). Computerized migraine diagnostic tools: A systematic review. *Therapeutic Advances in Chronic Disease*, *13*, 204062232110652.

[19] Zhu, B., Coppola, G., & Shoaran, M. (2019). Migraine classification using somatosensory evoked potentials. *Cephalalgia*, *39*(9), 1143–1155.

[20] Kaur, P., Kumar, Y. & Gupta, S. (2022). Artificial intelligence techniques for the recognition of multi-plate multi-vehicle tracking systems: A systematic review. *Arch Computat Methods Eng* 29, 4897–4914.

[21] M. Gosain, S. Gupta and S. Kaur, Machine and deep learning techniques to classify and predict thyroid diseases, *2022 3rd International Conference on Intelligent Engineering and Management (ICIEM)*, London, United Kingdom, 2022, pp. 675–680.

Computational Methods in Science and Technology Volume 1 – Kaur et al. (Eds)
© 2025 the Author(s), ISBN: 978-1-032-81760-6

Classification of lung cancer images using machine learning

Bhumi Goyal, Saksham Jha, Tanmay Sharma, Hardik Sharma & Nirbhay Kashyap
ASET, Amity University Noida

ABSTRACT: In Today's world there are many diseases but among them there is lung carcinoma which is a deadly disease and is more dangerous than other types of cancer. Lung carcinoma is another name for lung cancer. Uncontrollable cell development inside lung tissue is the root reason of lung cancer. Initial lung carcinoma identification is crucial for accurate diagnosis. Detecting of lung carcinoma initially is tough as lung cancer does not have any major symptoms at an early stage. However, initial identification of lung carcinoma contributes in maintaining overall survival rate of people. Consequently, by conducting numerous investigations, several researchers develop diverse approaches for identifying lung cancer. This research paper focus on the hybrid methodology for classification of lung cancer. The proposed methodology is first pre-processed using the Adaptive median filtering. Once the images are pre-processed then the processed images are sent for feature extraction. Feature are extracted using Wavelet-based partial Hadamard Transform, Local Ternary Pattern and Grey level co-occurrence Matrix is used. Then from those extracted features the necessary features are selected using Wrapper algorithm. After the features are selected the classification takes place using KNN, MLP and SVM. The input is characterized as benign and malignant.

1 INTRODUCTION

Uncontrolled growth of tissue is the cause of lung cancer. This condition is referred to as mutation. A mutation can be caused by environmental factors or inherited genetic flaws. While there are numerous other carcinogens, the chemicals in the smoke from cigarettes are the most frequent environmental offenders. New tissue often grows in the wounded spot, but a purposeful alteration might cause the new tissue to grow out of control and cause malignant cells to emerge. Although the majority of tumours in the lungs are benign, some may be early indicators of malignancies. Unusual growths in the lungs that may or may not be cancerous. According to a recent study, only about 19% of those with lung cancer patients make it to their fifth year of life.

Initial disease diagnosis through observation of symptoms and indications is a very difficult undertaking that necessitates several medical tests and analyses to reach an accurate diagnosis. It requires the use of diagnostic instruments to help medical professionals get a diagnosis with accuracy. The death rate is decreased by early disease detection and appropriate treatment. The WHO approximations that 7.6 million people globally lose their lives to lung cancer each year.

Symptoms including blood in the cough, fatigue, shortness of breath, fractures in bones, headaches blurred memory, pain in the joints, blood loss altered sound, brain disorder, variation in sputum colourr, and loss of weight are commonly used to diagnose lung cancer physically.

Lung cancer stages are identified, classified, and diagnosed using a variety of imaging modalities. A few of the commonly used methods include CT, MRIs, PET scans, and chest radiographs, also referred to as X-rays. These imaging modalities offer localised information

DOI: 10.1201/9781003501244-39

on the existence of a tumour, the disease's metastatic state, and even the body's lymph nodes. Additionally, a mixture of image modalities, including PET and CT, is used to obtain more detailed anatomical information about the area. Radiologists can detect disease by using CAD devices, which automatically recognize lung nodules. The section on image acquisition contains information on the various imaging modalities.

2 LITERATURE SURVEY

Tian (2021) produced an optimised Fuzzy C-Ordered Means (DSARFCM) utilising the search and rescue optimisation algorithm, as well as a novel clustering-based method for the automatic segmentation of lung tumour images. On the basis of enhanced capsule networks, the final categorization is made. Advantages of the proposed method include higher efficiency in accuracy, reliability, and convergence compared to other methods. Disadvantages mentioned in the literature include the sensitivity of certain methods to initial centroids and the need for improved classification models to enhance system accuracy.

Pankaj *et al.* [10] developed a model for lung carcinoma classification using SVM and NN. Proposed algorithm, named KASC, integrates SVM with FFBPNN to enhance classification accuracy. The advantage of this hybrid approach is demonstrated through the achievement of an overall accuracy is 98.08% for classification of lung cancer using more than 500 CT imaging data samples. a potential disadvantage could be the complexity of implementing and fine-tuning the hybrid algorithm, which may require specialized expertise and computational resources.

Venkatesan *et al.* (2024) proposed an efficient lung cancer detection technique the use of most appropriate SVM and improved swarm optimization, demonstrating high accuracy fees of ninetynine.35% and ninety-nine. Fifty-five% with more suitable performance and decreased jogging time. Their technique includes the usage of DLBP, Hybrid WPHT, adaptive median filtering, ultimate SVM, and stepped forward weight-primarily based beetle swarm optimization, leveraging datasets which includes LUNA-16 and benchmark lung databases.

Bushara *et al.* (2023) advanced an ensemble approach for lung most cancers detection and type using Computed Tomography (CT) pics. Their method utilizes VGG-CapsNet, a fusion of Visual Geometry Group (VGG) and Capsule Network, showcasing excessive accuracy, precision, and specificity in lung most cancers detection using datasets like LIDC-IDRI and Kaggle. While the examine emphasizes the superior overall performance of VGG-CapsNet over traditional Convolutional Neural Networks (CNNs), it does not explicitly point out any boundaries. The study reports high accuracy, precision, and specificity of VGG-CapsNet in detecting lung cancer using LIDC-IDRI and Kaggle datasets. The advantages include superior performance over traditional CNNs, while the limitations are not explicitly mentioned. The algorithms used include CapsNet, CNN-CapsNet, and VGG-CapsNet. The datasets used are LIDC-IDRI and Kaggle datasets.

Elnakib *et al.* (2020) developed an early lung cancer detection method utilising deep mastering optimization, with the I-ELCAP dataset. The algorithm includes function extraction from CNN fashions like VGG16, VGG19, and AlexNet, alongside Genetic Algorithm (GA) for function choice and classification using SVM, Decision Trees, and KNN. Limitations encompass the inadequacy of the take a look at set to generalize version performance due to the small dataset length.

Discovery Prognosis-Related Genes and Paths by Multi-Omics Analysis in Lung Carcinoma. Biomolecules. Asada *et al.* highlights the use of TCGA, GEO dataset, it used multiple techniques like multi-omics analysis, random forests, survival analysis models, advantages and limitation of this model was multi-omics approach provides a more complete picture of biological processes compared to focusing on single omics data. Improved accuracy, identifies potential pathways, Data integration challenges Combining data from different sources with varying formats and quality can be challenging, requiring careful normalization and preprocessing.

Shafi *et al.* (2022) offered a thorough examination of methods for detecting lung cancer, with a particular emphasis on the LUNA16 dataset and a range of approaches, including hybrid models, deep learning, and machine learning. The LUNA16 dataset, comprising 888 samples of CT scanned in DICOM format, is the one that was used. SVM, DT, RF, CNN, and CapsNet are among the ML, DL, and hybrid models used in the research to categorise nodules as cancerous or noncancerous. The suggested hybrid model outperforms state-of-the-art methods and yields faster and more accurate results than other existing approaches. It combines CNN for feature selection with SVM for classification. It achieves a 94% accuracy rate. But the paper also addresses the drawbacks of specific methods, such the requirement for a lot of training data for CNN, the risk of overfitting, and the max-pooling layer's ineffectiveness at retaining spatial info.

Pradhan *et al.* (2023) proposed an ensemble learning approach for accurate lung cancer diagnosis, leveraging features extracted from CNN, PCA, and t-SNE, and optimized using the BF-SSA algorithm. Model's advantages include improved accuracy and early disease detection, while its limitations involve challenges in combinatorial optimization and data scarcity. The dataset used for the anticipated lung carcinoma analysis model is collected from Kaggle and machine-learning databases, containing attributes related to lung carcinoma detection.

Liu *et al.* introduced a deep CNN version for number one lung nodule identification, utilising a non-overlapping patch-based totally method for lung segmentation. Improvements in structure and training samples are advised for stronger accuracy. An improvement in DCNN architecture and an increase in training samples are required to further improve lung segmentation accuracy.

The Table 1 shows the tabular literature survey of the research paper studied.

Table 1. Tabular literature survey of the research paper studied.

Authors Name	Dataset	Techniques	Advantage	Drawback
Kobayashi *et al.* 2020	TCGA, GEO	Multi-omics analysis, random forests, survival analysis models	Multi-omics approach provides a more complete picture of biological processes compared to focusing on single omics data. Improved accuracy, identifies potential pathways	Data integration challenges Combining data from different sources with varying formats and quality can be challenging, requiring careful normalization, and preprocessing.
Amer *et al.* 2020	I-ELCAP	extraction from CNN methods, in-cluding VGG16, VGG19, AlexNet, utilizing the Genetic Algorithm (GA) for feature selection, Classification with SVM, Decision Trees, and KNN	With the help of a pipeline of traditional techniques and feature descriptors obtained by deep learning, this work produces a CADe system.	Because of the very small dataset size, the test set considered is not large enough to generalize the model's performance.
Tang *et al.* 2018	LIDC-IDRI	3D Faster R-CNN with multi-scale, 3D CNN	Improved Accuracy, Endtoend 3D CNN, Effective false positive reduction	Computational cost, data dependence, interpretability challenges
Murugi *et al.* 2018	LIDC-IDRI	Residual U-Net, Direct nodule detection	Improved accuracy: The study demonstrated that RUN attained a sensitivity of 90.90% at 2 FP per scan, surpassing several remaining models. Direct prediction	The performance of RUN might be affected by the specific characteristics of the LIDCIDRI dataset, and its generalizability to other datasets or scan protocols needs further evaluation. Computational cost

(*continued*)

Table 1. Continued

Authors Name	Dataset	Techniques	Advantage	Drawback
Alam *et al.* 2019	LUNA16, LIDC	3D customized mixed link network, R-CNN, U-Net, (GBM)	Physiological symptoms, biomarkers were also taken into account. A unique multi-strategy 3D deep learning architecture is put forth.	The required training period is one week or so. Furthermore, by utilizing diverse features and robust 3D networks, the outcomes could be enhanced.
Montana *et al.* 2019	LIDC	Thresholding-based region, ReCTnet	The work's results indicate that LSTM layers help to improved combine structural data from adjacent slices, which eventually improves discriminating ability.	Given its deep architecture, segmentation shouldn't be required as an additional step as the model gets more complicated.
Srinivasan *et al.* 2020	LIDC dataset	FPSOCNN	The paper offers a thorough examination of several image features, including SIFT, HOG, and others, for the purpose of classifying lung cancer.	The article provides numerous feature combination examples that are utilized in conventional techniques to simplify the CNN model.
Bhattacharyya *et al.* 2021	LUNA16	3D AlexNet with lightweight Architecture	The technique makes use of 3D AlexNet, which has a lightweight architecture and suggests that the model's complexity is lower than that of other deep learning models.	The authors can expand on this work by classifying detected malignant nodules according to stage.

3 METHODOLOGY

The methodology of the model can be seen in the Figure 1 shown below.

In the model we have performed the tasks according to this flow chart diagram. At first we will take input data that is the CT scans or MRIs images of the patients lungs and then send the input images to further processes and the further process is pre-processing and after this process is the Feature Extraction process. After these processes we went forward towards the feature selection process and after that we move towards the classification for which we have 3 machine learning algorithms that are KNN, MLP and SVM. After the comparison between these three algorithms, we the one which gives best results we are gonna use the most effective and accurate algorithm. Figure 1 shows the diagram of the proposed methodology.

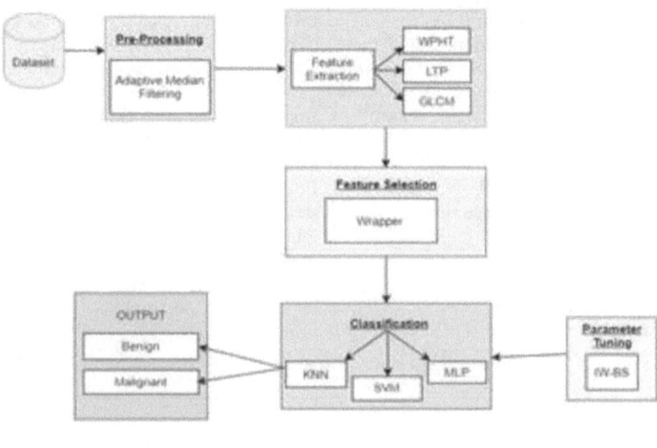

Figure 1. The proposed methodology.

3.1 *Dataset*

Creating a ML model for the classification of lung carcinoma images requires a high-quality and diverse dataset. The dataset should include a variety of lung cancer images captured through different imaging modalities such as X-rays, CT scans, or MRIs. The data sources for classification of lung cancer pictures can be taken from hospitals and medical institutions by collaborating with healthcare institutions to take lung cancer images from lung cancer patients. The dataset which was used for this was Lung cancer dataset (IQ-OTH/NCCD).

3.2 *Pre-processing*

In this model we worked with the Adaptive Median Filtering which comes under noise removal technique. Adaptive median filtering is a noise elimination approach utilized in photo processing to cast off impulsive noise, together with salt-and-pepper noise, from scans. This technique is particularly beneficial when the noise impacts handiest sure pixels in a scan. Adaptive median filtering is a treasured preprocessing technique within the class of lung cancer snap shots. It enables in noise elimination even as maintaining important image details, in the long run contributing to the accuracy and reliability of system mastering-based totally class structures for lung most cancers detection and evaluation.

3.3 *Feature extraction*

In this model we have worked with the Wavelet-based partial Hadamard transform (WPHT), LTP and GLCM. In general, during feature extraction, the DWT performed well with few incorrect findings. The partial Hadamard Transform, which is based on hybrid DWT, is used to extract the features. These deduced characteristics can be used for scaling, rotation, and translation. In addition to using two-level DWT to analyze the segments, a partial Hadamard transform is then utilized to extract the multidirectional characteristics from the low-frequency subband. In addition, DWT can be used to sequentially apply low- and high-pass filters on an image to extract its features. The LTP Local Ternary pattern is also used with WPHT and GLCM as it is a texture descriptor and it is a higher version of the LBP Local Binary Pattern as the LTP increases the discriminatory power of LBP by including three possible samples: 0, 1, and 2) instead of two samples (two possible sample values: 0, 1, and 2) will be added. Apart from WPHT and LTP the GLCM technique is also used. GLCM is a broadly used surface analysis method in image processing. Features have gained recognition in the field of lung cancer categorization study, especially when combined with machine learning algorithms. Their power resides in their ability to use spatial correlations between picture intensities to capture small texture differences within lung nodules that are frequently imperceptible to the unaided eye. After that, the potentially useful data is encoded as features, which provide adaptability and interpretability for various categorization tasks.

3.4 *Feature selection*

In this proposed model we have used Wrapper algorithm. Wrapper algorithms are a class of characteristic choice techniques in system getting to know in which the choice of features is handled as a search trouble. These algorithms compare special subsets of capabilities using a selected machine getting to know model and select the subset that outcomes within the pleasant overall performance. wrapper algorithms may be employed to select the most relevant features for constructing an accurate and green class model. Popular wrapper algorithms consist of recursive function elimination (RFE), forward choice, backward elimination, and genetic algorithms. The choice of the wrapper algorithm relies upon at the dataset.

3.5 *Classification*

SVM, MLP, and KNN classification techniques are used in the model. K-Nearest Neighbours (kNN) presents itself as a compelling alternative for lung cancer classification research, providing a unique methodology with unique advantages and disadvantages. Support Vector Machines (SVMs) have arose as a tough competitor in the field of lung cancer categorization research, demonstrating efficacy in a multitude of tasks. They are especially appealing because of their high accuracy (sometimes above 90% in nodule classification), their capacity to manage high-dimensional data efficiently, and their degree of interpretability. Multilayer Perceptron (MLP), can learn intricate correlations from a variety of input types, they present a viable method for classifying lung cancer. They are particularly good at working with numerical, image, and even molecular data, and they can also provide some interpretability via network analysis.

4 RESULT

The experimental outcomes of the proposed method for lung carcinoma Classification are examined in this unit. To match performance of the suggested method, further modern techniques are applied. The Colaboratory platform is used to implement the suggested technique, and the accuracy, running time, precision, recall, F-score, area under the curve, and ROC are used to calculate the methodology's performance.

Table 2 can be referred for the performance metrics of the algorithms by applying Feature extraction and selection. The overall Accuracy, Recall, Precision and F1-Score is determined in the proposed model after the evaluation the Accuracy came out to be 99%, recall is 99%, Precision is 99%, and F1-Score came out to be 99%.

Table 2. The performance metrics of the algorithms.

Algorithm	Accuracy	Recall	Precision	F1-Score	AUC
KNN	97.2%	97.2%	97.2%	97.2%	97.4%
MLP	98.5%	98.5%	98.5%	98.5%	99.5%
SVM	99.1%	99.1%	99.1%	99.1%	99.3%

5 CONCLUSION

Additionally, the application of machine learning techniques in the classification of lung cancer images, in conclusion, constitutes a paradigm shift in medical diagnosis. Machine learning models have shown tremendous potential in lung cancer early detection, accurate diagnosis, and personalised therapy planning due to their quick analysis of large datasets. By making it possible for prompt treatments, customised medicines, and increased survival rates, the results hold the prospect of improving patient outcomes. Additionally, the proposed method adds much value in the field of early diagnostics as this proposed method achieves great accuracy with the help of hybrid algorithms used in the model.

REFERENCES

Ajai, A. K.; Anitha, A. (2022). Clustering based lung lobe segmentation and optimization based lung cancer classification using ct images. *Biomedical Signal Processing and Control*, 78, 103986. https://doi.org/10.1016/j.bspc.2022.103986.

Alyasriy, hamdalla; AL-Huseiny, Muayed. (2021). *The IQ-OTHNCCD Lung Cancer Dataset*, Mendeley Data, V2, doi: 10.17632/bhmdr45bh2.2

Asada, K., Kobayashi, K., Joutard, S., Tubaki, M., Takahashi, S., Takasawa, K., Komatsu, M., Kaneko, S., Sese, J., & Hamamoto, R. (2020). Uncovering prognosis-related genes and pathways by multi-omics analysis in lung cancer. In *Biomolecules* (Vol. 10, Issue 4, p. 524). MDPI AG. https://doi.org/10.3390/biom10040524.

Asuntha, A., & Srinivasan, A. (2020). Deep learning for lung Cancer detection and classification. In *Multimedia Tools and Applications* (Vol. 79, Issues 11–12, pp. 7731–7762). Springer Science and Business Media LLC. https://doi.org/10.1007/s11042019-08394-3.

Bushara, A.R., Vinod Kumar, R.S., & Kumar, S.S. (2023). An ensemble method for the detection and classification of lung cancer using Computed Tomography images utilizing a capsule network with Visual Geometry Group. *Biomedical Signal Processing and Control*, 85, 104930. DOI: 10.1016/j.bspc.2023.104930.

Cheng ES, Weber M, Steinberg J, et al. (2021). Lung cancer risk in never-smokers: An overview of environmental and genetic factors, *Chin. J. Cancer Res.* 33 (5) 548.

Elnakib, A., M. Amer, H., & E.Z. Abou-Chadi, F. (2020). Early lung cancer detection using deep learning optimization. In *International Journal of Online and Biomedical Engineering (iJOE)* (Vol. 16, Issue 06, p. 82). International Association of Online Engineering (IAOE). https://doi.org/10.3991/ijoe.v16i06.13657.

Foox J, Tighe SW, Nicolet CM, et al. (2021). Performance assessment of DNA sequencing platforms in the ABRF Next-generation sequencing study, *Nat. Biotechnol.* 39 (9) 1129–1140.

Lan, T., Li, Y., Murugi, J. K., Ding, Y., & Qin, Z. (2018). RUN:Residual u-net for computer-aided detection of pulmonary nodules without candidate selection (Version 1). *arXiv*. https://doi.org/10.48550/ARXIV.1805.11856.

Liu, C., & Pang, M. (2020). Extracting lungs from CT images via deep convolutional neural network based segmentation and two-pass contour refinement. In *Journal of Digital Imaging* (Vol. 33, Issue 6, pp. 1465–1478). *Springer Science and Business Media LLC.* https://doi.org/10.1007/s10278-020-00388-0.

Manoharan, S. (2020). Early diagnosis of lung cancer with a probability of malignancy calculation and automatic segmentation of lung CT scan images. *Journal of Innovative Image Processing (JIIP)*, *2*(04), 175–186.

Nanglia, P., Kumar, S., Mahajan, A. N., Singh, P., & Rathee, D. (2021). A hybrid algorithm for lung cancer classification using SVM and Neural Networks. In *ICT Express* (Vol. 7, Issue 3, pp. 335–341). Elsevier BV. https://doi.org/10.1016/j.icte.2020.06.007.

Nasrullah, N., Sang, J., Alam, M. S., Mateen, M., Cai, B., & Hu, H. (2019). Automated lung nodule detection and classification using deep learning combined with multiple strategies. In *Sensors* (Vol. 19, Issue 17, p. 3722). MDPI AG. https://doi.org/10.3390/s19173722.

Neal Joshua, E. S., Bhattacharyya, D., Chakkravarthy, M., & Byun, Y.-C. (2021). 3D CNN with visual insights for early detection of lung cancer using gradient-weighted class activation. In H. Gritli (Ed.), *Journal of Healthcare Engineering* (Vol. 2021, pp. 1–11). Hindawi Limited. https://doi.org/10.1155/2021/6695518.

Pradhan, K. S., Chawla, P., & Tiwari, R. (2023). HRDEL: High ranking deep ensemble learning-based lung cancer diagnosis model. *In Expert Systems with Applications* (Vol. 213, p. 118956). Elsevier BV. https://doi.org/10.1016/j.eswa.2022.118956.

Shakeel PM, Burhanuddin MA, Desa MI, (April 2020). Automatic lung cancer detection from CT image using improved deep neural network and ensemble classifier, *Neural Computing and Applications*, pp.1–4.

Tang H., Kim D. R. and Xie X., Automated pulmonary nodule detection using 3D deep convolutional neural networks, *2018 IEEE 15th International Symposium on Biomedical Imaging (ISBI 2018)*, Washington, DC, USA, 2018, pp. 523–526, doi: 10.1109/ISBI.2018.8363630.

Tian, Q., Wu, Y., Ren, X., & Razmjooy, N. (2021). A new clustering-based technique for automatic segmentation of lung tumor images and an optimized Fuzzy C-Ordered Means (DSARFCM) according to the new version of the search and rescue optimization algorithm for lung cancer diagnosis. *Biomedical Signal Processing and Control*, 68, 1027612–1027618. https://doi.org/10.1016/j.bspc.2021.102761.

Uplaonkar, D. S., appa, V., & Patil, N. (2021). An efficient discrete wavelet transform based partial hadamard feature extraction and hybrid neural network based monarch butterfly optimization for liver tumor classification. In *Engineered Science. Engineered Science Publisher*. https://doi.org/10.30919/es8d594.

Venkatesan, N., Pasupathy, S., & Gobinathan, B. (2024). An efficient lung cancer detection using optimal SVM and improved weight-based beetle swarm optimization. In *Biomedical Signal Processing and Control* (Vol. 88, p. 105373). Elsevier BV. https://doi.org/10.1016/j.bspc.2023.105373.

Westeel V, Foucher P, Scherpereel A, et al. (2022). Chest CT scan plus x-ray versus chest x-ray for the follow-up of completely resected non-small-cell lung cancer (IFCT-0302): A multicentre, open-label, randomized, phase 3 trial, *Lancet Oncol.* 23 (9) 1180–1188.

Ypsilantis, P.-P., & Montana, G. (2016). Recurrent convolutional networks for pulmonary nodule detection in ct imaging (Version 2). *arXiv*. https://doi.org/10.48550/ARXIV.1609.09143.

Computational Methods in Science and Technology Volume 1 – Kaur et al. (Eds)
© 2025 the Author(s), ISBN: 978-1-032-81760-6

College bus tracker andriod application using java and google firbase

K.L. Anusha
Department of Computer Science and Engineering (Data Science), Associate Professor at Vignan Institute of Technology and Science, Hyderabad, Telangana State, India

K. Upendhar Reddy, S. Rakesh Paul & B. Sai Prakash
Department of Computer Science and Engineering (Data Science), Student at Vignan Institute of Technology and Science, Hyderabad, Telangana State, India

ABSTRACT: A university bus monitoring system is being developed to enhance campus transportation efficiency and enhance the overall campus experience. The system uses GPS technology and a mobile application to provide accurate information about bus routes, schedules, and arrival times. It uses cloud-based infrastructure for seamless communication between the bus and central management, facilitating dynamic course optimization and service response changes. Key features include intuitive user interfaces, configurable alert mechanisms for delays or route deviations, and comprehensive reporting capabilities. The system also features robust security measures to protect sensitive information and ensure user privacy. Implementing this system can reduce waiting times, improve schedule optimization, enhance productivity, and improve resource allocation for managers.

Keywords: College bus monitoring gadget, site visitors control, GPS generation, cellular application, path optimization, campus logistics

1 INTRODUCTION

Transportation is crucial for the daily operations of educational institutions, with campuses often covering large areas and serving a large population of students, faculty, and staff. Efficient control of transportation is essential for smooth logistics and improving the overall experience of the campus community. Traditional methods of managing bus systems often suffer from inefficiencies such as loss of real-time records, unpredictable schedules, and limited communication channels. To address these issues, a College Bus Tracker System has emerged as a promising solution. This system uses modern technologies like GPS monitoring and cell programs to provide real-time visibility of college buses, allowing users to access accurate statistics about bus routes, schedules, and anticipated arrival times. This system aims to streamline transportation logistics, reduce wait times, and enhance comfort for students, staff, and the general community. The paper explores the design, implementation, and potential benefits of a College Bus Tracker System for educational institutions, focusing on its potential to revolutionize transportation control, improve efficiency, reliability, and customer satisfaction. It also discusses the challenges and opportunities associated with implementing such a system and recommends strategies for overcoming barriers to adoption. The adoption of a College Bus Tracker System could transform transportation on university campuses, enhancing performance, accessibility, and enjoyment for all stakeholders.

 DOI: 10.1201/9781003501244-40

2 RELATED WORK

College Bus Tracker systems are increasingly being used in educational institutions worldwide to optimize transportation logistics and enhance campus experience. These systems use technology such as GPS monitoring, mobile applications, and web-based systems to provide real-time monitoring and control of college bus fleets. GPS-based tracking systems display the location of university buses in real-time, allowing users to plan their commutes. Mobile applications provide real-time bus monitoring, direction maps, estimated arrival times, and notifications for delays or route changes. These apps can be downloaded onto smartphones for easy access to bus-related information. Web-based platforms allow users to view bus routes, track bus locations in real-time, and receive notifications about arrivals and departures. These systems provide a convenient way for users to access transportation information without the need for additional software installation. Overall, these systems aim to improve campus transportation management and enhance student experience.

3 PROPOSED WORK

Bus passes are widely used for college transportation, allowing students to easily access their campus. However, this method is cumbersome for students and administrative staff due to the need for separate software forms and paper records. This can lead to loss of information and reliance on manual labor. An automated system for bus skip management is proposed to address these issues and reduce administrative burden. This system uses database generation to streamline software for students and simplifies file-keeping for administrative staff. Students can now submit their utility information electronically, eliminating the need for paper paperwork and manual records access. Administrative staff can now evaluate and process applications electronically, enhancing accuracy and transparency while reducing administrative overhead. Once an application is approved, students are issued a bus pass ID card, allowing them to access college buses at different pick-up points. This eliminates the need for manual verification of eligibility, streamlining the boarding process for both students and bus drivers. The proposed automated bus pass system offers numerous advantages, including increased efficiency, decreased administrative burden, and improved student comfort. By implementing this technology, schools can enhance their students' transportation experience and optimize administrative strategies.

4 SYSTEM ARCHITECTURE

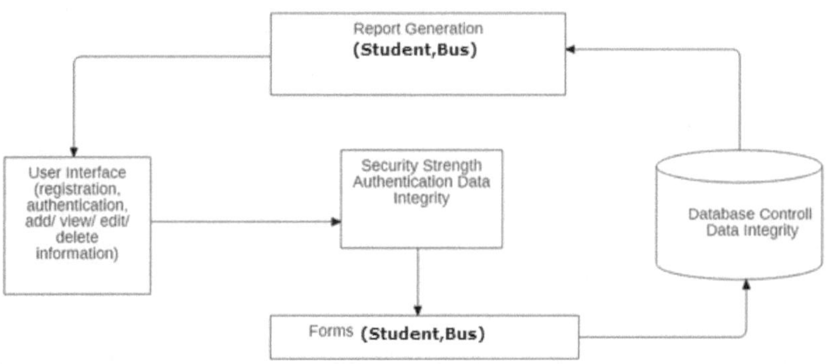

Figure 1. Working of the application.

5 TEST CASES MODEL BUILDING

Table 1. Test cases for running the app.

Test ID	Test Name	Inputs	Process	Excepted Output	Actual Output	Status
1	Login Test	Username, Password	Validate User-name and pass-word on database	Need to redirect to user home page	It's Redir-ected to user home	Success
2	Registration Test	Username, Password, email, mobile number, email etc.	insert the users into database	Need to insert the user details into da-tabase	It's inserted	Success
3	Add Bus routes	Route details	Insert the files into the database	Need to insert the files into the data-base	It's inserted	Success
4	Track bus details	Bus number	Grant the access to the user to ac-cess the files	Need to grant per-mission to user	Inserted successfully	Success
5	View track bus details	Bus number	Insert the files into the database	Need to send the details to mail id	Inserted successfully	success
6	View Bus details	Bus Id	Insert the file into the database	Need to send the details to database	Inserted successfully	Success

6 RESULTS

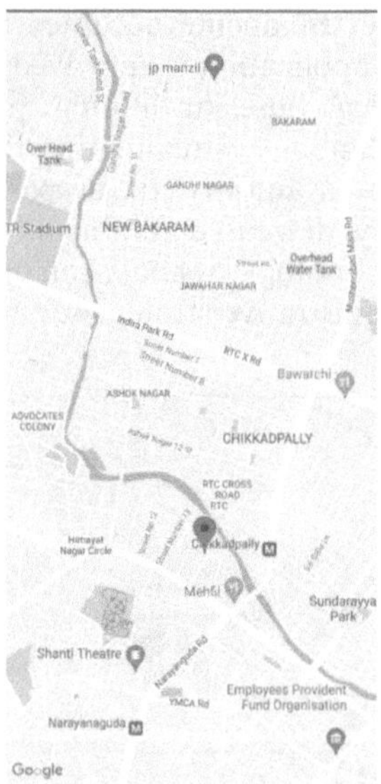

Figure 2. Bus Location on Maps. This is the final output, this is how the student will be able to track the location of the Bus.

7 FUTURE SCOPE

College Bus Tracker systems are poised for significant innovation and advancement. Future systems will incorporate emerging technologies like AI, machine learning, and IoT to enhance performance and efficiency. AI and ML algorithms will analyze transportation data, anticipate demand, and optimize direction planning in real-time. IoT-enabled sensors will provide additional data insights, enabling better decision-making. The focus is on improving user experience across all touchpoints, with future developments focusing on refining interfaces, optimizing mobile apps, and incorporating personalized functions. Augmented reality and virtual reality will also be explored for innovative ways to engage with transportation information. Multimodal transportation integration will be a priority, integrating with public transit, ridesharing, cycling infrastructure, and pedestrian pathways to promote sustainable and multimodal transportation practices.

8 CONCLUSION

The improvement and implementation of College Bus Tracker systems constitute a good sized advancement in transportation management inside academic institutions. Throughout this paper, we've got explored the key functions, functionalities, and capacity advantages of such structures, highlighting their significance in streamlining logistics and improving the overall campus enjoy for college students, college, and team of workers. College Bus Tracker systems offer actual-time monitoring competencies, course optimization algorithms, and consumer-friendly interfaces that empower users to access correct bus-related facts without difficulty. By leveraging GPS generation, mobile packages, and net-primarily based structures, those structures provide unheard of visibility into the place and standing of university buses, permitting users to plot their commutes more effectively and decrease wait instances. In conclusion, College Bus Tracker structures constitute a transformative solution for reinforcing transportation management inside instructional institutions. By leveraging generation, statistics-driven insights, and person-centric design standards, these systems empower users, sell sustainability, and foster network engagement. As educational establishments keep to evolve and develop, the significance of green and dependable transportation services facilitated by means of College Bus Tracker systems will simplest hold to increase, ensuring a unbroken and enjoyable campus experience for all.

REFERENCES

[1] Daza, L. and Misra, S., (2017). Past the web of things: everything interconnected: innovation, communications and computing [book survey]. *IEEE Remote Communications*, 24(6), pp.10–11.

[2] Turnquist, M. A., (1974). Methodologies for moving forward unwavering quality of transport travel benefit. *Computers and Operations Inquire About*, 1, pp. 201–211.

[3] Booysen, M. J., Gilmore, J. S., Zeadally, S. and Van Rooyen, G. J. (2012). Machine-to-machine (M2M) communications in vehicular systems. *KSII Exchanges on Web and Data Frameworks (TIIS)*, 6 (2), pp.529–546.

[4] Boswarthick, D., Elloumi, O. and Hersent and O. eds., (2012). *M2M communications: a frameworks approach. John Wiley & Sons.*

[5] Bezzina, D. and Sayer, J., (2015). *Security Pilot Show Arrangement: Test Conductor Group Report Usdot Report No.* speck hs 812 171.

[6] Padilla, V. S., Collaguazo, A., Cabrera, J. J. and Pesantez, J., (2020, June). M2M innovation for transport armada administration. case ponder: A college transportation framework. In *2020 IEEE Innovation & Designing Administration Conference (TEMSCON)* (pp. 1–5). IEEE.

[7] Jornod, G., El Assaad, A., Kwoczek, A. and Kürner, T., (2019, June). Bundle inter-reception time modeling for high-density platooning in shifting encompassing activity thickness. In *2019 European Conference on Systems and Communications (EuCNC)* (pp. 187–192). IEEE.

[8] Guerrero-Ibanez, J.A., Zeadally, S. and Contreras-Castillo, J., (2015). Integration challenges of brilliantly transportation frameworks with associated vehicle, cloud computing, and web of things innovations. *IEEE Remote Communications*, 22(6), pp.122–128.

Computational Methods in Science and Technology Volume 1 – Kaur et al. (Eds)
 ISBN: 978-1-032-81760-6

Lightweight encryption techniques: Types, challenges and security issues

T.C. Swetha Priya
Research Scholar, Department of CSE, JNTUH University College of Engineering, Science and Technology, Hyderabad, Telangana, India

R. Sridevi
Professor, Department of CSE, JNTUH University College of Engineering, Science and Technology, Hyderabad, Telangana, India

K.S. Sadasiva Rao
Professor, Department of CSE, Sri Indu College of Engineering and Technology, Hyderabad, Telangana, India

ABSTRACT: Due to the deployment of Wireless Sensor Networks (WSN) in hostile environments, they are prone to a different set of attacks. Securing the data that is being transmitted over the network in such environments is one of the crucial job. If the data is sent without encryption, then the attacker may misuse the data causing hindrance to Data Confidentiality. So, a good lightweight encryption mechanism must be designed such that it uses low energy and low memory. This paper presents various lightweight encryption techniques that are used in Wireless Sensor Networks.

Keywords: Encryption, key, Nonce, S-Box, LiCi, Joltik

1 INTRODUCTION

A WSN is a collection of sensor nodes with restrained set of resources that are deployed in unfocused environments. Huge amount of data will be transmitted through insecure communication channel over the network. This data communication is done over an insecure channel that is open to malicious attackers. Data Security has become one of the most important issues that is being faced by a lot of networks. Due to advanced technologies, the existence of malicious attackers had been slightly large these days. Confidentiality, Authentication, Integrity and privacy are the crucial security requirements that must be preserved in a network to ensure secure environment. Confidentiality service can be achieved by employing encryption to the data that is being transmitted through communication channel. Confidentiality always aims at disclosing data to authorized users. This can be achieved by using Encryption techniques. Generally, the Encryption techniques are classified into 2 types. They are secret key and public key Encryption. In Symmetric Encryption, same key is used for both encryption and decryption whereas in Asymmetric Encryption, public key and private key namely two keys are used for encryption and decryption. If public key is utilized for encrypting the data, then the private key is utilized for decryption and vice-versa. This paper provides a study of all the lightweight cryptographic techniques that can be implemented in Wireless Sensor Networks to provide data security.

 DOI: 10.1201/9781003501244-41

2 LITERATURE SURVEY

2.1 *Attribute based encryption*

Attribute based Encryption (Bethencourt, Sahai and Waters 2007; Naru, Saini and Sharma 2017; Qiao, Liang, Davis and Jiang 2014; Rasori, Pericle and Gianluca 2019) is a public key cryptographic technique where the key is generated from the user's attributes and cipher text key policies. The decryption process of the cipher text is possible only if there is a match between the arrangement of users' key attributes and the cipher text attributes. The key feature of this encryption technique is that it maintains both forward and backward secrecy. This encryption is classified into two types : Cipher Text Policy based Attribute Encryption and Key Policy based Attribute Encryption. In Cipher Text Policy based technique, the access trees are used to rearrange the data in a reverse manner such that it is irreversible. It produces a secret key from a set of user attributes. In a Key Policy based technique, the secrecy of the key is maintained based on attributes of a particular user that are characterized at the time of creation of access tree.

2.2 *Identity based encryption*

Identity based Encryption (Boldyreva *et al.* 2008; Li *et al.* 2013) belongs to the category of public Key Cryptography techniques where the public key is generated based on unique information about the identity of user. The sender who has access to the network can encrypt the data by generating a key from text-value pair of identity information of user. The receiver obtains the key pattern from a trusted central server or a public key generator that produces users' secret keys. This method does not require any certificate mechanism for keys. Automatic Key Expiry enables retrieving messages unreadable after certain expiration time and hence no revocation of keys is required. But a central key distribution server causes a single point of failure.

2.3 *Identity based broadcast encryption*

This method protects data privacy by encrypting the data over broadcast channel in such a way that only privileged set of users will be able to decrypt the data (Aijun and Wei 2019). In this method, an entity is employed for generation and distribution of keys for decryption. A sender will encrypt data that is sent to a set of receivers by adaptively choosing their own public keys that are derived from identity information such as their email or IP address. So, it does not require Public Key Infrastructure. Based on these public keys, a third party Private Key Generator generates corresponding secret keys associated with user's public identities. So, the information will be decrypted only by those who have a valid secret private key among the receivers. Access to public parameters of the system is provided to the sender who can encrypt message using receivers' identity as public key and the intended receiver that obtains decryption key from private key generator can decrypt the data.

2.4 *Elliptic Curve based lightweight Cryptography (ECC)*

Elliptic curve cryptography (Banerjee and Patil 2020; Lara-Nino *et al.* 2020; Patel and Doshi 2020; Sowjanya K *et al.* 2019) is a public key cryptography technique that is based on elliptic curves over gaussian field G. ECC is one of the strong lightweight encryption techniques because of low memory consumption. It is a one-way function as it is very difficult to compute its inverse. The main advantage of this technique is that it uses shorter keys which includes less memory requirements and fast mathematical computations which make them suitable for less constrained devices. The strength of it depends on computation of set of points that satisfies equation. There will be different set of curves which can be selected for providing different security provisions. The key generation in ECC is done at an affordable cost and provides strength as that is done in general encryption and decryption processes.

2.5 *Signcryption*

It is a lightweight encryption technique based on the public key Infrastructure. This method [Ferrari *et al.* 2018; Malone-Lee 2002) performs encryption and digital signature which helps in ensuring both Confidentiality and Integrity. First the data is digitally signed and then encryption technique is applied. But the computation is expensive and provides less efficiency although it provides confidentiality and Integrity services.

2.6 *Tiny Encryption Algorithm (TEA)*

This technique is cryptographically strong, simple and fastest lightweight encryption. This follows Feistel cipher structure. It performs encryption of 64 bits data at a time using a 128-bit key with 32 rounds. It does not involve any S-boxes, so the data space required is also tiny. TEA (Abdul *et al.* 2023; Adiwiganda *et al.* 2019) has some weaknesses. One of the disadvantages is the use of equivalent keys where each key is equivalent to others, so the effective key size is only 126 bits. That's why, this algorithm is weak if used for creating a cryptographic hash. It is resistant against differential cryptanalysis and accomplishes complete diffusion property.

2.7 *DNA computing based lightweight algorithm*

This algorithm is based on encrypting the data based on DNA sequences (Gamil and Nadhem 2023) which are very random. This idea creates a strong secret key based on DNA which is difficult for attackers to break it. The idea is that binary code is assigned to DNA Bases: A, C, G, T. It uses a Feistel structure and SP networks. It encrypts the data with n-bit block size and requires n-bit key. First the data is collected from users in network then convert it into text format and choose DNA format for generating a secret key. Then, using this key the data is encrypted into two bytes at a time. Then 16 bits secret key is extracted from DNA tape letters and apply XOR on each data segment and DNA secret key and then a series of rules are applied based on DNA tape to obtained XOR result and transpose them. Repeat the same process until entire data is encrypted. This approach is lightweight because of its less complexity, high throughput, less memory, and immunity against differential and linear attacks.

2.8 *Present*

PRESENT (Bogdanov *et al.* 2007; Deepsikha Sharma *et al.* 2023, Kubba & Hoomod,2020; Latif *et al.* 2020) is a block cipher developed for implementing lightweight cryptography. It follows substitution-permutation network to produce unreadable text by taking plaintext and key as input and follows 16 rounds by applying several substitutions and permutations combination. This works for 80-bits and 128-bits. Each round introduces a round key that is used for performing XOR Operation with 64-bit block size data. This method is simple and suitable for constrained environments with less number of resources. Although it is simple, the computational overhead is still present in this method.

2.9 *PRINCE algorithm*

PRINCE (Victor *et al.* 2023; Rani *et al.* 2016; Abdullah and Obeid 2021) is the first lightweight block cipher that is dependent on Substitution-Permutation Network. This algorithm takes 64-bit plaintext as input along with 128-bit key. It takes 12 rounds where each round constitutes 4 main steps: key-addition by using XOR with 64-bit subkey, substitution layer for replacing using 4-bit S-Box, linear Matrix multiplication is done by multiplying 16x16 matrix and Shifting rows by permuting 16 nibbles and a round constant addition is performed using XOR operation. The 128-bit key is first split into 2 identical parts each of 64-bit size. The first one is used as whitening key and second part will be used directly at each round. At the decryption side, we can use the key that is derived from second part of the key. This is how computational overhead can be minimized. PRINCE cipher takes six to seven times less space compared to that of PRESENT with 80-bit key size and uses fourteen to fifteen times less area than that of AES with 128-bit key size.

2.10 *Hash-based Authenticated Nonce-Misuse Resistant Encryption (HANMRE)*

HANMRE (Tran, Seok and Lee 2020) is adaptable to lightweight leak resistant authenticated encryption with associated data (AEAD) scheme. It is designed for achieving side-channel security and to be suitable for restricted environments w ith limited resources. The strength of this scheme is that it ensures strong security against the adversaries for authenticated encryption. It makes use of unique Initialization vectors which are used in various processes. In this method, the encryption takes four entities as input: Data, Key, Nonce and Plaintext and gives Ciphertext as output. The sponge uses a state with a quite large size that constantly progresses in the operational functions inside the scheme, which allows us to give resistance to information leakage.

2.11 *Improved Lightweight Encryption Algorithm (ILEA)*

The Improved Lightweight Encryption Algorithm (Jha *et al.* 2020) is based on the concept of PRINCE cipher. This has two additional layers of block balanced Mixing. It works with Python and C++ languages efficiently. This additional feature is used to achieve diffusion property of data blocks. This mixing helps in showing difference between original plain text and encrypted text. These additional layers are added at the beginning of main PRINCE algorithm. ILEA consists of 3 stages primarily: PRINCE and two Block balanced Mixing layers. In encryption phase, the plain text is divided into 64-bit blocks and converted into hexadecimal form and then given to the Block Balanced Mixing phase for permutation of bits and then apply PRINCE algorithm and then the output is given to another Block Balanced Mixing round to obtain the final ciphertext. Then decryption is applied in reverse manner to obtain original text. This is suitable for devices with less resources and require high speed performance.

2.12 *LISP-XK*

Lightweight Signcryption Protocol with Keccak (LISP-XK) (Devgun and Thomas 2021; Kim *et al.* 2021) was designed for resource constrained devices. It consists of 3 layers: Perception Layer, Network layer, and Application Layer. In the first layer, the data is collected from nodes and then given to the Network layer where the data is processed and security controls will be added to provide security. Then, the second layer is used as an interface to store data from the nodes. In this method, Signcryption technique is used where both signing and encryption of the data can be done in a single step. Then, at the receiver side first decryption of data is taken place followed by finding its hash and then compare it with unsigned data. If both hashes match, then message is authenticated otherwise it will be discarded.

2.13 *DBST*

This is a lightweight block cipher that is based on dynamic S-Boxes (Yan, Li and Guo 2023). It uses 128-bit block data and key size of 64-bits. It adapts generalized Feistel variant structure for encryption and decryption process. Here the S-Boxes will be dynamically associated with Key. It consists of 4 stages: Byte Substitution, Shift Rows, Column Mixing and Round Key Addition. The round function designed in this algorithm achieves diffusion property. This method includes two concepts: (1) novel generalized Feistel variant structure, and (2) dynamic S-box. It uses an amalgam of the bit-slice method and subkeys to construct a novel dynamic S-box structure. The decryption process is the reverse of encryption except with the use of keys in reverse manner. It achieves high throughput, good avalanche effect and low hardware resources.

2.14 *LiCi*

LiCi (Patil, Bansod and Kant 2017) is a lightweight cipher that is based on Feistel structure. It operates on a plaintext of 64-bits and a key size of 128-bits. It is considered as a balanced Feistel cipher network that is run for 31 rounds. It contains 4-bit input and a 4-bit output S-Box. From 128-bit key, we have to extract the left 32-bits and consider as first round key and the right 32-bits as second round key. Then, by using the Key-Scheduling algorithm, we can update the round keys periodically. The left most 32-bits of plain text has to be passed through S-Box and substitution is performed by producing a 32-bit output. Then the rightmost 32-bits are XORed with second round key to obtain 32-bit output. Then left circular shift has to be applied to output circularly. The obtained output is then XORed with S-Box output and leftmost 32-bits of round key and the output obtained is fed to right circular shift block which will shift the output by 7 positions to the right. These operations must be repeated for 32 rounds to produce cipher text. This method is resistive against basic and advance attacks and is suitable for less memory restrained nodes.

2.15 *Piccolo*

Piccolo (Mhaouch, Elhamzi and Atri 2020) is a block cipher that works in iterative manner and is based on the Feistel structure. It takes 64-bits data as plaintext and works for two different key sizes i.e., 80-bits and 128-bits. This algorithm runs for 25 rounds if the key-size is 80-bits and runs for 31 rounds if the key size is 128-bits. A mix key function is implemented at the beginning and end of the algorithm resulting in additional two rounds. This results in increase in total number of rounds to 27 for Piccolo-80 and 33 for Piccolo-128. After n rounds, the plaintext is transformed into a ciphertext of 64-bits after completion of 25 rounds. It uses 4-bit input-output ports. To store the plain text and key bits, this method uses registers. It takes 16 clock cycles to generate output. This method uses two Feistel schema structures where each schema consisting of 32-bit input and output data consisting of XOR and F functions. The XOR is used for addition of Output Function F with round key. The F function uses 16-bit input and gives 16-bits data as output which comprises of 2 S-Box layers separated by diffusion matrix. The first S-Box layer will be multiplied by a diffusion matrix. For loading the data and key, 32 clock cycles are consumed. Whereas sixteen clock cycles are taken for each round and sixteen clock cycles for generating output. This method ensures high security, compact hardware implementation and suitable for extremely constrained environments.

2.16 *JOLTIK*

JOLTIK (Duan and Peng 2020; Deepsikha Sharma *et al.* 2023; Li, M. and Chen 2021; Zainab AlJabri *et al.* 2023) is a lightweight authenticated encryption algorithm which is tweakable to 64 bits cipher and four sets of key-tag pairs such as 64-64, 80-112, 96-96 and 128-64 are used to achieve 64 bits security in terms of confidentiality and authenticity. This algorithm consists of two parts: 1. Encryption and 2. Verification. The encryption takes variable size message, fixed size nonce, a k-bit key, variable AD and a fixed nonce to generate m-bits cipher text and authentication tag. These four sets are mathematically strong in providing security even though the nonce is reused. It is efficient in less constrained environments.

2.17 *SILC*

SILC (Jahanbani *et al.* 2020; Rao *et al.* 2021) is an authenticated block cipher technique that comes with two variants: SILC-AES and SILC-LED. SILC-AES encrypts the message of 64/128 bits by using a key of size 128 bits along with either a 64/96-bit nonce and a tag of 64-bit size. This method works well on standard hardware and exhibits a low transmission rate that is most suitable for low-rate applications if the throughput is not of concern. SILC-LED encrypts data by using a message of 64-bit size, key of 80-bits size, a nonce of 48-bits and a tag of 32-bits size. This method is chosen if low data transmission rate is required. This method is suitable for low-power devices.

2.18 *AEGIS*

AEGIS (S. Ullah *et al.* 2022; Wu and Preneel 2014) is one of the authenticated encryption with associated data scheme that is suitable to protect data packets over a network. AEGIS-128 will use eight round functions for processing blocks of 128-bits message. It uses 5 round functions of AES algorithm to process 128-bit message block. AEGIS-256 takes 256-bit key and an initialization vector of 256-bits along with a 128-bit tag. AES-256 runs six round functions of AES algorithm. The initialization phase takes atleast 10-steps which makes it strong. For long messages, AES-128 is slower than AES-256. It is the fastest authenticated encryption algorithm. It is freely available algorithm.

2.19 *COLM*

COLM (Ibrahim *et al.* 2023; Jahanbani *et al.* 2020; Zainab AlJabri *et al.* 2023) is a block cipher that works on the Authenticated encryption scheme. It follows a block cipher based Encrypt-Linear mix-Encrypt mode which is designed to achieve resistance against misuse to secure against adversaries. It uses AES-128 with a key and a 128-bits length state matrix. This method achieves confidentiality and Integrity even the nonce value is repeated and also resist from some types of differential attacks.

2.20 *ACORN*

ACORN-128 (Wu *et al.* 2016; William *et al.* 2017) uses an input of 128-bit key size and an initialization vector of 128-bit. The data and the plaintext size associated with this is less than 264 bits. The length of authentication tag is less than or equal to 128 bits. The size of state matrix in this algorithm is 293 bits. It uses six Linear Feedback Shift Registers that are being concatenated in this method. It uses three functions: keystream bit generation from the state, overall feedback bit computation, and the state update function. The initialization phase consists of loading of keys and IV into the state, and running the cipher for 1792 steps. The keystream bit is used to update the state. After initialization, the associated data is used to update the state. After processing, at each step of the encryption one bit of plaintext is used to update the state and is encrypted to cipher text. Even when there is no plaintext to be encrypted, the cipher is run for 256 steps. There are 3 versions in ACORN: v1, v2 and v3. All three versions have common characteristics: parallelizable, online and inverse-free. ACORN v2 is insecure with Non-linear feedback function. To overcome this drawback, in ACORN v3 a filtering function that provides security is employed. This algorithm is resistant to side channel attacks.

3 CONCLUSION

This paper presents a brief study on various Lightweight Encryption techniques. This study reveals that Lightweight Encryption techniques has their own advantages and limitations. These encryption techniques have to be applied based on the requirement of application that has to be designed. This paper will be a useful resource to all the researchers working in the domain of security and lightweight encryption.

ACKNOWLEDGEMENT

I would like to thank my supervisor Dr. R. Sridevi, Professor, Department of CSE, JNTU Hyderabad for providing her valuable timely suggestions in the preparation of this paper and in guiding me in all aspects to carry out my research.

REFERENCES

Abdul, A., Narayana, G., Kishore, R. S., Srikanth, B., Kumar, K. K., and Indira, D. (2023). LWC: efficient lightweight block ciphers for providing security to constrained devices a solution For IOT devices. *Journal of Theoretical and Applied Information Technology*, 101(7).

Abdullah, A. A., and Obeid, N. R. (2021, March). Efficient implementation for PRINCE algorithm in FPGA based on the BB84 protocol. *In Journal of Physics: Conference Series* (Vol. 1818, No. 1, p. 012216). IOP Publishing.

Adiwiganda, M. R., Ariyanto, E., Yasirandi, R., Suwastika, N. A., and Setyoko, Y. A. (2019, November). Adopting tiny encryption algorithm for patient healthcare record on smart card. *In 2019 International Conference of Computer Science and Information Technology (ICoSNIKOM)* (pp. 1–5). IEEE.

Banerjee, S., and Patil, A. (2020). ECC based encryption algorithm for lightweight cryptography. *In Intelligent Systems Design and Applications: 18th International Conference on Intelligent Systems Design and Applications (ISDA 2018)* held in Vellore, India, December 6-8, 2018, Volume 1 (pp. 600–609). Springer International Publishing.

Bethencourt, J., Sahai, A., & Waters, B. (2007, May). Ciphertext-policy attribute-based encryption. *In 2007 IEEE Symposium on Security and Privacy (SP'07)* (pp. 321–334). IEEE.

Bogdanov, A., Knudsen, L. R., Leander, G., Paar, C., Poschmann, A., Robshaw, M. J., and Vikkelsoe, C. (2007). PRESENT: An ultra-lightweight block cipher. *In Cryptographic Hardware and Embedded Systems-CHES 2007: 9th International Workshop*, Vienna, Austria, September 10–13, 2007. Proceedings 9 (pp. 450–466). Springer Berlin Heidelberg.

Boldyreva, A., Goyal, V., and Kumar, V. (2008, October). Identity-based encryption with efficient revocation. *In Proceedings of the 15th ACM conference on Computer and communications Security* (pp. 417–426).

Deepsikha Sharma, Jain Singh, Niharika Namdev, Suman Kumar Jha, New ultra light-weight cryptography encryption and decryptions for the internet of things, *European Chemical Bulletin*, 2023, Volume 12(Special Issue 12), pp 266–276 DOI:10.48047/ecb/2023.12.si12.019.

Devgun, T., and Thomas, R. LiSP-XK: extended light-weight signcryption for IoT in resource-constrained environments.

Duan, T., and Peng, Y. J. (2020, August). Interactive symmetric context encryption in framework of asynchronous data exchange. *In 2020 Chinese Control And Decision Conference (CCDC)* (pp. 3665–3670). IEEE.

Ferrari, N., Gebremichael, T., Jennehag, U., and Gidlund, M. (2018, October). Lightweight group-key establishment protocol for IoT devices: Implementation and performance analyses. *In 2018 Fifth International Conference on Internet of Things: Systems, Management and Security* (pp. 31–37). IEEE.

Gamil R. S. Qaid and Nadhem Sultan Ebrahim, A lightweight cryptographic algorithm based on DNA computing for IoT devices, *Security and Communication Networks*, vol.2023 Article ID 9967129, 12 pages, 2023.

Ge, Aijun and Wei, Puwen. (2019). *Identity-based broadcast encryption with efficient revocation.* 10.1007/978-3-030-17253-4_14.

Ibrahim, Nahla and Agbinya, Johnson. (2023). design of a lightweight cryptographic scheme for design of a lightweight cryptographic scheme for resource-constrained internet of things devices. *Applied Sciences.* 13. 1–17. 10.3390/app13074398.

Jahanbani, M., Bagheri, N., & Norozi, Z. (2020). Lightweight implementation of SILC, CLOC, AES-JAMBU and COLM authenticated ciphers. *Microprocessors and Microsystems*, 72, 102925.

Jha, P., Zorkta, H. Y., Allawi, D., and Al-Nakkar, M. R. (2020, June). Improved lightweight encryption algorithm (ILEA). *In 2020 International Conference for Emerging Technology (INCET)* (pp. 1–4). IEEE.
Kar, J., Naik, K., and Abdelkader, T. (2020). A secure and lightweight protocol for message authentication in wireless sensor networks. *IEEE Systems Journal*, 15(3), 3808–3819.
Kim, T. H., Kumar, G., Saha, R., Buchanan, W. J., Devgun, T., and Thomas, R. (2021). LiSP-XK: extended light-weight signcryption for IoT in resource-constrained environments. *IEEE Access*, 9, 100972–100980.
Kubba, Z. M. J., and Hoomod, H. K. (2020, November). Modified present encryption algorithm based on new 5D chaotic system. *In IOP Conference Series: Materials Science and Engineering* (Vol. 928, No. 3, p. 032023). IOP Publishing.
Lara-Nino, C. A., Diaz-Perez, A., and Morales-Sandoval, M. (2018). Elliptic curve lightweight cryptography: A survey. *IEEE Access*, 6, 72514–72550.
Latif, M. A., Ahmad, M. B. and Khan, M. K. (2020, October). A review on key management and lightweight cryptography for IoT. *In 2020 Global Conference on Wireless and Optical Technologies (GCWOT)* (pp. 1–7). IEEE.
Li, M., and Chen, S. (2021). Improved meet-in-the-middle attacks on reduced-round Joltik-BC. *IET Information Security*, 15(3), 247–255.
Li, Y., Du, L., Zhao, G., and Guo, J. (2013, August). A lightweight identity-based authentication protocol. *In 2013 IEEE International Conference on Signal Processing, Communication and Computing (ICSPCC 2013)* (pp. 1–4). IEEE.
Malone-Lee, John. (2002). *Identity-Based Signcryption. IACR Cryptology ePrint Archive*. 2002. 10.1007/978-3-540-89411-7_10.
Mhaouch, A., Elhamzi, W., and Atri, M. (2020, September). Lightweight hardware architectures for the piccolo block cipher in FPGA. *In 2020 5th International Conference on Advanced Technologies for Signal and Image Processing (ATSIP)* (pp. 1–4). IEEE.
Naru, E. R., Saini, H., and Sharma, M. (2017, February). A recent review on lightweight cryptography in IoT. *In 2017 International Conference on I-SMAC (IoT In Social, Mobile, Analytics and Cloud) (I-SMAC)* (pp. 887–890). IEEE.
Patel, C., and Doshi, N. (2020). Secure lightweight key exchange using ECC for user-gateway paradigm. *IEEE Transactions on Computers*, 70(11), 1789–1803.
Patil, J., Bansod, G., and Kant, K. S. (2017, February). LiCi: A new ultra-lightweight block cipher. *In 2017 International Conference on Emerging Trends & Innovation in ICT (ICEI)* (pp. 40–45). IEEE.
Qiao Z., Liang S., Davis S. and Jiang H. (2014). Survey of attribute based encryption, *15th IEEE/ACIS International Conference on Software Engineering, Artificial Intelligence, Networking and Parallel/ Distributed Computing (SNPD)*, Las Vegas, NV, USA, pp. 1–6, doi:10.1109/SNPD.2014.6888687.
Rani, D. J., and Roslin, S. E. (2016, November). Light weight cryptographic algorithms for medical internet of things (IoT)-a review. *In 2016 Online International Conference on Green engineering and Technologies (IC-GET)* (pp. 1–6). IEEE.
Rao, Vidya and Prema, K (2021). A review on lightweight cryptography for Internet-of-Things based applications. *Journal of Ambient Intelligence and Humanized Computing*. 12. 1–23. 10.1007/s12652-020-02672-x.
Rasori, Marco & Perazzo, Pericle & Dini, Gianluca. (2019). A lightweight and scalable attribute-based encryption system for smart cities. *Computer Communications*. 149. 10.1016/j.comcom.2019.10.005.
Sowjanya, K., Dasgupta, M., Ray, S., & Obaidat, M. S. (2019). An efficient elliptic curve cryptography-based without pairing KPABE for internet of things. *IEEE Systems Journal*, 14(2), 2154–2163.
Thakor, V. A., Razzaque, M. A., and Khandaker, M. R. (2021). Lightweight cryptography algorithms for resource-constrained IoT devices: A review, comparison and research opportunities. *IEEE Access*, 9, 28177–28193.
Tran, S. D. P., Seok, B., and Lee, C. (2020). HANMRE-An authenticated encryption secure against side-channel attacks for nonce-misuse and lightweight approaches. *Applied Soft Computing*, 97, 106663.
Ullah S., Radzi R. Z., Yazdani T. M., Alshehri A. and Khan I., (2022). Types of lightweight cryptographies in current developments for resource constrained machine type communication devices: challenges and opportunities, *in IEEE Access*, vol. 10, pp. 35589–35604, doi:10.1109/ACCESS.2022.3160000.
Victor, M., Praveenraj, D. D. W., Sasirekha, R., Alkhayyat, A. and Shakhzoda, A. (2023). Cryptography: advances in secure communication and data protection. *In E3S Web of Conferences* (Vol. 399, p. 07010). EDP Sciences.
William J. Buchanan, Shancang Li and Rameez Asif (2017) Lightweight cryptography methods, *Journal of Cyber Security Technology*, 1:3–4, 187–201, DOI:10.1080/23742917.2017.1384917.
Wu, H. (2016). *Acorn v3. Submission to CAESAR Competition*.
Wu, H., and Preneel, B. (2014). AEGIS: A fast authenticated encryption algorithm. *In Selected Areas in Cryptography–SAC 2013: 20th International Conference*, Burnaby, BC, Canada, August 14–16, 2013, Revised Selected Papers 20. 185–201. Springer Berlin Heidelberg.
Yan, L., Li, L. and Guo, Y. (2023). DBST: a lightweight block cipher based on dynamic S-box. *Front. Comput. Sci.* 17, 173805. https://doi.org/10.1007/s11704-022-1677-5
Zainab AlJabri, Jemal Abawajy, Shamsul Huda, and Chin-Ling Chen. (2023). A comprehensive review of lightweight authenticated encryption for IoT devices. *Wirel. Commun. Mob. Comput.* 2023 (2023). https://doi.org/10.1155/2023/9071969

Computational Methods in Science and Technology Volume 1 – Kaur et al. (Eds)
© 2025 the Author(s), ISBN: 978-1-032-81760-6

Improved hierarchical neural network architecture for house price prediction

Pandiyanathan Murugesan, Md Shamsul Haque Ansari & Nafees Akhter Farooqui
Department of Computer Science & Engineering, Koneru Lakshmaiah Education Foundation, Vaddeswaram, Andhra Pradesh, India

ABSTRACT: In the current environment, the value of real estate properties, in particular land and homes, might be considered substantial. When it comes to forecasting house prices, one of the most important factors to consider is the relationship between current housing costs and the state of the economy. The trends in house prices are not only relevant to buyers and sellers, but they also act as indications of the economic conditions that are now prevalent. The price of homes is determined by several factors, some of which include the size of the land in square feet, the location of the property, proximity to sources of water and transportation, and the presence or absence of amenities. The cost per square foot rises proportionally with amenities. Property prices are often estimated using "ma-chine learning" based on certain parameters. Neural networks with three or five layers estimate a house's price in this way. Price prediction and recognition/measurement precision are the main goals employing several ANN layers.

Keywords: House Price Prediction, Machine Learning, Neural Networks, Proximity.

1 INTRODUCTION

House price prediction helps with property assessment, finance, and urban planning. Buyers, sellers, governments, financial institutions, and the economy need accurate property price estimates. Artificial intelligence and machine learning improve house price predictions. Discovering and using hierarchical neural network concepts is exciting. Simple feedforward neural networks or linear regression algorithms forecast house values. These models make one scalar prediction based on the given input features. Some of these methods have worked, but they fail to convey the complexity and hierarchical structure of the real estate market, where many factors affect property values locally and globally. Deep learning reveals complex, multilayer component interactions in hierarchical neural network topologies. This research examines the increasing prevalence of hierarchical neural network designs to predict property prices. Hierarchical systems' core principles and ability to depict complicated interactions between numerous components are examined in this study and analyse housing data modelling benefits of geographical and temporal hierarchy. This approach emphasises these algorithms' ability to grasp housing price variables more thoroughly, perhaps improving projections. The economy is key to home price estimates. The lack of appropriate metrics makes housing expenses difficult to measure. Real estate deals require evaluation. Housing price changes affect buyers and sellers and disclose economic conditions (Sippel *et al.* 2017). So, objective house price estimates assist buyers and sellers decide. Several machine learning methods forecast house values. For this, neural networks will be deployed. Comparing three- and five-layer neural

 DOI: 10.1201/9781003501244-42

networks (Farooqui *et al.*, n.d.). House costs depend on many things, including bedrooms and baths. In addition, support vector regression parameter combinations can greatly affect predictions (Farooqui and Ritika 2020).

2 RELATED WORK

Professional evaluations determine property values in the competitive real estate market. Realtors, appraisers, developers, investors, lenders, and buyers need appraisals. Accurate appraisals help buyers and sellers negotiate, mortgage holders lend and investigate, and governments build cities. House price prediction is a growing trend in real estate and finance (Hacıevliyagil *et al.* 2022). Previously, linear regression models and simpler machine learning methods were used. These approaches are unstable and only moderately accurate in ideal settings since real estate markets are complicated and non-linear (Battifarano and Qian 2019). CNNs are prominent hierarchical neural networks in computer vision (Khan *et al.* 2020). CNNs' success has motivated scientists to predict real estate prices with hierarchical neural networks (Zhou 2020). The worth of a property depends on market, demographic, and ecological factors (Nelson and Frost 2023). Hierarchical neural network de-signs can store temporal hierarchies and explain how market fluctuations affect property prices using prior market data (Feng *et al.* 2019). Hierarchical neural networks forecast home values, changing reality. Classic linear models and neural networks struggled to explain the housing market's complexity. Multilayers of connected neurons automatically extract hierarchical characteristics from incoming data in these models. Hierarchical neural networks' real estate market flexibility is prospective. They can reveal hidden patterns and feature links to better illustrate local and global property value influences. This aids predictions and informs buyers, sellers, investors, city planners, and regulators (Law *et al.* 2019). By the 1990s, print adverts may have been hosted online due to their popularity (Twenge *et al.* 2019).

Selim (2009) projected property values with LR/ANN. Ad frequency and Google search data predicted property values. Hedonic regression is statistical, but ANN with input, hidden, and output layers predicts prices better. Many factors also affect-ed property prices (Selim 2009). Water, pool, home or building type, square footage, bedrooms, and neighbourhood were most important. SVR, Decision Trees, and RF solved this problem. C4.5 decision tree, RIP-PER, Nave Bayes, and AdaBoost projected pricing using Multiple Listing Service data, past mortgage rates, and local school ratings (Baldominos *et al.* 2018). The authors compare RF, SVD, Ridge, Lasso, and XGB housing market price prediction algorithms (Markovics and Mayer 2022). Large datasets suit deep learning architectures, making them popular. Sadhwani *et al.* (2021) suggested assessing mortgage risk from 120 million 1995–2014 US loan records. Poursaeed *et al.* (2018) explore how interior and exterior property attributes affect market value using ML. Property pricing studies use multiple data modelling methods. These models are econometrics. Statistics or math are emphasised in modelling. ARIMA uses machine learning, linear regression, ANN, hedonic price model, and fuzzy logic. Many real estate experts have investigated these models' uses (Abidoye 2017). AI models use neural networks. Successful real estate strategies are now used elsewhere. This modelling method predicts real estate prices well (Potrawa and Tetereva 2022).

3 METHODOLOGY

In this section, examine the dataset, perform preprocessing tasks, construct a neural network, create a model, conduct model testing, and generate predictions.

3.1 *Dataset description*

In computing and business, "data" is computer-readable information. Dataset should highlight all pertinent data. Data is modified to suit our model after collection (Costa-jussà *et al.* 2022). Transforming text, images, videos, and music into vectors and tensors simplifies linear algebra. This data needs normalisation, standardisation, and cleansing to work. "House Prices: Advanced Regression Techniques" data is used to address the problem. That has 80 attributes.

3.2 *Data analysis and feature selection*

Extraction of features reduces the resources needed to display enormous datasets. Complex data processing is difficult due to variables (Bhattarai *et al.* 2019). Multiple variable data analysis needs a lot of memory and computational capacity and can overfit classification algorithms on training samples, resulting in poor generalisation. Feature extraction uses variables to prevent these issues while maintaining data accuracy. Many machine learning researchers believe model construction requires optimised feature extraction (Yu *et al.* 2022). Business, science, and social sciences use data analysis techniques, each with their own titles. Modern firms use data analysis to make more informed decisions and improve operational efficiency. Data analysis and feature selection are needed for accurate forecasts. Application of the correlation function is the most efficient technique to identify which characteristics affect property values (Chou *et al.* 2022). Characteristics with a property price correlation greater than 0.1 are more likely to predict the outcome. Create a model with these traits to improve forecasts (Joloudari *et al.* 2019). Missing values in the dataset and appropriate scaling require additional work. Each feature's median value estimates missing values, and the Min-Max Scaler algorithm yields the best results (Koc and Gurgun 2022). After dataset scaling, each home area's needed features are determined.

3.3 *Model selection and methods*

Model selection involves the amalgamation of data and prior knowledge to make a choice among a set of statistical models denoted as "Mγ" from the collection of models denoted as "M." When constructing a model, determinations are made regarding the incorporation or omission of covariates, along with managing uncertainty surrounding the encoding of these covariates (Zhou *et al.* 2023). Design matrix "Xγ" for any given model "Mγ" are based both on the prior hypotheses and the data. With many potential covariates, these decisions become difficult. Some algorithms will select variables to be included in the model, but only return a single "best" model.

3.4 *Artificial neural networks*

ANN algorithms resemble the brain's computational processes to model complex patterns and predictive problems (Chung and Shin 2020). Our brain's billions of neurons send electrical signals to process information. Neuron dendrites receive external stimuli. The cell body processes the information into an output before sending it along the axon to the next neuron. Signal intensity affects whether the next neuron accepts the signal. Figure 1 is a diagram that illustrates the architecture of artificial neural networks. Here, w1, w2, and w3 represent the relative strengths of the input signals.

As shown below, an ANN simplifies brain neuron function. Input, hidden, and output layers make up a network. MLP, which stands for "multi-layer perceptron," refers to this setup's numerous layers.

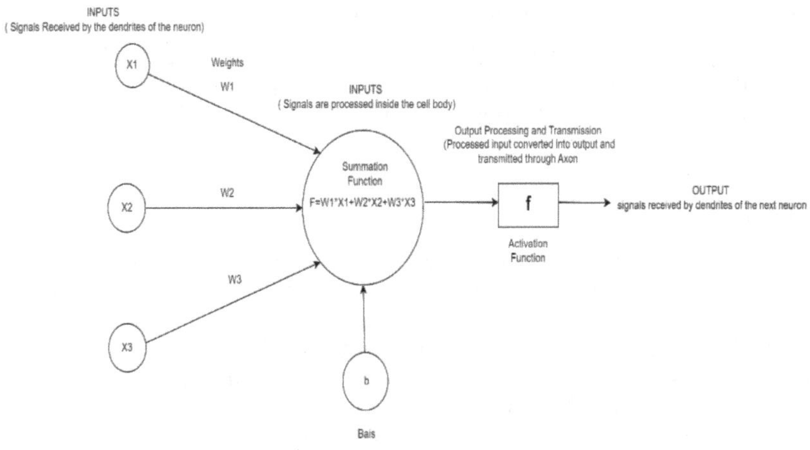

Figure 1. Architecture of artificial neural network (Chung and Shin 2020).

The hidden layer acts as a "distillation layer," extracting important patterns from inputs and passing them on to the next layer. This method enhances network performance by segregating important from irrelevant data. contained in the inputs. Its activation function holds dual significance. captures non-linear relationships within the inputs. It transforms the input into a more valuable and meaningful output.

3.5 *Construction of neural networks*

This paper addresses the problem through the application of two neural network solutions: 3-layer neural network and a 5-layer neural network, both implemented using Keras.

3.6 *Three layered neural network*

Three-layer Neural network consists of one input layer, hidden layer and one output layer. Input layer has 100 neurons, RELU functions as activation and input dimension of 24, the hidden layer has 50 neurons and the activation RELU and output layer has one neuron linear activation. The model is compiled with the "Mean squared error" and optimizer "Adam" is the loss function. Both equations are represented in equation1 and equation 2, respectively.

$$\frac{1}{n}\sum_{i=1}^{n}\left(Y_i - \vec{Y}_i\right)^2 \tag{1}$$

where: n – numbers of data points
Y_i – observed values
$\vec{Y}_i$ – predicted values

$$\theta_{t+1} = \theta_t - \frac{\eta \;\widehat{m}_t}{\sqrt{\widehat{v}_t} + \varepsilon} \tag{2}$$

where: $\widehat{m}_t = \frac{\widehat{m_t}}{1-\beta_1^t}$

$\widehat{v}_t = \frac{\widehat{v}_t}{1-\beta_2^t}$

Predictions with 3-layer neural network are as follows:

Predictions:[[0.20121609] [0.23529924] [0.2953675] [0.29329547] [0.25579336]]
Actual values: [[0.224208] [0.248687] [0.243157] [0.237507] [0.199642]]
Model Info: Batch Size: 32, No of Epochs: 30, Loss Function: MSE
Optimizer: Adam

3.7 *Five layered neural network*

The architecture of the five-layer neural network comprises an input layer with 300 neurons, three hidden layers with 100, 50, and 10 neurons respectively, and an output layer with a single neuron. The activation function utilized for the input layer is ReLU, with an input dimension of 24. Similarly, the activation function for the hidden layers is ReLU, while the output layer employs linear activation. The model is compiled with loss function "mean squared error" and optimizer "Adam".

Predictions with 5-layer neural network are as follows:

Predictions: [0.18765602] [0.21876758] [0.27828076] [0.28479898] [0.25641984]]
Actual *values:* [[0.224208] [0.248687] [0.243157] [0.237507] [0.199642]]
Model Info: Batch Size: 32, No of Epochs: 30, Loss Function: MSE
Optimizer: Adam

4 CONCLUSION

The present investigation predicts property values using three- and five-layer neural networks. This research includes 24 house-price-correlated features. Five-layer neural networks are more accurate than three-layer neural networks. A direct correlation was observed between property prices and 24 criteria. Many factors affect the housing market, including properties, neighbourhoods, and economic indicators. The research aims to increase model predictability by increasing real estate market representation. Extensive testing and analysis conducted in this research demonstrate that five-layer neural networks outperform three-layer ones in predicting home values. More intricate dataset patterns and dependencies with more layers in the neural net-work design could lead to more accurate and dependable predictions. The depth and complexity of the five-layer network enable it to adapt to shifting trends in the real estate market, im-proving forecasting accuracy. The findings show that neural network architecture, feature se-lection, and data preparation are essential for prediction model effectiveness in this domain. These findings can be used by academics, real estate agents, property owners, investors, and policymakers who are looking for more precise methods for managing and accessing the housing market.

REFERENCES

Abidoye, R. B. (2017). *Towards Property Valuation Accuracy: A Comparison of Hedonic Pricing Model and Artifiical Neural Network.*

Baldominos, A., Blanco, I., Moreno, A. J., Iturrarte, R., Bernárdez, Ó., and Afonso, C. (2018). Identifying real estate opportunities using machine learning. *Applied Sciences, 8*(11), 2321.

Battifarano, M., and Qian, Z. S. (2019). Predicting real-time surge pricing of ride-sourcing companies. *Transportation Research Part C: Emerging Technologies*, 107, 444–462.

Bhattarai, B. P., Paudyal, S., Luo, Y., Mohanpurkar, M., Cheung, K., Tonkoski, R., Hovsapian, R., Myers, K. S., Zhang, R., and Zhao, P. (2019). Big data analytics in smart grids: state-of-the-art, challenges, opportunities, and future directions. *IET Smart Grid*, 2(2), 141–154.

Chou, J.-S., Fleshman, D.-B., and Truong, D.-N. (2022). Comparison of machine learning models to provide preliminary forecasts of real estate prices. *Journal of Housing and the Built Environment*, 37(4), 2079–2114.

Chung, H., and Shin K. (2020). Genetic algorithm-optimized multi-channel convolutional neural network for stock market prediction. *Neural Computing and Applications*, 32, 7897–7914.

Costa-jussà, M. R., Cross, J., Çelebi, O., Elbayad, M., Heafield, K., Heffernan, K., Kalbassi, E., Lam, J., Licht, D., & Maillard, J. (2022). No language left behind: Scaling human-centered machine translation. *ArXiv Preprint ArXiv:2207.04672*.

Farooqui, N. A., & Ritika. (2020). A machine learning approach to simulating farmers' crop choices for drought prone areas. *Proceedings of ICETIT 2019: Emerging Trends in Information Technology*, 472–481.

Farooqui, N. A., Mishra, A. K., & Mehra, R. (n.d.). Automatic crop disease recognition by improved abnormality segmentation along with heuristic-based concatenated deep learning model. *Intelligent Decision Technologies, Preprint*, 1–23.

Feng, F., He, X., Wang, X., Luo, C., Liu, Y., and Chua, T.-S. (2019). Temporal relational ranking for stock prediction. *ACM Transactions on Information Systems (TOIS)*, 37(2), 1–30.

Hacıevliyagil, N., Drachal, K., & Eksi, I. H. (2022). Predicting house prices using DMA method: evidence from Turkey. *Economies*, 10(3), 64.

Joloudari, J. H., Saadatfar, H., Dehzangi, A., and Shamshirband, S. (2019). Computer-aided decision-making for predicting liver disease using PSO-based optimized SVM with feature selection. *Informatics in Medicine Unlocked*, 17, 100255.

Khan, A., Sohail, A., Zahoora, U., and Qureshi, A. S. (2020). A survey of the recent architectures of deep convolutional neural networks. *Artificial Intelligence Review*, 53, 5455–5516.

Koc, K., and Gurgun, A. P. (2022). Scenario-based automated data preprocessing to predict severity of construction accidents. *Automation in Construction*, 140, 104351.

Law, S., Paige, B., and Russell, C. (2019). Take a look around: using street view and satellite images to estimate house prices. *ACM Transactions on Intelligent Systems and Technology (TIST)*, 10(5), 1–19.

Markovics, D., and Mayer, M. J. (2022). Comparison of machine learning methods for photovoltaic power forecasting based on numerical weather prediction. *Renewable and Sustainable Energy Reviews*, 161, 112364.

Nelson, P. B., and Frost, W. (2023). Migration responses to the covid-19 pandemic: a case study of New England showing movements down the urban hierarchy and ensuing impacts on real estate markets. *The Professional Geographer*, 75(3), 415–429.

Potrawa, T., and Tetereva, A. (2022). How much is the view from the window worth? Machine learning-driven hedonic pricing model of the real estate market. *Journal of Business Research*, 144, 50–65.

Poursaeed, O., Matera, T., and Belongie, S. (2018). Vision-based real estate price estimation. *Machine Vision and Applications*, 29(4), 667–676.

Sadhwani, A., Giesecke, K., and Sirignano, J. (2021). Deep learning for mortgage risk. *Journal of Financial Econometrics*, 19(2), 313–368.

Selim, H. (2009). Determinants of house prices in Turkey: Hedonic regression versus artificial neural network. *Expert Systems with Applications*, 36(2), 2843–2852.

Sippel, S. R., Larder, N., and Lawrence, G. (2017). Grounding the financialization of farmland: perspectives on financial actors as new land owners in rural Australia. *Agriculture and Human Values*, 34, 251–265.

Twenge, J. M., Martin, G. N. and Spitzberg, B. H. (2019). Trends in US Adolescents' media use, 1976–2016: The rise of digital media, the decline of TV, and the (near) demise of print. *Psychology of Popular Media Culture*, *8*(4), 329.

Yu, G., Tabatabaei, M., Mezei, J., Zhong, Q., Chen, S., Li, Z., Li, J., Shu, L., and Shu, Q. (2022). Improving chronic disease management for children with knowledge graphs and artificial intelligence. *Expert Systems with Applications*, 201, 117026.

Zhou, C. Y., Talmi, D., Daw, N., and Mattar, M. G. (2023). *Episodic Retrieval for Model-Based Evaluation in Sequential Decision Tasks*.

Zhou, X. (2020). The usage of artificial intelligence in the commodity house price evaluation model. *Journal of Ambient Intelligence and Humanized Computing*, 1–8.

Computational Methods in Science and Technology Volume 1 – Kaur et al. (Eds)

Analyzing COVID-19 sentiments on Twitter: A machine learning approach

V. Aishwarya, K. Sanmuga Priyan, R. Vyshnavi, M. Azhagiri & S. Sathya Priya
Student, Chennai, SRM Institute of Science and Technology

ABSTRACT: Throughout the outbreak, COVID-19 has been a hot issue on Twitter, providing a forum for people to voice their ideas and perspectives. Unfortunately, there isn't a thorough rundown of sentiment analysis in Twitter discussions around COVID-19. By analyzing sentiment analysis utilizing COVID-19 Twitter data from both a machine learning and a behavioral and social science standpoint, this article fills this gap. Forty papers examining public attitudes were found in a study of the literature conducted between October 2019 and January 2022. Higher accuracy is shown by ensemble models, especially those that use BERT and RoBERTa. These models do not provide exact outcomes, despite their theoretical promise. Based on the COVIDSENTI dataset, the research emphasizes the need for government action on social media to counter false information. It highlights that, despite the continual examination of datasets, proactive public health actions are crucial in lowering unfavorable sentiments during a pandemic.

Keywords: Sentiment Analysis, Twitter API, Random Forest, COVID–19

1 INTRODUCTION

Unprecedented uncertainty brought about by the COVID-19 epidemic has caused a dramatic change in human behavior. These difficult circumstances have made social media platforms—especially Twitter—essential for emotional expression, information sharing, and communication. Twitter is an essential platform for delivering critical COVID-19 updates and facilitating real-time collaboration. Comprehending the array of perspectives articulated on Twitter is vital in comprehending the dynamics of public opinion, providing direction for policy development, and augmenting outreach tactics.

Twitter has developed into a dynamic forum where users participate in discussions in real time, commenting on and influencing public discourse around the epidemic. This work uses sophisticated machine learning techniques to explore the emotional terrain seen in large datasets of COVID-19-related tweets. The research aims to chronicle dominant perspectives and provide insights into the changing public perceptions of the epidemic via sentiment patterns and repeating themes throughout time. As the pandemic story develops, these findings may help communicators, policymakers, and decision-makers navigate the convoluted realm of COVID-19 arguments on Twitter.

Because Twitter is real-time and allows for the recording of strong emotions, it is an invaluable resource for monitoring and assessing the growing discourse around the epidemic. This study intends to demonstrate the enormous influence of social media, notably Twitter, in molding the narrative around COVID-19, in recognition of Twitter's position as a real-time indicator of public opinion. The research highlights the significance of social media in creating and sharing knowledge amid global crises by illuminating the collective awareness shown in digital environments.

 DOI: 10.1201/9781003501244-43

2 LITERATURE REVIEW

Braig *et al.* (2023) [1] examine Twitter attitude towards COVID-19. Using over 120 million messages between January and May 2020, sentiment analysis was found to be a useful approach for analysing public impressions of containment efforts, vaccine readiness, and government trust. Machuca *et al.* (2021) [2] study the massive effects of pandemics on society, focusing on COVID-19. The research emphasises the importance of understanding people's viewpoints in difficult situations and web scraping's ability to reveal implicit public opinion trends. It also considers how the epidemic is changing behaviour, mental health, cultural norms, and economic inequality. Usman *et al.* (2021) [3] emphasise the dynamic nature of data concerns and its usage in social media, healthcare, and calamities like the COVID-19 pandemic. It achieves this by reviewing machine learning, sentiment analysis, and textual analytics literature. Andrea and others, [4] Social media, which uses word embedding and emotion analysis, was essential for spreading information and assessing public opinion during these troubling times. Global lockdowns and travel restrictions followed. Pedrosa and colleagues [5]. The epidemic has increased mental health difficulties in certain demographic groups. Global emotional distress reduction techniques and remote support networks have been established to address this. Muvazima *et al.* (2020) [6] use Twitter sentiment analysis and exploratory data analysis of COVID-19 cases and fatalities to analyse public opinion of distant employment and online education during these tough times. Salvador *et al.* (2023) [7] compared multilingual and Spanish-only BERT models with SVM, Decision Trees, Logistic Regression, and Naive Bayes to assess public opinion of the epidemic's severity and government reaction. Dharmendra Associates [8] Use five machine learning classifiers and propose SATD to improve F1-score, recall, and accuracy. Test classifier robustness against overfitting, outliers, and handling many features using the twitter dataset. Imran's crew [9] Six models (RF, XGBoost, SVC, ETC, DT, LSTM), a new feature extraction methodology, and feature engineering methods (BoW, TF-IDF) were utilised to highlight the importance of diverse models and feature engineering for sentiment classification. The assessment used IEEE data port sentiment ratings. Deep neural networks and NLP were used by Rustam *et al.* (2021) [10] to classify emotions in two stages in Sweden and Denmark.

3 RELATED WORKS

Social networking sites like Twitter have completely changed the way people communicate by giving them a quick and easy way to share their thoughts and feelings. Sentiment analysis is the process of identifying the emotional undertones in textual data. It has broad applications in a variety of domains, such as politics, public opinion tracking, marketing, and customer service. Preprocessing is required to get raw Twitter text data ready for sentiment analysis. Tokenization, stop-word reduction, and translating emoticons and emojis into text counterparts are some of these procedures; they all help to expedite the analytical phase that follows.

Sentiment representation in Twitter datasets is made more difficult by a variety of features, including text content, location data, and user information. Using the Random Forest algorithm to classify tweets into positive, negative, and neutral emotions works well. Through the process of building decision trees during training and combining their predictions, this approach enables more sophisticated analysis and makes it possible to discern subtle emotional patterns in Twitter conversations. This method is very useful for identifying nuanced feelings, which makes it an essential tool for academics and companies looking to get a better understanding of the dynamics of online sentiment. The first round of data extraction using the Twitter API guarantees a large and varied dataset for examination. In order to curate relevant information that reflects the real-time nature of Twitter debates, filtering factors like language requirements and hashtags are helpful. Further preparation operations, such as normalization to normalize text and remove superfluous characters, are essential to improving the dataset's quality and applicability for sentiment analysis. When determining the semantic meaning of tweets and grading them for sentiment, the Sentiments npm package is essential. The basis for deciphering the emotional tone ingrained in Twitter data is these ratings. After preprocessing, relevant characteristics (text content, user data) are extracted from tweets to provide a comprehensive understanding of emotional expression in the dataset.

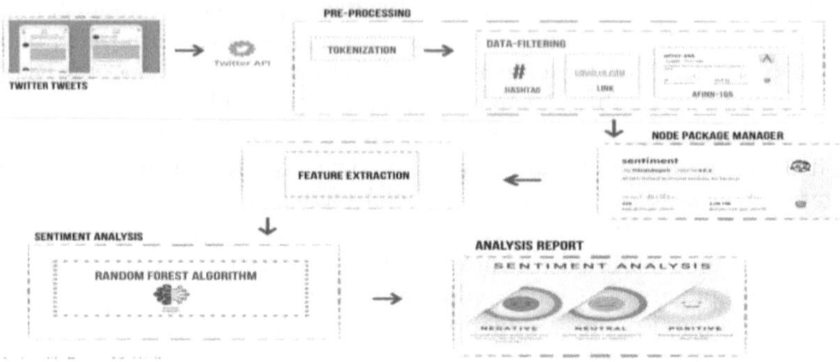

Figure 1. Proposed architecture of sentiment analysis.

A strong sentiment analysis model is built using the Random Forest classification technique. This model efficiently classifies tweets into positive, negative, and neutral feelings by using features gathered from preprocessed tweets and matching sentiment labels. Evaluation criteria that guarantee the validity and effectiveness of sentiment classification include accuracy, precision, recall, and F1-score.

The last stage involves creating a website using React.JS that presents sentiment analysis results in an approachable and user-friendly way. The use of interactive visualizations allows users to investigate and evaluate sentiment patterns, thereby improving the usability and accessibility of sentiment analysis results for a wider range of users. The use of visualization tools enables a more captivating and easily understood investigation of sentiment patterns in Twitter conversations.

4 EMPIRICAL RESULTS

When sentiment analysis of Twitter data was conducted, the Random Forest classifier produced noteworthy results in terms of accuracy, precision, recall, and F1-score, among other important performance measures. This accomplishment may be ascribed to a methodical strategy to data extraction that uses hashtags and language filters to generate a diversified dataset. Thorough preparation methods, such as tokenization and stop-word elimination, improved the raw text even more and made it easier for the classifier to identify the nuanced emotions tweeters convey.

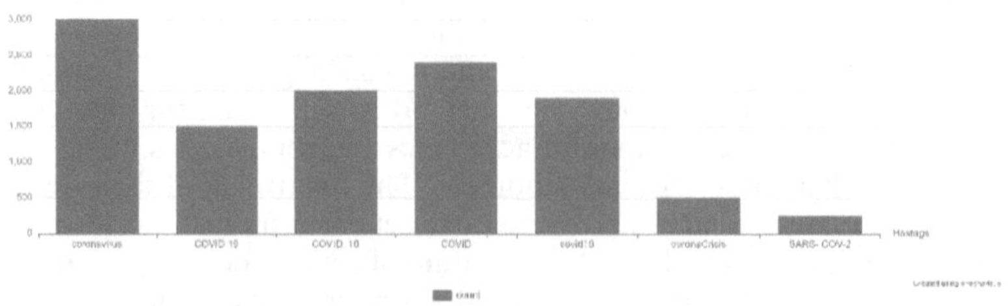

Figure 2. Keyword used in twitter tweet using Hashtag.

Figure 2 illustrates a visual representation or a word cloud is likely presented to depict the prominent keywords extracted from Twitter tweets related to COVID-19. A word cloud visually emphasizes words based on their frequency in the dataset, with more frequently occurring words appearing larger and bolder. This approach offers a quick and intuitive overview of the most commonly used terms in tweets associated with the hashtag #COVID19.

By including the Sentiments npm package for sentiment analysis and feature extraction, the classifier's performance was much improved. By exploring each tweet's semantic meaning, this integration allowed the computer to precisely identify the sentiment ratings—positive, negative, or neutral—associated with it. The classifier's overall performance in sentiment analysis was boosted by the incorporation of a variety of input characteristics, including textual content, user information, and geographic data, in the feature extraction process. This gave the classifier a thorough knowledge of sentiment expression.

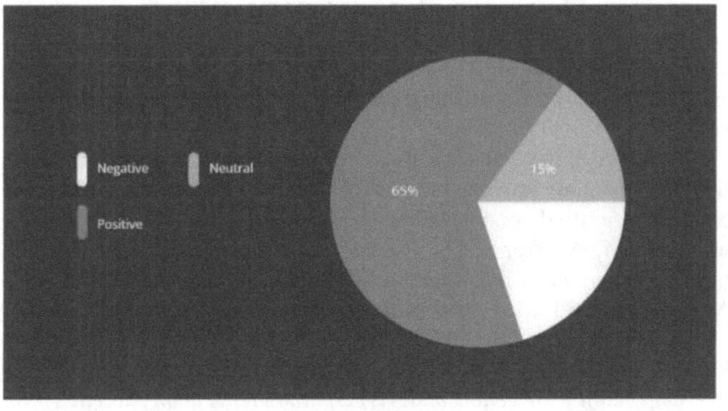

Figure 3. Training performance metrics over sentiments.

Figure 3 displays the evaluation metrics, which included F1-score, recall, accuracy, and precision, which functioned as quantitative measures of the classifier's effectiveness. The remarkable outcomes in these measures confirm that the Random Forest classifier can accurately classify tweets according to sentiment, demonstrating its effectiveness in managing the intricacies of Twitter data.

5 CONCLUSION

The results of this study highlight how well Random Forest categorization works when used for sentiment analysis of Twitter data. A thorough framework for examining public sentiment trends on social media platforms is presented by the methodical approach described in this study, which includes data extraction, preprocessing, sentiment analysis, and feature extraction. Additionally, a user-friendly web interface for result visualization is developed. This work advances our knowledge of sentiment analysis techniques and provides useful applications in a variety of fields, including opinion mining, brand monitoring, and market research.

REFERENCES

[1] Niklas Braig, Alina Benz, Soeren Voth, Johannes Breitenbach and Ricardo Buettner. (2023). Machine learning techniques for sentiment analysis of COVID-19-related twitter data. *IEEE Access*.

[2] *Twitter sentiment analysis on coronavirus: machine learning approach* to cite this article: Cristian R. Machuca *et al.* 2021.

[3] Naseem U, Razzak I, Khushi M, Eklund PW, Kim J. (2021). Covidsenti: A large-scale benchmark Twitter data set for COVID-19 sentiment analysis. *IEEE Trans Comput Soc Syst.* 8:1003–15. doi: 10.1109/TCSS.2021.3051189.

[4] Sciandra A. (2020). COVID-19 outbreak through tweeters' words: monitoring Italian social media communication about COVID-19 with text mining and word embeddings. In: 2020 *IEEE Symposium on Computers and Communications (ISCC)*. Rennes: IEEE.

[5] Pedrosa AL, Bitencourt L, Fróes ACF, Cazumbá MLB, Campos RGB, de Brito SBCS, *et al.* (2020). Emotional, behavioral, and psychological impact of the COVID-19 pandemic. *Front Psychol.* 11:566212. doi: 10.3389/fpsyg.2020.566212.

[6] Mansoor M, Gurumurthy K, Prasad V. (2020). Global sentiment analysis of COVID-19 tweets over time. *arXiv preprint* arXiv:201014234.

[7] Salvador Contreras Hernández1· María Patricia Tzili Cruz1· José Martín Espínola Sánchez1 · Angélica Pérez Tzili2 Deep Learning Model for COVID-19 Sentiment Analysis on Twitter (2023).

[8] Dangi, D., Dixit, D.K. and Bhagat. (2022). A. Sentiment analysis of COVID-19 social media data through machine learning. *Multimed Tools Appl* 81, 42261–42283. https://doi.org/10.1007/s11042-022-13492-w.

[9] Imran AS, Daudpota SM, Kastrati Z, Batra R. (2020). Cross-cultural polarity and emotion detection using sentiment analysis and deep learning on COVID-19 related tweets. *IEEE Access.* 8:181074–90. doi: 10.1109/ACCESS.2020.3027350.

[10] Rustam F, Khalid M, Aslam W, Rupapara V, Mehmood A, Choi GS. (2021). A performance comparison of supervised machine learning models for Covid-19 tweets sentiment analysis. *PLoS ONE.* 16: e0245909. doi: 10.1371/journal.pone.0245909.

[11] Desai M. and Mehta M. A. (Apr. 2016). Techniques for sentiment analysis of Twitter data: A comprehensive survey, in *Proc. Int. Conf. Comput., Commun. Autom. (ICCCA)*, pp. 149154.

[12] Giachanou A. and Crestani F., Like it or not: A survey of Twitter sentiment analysis methods, *ACM Comput. Surv.*, vol. 49, no. 2, p. 28, 2016.

[13] Lohani, Prakash Bhanu, Dagur., Arvind, Shukla and Dhirendra, *An Efficient Approach for Diabetes Classification Using Feature Selection and Hyperparameter Tuning,Recent Advances in Electrical & Electronic Engineering*, volume 17, issue, pages 1–15, year 2024

[14] Kushwaha, Kumar Pradeep, Dagur, Arvind, Shukla and Dhirendra. (2024). *HSLE: A Hybrid Ensemble Classifier for Prediction of Heart Disease, Recent Advances in Electrical & Electronic Engineering*, volume 17, issue, pages 1–12

[15] Aatif Jamshed, Anshu Kumar Dwivedi, Pawan Singh Mehra, Arvind Dagur. (2024). Internet-of-things-enabled networks: vision challenges and smart applications, *Advances in Computing Communications*, 2024, Vol 6, 1–17

[16] Ji X., Chun S. A., Wei Z., and Geller J. (Dec. 2015). Twitter sentiment classification for measuring public health concerns, *Social Netw. Anal. Mining*, vol. 5, no. 1, pp. 125.

[17] Cambria E. (Mar. 2016). Affective computing and sentiment analysis, *IEEE Intell. Syst.*, vol. 31, no. 2, pp. 102107.

[18] Kaila R. P. and Prasad A. K. (2020). Informational ow on Twitter Corona virus outbreak topic modeling approach, *Int. J. Adv. Res. Eng. Technol. (IJARET)*, vol. 11, no. 3, pp. 128134.

[19] Kaliyar, R.K., Goswami, A., Narang, P. (2021). FakeBERT: fake news detection in social media with a BERT-based deep learning approach. *Multimed. Tools Appl.*. https://doi.org/10.1007/ s11042-020-10183-2

Computational Methods in Science and Technology Volume 1 – Kaur et al. (Eds)
© 2025 the Author(s), ISBN: 978-1-032-81760-6

Improving the quality of Indian celebrity data for machine learning applications

Jyoti Kesarwani & Himanshu Rai
United University, Prayagraj

ABSTRACT: This paper presents a comparative study of various convolutional neural network (CNN) models like Inception, ResNet, and VGG16 for face recognition on the Indian Celebrity Dataset. The work highlights the importance of an Indian celebrity face dataset to improve facial recognition performance globally, given the massive and multicultural population in India. The paper discusses different data preprocessing techniques applied to the Indian Celebrity Dataset including data cleaning, reduction, and transformation methods like normalization. It investigates how these preprocessing steps can enhance the data quality and model performance. The CNN models are evaluated on tasks such as facial attribute prediction and face recognition using the preprocessed Indian Celebrity Dataset. The paper aims to provide insights into effectively utilizing CNNs and data preprocessing for face analysis on a diverse Indian celebrity face dataset.

Keywords: CelebA Dataset, Data Preprocessing, Normalization, VGG16

1 INTRODUCTION

A significant number of humanities are becoming victim to image Forging in this day of technology. Many people alter images using technology and present them to the court as proof to deceive them. Therefore, any images published on social media should be appropriately classified as authentic or illegal in order to stop this. Social media can be an excellent tool for networking, information sharing, and socialization, but it can also unintentionally propagate false propaganda, mislead people, and even cause panic if care fails to be taken. A celebrity is a well-known and well-recognized person who attracts a lot of media and public attention in addition to being well-known in society. A model, athlete, or movie star can all be termed celebrities [1]. The movie reflects celebrity feelings, lifestyles, and some wonderful scenes [2]. Images or movies that have been digitally edited or changed are referred to as fake images of Indian Bollywood celebrities. These images may have been produced with harmful intent or humor. Fake image may show famous people doing things they haven't done, or hanging out with people they haven't met. Through social media, websites, and other online platforms, these images may circulate and result in false information, harm to the celebrities' reputations, and invasions of their privacy [3]. Certain technologies that make it possible to create or manipulate realistic-looking visuals. The process of assigning labels or classifications to facial images according to specific features or attributes is sometimes referred to as "face classification" [12]. This technology has considerably replaced approaches for identifying fake images, and even experts frequently are unable to tell real digital images from fakes just by looking at them. Hence, fraudulent individuals can effortlessly release altered visual content to trick their audience (Shen *et al.* 2019). Deepfakes are created using Generative Adversarial Networks (GANs). Changing the features of well-known politicians, celebrities, and people over explicit images and videos is known as deepfake [14].

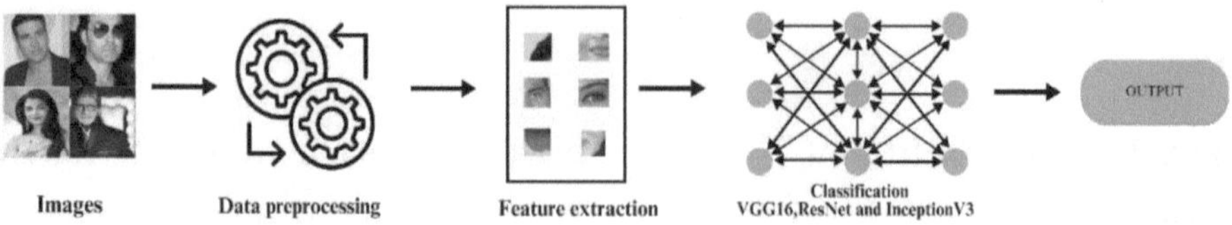

Figure 1. Process of classification from images.

Indian celebrities are famous not just in India but also in a number of countries where there is a huge Indian community. India offers a massive and multicultural population. As a result, adding Indian superstars to the dataset makes the model more applicable and relevant globally. The main reason for the need for an Indian celebrity dataset is to identifying modified or fraudulent Indian celebrity images. A gallery of 100 Indian superstars' faces, with a variety of expressions, positions, and lighting settings, is in Bollywood Celebrity Faces dataset. Numerous sources, including images of celebrities and images from the internet, were used to compile these pictures. Some features in this dataset are applied to each image, including laughing, posing with glasses, young, male and female. Tasks like attribute-based face recognition and facial attribute prediction benefit greatly from the information these annotations include.

Figure 2. Representations of various facial attributes [22].

Data preparation attempts to minimize data size, identify relationships between data, organize data, eliminate anomalies, and extract features from data [8]. Cleaning, integration, sampling, normalization, transformation, and reduction are just a few of the techniques that are included [4]. Remove missing values and eliminate mistakes when cleaning data. Feature scaling, one-hot encoding, and dimension reduction are examples of data transformation. To balance the class distribution, the majority class is undersampled and the minority class is oversampled during the data sampling procedure. The data sampling stage uses methods as SMOTE (Synthetic Minority Over-sampling Technique) for balancing the class distribution [5].

This study takes a different approach, focusing on the development and use of a database of Indian celebrities. Although current facial profiles such as CelebA and LFWA are commonly used, they fail to adequately capture the diversity of the Indian population. This study addresses the gaps in existing face recognition research by maintaining data storage specifically for Indian superstars, in order to improve global applications, including face analysis models. Recent CNN frameworks like Inception, ResNet, VGG16, and others can be compared on pre-processed Indian celebrity face datasets to get an understanding of how well these models work and how suitable they are for face recognition and positive prediction for Indian Celebrities.

The remaining portion of the paper is structured as follows: In Section 2, the different datasets are explained briefly. In Section 3, various data preprocessing techniques on an Indian celebrity dataset are briefly described along with their outcomes. In this paper discuss various CNN architectures, such as Inception, ResNet, and VGG16 using an Indian celebrity dataset in Section 4. Finally, Section 5 provides some conclusions about this work.

2 LITERATURE REVIEW

With over 200K images of celebrities and 40 attribute annotations, the CelebFaces Attributes collection (CelebA), is a complete face attributes collection. This dataset contains images with a wide range of expressions. CelebA has 202,599 facial images, 5 marker locations, 40 basic feature annotations in each image, and 10,177 identities [7]. The dataset can be used as training and test sets for facial synthesizing, landmark localization, face enhancing, identification of faces, recognition of faces, and face attribute recognition.

In machine learning, one of the most prominent datasets is LFWA (Labeled Faces in the Wild Attribute). Categorizing a face image, such as a hat, goggle, or beard. The LFWA dataset is used in face recognition, feature categorization, and facial analysis applications [6]. LFWA is a well-known unconstrained face attribute dataset, comprising 13,143 facial photos representing 5,749 people. For each face image, there are forty attribute annotations.

Age, ethnicity, and image background are among a number of factors present in Flickr-Faces-HQ (FFHQ), a collection of 70,000 high-quality PNG images with a 1024 x 1024 size [9]. Accessories including hats, sunglasses, eyeglasses, and other items are also well covered. Generative adversarial networks (GANs) and other deep learning approaches have created FFHQ a benchmark dataset for tasks like face generation, style transfer, and facial feature manipulation [10].

Introducing a different benchmark for the identification of large-scale scenes is the aim of the Large-scale Scene Understanding (LSUN) dataset. approximately 120,000 and 3,000,000 images are available [11]. Figure 3 depicts the number of publications for the LFWA, FFHQ, LSUN, and Celeba datasets throughout the years. The y axis shows the number of papers, while the x axis shows the year.

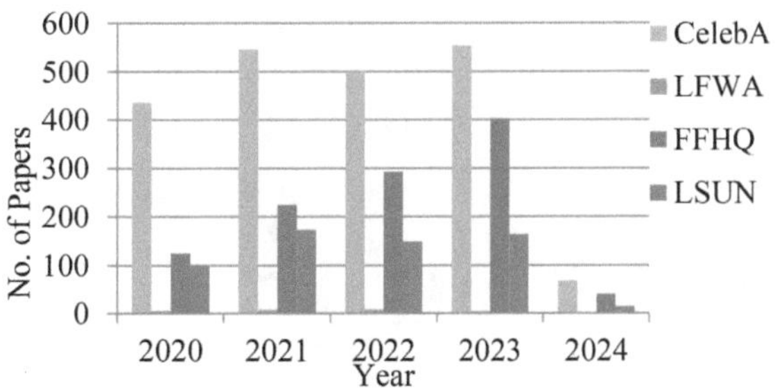

Figure 3. A list of all the papers published on various datasets in each year [8].

3 DATA

Data from the real world is frequently noisy, lacking, or unreliable. By removing missing values, identifying outliers, and eliminating errors called preprocessing. Data preprocessing enhances data quality and guarantees that models are trained on accurate and dependable data [17][18]. Preprocessing approaches lessen the likelihood of both overfitting and underfitting the model by adequately cleaning and modifying the data. Several techniques emerge for preparing data, such as data cleaning, data reduction, imputation, and transformation [20]. The purpose of data cleaning is to eliminate noise and fix errors in unprocessed business information. Data integration, like combines information from several sources into an organized form of storage [13]. Through aggregation, feature elimination of redundant information, or clustering, data reduction removes certain samples or the irrelevant features. Transforming data in order to fit a specific range is known as data transformation such as normalization [15]. Data processing encompasses a number of methods such as:

3.1 *Data collection*

The first stage of the data preprocessing is data collection, during which raw data is obtained from a variety of sources. It involves collecting relevant information with the data required for an activity or research.

3.2 *Data cleaning*

Finding inaccurate or noisy data and improving it or eliminating it from the dataset is the process of data cleaning. In addition, it replaces any noisy data that is erroneous, incomplete, or unnecessary. When information is aggregated from different places, such as scrape information, or information from several clients. There is a chance that duplicate data will be generated. Mislabeled categories or classes are the cause of the inconsistencies. This process includes handling missing values, removing duplicate data, and eliminating data that is unnecessary. An Indian celebrity dataset includes images that are connected to each celebrity. This involves investigating several variables, including noise present, lighting, resolution, and clarity of the images. It could be necessary to eliminate or improve images with poor quality such as those with a lot of noise or low resolution using preprocessing techniques. It is required to standardize image sizes and formats to guarantee uniformity across the collection.

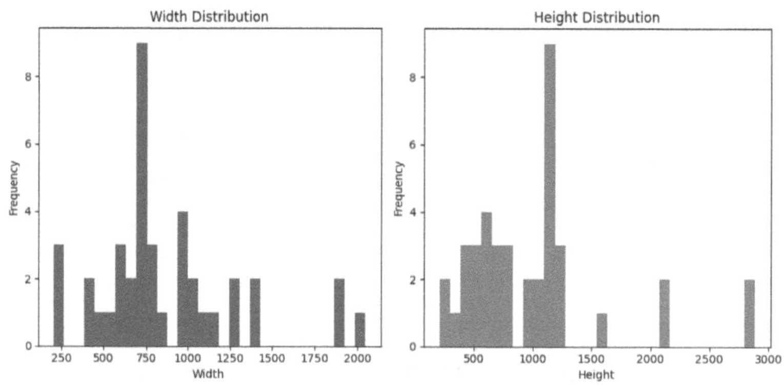

Figure 4. Height and width distribution of indian celebrity dataset.

3.3 *Data transformation*

The methods used to modify data into the structure required by machine learning algorithms are commonly referred to as data transformation [16]. Normalization and data aggregation represent two essential methods for data transformation. Normalization, which involves scaling feature data to a specific range, such as 0 to 1 or -1 to 1, is typically necessary when a dataset contains features with different scales. When features with values on an even bigger scale have the potential to reduce the effectiveness of another feature that is similarly important but on a lesser scale. This will have an adverse impact on the data mining model's accuracy performance.

3.4 *Data reduction*

A reduced representation of the data set that is significantly reduced in volume and aims to maintain almost all of the original data integrity can be obtained by applying data reduction techniques. Furthermore, feature selection is a preprocessing technique that indicates to

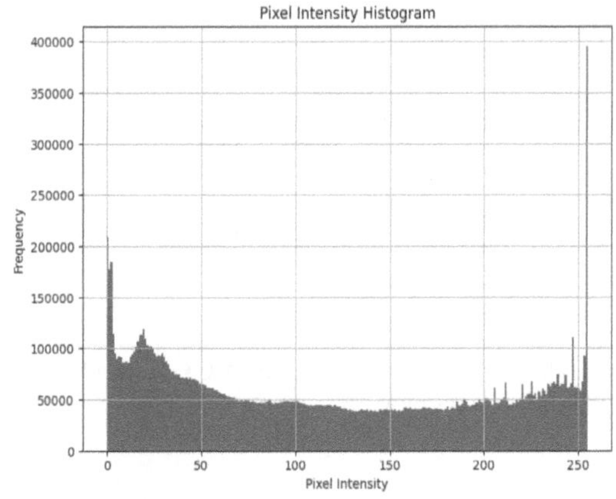

Figure 5. Pixel intensity histogram on indian celebrity dataset.

selection of main features. Reducing the number of columns in a dataset, or the number of features, is essentially how this is done (Alasadi *et al.* 2017). Reducing the features without compromising the quality of the dataset results in an increase in the model precision and prediction value, along with a decrease in learning time as well as memory needs. Several feature selection techniques are available to provide these benefits [19]. Principal Component Analysis (PCA) and t-Distributed Stochastic Neighbor Embedding (t-SNE) are two techniques that can be used to convert high-dimensional picture data in an Indian celebrity dataset into a lower-dimensional space.

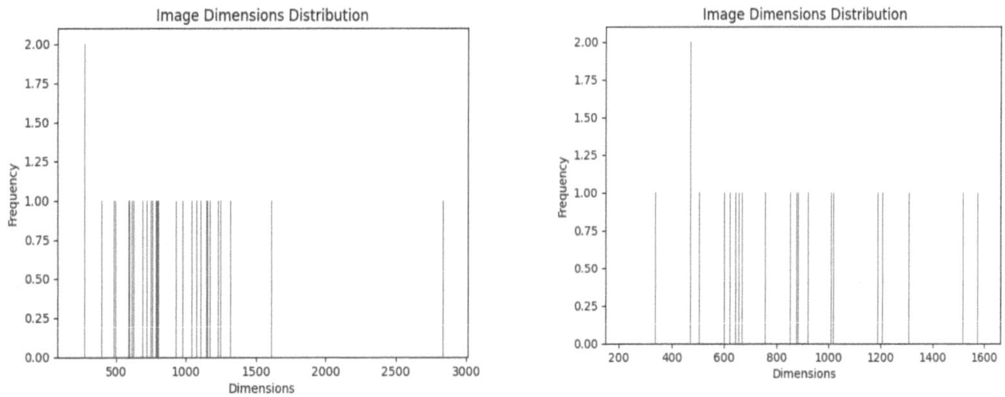

Figure 6. Image dimension distribution on Indian celebrity dataset.

4 DATA MODELLING

Several deep learning models for generative modeling, attribute prediction, and face recognition have utilized the use of the Indian Celebrity dataset [22]. Convolutional Neural Networks (CNNs) have become popular in image classification applications, such as facial recognition and attribute prediction. The Indian Celebrity dataset to train CNN architectures including VGG [21], ResNet, and Inception for tasks like predicting features like hair color, the availability of glasses [15] etc.

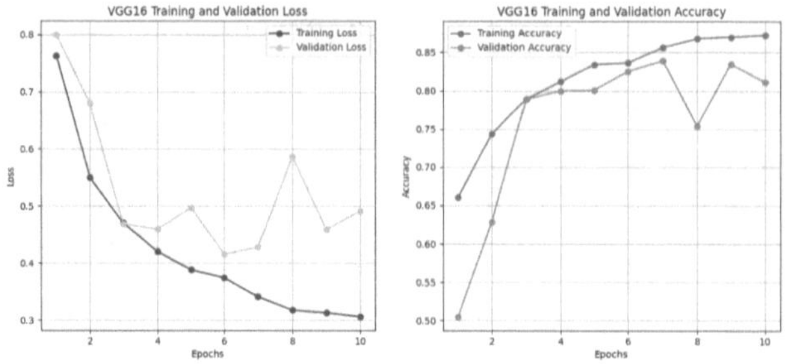

Figure 7. Performance of Indian celebrity dataset using VGG16 model (a) Training and validation loss and (b) Training and validation accuracy.

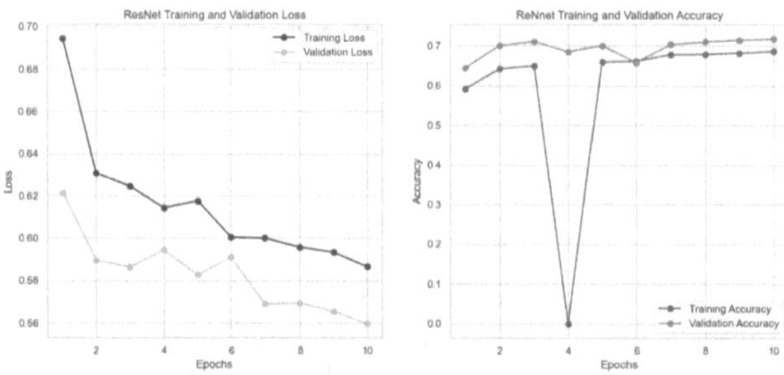

Figure 8. Performance of Indian celebrity dataset using ResNet model (a) training and validation loss and (b) training and validation accuracy.

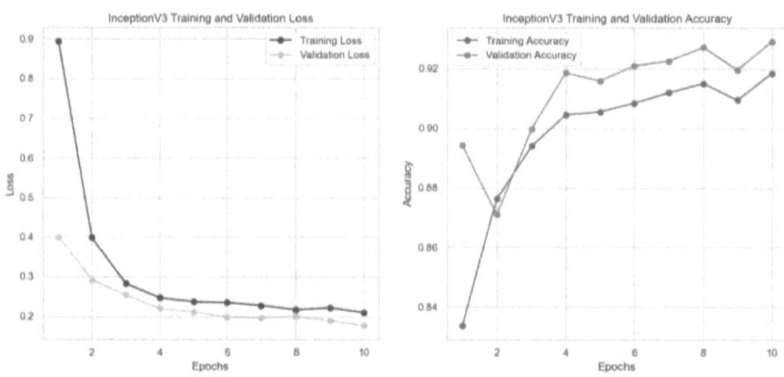

Figure 9. Performance of Indian celebrity dataset using inceptionV3 model (a) training and validation loss and (b) training and validation accuracy.

5 CONCLUSION

This study investigated the use of pretrained convolutional neural network (CNN) models such as Inception, ResNet, and VGG16 for face recognition and feature prediction tasks on Indian celebrity datasets extensive data management techniques were used to optimize the well-known Indian dataset. Data cleaning techniques were used to remove noise, duplicate images and inconsistencies, while data reduction techniques such as PCA and t-SNE helped to reduce multidimensional image data, as the higher theory of Normalization, is important in data transformation. Using a pre-processed Indian celebrity dataset, comparisons and trained testing of determined CNN architectures for face attribute calculation and face recognition operations offered useful details about how well such initiatives perform and how suitable CNN models are across India.

In summary, the results of the study emphasize the significance of data pre-processing and the CNN model's ability to perform facial recognition testing modified for the Indian environment. These developments support the development of a reliable and face recognition system in India that can accommodate the country's complex community. To boost the proposed method's broad applicability, future research should look at ensemble models, transfer learning techniques, and enlarge the dataset to include additional prominent Indian groups.

REFERENCES

[1] Eason G., Noble B., and Sneddon I. N., (April 1955). On certain integrals of Lipschitz-Hankel type involving products of Bessel functions, *Phil. Trans. Roy. Soc. London*, vol. A247, pp. 529–551.

[2] Halder, D., and Jaishankar, K. (2016). *Celebrities and Cyber Crimes: An Analysis of the Victimization of Female Film Stars on the Internet*. Available at SSRN 3049543.

[3] Shen, C., Kasra, M., Pan, W., Bassett, G. A., Malloch, Y., and O'Brien, J. F. (2019). Fake images: The effects of source, intermediary, and digital media literacy on contextual assessment of image credibility online. *New Media & Society*, 21(2), 438–463.

[4] Sutha, K. and Tamilselvi J. J. (2015). A review of feature selection algorithms for data mining techniques. *International Journal on Computer Science and Engineering*, 7(6), 63.

[5] Zhang, N., Paluri, M., Ranzato, M. A., Darrell, T. and Bourdev, L. (2014). Panda: Pose aligned networks for deep attribute modeling. *In Proceedings of the IEEE Conference on Computer Vision and Pattern Recognition* (pp. 1637–1644).

[6] Liu, Z., Luo, P., Wang, X., & Tang, X. (2015). Deep learning face attributes in the wild. *In Proceedings of the IEEE International Conference on Computer Vision* (pp. 3730–3738).

[7] Zhang, Y., Yin, Z., Shao, J., Liu, Z., Yang, S., Xiong, Y., ... and Hao, Z. (2021). *Celeba-spoof challenge 2020 on face anti-spoofing: methods and results. arXiv preprint* arXiv:2102.12642.

[8] Zhang, H., Chen, W., Tian, J., Wang, Y., and Jin, Y. (2018). Show, attend and translate: unpaired multi-domain image-to-image translation with visual attention. *ArXiv*, abs/1811.07483.

[9] Karras, T., Laine, S., and Aila, T. (2019). A style-based generator architecture for generative adversarial networks. *In Proceedings of the IEEE/CVF Conference on Computer Vision and Pattern Recognition* (pp. 4401–4410).

[10] Sauer, A., Schwarz, K., and Geiger, A. (2022, July). Stylegan-xl: scaling stylegan to large diverse datasets. *In ACM SIGGRAPH 2022 Conference Proceedings* (pp. 1–10).

[11] Yu, F., Seff, A., Zhang, Y., Song, S., Funkhouser, T., and Xiao, J. (2015). Lsun: Construction of a large-scale image dataset using deep learning with humans in the loop. *arXiv preprint* arXiv:1506.03365.

[12] Ehrlich, M., Shields, T. J., Almaev, T., and Amer, M. R. (2016). Facial attributes classification using multi-task representation learning. *In Proceedings of the IEEE Conference on Computer Vision and Pattern Recognition Workshops* (pp. 47–55).

[13] Singhal, S., Ali, S., Shukla, D.K. and ...Rathod, V., Awasthy, M., Enhanced-honey bee based load balancing algorithm for cloud environment, *Artificial Intelligence, Blockchain, Computing and Security*: Volume 1, 2023, 1, pp. 951–956.

[14] Kushwaha, P.K., Dagur, A. and Shukla, D., (2023). GA-ESE: A high-performance heart disease classification using hybrid machine learning approach, *Artificial Intelligence, Blockchain, Computing and Security*: Volume 2(2), pp. 82–89.

[15] Sharma, R.K. and Dagur, A. (2023). Various methods to classify the polarity of text based customer reviews using sentiment analysis, *Artificial Intelligence, Blockchain, Computing and Security*: Volume 2 (2), pp. 107–114.

[16] Sharma, R.K., Dagur, A. (2023). Analyzing the performance of different classifier for detecting polarity of customer reviews, *International Journal on Recent and Innovation Trends in Computing and Communication*, 11(7), pp. 142–148.

[17] Maharana, K., Mondal, S., and Nemade, B. (2022). A review: Data pre-processing and data augmentation techniques. *Global Transitions Proceedings*, 3(1), 91–99.

[18] Alasadi, S. A., and Bhaya, W. S. (2017). Review of data preprocessing techniques in data mining. *Journal of Engineering and Applied Sciences*, 12(16), 4102–4107.

[19] García, S., Luengo, J., & Herrera, F. (2015). *Data preprocessing in data mining* (Vol. 72, pp. 59–139). Cham, Switzerland: Springer International Publishing.

[20] Çetin, V., & YILDIZ, O. (2022). A comprehensive review on data preprocessing techniques in data analysis. *Pamukkale Üniversitesi Mühendislik Bilimleri Dergisi*, 28(2), 299–312.

[21] Jha, R., Jain, P., Tayal, S., and Sharma, A. (2021). Landmark recognition using VGG16 training. *Smart and Sustainable Intelligent Systems*, 17–39.

[22] Abdullah, I. A. (2021). Face recognition using face embedding method. *Turkish Journal of Computer and Mathematics Education (TURCOMAT)*, 12(10), 3383–3394.

Computational Methods in Science and Technology Volume 1 – Kaur et al. (Eds)

Advancements in Bat algorithm

Arvinder Kaur
Associate Professor, Department of Information Technology, Chandigarh Engineering College-CGC, Landran, Mohali (Punjab)

Yugal Kumar & Jagpreet Sidhu
Associate Professor, Computer Engineering/STME, NMIMS Chandigarh Campus

ABSTRACT: The modern times have led to the adoption of distinctive meta-heuristic procedures for solving distinct class of optimization-problems. The meta-heuristics procedures have benefit above conventional algorithms because of strong capabilities of exploration and exploitation. In a similar vein, a Bat algorithm was created based on the echolocation habits of bats to address optimisation difficulties. Several advancements have been made in bat algorithm since then to achieve the state-of art results. The purpose of the study is to investigate bat algorithm for the various advancements done. Henceforth, the study presents the miniature analysis on advancements in bat algorithm. The bat algorithm applications in different areas are also investigated.

Keywords: Metaheuristic, Echolocation, Bat

1 INTRODUCTION

Many metaheuristics procedures have been developed that are motivated from natural phenomena. Such procedures have the inspiration from insect's behaviour, chemical reaction, intelligence of swarm's etc. (Glover and Kochenberger 2006). These metaheuristics are adopted to resolve the different optimization problems. The distinct trait of each metaheuristic helps in achieving the optimal solution. Inspired from the echolocation behaviour of bat, a metaheuristic Bat algorithm was developed (Yang 2010). In recent times, the bat algorithm has gained attention for solving different optimization problems. The bats use their echo-location feature to avoid barriers and sense prey. When searching for a prey, the microbats emit short pulses. The loudness and emission pulse rate depends on the distance to the prey or obstacle. If microbat is near to prey/obstacle the emission pulse rate increase and there is decrease in the loudness. The frequency is further tuned for the microbats thru pulse rate and loudness and influences echolocation wavelength that helps to detect objects.

Equations 1 and 2 are used to initialise the loudness and emission rate of the bat algorithm.

$$L_j^{t+1} = s\left(L_j^t\right) \tag{1}$$

$$per_j^{t+1} = [1 - \exp(-s)] \tag{2}$$

where L_j^t is loudness, pulse emission rates denoted by per_j^t and (s) represents user defined variable with values between [0-1]. Equation 3–4 are used to compute the frequency and velocity of bats.

DOI: 10.1201/9781003501244-45

$$f_j^t = f_{\min}^t + \quad (f_{\max}^t - f_{\min}^t) rand() \tag{3}$$

$$v_j^t = \quad v_j^{t-1} + \left(x_j^t - x_*\right) f_j^t \tag{4}$$

where $f_{\min}^t$ and $f_{\max}^t$ indicates lowest frequency and highest frequency on time-stamp (t). rand() denotes random function, v_j^{t-1} means starting velocity, x_j^t denotes existing position; x_* is best existing position. Equation 5 and Equation 6 are used for updating positions of bats.

$$x_i^{t+1} = X_{new} + v_i^t \tag{5}$$

$$X_{new} = x_i^t + randi[-1, 1] A_i^t \tag{6}$$

Where X_{new} signifies new- position and x_i^{t+1} denotes final updated- position.

The key purpose of this study is to give a review on the advancements in bat algorithm. The study also investigates the applications of the Bat algorithm in various research areas. This research paper is organized as: Section 2 briefs the advancements in bat algorithm; Section 3 converses the different applications of bat algorithm and at the end Section 4 presents the conclusion.

2 ADVANCEMENTS IN BAT ALGORITHM

This section discusses the related works of bat algorithm and the various improvements and advancements done in bat algorithm.

2.1 *Candidate solution based*

Aboubi et al. introduced BAT-CLARA, a robust optimisation technique (Aboubi *et al.* 2016) using k-medoids with bat algorithm.

Binary version of the bat algorithm was given (Mirjalili *et al.* 2014). It utilizes hypercube in binary bat algorithm and also employs transform function to map velocity.

Locus-based adjacency schemes has been applied (Hassan *et al.* 2015). The authors proposed discrete bat algorithm to overcome from local optima.

A chaotic bat algorithm was proposed and employed to measure the arbitrariness using chaotic maps at each iterative search steps (Gandomi and Yang 2014).

To progress the performance of bat algorithm, an improved bat algorithm with extremal optimization (EO) algorithm, called as IBA-EO was given (Chen *et al.* 2021). The authors have improved update strategy to enhance global search and obtain optimal solution. Further, to balance exploration and exploitation, authors have added (i) Boltzmann selection, (ii) a monitor mechanism.

Table 1. Candidate solution-based advancements in bat algorithm.

Author, Year	Advancement Named as	Technique Used
Chen *et al.* 2021	IBA-EO	Boltzmann selection and monitor mechanism
Aboubi *et al.* 2016	BAT-CLARA	k-medoids
Hassan *et al.* 2015	Discrete Bat Algorithm	Locus based encoding scheme
Mirjalili *et al.* 2014	Binary-Bat algorithm	Transform Function
Gandomi and Yang 2014	Chaotic-Bat Algorithm	Chaotic-Maps

2.2 *Operator based*

Pang *et al.* (2024) proposed hybrid binary bat algorithm. The proposed algorithm uses the improved crossover-operator, chaotic map and Lambda iteration to handle the different issues of bat algorithm for unit commitment problem.

The differential operator and mean based population-diversity techniques have been adopted to achieve global search in modified bat algorithm and named proposed approach as modified DBA (dos Santos and Askarzadeh 2016).

An improved discrete bat algorithm (IDBA) was developed to calculate velocity (Osaba *et al.* 2016). The proposed algorithm utilizes the random number alongwith hamming distance for calculating velocity. The successor operator has been utilized to generate new solutions.

The differential operator with Levy flight were incorporated into the bat algorithm and dubbed the DLBA (Xie *et al.* 2013). The proposed algorithm has helped in self-adaptability and handling premature convergence.

Additionally, the Bat algorithm employed Gaussian walk and used the Standard Gaussian distribution in the algorithm to aid enhance local search capability (Cai *et al.* 2014).

Table 2. Operator based advancements in bat algorithm.

Author, Year	Advancement Named as	Technique Used
Pang *et al.* 2024	Binary bat algorithm	crossover operator based on exponential-logic-modulo map, Chaotic map, Lambda iteration
dos Santos and Askarzadeh 2016	Modified Differential BA	Differential operator & diversity mean
Osaba *et al.* 2016	Improved DBA	Hamming distance, Random number
Cai *et al.* 2014	Bat Algo with Gaussian walk	Gaussian distribution
Xie *et al.* 2013	Differential Operator & Lévy-flight trajectory	Differential Operator & Lévy-flights trajectory

2.3 *Strategic advancements*

Modern Computerized Bat Algorithm (MCBA) approach was proposed (Bangyal *et al.* 2022). To improve the local search ability, Torus walk based technique has been proposed instead of the random walk. Applying torus walk has helped in improving the diversity and convergence speed of the algorithm.

An improved BAT algorithm (IBA) was introduced for improving global search (Huang and Ma 2020).

Additionally, a Two-stage bat algorithm (TSBA) was introduced (Zeng *et al.* 2017). The algorithm inspired after artificial bee colony and bat algorithm.

The proposed algorithm utilized an operator to improve the global search and also combined gaussian functions with different variants to improve local optima. The proposed IBA approach additionally uses adaptive weight to update the velocity equation, balancing exploration and exploitation.Foraging habitations was integrated with doppler-effect into bat algorithm, named Novel Bat Algorithm (NBA) (Meng *et al.* 2015).

Table 3. Strategic advancements in bat algorithm.

Author, Year	Advancement Named as	Technique Used
Bangyal *et al.* 2022	MCBA	Torus Walk
Huang and Ma 2020	IBA	Gaussian Functions
Zeng *et al.* 2017	TSBA	Trade-off strategy
Meng *et al.* 2015	NBA	Doppler Effect, Foraging Habitats

2.4 *Hybridization*

Hybrid bat algorithm Artificial Bee Colony and Bat Algorithm (ABC-BA) was given by Ahanger *et al.* (2023). The proposed work also helps in improving candidate solution through bat agent in ABC and further, improves the convergence rate by introducing elitist strategy in bat algorithm. Wang *et al.* (2016) enhanced the bat algorithm for local search ability and variety using a variable neighbourhood search technique.

The idea of self-adaptation and hybrid self-adaptive in bat algorithm was introduced by Fister *et al.* (2014). Arunarani *et al.* (2017) combined firefly algorithm with bat algorithm and proposed FFBAT. Neelima *et al.* (2018) employed Random walk property and proposed artificial bee colony bat. A hybrid metaheuristic known as the enhanced Binary Bat Algorithm was presented to preserve equilibrium between the local and global search ability of BA (Li and Le 2019).

Table 4. Hybridization in bat algorithm.

Author, Year	Advancement Named as	Technique Used
Ahanger *et al.* 2023	Aritificial Bee Colony and Bat Algorithm (ABC-BA)	Used Bat agent for improvement in ABC and the Elitist strategy in BA
Li and Le 2019	Improved Binary Bat thru Cross- Entropy	Cross- Entropy
Neelima *et al.* 2018	Hybrid ABC-BAT	Random-walk
Arunarani *et al.* 2017	Firefly and bat algorithms	Multi-objective functions
Wang *et al.* 2016	Variable Neighbourhood Bat Algorithm	Variable Neighbourhood-search
Fister *et al.* 2014	Self-Adaptive & Hybrid Self-Adaptive	Self-adaptive mechanism, Differential evolution strategy &crossover CR

3 APPLICATIONS OF BAT ALGORITHM

The bat algorithm unearths optimal solutions in number of real-world disciplines. Figure 1 depicts these varied fields' specifics. Bat algorithm has been widely adopted for solving data mining problems such as clustering etc. Different variations of Bat algorithm have been developed for solving the partitional clustering problems (Kumar and Kaur 2022). BAT-CLARA is developed for resolving clustering problem (Aboubi *et al.* 2016). Crop classification, Fish classification have been done through bat algorithm (Senthilnath *et al.* 2016; Fouad *et al.* 2016).

Further, Anam *et al.* (2023) employed K-means algorithm and optimized Bat algorithm for diabetes-mellitus diagnosis. The bat method has been used to analyse redundancy in online-offline streaming feature selection, improving classification accuracy (Eskandari and Seifaddini 2023). Bat algorithm is also applied for improving web service (Dahan 2023).

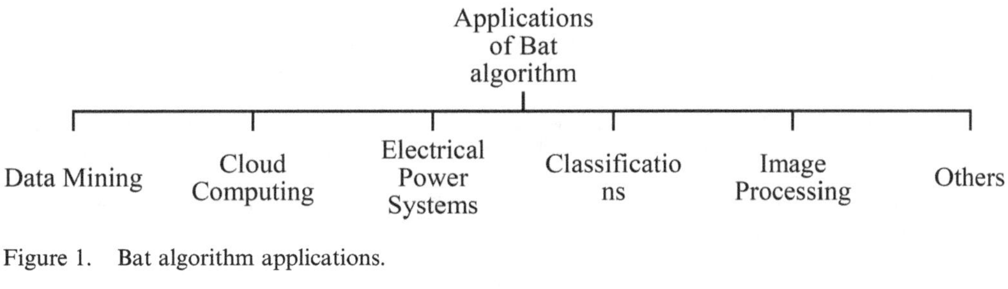

Figure 1. Bat algorithm applications.

Further, intrusion detection system also uses Bat algorithm due to its fast convergence (Priya and Kumar 2023).

Kaur *et al.* (2023) employed the bat algorithm to balance the load on the virtual machines in cloud environment. For cloud computing resource scheduling, enhanced bat method with time-varying wavelet perturbations has been employed (Yu *et al.* 2023). Energy management has been done through Bat algorithm for micro grid that helps in minimizing cost of renewable power sources (Baziar *et al.* 2013). A bat inspired algorithm, is developed for detecting brain tumors in MRI images (Chawla *et al.* 2022). A proficient bat algorithm is suggested for the routing of electric automobiles (Vani *et al.* 2023). Rashid *et al.* (2021) employed improved Bat algorithm in solving job scheduling problems.

4 CONCLUSION

The improvements made to the bat algorithm are the main topic of this research. It is evident that the bat algorithm has included a number of advancements to make it more trustworthy, capable, and useful. Furthermore, it appears that other scholars have proposed different strategies to solve the local optima issue in the bat algorithm. Exploration and exploitation processes are balanced by the changes made to the solution search equations and control settings. It is further noticed that bat algorithm has vast areas of application for solving global optimization problems.

REFERENCES

Aboubi Y Drias H. and Kamel N. (2016). BAT-CLARA: BAT-inspired algorithm for clustering LARge applications. *IFAC-PapersOnLine*, 49(12), 243–248.

Ahanger, T. A., Dahan, F., and Tariq, U. (2023). Hybridizing artificial bee colony with bat algorithm for web service composition. *Computer Systems Science & Engineering*, 46(2).

Fouad, M. M., Zawbaa, H. M., Gaber, T., Snasel, V., and Hassanien, A. E. (2016). A fish detection approach based on BAT algorithm. In *The 1st International Conference on Advanced Intelligent System and Informatics (AISI2015)*, November 28–30, 2015, Beni Suef, Egypt (pp. 273–283). Springer International Publishing.

Anam, S., Fitriah, Z., Hidayat, N., and Maulana, M. H. A. A. (2023). Classification model for diabetes mellitus diagnosis based on K-Means clustering algorithm optimized with bat algorithm. *IJACSA) International Journal of Advanced Computer Science and Applications*, 14(1).

Arunarani A R Manjula D and Sugumaran V 2017 FFBAT: A security and cost-aware workflow scheduling approach combining firefly and bat algorithms. *Concurrency and Computation: Practice and Experience*, 29 (24), e4295.

Bangyal, W. H., Hameed, A., Ahmad, J., Nisar, K., Haque, M. R., Ibrahim, A. A. A., and Etengu, R. (2022). New modified controlled bat algorithm for numerical optimization problem. *Computers, Materials & Continua*, 70(2).

Baziar A Kavoosi-Fard A and Zare J. (2013). A novel self adaptive modification approach based on bat algorithm for optimal management of renewable MG. *Journal of Intelligent Learning Systems and Applications*, 5(01), 11.

Cai X Wang L Kang Q and Wu Q. (2014). Bat algorithm with Gaussian walk. *International Journal of Bio-Inspired Computation*, 6(3), 166–174.

Chawla, R., Beram, S. M., Murthy, C. R., Thiruvenkadam, T., Bhavani, N. P. G., Saravanakumar, R., and Sathishkumar, P. J. (2022). Brain tumor recognition using an integrated bat algorithm with a convolutional neural network approach. *Measurement: Sensors*, 24, 100426.

Chen, M. R., Huang, Y. Y., Zeng, G. Q., Lu, K. D. and Yang, L. Q. (2021). An improved bat algorithm hybridized with extremal optimization and Boltzmann selection. *Expert Systems with Applications*, 175, 114812.

Dahan, F. (2023). Neighborhood Search Based Improved Bat Algorithm for Web Service Composition. *Computer Systems Science & Engineering*, 45(2).

dos Santos Coelho L & Askarzadeh A. (2016). An enhanced bat algorithm approach for reducing electrical power consumption of air conditioning systems based on differential operator. *Applied Thermal Engineering*, 99, 834–840.

Eskandari, S., and Seifaddini, M. (2023). Online and offline streaming feature selection methods with bat algorithm for redundancy analysis. *Pattern Recognition*, 133, 109007.

Fister I. Fong S. and Brest J. (2014). A novel hybrid self-adaptive bat algorithm. *The Scientific World Journal*, 2014.

Gandomi A. H. and Yang X. S. (2014). Chaotic bat algorithm. *Journal of Computational Science*, 5(2), 224–232.

Glover, F. W., and Kochenberger, G. A. (Eds.). (2006). *Handbook of Metaheuristics* (Vol. 57). Springer Science & Business Media.

Hassan E. A., Hafez A. I., Hassanien A. E. and Fahmy A. A. (2015). June A discrete bat algorithm for the community detection problem. *In International Conference on Hybrid Artificial Intelligence Systems* (pp. 188–199). Springer, Cham.

Huang, J., and Ma, Y. (2020). Bat algorithm based on an integration strategy and Gaussian distribution. *Mathematical Problems in Engineering*, 2020, 1–22.

Kaur, A., Kamboj, S., Kaur, B., and Hrisheekesha, P. N. (2021, December). Hybrid approach for virtual machine optimization using BAT algorithm in cloud. *In 2021 2nd International Conference on Computational Methods in Science & Technology (ICCMST)* (pp. 57–61). IEEE.

Kumar, Y., and Kaur, A. (2022). Variants of bat algorithm for solving partitional clustering problems. *Engineering with Computers*, 38(Suppl 3), 1973–1999.

Li G and Le C. (2019). April hybrid binary bat algorithm with cross-entropy method for feature selection. *In 2019 4th International Conference on Control and Robotics Engineering (ICCRE)* (pp. 165–169). IEEE.

Meng X B Gao X Z Liu Y and Zhang H. (2015). A novel bat algorithm with habitat selection and Doppler effect in echoes for optimization. *Expert Systems with Applications*, **42**(17–18), 6350–6364.

Mirjalili S. Mirjalili S. M. and Yang X. S. (2014). Binary bat algorithm. *Neural Computing and Applications*, 25(3–4), 663–681.

Neelima S. Satyanarayana N. and Murthy P. K. (2018). Minimizing frequent itemsets using hybrid ABCBAT algorithm. *In Data Engineering and Intelligent Computing* (pp. 91–97). Springer, Singapore.

Osaba E Yang X S Diaz F Lopez-Garcia P and Carballedo R. (2016). An improved discrete bat algorithm for symmetric and asymmetric traveling salesman problems. *Engineering Applications of Artificial Intelligence*, 48, 59–71.

Pang, A., Liang, H., Lin, C. and Yao, L. (2024). A binary bat algorithm with improved crossover operators and Cauchy mutation for unit commitment problem. *The Journal of Supercomputing*, 1–32.

Priya, S., and Kumar, K. P. M. (2023). Binary bat algorithm based feature selection with deep reinforcement learning technique for intrusion detection system. *Soft Computing*, 27(15), 10777–10788.

Rashid, T. A., Shekho Toghramchi, C. I., Sindi, H., Alsadoon, A., Bačanin, N., Umar, S. U., and Mohammadi, M. (2021). An improved BAT algorithm for solving job scheduling problems in hotels and restaurants. *Artificial Intelligence: Theory and Applications*, 155–171.

Senthilnath, J., Kulkarni, S., Benediktsson, J. A. and Yang, X. S. (2016). A novel approach for multispectral satellite image classification based on the bat algorithm. *IEEE Geoscience and Remote Sensing Letters*, 13 (4), 599–603.

Vani, B. V., Kishan, D., Ahmad, M. W. and Reddy, C. R. P. (2023). An efficient optimization algorithm for electric vehicle routing problem. *IET Power Electronics*.

Wang G. G., Lu M. and Zhao X. J. (2016). July An improved bat algorithm with variable neighborhood search for global optimization. *In Evolutionary Computation (CEC)*, 2016 IEEE Congress on (pp. 1773–1778). IEEE.

Xie J., Zhou Y. and Chen H. (2013). A novel bat algorithm based on differential operator and Lévy flights trajectory. *Computational Intelligence and Neuroscience*, 2013.

Yang, X. S. (2010). A new metaheuristic bat-inspired algorithm. *In Nature Inspired Cooperative Strategies for Optimization (NICSO 2010)* (pp. 65–74). Berlin, Heidelberg: Springer Berlin Heidelberg.

Yu, F., Chen, M. and Yu, B. (2023). An improved bat algorithm with time-varying wavelet perturbations for cloud computing resources scheduling. *International Journal of Cognitive Informatics and Natural Intelligence (IJCINI)*, 17(1), 1–16.

Zhang, B., Yuan, H., Sun, L., Shi, J., Ma, Z. and Zhou, L. (2017). A two-stage framework for bat algorithm. *Neural Computing and Applications*, 28, 2605–2619.

Computational Methods in Science and Technology Volume 1 – Kaur et al. (Eds)

Multimodal fusion for early chronic kidney disease detection

S. Aryan
Student, Galgotias University, Greater Noida, India

Pooja
Galgotias University, Greater Noida, India

K. Gaurav
Bennett University, Greater Noida, India

ABSTRACT: Chronic Kidney Disease (CKD) poses a substantial universal health concern, particularly in developing and underdeveloped nations. It manifests as a progressive deterioration of kidney function over a period of time. In many regions, hypertension and diabetes stand as primary causative factors for CKD. Therefore, early detection of CKD is imperative to impede further kidney deterioration. However, current diagnostic methods and models often lack precision in early CKD detection. Hence, we introduce a novel model designed to extract insights from both medical images and clinical records. The proposed model is amalgamation of three ML techniques, i.e., Convolutional Neural Networks (CNNs) algorithm that is employed for analyzing images, Recurrent Neural Networks (RNNs) that is applicable for interpreting clinical data, and a Support Vector Machine (SVM) that is capable of classification and predictive purposes. The experiments results show that our model generates better results as compared to CNN and RNN ML algorithms.

Keywords: CKD, CNN, RNN, SVM, Clinical data

1 INTRODUCTION

CKD is categorized by damaged kidney function, leading to difficulties in adequately filtering blood. The kidneys' fundamental role is to remove excess water and waste from the bloodstream, expelling them through urine. Typical factors contributing to CKD comprise diabetes, hypertension, glomerulonephritis, and associated ailments. In 2013, approximately one million fatalities were attributed to chronic kidney disease (Raghupati and Raghupati 2018; Anderson and Horvath 2001; Nugent 2010). Approximately 68% of chronic diseases stem from Hypertension and Diabetes (Jana and Chattopadhyay 2022). CKD is an extensive health challenge that often progresses silently, resulting in significant kidney damage before symptoms manifest. Detecting CKD in its early stages is crucial for effective management and patient well-being. The condition persists as damage gradually accumulates over a prolonged duration. While CKD is a global concern, it affects individuals differently based on various factors, including underlying health conditions, age, and gender. Symptoms such as back pain, stomach discomfort and more may indicate CKD, but they often appear when the disease has advanced. Notably, CKD is often asymptomatic, and people may remain unaware of its presence until it reaches an advanced state.

In 2013, approximately one million individuals lost their lives due to CKD (The daily star 2018). The burden of CKD is notably higher in the developing economies, with 387.5 million

 DOI: 10.1201/9781003501244-46

CKD patients worldwide being found in low-to middle-income nations (Prothomalo 2018; Pooja & Chauhan 2022). These statistics highlight a significant occurrence of CKD in developing nations, with the number of affected individuals steadily increasing. Efforts have been made towards early diagnosis for timely intervention, particularly focusing on machine learning prediction models. The key and most effective factor for assessing kidney function is glomerular filtration rate (GFR), which essentially assesses the kidney's capacity for blood filtration. GFR is widely regarded as the best accurate gauge of renal function and is typically estimated (eGFR) through a creatinine blood test. The eGFR value is given in $1.73m^2$/mL/min or milliliters per minute per $1.73m^2$. Based on eGFR, renal function is typically segmented into five stages, as depicted in table 1 (Webster *et al.* 2017).

Table 1. Phases of CKD.

Phase	Explanation	GFR (mL/min/$1.73m^2$)
1	Normal	$\geq$90
2	Mild CKD	60-89
3	Moderate CKD	30-59
4	Severe CKD	15-29
5	End Stage CKD	<15

This research project on Multi-Modal CKD Detection pursues innovative approach to CKD diagnosis, its imperative to address the challenges and limitations of current diagnostic methods. The fusion of clinical data and medical imaging through advanced machine learning models, such as CNNs and RNNs, offer a promising direction. This approach can improve early detection and intervention, potentially leading to better patient outcomes. Additionally, data mining techniques applied to vast patient medical and treatment data can play a crucial role in uncovering hidden insights that can aid in diagnosing and managing CKD more effectively. By integrating these advanced methods, the research strives to revolutionize CKD diagnosis and enhance patient care.

The remainder of the paper is arranged as follows, the overview of the related work is found in section 2. In section3, the suggested model and dataset details are presented. Section 4 includes results and experiments. Finally, the paper concludes in section 5, followed by future directions for research.

2 LITERATURE REVIEW

Polat *et al.* (2017) undertook an extensive study that encompassed a variety of classifiers, including K-nearest neighbours (kNN), naïve Bayes (NB), SVM, preference tables, random forest (RF), and J48. They investigated a wide range of possible therapies for persistent kidney disease by utilising algorithms such as K-means and Apriori. In their model, they made a diagnostic test using decision tress (DT) along with SVM, NB, and KNN calculations to detect CKD. Ani *et al.* (2016). expanded the range of how their methods could be used by trying several classification approaches, like DT, naïve NB, and linear discriminant analysis (LDA), as well as others like randomized subspace, k-nearest neighbours (KNN), and backpropagation network (BPN). They used DTs and NB methods to lower the chances of death related to CKD.

Charleonnan *et al.* (2016) looked into four different machine learning methods in their research: KNN, SVM, Logistic Regression (LR), and DT. They used information from patients with CKD to build models that could predict who might have CKD. Then, they compared how well each method worked to determine which one is most effective in predicting CKD. Almasoud and Ward (2019) investigated if machine learning methods could guess if someone had CKD using only a few features. Authors applied ANOVA, Pearson's correlation, and Cramer's V to get rid of features they did not need, and they utilised a 10-fold cross-validation approach for training and testing their model. The gradient boosting classifier showed it could guess CKD correctly 99.1% of the time, showing how important features like haemoglobin are for spotting CKD.

Qin *et al.* (2019) applied ML algorithms to diagnose CKD. They used a data from the UCI machine learning repository, which had some missing information. To deal with the missing information, they used a technique called KNN imputation. Then, they tried out various ML methods to proposed their own model. Among them, the random forest (RF) technique had the best accuracy. Sinha and Agarwal (2018) and Pooja & Chauhan (2017) proposed a model to evaluate the effectiveness of two data mining techniques. The author utilized KNN and SVM methods to assess accuracy and precision. Notably, the study revealed that the KNN classifier outperformed the SVM classifier in terms of both precision as well as accuracy metrics.

3 RESEARCH GAP

The existing studies on CKD diagnosis and prediction revolve around several critical areas. These include the need for a more holistic approach that integrates multi-modal data, standardized and comprehensive datasets for CKD research, a focus on early detection and preventive strategies, clinical validation and real-world applicability of machine learning models, increased interdisciplinary collaboration between data scientists and healthcare professionals. Few other problems that also need to be addressed are listed below,

(1) Limited and imbalanced datasets: Many studies suffer from small and imbalanced datasets. The imbalance in the distribution of CKD and non-CKD cases can result in models that are biased and perform well on the majority but poorly on the minority class.
(2) Data quality and missing values: In practical healthcare datasets, the occurrence of missing values and data quality concerns is a common phenomenon. Handling missing data appropriately is crucial for building robust models. Some studies do not adequately address these challenges, leading to potential biases in the models.
(3) Lack of standardized evaluation metrics: There is frequent disagreement over the assessment metrics that are used to rate the effectiveness of CKD detection. Various studies may employ distinct metrics, making it difficult to compare outcomes and generalise the results.

Addressing these challenges holds the potential to advance CKD diagnosis and prediction, improving patient care and outcomes.

4 PROPOSED MODEL

The proposed model in our research presents an integrated approach to CKD diagnosis and prediction, it is implemented in python with all the necessary libraries downloaded. The CKD dataset has been downloaded from multiple sources, as detailed in the following section. There are two groups CKD and non-CKD represented by 1 and 0 respectively. This model combines the strengths of machine learning, clinical data and medical imaging data, using CNN, RNN, and SVM as a classifier aiming to enhance the accuracy of early CKD detection by combining these diverse data sources and leveraging advanced algorithms. Figure 1. represents the flowchart of our suggested method.

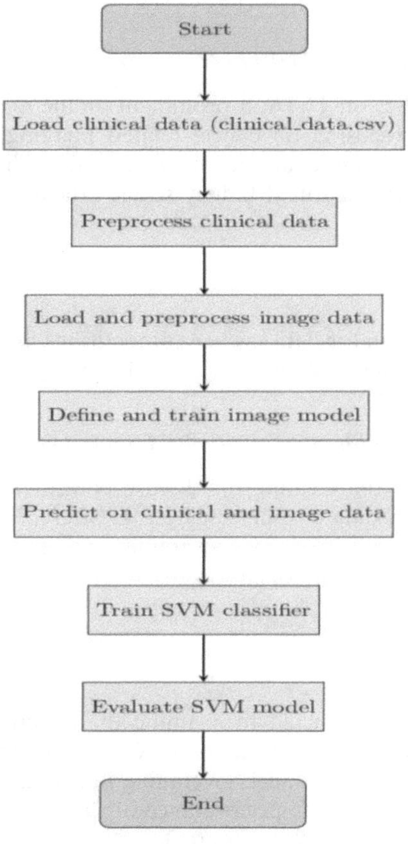

Figure 1. Flow chart of recommended model.

The proposed model begins by loading clinical data from a CSV file and preprocessing it by handling missing values, encoding categorical data, performing one-hot encoding. Then the model proceeds to load and preprocess image data using data generators and image resizing. After all the preprocessing steps the model defines and trains a Convolutional Neural Network (CNN) for image data and then uses the trained model to predict on testing images. The predictions from the image model are combined with a subset of clinical data and standardized. The combined and standardized data is then used to train a SVM classifier. Finally, the trained SVM model is valuated on testing data, and its accuracy is calculated. Machine learning techniques, integrating multimodal data sources for enhanced diagnostic accuracy.

To construct the sample for this study, NIFTY-500 indexed companies are selected from the National Stock exchange (NSE), India. The selected sample size for the both the pre-crisis and post-crisis period is decided according to the data availability of the variables (ownership holdings, stock return, and company-specific). Balanced panel dataset of 316 listed companies is selected for the pre-crisis period while the post-crisis period comprises 404 balanced panel datasets of listed companies. The dataset related to the ownership holdings, stock return, company-specific data are extracted from the Centre for Monitoring Indian Economy (CMIE) database.

5 DATASET

5.1 *The Cancer Imaging Archive (TCIA)*

Our research extensively relies on TCIA, a prominent online repository for a wide range of medical images, primarily focusing on cancer-related imaging. TCIA hosts several radiological images that align with the requirements of our study. This resource provided us with an invaluable collection of de-identified kidney scans, crucial for the image-based component of our model.

5.2 *Kaggle*

Kaggle, a well-known platform for data science competitions and datasets, offered resources for our clinical data requirements. Website: Kaggle Datasets

5.3 *NHANES (National Health and Nutrition Examination Survey)*

NHANES, administered by the Centers for disease control and prevention (CDC), also contributed valuable clinical data to our research. It offered insights into the health and nutrition data collected through interviews, examinations, and laboratory tests, enhancing the richness of our clinical data component. Website: NHANES

6 RESULTS

A thorough assessment was conducted to examine the effectiveness and potential of the multimodal fusion model in the early detection of CKD. The manner in which predictions are generated by the trained CNN model, leveraging image data, can be observed through the accompanying Figure 2. Also, the performance of the trained RNN model used to predict on clinical data can be visualized in Figure 3.

Now, the SVM classifier is trained using the results obtained from the individually trained CNN and RNN models to make predictions on the given set of data. The performance of the combined model can be visualized in Figure 4.

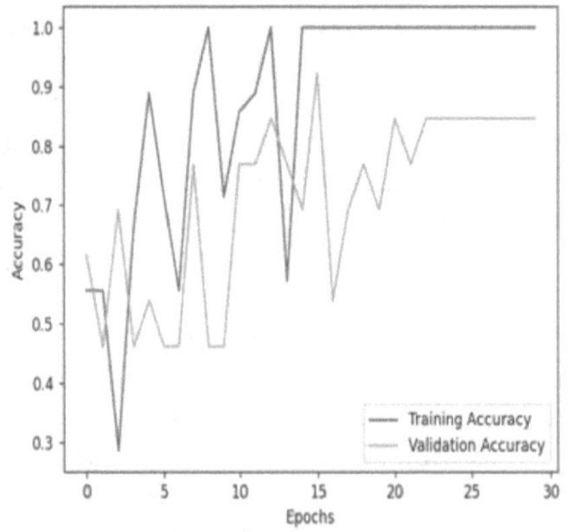

Figure 2. Performance of the CNN model.

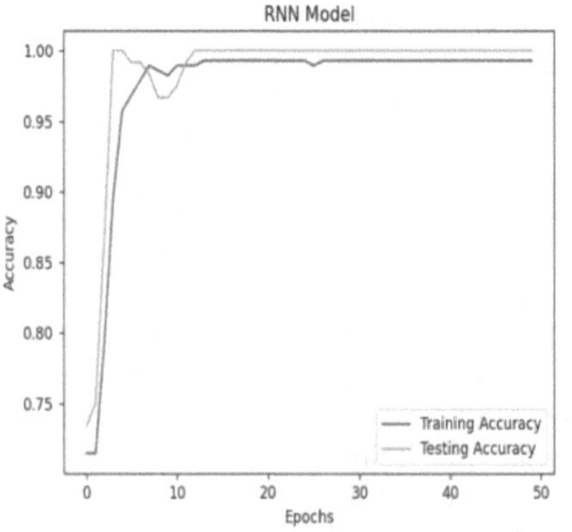

Figure 3. Performance of the RNN model.

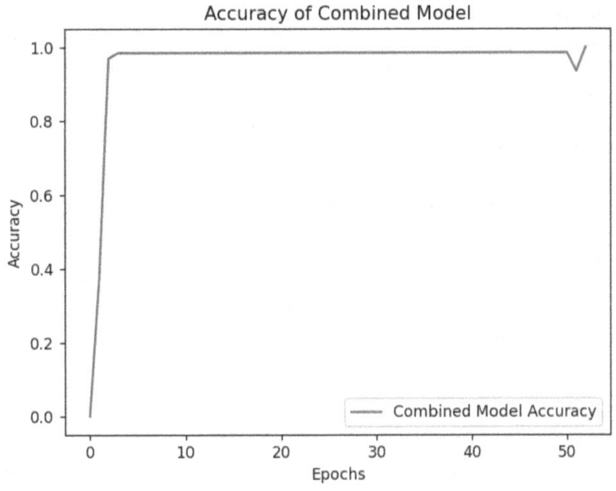

Figure 4. Performance of the combined CNN, RNN, SVM model.

7 CONCLUSION

In conclusion, the multimodal fusion model developed for early CKD detection demonstrates promising accuracy of '99.5%' and with a F1 score of '1' underscores the potential of integrating CNN analysis of medical images with RNN processing of sequential clinical data. The combined approach, complemented by SVMs classification, establishes a robust foundation for accurate CKD diagnosis. Sensitivity analyses highlight the unique contributions of each modality, emphasizing the significance of holistic information fusion. Despite these advancements, acknowledging limitations, such as data quality and model interpretability challenges, is crucial. Future work should explore advanced model architectures, incorporate additional modalities like genomics, and focus on clinical validation and integration. Addressing interpretability concerns and conducting longitudinal studies for predictive modeling are vital for bridging the gap between research findings and practical applications in clinical settings. This research marks a significant stride towards enhanced diagnostic methodologies for CKD, with a commitment to continuous refinement and validation in real-world healthcare scenarios.

REFERENCES

Almasoud, M. and Ward, T. E. (2019). *Detection of Chronic Kidney Disease using Machine Learning Algorithms with Least Number of Predictors.*

Anderson, G., and Horvath, J. (2001). The growing burden of chronic disease in America. *Public Health Reports*, 11(3), 263–270.

Ani R., Sasi G., Sankar U.R., Deepa O. (2016). Decision support system for diagnosis and prediction of chronic renal failure using random subspace classification. *2016 International Conference on Advances in Computing, Communications and Informatics (ICACCI)*. IEEE; 2016.p. 1287–1292.

Arora, M., Sharma, E.A., (2016). Chronic kidney disease detection by analyzing medical datasets in weka. *International Journal of Computer of machine learning algorithms. New advances in machine learning* 3, 19–48.

Charleonnan A., Fufaung T., Niyomwong T., Chokchueypattanakit W., Suwannawach S., Ninchawee N. (2016). Predictive analytics for chronic kidney disease using machine learning techniques. *Management and Innovation Technology International Conference (MITicon)*. IEEE; 2016.pp. MIT–80.

Islam, M. A., Majumder, M. Z. and Hussein, M. A. (2023). Chronic kidney diesase prediction based on machine learning algorithms. https://doi.org/10.1016/j.jpi.2023.100189.

Jana, A. and Chattopadhyay, A. (2022). Prevalence and potential determinants of chronic disease among elderly in India: Rural-urban perspectives. *National Library of Medicine*. https://pubmed.ncbi.nlm.nih.gov/35275937/#:~:text=About%2021%25%20of%20the%20elderly,68%25%20of%20all%20chronic%20diseases.

Momani, A. R., Mustafa,A., Zeidan, G. R., Alquran, H., Mustafa, W. A., and Alkhayyat, A. (2022). Chronic kidney disease detection using machine learning technique. *5th International Conference on Engineering Technology and its Applications (IICETA)*, Al-Najaf, Iraq, 2022, pp. 153–158, doi:10.1109/IICETA54559.2022.9888564.

Nugent, R. (2010). Chronic diseases in developing countries: Health and economic burdens. *Annals of the New York Academy of Sciences*. 1136(1), 70–79.

Polat, H., Danaei, H.M. and Cetin, A. (2017). Diagnosis of chronic kidney disease *Based on Support Vector Machine by Feature Selection Methods*. 41:1–11.

Pooja, Chauhan, R. K. (2022). Triple phase hybrid cryptography technique in wireless sensor network. *International Journal of Computers and Applications (Taylor & Francis)*. ISSN No.: 1206-212X. January 2022. Doi: 10.1080/1206212X.2019.1710342.

Pooja, Chauhan,R. K. (2017). A survey on comparisons of cryptographic algorithms using certain parameters in WSN. *International Journal of Electrical and Computer Engineering (IJECE)*, Vol.7, No.4, pp. 2232–2240, ISSN No.: 2088-8708, August. DOI: http://doi.org/10.11591/ijece.v7i4.pp2232-2240,

Prothomalo. (2018). Women more afected by kidney diseases. https://en.prothomalo.com/bangladesh/Womenmore-affected-by-kidney-diseases.

Qin J, Chen L, Liu Y, Liu C, Feng C, Chen B. (2019). A machine learning methodology for diagnosing chronic kidney disease. *IEEE Access 2019*;8:20991–21002.

Raghupati,W. and Raghupathi,V. (2018). An empirical study of chronic diseases in the United States: a visual analytics approach to public health. *International journal of Environmental Research and Public Health*. 15 (1), 431–437.

Salekin, A. and Stankovic, J. (2016). Detection of chronic kidney disease and selecting important predictive attributes. *2016 IEEE International Conference on Healthcare Informatics (ICHI)*, Chicago, IL, USA. pp. 262–270, doi:10.1109/ICHI.2016.36.

Sinha, A. D. and Agarwal, R. (2018). *Clinical Pharmocology of Antihypertensive Therapy for the Treatment of Hypertension in CKD*. 10.2215.

Thedailystar. (2019). Over 35,000 develop kidney failure in Bangladesh every year. https://www.thedailystar.net/city/news/18m-kidney-patients-bangladesh-every-year-1703665.

Tusar, M. T. H. K., Islam, M. T., and Raju, F. I. (2022). Detecting Chronic Kidney Disease(CKD) at the initial stage: A novel hybrid feature-selection method and robust data preparation pipeline for different ML techniques. *5th International Conference on Computing and Informatics (ICCI)*, New Cairo, Cairo, Egypt, 2022, pp. 400–407, doi:10.1109/ICCI54321.2022.9756094.

Venkatesan, V. K.; Ramakrishna, M. T.; Izonin, I.; Tkachenko, R.; Havryliuk, M. (2023). Efficient data preprocessing with ensemble machine learning technique for the early detection of chronic kidney disease. *Appl. Sci.* 2023, 13, 2885. https://doi.org/10.3390/app13052885.

Webster, A.C., Nagler, E.V., Morton, R.L., Masson, P. (2017). Chronic kidney disease. *Lancet*, 389, 1238–1252.

Computational Methods in Science and Technology Volume 1 – Kaur et al. (Eds)
© 2025 the Author(s), ISBN: 978-1-032-81760-6

Development and evaluation of a novel medicine expiry detection application: An innovative approach to ensuring medication safety

Pooja
Assistant Professor, Galgotias University, Greater Noida, India

Yogita Sangwan & Sabhya Rani
Assistant Professor, DCSA, Kurukshetra University, Kurukshetra, India

ABSTRACT: Medicines are the part of our daily lives and sometimes more crucial than water, as we can live without a water for some hours but can't handle cervical pain, high blood pressure and so on. In this fast pace life, either we forget to take medications on time or mistakenly took expired medicines or somehow damaged the packaging of medications. If medicines are giving life to human, they have some negative repercussions as well. Therefore, it is must to handle drugs very safely. In today's era, everyone is busy with mobile phones, why don't we give the solution to this problem on this phone itself. In this research paper, we have discovered such a mobile application, called Medicine Expiry Detection Application (MEDA), which will inform the patients about the expiry date of medicines or will notify them about taking medicines on time. When all the users tried it, its performance proved to be quite good.

Keywords: Medicines, Mobile application, QR code

1 INTRODUCTION

The widespread use of medications globally has sparked worries about the safe and efficient use of pharmaceutical products. It is essential to ensure that drugs are still within their expiration date to maintain their effectiveness and minimize the potential health hazards linked to expired medications (Ventola, 2014). Expired medications may lose their potency, develop harmful byproducts, or lead to antibiotic resistance when used inappropriately (Alsultan & Mayet, 2016; Fallah & Yasini, 2017). Despite this, monitoring and managing medication expiry dates remain a challenging task for patients, particularly those with multiple medications or chronic illnesses (Lim *et al.* 2016; Saghaeiannejad-Isfahani S.*et al.*, 2017; Pooja & Chauhan, 2017).

Therefore, it is very much important to know the expiry date of the medicine. There is a QR code framework these days in pretty much every item whether it is medication, food, or beverages. It addresses heaps of data about an item when the code is scanned. You simply need your mobile to examine it and get all data soon. Yet, consider the possibility that just in a few cases there will not be any QR code or any expiry date an exceptionally restricted measure of data will be there behind the covering if you just buy two or their capsules. Recent advancements in technology have paved the way for innovative solutions to tackle the problem of medication expiry management. Smartphone applications, in particular, have demonstrated significant potential in improving healthcare outcomes and patient self-management (Kao & Liebovitz, 2017; Pooja & Chauhan, 2015).

After all these, authors have suggested a new mobile application, called MEDA, in this paper which uses OCR technology with the help of which the expiry date of medicines can be easily detected. MEDA is a user friendly mobile app which can be easily used by everyone from children to old people and women.

2 LITERATURE SURVEY

In this section we will brief about the research done in the last 5 years.

In paper (AL-Dhaheri *et al.* 2018) the authors have done an exhaustive survey of papers related to the mobile apps of medication safety. The authors compared the apps on the basis of their ability for detecting expiry dates, unique features, their advantages and their disadvantages. In paper (Sahoo & Sahoo, 2020; Kaim *et al.* 2022), researchers created a novel smart medication box that uses RFID technology. The features of this box are checking expired medicines and sends alerts to patients. Study (Patel & Ahmed, 2019), shows that the authors examined most of the applications related to medication safety. After that they explained about the applications' unique features, OCR technology and user friendly interfaces.

Research in paper (Dang & Dang, 2021; Garg *et al.*, 2023), shows that the authors made use of state-of-art technology such as IoT and RFID to invent a smart and powerful medication storage system. IoT technology was used for the collection of user's information and RFID tags and sensors were used to track expiry dates so that app immediately alerts the users about expiry of medicines. Similarly, same study was conducted by paper (Kumar & Kumar, 2018; Singh *et al.*, 2024), in which authors used latest technologies to build a portfolio management system that was based on IoT technology to keep tracks of all expired medicines, send alerts to users and preserve a central inventory which is based on real time information. Then in paper (Daud & Zulkifi, 2019, Chaudhary *et al.* 2024), authors discovered a mobile application for medication safety by using IoT technology. Researchers described the system's architecture and evaluate its performance, emphasizing its potential for healthcare applications. In paper (Albrecht *et al.* 2019), authors investigated the potential of app-based tracking in monitoring blood pressure among patients undergoing chemotherapy, highlighting the potential of mobile health applications in improving patient health management and monitoring. The authors (Kooij *et al.*, 2018; Pooja & Chauhan, 2022) systematically reviewed the effectiveness of IT-supported shared care in managing chronic diseases, emphasizing the potential of technology-based interventions to improve healthcare management and patient-provider collaboration.

3 RESEARCH GAP

However, the primary distinction between proposed work and existing applications lies in the fact that all the aforementioned applications require users to add medications themselves, either by scanning barcodes or entering serial numbers. This can lead to issues if they forget to do so and end up consuming expired medications. We address this problem by recording the medication details at the time of purchase, eliminating the possibility of overlooking crucial information.

Moreover, some enhanced features in proposed application have been incorporated such as stock management, detailed information, notes, calendar, expiry tracking, alerts, and consultation options. These added functionalities aim to assist users in various ways and improve their overall experience.

4 PROPOSED MODEL

The proposed application known as 'Medication Expiry Detection Application' (MEDA) will give the subtleties of the drugs i.e. when the medication will lapse, what all drugs are accessible to you and so on. Essentially, there will be two ends of the application: one at the charging counter (pharmacist end) and other for the client (User end).

So, basically it will work as follow and depicted in Figures 1 and 2:

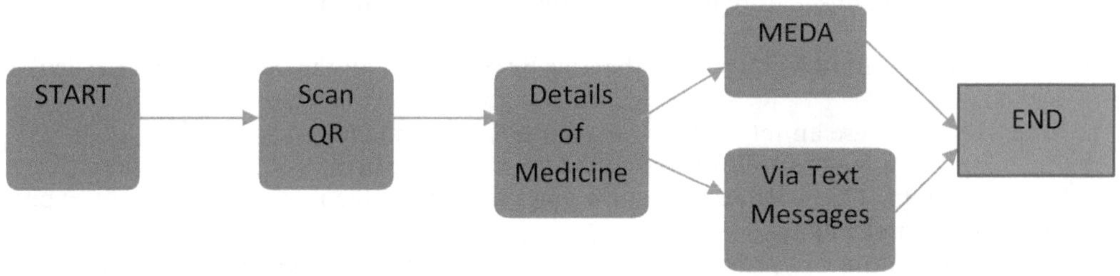

Figure 1. Flowchart of MEDA (Pharmacist end).

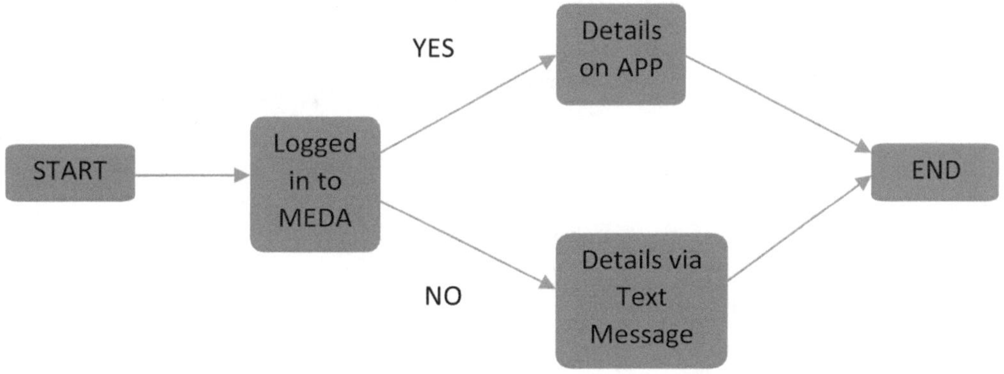

Figure 2. Flowchart of MEDA (user end).

STEP 1: As soon as user buys any medicine from any pharmacy, at time of billing he needs to provide a phone number, pharmacist scan all the medicines using QR code. After scanning, all the details of the medicine will automatically feed on an application (Pharmacist End).

STEP 2: If user have an application, he will receive all the information on his app otherwise, he will receive text messages regarding medication on his registered mobile number (User End).

5 METHODOLOGY

The main feature of MEDA is it prompt a message about the medicine whenever it will be expiring or 6-7 days before expiring to change your medicine or avoid consuming it. When a user buys a medicine, pharmacist scans a QR code at the time of billing and automatically sends the details to user registered mobile umber, such as name of medicine, for what cause it is used, Manufacturing date, Expiry date, company in which that medicine is produced, drug contains in the medicine, price of the medicine etc. These detail will be feed in the database of application for the future use of the user.

On other hand, user will be using this application in various aspects, user will receive alert message few days before the medicine get expired and to change that medicine i.e. Buy the new one and remove that from your stock.

Other than this we will enhance much more feature for the user end to facilitate them in various aspects as follows:

(1) Stock: - By using this feature user will be able to check all the medicines that are in stock with them.
(2) Details: -This feature will facilitate the user by providing all the details of the particular medicine.
(3) Note: -User can use an interface to create a note for remembering in future.
(4) Calendar: -This feature will remind every day which medicine you have to take.
(5) Expiry Region: -This interface will show you the name of the medicines sorted by expiry dates from closer to later.
(6) Alert: -This feature will provide you a facility to set alarms for remembering the various things.
(7) Consult: -This feature will provide you the list of medicines you have to take at time of illness. Example-at the time of fever what all medicines are available in your stock.

Below are the screenshots of MEDA application in Figures 3 and 4.

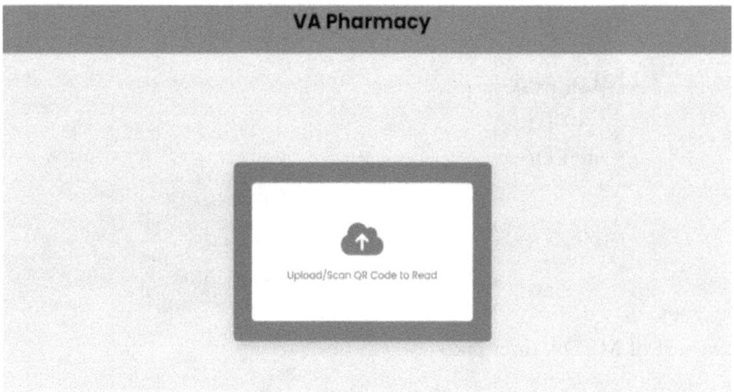

Figure 3. QR code scanner at pharmacist end.

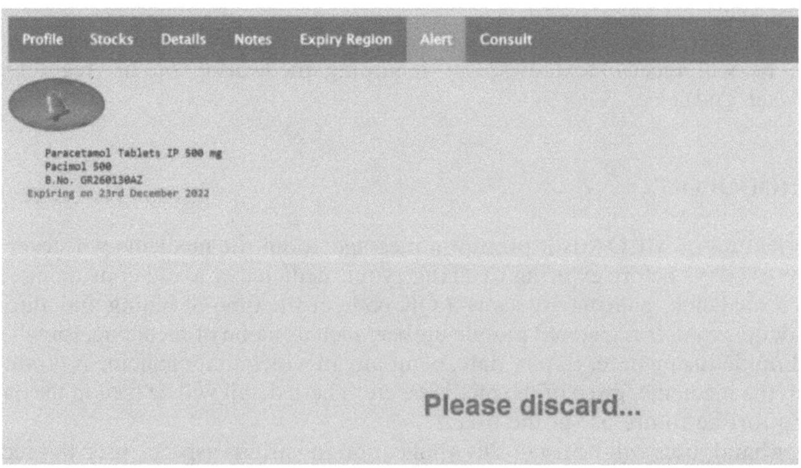

Figure 4. Notification of expired medicine at user end.

6 CONCLUSION AND FUTURE WORK

Medication expiry recognition application helps in really taking a look at every one of the subtleties of medication in an extremely simple way. It will caution individuals about their expiry medication, by simply taking a look at the versatile application subsequent to checking the OR code of medication at the hour of purchasing medication to get every one of the connected information in our application. It helps to diminish the pace of individuals consuming lapsed medication and getting different medical problems.

REFERENCES

Albrecht, U. V., Pramann, O., & von Jan, U. (2019). App-based tracking of self- logged blood pressure data among patients undergoing chemotherapy: A cross- sectional study. *Journal of Medical Internet Research*, 21(9), e15360.

Al-Dhaheri, M., Al-Hosani, F., & Al-Zaabi, M. (2018). Mobile applications for medication management: Review and analysis. *Journal of Medical Systems*, 42(9), 164.

Alsultan, M. S., & Mayet, A. Y. (2016). Dispensing expired medications: A pharmacist's view. *International Journal of Health Sciences*, 10(3), 425–427.

Chaudhary, M. Pooja and Chandwani G. (2024). Unveiling insights: Vision-based data mining analysis of webpages in transition from Html 3 To Html 5. *IC2PCT-2024*, IEEE, February-2024. pp. 1646–1651, doi: 10.1109/IC2PCT60090.2024.10486413.

Dang, L., & Dang, L. H. (2021). IoT-based smart medication storage system for detecting expired medicine. *Journal of Sensors*, 2021, 1–8.

Daud, N. A., & Zulkifli, A. N. (2019). An IoT-based medication management system for healthcare applications. 2019 *IEEE 6th International Conference on Engineering Technologies and Applied Sciences (ICETAS)*, Bangkok, Thailand, 2019, 1–6.

Fallah M. and Yasini M.(2017). A medication reminder mobile app: Does it work for different age ranges. *Stud Health Technol Inform*. 235, pp. 68–72. PMID: 28423757.

Garg, S., Pooja, Chaoudhary, S.L., Dixit, R. & Govil, N. (2023) "Implementing and Analyzing Behaviour of ML Based Tamil Songs Recommendation System", *International Conference on Sustainable Computing and Data Communication Systems (ICSCDS-2023)-IEEE*, March-2023. pp. 381–386, doi: 10.1109/ICSCDS56580.2023.10104744.

Kaim, V.K., Misra, D.K., Sharma, S., and Kumar, V. (2022). MedCue (A Medicine Reminder Android Application). *2022 International Conference on Machine Learning, Big Data, Cloud and Parallel Computing (COM-IT-CON)*, Faridabad, India, 2022, pp. 485–488, doi: 10.1109/COM-IT-CON54601.2022.9850754.

Kao, C. K., & Liebovitz, D. M. (2017). *Consumer Mobile Health Apps: Current State, Barriers, and Future Directions*. PM&R, 9(5S), S106–S115.

Kooij, L., Groen, W. G., & van Harten, W. H. (2018). The effectiveness of information technology-supported shared care for patients with chronic disease: A systematic review. *Journal of Medical Internet Research*, 20 (6), e221.

Kumar, P., & Kumar, N. (2018). Smart medicine inventory management using IoT. *International Journal of Computer Sciences and Engineering*, 6(4), 940–944.

Lim, S. S., Smith, E., Wiecek, E., & Blyth, F. M. (2016). Medicine storage practices in Australian homes. *Australian Health Review*, 40(4), 391–397.

Patel, R., & Ahmed, K. (2019). Development of a mobile application for medicine expiry management. *International Journal of Computer Applications*, 178(39), 18–23.

Pooja and Chauhan, R. K. (2015). An assessment based approach to detect black hole attack in MANET. *International Conference on Computing, Communication and Automation (ICCCA-2015)*. IEEE, April-2015. pp. 552–557, doi: 10.1109/CCAA.2015.7148439.

Pooja, and Chauhan, R.K. (2017). A survey on comparisons of cryptographic algorithms using certain parameters in WSN. *International Journal of Electrical and Computer Engineering (IJECE)*, Vol.7, No.4, pp. 2232–2240, ISSN No.: 2088–8708, August. DOI:http://doi.org/10.11591/ijece.v7i4.pp. 2232–2240.

Pooja, and Chauhan, R.K. (2022). Triple phase hybrid cryptography technique in wireless sensor network. *International Journal of Computers and Applications (Taylor & Francis)*. ISSN No.: 1206-212X. January 2022. Doi:10.1080/1206212X.2019.1710342.

Saghaeiannejad-Isfahani S, Ehteshami A, Savari E, Samimi A. (2017). Developing the medication reminder mobile application "Seeb". *Acta Inform Med.* 2017 Jun;25(2):108–111. doi:10.5455/aim.2017.25.108-111. PMID: 28883675; PMCID: PMC5544449.

Sahoo, S. K., & Sahoo, S. (2020). A smart medicine box for monitoring medication adherence. *Journal of Ambient Intelligence and Humanized Computing*, 11(4), 4573–4584.

Singh, S., Saini, N., Pooja & Singh, D. (2024). IoB based intelligent healthcare system for disease diagnosis in humans. *ICDAM-2023*, Springer, LONDON, June-2023. Lecture Notes in Networks and Systems, vol 785. Springer, Singapore. https://doi.org/10.1007/978-981-99-6544-1_43.

Ventola, C. L. (2014). Mobile devices and apps for health care professionals: Uses and benefits. *Pharmacy and Therapeutics*, 39(5), 356–364.

Computational Methods in Science and Technology Volume 1 – Kaur et al. (Eds)

Enhancing driver safety: A comprehensive evaluation of the Guardian Drive Vigilant System using machine learning

Pooja
Assistant Professor, Galgotias University, Greater Noida, India

Yogita Sangwan & Sabhya Rani
Assistant Professor, DCSA, Kurukshetra University, Kurukshetra, India

ABSTRACT: Laziness and Sleepless of drivers are most common reasons for accident. Year after year the graphs of road accidents are increasing. In this paper, authors have come up with a solution of drivers' fatigue using a novel model called Guardian Drive Vigilant System (GDVS) which will help to reduce road accidents by alerting the driver about the situation and it is based on real time visual information and Machine Learning technique. Authors presents a mechanism to discover, track and determine both driver face and eyes for computing the PERCLOS (a measure related with slow eye closure). The Machine Learning algorithm which is used in this model is Support Vector Machine (SVM) that detects the drivers' facial expression and analyze them to create alertness to the driver.

Keywords: Laziness detection, Fatigue detection, Drowsiness detection, GDVS, PERCLOS, SVM

1 INTRODUCTION

At present, transportation systems constitute a vital aspect of human endeavors. Over recent years, there has been a significant surge in road accidents attributed to driver fatigue (Kiranmayee 2021). Anyone can fall prey to drowsiness while driving, whether due to inadequate sleep, physical discomfort, or extended journeys.

Numerous investigations have revealed that sleep deprivation can impair driving abilities to a similar extent as alcohol intoxication. Approximately 20% of individuals acknowledge experiencing episodes of nodding off while driving, with 40% admitting to such occurrences at least once in their driving history (Alshaqaqi *et al.* 2013; Pooja and Chauhan, R. K. 2015)). Statistics reveal that in India 40% of highways accidents or in nearby area are due to the driver's fatigue, and also more than 50% of fatalities on highways, involving more than two-vehicles, are due to the drunk driving (Manikandan *et al.* 2018). Moreover, approximately 65% of highway accidents, involving single-car, are drunk driving cases (Kiran *et al.* 2020). According to European Union Research Survey, 20% of road accidents are caused by driver's fatigue. (Fatima *et al.* 2020).

Given all this data, we need a system that responds immediately to driver fatigue. In this paper, the authors have developed one such system named Guardian Drive Vigilant System (GDVS) which is based on the algorithm of PERCLOS and Machine Learning, which identifies the driver's fatigue and issues timely alerts.

DOI: 10.1201/9781003501244-48

The remainder of the paper is structured as follows: Section 2 explores Related Work, providing an overview of research conducted in this domain. Section 3 illustrates the Proposed Model through a block diagram. Section 4 concludes the paper, offering insights into future directions aimed at enhancing the proposed model.

2 LITERATURE REVIEW

In paper (Patel *et al.* 2010; Garg *et al.* 2023), authors introduced a model for driver's fatigue employing a color tool which was designed by JVC, a video recording technique. They used some sort of skin segmentation module for face detection relying on Gaussian distribution, followed by iris-based eye detection using binary images. Another research (Lenskiy and Lee 2012) presented a pioneering algorithm for segmenting skin color, which utilizes a Neural Network (NN) and a texture segmentation algorithm that depends on SURF features to detect eyes. In their study, (Jo, Lee and Jung 2011) utilized near-infrared (NIR) LEDs equipped with a narrow band-pass filter to illuminate the driver's face. First of all, this model fetches all the images and applies eye detection algorithm on them. Not only this, it then further filters the images by performing template matching and bob detection. In the last iteration of this model, authors have included the most unique feature that is eye validation which is done with the help of SVM, PCA and LDA. Researchers of paper (Jayanthi and Bommy 2012) did face detection using a skin color algorithm, followed by eye tracking. They used dynamic template matching, which is excellent, and templates were trained by Artificial Neural Network (ANN) machine learning algorithm. In study, (Batchu and Kumar 2015), authors developed a new system to check drivers' fatigue that can detect Face and Eyes with the help of a simple HARR classifier in OpenCV.

A drowsiness detection system was introduced by (Suryaprasad *et al.* 2013; Pooja and Chauhan, R. K. 2022), which consisted of five modules that spanned from video acquisition to fatigue detection. The first step of this model is pn-point the regions of interest, after that face and eyes were detected by in-built method HARR cascade samples. In their study, (Karchani, Mazloumi, and Saraji 2015) put forth a design for a drowsiness detection system that utilized a virtual-reality driving simulator. The study involved 20 male bus drivers who work in urban areas and had perfect eyesight, i.e. they haven't use spectacles. During the research, pictures of the drivers' faces were captured using a CC Camera positioned in front of them. In paper (Kong *et al.* 2015; Chaudhary *et al.* 2024) pioneered a new technology to detect a driver's drowsiness by capturing images from two directions, such as front view and side view of the driver's face using Viola-Jones Adaboost algorithm.

Authors of paper (Bharambe and Mahajan 2015) created a system to detect driver's drowsiness that was based on real-time information. Using the Viola-Jones classifier and webcam, they recorded video of 8 drivers in different lighting conditions. Study by (Shen and Xu 2015; Pooja and Chauhan, R. K. 2017) invented a model that consists of four small devices namely image detection, image processing, fatigue detection and display. They have used an infrared camera which is not affected by lighting. Authors of (Saradadevi and Bajaj 2008) used the Viola-Jones framework to create a system that detected faces and mouth regions. Not only this, they also incorporated SVM algorithm which was useful in detecting yawning. Another model by (Abtahi 2012; Singh *et al.* 2024) invented a unique technique that combines three things namely color segmentation, active contour modeling and Viola-Jones, to detect driver's fatigue by analyzing yawning behavior. In paper (Sigari, Fathy, and Soryani 2013) put forward a flexible expert system based on fuzzy logic. The system utilizes a blend of symptoms related to the eye region, such as the distance between eyelids and eye closure, as well as factors associated with the face region, such as head rotation, to detect drowsiness.

3 PROPOSED SYSTEM

In this section, proposed model is demonstrated i.e., Guardian Drive Vigilant System (GDVS) with the help of flowchart. GDVS has the major steps to be followed and these are face detection, eyes localization, identification of eyes states, detect yawning on mouth, apply SVM algorithm and lastly identify driver state. Block diagram of GDVS is depicted in Figure 1.

3.1 *Face detection*

In order to depict the symmetry of a digital image, 1-D signal known as an accumulator vector is utilized. This vector has a size equivalent to the width of the image and provides us with the value that corresponds to the position of the vertical axis of symmetry for objects within the image. The conventional method for computing the symmetry signal involves incrementing the value in the accumulator vector at the midpoint between every two white pixels that fall on the same line.

3.2 *Eye localization*

Since eyes are a distinct feature on the face, our research has focused on the region between the mouth and forehead. We have confidently identified the eyes within this area based on their symmetrical characteristics. To begin, we vertically measure the eROI by marking a rectangle with the already computed height of the eye and face. After that assess the symmetry of the eye region, taking into account both the sides. So that value of height is associated with the centroid of the eyes.

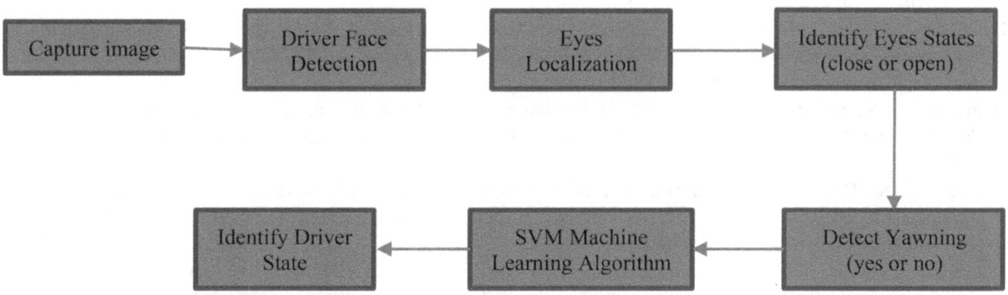

Figure 1. Block diagram of GDVS.

3.3 *Eye states*

The main aim of determination of eyes' states is to check whether the eyes are open or closed. We utilized HTC (Hough Transform of Circle) to detect iris. In this process, HTC was applied to the eye to identify the circle with specific characteristics. The largest number in HTC is selected for every rays. Next, '&&' operation is used between the edge image and the circle obtained through HTC to measure the level of interaction with 'S'.

$$\text{State eye} = \{\text{open if } S \geq Th; \text{ Close if } S < Th\} \tag{1}$$

3.4 *Detect yawning*

After identify the eyes states, we detect whether the driver is yawning or not. It follows the simple procedure, just check the driver's mouth is open for more than 5 seconds. Here 'Th' is the threshold that is 5 second.

$$\text{Yawning} = \{\text{yes if } \text{mouth}_{\text{open}} \geq \text{Th}; \text{ no if } \text{mouth}_{\text{open}} < \text{Th}\} \quad (2)$$

3.5 *Support vector machine algorithm*

SVM is categorized as supervised learning model used for classification and pattern recognition in data analysis. SVM operates on the belief of maximum margin, representing samples as points and mapping them in a way that creates a clear gap between different classes. This separation allows new test samples to be accurately classified based on their placement relative to the boundary. SVM relies on two datasets: a training dataset and a testing dataset.

4 CONCLUSION AND PERSPECTIVE

In this paper, authors have proposed a system for drivers' drowsiness with a far sighted view of alarming the driver whenever he feels sleepy. Under this system, it is able to find out the state of driver whether he is filling sleepy or not with help of camera. As it has discussed earlier that face and detection of eye is totally based on symmetrical data of face. Further the results are more satisfying well authors have used other techniques related to calculation of symmetry.

REFERENCES

Abtahi, S. (2012). *Driver Drowsiness Monitoring Based on Yawning Detection.* http://www.ruor.uottawa.ca/fr/handle/10393/23295.

Alshaqaqi, B., Baquhaizel, A. S., Md. Ouis, E. A., Boumehed, M., Ouamri, A. and Keche, M. (2013). Driver drowsiness detection system, *8th International Workshop on Systems, Signal Processing and their Applications (WoSSPA)*, pp: 151–155.

Batchu, S., and S. P. Kumar. (2015). Driver drowsiness detection to reduce the major road accidents in automotive vehicles. *International Research Journal of Engineering and Technology (IRJET)* 2 (1). http://www.irjet.net/archives/V2/i1/Irjet-v2i163.pdf.

Bharambe, S. S., and P. M. Mahajan. (2015). Implementation of real time driver drowsiness detection system. *International Journal of Science and Research (IJSR)* 4 (1). http://www.ijsr.net/archive/v4i1/SUB15742.pdf.

Chaudhary, M. Pooja and Chandwani G. (2024). *Unveiling Insights: Vision-Based Data Mining Analysis of Webpages In Transition From Html 3 To Html 5.* IC2PCT-2024, IEEE, February-2024. pp. 1646–1651, doi: 10.1109/IC2PCT60090.2024.10486413.

Fatima, B., Shahid, A. R., Ziauddin, S., Safi, A. A. and Ramzan, H. (2020). Driver fatigue detection using viola jones and principal component analysis, *Applied Artificial Intelligence*, 34:6, 456–483, DOI: 10.1080/08839514.2020.1723875.

Garg, S., Pooja, Chaoudhary, S.L., Dixit, R. and Govil, N. (2023) "Implementing and analyzing behaviour of ML based tamil songs recommendation system", *International Conference on Sustainable Computing and Data Communication Systems (ICSCDS-2023)*-IEEE, March-2023. pp. 381–386, doi: 10.1109/ICSCDS56580.2023.10104744.

Jayanthi, D., and M. Bommy. (2012). Vision-based real-time driver fatigue detection system for efficient vehicle control. *International Journal of Engineering and Advanced.* http://scholar.google.com.pk/scholar?q=Vision-based+Real-time+Driver+Fatigue+Detection+System+for+Efficient+Vehicle+Control&btnG=&hl=en≈sdt=0%2C5≈ylo=2011#0.

Jo, J., S. J. Lee, and H. G. Jung. (2011). Vision-based method for detecting driver drowsiness and distraction in driver monitoring system. *Optical Engineering* 50:127202. http://opticalengineering.spiedigitallibrary.org/article.aspx?articleid=1158525.

Karchani, M., A. Mazloumi, and G. N. Saraji. (2015). *The Steps of Proposed Drowsiness Detection System Design Based on Image Processing in Simulator Driving.* http://irjabs.com/files_site/paperlist/r_2607_150608134844.pdf.

Kiran, V.B.N., Raksha, R., Rahman, A., Varsh, K. N. and Nagamani, N. P. (2020). Driver drowsiness detection. Volume 8, Issue 15. *International Journal of Engineering Research & Technology (IJERT)*, Special Issue, Volume 8, Issue 15, pp: 33–35.

Kiranmayee, V., Pandey, V., Jain, R., and Ramaswamy, T. (2021). Driver drowsiness detection system, *International Journal for Research in Applied Science & Engineering Technology*, Volume 9 Issue VII July 2021, pp: 1226–1231.

Kong, W., L. Zhou, Y. Wang, J. Zhang, J. Liu, and S. Gao. 2015. A system of driving fatigue detection based on machine vision and its application on smart device. *Journal of Sensors* 2015:1–11. http://www.hindawi.com/journals/js/aa/548602/.

Lenskiy, A. A., and J.-S. Lee. 2012. Driver's eye blinking detection using novel color and texture segmentation algorithms. *International Journal of Control, Automation and Systems* 10 (2):317–27. doi:10.1007/s12555-012-0212-0.

Manikandan, R., Abilash, S., Agilakalanchian, C. and Tamilselvan, P. (2018). Driver drowsiness detection system using open computer vision, *International Journal of Current Research and Modern Education (IJCRME)*, Volume 3, Issue 1.

Patel, M., S. Lal, D. Kavanagh, and P. Rossiter. (2010). Fatigue detection using computer vision. *International Journal of Electronics and Telecommunications* 56 (4):457–61. doi:10.2478/v10177-010-0062-8.

Pooja and Chauhan, R. K. (2015). An assessment based approach to detect black hole attack in MANET. *International Conference on Computing, Communication and Automation (ICCCA-2015)*. IEEE, April-2015. pp. 552–557, doi: 10.1109/CCAA.2015.7148439.

Pooja, and Chauhan, R. K. (2017). A survey on comparisons of cryptographic algorithms using certain parameters in WSN. *International Journal of Electrical and Computer Engineering (IJECE)*, Vol.7, No.4, pp. 2232–2240, ISSN No.: 2088–8708, August. DOI: http://doi.org/10.11591/ijece.v7i4.pp2232-2240.

Pooja, and Chauhan, R. K. (2022). Triple phase hybrid cryptography technique in wireless sensor network. *International Journal of Computers and Applications (Taylor & Francis)*. ISSN No.: 1206–212X. January 2022. Doi: 10.1080/1206212X.2019.1710342.

Saradadevi, M., and P. Bajaj. (2008). Driver fatigue detection using mouth and yawning analysis. *Journal of Computer Science and Network*. http://paper.ijcsns.org/07_book/200806/20080624.pdf.

Shen, H. M., and M. H. Xu. (2015). Design and implementation of embedded driver fatigue monitor system. 2015 *International Conference on Artificial*. http://www.atlantis-press.com/php/download_paper.php?id=22072.

Sigari, M. H., M. Fathy, and M. Soryani. (2013). A driver face monitoring system for fatigue and distraction detection. *International Journal of Vehicular* 2013:1–11. http://www.hindawi.com/journals/ijvt/aip/263983/.

Singh, S., Saini, N., Pooja and Singh, D. (2024). *IoB Based Intelligent Healthcare System for Disease Diagnosis in Humans. ICDAM-2023*, Springer, LONDON, June-2023. Lecture Notes in Networks and Systems, vol 785. Springer, Singapore. https://doi.org/10.1007/978-981-99-6544-1_43.

Suryaprasad, J., Dr., D. Sandesh, V. Saraswathi, D. Swathi, and S. Manjunath. (2013). Real time drowsy driver detection using haarcascade samples. In *Computer Science & Information Technology (CS & IT)*, 45–54. Academy & Industry Research Collaboration Center (AIRCC). doi:10.5121/csit.2013.3805.

Computational Methods in Science and Technology Volume 1 – Kaur et al. (Eds)
© 2025 the Author(s), ISBN: 978-1-032-81760-6

An empirical study on text based sentimental analysis of using various machine learning classifiers on newspaper

Harjender Singh
Research Scholar, GLA University Mathura, U.P, India

Hitendera Garg
GLA University Mathura, U.P, India

Ajay Kumar Phogat
Maharaja Surajmal Institute, Delhi, India

ABSTRACT: Sentiment classification is a way to evaluate or analyse the important subjective information from the text. Analyzing the sentiments in newspaper is crucial for finding the pulse of public opinion i.e. parsing sentiments within newspaper is a complex task. We prioritize models that not only make accurate predictions but also provide clear explanations for their decisions. By comparing different models, we identify approaches that offer both accuracy and transparency and giving more insightful analyses of newspaper content. Our study involves a diverse collection of newspaper articles, and through rigorous evaluation, we highlight models that provide a balance between accuracy and interpretability. Text analysis methods for sentiment analysis typically work at a specific level, such as the pharse level, sentence or document level. This research paper aims to explore the application of TfidfVectorizer and counter vectorization techniques including Random Forest Classifier, Logistic Regression and Naïve Bayes methods, these techniques are evaluated on text based sentence-level sentiment analysis on newspaper headlines. In this paper the dataset used by me the target label is not provided, which should indicate whether the headlines are positive, negative, or neutral. To create the specified column, we can calculate the positive, negative, neutral, and compound scores for each headline. Based on the compound score, we will generate a 'Sentiment' column, which will serve as our target column during model training.

Keywords: phrase level, Random Forest Classifier, Logistic Regression and Naive Bayes

1 INTRODUCTION

Sentiment analysis refers to a type of classification wherein a dataset of text, such as a message, is analyzed using machine learning and deep learning algorithms to determine if the sentiment that comprises the dataset is positive or negative[1][2]. The process of interpreting data in the province of text mining and natural language processing (NLP) is known as analysis. Sentiment analysis is widely used to track consumer sentiment on social media, as well as to analyze brand and campaign performance. Our research paper focuses on sentence-level sentiment analysis of newspaper headlines using TfidfVectorizer and counter vectorization. Analyzing the sentiment expressed in newspaper headlines is a crucial task in understanding the emotional tone and overarching themes of news articles. The analysis typically involves pre-processing the text, which may include tasks like removing special characters, handling capitalization, and eliminating stopwords. Next, the headlines are

 DOI: 10.1201/9781003501244-49

passed through a sentiment analysis model, which can be based on rule-based systems, machine learning algorithms. These models assign a sentiment score or label to each headline based on the sentence. The result of this study have the potential to increase the sentiment and its implication in the context of news headlines [3][4][5].

Natural Language Processing (NLP) is the process of processing text input in any language to produce useful outcomes that may be applied to many situations. Sentiment analysis is a field of NLP where vast amounts of text are analyzed to extract sentiments and emotions. If we use sentiment analysis to the language above, our model should identify that consumers wish to express good things about the product, such as "The product quality of this Oven is great". The objective is to generate perspectives from the document and social media platforms in order to evaluate the attitudes of different user categories. These days, sentiment analysis has numerous applications in sociology, politics, economics, and other disciplines. Pre-processing the data is a crucial step [6][7][8].

2 LITERATURE REVIEW

Existing research on sentence-level sentiment analysis in the context of newspaper headlines has been explored. Previous studies have investigated a variety of techniques, including machine learning algorithms and feature representation approaches. This section presents a comparative examination of previous sentiment analysis research conducted with different sentiment classification methods. After reviewing the literature on numerous studies and models developed using the Amazon dataset, we discover that many different machine learning models have been attempted to be constructed by researchers [9] [10] [18].

Kharde, V[10] *et al.* (2016) presented the process of the opinion mining using machine learning and lexicon-based approaches, they done a survey and comparative study on the existing techniques. For this purpose, they utilize the cross domain and cross-lingual methods together for the evaluation of metrics [16]. As per obtained results, it is predicted that highest accuracy is present in SVM and naive Bayes classifier subpart of machine learning methods. Due to this accuracy, they are regarded as the baseline learning methods. In some cases, lexicon-based methods are very effective as it requires human labelled document efforts. Hence, the accuracy of the obtained result is more if the data is cleaner. Better sentiment accuracy is provided by the bigram model as compared to existing methods.

Islam, M. [11]: Determining the polarity of a news story is the primary goal of this work. Both the administrator and the user will find the algorithm's results useful in deciding whether to display or conceal the output. Theoretically, the algorithm should perform well across all areas under consideration; in practice, however, it displays fair results while displaying both positive and negative news. Due of human acclimatization to knowledge, this task has enormous significance. But not everyone needs to know everything. For instance, violent news is not suitable for a child's mental development. Large corporations can also benefit from this by using analysis to find the advantages and disadvantages of reviews of their products.

Htait, A. [12]: This study examines the relationship between sentence classification in lengthy book search queries and sentiment evaluation. Additionally, it looks into how book reviews and book search sentences are written similarly. In order to conclude this study, a language model derived from book reviews and a semi-supervised method for sentiment intensity prediction are provided.

Başarslan, M [13]: In this study, customer feedback on websites in two different industries was made public via sentiment analysis. The TripAdvisor dataset comprises reviews left by users on hotels, both positive and negative. The dataset on Rotten Tomatoes contains both positive (fresh) and negative (rotten) user reviews for movies. F-measures and accuracy are used in studies as performance metrics. The findings imply that group learning techniques perform better than individual machine learning algorithms.

Kumari, S.[14] : In this work, a finer-grained approach to sentiment classification—that is, classifying sentences as positive, negative, or neutral based on their polarity—is investigated. This study examines popular Sentiment Analysis techniques, such as Maximum Entropy, SVM, Naïve Bayes, K-Nearest Neighbor, Random Forest, and Voted Perceptrons, for evaluating movie reviews.

Kumar, S [15]: The purpose of tweet sentiment analysis is to identify the positive, negative, and neutral sentiment components in tweeter data. Sentiment analysis may assist any corporation in determining people's thoughts of their company and products. We performed sentiment analysis on the Twitter dataset. Our algorithm accepts an input tweet, a sentiment, and outputs selected text that begins and ends with the input tweet. To discover sentiment positions in tweets, we use the BERT (Bidirectional Encoder Representations from Transformers) model with one additional layer. The extra layer is utilized to determine the sentiment of a tweet. Our model can reach 70% accuracy on the validation dataset.

Shirodkar, V. [16]: The authors of this research present a method for stock price prediction that combines deep learning and sentiment analysis. The presented approach primarily employs Artificial Neural Networks (ANN) in conjunction with Natural Language Processing (NLP) and Information Retrieval (IR) as its fundamental components. The predictions are based upon two things which are historical stock data and financial news that we come across daily. Financial news includes news articles and datasets which are available on Yahoo Finance and could be obtained with the help of API(s). The Paper illustrates prediction of stock prices based on individual sentimental ratings of stocks extracted by applying sentimental analysis on news from various data sources. The sentimental analysis is specifically on the news that belongs to that companies' stocks rather than determining a general stock market sentiment.

Sivananda, D. [17]: The sentiments expressed in tweets or reviews that are published on Twitter are analyzed by the author in this study using particular keywords found in the tweets. We then determine if the tweets present a good or negative perspective. The ideas conveyed through tweets are predicated on a particular collection of tweets. The author will use three machine learning classifiers—Random Forest, Naive Bayes, and Support Vector Machine—to assess the accuracy of several domains, including precision, accuracy, etc. (SVM).

3 PROPOSED MODEL & RESULTS

3.1 *Machine learning based approach*

It applies machine learning algorithms for text classification. Sentiment classification using machine learning can be divided into two categories: supervised and unsupervised techniques of learning.

3.2 *Models used*

3.2.1 *Logistic regression*

This technique is used to define the dependent variable from the group of separate variables. It is based on estimating probabilities using a logistic regression formula.

3.2.2 *Random forest*

The Random Forest technique is used for both classification and regression. In this method, a decision tree is built using a data sample, and predictions are derived from each of them before voting to determine the best option

3.2.3 *Naive Bayesian classifier*

The technique is based on the Bayes theorem. It predicts the membership probability, which is the likelihood that a given data point belongs to a specific class. The data will be classified based on the class with the highest likelihood.

3.3 *Preprocessing*

Pre-processing starts with converting text documents into a word format that can be read. The papers are prepared for the next stage of text categorization due to a number of factors. Usually, the following steps are performed: Tokenization: After being treated as a string, a document is processed into a list of tokens. Getting rid of stop words: Because stop words are often used, they should be eliminated.

3.4 *Data collection*

The information will be utilized after it was gathered from various sources in various formats. Finding and replacing coarse data that is inaccurate, imprecise, incomplete, or superfluous is known as data cleaning; a sentence should be consistent following cleansing. The newspaper headlines for analysis will be collected from the Times of India (TOI) dataset available on Kaggle. The dataset covers a specific time period and headlines will be selected based on relevant categories or keywords. This approach ensures a diverse and representative sample for sentiment analysis. According to the Dataset description this news dataset is a persistent historical archive of notable events in the Indian subcontinent from start-2001 to q1- 2022, recorded in real-time by the journalists of India. It includes almost 3.6 million Times of India-published events. Most of the data is focusing on Indian local news including national, city level and entertainment. The dataset consists of the following columns:

Column Name	Description
S_No	Serial number or unique identifier for each headline
URL	The URL source of the headline.
Date	The date when the headline was published
Headline	The text of the newspaper headline.
Headline Link	The hyperlink associated with the headline, if available.
Positive	The count or score indicating the positive sentiment associated with the headline.
Negative	The count or score indicating the negative sentiment associated with the headline.
Neutral	The count or score indicating the neutral sentiment associated with the headline.
Compound	The compound score, which is a combined measure of sentiment polarity positive/ negative) and intensity.

After labelling we have the following results:
Neutral 82057
Negative 57211
Positive 37682

3.5 *Model training using count vectorizer*

selecting the right value for n in n-grams is crucial because it determines the size and type of language patterns our Vectorizer will capture. Choosing an appropriate n value can have a significant impact on the quality and relevance of the extracted features, which in turn can affect the performance of your models.

Table 1. Logistic regression algorithm.

Sr. No	Value for n in n-grams	Cross-validation scores	Mean CV accuracy	Test accuracy
1	N=1	[0.56099887 0.5584911 0.55937412 0.55902091 0.55785533]	0.5591480644249788	0.5696241876236225
2	N=2	[0.70263493 0.70436564 0.70496609 0.70217576 0.70330602]	0.7034896863520769	0.717745125741735
3	N=3	[0.56099887 0.5584911 0.55937412 0.55902091 0.55785533]	0.5591480644249788	0.5696241876236225

We can see for n = 1 , we have the best accuracy.

Accuracy score : 0.8924554959027974

Sr. No	Sentiments	Precision	Recall	F1 Score	Support
1	Negative (-)	0.87	0.90	0.88	11078
2	Neutral (0)	0.92	0.90	0.91	16882
3	Positive (1)	0.86	0.87	0.87	7430
4	accuracy			0.89	35390
5	macro avg	0.88	0.89	0.89	35390
6	weighted avg	0.89	0.89	0.89	35390

Overall Logistic Regression is giving good results and took around **15 seconds** to train.

Accuracy score : 0.8964679287934445

Sr. No	Sentiments	Precision	Recall	F1 Score	Support
1	Negative (-)	0.87	0.90	0.89	11160
.2	Neutral (0)	0.93	0.90	0.92	16857
3	Positive (1)	0.86	0.87	0.86	7373
4	accuracy			0.90	35390
5	macro avg	0.89	0.89	0.89	35390
6	weighted avg	0.90	0.90	0.90	35390

Overall RandomForest Classifier is giving slight better results than Logistic Regression and took around **23 minutes** to train. But, we should prefer Logistic Regression over RandomForest considering how time intensive RandomForest was.

Accuracy score : 0.8018084204577565

Sr. No	Sentiments	Precision	Recall	F1 Score	Support
1	Negative (-)	0.81	0.77	0.79	12194
2	Neutral (0)	0.82	0.84	0.83	16021
3	Positive (1)	0.74	0.77	0.75	7175
4	accuracy			0.80	35390
5	macro avg	0.79	0.79	0.79	35390
6	weighted avg	0.80	0.80	0.80	35390

Although it took **no time**(almost 0s) to train Multinomial Naive Bayes but it is giving the worst results among the three algorithms

3.6 *Model training using TfidfVectorizer*

Based on these observations, we can conclude that the model using CountVectorizer outperforms the model using TfidfVectorizer in terms of accuracy score, with a slight decrease of 4% in accuracy for the TfidfVectorizer approach.

Table 2. Classification report of logistic regression, randomforest classifier & Navie Bayes.

Accuracy score : 0.8763775077705567

Sr. No	Sentiments	Precision	Recall	F1 Score	Support
1	Negative (-)	0.84	0.90	0.87	10709
2	Neutral (0)	0.93	0.86	0.90	17642
3	Positive (1)	0.81	0.87	0.84	7039
4	accuracy			0.88	35390
5	macro avg	0.86	0.88	0.87	35390
6	weighted avg	0.88	0.8	0.88	35390

Overall Logistic Regression is giving good results and took around **13 seconds** to train.

Accuracy score : 0.897259112743713

Sr. No	Sentiments	Precision	Recall	F1 Score	Support
1	Negative (-)	0.88	0.90	0.89	11223
2	Neutral (0)	0.93	0.91	0.92	16690
3	Positive (1)	0.86	0.87	0.86	7477
4	accuracy			0.90	35390
5	macro avg	0.89	0.89	0.89	35390
6	weighted avg	0.90	0.90	0.90	35390

Overall RandomForest Classifier is giving slight better results than Logistic Regression and took around **21 minutes** to train.But, we should prefer Logistic Regression over RandomForest considering how time intensive RandomForest was.

Accuracy score : 0.7687199773947443

Sr. No	Sentiments	Precision	Recall	F1 Score	Support
1	Negative (-)	0.77	0.78	0.77	11322
2	Neutral (0)	0.90	0.74	0.81	19885
3	Positive (1)	0.48	0.86	0.62	4183
4	accuracy			0.77	35390
5	macro avg	0.72	0.79	0.74	35390
6	weighted avg	0.81	0.77	0.78	35390

Although it took **no time**(almost 0s) to train Multinomial Naive Bayes but it is giving the worst results among the three algorithms.

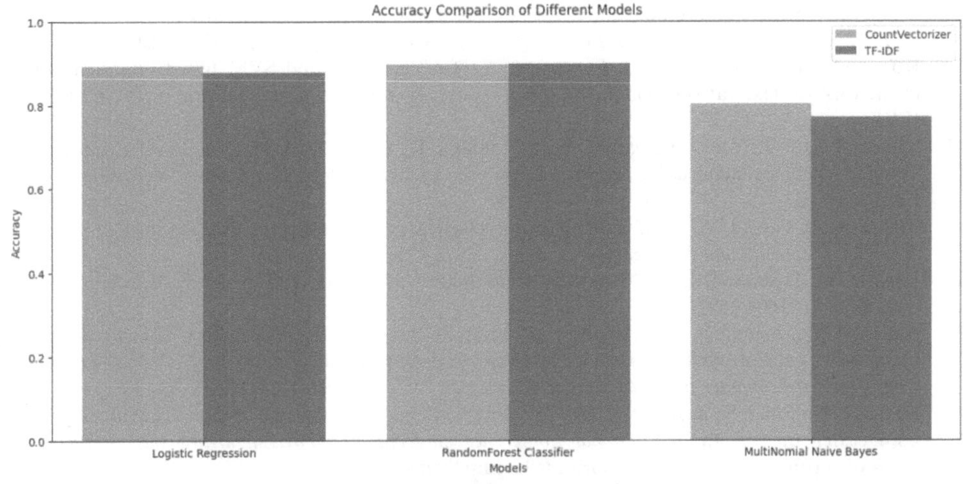

Figure 1. Accuracy comparison of different models using TfidfVectorizer.

4 CONCLUSION

The study briefly discussed the various machine learning based techniques for sentiment analysis from 2012 to 2020 also gives a brief introduction about the application of sentiment analysis in various fields. The analysis and extensive study of the literature gives an insight in the field of sentiment analysis. Through logistic regression, it may leverage a simple yet effective linear classifier, providing interpretability through feature coefficients, which aids in understanding the influential words in sentiment determination. Random forest, with its ensemble learning strategy, offers robustness to noise and over fitting, enhancing the model's generalization capabilities. Naive Bayes, despite its simplistic assumptions, demonstrates competitive performance, particularly in scenarios with limited training data, owing to its probabilistic nature. Both count vectorization and TF-IDF vectorization prove instrumental in converting text into numerical representations, capturing the essence of the headlines' sentiment while mitigating the curse of dimensionality. The performance of the sentiment analysis model can be further enhanced through techniques such as feature engineering, hyper parameter tuning, or utilizing more sophisticated algorithms like Support Vector Machines (SVM) or neural networks.

REFERENCES

[1] Poornima, A., & Priya, K. S. (2020, March). A comparative sentiment analysis of sentence embedding using machine learning techniques. *In 2020 6th International Conference on Advanced Computing and Communication Systems (ICACCS)* (pp. 493–496). IEEE.

[2] Mestan Fırat C¸ eliktu˘g Computer Engineering Bilkent University, Gazi University Twitter Sentiment Analysis, 3-Way Classification: Positive, Negative or Neutral? *2018 IEEE International Conference on Big Data* (Big Data)

[3] Juyal, P., & Kundaliya, A. (2023, July). A Comparative Study of Hybrid Deep Sentimental Analysis Learning Techniques with CNN and SVM. In *2023 IEEE World Conference on Applied Intelligence and Computing (AIC)* (pp. 596–600). IEEE.

[4] Sharma, R., & Pathak, P. (2021, August). A comparative study on various approaches of sentimental analysis. In *2021 International Conference on Simulation, Automation & Smart Manufacturing (SASM)* (pp. 1–4). IEEE.

[5] Kumar, T. S., Pandian, K. A., Aara, S. T., & Pandian, K. N. (2021, August). A reliable technique for sentiment analysis on tweets via machine learning and bert. In *2021 Asian Conference on Innovation in Technology (ASIANCON)* (pp. 1–5). IEEE.

[6] Chavan, S., Balasubramanian, J., Puro, J., Naik, M., & Nimkar, A. V. (2020, July). Similarity analysis of legal documents using content and network based approach. In *2020 11th International Conference on Computing, Communication and Networking Technologies (ICCCNT)* (pp. 1–7). IEEE.

[7] Kumar, V., & Subba, B. (2020, February). A TfidfVectorizer and SVM based sentiment analysis framework for text data corpus. In *2020 National Conference on Communications (NCC)* (pp. 1–6). IEEE.

[8] kumari Bygani, J., Venkateshwaralu, Y., & Ramana, K. V. (2023). A Sentence Level Classification of Telugu News Document using Sentiment Analysis. In *E3S Web of Conferences* (Vol. 391, p. 01037). EDP Sciences.

[9] Mann, S., & Phogat, A. K. (2020). Dynamic construction of lattice of cuboids in data warehouse. *Journal of Statistics and Management Systems, 23*(6), 971–982.

[10] Kharde, V., & Sonawane, P. (2016). Sentiment analysis of twitter data: a survey of techniques. *arXiv preprint arXiv*:1601.06971.

[11] Islam, M. U., Ashraf, F. B., Abir, A. I., & Mottalib, M. A. (2017, December). Polarity detection of online news articles based on sentence structure and dynamic dictionary. In *2017 20th International Conference of Computer and Information Technology (ICCIT)* (pp. 1–5). IEEE.

[12] Htait, A., Fournier, S., & Bellot, P. (2019, April). Sentiment analysis and sentence classification in long book-search queries. In *International Conference on Computational Linguistics and Intelligent Text Processing* (pp. 248–259). Cham: Springer Nature Switzerland.

[13] Başarslan, M. S., & Kayaalp, F. (2023). Sentiment analysis with ensemble and machine learning methods in multi-domain datasets. *Turkish Journal of Engineering*, 7(2), 141–148.

[14] Kumari, S. P. S. G. S., & Yadav, P. Sentiment Analysis Techniques and Approaches.

[15] Kumar, S. (2023). *Real Time Twitter Sentimental Analysis*, IJCRT. Volume 11, Issue 3 March 2023 | ISSN: 2320-2882

[16] Shirodkar, V. D., Kadri, N. J., Yadav, V. V., & Patil, P. V. Stock prices prediction using news categorization, *Sentimental Analysis And Deep Learning.*

[17] Sivananda, D., Asmai, S. A., Abidin, Z. Z., Abas, Z. A., & Ahmad, S. (2021). Sentiment analysis dashboard using twitter data in power BI. *Manuscript Editor*, 2021, 61.

[18] Phogat, A. K., & Mann, S. (2022). Optimal data cube materialization in hyper lattice structure in data warehouse environment. *Journal Of Algebraic Statistics*, 13(1), 149–158.

Computational Methods in Science and Technology Volume 1 – Kaur et al. (Eds)
 ISBN: 978-1-032-81760-6

Unraveling herding biases in the Indian equity market

Pooja Gupta & Neeraj
Assistant Professor, Geeta University, Panipat.

Monisha
Professor, Tecnia Institute of Advanced Studies, Delhi

Pooja Sharma
Associate Professor, Tecnia Institute of Advanced Studies, Delhi

ABSTRACT: The research emphasis on the presence of Herding Bias in Indian Equity Market. CNX Nifty has been taken as substitute variable for the study. To study herding behaviour, CSAD model is used to discern the presence of herding behaviour during extreme market movements in CNX Nifty during 2009-2022. Findings of the research depicted non-presence of herding behavior in CNX NIFTY during bull and bearish market respectively.

Keywords: Herding, Stock Return, Behavioural Bias, Behavioural Finance, CSAD, CSSD

1 INTRODUCTION

History is filled with the examples of irrational behaviour of investors while making investment decision like the Stock Market Crash of 2008, one of the greatest crashes ever happened in Indian Equity Market. Thus, behavioural finance came into picture which focuses on the reason behind investor investment decision. It is a mechanism which acknowledges the market developments in tandem with the behavioural patterns among investors. Various behavioural biases have been observed among the investors while taking investment decision. Herding is one of the biases which have been identified in investors and resists EMH assumption that investors behave rationally. Herding can be defined as mimicking the action of crowd by investors while taking investment decision. Investor takes decisions on other investor's judgement which prompts to mispricing of securities. Herding behaviour can shift the prices from their based value (Tan *et al.* 2008). It leads to shift in the financial market dynamics rendering the entire financial market fragile. The herding behaviour's presence in the market is indicative of market inefficiencies (Nofsinger and Sias 1999) and the market fluctuation are completely based on the investors. Therefore, herding behaviour exists only when investors chase the crowd instead of making rational decisions.

2 LITERATURE REVIEW

Herding as important bias exists on account of the way that a few investors overlook their very own information and follows others for decision making (Allsopp and Hey 2000). It can be intentional or unintentional: following the crowd's decision enthusiastically is intentional herding whereas when it is accidental at that point it is unintentional herding (Froot *et al.* 1992). Intentional herding can be judicious if the investors pursue the crowd on the premise that others are better informed over them and irrational if investors indiscriminately pursue

 DOI: 10.1201/9781003501244-50

the crowd's investment decision (Banerjee 1992; Bikhchandani *et al.* 1992). During bubble times, herding is due to the greed for money making and at the time of market crashes, they again herd due to the fear of losing money (Puchter 1999). Christie and Huang (1995) developed a model CSSD and opined that at the time of abnormal market movement, investors are more likely to conquer their personal beliefs. Chang *et al.* (2000) believed that the behavior of the herd differs between the biased opinion and the decision to make the wrong decision, and it increases the volatility in the returns and causes instability in the financial systems and make some modification in CSSD model and introduced CSAD model. Prosad *et al.* (2012) found that the outcome of the investigation supports the fact that Indian markets are efficient because herding bias does not exist. However, the herding is presence strongly in the bull phase but not in bearish stage of market individually. Ganesh *et al.* (2016) opined that the Indian Equity Market as whole is not affected by industry herding behavior. Jose *et al.* (2018) concluded that herding behavior does not prevail in Indian Equity Market and the investors make rational investment decisions.

2.1 *Research gap*

In September 2008 the Indian Equity Markets observed a consistent dip in CNX NIFTY. Because of the behavioral attitudes most investors suffered during this crisis (Adam 2010). The behavior of investor in the stock market depends upon elements like investment capacity, benchmark performance, other investor's behavior, presence of volatility in the equity market (Chang *et al.* 2000). Some investors invest with planning and a motive of high return. (Lin and Zhang 2012) suggested that behavioral biases are committed due to absence of technical expertise and lack of confidence among investors while making better decisions about investments. To have an in-depth insight that how investors' make decision, there is a necessity to examine the influence of herding on the individual investors' decision making at Indian Equity Market.

2.2 *Research objective*

- To discover that herding behavior exists in CNX Nifty from 2009 to 2022.
- To discover the herding bias in bull and bear conditions of market in CNX Nifty from 2009 to 2022.

3 RESEARCH METHODOLOGY

The present research used daily observation gathered from the website www.nseindia.com. The data is represented with the help of graph, tables etc. Unit root test is used to detect the stationarity of all the present variables. To study the herding behaviour, CSAD model methodology developed by Chang *et al.* (2000). CSAD approach uses OLS regression model to demonstrate presence of herding in CNX Nifty.

3.1 *Research model*

The present research utilized the following Ordinary Least Square Regression Model to identify the herding bias in NSE S&P CNX Nifty 50.

$$CSAD_t = \beta_0 + \beta_1 |r_{mt}| + \beta_2 r_{mt}^2 + e_t \quad (1)$$

Where $\beta_0, \beta_1, \beta_2$ are the coefficient of OLS Regression Model and e_t expressed as error term. $|r_{mt}|$ is the absolute market term. Under the CSAD model, if the coefficient r^2_{mt} ($\beta 2$)

demonstrates the significant negative value then it tends to be seen during existence of herding in the market. The present research used the following Ordinary Least Square Regression Model to identify the presence of herding bias in extreme phase (bull and bear) of the market respectively in NSE S&P CNX Nifty 50

$$CSAD_t^{bull} = \beta_0^{bull} + \beta 1^{bul}|r_{mt}^{bull}| + \beta 2^{bull} r^2_{mt}{}^{bull} + e_t \quad (2)$$

$$CSAD_t^{bear} = \beta_0^{bear} + \beta 1^{bear}|R_{mt}^{bear}| + \beta 2^{bear} R^2_{mt}{}^{bear} + e_t \quad (3)$$

The bullish and bearish trend is analyzed by the daily market return. If $r_{mt} > 0$, the market represents bullish trend and if $r_{mt} <= 0$, the market represents bearish trend.

3.2 *Research variables*

To address the aim of the study, two variables namely, $CSAD_t$ and R_{mt} have been selected from CNX Niftygiven by Chang *et al.* (2000)

$CSAD_t$ is the dependent variable

$$CSA = 1/n \sum_i^n = 1 \sum_i^n = 1|R_{it} - R_{mt}|$$

R_{mt} is the independent variable$R_{mt} = ((CV_t - CV_{t-1}) / CV_{t-1})*100$

3.4 *Research hypothesis*

H1: Herding behavior exists in CNX Nifty from 2009 to 2022.

4 DATA ANALYSIS & DATA INTERPRETATION

4.1 *Analysis through graph*

Figure 1 depicts that the association between CSAD and return of the market to be linearly positive, because all the values lie in the positive horizon.

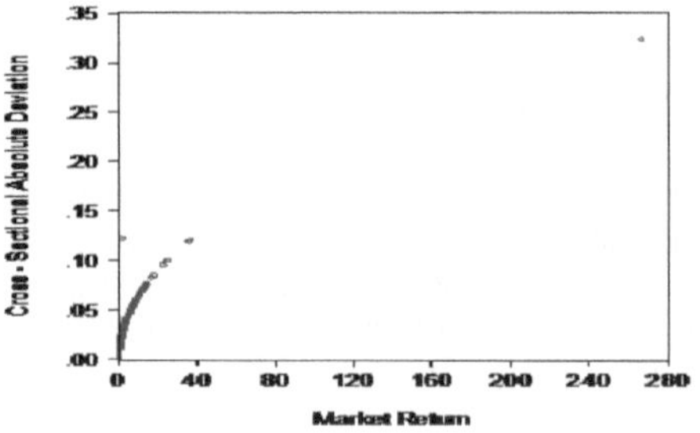

Figure 1. Represents the association between CSAD and return of the market for the duration 1st April, 2009 to 31st March, 2022.
Source: Compiled by researcher from E-views

4.2 *Descriptive statistics*

It is observed that CSAD is 00.399532 and S.D is 00.544432 while average return is 0.060394 and standard deviation is 1.103617 on daily basis. The high variation in Standard deviation in both the series may be because of abnormal variation caused by uncertainties in the whole

Table 1. Descriptive statistics.

Situation	Entire Period	Entire Period	Up Market	Bull Market	Down Market	Bear Market
Variable	CSADt	rmt	CSADt up	rmtbull	CSADtdown	rmt bear
Mean	00.399532	00.060394	00.015596	00.782982	00.015188	00.766129
Median	00.279099	00.056400	00.011410	00.577124	00.011159	00.563130
Maximum	10.21394	17.74407	00.323659	16.33432	00.120492	6.097258
Minimum	0.000305	−5.915096	8.76E-07	0.000632	2.54E-06	0.000000
Std. Dev.	0.544432	1.103617	0.016494	0.818729	0.014332	0.723094
Skewness	8.875463	1.591724	6.124748	6.233714	1.977777	1.976234
Kurtosis	135.7374	30.78126	97.42209	102.8070	9.794101	9.799621
Jarque-Bera	1852465.	80767.15	495962.8	553477.6	3002.752	3005.212
Probability	0.000000	0.000000	0.000000	0.000000	0.000000	0.000000
No. of observ.	2479	2479	1313	1313	1166	1166

Source: Data: Prowess, Calculations: E-Views.

market. Kurtosis is larger than 3 depicting that the return of the market and return dispersion (CSAD) are abnormal in nature. The result of kurtosis is 30.78126 for market return and 135.7374 for CSAD. The series shown values higher and significantly different from 3, it means the series depicts leptokurtic position, The Jarque-Bere test analysis are significant for both CSAD and market return series forcing to reject null hypothesis. The skewness and kurtosis have been showing larger values may be due to the occurrence of outliers. Hence primary data has been utilized. The present objective is studied by dividing the variables in two more phases.

Table 2. ADF test statistics.

	CSADt	r_{mt}
ADF test statistic	−47.9456**	−8.66862**
Test critical values:	0.0001	0.0000

*Note: The Null hypothesis is: Ho= Variable is non-stationary. The lag order for ADF test is selected by SIC. ** Significant level at 1%.*

4.3 *Unit root tests statistics*

Table 2 predicts that the test is supporting all the two variables and there is no trend in these variables. Therefore, we reject null hypothesis. It states that the $CSAD_t$ and r_{mt} series do not contain unit root and thus all series are stationary.

4.4 *Presence of herding behavior in NSE S&PCNX NIFTY50*

From the Table 3 it can be inferred that $\beta 2$ is -0. 000731. $\beta 2$is observed as the primary index to find out the presence of herding behaviour in CNX Nifty. As referred by Chang *et al.* (2000) model, if the coefficient of $\beta 2$ is of negative value but significant, then it shows the presence of herding behavior in the stock market. But in this study $\beta 2$ is negative but insignificant indicating that herding does not prevails in the CNX Nifty during this time period.

Table 3. Herding behavior for the complete market period.

Variables Coefficient	Coefficient	Std. Error	t-Statistic	Prob.
B0	0.338115	0.016882	20.0285	0.00000
+++β1	0.080305	0.018811	4.269143	0.00000
β2	−0.000731	0.002214	−0.330171	0.7413
R-squared	0.012141			
Adjusted R-squared	0.011343			
F-statistic	15.21566			
Prob(F-statistic)	0.00000			

Source: Data: Prowess, Calculations: E-Views.

4.5 *Presence of Herding Behavior in NSE S&P CNX NIFTY50 during Bull and Bearish market respectively.*

Table 4. Presence of herding behavior.

	Regression for Bull Market					Regression for Bear Market			
Co-efficient of Variable	Coefficient	Std. Error	t-Statistic	Prob.	Variable	Coefficient	Std. Error	t-Statistic	Prob.
β0	−6.98E-06	0.000115	−0.060493	0.9518	β0	7.25E-06	6.62E-06	1.095106	0.2737
β1	0.019942	0.000129	154.8603	0.0000	β1	0.01981	1.09E-05	1816.939	0.0000
β2	−8.95E-06	1.38E-05	−0.646649	0.518	β2	3.18E-06	3.24E-06	0.982325	0.3261
R-squared	0.973949				R-squared	0.999921			
Adjusted R-squared	0.973909				Adjusted R-squared	0.99992			
F-statistic	24487.7				F-statistic	7316497			
Prob (F-statistic)	0.00000				Prob (F-statistic)	0.0000			

Source: Data: Prowess, Calculations: E-Views

From Table 4 it is observed that the coefficient of market return (β2) in bull and bearish market is -8.95E-06, 3.18E-06 respectively. The herding does not exist in both the phases of market as the β2 in bull market is negative and in bearish market is positive, but in both the markets β2 is not significant. Hence, we can say that asymmetry of herding behavior is not observed in CNX NIFTY at times of Bull and Bearish market respectively.

5 CONCLUSION

The present study is conducted to detect the herding behavior in CNX Nifty during 2009-2022. The result confirms that herding behavior is not exhibited in CNX Nifty during 2009-2022. And, the study depicted non- presence of herding behavior in CNX NIFTY during bull and bearish market respectively. Thus, it can be inferred from the present research that the investors are rational and their judgments or decisions are not based on the crowd while taking investment decisions. The reason for this asymmetry could be reinforcement learning of past events that have impact on future events. The other reason could be that investors do not panic and engage into herding.

5.1 *Implications of the study*

This study offers valuable policy implications for brokers, investors, policy makers, academicians, and regulators to improve the effectiveness of decision making by the investors. The Managerial Implication is that the investors must have knowledge of these propensities and the biases that are associated with them, due care should be taken while making investment decisions. The investor must be more conscious of his/her own heuristics and frame dependencies, which is sowed by experience and personality.

5.2 *Future directions*

Behavioral finance has vast area that can be studied. There is a scope of deeper research on untapped biases like greed, fear, optimism, anchoring. This would be fascinating to examine the endowment of presently studied bias during the economic recession for developing capital markets collating with developed capital markets in recent years.

REFERENCES

Banerjee, Abhijit, V. (1992). A simple model of herd behavior. *The Quarterly Journal of Economics*, 107(3), 797–817.

Chang, E. C., Cheng, J. W. and Khorana, A. (2000). An examination of herd behavior in equity markets: An international perspective. *Journal of Banking and Finance*, 24(10), 1651–1679.

Christie, W. G., and Huang, R. D. (1995). Following the pied piper: do individual returns herd around the market? *Financial Analysts Journal*, 51(4), 31–37.

Froot, Kenneth A., and Emil M. Dabora, (1999). How are stock prices affected by the location of trade?. *Journal of Financial Economics* 53, 189–216.

Ganesh J and Shakthi V. (2013). A study on the individual investor behavior with special reference TO Geojit BNP Paribas financial service Ltd Coimbatore. *International Journal of Social Sciences Research*, 1(2), 1–13.

Jose, B., Varghese, J. and Surendram, A. (2018). Investors' Herding In Indian Equity Market; An Empirical Analysis. *IOSR Journal of Business and Management (IOSR-JBM)*, 76–86.

Kumar, A., and Bharti. (2015). Herding in indian equity markets: An evidence from information technology sector. *Journal of Economics and Finance*, 1–7.

Nofsinger, J. R., and Sias, R. W. (1999). Herding and feedback trading by institutional and individual investors. *The Journal of Finance*, 6.

Prosad, J. M., Kapoor, S., and Sengupta, J. (2012). An examination of herd behavior: An empirical evidence from indian equity market. *International Journal of Trade, Economics and Finance*, 3(2).

Tan, L., Chiang, T. C., Mason, J. R., and Nelling, E. (2008). Herding behavior in Chinese stock market: An examination of A and B shares. *Pacific-Basin Finance Journal*, 16 (1/2), 61–77.

Computational Methods in Science and Technology Volume 1 – Kaur et al. (Eds)
 ISBN: 978-1-032-81760-6

Revolutionizing healthcare: Unleashing the power of machine learning in remote diagnosis and health management

Parth Puneet, Prakhar Shukla & Gaurav Parashar
KIET Group of Institutions, Delhi-NCR, Ghaziabad, India

ABSTRACT: The medical industry has increasingly used artificial intelligence (AI) as technology progresses. The primary advantage of machine learning comes in its capacity to derive significant insights from the immense volumes of health data gathered daily. The progress in the Internet of Things (IoT), machine learning, big data, and high-performance computing has created exceptional possibilities to enhance public health by maximizing scarce human resources. This work presents a diabetes prediction model that categorizes diabetes based on common parameters such as blood sugar, body mass index (BMI), age, insulin, and other relevant factors, as well as external factors that contribute to the development of diabetes. Our analysis utilizes the PIMA India and Sylhet datasets. Machine learning techniques, such as the Support Vector Machine (SVM), Naive Bayes classifier (NB), and decision tree, are used in the study.

Keywords: Healthcare, Artificial Intelligence, Machine Learning, Diabetes Prediction

1 INTRODUCTION

A strong state of well-being lays the foundation for a satisfying life, unlocking pathways to joy and success. The health index of a nation serves as a reflection, indicating its economic strength, scientific achievements, defense capabilities ties, and social unity. While health is traditionally defined as the “absence of diseases,” its true essence lies in the timely identification and anticipation of illnesses, followed by prompt and suitable interventions. This necessity becomes even more pronounced in a nation as diverse as India, where a mosaic of demographics, climate variations, and socio-cultural intricacies converge. Extending our focus to the national level, the health index becomes apparent as a complex indicator of a country’s general resilience and vibrancy. A country’s economic strength is reflected in the health index, which acts as a mirror to show patterns of wealth or inequality. A thriving economy makes investments in public health programs, healthcare infrastructure, and research possible, all enhancing well-being overall.

Diabetes is a prevalent, lifelong condition that affects individuals of all ages. It occurs when the level of glucose in the blood becomes too high. According to the World Health Organization (WHO)[1], the number of people diagnosed with diabetes surged from 108 million in 1980 to 422 million in 2014, highlighting a significant global health issue. In 2019 alone, an estimated 2 million deaths were attributed to diabetes and related kidney diseases. For Type 1 Diabetes[2], risk factors include a family history of the condition and age, although currently, there’s no known method for preventing Type 1 Diabetes. In the case of

[1]https://www.who.int/news-room/fact-sheets/detail/diabetes
[2]https://www.cdc.gov/diabetes/basics/risk-factors.html

 DOI: 10.1201/9781003501244-51

Type 2 Diabetes, factors primarily include being overweight, being aged 45 years or older, having prediabetes, and having a family member with Type 2 Diabetes.

2 LITERATURE REVIEW

The experiment's primary goal is to design and implement a model for diabetes prediction that utilizes machine learning techniques successfully. Soni *et al.* (2020) [1] suggested ensemble learning and the use of classification techniques SVM, Random Forest (RF), Decision Tree (DT), Logistic Regression (LR), Gradient Boosting classifiers (GBC), and K-nearest neighbor (KNN). Furthermore, 77% classification accuracy was attained.

In their research, Xue *et al.* (2020) [2] use the NB as a classification technique. SVM achieved 96.54% accuracy. Tigga *et al.* (2020) [3] in the paper, implement six machine-learning classification strategies and compare the results with independent metrics. The tests were conducted on a dataset collected through online and offline surveys of 18 questions. The same statistics were also linked to the PIMA dataset (PIMA). Research results show that irregular timberland has an accuracy of 94.10%. Thomas and others [4] propose DT for diabetes prediction. For their research, they used the PIMA dataset. The accuracy of DT algorithms is examined and assessed in this work. The test results demonstrated that the devised system was effective, with an 87%. At small sample numbers, DT scales quickly and yields inaccurate response predictions. You *et al.* (2020) [5] conducted a study using two-class SVM and two-class DT to search for important diabetic factors in the PIMA dataset. It was found, through correlation analysis, that examining only the potential patient's glucose levels, BMI, and age is more efficient than conducting a comprehensive medical examination, which is time-consuming. The experiment with these features using SVM resulted in an accuracy of 70.4 %.

In the study, Rajni *et al.* [6] propose a procedure using the Bayes hypothesis for diabetes prediction utilizing machine learning methods to extract the necessary data so that the problem can be solved with high accuracy. Here, the mean is calculated to handle missing data and probability is used for calculating yes (positive) and no (negative) values. In comparison to previous controlled techniques, the suggested method enhances accuracy on the Pima dataset by 72.9%. In the study conducted by Sisodia D *et al.* (2018) [7], the Naive Bayes (NB), SVM, and DT algorithms are used among which The NB algorithm is considered the best machine learning method for this test because it has higher accuracy than other classification algorithms, with an accuracy of 76.30%. The study carried out by Mujumdar A *et al.* (2019) [8] suggests the use of numerous classification and common machine learning methods that yield the most precise results. Various classification methods include methods like LR, GBC, LDA, AdaBoost Classifier (ABC), Gaussian NB, Extra Trees Classifier (ETC), Bagging, RF, DT, Perceptron, SVC, and KNN on the PIMA dataset. Out of all, GBC, ETC, and ABC gave the highest accuracy of 77%.

In the model, author Saru S *et al.* (2019) [9] employ ensemble methods by using various classifiers like DT, KNN, and NB on the PIMA dataset. Here, the highest prediction result of 78% was obtained using DT without bootstrap. Authors of [10] paper used seven machine-learning algorithms used on the Pima Dataset for prediction namely DT, KNN, RF, NB, AB, LR, and SVM. For some criteria, including precision, accuracy, recall, and F measure, all models exhibit good results. Each model has an accuracy of 70%. For the train/test and test phases, LR and SVM yielded accuracies of 77% and 78%, respectively. Another neural network approach was used to predict diabetes in which the hidden layers in the network model were 1, 2, and 3, and occasionally 200, 400, and 800. In 400 epochs, the accuracy of the second hidden layer was 88.6%, the highest of all the models used in PIDD. The research conducted by Zou *et al.* (2018) [11] uses two datasets for better diabetes prediction. The first dataset is from a hospital physical examination in Luzhou, China and the second dataset is Pima Indians. When RF is used, the accuracy for the Luzhou dataset is

80.84%, and for the Pima Indians dataset, the accuracy is 77.21%. Talha Mahboob and others [12] propose ANN for diabetes prediction. The accuracy of the ANN approach was 75.7%, the RF method was 74.7%, and the K-means clustering method was 73.6%. ANN outperforms other methods. Authors of paper [13] found that NB is better than the DT method J48. The results predicted that NB got 76.3021% accuracy followed by J48 with an accuracy of 73.8281%.

N. Gupta *et al.* (2013) [14] uses various classification techniques namely Multilayer Perception(MLP), J48, JRIP, and Bayes Network on Pima Indians dataset. The paper determined that J48 has the highest accuracy of 81.33%. A. Iyer *et al.* (2015) [15] proposed DT and NB algorithm for Diabetes Prediction on the PIMA dataset. Based on the cross-validation technique, the DT achieves an accuracy of 74.8698%. Based on the percentage spilt(70:30) technique, J48 gives 76.9565% accuracy and NB achieves an accuracy of 76.5652%. The paper [16] predicted diabetes in females by taking the PIMA dataset. Various supervised learning algorithms have been used such as SVM, KNN, NB, RF, CT, NN, AB, and LR. When 10-fold cross-validation is applied, LR outperforms other techniques. The AUC for LR is 0.825. Aziz Perdana and others [17], in their research paper have used the KNN method for classification. By employing KNN with a value of k = 22, they achieved the highest accuracy of 83.12%. R. Sivanesan *et al.* (2017) [18] analyzed the performance of the J48 Decision Tree with different metrics. Based on the training set, the accuracy of correctly classified was 84.11%. After using 10-fold cross-validation, the accuracy of correctly classified was 73.82%.

3 METHODOLOGY

In our study, we use two diabetes-related datasets from UCI and Kaggle. These datasets contained different numbers of features and multiple inconsistencies. We use the following steps to ascertain the best performance of the models.

(1) Load and preprocess the data. Remove inconsistencies like missing values, outliers, inconsistent data, and duplicate rows. Perform feature scaling and normalization. Split the data into training and test sets.
(2) Select the models and perform hyperparameter tuning to achieve the best results.
(3) Finalize the model based on the performance metric.

Dataset Collection The module covers gathering data and preprocessing it for further analysis. For the study, two datasets were identified from the UCI repository.

Dataset 1: Dataset-1[3] was gathered by doctors from patients from Sylhet Diabetes Hospital patients in Sylhet, Bangladesh. The dataset contains attributes namely age, gender, polyuria, polydipsia, sudden weight loss, weakness, polyphagia, genital thrush, visual blurring, itching, irritability, delayed healing, partial paresis, muscle stiffness, alopecia, obesity, and class. The dataset contains 520 entries.

Dataset 2: For the research, dataset-2[4] is the Pima Indians Diabetes dataset from kaggle. The dataset was obtained from the National Institute of Diabetes and Digestive and Kidney Diseases. Based on specific diagnostic metrics included in the collection, the dataset aims to predict diagnostically whether or not a patient has diabetes. These examples were chosen from a larger database under several restrictions. Specifically, all of the patients in this facility are Pima Indian women who are at least 21 years old. The dataset contains attributes: pregnancies, glucose, blood pressure, skin thickness, insulin, bmi, diabetes pedigree function, age, and outcome. This dataset has 768 entries, of which 268 are positive and the remaining 500 are negative for diabetes.

[3]https://archive.ics.uci.edu/dataset/529/early+stage+diabetes+risk+prediction+dataset
[4]https://www.kaggle.com/datasets/uciml/pima-indians-diabetes-database

Data Pre-Processing: The data pre-processing method improves results by handling superfluous data in the dataset in an efficient manner. But dealing with missing information is a major problem, especially regarding important variables like age, blood pressure, skinfold thickness, insulin, bmi, and blood sugar level. Since the values of these attributes must not have zero, therefore we used the imputation technique to remedy and preserve the integrity of the dataset. Our goal is to maintain data consistency and correctness by expanding the dataset to include all values, that enable more precise and significant analysis and forecasts. Model Selection The choice of a suitable machine-learning technique is crucial during this stage of creating a diabetes prediction model. Used different machine-learning methods, each with unique advantages and disadvantages, to accurately predict diabetes. SVM, DT, and NB are a few of the techniques used. The goal is to optimize both predictive accuracy and robustness while maintaining the consistency of the model's predictions by experimenting with different approaches.

3.1 *Support vector classifier*

SVM is a supervised machine learning algorithm that is used for both classification and regression tasks. For learning instances, the maximal reserve hyperplane serves as its crucial limit. SVM uses the hinge loss function to calculate the empirical risk and adds a solution system regularization term to optimize the structural risk. SVM can do non-linear classification thanks to the kernel method, one of the most often used kernel learning strategies.

3.2 *Decision trees*

A non-parametric supervised learning technique for regression and classification is called a DT. The goal is to create a model that can forecast the value of a target variable using fundamental decision rules inferred from the data attributes.

3.3 *Naïve Bayes*

NB uses a set of algorithms based upon Bayes' theorem where the probability of previously occurred events is used to calculate the probability of posterior events. It is called naive as it takes the assumptions of features to be independent of each other. The classification model assigns class labels, represented by feature values, and the class labels are extracted from the dataset. For an item to be classified, it finds the maximum probability of each occurrence under the conditions that the item is considered as the item to be classified.

Table 1. Distinctive paper examination.

Author	Algorithm	Dataset	Performance Metric
Soni [1]	RF	PIMA	77% classification accuracy
Xue [2]	SVM	Sylhet	Accuracy of 96.54%
Tigga [3]	RF	PIMA	Accuracy of 94%
Thomas [4]	Decision Tree	PIMA	Accuracy of 94%
S.You [5]	SVM	PIMA	Accuracy of 70.4%
Rajni [6]	RB-Bayes	PIMA	Accuracy of 72.9%
Sisodia [7]	Naive Bayes	PIMA	Accuracy of 76.30
Mujumdar [8]	Random Forest	PIMA	77% classification accuracy
Saru [9]	Decision Tree	PIMA	Accuracy of 78%
Khanam [10]	SVM	PIMA	Accuracy 77%-78%

The Table 1 summarizes several studies using different machine learning algorithms to predict diabetes. It includes the author's name, the algorithm used, the dataset used, and its performance metrics, expressed as a percentage of accuracy. Algorithms exhibiting varying accuracy degrees on datasets include RF, SVM, DT, and NB. This highlights the significance of choosing algorithms depending on certain attributes. The table provides a quick comparative overview of the prediction performance of various algorithms in different papers.

4 RESULTS & CONCLUSION

Among the various machine learning algorithms evaluated in our study, DT emerged as the most accurate model, exhibiting a diabetes prediction accuracy of 98%, refer Table 2. We utilized various machine learning algorithms such as SVM, DT, and NB on the two datasets. The dataset from the Pima Indian dataset contains fewer parameters. Consequently, due to this limitation, the model's performance is not satisfactory. Conversely, when dealing with the Sylhet dataset, which encompasses more parameters and accounts for both genders, the model achieved higher accuracies. The highest accuracy defines the effectiveness of DT in handling the intricacies of datasets and making precise predictions. The study compared the accuracy of a machine learning algorithm using two different datasets. This comparative analysis revealed that the developed model significantly enhanced the accuracy and precision of diabetes prediction across various datasets.

Table 2. Comparison between accuracy of PIMA dataset and sylhet dataset.

Algorithm	PIMA Dataset	Sylhet Dataset
SVM	77.2%	93.2%
Decision Tree	74%	98%
Naive Bayes	77.2%	90.3%

The model's adaptability to different data sources underscores its relevance to real-world situations. Additionally, the study lays the groundwork for further extensions. The natural progression involves examining the risk of mortality within the next few years among individuals with diabetes. Expanding the analysis to include predictors of diabetes in individuals without the condition, can furnish valuable insights for health interventions. This broadening of the research scope holds promise for advancing diabetes prediction and prevention strategies.

REFERENCES

[1] Soni M. and Varma S. (2020). Diabetes prediction using machine learning techniques. *International Journal of Engineering Research & Technology (Ijert)* Volume. 9.

[2] Xue J., Min F. and Ma F. (2020). Research on diabetes prediction method based on machine learning. *In: Journal of Physics: Conference Series*; Vol. 1684; IOP Publishing; p. 012062.

[3] Tigga N. P. and Garg S. (2020). Prediction of type 2 diabetes using machine learning classification methods. *Procedia Computer Science.* 167:706–716.

[4] Thomas J., Joseph A., Johnson I., *et al.* (2019). Machine learning approach for diabetes prediction. *International Journal of Information.* 8(2).

[5] You S., Kang M. (2020). A study on methods to prevent pima indians diabetes using svm. *Korean Journal of Artificial Intelligence.* 8(2):7–10.

[6] Rajni R. and Amandeep A. (2019). Rb-bayesalgorithm for the predictionof diabetic in pima indian dataset. *International Journal of Electrical and Computer Engineering.* 9(6):4866.

[7] Sisodia D. and Sisodia D. S. (2018). Prediction of diabetes using classification algorithms. *Procedia Computer Science.* 132:1578–1585.
[8] Mujumdar A. and Vaidehi V. (2019). Diabetes prediction using machine learning algorithms. *Procedia Computer Science.* 165:292–299.
[9] Saru S. and Subashree S. (2019). Analysis and prediction of diabetes using machine learning. *International Journal of Emerging Technology and Innovative Engineering.* 5(4).
[10] Khanam J. J. and Foo S. Y. (2021). A comparison of machine learning algorithms for diabetes prediction. *Ict Express.* 7(4):432–439.
[11] Zou Q., Qu K., Luo Y., *et al.* (2018). Predicting diabetes mellitus with machine learning techniques. *Frontiers in Genetics.* 9:515.
[12] Alam T. M., Iqbal M. A., Ali Y., *et al.* (2019). A model for early prediction of diabetes. *Informatics in Medicine Unlocked.* 16:100204.
[13] Diwani S. A. and Sam A. (2014). Diabetes forecasting using supervised learning techniques. *Adv Comput Sci an Int J.* 3:10–18.
[14] Gupta N., Rawal A., Narasimhan V., *et al.* (2013). Accuracy, sensitivity and specificity measurement of various classification techniques on healthcare data. *IOSR Journal of Computer Engineering (IOSR-JCE).* 11(5):70–73.
[15] Iyer A., Jeyalatha S. and Sumbaly R. (2015). Diagnosis of diabetes using classification mining techniques. *arXiv preprint* arXiv:150203774.
[16] Bhoi S. K., *et al.* (2021). Prediction of diabetes in females of pima indian heritage: a complete supervised learning approach. *Turkish Journal of Computer and Mathematics Education (TURCOMAT).* 12 (10):3074–3084.
[17] Perdana A., Hermawan A. and Avianto D. (2023). Analyze important features of pima indian database for diabetes prediction using knn. *Jurnal Sisfokom (Sistem Informasi dan Komputer).* 12(1):70–75.
[18] Sivanesan R. and Dhivya KDR. (2017). A review on diabetes mellitus diagnoses using classification on pima indian diabetes data set. *International Journal of Advance Research in Computer Science and Management Studies.*;5(1).

Computational Methods in Science and Technology Volume 1 – Kaur et al. (Eds)
 ISBN: 978-1-032-81760-6

Machine learning and data analytics driven model for income tax rebate estimation: A neoteric and benign approach

Viney Sharma & Achal Kumar
Department of Computer Science & Engineering, Sharda University Agra India

Archana Singh
Department of Computer Science & Engineering, Anand Engineering College Agra India

Manish Kumar
Department of Computer Science & Engineering, Chandigarh Engineering College, Landran India

ABSTRACT: The provision of rebate to needy/underprivileged sections of society has been in practice since long in government organizations. The efficacy of such provisions lies in the fact that whether this rebate reaches people for whom it is intended. The biggest problem in awarding relief lies in the fact that people may not disclose their actual income. This paper conducts a study which strives for proposing a model which helps in assessing the financial status of any person given certain information about that person. This paper aims at proposing a Machine learning and Data Analytics driven model that predict the financial status of a person. The striking feature of this study is that results are applicable not only to income tax rebate problem but with minor customization, the model can be extended to work in other problems also where financial status of a person matters.

1 INTRODUCTION

A huge amount of information is produced today. It can be created by human, machine or by combination of both. Accurate information helps us make better decision. Data Analysis defined as science of data has helped us in going into insights of data.

1.1 *Classification of data analytics*

There are four main types of data analysis, depending on the level of work and the type of analysis to be performed.

1.1.1 *Descriptive analysis*

Descriptive analysis indicates analysis of data and business IQ. It is designed to provide an overview or "summary view" of facts and information in an easy-to-understand format [8]. This can inform or prepare data for further analysis. It collects raw data and converts it into human readable format. These can describe past events.

1.1.2 *Diagnostic analysis*

"Why did it happen?" it is an advanced analysis that examines data or content to answer the question. It helps analysts dig deeper into the problem to find the root cause of the problem. In a business environment, descriptive analysis tools and analytical tools go hand in hand. The technology it uses include: data discovery, data mining, social networking and more [9].

 DOI: 10.1201/9781003501244-52

1.1.3 *Predictive analytics*

Predictive analysis helps predict events based on current events. Predicting the probability of future events or predicting the time period in which an event occurs can be determined with the help of statistical models. We identify distinct variables to predict trends in this type of analysis [10].

1.1.4 *Prescriptive analysis*

Consists of a set of methods showing how best to study the data. It tells us what decisions we need to make to achieve good results. The purpose of prescriptive analysis is to achieve: quality improvement, service improvement, cost reduction and productivity improvement [11,12].

2 LITERATURE SURVEY

The use of Machine Learning (ML) technology is a new trend in the current business world offering financial researchers with new research and analysis across the region and increasing the predictive power of the old structure. Hall (2018), Saltzman & Yung (2018) demonstrated the present use of ML in financial research to conduct economic research in various domains of the economy[1,6]. New works in the subdomain of financial markets is using this ML to explore various possibilities in share market. Hanh and Viviane (2017) and Barzarbash (2019) demonstrated the feasibility of this approach in the subfield of forecasting oraganizational outcomes for financial institutions [3,5]. They used similar ML techniques such as SVM, ANN and Random Forest algorithms to better predict the company credit score. Matz et al. (2019) also used ML techniques to mine Facebook profiles with basic demographic information to predict revenue [4]. Sharma et al. (2016) illustrated the use of ML techniques in VM allocation in cloud computing environment [2]. Sharma et al.(2020) demonstrated the use of various bin packing techniques for VM allocation in cloud datacenters[7].

3 PROBLEM FORMULATION

This study works on a large data set of people across the world, taken from Kaggle with 31978 samples and 13 features. In numerous situations, organizations need to award rebate, subsidies, perks, and bonuses to some entities categorically. Quite often, a salary of a person is taken as deciding factor to be classified him/her in a particular category; but accurate income data is one of the hardest pieces of data to obtain across the world. In this study we have taken data from a Kaggle data source, supposed to be well authenticated. We propose a model which could be used for deciding whether subsidy/concession should be awarded to a person or not which will depend upon his financial health.

Table 1. Snapshot of data set.

Age	Job Type	Ed Type	Marital Status	Occu_pation	Relation_ ship	Race	Gen_ der	Capital Gain	Capital Loss	Hours per Week	Native Country	Fin Status
45	Private	HS-grad	Separated	Typist	Not-in-family	White	F	0	0	28	US	HIG
40	Private	12th	Married	other	Spouse	Black	M	0	0	40	Jamaica	LIG/ MIG
44	Private	Some-college	Married	Prof-Emeritus	Spouse	White	M	0	0	40	US	HIG
27	Private	9th	Unmarried	Craftman	Relative	White	M	0	0	40	Mexico	HIG
33	Private	Assoc-admin	Married	Craftman	Spouse	White	M	0	0	45	Portugal	LIG/ MIG
23	Private	5th-6th	Unmarried	Other	child	White	M	0	0	40	El-Salvador	LIG/ MIG
23	Local-gov	HS-grad	Unmarried	Farming and fishing	Not-in-family	White	M	0	0	40	Poland	LIG/ MIG

4 METHODOLOGY

Machine Learning and Data Analytics are very promising approaches for taking insights of data and model formulation in case of making prediction. This study aims at two main objectives:

(1) To determine at first instance, how various attributes are correlated/uncorrelated with each other. If they are correlated, whether they have a positive correlation or negative correlation among them. The examination of correlation could help answer any intermediate queries if asked by the organization employing this model.
(2) To formulate a model for future data sets so that when unseen data arrives, it could easily be classified for the appropriate category

4.1 *Logistic regression*

This model is used for classification task when output/label is discrete quantity as in whether the house will be sold or not. So, suppose instead of predicting a real number we are interested in predicting some kind of a discrete quantity. Let us represent the real number that we got out of linear regression using z; let us say z is $b + w_1x_1 + w_2x_2 \ldots w_mx_m$ given m features. We can perform the multiplication of the parameter with the feature value and then we add all these things together this is called as a linear combination. Our job is to take this particular real number and convert that into a discrete quantity between 0 and 1. So, what we can possibly do is, we can create a linear combination of the features and the parameters and we get this intermediate representation which is z which is the real number and we want to convert this real number into a discrete quantity between 0 to 1. For that, we have a special function called logistic function. [13,14]. So, what happens is this particular function takes the real number and squashes it between 0 and 1; it crosses the y-axis exactly at 0.5 and this S shape function is called as a sigmoid function [15,16].

The sigmoid function is represented by $\sigma(z)$ and is

$$\sigma(z) = 1/(1 + e^{-(b + w1x1 + w2x2 \ldots wmxm)})$$

4.2 *k-Nearest Neighbor*

This algorithm is based upon the concept of distance of newly arrived data point from the developed model. It is instance based model. This algorithm assumes all instances correspond to points in the n-dimensional space. Sharma et al. describes the distance is calculated as standard Euclidean distance [17,18]. Let us assume an instance is denoted by collection of features as below.

$$< \mathbf{a_0(x), a_1(x), a_2(x) \ldots .. a_n(x)} >$$

where $a_r(x)$ gives the value of the rth feature of instance x. The distance between two instances x_i and x_j is defined to be $d(x_i, x_j)$, where

$$\mathbf{d(x_i, x_j) = \sum_{r=1}^{n} (a_r(x_i) - a_r(x_j))^{1/2}}$$

In KNN learning the objective function can be continuous or non-continuous. If we choose $k = 1$, then the 1-NN algorithm assigns the value of the point nearest to point in question.

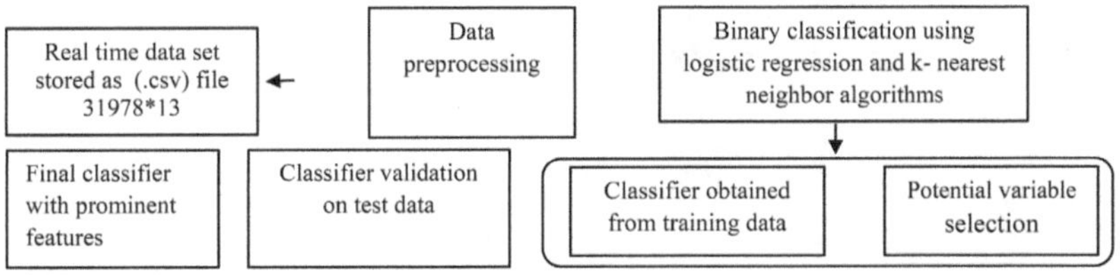

Figure 1. Flow chart in proposed framework.

5 DATA PREPROCESSING AND EXPLORATORY DATA ANALYSIS

Data preprocessing is crucial step in data analytics which has vital impact on exploratory data analysis and machine learning model formation. In the data set, null values have been identified and have been substituted with appropriate values. We have done exploratory analysis for having deeper insights into data. Python has been used as it has powerful tools and libraries for data analytics. Data analysis of give data set has been given in the following table which shows mean value, minimum value and maximum value and standard deviation for numerical attributes. 25% of samples are under 28 years of age f the samples are under 50.

Table 2. Exploratory data analysis.

Attributes Parameters	age	capitalgain	capitalloss	hoursperweek
Mean	38.58	1064.36	86.74	40.42
Standard Deviation	13.66	7298.59	401.59	12.34
Min	17	0	0	1
Max	90	99999	4356	99
25%	28	0	0	40
50%	37	0	0	40
75%	48	0	0	45

Data Analysis shows that 25% of samples belong to HIG while 75% belong to LIG/MIG. Gender wise classification is given below in table.

Table 3. Genderwise income group.

Income Group Gender	LIG/MIG	HIG
Female	0.89	0.11
Male	0.69	0.31

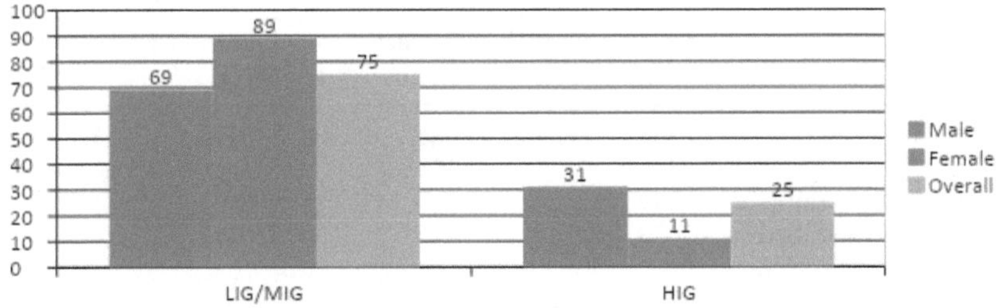

Figure 2. Genderwise income group bar chart (Data in percentage).

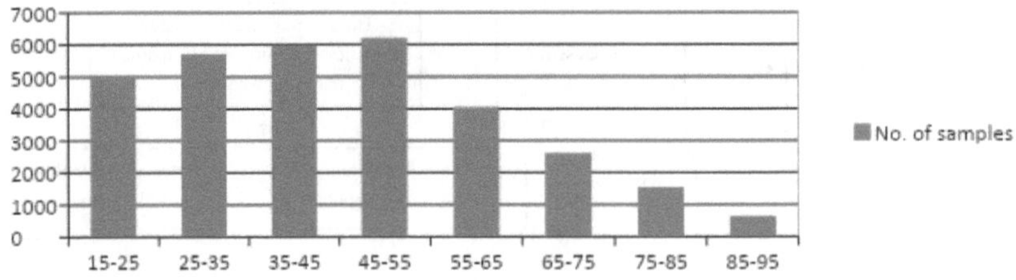

Figure 3. Number of samples with respect to age group.

The above graph shows that maximum person belong to age group 25-55 which are prominent candidates for income tax

6 MODEL FORMATION AND TESTING

Machine learning model for financial status classification has been formed using logistic regression and k-nearest neighbor algorithm. We have done coding in python for model formation and testing. 70% of samples have been kept for training purpose and 30% for testing purpose. In logistic regression, one hot encoding is used to represent categorical attributes, thus converting 9 categorical attributes into 94 attributes. The value of regression intercept has been obtained as 2.83875548. The value of some of coefficients for 94 categorical variables) have been obtained as follows.

Table 4. Values of coefficients for 94 categorical variables obtained after one hot encoding.

1	6.0e-02	20	-1.2e-02	39	-5.3e-03	58	3.5e-02	77	-1.1e+00
2	-4.4e-02	21	-1.8e-01	40	-5.8e-02	59	-1.1e+00	78	-1.7e-01
3	-7.6e-01	22	2.3e-02	41	9.0e-02	60	5.8e-01	79	-2.8e-02
4	-2.2e-01	23	-5.3e-02	42	6.3e-04	61	-4.1e-01	80	-7.4e-01
5	-1.0e+00	24	3.1e-04	43	-5.3e-01	62	-9.2e-03	81	-2.3e-02
6	8.5e-01	25	-3.1e-02	44	-7.9e-01	63	-1.9e-01	82	7.8e-01
7	-6.8e-01	26	1.1e-02	45	-1.0e-01	64	-1.5e-01	83	-4.3e-01
8	-7.4e-02	27	-2.1e-01	46	-3.4e-01	65	-7.7e-02	84	-8.7e-01
9	-5.0e-02	28	-7.1e-01	47	1.1e+00	66	-1.7e-01	85	-4.7e-02
10	-8.2e-02	29	3.7e-02	48	-9.5e-01	67	5.2e-01	86	-3.4e-02
11	-2.3e-02	30	-2.9e-02	49	-1.9e-01	68	7.0e-01	87	-8.7e-01
12	-4.5e-02	31	-5.1e-01	50	-1.2e-01	69	-6.8e-01	88	6.3e-01

The confusion matrix for obtained results for test data is given below. The accuracy is 84.241%. The number of misclassified samples is 1426 out of 9049 samples in test data set. These are results with logistic regression algorithm. Now model is formulated and tested with KNN algorithm. The accuracy score of model with KNN algorithm is given below.

Table 5. Accuracy score.

	Logistic Regression		k-Nearest Neighbor	
	Predicted LIG/ MIG	Predicted HIG	Predicted LIG/ MIG	Predicted HIG
Actual LIG/MIG	6338	485	6181	642
Actual HIG	941	1285	814	1412
Number of test data set samples	9049		9049	
Number of correctly classified samples	7623		7591	
Number of misclassified samples	1426		1458	
Accuracy Score	84.2419		83.90982	

The KNN algorithm has been run for various values of n and result obtained is given below

Table 6. K-NN for various values of N.

Value of n	Number of misclassified samples	Value of n	Number of misclassified samples	Value of n	Number of misclassified samples
1	1763	6	1445	11	1763
2	1509	7	1461	12	1509
3	1510	8	1450	13	1510
4	1431	9	1459	14	1439
5	1456	10	1446		

7 RESULTS & DISCUSSIONS

The paper has aimed at training and testing model for predicting the financial status of a person using two algorithms-logistic regression and k-nearest neighbor. Logistic regression has given slightly better performance with 84.2419% accuracy than k-nearest neighbor with 83.90982%. K-nearest neighbor algorithm has been run for n = 1,2,3 … ..14 and n = 4 has given best results. The proposed model could be applied in various situations where some decision is to be taken on the basis of financial status of a person. This model is versatile as it could be utilized in selecting candidates for award of subsidy, scholarship, concession etc.

8 FUTURE SCOPE

Data analysis among various features can be done to have more insights in data. Depending upon analysis, features having very less correlation with output can be dropped for fast results. Data can be subjected to artificial neural networks to get deep learning based model and then comparison can be made between machine learning and deep learning model results.

REFERENCES

[1] Hall, Aaron Smalter. (2018). Machine learning approaches to macroeconomic forecasting. *The Federal Reserve Bank of Kansas City Economic Review* 103, no. 4. https://doi.org/10.18651/er/4q18smalterhall

[2] Sharma, V. and Srivastava, G. M. S. (2016). Evolution and present status of cloud computing: a comprehensive analysis. *Int. J. Business Information Systems*, 22(2),123–142

[3] Le, Hong Hanh, and Jean-Laurent Viviani. (2017). Predicting bank failure: An improvement by implementing a machine-learning approach to classical financial ratios. *Research in International Business and Finance* 44 (February 17, 2017): 16–25. https://doi.org/10.1016/j.ribaf.2017.07.104

[4] Matz, Sandra C., Jochen I. Menges, David J. Stillwell, and Andrew H. Schwartz. (March 28, 2019). Predicting Individual-Level Income From Facebook Profiles. *Plos One* 14, no. 3: 1–13.

[5] Barzarbash, Majid. (2019). Fintech in financial inclusion: Machine learning applications in assessing credit risk. *IMF: Money and Capital Department.*

[6] Saltzman, Bennett, and Julieta Yung. (July 11, 2018). A machine learning approach to identifying different types of uncertainty. *Economics Letters* 171: 58–62. https://doi.org/10.1016/j.econlet.2018.07.003.

[7] Sharma, V. and Srivastava, G.M.S. (2020). Correlation-based heuristics and evaluation of existing greedy heuristics for VM allocations in cloud datacenters. *IJITCC, Inderscience Publishers.* 3(4). pp. 276–318

[8] Mishra, A. K., malik, S., Sharma, V., Paliwal, S. and Ray, K. (2023). Integrated linear regression and random forest framework for e-commerce price prediction of pre-owned vehicle. *Proceedings of International Conference on Trends in Electronics and Health Informatics. Tehi* 2022. pp 107–116.

[9] Sharma, V and Srivastava GMS. (2013). Energy efficient architectural framework for virtual machine management in IaaS clouds. *Proceedings of Sixth International Conference on Contemporary Computing (IC3)*. pp 369–374.

[10] Bambil, D., Pistori, H., Bao, F., Weber, V., Alves, F. M., Gonçalves, E. G. ...& Bortolotto, I. M. (2020). Plant species identification using color learning resources, shape, texture, through machine learning and artificial neural networks. *Environment Systems and Decisions* 40(4).pp 480–484.

[11] Kaur, S., & Kaur, P. (2019). Plant species identification based on plant leaf using computer vision and machine learning techniques. *Journal of Multimedia Information System*, 6(2), 49–60.

[12] Sharma V. (2013). *Virtual Machine Management in Cloud Computing Environment, PhD. Thesis.*

[13] Shibani, K., Sendhil Kumar, K. S., & Siva Shanmugam, G. (2020). An effective approach for plant monitoring, classification and prediction using iot and machine learning. *In Artificial Intelligence and Evolutionary Computations in Engineering Systems*. pp 143–154.Springer Singapore.

[14] Sharma, V. (2010). Ipv6 and ipv4 security challenge analysis and best-practice scenario. Int. *J. of Advanced of Networking and Applications* 1 (04), pp 258–269.

[15] Soltis, P. S., Nelson, G., Zare, A., & Meineke, E. K. (2020). Plants meet machines: Prospects in machine learning For plant biology. *Applicatiopns in Plant Sciences*. 8(6).

[16] Bojamma, A. M., & Shastry, C. (2021). A study on the machine learning techniques for automated plant species identification:current trends and challenges. *International Journal of Information Technology*.13(3).pp 989–995.

[17] Sharma, V, Mishra AK and Paliwal S. (2023). Machine Learning Framework for Recognition and Classification of Plant Species: A Study using Digital Database. *ICDCN '23: Proceedings of the 24th International Conference on Distributed Computing and Networking January 2023*. pp 407–413. https://doi.org/10.1145/3571306.3571444

[18] Wagle, S. A. (2021). Comparison of plant leaf classification using modified alexnet and support vector machine. *Traitement du Signal*, 38(1).

Computational Methods in Science and Technology Volume 1 – Kaur et al. (Eds)
© 2025 the Author(s), ISBN: 978-1-032-81760-6

CNN-based fundus image analysis for improved glaucoma detection

Shajila Beegam M.K. & Mala Kalra
National Institute of Technical Teachers Training & Research, Chandigarh, India

ABSTRACT: Glaucoma stands as a primary contributor to visual impairment worldwide, emphasizing the need for precise and early detection. Fundus images offer a window into the health of the retina. However, clinical diagnosis relies on trained specialists, making it resource-intensive. Researchers are harnessing the power of Convolutional Neural Networks (CNNs) to analyze fundus images, automatically identifying patterns indicative of glaucoma. Latest advancements in deep learning (DL) are streamlining glaucoma detection through the extraction of pertinent traits from retinal fundus images. This work outlines a CNN-based model designed to recognise glaucoma by analyzing fundus images. The model utilises an integrated heterogenous dataset for training and testing to ensure its robustness and effectiveness. This study gives greater accuracy and sensitivity in comparison to existing studies thereby facilitating precise glaucoma detection and intervention to reduce vision loss.

Keywords: Glaucoma, Convolutional Neural Network, Automated System, Deep Learning, Fundus image

1 INTRODUCTION

Glaucoma, the second dominant factor for permanent eye nerve damage globally, might cause total vision loss if not identified and addressed in its initial phase. Because of its asymptomatic nature, it is frequently recognized as the stealthy thief of sight (Abdull *et al.*, 2016). Studies highlight its significant impact, with over 60 million cases worldwide, affecting about 1.69% with "Open-Angle Glaucoma" and 0.69% with "Angle-Closure Glaucoma" (Tham *et al.*, 2014). Anticipated estimates indicate that by 2025, the global number of individuals affected by glaucoma will reach 80 million and by 2040, 111.8 million (Tham *et al.*, 2014). This increasing predominance of glaucoma necessitates better diagnostic methods.

IOP checking and visual field testing are two time-consuming, conventional methods for screening glaucoma but they do not give accurate result always. The "cup-to-disc ratio (CDR)" is a standard method for grading this disease. An affordable modality to evaluate glaucoma (Abramoff *et al.*, 2010) is fundus imaging, but it is subjective. To deal with this problem, automated computer-aided evaluation frameworks are developed to diagnose this disorder from fundus photos. There are two methods for doing this. The first one uses segmentation results(Joshi *et al.*, 2011) to calculate clinical measures such as CDR. The second approach make use of deep learning to extract features from the source image (Maheshwari *et al.*, 2017). Compared to typical handcrafted features, the second approach that uses deep learning (CNN in particular), demonstrates excellent performance.

This study's main goal is to build an automated system based on a 16-layer CNN for assessing glaucoma utilizing fundus images. The succeeding segments of the research are arranged as follows: Literature survey is done in Section 2. Proposed methodology is

DOI: 10.1201/9781003501244-53

portrayed in Section 3. Experimental setup with datasets, performance measures and results obtained are elaborated in Section 4. Conclusion is given in Section 5.

2 LITERATURE REVIEW

Numerous studies have employed CNNs for glaucoma detection. In specific investigations (Al-Bander *et al.*, 2017; Li *et al.*, 2020; Raghavendra *et al.*, 2018; Raja *et al.*, 2021), researchers have developed custom CNN models from scratch to recognize glaucoma in fundus data. Moreover, transfer-learning techniques utilizing pre-trained CNN architectures have been looked into for glaucoma identification in other studies (Christopher *et al.*, 2018; Gheisari *et al.*, 2021; Kashyap *et al.*, 2022; Pascal *et al.*, 2022; Sun *et al.*, 2023; Velpula & Sharma, 2023).

(Al-Bander *et al.*, 2017) devised a model combining CNN and SVM for glaucoma classification utilizing the RIM-ONE dataset. Their method produced 88.2 percent accuracy, 90.8 percent specificity, and 85 percent sensitivity. (Raghavendra *et al.*, 2018) constructed a deep convolutional neural network comprising 18 layers using a private dataset containing 589 healthy and 837 glaucoma images. Their model yielded a sensitivity of 0.98, specificity of 0.983, and accuracy of 0.9813. (Li *et al.*, 2020) devised an attention-guided CNN to detect glaucoma, obtaining 96.2% accuracy, 95.4% sensitivity, and 96.7% specificity with the LAG dataset during validation. Testing on RIM-ONE dataset gave 85.2% accuracy, 84.8% sensitivity, and 85.5% specificity. (Raja *et al.*, 2021) developed a clinically validated hybrid DL system for grading the progression of glaucoma in retinal ganglion cells (RGC), achieving a higher F1 score (0.9577) and accuracy (0.9117).

(Christopher *et al.*, 2018) used pre-trained models like Inception v3, VGG16, and ResNet50 on a private dataset, achieving a sensitivity of 92, specificity of 93, and AUC of 97 percent. (Gómez-Valverde *et al.*, 2019) developed an automatic glaucoma classifier employing fundus images and CNNs with transfer learning on a combined public and private dataset, achieving the best outcomes as a sensitivity of 87.01%, specificity of 89.01%, and AUC of 94.20%. (Gheisari *et al.*, 2021) formulated a "CNN and recurrent neural network (RNN)" combination for improved glaucoma identification integrating VGG16 and LSTM to achieve 95% sensitivity, 96% specificity, 99% AUC, and 96.19% F-measure. (Kashyap *et al.*, 2022) used a glaucoma dataset, DenseNet-201 for extracting features and deep CNN (DCNN) for classification, achieving 0.9882 accuracies during training and 0.969% during testing, surpassing existing methods. Pascal and colleagues (Pascal *et al.*, 2022) devised a multi-task DL structure, combining U-Net and VGG-16 networks, for glaucoma diagnosis, outperforming other models with 3.5 times fewer parameters on REFUGE dataset. ResFPN-Net, a convolutional neural network, was developed by (Sun *et al.*, 2023) for glaucoma identification, achieving an AUC of 0.8947 for Drishti-GS and 0.7964 for RIM-ONE datasets. (Velpula & Sharma, 2023) developed a glaucoma classification method employing pre-trained CNNs and a voting-based fusion approach. Their model achieved high accuracy across various datasets, with an AUC of 1.0 and 99.57% on ACRIMA and HVD.

Currently, available frameworks in the study have relied on either private datasets or pre-trained architectures, making it challenging to replicate their outputs due to the non-availability of datasets and the need for a substantial quantity of images to assess pre-trained model effectiveness. This necessitates a CNN architecture for glaucoma classification employing publicly accessible datasets. The study's primary aim is to create a computationally efficient glaucoma classifier for fundus images based on CNN.

3 PROPOSED METHODOLOGY

To classify glaucoma in fundus images, this work uses a CNN architecture, which is illustrated in Figure 1.

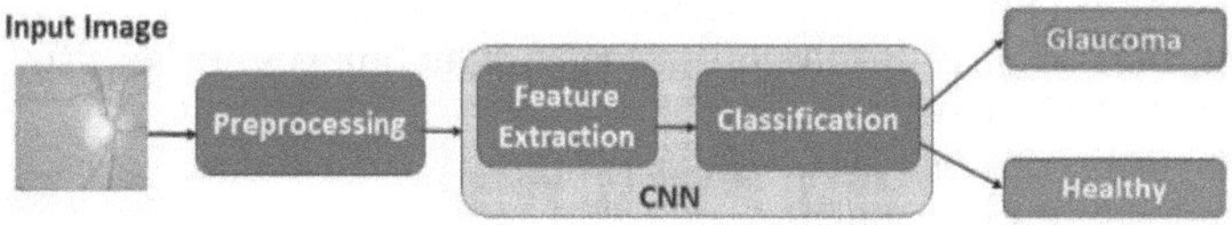

Figure 1. Proposed methodology.

The method comprises three primary stages: preprocessing, feature extraction, and classification. Initially, to standardize inputs of varying sizes, all images are resized to 256x256 pixels during preprocessing. Ten percent of the data goes for testing, twenty percent for validation and seventy percent is utilised for training. Various augmentation methods including rotation, shifting, shearing, zooming, and horizontal flipping, are employed to enrich the training image dataset. Subsequently, the images subjected to preprocessing are fed into the proposed CNN model for feature extraction and classification.

The suggested CNN architecture utilizing Softmax classifier (CNN-Soft) comprises 16 layers, including convolutional, pooling, dense, and softmax layers and is depicted in Figure 2. Batch normalization (BN) is employed in every convolutional layer to expedite training and enhance network initialization. The softmax layer is employed for binary classification, distinguishing between glaucoma and non-glaucoma cases. Binary cross-entropy loss function and ReLU activation function are also utilized.

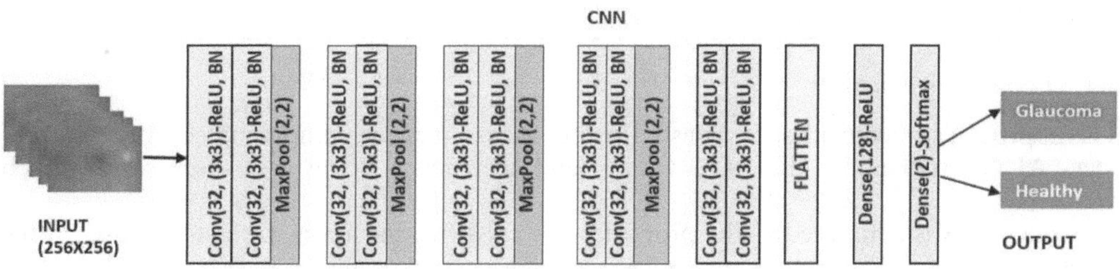

Figure 2. Proposed CNN-Soft architecture.

Utilizing this architecture ensures superior performance measures, even when trained from scratch. Carefully selecting the convolutional layers's stride, padding, filter dimension and count improves the model's effectiveness. These hyperparameters reduce computational complexity and allow for more local feature extraction. A 3×3 kernel size and 32 filter size are used in all layers. Proper selection of stride and padding ensures that the output dimension remains the same as the input dimension, preventing pixel information loss.

4 EXPERIMENTAL SETUP AND RESULTS

TensorFlow and Python Jupyter Notebook are used for CNN model development and training, with Google Colab's GPU and TPU resources accelerating experimentation and refinement.

4.1 *Datasets*

The experimentation is carried out on fundus images, as detailed in Table 1 and described below.

The High-Resolution Fundus (HRF) (*High-Resolution Fundus (HRF) Image Database*, n. d.) dataset comprises 45 high-resolution retinal fundus images. It consists of 15 healthy, 15 glaucomatous and 15 diabetic images.

The sjchoi86-HRF (*GitHub-Cvblab/Retina_dataset.*, n.d.) dataset consists of 101 glaucoma images and 300 healthy images.

The Drishti-GS1 (Sivaswamy *et al.*, 2014) dataset comprises 70 glaucoma images and 31 healthy images collaboratively developed by the Aravind Eye Hospital in India and the International Institute of Information Technology in Hyderabad, India.

The ACRIMA dataset (Diaz-Pinto *et al.*, 2019), a publicly available resource for glaucoma classification, includes 705 fundus images of the posterior segment of the eye, with 396 showing glaucoma signs and 309 normal eyes developed as part of the ACRIMA project.

Table 1. Description of datasets employed for this work.

Dataset	Glaucoma	Healthy	Total	Resolution
HRF	15	15	30	3504×2336
sjchoi86-HRF	101	300	401	2048 × 1536
Drishti-GS1	70	31	101	2896 × 1944
ACRIMA	396	309	705	2048 × 1536
Total	582	655	1237	

4.2 *Performance metrics*

The approach's effectiveness is assessed using diverse metrics comprising sensitivity, accuracy, AUC (Area under the ROC Curve) and specificity. Below is a discussion of each metric:

Accuracy: It is computed as the proportion of classified instances to the total cases in the dataset.

$$\text{Accuracy} = (\text{TN} + \text{TP})/(\text{TP} + \text{TN} + \text{FN} + \text{FP}) \tag{1}$$

Sensitivity: It denotes the proportion of actual positive instances accurately identified by the model to detect positive (glaucoma) cases correctly.

$$\text{Sensitivity} = \text{TP}/(\text{FN} + \text{TP}) \tag{2}$$

Specificity: The proportion of actual negative instances accurately identified by the model to correctly reject negative (non-glaucoma) cases is specificity.

$$\text{Specificity} = \text{TN}/(\text{TN} + \text{FP}) \tag{3}$$

AUC: This metric is employed in glaucoma detection to assess a classifier model's diagnostic ability. The ROC curve depicts the true positive rate and the false positive rate for different threshold levels. A higher AUC implies better discrimination between diseased and healthy eyes, crucial for accurate diagnosis and early intervention.

4.3 *Results and discussion*

To optimize the model's effectiveness, hyperparameters are adjusted during training. Initially, experimentation with multiple learning rates: 0.001, 0.0001, and 0.00001 are done. Among these options, a learning rate of 0.001 consistently yielded the most favourable results for this model. The proposed CNN architecture is trained over 50 epochs, exploring various batch sizes including 8, 16, 32, 64, and 128. Batch sizes significantly influence learning dynamics, directly affecting the model's accuracy. Through thorough experimentation, a batch size of 32 achieved the highest accuracy, reaching 89.69%. This meticulous tuning process significantly enhanced the model's performance for glaucoma detection. The proposed model is evaluated with the reserved ten percent of data for testing, comprising 124 fundus images, including 66 healthy and 58 glaucoma images. Figures 3 and 4 depicts the confusion-matrix of the model and the performance comparison graphs with existing studies respectively.

Three alternative models are compared with the suggested CNN-Soft model: the CNN +SVM model (Al-Bander *et al.*, 2017), the attention-guided CNN (Li *et al.*, 2020), and both the M-Net and DENet models by (Fu *et al.*, 2019). Table 2 portrays the comparison of the results.

Table 2. Comparative analysis of the suggested approach with existing methods.

Reference	Method	Sensitivity	Accuracy	AUC	Specificity
(Al-Bander *et al.*, 2017)	CNN+SVM	0.85	0.882	0.9102	0.908
(Li *et al.*, 2020)	AG-CNN	0.848	0.852	0.916	0.855
(Fu *et al.*, 2019)	MNet	0.7609	0.8157	0.8998	0.8706
(Fu *et al.*, 2019)	DENet	0.8478	0.8429	0.9183	0.838
CNN-Soft	CNN with Softmax classifier	**1.0**	**0.8969**	0.9044	0.8088

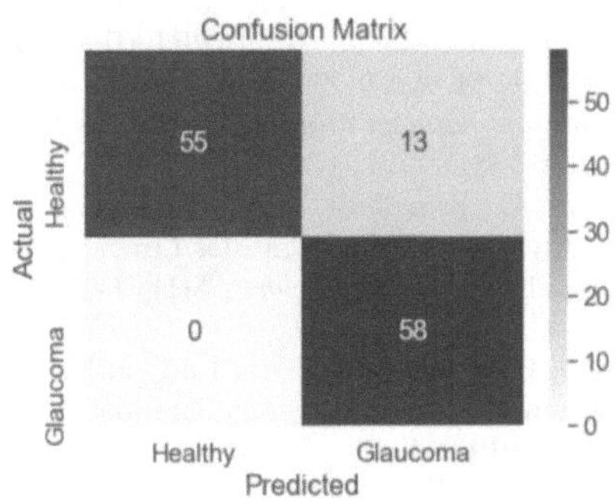

Figure 3. Confusion matrix of CNN-soft model.

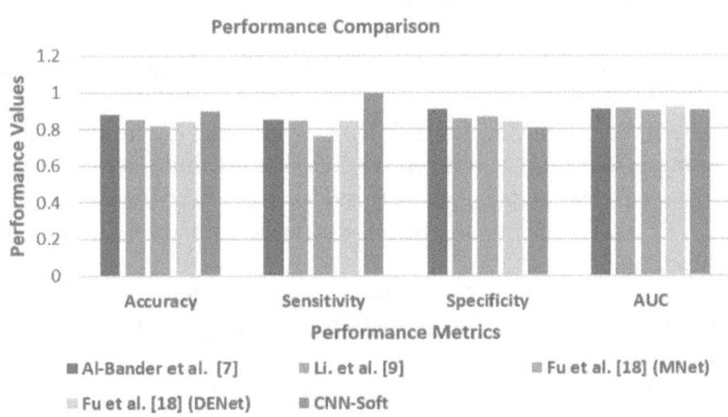

Figure 4. Performance comparison graph.

5 CONCLUSION

A comprehensive DL model is designed to detect glaucoma in fundus data. The design is executed on a combined dataset representing a diverse dataset. Seventy percent of the dataset is utilised for training, 20 percent for validation and 10 percent for testing. Different pre-processing techniques, like rotation, shifting, shearing, zooming, horizontal flipping etc., are done to enrich the training dataset. Several hyperparameters, including epochs, batch size and learning rate, were tried to optimize the model's effectiveness.

This study achieves a significant milestone by accurately detecting glaucoma in patients' eyes, showcasing the model's reliability in predicting disease occurrence. However, a limitation of this work is its reliance on labeled data for training. Deep learning models demand extensive datasets to learn features and make precise predictions, and annotating retinal images is a resource-intensive process in terms of both cost and time. To mitigate this challenge, collaborative efforts between eye specialists and machine learning experts can be instrumental in sharing resources. Despite these limitations, this study demonstrates that the suggested CNN-Soft model is capable of correctly identifying glaucomatous cases.

REFERENCES

Abdull, M. M., Chandler, C., & Gilbert, C. (2016). Glaucoma, "the silent thief of sight": Patients' perspectives and health seeking behaviour in Bauchi, northern Nigeria. *BMC Ophthalmology*, 16(1), 1–9. https://doi.org/10.1186/s12886-016-0220-6

Abramoff, M. D., Garvin, M. K., & Sonka, M. (2010). Retinal imaging and image analysis. *IEEE Reviews in Biomedical Engineering*, *3*, 169–208. https://doi.org/10.1109/RBME.2010.2084567

Al-Bander, B., Al-Nuaimy, W., Al-Taee, M. A., & Zheng, Y. (2017). Automated glaucoma diagnosis using deep learning approach. *2017 14th International Multi-Conference on Systems, Signals and Devices (SSD), Jan*, 207–210. https://doi.org/10.1109/SSD.2017.8166974

Christopher, M., Belghith, A., Bowd, C., Proudfoot, J. A., Goldbaum, M. H., & Weinreb, R. N. (2018). Performance of Deep Learning Architectures and Transfer Learning for Detecting Glaucomatous Optic Neuropathy in Fundus Photographs. *Scientific Reports*, 8(1), 1–13. https://doi.org/10.1038/s41598-018-35044-9

Diaz-Pinto, A., Morales, S., Naranjo, V., Köhler, T., Mossi, J. M., & Navea, A. (2019). CNNs for automatic glaucoma assessment using fundus images: An extensive validation. *BioMedical Engineering Online*, 18(1), 1–19. https://doi.org/10.1186/s12938-019-0649-y

Fu, H., Cheng, J., Xu, Y., & Liu, J. (2019). Glaucoma Detection Based on Deep Learning Network in Fundus Image. *Advances in Computer Vision and Pattern Recognition*, *September*, 119–137. https://doi.org/10.1007/978-3-030-13969-8_6

Gheisari, S., Shariflou, S., Phu, J., Kennedy, P. J., Agar, A., Kalloniatis, M., & Mojtaba Golzan, S. (2021). A combined convolutional and recurrent neural network for enhanced glaucoma detection. *Scientific Reports*, 11(1), 1–11. https://doi.org/10.1038/s41598-021-81554-4

GitHub-cvblab/retina_dataset. (n.d.). Retrieved January 4, 2023, from https://github.com/cvblab/retina_dataset

Gómez-Valverde, J. J., Antón, A., Fatti, G., Liefers, B., Herranz, A., & Santos, A. (2019). Automatic glaucoma classification using color fundus images based on convolutional neural networks and transfer learning. *Biomedical Optics Express*, 10(2), 892. https://doi.org/10.1364/boe.10.000892

High-Resolution Fundus (HRF) Image Database. (n.d.). Retrieved January 4, 2023, from https://www5.cs.fau.de/research/data/fundus-images/

Joshi, G. D., Sivaswamy, J., & Krishnadas, S. R. (2011). Optic disk and cup segmentation from monocular color retinal images for glaucoma assessment. *IEEE Transactions on Medical Imaging*, 30(6), 1192–1205. https://doi.org/10.1109/TMI.2011.2106509

Kashyap, R., Nair, R., Gangadharan, S. M. P., Botto-Tobar, M., Farooq, S., & Rizwan, A. (2022). Glaucoma Detection and Classification Using Improved U-Net Deep Learning Model. *Healthcare (Switzerland)*, 10(12). https://doi.org/10.3390/healthcare10122497

Li, L., Xu, M., Liu, H., Li, Y., Wang, X., & Jiang, L. (2020). A Large-Scale Database and a CNN Model for Attention-Based Glaucoma Detection. *IEEE Transactions on Medical Imaging*, 39(2), 413–424. https://doi.org/10.1109/TMI.2019.2927226

Maheshwari, S., Pachori, R. B., & Acharya, U. R. (2017). Automated Diagnosis of Glaucoma Using Empirical Wavelet Transform and Correntropy Features Extracted from Fundus Images. *IEEE Journal of Biomedical and Health Informatics*, 21(3), 803–813. https://doi.org/10.1109/JBHI.2016.2544961

Pascal, L., Perdomo, O. J., Bost, X., Huet, B., Otálora, S., & Zuluaga, M. A. (2022). Multi-task deep learning for glaucoma detection from color fundus images. *Scientific Reports*, 12(1), 6–15. https://doi.org/10.1038/s41598-022-16262-8

Raghavendra, U., Fujita, H., Bhandary, S. V., Gudigar, A., Tan, J. H., & Acharya, U. R. (2018). Deep convolution neural network for accurate diagnosis of glaucoma using digital fundus images. *Information Sciences*, 441, 41–49. https://doi.org/10.1016/j.ins.2018.01.051

Raja, H., Hassan, T., & Akram, M. U. (2021). Clinically Verified Hybrid Deep Learning System for Retinal Ganglion Cells Aware Grading of Glaucomatous Progression. *IEEE TRANSACTIONS ON BIOMEDICAL ENGINEERING*, 68(7), 2140–2151. https://doi.org/10.1109/TBME.2020.3030085

Sivaswamy, J., Gopal, S. R. K., Joshi, D., Jain, M., Syed, U., & Hospital, A. E. (2014). *DRISHTI-GS: Retinal Image Dataset For Optic Nerve Head (ONH) Segmentation, IIIT, Hyderabad, India*. 53–56.

Sun, G., Zhang, Z., Zhang, J., Zhu, M., Zhu, X. rong, & Yang, J. K. (2023). Joint optic disc and cup segmentation based on multi-scale feature analysis and attention pyramid architecture for glaucoma screening. *Neural Computing and Applications*, 35(22), 16129–16142. https://doi.org/10.1007/s00521-021-06554-x

Tham, Y. C., Li, X., Wong, T. Y., Quigley, H. A., Aung, T., & Cheng, C. Y. (2014). Global prevalence of glaucoma and projections of glaucoma burden through 2040: A systematic review and meta-analysis. *Ophthalmology*, 121(11), 2081–2090. https://doi.org/10.1016/j.ophtha.2014.05.013

Velpula, V. K., & Sharma, L. D. (2023). Multi-stage glaucoma classification using pre-trained convolutional neural networks and voting-based classifier fusion. *Frontiers in Physiology*, 14(June), 1–17. https://doi.org/10.3389/fphys.2023.1175881

Computational Methods in Science and Technology Volume 1 – Kaur et al. (Eds)
 ISBN: 978-1-032-81760-6

Innergram: A social media analysis platform

P. Rout, S. Khetan, S. Ahkam, M. Raghuvanshi, P. Pargai & K. Bhardwaj
Bennett University, Greater Noida, Uttar Pradesh, India

ABSTRACT: The remarkable development of social media applications has catalysed a significant change in computerized cooperations, especially obvious on account of Instagram. As clients draw in with this application, they produce tremendous and varied information, requiring progressed devices for analysis. There are different devices for dissecting other social media stages however no such application for analysing information from Instagram. This paper presents Innergram, a particular social media investigation application planned and made explicitly for Instagram. Innergram offers a complete set-up of highlights, including addiction assessment, follower growth analysis, keyword usage tracking, identification of inactive accounts, post-specific analytics, and user engagement assessment. Through fastidious information pre-handling and algorithmic methodologies, Innergram explores the intricacies of Instagram information, giving clients nuanced experiences into their web-based presence. With a promise to client security and continuous refinement, Innergram arises as a strong and easy to understand instrument for people, organizations, and advertisers trying to enhance their Instagram investigation experience. Its imaginative way to deal with information examination and devotion to client strengthening position Innergram as a significant resource in the steadily developing scene of social media investigation.

1 INTRODUCTION

In the present computer-oriented period, social media has changed past its unique reason as a straightforward communication device, turning into a strong power that shapes worldwide collaborations. Individuals worldwide go to social media for bunch reasons - to associate with friends and family, share individual experiences, and participate in virtual networks. This far reaching utilization highlights the different jobs social media plays in correspondence, self-articulation, and data sharing.

The data generated within these spaces becomes increasingly vast, intricate and sophisticated as social media platforms continue to attract growing user populations. Grasping client ways of behaving commitment patterns and the advancing elements of online networks (Liu, 2010) in the midst of this downpour of data presents imposing challenges. To address these intricacies modern social media examination instruments have become fundamental.

In the unique scene of social media examination, the development of complex apparatuses has become key for translating the intricacies inborn in the tremendous datasets produced by platforms like Instagram. While numerous examination apparatuses are accessible for platforms like Spotify and Discord to interpret with bits of knowledge into music and correspondence communicable (Cornelio and Roig, 2020) ways of behaving, the complexities of Instagram information has remained neglected to a great extent. This research paper presents a far reaching Nobel social media examination tool, made to explore this many-sided landscape (Smith, 2015) and distil significant bits of knowledge from the abundance of accessible information to user's need, such as their activity, top keywords, most interactions and all.

 DOI: 10.1201/9781003501244-54

In this paper, "Innergram" is proposed, a committed social media examination tool specifically created for Instagram users. Not at all like its ancestors, Innergram remains at the very front of development, offering custom-made answers for exploring different substances, keywords, frequency and measurement (Schober, Pasek, Guggenheim, Lampe and Conrad, 2016) user cooperations exceptional to Instagram. Diving into the complexities of Innergram, it becomes obvious that this instrument is ready to address the advancing difficulties of understanding and tackling the force of social media information. By furnishing users with a nuanced comprehension of their web-based (Batrinca and Treleaven, 2015) presence Innergram not just makes up for a significant shortcoming in the domain of social media examination and analysis, yet additionally fills in as an important asset for both individual and expert undertakings.

The turn of events and excursion of Innergram have been set apart by a guarantee to tending to the neglected requirements of Instagram user in the domain (Maynard and Bontcheva, 2016, May) of social media analysis. The difficulties illustrated prior highlight the complexities of working with Instagram information and Innergram imaginative elements have been intended to explore and conquer these obstacles. All through this excursion user criticism and commitment play had a significant impact in refining and improving the device. Looking forward, Innergram imagines expanding its capacities consolidating progressed calculations and refining its UI for a much more consistent experience. As part of a dedication to community collaboration, active consideration is being given to transitioning the Innergram into an open-source initiative (Younis, 2015). Thus, the aim is to foster a cooperative environment where designers and data enthusiasts can contribute to the evolution of Innergram, ensuring its adaptability to the ever-changing landscape of Instagram and social media analytics. The change to open source adjusts with the vision of making a device that enables users with experiences as well as welcomes an aggregate work to constantly improve its elements and functionalities as said by (Sapountzi and Psannis, 2018). Innergram embodies continuous investigation, fuelled by the anticipation of future possibilities driven by user engagement (Young, Soliz, Xu and Young, 2020) and a dedication to transparency and community collaboration.

The structure of the paper is as follows. Section 2 delves into the comprehensive approach employed in developing Innergram, detailing the extraction and analysis of Instagram data, as well as the intricate algorithms utilized for features like Addiction Counter, Follower Growth Graph, Top Keyword Usage, Inactive Accounts identification, Post-Specific Analytics, and User Engagement Level Assessment. The section starts with the Data preprocessing segment outlining the obstacles encountered during the tool's development, from designing a user-friendly interface to addressing data inconsistencies and limitations. In section 3 the conclusion is provided, encapsulating the promise and potential of Innergram, highlighting its role as a user-friendly Instagram analysis tool and providing insights into follower engagement and addiction frequency.

2 METHODOLOGY

The proposed analytical (Rogers, 2021) tool starts with getting the data from Instagram and then performing multiple operations on the data to find insightful operations. Figure 1 provides a glimpse into the application, showcasing its capacity to furnish users with nuanced insights into their daily Instagram activities.

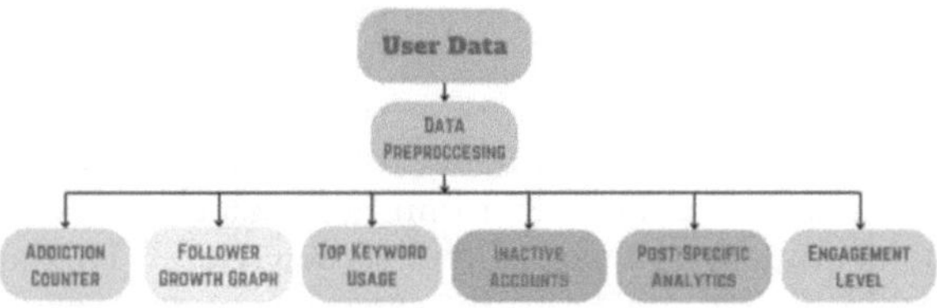

Figure 1. Operations performed on user data. In first step data processing is done to generate usable data and then six different operations are performed.

In first step data processing is done to generate usable data and then six different operations are performed.

2.1 *Data pre-processing*

The development of an Instagram data analysis tool presented several data pre-processing steps that demanded careful consideration and innovative solutions. These were categorized into distinct phases of the tool's workflow-

2.1.1 *Uploaded the zip file on a webpage*

Developed a mechanism for users to upload Instagram data packaged in a zip file onto a webpage. Ensured seamless integration with backend processing for efficient data submission.

2.1.2 *Extracted it in-memory and performed search operations*

Implemented algorithms for in-memory extraction of data and search operations and optimized algorithms to efficiently locate files based on relative paths while considering both search speed and memory utilization.

2.1.3 *Worked with bad data structuring from instagram*

Employed robust data pre-processing techniques to handle inconsistencies and irregularities in Instagram's data structuring and developed mechanisms to adapt to diverse and erratic data structures encountered in Instagram datasets.

2.1.4 *Made assumptions due to lack of data*

Incorporated assumptions into the analysis tool's algorithms to compensate for the scarcity and limitations of data provided by Instagram and considered the impact of these assumptions on the accuracy and reliability of generated insights.

2.1.5 *Displayed results to users in a fast and predictable manner*

Optimized the output generation process to present analysed results to users quickly and predictably and minimized latency and ensured a responsive user interface for an enhanced user experience.

These steps outlined the data pre-processing tasks involved in developing an Instagram data analysis tool, addressing various aspects of handling, and preparing the data for analysis.

2.1.6 *Dataset explanation*

Extracted from zip files and organized into JSON-formatted subdirectories, this dataset offers a comprehensive (Nanda & Kumar, 2021) analysis of Instagram user activities. Encompassing ad engagement, content consumption and it's marketing (Berg & Sterner, 2015), social connections, and personal preferences, the exploration provides nuanced insights into user behaviour. While some data categories exhibit gaps, the richness in others contributes valuable perspectives for future research on digital privacy (Killekar, Shah & Kolge, 2013), user engagement, and platform transparency.

2.2 *Addiction counter / status*

The Addiction Counter employs a stack or queue as its data structure to record timestamps of user interactions (Penni, 2017). The algorithm here calculates both the frequency and duration of these interactions to evaluate addictive behaviour. This approach provides a quantitative (Ahmed, 2018) and time-sensitive analysis of user engagement to identify potentially problematic patterns.

2.2.1 *Feature extraction*

Define a set of features that could indicate addiction to the platform. These features include:

- How often the platform is used (like viewing posts, giving likes or commenting)
- Engagement level (e.g., number of followers, frequency of interactions)
- Content consumption habits (e.g., types of posts viewed, time spent on the platform)
- Dependency indicators (e.g., excessive use of certain features like search history)

2.2.2 *Scoring system*

Define a scoring system where each feature contributes to the overall addiction score.

- Assign weights to each feature based on its perceived importance in indicating addiction.
- Normalize scores to ensure consistency across different features.

2.2.3 *Calculating addiction score*

- Combine the scores using the defined weights to calculate the overall addiction score. Normalize the overall score to fit within the 1 to 5 scale.

2.2.4 *Interpretation*

- Define thresholds on the addiction score to categorize users into different levels of addiction (e.g., 1-2: Low addiction, 3: Moderate addiction, 4-5: High addiction).
- Provide explanations for each level of addiction based on the user's data patterns (e.g., "Low addiction: User engages with Instagram occasionally and shows balanced usage patterns").

2.3 *Follower growth graph*

The Follower Growth Graph utilizes a graph data structure to represent followers and their connections evolving over time. This approach allows for a dynamic representation of the network, offering insights into how a user's followers expand and connect throughout different periods.

In Figures 2 and 3, first there is an increase in the number of followers and then there is a sudden drop.

Figure 2. Increase in the number of followers.

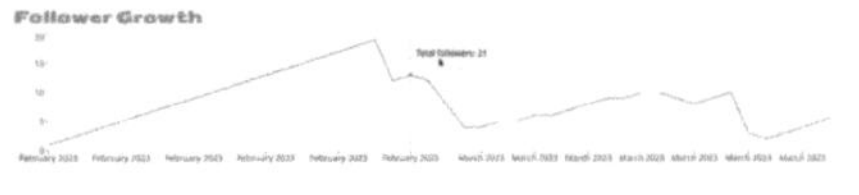

Figure 3. Decrease in the number of followers.

(1) Sort followers by their follow dates.
(2) Store data for the current window: formatted time, unique follower count, and cumulative follower count
(3) Return the result: an array containing data for each time window.

2.4 *Top keyword usage*

For tracking top keyword usage, a Trie data structure is employed for efficient keyword storage. The algorithm involves tokenizing and pre-processing text data, breaking it down into individual keywords. These keywords are then counted using the Trie data structure, facilitating a quick and space-efficient means of identifying and quantifying the frequency of specific terms in the overall dataset.

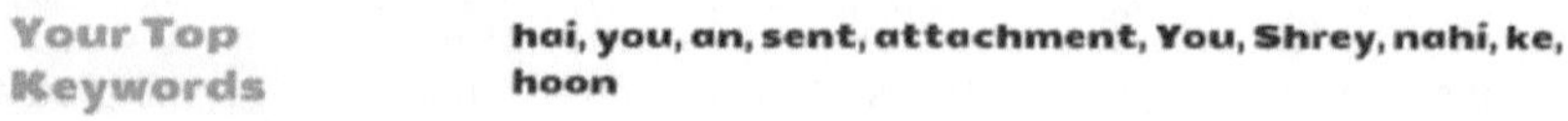

Figure 4. Top keywords for user 1.

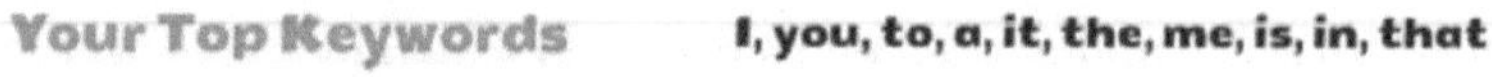

Figure 5. Top keywords for user 2.

Figures 4 and 5 shows the top keywords used by both the users frequently.

(1) Extract messages: Iterate through each chat and accumulate messages from the specified sender.
(2) Tokenize and count words: Split accumulated messages into words and count occurrences of each word.
(3) Sort by occurrence count: Sort words by their occurrence count in descending order.

2.5 *Inactive accounts*

Identifying inactive accounts involves using a linked list or a database to store account activity data. This algorithm is designed to define criteria for inactivity, such as a lack of posts or interactions over a specified period. Periodic scans of the stored data are performed to flag accounts that meet the defined criteria, allowing for the identification and management of inactive accounts. Figure 6 and Figure 7 gives different examples of 2 different Instagram accounts with the followers they have engaged/interacted least with.

Figure 6. List of least engaging accounts for user 1.

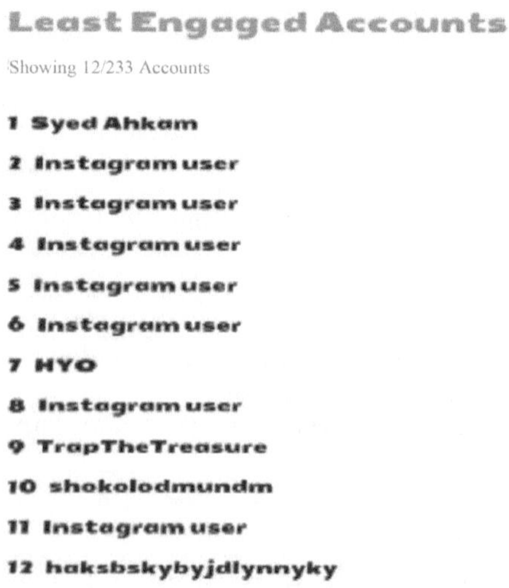

Figure 7. List of least engaging accounts for user 2.

2.6 *Post-specific analytics*

Post-Specific Analytics depends on putting away post information in a data set. The calculation investigates individual post measurements, including likes, remarks, offers, and commitment rate, to evaluate the general execution of each post. This approach gives a nitty gritty comprehension of the effect and reach of individual posts, helping content makers and help content creators in refining their procedures and streamlining future content (Malik, Berggren & Al-Busaidi, 2022).

2.7 *User engagement level assessment*

The assessment of User Engagement Level is a complex investigation, talking into account the length of user discussions and the number of times of remarks and likes on posts (Chang, 2016). This algorithmic exploration goes past a twofold engagement assessment, consolidating logical elements and behaviour subtleties. By taking into account the dimensions of user connections, the framework gives a nuanced assessment of engagement habits. The assurance of a fixation score is certainly not a inflexible yield yet a dynamic and versatile interaction, guaranteeing a customized comprehension of user behaviour in the application. This approach makes preparations for the traps of a one-layered assessment, perceiving the different ways users draw in with the application.

3 CONCLUSION

To conclude, Innergram is a promising and user friendly tool for social media analysis, specifically for Instagram. Innergram provide people with the power to comprehensively understand and enhance their social media presence. Highly valuing user security is evident through the commitment to data privacy, achieved by storing user-provided data temporarily. Innergram, despite being in the production phase, anticipates user adoption and feedback to enhance its offerings even further. Some potential future features includes recommendations and consensus building for annotations, highlight a commitment to continuous improvement. With a diverse target audience (Stavrakantonakis, Gagiu, Kasper, Toma & Thalhammer, 2012), from individuals to businesses and marketers, Innergram stands out by offering its valuable services entirely free of charge, making it an accessible and powerful tool for anyone seeking to elevate their Instagram analytics experience.

REFERENCES

Ahmed, W. (2018). Using social media data for research: An overview of tools. *Journal of Communication Technology*, 1(1), 77–94.

Batrinca, B., & Treleaven, P. C. (2015). Social media analytics: a survey of techniques, tools, and platforms. *Ai & Society*, 30, 89–116.

Berg, L., & Sterner, L. (2015). Marketing on Instagram: A qualitative study on how companies make use of Instagram as a marketing tool.

Chang, S. (2016). *Instagram post data analysis. arXiv preprint arXiv*:1610.02445.

Cornelio, G. S., & Roig, A. (2020). Mixed methods on Instagram research: Methodological challenges in data analysis and visualization. *Convergence*, 26(5–6), 1125–1143.

Killekar, O., Shah, H., & Kolge, A. (2013). Social media metrics, tools & analytics. In *PRIMA* (Vol. 3, No. 2, p. 35). Publishing India Group.

Liu, Y. (2010). Social media tools as a learning resource. *Journal of Educational Technology Development and Exchange (JETDE)*, 3(1), 8.

Malik, A., Berggren, W., & Al-Busaidi, A. S. (2022). Instagram as a research tool for examining tobacco-related content: A methodological review. *Technology in Society*, 70, 102008.

Maynard, D. G., & Bontcheva, K. (2016, May). Challenges of evaluating sentiment analysis tools on social media. In *Proceedings of the Tenth International Conference On Language Resources and Evaluation (LREC 2016)* (pp. 1142–1148). LREC

Nanda, P., & Kumar, V. (2021). Social media analytics: tools, techniques and present day practices. *International Journal of Services Operations and Informatics*, 11(4), 422–436.

Penni, J. (2017). The future of online social networks (OSN): A measurement analysis using social media tools and application. *Telematics and Informatics*, 34(5), 498–517.

Rogers, R. (2021). Visual media analysis for Instagram and other online platforms. *Big Data & Society*, 8(1), 20539517211022370.

Sapountzi, A., & Psannis, K. E. (2018). Social networking data analysis tools & challenges. *Future Generation Computer Systems*, 86, 893–913.

Schober, M. F., Pasek, J., Guggenheim, L., Lampe, C., & Conrad, F. G. (2016). Social media analyses for social measurement. *Public Opinion Quarterly*, 80(1), 180–211.

Smith, M. A. (2015). Catalyzing social media scholarship with open tools and data. *Journal of Contemporary Eastern Asia*, 14(2), 87–96.

Stavrakantonakis, I., Gagiu, A. E., Kasper, H., Toma, I., & Thalhammer, A. (2012). An approach for evaluation of social media monitoring tools. *Common Value Management*, 52(1), 52–64.

Young, L. E., Soliz, S., Xu, J. J., & Young, S. D. (2020). A review of social media analytic tools and their applications to evaluate activity and engagement in online sexual health interventions. *Preventive Medicine Reports*, 19, 101158.

Younis, E. M. (2015). Sentiment analysis and text mining for social media microblogs using open source tools: an empirical study. *International Journal of Computer Applications*, 112(5).

Computational Methods in Science and Technology Volume 1 – Kaur et al. (Eds)

Optimizing resource allocation in big data environments through a minimum-cost policy

Rohit Kumar Verma
PhD Research Scholar, Department of Computer Science, Himachal Pradesh University, Shimla, India

Sukhvir Singh
Assistant Professor, Department of Computer Science, Himachal Pradesh University, Shimla, India

ABSTRACT: When it comes to big data, the smartest thing is often allocating resources through cheapest channel. The latter involves assessing the costs and selecting the cheapest alternatives so as to make informed resource allocation decisions regarding, among others, processing capa bility, storage and network bandwidth. This way businesses can employ Big Data analytics using efficient algorithms for optimum use of resources. To solve the problem of achieving higher ac curacy in cloud computing environmentsfor information management system in terms of resource allocation, authors develop a minimum cost policy approach. It uses big data association mining techniques to come up with an allocation algorithm of resources for an information management system. In this Information Management System, Resource Distribution Data is pooled and re structured based on Keywords and Semantic Connection Features. In the subsequent context of cloud computing, the association rule features are extracted from information management system resources. For example, these derived association rules enable the internal resource allocation of the information management systems by creating a big data training set. The allocation process is enhanced even further with adaptive scheduling and big data association mining, enabling dy namic resource allocation control and optimization. The simulation results confirm the effective ness of this approach in terms of improved information management and resource utilization ef ficiency, which leads to interference mitigation and accurate resource allocation.

Keywords: Resource Allocation, Minimum Cost, Trust Management, Big Data

1 INTRODUCTION

In the modern era, a new emphasis has been placed on the management and optimization of re source allocation due to the use of Big Data. On performance and cost-efficiency, the employment of large-scale computing resources for developing Big Data applications is very significant (As sunção 2015). In this case problem can be resolved by minimum-cost approach resource alloca tion. Resource allocation in big data world is one of the important tasks that must be done with great care. This helps in managing the huge amounts of data through careful allocation of compute resources such as storage and memory. In terms of allocating resources this way, it is one of the most effective methods that can be used to achieve this objective i.e. minimizing costs. A mini mum cost policy is guided by systematic procedures aimed at reducing costs while providing sufficient resources for big data applications (Sharma 2024). It takes a variety of factors into account like resource provisioning costs, workload dynamics and data processing requirements among others. Organizations are able to balance access with cost effectiveness thus optimizing their big data processes when they implement a

DOI: 10.1201/9781003501244-55

cost-effective approach. This way could help to cut expenses as well as improve performance and scalability of big data applications under present conditions in its data-driven environment. Utilizing big data and cloud computing go hand in hand. Big data utilization enables users to take use of commodity computing to manage dispersed queries over several datasets and deliver fast result sets. The underlying engine is powered by cloud computing and utilizes the Hadoop platform, a type of platform for distributed data pro cessing (Modica *et al.* 2022). The usage of CC for big data applications is seen in Figure 1.

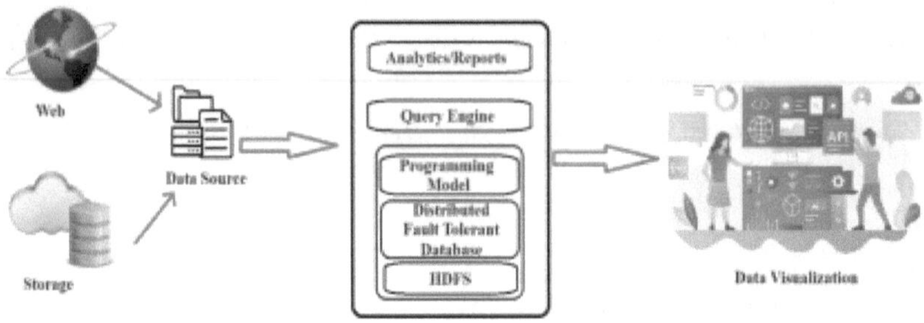

Figure 1. Usage of cloud computing in big data.

The focus has mostly been on making it easier to provision resources and services as needed to users and tenants, often dynamically and on demand (Sharma and Mangat 2015). However, there hasn't been as much focus on dealing with the dynamic changes in workloads and jobs, as well as application flows. The ambiguity around the precise resource required is one of the main obstacles in this respect. It's usually not known upfront, which makes it difficult to provision cloud services, whether it's reservation-based or on-demand. The challenge lies in the fact that the exact number of resources can't be set in advance due to the variability of workloads and usage patterns (Dargie *et al.* 2017; Hanani *et al.* 2017). Thus, in order to guarantee the lowest possible makespan, lowest possible prices, and lowest possible CPU utilization while minimizing resource consumption, large data services must be scheduled in a cloud-based environment.

1.1 *Background*

Managing and allocating resources to efficiently analyse and analyse massive data volumes has become crucial in the big data age. With the rapid increase in data volumes, resource allocation optimization in Big Data environments is getting a lot of attention. Resource allocation in the realm of Big Data is a pivotal challenge. It involves the proper allocation of computational resources, including processing and storage, to manage large datasets efficiently. One way to allo cate resources in this way is to have a minimum cost policy. Basically, a minimum cost policy is a systematic approach to reduce expenses while ensuring sufficient resources are available to meet the needs of Big Data apps (Liu *et al.* 2021; Wu *et al.* 2018).

2 WHY RESOURCE ALLOCATION IS A CONSTRAINT IN BIG DATA?

The allocation of resources in the context of Big Data (Ben Ayed *et al.* 2014; Hajjaji *et al.* 2021) is a limiting factor due to the following fundamental elements:

(1) Data Volume: The processing of massive amounts of data, which demands significant compu ting and storage resources, is referred to as "big data."

(2) Data Variety: Big Data is made up of different types of data, each of which needs different ways to be processed and analyzed.
(3) Data Velocity: The pace at which data is being added to Big Data systems is known as data velocity, and in order to keep up with changes, dynamic resource allocation is needed.
(4) Complex Analytics: Complex Analytics refers to the use of high-throughput algorithms and machine-learning models in big data analytics.
(5) Data Locality: In order to optimize performance, it is essential to allocate resources close to the data source.
(6) Dynamic Workloads: The demand for resources in a Big Data environment can fluctuate rap idly as a result of user inquiries and processing operations.
(7) Cost Efficiency: In order to effectively manage costs and avoid over/under provisioning, it is essential to allocate resources in an efficient manner.
(8) Resource Competition: The concept of resource competition involves the allocation of re sources to multiple users in a space that is shared by multiple tenants.

Big data scheduling architecture is composed of three distinct components: a data node, a name node, and a user. A big data computing unit is referred to as a name node (nm) and consists of multiple data nodes (dn). Data nodes are the computational hardware processors that receive the task when it is requested from the user. To accomplish a task, the data nodes must not be over utilized or classified as under-utilized systems in order to maintain a reasonable load on the data nodes over time. The load is managed by the name node, which also shifts the burden from one data node to another. This entire process consumes a lot of power which is defined by eq (1) and eq (2) as follows:

$$Pc = \int_{i=1}^{n} \int_{t=0}^{T} P \ tdt \tag{1}$$

$$Pc = \int_{i=1}^{itr} \oint_{j=1}^{data_{node_{count}}} \int_{k=1}^{user\_count} Pc_{storage} + Pc_{processing} + Pc_{migration} \tag{2}$$

3 RESOURCE ALLOCATION BY TRUST MANAGEMENT

There are a number of trust issues when aggregating and analyzing users' data because users are mobile in big data environments. When users request services, sometimes they are not receiving services that meet their needs, which is a breach of service level agreements between service providers and users. In any event, if the service provider does not adhere to the minimum require ments specified in the agreement, the penalty shall be imposed on the user in accordance with the terms of the agreement. Service Level Agreements (SLAs) are contracts that, in the case of a loss, obligate the company to pay as specified in the agreement, much like insurance plans. The name node is in charge of moving the load from the single data node to the other data node in the context of large data settings. Every time a user is moved, a certain amount of computational time is added to the total completion time, resulting in an SLA violation.

4 COMPARATIVE STUDY

In practice, the optimization method to use within GRA-BD will depend on the nature of the challenge, the trade-offs that need to be made between exploration and utilization, and the capac ity to adjust to dynamic workloads and resource limitations in dynamic big data environments. Hybrid approaches that leverage the power of these algorithms can also be considered to strike a balance between resource efficiency and flexibility in dynamic Big data environments. The review delves further into different optimization approaches in the context of cloud compu ting for big data. Numerous studies as mentioned in (Gibet *et al.* 2018; Manivannane and Chid ambaram 2019; Rjoub *et al.* 2020) have been conducted to address the intricate issues of resource scheduling for big data in cloud systems. Accordingly, recent research contributions have been taken into account, including the work of S. Talwani *et al.* 2022 (Talwani *et al.* 2022). Their study presented a solution to cloud computing that optimizes virtual machine migration and allo cation through machine learning. By using well-known machine learning frameworks and lan guages, the study showed increased efficiency and accuracy in virtual machine administration decision-making. Recognizing the difficulties in using representative data for model training and the possible effects of changing workload patterns, the study emphasized improvements in virtual machine management performance. Furthermore, the research conducted by (Liu *et al.* 2013; Wang *et al.* 2023) adds valuable insights into the optimization methodologies used in big-data cloud computing.

Criteria	Simulated Annealing	Artificial Bee Colony	Genetic Algorithms	Firefly Algorithm
Search Space Exploration	Good	Moderate	Excellent	Good
Adaptability	Good	Excellent	Moderate	Excellent
Complexity	Moderate	Moderate	High	Moderate
Convergence Speed	Moderate	Good	Moderate	Good
Dynamic Adaptation	Possible	Yes	Possible	Yes
Parameter Tuning	Required	Required	Required	Required

Figure 2. Algorithmic comparison.

Together, these efforts add to the thorough examination and analysis of optimization strategies for the effective deployment and management of various applications in cloud settings. Many academics from around the world have provided insightful references for studies on data resource allocation and economic management information systems. (Liu *et al.* 2013) developed an improved ABC technique that used chaotic mapping with enhanced ergodicity during initialization and sufficient population unpredictability. Better convergence was achieved in the scout bee phase with the updated search equation. PSO inspired one of the variants, the gbest-guided ABC (GABC) algorithm, to improve the exploitation of ABC by using the global best solution (gBest).

By including the gbest factor in ABC's solution search equation, exploitation is enhanced while maintaining the balance between solution space exploration and exploitation. The authors Buyya *et al.* 2010 explore a number of difficulties and open questions around resource distribution in cloud computing settings. They highlight how much energy is used to host consumer apps on the cloud and offer a plan for managing these environments in an energy-efficient manner. They employ dynamic resource allocation and provisioning algorithms that consider the interdependencies between different types of data center infrastructures.

An analytical model centered on virtual machine migration is presented in (Kochut *et al.* 2007), with the goal of enhancing reaction times through well-informed migration decisions.

This model considers the features of infrastructure virtualization as well as the predictability of resource re quests. The authors ran simulations utilizing traces of data center resource utilization to verify the efficacy of their model.

5 EVALUATION PARAMETERS

The process of evaluation involves a critical analysis of the services that the service providers offer. In this kind of assessment procedure, parameters are crucial. In order to assist the users, they evaluate the services and choose the best option, with the suppliers providing the best possi ble quality of service. There are two categories of evaluation parameters: quantitative and quali tative.

5.1 *Quantitative parameters*

Parameters, which can be made up on the basis of countability/quantity will be termed as Quan titative parameters. These quantitative factors are measured in machine learning using a perfor mance measurement tool called the confusion matrix. The confusion matrix illustrates the classi fication model's accuracy by evaluating model performance, spotting misclassifications, and enhancing predicted accuracy. The confusion matrix displays the quantity of true positives, true negatives, false positives, and false negatives.

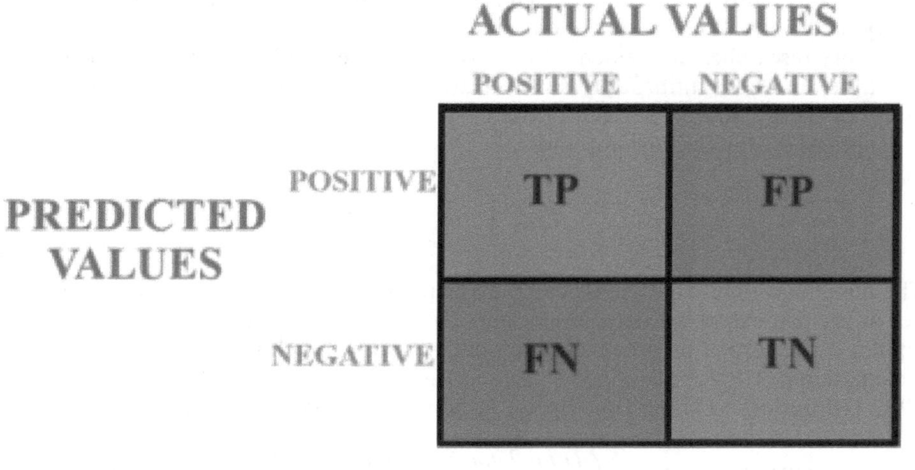

Figure 3. Confusion matrix.

True positives are the number of samples that were correctly anticipated to be positive. True neg atives are the number of samples that were correctly predicted to be negative. A false positive is the quantity of samples that were incorrectly projected as positive. False negatives are among the samples that were incorrectly predicted to be negative. Confusion matrix may be used to compute many metrics, such as F1 score, recall, accuracy, and precision.

The accuracy measure shows the percentage of correctly anticipated situations with successful outcomes. Calculation of Precision.

$$P = \frac{TP}{TP + FP} \tag{3}$$

The percentage of actual positive cases that proposed model correctly predicted is known as recall. Calculation of Recall:

$$R = \frac{TP}{TP + FN} \tag{4}$$

Accuracy indicates how many of observed predictions came true out of all of them. Calculation of Accuracy:

$$A = \frac{TP + TN}{TP + TN + FN + FP} \tag{5}$$

F1 score is a combination of recall and precision. The efficiency with which proposed models achieve that trade-off between recall and precision can be determined using F1 score. Calculation of F1 score:

$$F1 = 2 * \frac{Precision \times Recall}{precision + Recall} \tag{6}$$

5.2 *Qualitative parameters*

The subjective quality of anything, which is usually based on user experience, such as service usability, is the basis for qualitative criteria.

For every researcher, it is essential to understand the quality of service—such as power usage, time, and CPU utilization is crucial for employing qualitative methodologies. The quantity of energy utilized per unit of time is known as power consumption.

Calculation of Power Consumption:

$$Pc = \int_{i=1}^{n} \int_{t=0}^{T} P \ tdt \tag{7}$$

The maximum time a data node can take to perform a job is indicated by the time parameter in the context of big data processing. The phrase "CPU utilization" refers to how a computer makes use of its processing capability, or how much work a CPU can do.

Calculation of CPU Utilization:

CPU Utilization = (Total Time Spent on Non-Idle Tasks / Total Time) x 100

$$CPU\ Utilization = \left(\frac{Total\ Time\ Spent\ on\ Non - Idle\ Tasks}{Total\ Time}\right) * 100 \tag{8}$$

6 CONCLUSION

To sum up, the big data computing era of today provides an abundance of resources that are customized to meet the demands of users. Efficient resource management is critical in many do mains, highlighting the need of optimization. Using the minimum cost policy turns out to be a wise move, especially in situations where there isn't a conventional resource allocation mecha nism in place. This paper introduces and evaluates the MCMF strategy, a complex approach to address dynamic resource allocation issues in cloud computing environments. For cloud service providers, dynamic pricing adjustments ensure profitability and sustainable revenue development by using a basic autoregressive algorithm for predicting changing needs. This thorough investi gation emphasizes how crucial it is to employ advanced resource management techniques in order to successfully negotiate the complexity of contemporary computing environments.

REFERENCES

Assunção, M. D., Calheiros, R. N., Bianchi, S., Netto, M. A. S. and Buyya, R. (2015). Big Data computing and clouds: Trends and future directions. *Journal of Parallel and Distributed Computing* 79–80: 3(15). https://doi.org/10.1016/J.JPDC.2014.08.003

Ben Ayed, A., Ben Halima, M. and Alimi, A. M. (2014). Survey on clustering methods: Towards fuzzy clustering for big data. *6th International Conference on Soft Computing and Pattern Recognition, SoCPaR* 2014: 331–336. https://doi.org/10.1109/SOCPAR.2014.7008028

Buyya, R., Beloglazov, A. and Abawajy, J. (2010). Energy-efficient management of data center resources for cloud computing: A vision, architecture elements, and open challenges. *In: Proceedings of the 2010 International Conference on Parallel and Distributed Processing Techniques and Applications (PDPTA 2010)*. p. 12.

Dargie, W. and Eichhorn, F. (2017). Dynamic power management in data centres. *Advances in Computer Com munications and Networks From Green, Mobile, Pervasive Networking to Big Data Computing*: 37–62. https://doi.org/10.1201/9781003337096-3

Di Modica, G. and Tomarchio, O. (2022). A hierarchical hadoop framework to process geo-distributed big data. *Big Data and Cognitive Computing* 6(5): 6, 5. https://doi.org/10.3390/BDCC6010005

GIBET TANI, H., EL Amrani, C. (2018). Smarter round robin scheduling algorithm for cloud computing and big data. *Journal of Data Mining & Digital Humanities*. https://doi.org/10.46298/jdmdh.3104

Hajjaji, Y., Boulila, W., Farah, I. R., Romdhani, I. and Hussain, A. (2021). Big data and IoT-based applications in smart environments: A systematic review. *Computer Science Review* 39: 100318. https://doi.org/10.1016/J.COSREV.2020.100318

Hanani, A., Rahmani, A. M. and Sahafi, A. (2017). A multi-parameter scheduling method of dynamic workloads for big data calculation in cloud computing. *Journal of Supercomputing* **73**: 4796–4822. https://doi.org/10.1007/S11227-017-2050-6/TABLES/8

Kochut, A. and Beaty, K. (2007). On strategies for dynamic resource management in virtualized server environments. *In: 2007 15th International Symposium on Modeling, Analysis, and Simulation of Computer and Telecommunication Systems*. Istanbul, Turkey: 193–200.

Liu, B., Li, J., Lin, W., Bai, W., Li, P. and Gao, Q. (2021). K-PSO: An improved PSO-based container scheduling algorithm for big data applications. *International Journal of Network Management* **31**: e2092. https://doi.org/10.1002/NEM.2092

Liu, H., Gao, L., Kong, X. and Zheng, S. (2013). An improved artificial bee colony algorithm. *In: 25th Chinese Control and Decision Conference (CCDC)*. Guiyang, China: 401–404.

Manivannane, K. and Chidambaram, M. (2019). A cloud resource scheduling framework for big data stream and analytics in cloud environment. *Proceedings of the 4th International Conference on Communication and Electronics Systems*, ICCES (2019): 1638–1642. https://doi.org/10.1109/ICCES45898.2019.9002225

Rjoub, G., Bentahar, J. and Wahab, O. A. (2020). BigTrustScheduling: Trust-aware big data task scheduling ap proach in cloud computing environments. *Future Generation Computer Systems* **110**: 1079–1097. https://doi.org/10.1016/j.future.2019.11.019

Sharma, M., Pandey, P. and Mital, S. (2024). Allocating resources in big data using a minimum-cost policy, *Recent Applications in Engineering and Technology*, **4**(2): 14–19.

Sharma, S. and Mangat, V. (2015). Technology and trends to handle big data: Survey. *International Conference on Advanced Computing and Communication Technologies*, ACCT 2015-April: 266–271. https://doi.org/10.1109/ACCT.2015.121

Talwani, S., Singla, J., Mathur, G., Malik, N., Jhanjhi, N. Z., Masud, M. and Aljahdali, S. (2022). Machine learning-based approach for virtual machine allocation and migration. *Electronics* **11**: 3249

Wang, X., Hu, X., Fan, W., Wang and R. (2023). Efficient data persistence and data division for distributed computing in cloud data center networks. *The Journal of Supercomputing* 79: 16300–16327. https://doi.org/10.1007/S11227-023-05276-2

Wu, W. T., Lin, W. W., Hsu, C. H. and He, L. G. (2018). Energy-efficient hadoop for big data analytics and com puting: A systematic review and research insights. *Future Generation Computer Systems* **86**: 1351–1367. https://doi.org/10.1016/J.FUTURE.2017.11.010

Computational Methods in Science and Technology Volume 1 – Kaur et al. (Eds)
 ISBN: 978-1-032-81760-6

ResNet-50: A breakthrough in plant disease detection accuracy for agricultural empowerment

Shadab Khan, Anil Kumar Yadav & Shiv Shankar Prasad Shukla
School of Computing Science and Engineering, VIT Bhopal University, Bhopal-Indore Highway, Kothrikalan, Sehore Madhya Pradesh

Vikas Shende
Madhya Pradesh Council of Science & Technology, MPCST, Bhopal Madhya Pradesh

ABSTRACT: Agriculture is integral to the economy of every nation, providing essential crops for consumption. Identifying plant diseases is imperative for sustaining agricultural growth. Early detection of these diseases is vital to ensure a thriving and productive agricultural sector, preventing both financial losses and resource depletion. Crop farmers experience significant setbacks annually due to the impact of various diseases on their plants. Deep learning technology presents a promising solution to mitigate crop failure by enabling timely detection of diseases in plant leaves. In this research, the performance of Convolution Neural Network (CNN), Visual Geometry Group (VGG-16), DenseNet-121, and Residual Network (ResNet-50) models is evaluated using a new plant disease dataset containing 87,000 RGB images to identify crop infections. The accuracy rates obtained were 94.60% for CNN, 92.49% for VGG-16, 95.73% for DenseNet, and 98.67% for ResNet-50, demonstrating ResNet-50 superior performance. This application aims to aid farmers in diagnosing plant diseases simply looking at images of plant leave. Leveraging the ResNet-50 transfer learning model, the application distinguishes between both diseased and healthy leaves, accurately identifying the specific disease present. The objective is to empower farmers to conserve resources and minimize economic losses through early disease detection and the implementation of appropriate treatment measures.

Keywords: Plant Disease Identification, Resnet-50, Convolutional Neural Networks, New Plant Disease Dataset, Deep Learning Technology

1 INTRODUCTION

Agriculture stands as a cornerstone of national economies, ensuring the production of vital crops for sustenance. The concurrent agricultural and industrial revolutions underscore its importance [1]. However, the challenge of food insecurity, primarily driven by crop diseases, poses a significant threat to global food security [2]. Plant diseases not only endanger food production but also have far-reaching implications for economies and food safety. Crop health, therefore, is paramount for ensuring both economic stability and food security, with the condition of leaves serving as a key indicator of crop health. Hence, understanding various plant diseases through the study of leaf symptoms becomes essential [3]. Diseases affecting vegetable plants, such as potatoes, tomatoes, or peppers, can lead to substantial financial losses for farmers each year. For instance, blight, whether early or late, caused by fungi or specific bacteria, respectively, poses significant threats to crops. Early identification of these diseases and timely intervention can mitigate waste and expenses, especially given the projected global population surpassing 9 billion in the next 25 years, necessitating a 70%

 DOI: 10.1201/9781003501244-56

rise in food output [4]. Crop diseases pose frightening challenges in many countries, particularly agronomic ones, with potato diseases being particularly alarming due to potatoes' widespread consumption. However, diseases affecting peppers and tomatoes are also of considerable worry. To battle potato crop illnesses, agricultural scientists have been investigating deep learning with the belief that "It's better to prevent than to cure." Advancements in computing have made real-time image processing and machine learning techniques increasingly promising.

Deep learning technology holds immense significance and popularity for its contributions to enhancing daily life and its potential for future advancements [5]. The quality and yield of crops, including leaves, fruits, stems, and roots, are significantly plant-related diseases, leading to global shortages in vegetable consumption [6]. Crop diseases result in a 16% annual reduction in crop yield, underscoring the urgency of implementing innovative solutions like smart farming, which leverages deep learning for novel algorithms and equipment in agriculture [7]. Machine learning addresses critical agricultural challenges, including feature extraction, transformation, identification of patterns, and image classification [8]. In this research, the development of a convolutional neural network (CNN)-based approach is used to detect crop infections and identify disease types. By implementing this system in real-time, we aim to mitigate crop yield losses.

In India, farmers have historically been reliant on their personal expertise and knowledge to identify agricultural diseases due to limited access to agricultural experts, particularly in countryside regions. Seeking assistance from specialists is often impractical, as it can consume significant time, labor, and financial resources, thereby impeding widespread utilization, especially in developing countries. Addressing these obstacles has prompted extensive research into artificial intelligence (AI'S) the development. The disease detection procedures expedite by Machine learning and deep learning methodologies, thus mitigating crop damage to a considerable extent. For instance, [9] proposes a CNN-based model achieving a 98.023% accuracy in leaf disease detection. [10] Designed custom 15-layer CNN architecture for leaf disease detection, though with a lower accuracy of 93%. [11] Investigated using ResNet152 for deep feature extraction from corn plants using EfficientNetB0 and DenseNet121. InceptionV3, EfficientNetB0, and DenseNet121 achieving accuracies of 98.56%, 98.37%, 96.26%, 97.91%, and 97.82%, respectively, in classifying diseases. Inspired by these works, we propose a dense-optimized CNN for classifying various categories of corn leaves from the New Plant Disease dataset, achieving an accuracy of 98.06% [12]. Moreover, [13] developed a Model for detecting diseases in potato leaves using different deep learning architectures like VGG-16, DenseNet-121, and ResNet-50, coupled with Utilizing algorithms f machine learning (ML) such as support vector machine (SVM), CNN(convolutional neural network), K-Nearest Neighbour and Logistic Regression resulted in an accuracy of 97.8% for logistic regression and Visual Geometry Group(VGG19) and. Similarly, [14] they introduced a framework for classifying plant leaf diseases employing a molecule swarm optimization algorithm, achieving accuracy of 98.00%. However, most of these studies employed default deep learning models without tuning, resulting in suboptimal accuracies, highlighting the need for further research and improvement. Additionally, providing preliminary treatment recommendations can significantly reduce crop losses. This study proposes to address these challenges by developing a highly accurate deep learning-based detection framework and offering initial treatment suggestions to farmers.

This study explores the application of ResNet-50 for disease identification in plants, highlighting its advantages over traditional methods and its potential to significantly enhance agricultural sustainability and efficiency. Through empirical evaluations and case studies, it demonstrate the efficacy of ResNet in accurately diagnosing plant diseases and discuss its implications for future research and practical applications in agriculture.

The structure of the article proceeds in the manner described below: Within Section 2, delve into a thorough analysis of the body of extant research. The approach used for this research project is described in detail in Section 3. The subsequent section, Section 4, delves into the experiments conducted, accompanied by detailed analysis and discussions. Concluding remarks are presented in Section 5. Finally, Section 6 encompasses the reference list.

2 RELATED WORKS

This section examines relevant research as well as the research on deep learning-based detection of leaf disease. The use of AI has significantly increased recently, especially in the healthcare [15] and agriculture, with a notable emphasis on imaging applications [16,17]. Sanjiv Sannakki *et al.* (2013) [18] integrated methods for image processing and artificial intelligence to assist in diagnosis, whereas Monika Jhuria *et al.* (2013) [19] utilized image processing for fruit grading and disease identification. Kaiyi Wang *et al.* (2023) [20] showcased the application of artificial neural networks in categorizing diseases, such as plant illnesses and insect infestations, utilizing image processing and computer vision methods, notably through images captured by smart phones. Several approaches and models have been suggested and evaluated for identifying plant diseases using machine learning methods. [21] Concentrated on identifying leaves that are unhealthy or healthy through the utilization of image processing techniques and machine learning strategies, attaining high accuracy with the use of methods such as Support Vector Machine (SVM) classification and Grey Level Co Occurrence Matrix (GLCM) feature extraction. Similarly, [22] outlined a machine learning-based method that used decision tree approaches to detect rice leaf diseases with above 97% accuracy. [23] Emphasized detecting outliers in wheat leaves, addressing challenges in wheat disease identification through machine learning approaches. [24] Suggested the use of Convolutional Neural Networks (CNNs) to identify plant leaf diseases, achieving significant accuracy improvements. [25] Explored machine learning under supervision algorithms for identifying plant diseases in maize, with Random Forest (RF) exhibiting the highest accuracy. Image processing methods were applied to determine the illnesses of plant leaves. In [26] image analysis classification algorithms were employed for disease recognition and categorization. Deep learning, a significant advancement in machine learning, has revolutionized various applications, including computer vision. [27] Recommending making use of open-source algorithms and categorizing images for detecting diseases on plant leaves, focusing on creating a reliable system for disease identification. Similarly, research by Durmuş *et al.* (2017) [28] planned to use deep learning algorithms to recognize different illnesses in the leaves of tomato plants, achieving real-time detection capabilities. Despite the progress, several challenges remain; challenges include the necessity of manually crafting features in conventional machine learning models, susceptibility to over fitting with small datasets, and limited applicability. Moreover, many studies lack robust experimentation with large datasets and fail to provide practical applications for real-world disease prediction.

3 METHODOLOGY

Plant disease detection accuracy and efficiency are intended to be improved by the suggested methodology, which combines CNN and Residual Networks (ResNet) – Deep Learning architecture. Initially, when compared to a 20-layer CNN design, CNN's application in identifying plant leaf disease showed greater error rates. Upon investigation, this discrepancy was attributed to the vanishing/exploding gradient phenomenon. To overcome this challenge, the ResNet-50 architecture was considered due to its capability to mitigate issues related to the vanishing/exploding gradient. ResNet employs skip connections, which connect certain layers of the network while bypassing others. This unique approach allows for the creation of residual blocks, which are then stacked together to form ResNet. Unlike traditional networks, ResNet focus on learning the differences between layers rather than independently learning the mapping, enabling the training of deeper networks without encountering gradient-related problems.

In this framework the following Deep learning models are applied.

3.1 *VGG-16*

With its uniform design and 16 convolutional layers, VGGNet-16 is one of the most often used structures for disease identification in picture categorization. The main achievement of VGG-16 is increased Network strength and enhanced system capabilities. The algorithm for VGG-16 transfer learning consists of three parts: pooling layers, fully linked layers, and convolution layers.

3.2 *DenseNet-121*

Convolutional neural networks called DenseNet [22] have feed-forward coupling between every layer of the network. The combination of every prior layer's feature map serves as a layer's input. Specifically, in the event of an L-layered network. Given that every ConvNet makes extensive use of 3×3 filters, it is a highly well-liked photo prediction model [7].

3.3 *ResNet50*

The ResNet family's subset [39–50], the ResNet50 model is composed of 48 convolutional layers, 1 MaxPool layer, and 1 Average Pool layer. This is a popular ResNet framework for image classification. The ResNet50 design is divided into four primary phases. The first convolution stage it composed of three layers: a 1*1, 64 kernel, a 3X 3, 64 kernel, 1* 1,256 kernel in the end. There have been three replications of these three layers, making nine layers overall. Next, a 1*1,128 kernel is observed, followed by a 3*3,128 kernel and finally a 1*1,512 kernel. For the whole of twelve layers, this approach has been employed four times. After that two more kernels with 3*3,256 and 1*1, 1024; this is repeated si* cycles to provide 18 layers. The next kernel is 1*1, 256. Two further kernels, measuring 3*3,512 and 1*1, 2048, were then constructed by a 1*1,512 kernel. Using this procedure three times gave us nine layers in total.

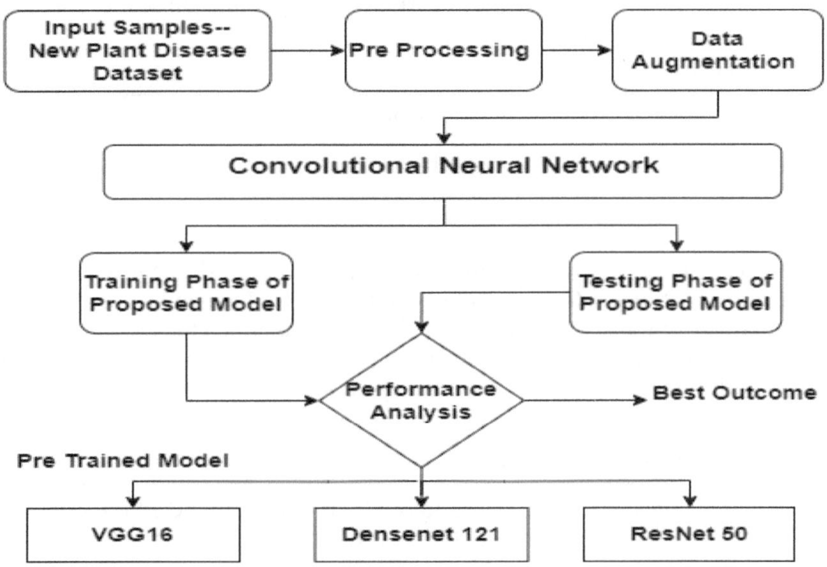

Figure 1. Depicts suggested model's general flow diagram representation.

The suggested solution uses a Raspberry Pi with a camera module to take pictures of plant leaves. These images undergo preprocessing to identify diseased leaves and classify them according to the specific disease present. Prior to training, the dataset is meticulously prepared by segregating images of diseased plants from healthy ones, facilitating more effective learning.

Next, subsets of the dataset are separated for testing and training, with the former comprising 80% of the data and the latter 20%. This section enables the performance of the trained model to be validated. This section validates the trained model's performance. The model, which forms the basis of the illness detection system, is then trained using the training data. When a new image of a plant leaf is captured using the Raspberry Pi, the trained model is employed alongside the CNN algorithm to determine if the leaf is sick and, if so, to identify the specific disease. It is anticipated that the system will achieve an accuracy rate of approximately 90% in disease detection. Overall, this suggested methodology utilizes advanced deep learning techniques and practical implementation using Raspberry Pi to improve plant disease detection efficiency and accuracy. By integrating CNN with ResNet architecture, the system intends to reduce challenges related with gradient issues and attain reliable disease recognition in real-world agricultural environment.

4 RESULTS AND DISCUSSIONS

For experimental setup Windows 10 workstation is used with an NVIDIA GeForce GTX graphics card. This workstation had an 8 GB RAM, Core i5 9th generation processor, and GDDR5 graphics memory. The software is implemented on Anaconda3 platform, leveraging libraries including Keras, OpenCV, NumPy, CuDNN, and Theano.CuDNN and CUMeM are specialized libraries developed by NVIDIA, specifically designed for optimizing memory efficiency and improving execution speed in deep learning applications. These libraries seamlessly integrate with the Theano backend. Additionally, OpenCV played a crucial role in this project development, providing support for both academic and commercial projects throughout a range of operating systems, such as Windows, Linux, and Mac OS. It also interfaces with popular programming languages such as Python, Java, iOS, and Android. Throughout the experiments, the evaluation of both training and testing accuracy. Making use of the matrix confusion displayed in Table 1, various deep learning models are investigated and assessed in terms of their work according to the subsequent standards.

Table 1. Matrix of confusion.

	True Positive	True Negative
Positive Prediction	TP	FP
Negative Prediction	FN	TN

In this instance, TP stands for True Positive, meaning that both the leaf's disease and the model's prediction of it were affected; True Negative (TN): The model ascertained leaf was healthy and unaffected; False Negative (FN): The model assumed the leaf was healthy but in reality, it was affected; and False Positive (FP): The model determined that the leaf was diseased but the instance was healthy. Accuracy shown in equation 1 used to express the percentage of test situations for which a correct forecast was made.

$$Accuracy = TP + TN/TP + FP + FN + TN \tag{1}$$

The equation 2 denotes Precision, the ratio of all favorably predicted leaves by the model to all accurately predicted disease-affected leaves. It has the following definition:

$$Precision = TP/TP + FP \tag{2}$$

In order to increase the CNN's learning speed utilizing transfer learning model uses the New Plant Disease dataset to train the models for study. For investigation purpose, the pre-trained models ResNet-50, VGG-16, and DenseNet-121 were chosen. Using offline augmentation techniques, the original dataset, which is available through a GitHub repository, was transformed into the dataset. This new collection, which is divided into 38 categories, has about 87,000 RGB photos that show both wholesome and damaged crop leaves. Preserving the initial directory hierarchy, the complete dataset has been separated into sets for training and validation using an 80/20 ratio. For prediction purposes, an additional directory with 33 test photos has been made.

Table 2. Accuracy results of experiment.

Model Name	Testing Accuracy	Precision
Simple CNN	94.60%	96.83%
VGG 16	92.49%	93.55%
DenseNet 121	95.73%	96.42%
ResNet 50	98.67%	98.96%

The summary results taking accuracy, precision, and into consideration are shown in Table 2. The chart also shows that, for any given model, ResNet50 outperforms all other models. In order to more accurately validate the experiment using ResNet50, a 5-fold cross-validation method on every experiment involving the ResNet50 model. Consequently, in such scenario, training examples use 80% of the data, while testing uses the remaining 20%. Accuracy values obtained were 94.60%, 92.49%, 95.73%, and 98.67%.Several deep transfer learning models have been proposed in this study for crop leaf detection. The suggested method for identifying multi-crop leaves is tested on large data sets. Three popular models for transfer learning VGG-16, DenseNet-121, and ResNet50—were tested to see how well they could accurately diagnose leaf diseases. Based on this analysis, ResNet50 performs better than the other two models. Though widely used image classification models, VGG-16 and DenseNet-121 find it difficult to deal with the problem of disappearing gradients. Their accuracy levels consequently fall short of allowable bounds. ResNet50, on the other hand, effectively resolves this problem and dramatically lowers testing mistakes in experimental circumstances. As a result, ResNet50 outperforms the previously described models in terms of performance.

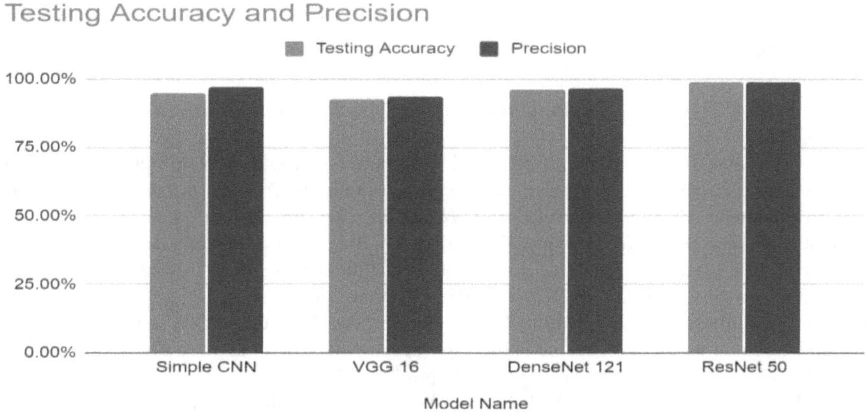

Figure 2. Comparison result.

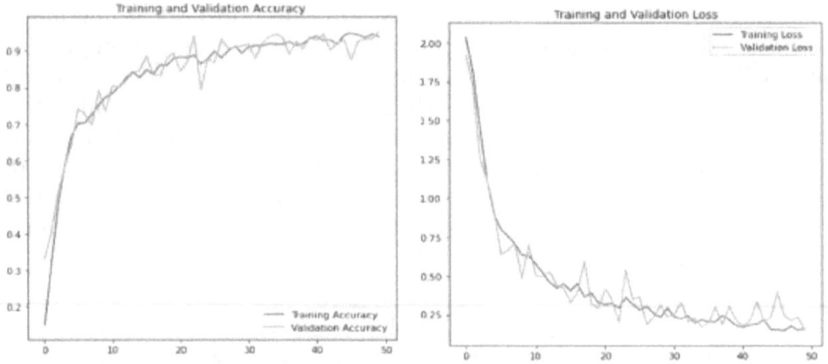

(a) Accuracy curve vs epoch.

(b) Loss curve vs epoch.

Figure 3. Accuracy and loss graph.

5 CONCLUSION

In global agriculture, identifying and classifying plant and leaf diseases poses significant hurdles. In order to accurately identify leaf illnesses and their causes, this study used architecture of a Convolutional Neural Network (CNN). Evaluation of several models using the New Plant Disease dataset is done, which is accessible to the public, including CNN, VGG-16, DenseNet-121, and ResNet-50 architectures. The investigations yielded encouraging outcomes, yields 98.60%, 92.49%, 95.73%, and 98.67% accuracy evaluations for the models CNN, VGG-16, DenseNet-121, and ResNet-50, in that sequence of particular notice was the ResNet-50 model's improved performance, which demonstrated improved accuracy in the effective identification of leaf diseases. Based on these findings, the combination of ResNet-50 model with an online tool is made to expedite the detection of pathogens in plant leaves, which improves the field of intelligent farming systems. This approach demonstrated improved efficacy in estimating symptom severity compared to prior studies on plant leaf disease detection. Overall the investigation underscores the efficacy of deep learning models, particularly ResNet-50, in precisely identifying and categorizing plant diseases, presenting implications for agriculture and crop management practices.

Funding: This paper is funded by the Madhya Pradesh Council of Science & Technology (MPCST), R&D Projects entitled "Multipurpose A1- powered drone for agricultural and safety measures in remote area" (File No. A/RD/RP-2/339/31.03.2023).

REFERENCES

[1] Bidwell, P. W. (1921). The Agricultural revolution in New England. *The American Historical Review 26 (4)*:683–702.

[2] Husin, Z. B., Shakaff, A. Y. B. M., Aziz, A. H. B. A. and Farook, R. B. S. M. (2012). Feasibility study on plant disease detection using image processing techniques. *Third International Conference on Intelligent Systems Modeling and Simulation, IEEE*: pp. 291–296.

[3] Talukder, M. A., Islam, M. M. and Pramanik, S. (2023). An Efficient Deep Learning Model to Categorize Brain Tumor Using reconstruction and Fine-Tuning expert systems with applications, *Expert Systems with Applications: Article 120534* Volume 230.

[4] Ahmed, N. R., Islam, M. M., Uddin, M. A., Akhter, A., Talukder, M. A. and Paul, B. K. (2021). Machine learning based diabetes prediction and development of smart web application. *International Journal of Cognitive Computing in Engineering*: pp. 229–241

[5] Uddin, M. A., Islam M. M., M. A. Talukder, Aryal, S. and Muntaha, M. (2023). Machine Learning Based Diabetes Detection Model for False Negative Reduction. *Biomedical Materials & Devices* 2(4):3: pp. 1–17

[6] Bah, M. D. A. Hafiane and R. Canals. (2018). Deep learning with unsupervised data labeling for weed detection in line crops in uav images. *Rem. Sens.*, 10 (11): p. 1690

[7] Talukder, M. A., Islam, M. M., Uddin, Akhter, M. A. A., Hasan, K. F. and Moni, M. A. (2022). Machine learning based lung and colon cancer detection using deep feature extraction. *Expert Systems with Applications*: pp. 1–36

[8] Talukder, M. A. Hasan, K. F. Islam, M. M. Uddin, M. A., Yousuf, M. A. and Moni, M. A. (2023). A dependable hybrid model for network intrusion detection. *Journal of Information Security and Applications* : pp. 1–44

[9] Jasim, M. A. and Al-Tuwaijari, J. M. (2020). Plant leaf diseases detection and classification using image processing and deep learning techniques. *International Conference on Software Testing IEEE*: pp. 259–265

[10] Pawar, S., Shedge, S., Panigrahi, N. W and Sayyad, S. (2022). Leaf disease detection of mutiple plants using deep learning. *International Conference on Machine Learning, Big Data, Cloud and Parallel Computing (COM-IT-CON), IEEE*: pp. 241–245

[11] Amin, H., Darwish, A., Hassanien, A. E. and M. Soliman. (2022). End-to-end deep learning model for corn leaf disease classification. *IEEE Access*, 10 : pp. 31103–31115.

[12] Waheed, A., Goyal, M., Gupta, D., Khanna, A., Hassanien, A. E. and Pandey, H. M. (2020). An optimized dense convolutional model for disease recognition and classification in corn leaf. *Comput. Electron. Agric., 175: Article* 105456

[13] Tiwari, D., Ashish, M., Gangwar, N., Sharma, A., Patel, S. and Bhardwaj, S. (2020. Potato leaf diseases detection using deep learning *4th (ICICCS), IEEE* : pp. 461–466

[14] Singh, V. (2019). Sunflower leaf diseases detection using image segmentation based on particle swarm optimization. *Artif. Intell. Agric.*: pp. 62–68

[15] Ahamed, M., Uddin, K., Islam, M. M., Uddin, Akhter, M. A., Acharjee, U. K. and Paul, B. K. (2023). An improved Resnet architecture to classify normal and conventional pneumonia cases from covid-19 instances utilizing chest x-ray images *.Diag nostics*, 13 (3) : p. 551

[16] Ahamed, K. U., Islam, M., Uddin, A., Akhter, A., Paul, B. K., Yousuf, M. A., Uddin, S. and Moni, M. A. A deep learning approach using effective preprocessing techniques to detect covid-19 from chest ct-scan and x-ray images Comput. *Biol. Med.: Article* 105014

[17] Ahmad, A., Saraswat, D. and El Gamal A. (2022). A survey on using DL techniques for plant disease diagnosis and recommendations for development of appropriate tools Smart Agric. *Technol. Article* 100083.

[18] Sannakki, S. S., Rajpurohit, V. S., Nargund, V. and Kulkarni, P. (2013). Diagnosis and classification of grape leaf diseases using neural networks *IEEE (2013)*: pp. 1–5.

[19] Jhuria M., Kumar A. and Borse R. (2013). Image processing for smart farming: detection of disease and fruit grading *IEEE Second (ICIIP-2013) IEEE.*New Plant Diseases Dataset (kaggle.com)

[20] Singhal, S., Ali, S., Shukla, D. K., Rathod, V. and Awasthy, M. (2023). Enhanced-honey bee based load balancing algorithm for cloud environment, *Artificial Intelligence, Blockchain, Computing and Security*: Volume 1, 1, pp. 951–956

[21] Kushwaha, P. K., Dagur, A. and Shukla, D. (2023). GA-ESE: A high-performance heart disease classification using hybrid machine learning approach, *Artificial Intelligence, Blockchain, Computing and Security*: Volume 2, 2, pp. 82–89

[22] Sharma, R. K. and Dagur, A. (2023). Various methods to classify the polarity of text based customer reviews using sentiment analysis, *Artificial Intelligence, Blockchain, Computing and Security*: Volume 2, 2, pp. 107–114

[23] Dixit, A. and Nema S. (2018). Wheat leaf disease detection using machine learning method. *A review Int. J. Comput. Sci. Mobile Comput.*, 7 (5) : pp. 124–129.

[24] Trivedi, J., Shamnani, Y.& Gajjar R. (2020). Plant leaf disease detection using machine learning. *Springer*: pp. 267–276.

[25] Panigrahi, K. P., Das, H., Sahoo, A. K. and Moharana, S. C. (2020). Maize leaf disease detection and classification using machine learning algorithms. *Progress in Computing, Analytics and Networking, Springer*: pp. 659–669.

[26] Prakash R. M., Saraswathy G., Ramalakshmi G., Mangaleswari K. & Kaviya T. (2017). Detection of leaf diseases and classification using digital image processing. *International Conference on Innovations in Information, Embedded and Communication Systems, IEEE* : pp. 1–4

[27] Ashok, S., Kishor, G., Rajesh, V., Suchitra, S., Sophia, S. G. and Pavithra B. (2020). Tomato leaf disease detection using deep learning techniques. *5th (ICCES), IEEE* :pp. 979–983.

[28] Durmus, H., Güneş, E. O. and Kırcı M. (2017). Disease detection on the leaves of the tomato plants by using deep learning. *6th International Conference on Agro-Geoinformatics, IEEE* : pp. 1–5.

Computational Methods in Science and Technology Volume 1 – Kaur et al. (Eds)
© 2025 the Author(s), ISBN: 978-1-032-81760-6

A survey of data-driven insights into air quality trends and prediction in India

K. Kabilan
Research Scholar, Department of Computer Science, VISTAS, Chennai

J. Jebathangam
Associate Professor, Department of Computer Science, VISTAS, Chennai

ABSTRACT: Air pollution is a major health and environmental issue in increasingly urbanizing countries like India. This study examines ML and DL implementations on Indian Air Quality Index (AQI) datasets from 2015 to 2020. The research seeks to identify patterns, anticipate trends, and guide air pollution mitigation decisions. The study carefully analyses AQI parameters, explores data, and uses powerful ML and DL algorithms to forecast AQI and identify pollution hotspots. Technical features include missing value imputation, feature engineering, model selection criteria, hyper parameter tuning procedures, and model performance assessment measures. Key results show that ML and DL models can accurately forecast air quality, identify pollution sources, and capture temporal and geographical fluctuations. The article suggests integrating more data sources, exploring advanced deep learning methods, and developing dynamic air quality control systems. This study uses data-driven insights and multidisciplinary techniques to improve our knowledge of air quality dynamics in India and create the framework for proactive air pollution mitigation and public health policies.

1 INTRODUCTION

In India, the combination of increasing industry, urbanization, and population increase has led to a significant decline in air quality, making air pollution a crucial environmental issue. The adverse effects of air pollution on human health, ecosystems, and general well-being have required the development of sophisticated techniques for monitoring, forecasting, and reducing its impact. The combination of machine learning (ML) and deep learning (DL) techniques has recently become a viable approach for tackling the complexity of air pollution analysis, especially in the Indian environment (Cowling *et al.* 2014; Prusty *et al.* 2022; Roy 2020).

Multiple research studies have explored the utilization of machine learning (ML) and deep learning (DL) in forecasting and examining air pollution levels in India. This demonstrates an increasing acknowledgment of the capabilities of these technologies in addressing environmental issues. This introduction utilizes findings from prominent research to emphasize the need of utilizing artificial intelligence to understand, forecast, and control air quality across the Indian subcontinent. The study conducted by (Chakraborty and Basu 2021) showcased the effectiveness of machine learning models in accurately forecasting air quality metrics using past data collected from several monitoring stations in Indian cities. The results emphasize the capacity of ML algorithms, such as support vector machines and random forests, to identify intricate patterns in the data and deliver precise predictions in the near run. (Nandi *et al.* 2024) investigated the use of deep learning methods, particularly convolutional neural networks (CNNs), to analyze air quality data from various sources, such as satellite images. The study showcased the capability of Convolutional Neural Networks (CNNs)(Alzubaidi *et al.* 2021) to effectively capture spatial patterns and detect areas with high pollution levels. This provides a fresh and comprehensive approach to monitoring air quality. Moreover, (Sharma *et al.* 2020) did study that specifically

 DOI: 10.1201/9781003501244-57

examined the temporal patterns of air pollution. They utilized recurrent neural networks (RNNs) to analyze time-series data. The study emphasized the need of taking into account temporal dependencies in air quality data, demonstrating the capability of Recurrent Neural Networks (RNNs) to capture long-term trends and improve forecast accuracy.

As this investigation examines the use of ML and DL to reduce air pollution in India, it is important to understand the urgency of these efforts to mitigate the significant and ubiquitous effects of air pollution on population and environmental health. The next sections will examine the changing landscape of AI applications in this vital domain. This review study synthesizes and builds on existing research efforts to provide a complete overview of current ML and DL methods for evaluating Indian air pollution datasets. By doing so, useful insights into the existing landscape, difficulties, and future perspectives for using AI to clean and improve India's air are offered.

2 MACHINE LEARNING APPROACHES

Machine learning (ML) techniques have emerged as powerful tools for predicting and analyzing air pollution levels in India. Several research papers have demonstrated the effectiveness of various ML models in discerning complex patterns, providing accurate predictions, and contributing to proactive air quality management.

2.1 *Traditional machine learning methods*

Conventional Machine Learning approaches: Conventional machine learning (ML) approaches have played a crucial role in enhancing the forecasting and examination of air pollution in India. These approaches use many methodologies such as regression models, classification algorithms, and clustering methods. These strategies have been successful in detecting patterns, forecasting air quality indices, and determining sources of pollution. The study conducted by (Maji *et al.* 2018) demonstrates the use of regression models to forecast air quality indices using historical data collected from monitoring stations in several cities in India. The research investigates the use of support vector machines and random forests, demonstrating their capacity to identify complex patterns in the data and provide precise predictions in the near term.

2.2 *Regression models for air quality prediction*

Regression models have become potent instruments for forecasting air quality indices, offering vital understanding into the intricate dynamics of air pollution in India. Multiple research projects have investigated the use of regression models to predict pollutant concentrations and evaluate the influence of contributing variables. The research conducted by (Sharma *et al.* 2020) centers on the use of regression models to forecast air quality metrics using historical data collected from many monitoring sites in Indian cities.

2.3 *Classification models for identifying pollution sources*

Classification models have shown their effectiveness in detecting and classifying pollution sources, providing significant information for focused treatments and regulatory actions. Multiple research projects have investigated the use of categorization models to identify distinct factors that contribute to air pollution in India. The study conducted by (Kaur and Pandey 2021) explores the use of machine learning techniques for categorizing pollutant origins.

2.4 *Clustering techniques for spatial analysis*

Clustering methods have shown efficacy in spatial analysis for air quality, facilitating the detection of pollution hotspots and trends across diverse locations. Several research papers have investigated the use of clustering techniques to improve spatial comprehension and provide important insights into the dispersion of air pollution in India. The study conducted by (Masood and Ahmad 2023) utilizes unsupervised learning techniques to cluster data and identify geographical patterns and clusters of pollution hotspots.

2.5 *Ensemble methods*

Ensemble approaches, which amalgamate the forecasts of numerous models, have become prominent in the field of air quality prediction and analysis. By harnessing the advantages of several models, ensemble approaches enhance accuracy and resilience. Multiple research papers have investigated the use of ensemble approaches to improve the forecasting and examination of air pollution in India (Samal *et al.* 2018).

2.6 *Random forests and gradient boosting*

Random Forests and Gradient Boosting are widely used ensemble methods that have shown to be effective in the field of air quality prediction and analysis. By harnessing the power of many decision trees, these techniques provide resilience and enhanced predictive capability. Multiple research projects have investigated the use of Random Forests and Gradient Boosting in the field of air pollution in India(Yu *et al.* 2016).

2.7 *Ensemble approaches for uncertainty estimation*

Ensemble methods are important for enhancing forecast accuracy and assessing uncertainty, which offers vital insights into the trustworthiness of air quality predictions. Uncertainty estimate is crucial for dealing with the inherent complexity and unpredictability of air pollution. Multiple research papers have investigated ensemble methods for estimating uncertainty in the field of air quality prediction (Mohan and Abraham 2024).

3 DEEP LEARNING APPROACHES

The popularity of deep learning in air quality prediction has increased due to its capacity to autonomously acquire nuanced patterns from complex data. Several researches have investigated the use of deep learning methods to improve the precision and comprehension of air pollution patterns in India. These studies demonstrate the wide range of uses of deep learning in predicting air quality, illustrating how different neural network structures may be used to various areas of air pollution study in India.

3.1 *Convolutional Neural Networks (CNNs)*

Convolutional Neural Networks (CNNs) are effective tools for analyzing geographical data and extracting features in the prediction of air quality. Multiple research projects have investigated the use of Convolutional Neural Networks (CNNs) to improve the comprehension and precision of air pollution dynamics, specifically in the setting of India(B. Zhang *et al.* 2022).

3.2 *Image-based air quality monitoring*

Researchers are using visual data to improve analysis and uncover anomalies in air quality monitoring, making the integration of image-based methodologies a cutting-edge development. Multiple research projects have investigated the use of image-based techniques to monitor air quality and detect abnormalities, providing a distinct viewpoint on evaluating pollution. The study conducted by (Chakma *et al.* 2017) centers on the use of Convolutional Neural Networks (CNNs) for the purpose of monitoring air quality using picture analysis.

3.3 *Recurrent Neural Networks (RNNs)*

RNNs are good at managing temporal dependencies and sequential patterns in air quality time series data. Several studies have used RNNs to examine temporal characteristics, forecast future air quality metrics, and comprehend pollution dynamics. RNNs for air quality time series analysis are studied (Nanduri and Sherry 2016). The work highlights RNNs' capacity to identify temporal dependencies in pollutant concentrations, enabling accurate forecasts and better understanding of air pollution dynamics. Long Short-Term Memory (LSTM), an RNN, is used to forecast air quality.

3.4 *Hybrid models*

Hybrid models, which integrate the advantages of many approaches, have become prominent in air quality prediction, providing a holistic approach that exploits the benefits of numerous methodologies. Multiple research projects have investigated the use of hybrid models for analyzing air pollution in India. The research conducted by centers on the integration of Convolutional Neural Networks (CNNs) with Long Short-Term Memory (LSTM) networks to develop a hybrid model for predicting air quality. The research demonstrates the advantageous effects of integrating geographical and temporal data, resulting in enhanced precision when forecasting pollution levels. Similarly, (Gu *et al.* 2022) investigates a hybrid approach that integrates machine learning and physical modeling to evaluate air quality.

3.5 *Combining CNNs and RNNs for comprehensive analysis*

The combination of Convolutional Neural Networks (CNNs) and Recurrent Neural Networks (RNNs) has emerged as a viable technique for accomplishing complete air quality prediction, particularly in circumstances with restricted data availability. Moreover, the use of transfer learning significantly improves the efficiency of models in situations with limited resources. Multiple research papers have investigated the effectiveness of these hybrid techniques in the context of analyzing air pollution. The research conducted by centers on the amalgamation of Convolutional Neural Networks (CNNs) and Recurrent Neural Networks (RNNs) to achieve a thorough examination of air quality (Z. Zhang *et al.* 2024).

4 IMPLEMENTATION AND DATA ANALYSIS

India's air quality necessitates a thorough analysis based on facts. This study rigorously analyses patterns and forecasts future changes based on the 2015-2020 AQI (Air Quality Index) dataset. The research progresses through numerous crucial stages, following a well-organized flowchart as shown in Figure 1 with a block diagram. Initially, the dataset is imported and each Air Quality Index (AQI) parameter, such as PM2.5, PM10, NO2, etc., is subjected to thorough examination. Data quality checks are conducted to detect and resolve missing values and outliers, while comprehensive documentation records specific parameter attributes, units, and any discrepancies. Subsequently, the process of exploratory data analysis reveals concealed narratives.

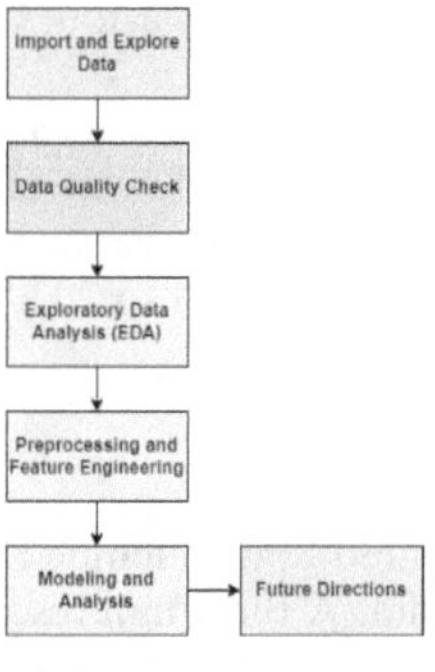

Figure 1. Block diagram of implementation in AQI prediction model.

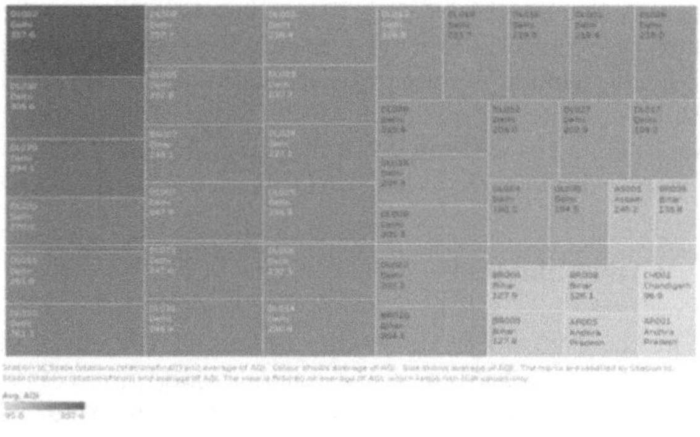

Figure 2. Overall tree map of Indian AQI impact on all the states.

Time series graphs, boxplots, and heatmaps show AQI changes across locations, pollutants, and time periods. Data preparation is essential for reliable analysis. Missing values are imputed, numerical properties are scaled or normalized, and pollutant ratios and meteorological indicators are constructed to increase data interpretation. Model selection and assessment are crucial for accurate forecasts and pattern recognition. AQI values are predicted using supervised learning algorithms like Random Forest, XGBoost, and LSTMs. Convolutional Neural Networks (CNNs) are also tested for satellite image utilization and forecasting. The efficiency of unsupervised learning approaches like K-means and DBSCAN in recognizing high-pollution locations is examined. Every model is painstakingly trained and evaluated on correctly separated datasets. MSE, R-squared, and accuracy are utilized to evaluate their AQI prediction and hotspot identification abilities.

4.1 *Dataset analysis*

Close investigation is needed to understand India's complex air quality. This study examines the 2015-2020 AQI (Air Quality Index) parameters dataset, concentrating on data from active monitoring stations around the nation. The dataset, perhaps from Kaggle, includes AQI readings and air quality data from active monitoring sites in several Indian cities between 2015 and 2020. While the list of cities may change, the information certainly covers air quality across varied geographic and climatic zones countrywide. Depending on the monitoring station and data collecting techniques, hourly readings to daily or monthly averages are predicted.

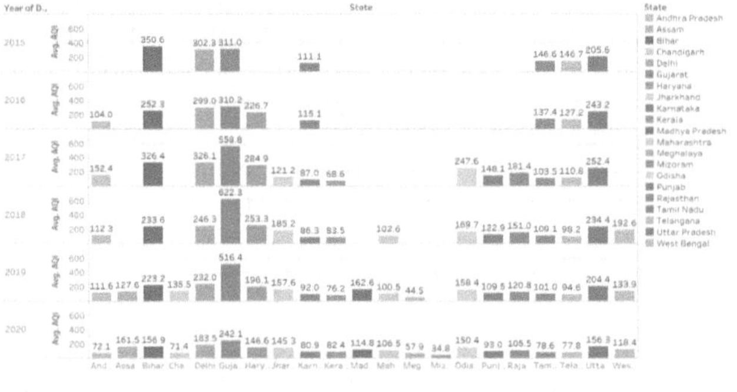

Figure 3. Impact of AQI data in overall India from the year 2015 to 2020.

Date, time, and measurement timestamps are important. The data came from this city or station. The overall Air Quality Index (AQI) shows air quality at a given time and area. Further air quality information is provided by PM2.5, PM10, NO2, SO2, CO, and O3 concentrations. Optional weather info: Temperature, humidity, wind speed, and precipitation can explain air quality influences when available. Data quality must be evaluated before analysis. Checks should identify and address: Values When data is lacking, utilize mean/median or interpolation. Outliers must be explored and maybe deleted for robust analysis.

Each pollutant's concentration determines its sub-index, and the highest sub-index determines the AQI. AQI rates air quality as Good, Satisfactory, Moderate, Poor, Very Poor, and Severe based on health concerns and suggested measures. Figure 2 shows how AQI altered India's tree map graph. Figure 3 displays the average AQI from monitoring centers across India from 2015 to 2020.

5 CONCLUSION

With the 2015–2020 Air Quality Index (AQI) dataset, this study evaluated how machine learning (ML) and deep learning (DL) can assess and forecast air quality in India. A thorough research of air pollution dynamics has provided the platform for educated decision-making and solutions to this essential issue. The investigation began with data preprocessing and exploratory data analysis to uncover AQI dataset properties and trends. For AQI prediction and pollution hotspot detection, Random Forest, XGBoost, LSTMs, CNNs, K-means, and DBSCAN were tried after careful model selection. This research emphasizes data-driven approaches to understanding and reducing air pollution in India. ML and DL models reliably predict air quality, identify pollution sources, and track temporal and geographical changes. Stakeholder confidence and decision-making insights depend on model interpretability. Many study and improvement possibilities remain despite this survey's improvements. First, models should contain socio-economic factors, traffic patterns, and industrial activities to anticipate and comprehend air quality changes. Deep learning research, notably employing satellite imagery and sensor data for spatial analysis, may improve air quality forecasts. Dynamic and adaptive air quality management systems employing real-time data streams and advanced ML algorithms may be researched. Such technologies might enable proactive air pollution control and public health measures. Finally, this study establishes the basis for Indian air quality forecast and management research and innovation. ML, DL, and trans-disciplinary techniques may help us clean up the environment for future generations.

REFERENCES

Alzubaidi, L., Zhang, J., Humaidi, A. J., Al-Dujaili, A., Duan, Y., Al-Shamma, O., Santamaría, J., Fadhel, M. A., Al-Amidie, M., and Farhan, L. (2021). Review of deep learning: Concepts, CNN architectures, challenges, applications, future directions. *Journal of Big Data*, 8(1), 53. https://doi.org/10.1186/s40537-021-00444-8

Chakma, A., Vizena, B., Cao, T., Lin, J. and Zhang, J. (2017). Image-based air quality analysis using deep convolutional neural network. *2017 IEEE International Conference on Image Processing (ICIP)*, 3949–3952. https://doi.org/10.1109/ICIP.2017.8297023

Chakraborty, J. and Basu, P. (2021). Air quality and environmental injustice in India: connecting particulate pollution to social disadvantages. *International Journal of Environmental Research and Public Health*, 18 (1), 304. https://doi.org/10.3390/ijerph18010304

Cowling, K., Dandona, R. and Dandona, L. (2014). Social determinants of health in India: Progress and inequities across states. *International Journal for Equity in Health*, 13(1), 88. https://doi.org/10.1186/s12939-014-0088-0

Gu, Y., Li, B. and Meng, Q. (2022). Hybrid interpretable predictive machine learning model for air pollution prediction. *Neurocomputing*, 468, 123–136. https://doi.org/10.1016/j.neucom.2021.09.051

Kaur, R., and Pandey, P. (2021). Air pollution, climate change, and human health in Indian cities: A brief review. *Frontiers in Sustainable Cities*, 3, 705131. https://doi.org/10.3389/frsc.2021.705131

Maji, K. J., Ye, W.-F., Arora, M. and Shiva Nagendra, S. M. (2018). PM2.5-related health and economic loss assessment for 338 Chinese cities. *Environment International*, 121, 392–403. https://doi.org/10.1016/j.envint.2018.09.024

Masood, A. and Ahmad, K. (2023). Data-driven predictive modeling of PM2.5 concentrations using machine learning and deep learning techniques: A case study of Delhi, India. *Environmental Monitoring and Assessment*, 195(1), 60. https://doi.org/10.1007/s10661-022-10603-w

Mohan, A. S. and Abraham, L. (2024). An ensemble deep learning approach for air quality estimation in Delhi, India. *Earth Science Informatics*. https://doi.org/10.1007/s12145-023-01210-5

Nandi, B. P., Singh, G., Jain, A. and Tayal, D. K. (2024). Evolution of neural network to deep learning in prediction of air, water pollution and its Indian context. *International Journal of Environmental Science and Technology*, 21(1), 1021–1036. https://doi.org/10.1007/s13762-023-04911-y

Nanduri, A. and Sherry, L. (2016). Anomaly detection in aircraft data using Recurrent Neural Networks (RNN). *2016 Integrated Communications Navigation and Surveillance (ICNS)*, 5C2-1-5C2-8. https://doi.org/10.1109/ICNSURV.2016.7486356

Prusty, R. K., Bairwa, M., Anwar, F., Mishra, V. K., Patel, K. K. and Mangal, D. K. (2022). Socio-biomedical predictors of child nutrition in India: An ecological analysis from a nationally representative Demographic and Health Survey, 2015–2016. *Journal of Health, Population and Nutrition*, 41(1), 1. https://doi.org/10.1186/s41043-021-00273-8

Roy, M. P. (2020). Socio-demographic factors of COPD mortality in India. *Clinical Epidemiology and Global Health*, 8(3), 939–942. https://doi.org/10.1016/j.cegh.2020.02.028

Samal, K., Babu, K. and Das, S. (2018). Spatio-temporal prediction of air quality using distance based interpolation and deep learning techniques. *EAI Endorsed Transactions on Smart Cities*, 168139. https://doi.org/10.4108/eai.15-1-2021.168139

Sharma, S., Zhang, M., Anshika, Gao, J., Zhang, H., and Kota, S. H. (2020). Effect of restricted emissions during COVID-19 on air quality in India. *Science of The Total Environment*, 728, 138878. https://doi.org/10.1016/j.scitotenv.2020.138878

Yu, R., Yang, Y., Yang, L., Han, G. and Move, O. (2016). RAQ–A random forest approach for predicting air quality in urban sensing systems. *Sensors*, 16 (1), 86. https://doi.org/10.3390/s16010086

Zhang, B., Rong, Y., Yong, R., Qin, D., Li, M., Zou, G. and Pan, J. (2022). Deep learning for air pollutant concentration prediction: A review. *Atmospheric Environment*, 290, 119347. https://doi.org/10.1016/j.atmosenv.2022.119347

Zhang, Z., Zhang, S., Chen, C. and Yuan, J. (2024). A systematic survey of air quality prediction based on deep learning. *Alexandria Engineering Journal*, 93, 128–141. https://doi.org/10.1016/j.aej.2024.03.031

Computational Methods in Science and Technology Volume 1 – Kaur et al. (Eds)
© 2025 the Author(s), ISBN: 978-1-032-81760-6

Breast cancer diagnosis using machine learning model

T.S. Arthi
Department of Computer Science and Engineering, Galgotias College of Engineering & Technology, Greater Noida, India

N. Partheeban
Department of Computer Science and Engineering, Galgotias University, Greater Noida, India

Bhoomika Singh, Aparajitha Singh & Nikhil Tripathi
Department of Computer Science and Engineering, Galgotias College of Engineering & Technology, Greater Noida, India

ABSTRACT: Early and accurate detection of breast cancer, particularly Invasive Ductal Carcinoma (IDC), is critical for improving patient outcomes. Traditional diagnostic methods like histopathology and mammography have limitations in terms of time, cost, and potential subjectivity. This study proposes a lightweight Machine Learning (ML) model for IDC classification using digital whole-mount breast cancer slide images. The model aims to achieve high accuracy while being suitable for deployment in resource- constrained environments. We evaluate our model on a publicly available breast histopathology dataset and compare its performance with other Deep Learning (DL) models, focusing on both accuracy and computational efficiency. The results demonstrate that our model achieves a high classification accuracy (around 89.3 percent) with minimal processing power and memory requirements. Furthermore, successful deployment on a Raspberry Pi embedded system validates its potential for real-world application in resource-limited settings, potentially leading to improved breast cancer diagnosis accessibility. Index Terms—Bi-directional Long Short-Term Memory (BiLSTM), Convolutional Neural Network (CNN), Deep Learning (DL), Invasive Ductal Carcinoma (IDC), Multilayer Perceptron (MLP), Raspberry Pi, Tiny Machine Learning (TinyML).

1 INTRODUCTION

Breast Cancer is responsible for the highest number of cancers diagnoses worldwide, with women being the most affected, accounting for approximately 11.7% of all cases (Sung *et al.*, 2021). Early detection of breast cancer significantly aids treatment. It allows for less invasive treatments, preserves more breast tissue, and improves survival rates. The most common type of breast cancer is Invasive Ductal Carcinoma (IDC), primarily affecting women in their 50s and older (Guan *et al.*, 2020). IDC is a type of breast cancer originating in the milk ducts, with the potential to infiltrate the adjacent breast tissues, affecting other parts of the body. In 2023, an estimated 297,790 new cases of invasive breast cancer are expected among U.S. women, while approximately 2,800 new cases are projected for men, with a man's lifetime risk at about 1 in 833 (Sharma and Purwar, 2022). Various methods for detecting Breast Cancer include breast exams, mammograms, ultrasounds, biopsies, (Magnetic Resonance Imaging) MRI scans and others, which can be time-consuming. However, Deep Learning (DL) can automate the process of detecting IDC for its early diagnosis, assisting medical professionals in making informed decisions. DL techniques have played a crucial role in the advancement of healthcare technologies over the last decade. DL can help

 DOI: 10.1201/9781003501244-58

improve the accuracy, efficiency, and scalability of breast cancer screening. The work done by (Krithiga and Geetha, 2021; Rajasekar *et al.*, 2023) highlighted methodologies, algorithms, and concepts related to histopathology image analysis. The merits and demerits of various methods and the performance of Machine Learning (ML) and DL techniques were evaluated. The techniques were applied to predict the recurrence rates of breast cancer. Moreover, an overview of the steps necessary to prepare tissue samples and analyze stained images was explained. Advanced methodologies and techniques within modern medical science are available for breast cancer identification.

The use of digital pathology is an emerging trend in modern medicine. The study conducted by (Aswathy and Jagannath, 2017) summarized the current applications of Digital Image Processing techniques on histopathological images for the detection of breast cancer as well as how such applications can be utilized in the future. (Gecer *et al.*, 2018) proposed a computer aided diagnosis system for breast histopathology using DL. Visualizations were presented to help understand discriminative features for breast cancer diagnosis (Pati *et al.*, 2023). These pieces of work motivated us to work in this domain. However, despite significant advances in breast cancer detection using high computational architectures and advanced ML models, a crucial aspect has been relatively unexplored i.e. utilizing low computational architecture (TinyML) that can be implemented on embedded hardware. TinyML represents a transformative approach for deploying ML models on resource-constrained devices, opening new possibilities in diverse fields. This is particularly relevant in healthcare, as demonstrated by (Tsoukas *et al.*, 2021), where TinyML models have been instrumental in developing portable and low-cost tools, making healthcare more accessible in underserved regions.

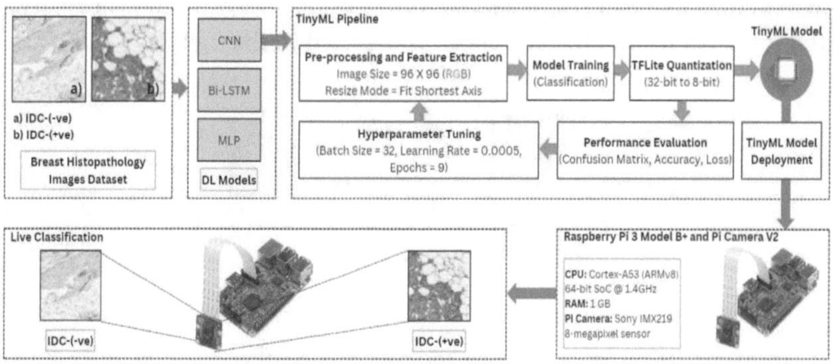

Figure 1. Methodology for research and TinyML workflow in creating smart edge solutions for IDC classification.

This research paper develops the low-computational TinyML model architecture, designed for low-end embedded hardware like Raspberry Pi. First, we utilized DL models such as Multilayer Perceptron (MLP) (Popescu *et al.*, 2009), Bi-directional Long Short Term Memory (Bi-LSTM) (Schuster and Paliwal, 1997), and Convolutional Neural Network (CNN) (O'Shea and Nash, 2015), and then evaluated their ac-curacy and performance on the Breast Histopathology Images dataset (Janowczyk and Madabhushi, 2016; Paul Mooney, n.d.) and based on it, the better DL model was identified so that its neural network architecture can be replicated for creating the TinyML model to utilize its advantages like low computational complexity, limited processing power and memory. After identifying the best DL model, the TinyML model is created, trained, and tested on the same dataset as shown in the TinyML pipeline block in Figure.1. The model is then deployed on embedded hardware like Raspberry Pi to incorporate our model into everyday devices, ensuring its usability and relevance in practical healthcare settings. The comprehensive block diagram illustrating our proposed work is shown in Figure 1.

2 METHODOLOGY

The proposed approach involves four stages: (i) Dataset, (ii) Training and testing for different conventional DL models, (iii) Training and testing for the proposed low computational model for embedded hardware, and (iv) Deployment of the low computational model on embedded hardware. Python 3.11.6 is utilized in creating, training, and testing the DL models, enabling comprehensive comparisons with our TinyML model. A. Dataset The initial step involves utilizing a comprehensive image dataset comprising IDC image patches. We utilized the preprocessed Breast Histopathology Images dataset for this purpose. Originally the dataset comprised 162 whole-mount slide images of Breast Cancer specimens scanned at 40× magnification. The dataset generates 277,524 image patches of size (50×50), of which 198,738 image patches are IDC-(- ve) and 78,786 are IDC-(+ve). B. Training and Testing for different conventional DL models. In this step, a DL model is trained using the IDC image patches to categorize them into two classes, as mentioned in Section II-A. Different conventional DL models like MLP, BiLSTM and CNN were tested to identify the best model for the given dataset. CNN performed the best. The neural network architecture for CNN is shown in Figure 2. Different image categories are shown in Figure 2. CNN Neural Network Architecture for IDC Classification based on distinctive patterns and features recognized by CNN. The training process aims to optimize the model's parameters to minimize the loss function. A batch size of 32 was used to train for 9 epochs. The model summary of the created CNN model is given in Figure 3. The total training time for the CNN model on the training dataset is 285.029 seconds. The resulting accuracy on the training and validation datasets for the CNN model was found to be 90.91% and 89.71% respectively. The accuracy and loss curves for the training and validation dataset have been depicted in Figure 4. The variation of accuracy and loss with respect to each epoch is depicted in the figures.

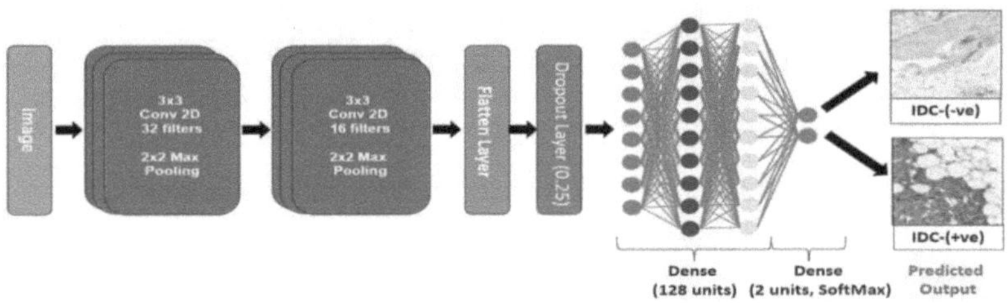

Figure 2. CNN Neural Network architecture for IDC classification

3 PROPOSED LOW COMPUTATIONAL MODEL FOR EMBEDDED HARDWARE

This approach requires significant memory resources and imposes high computational complexity, making it impractical for real-world applications. To improve efficiency in terms of memory consumption and computational complexity.

Layer (type)	Output Shape	Param #
conv2d (Conv2D)	(None, 48, 48, 32)	896
max_pooling2d (MaxPooling2D)	(None, 24, 24, 32)	0
conv2d_1 (Conv2D)	(None, 22, 22, 16)	4624
max_pooling2d_1 (MaxPooling2D)	(None, 11, 11, 16)	0
flatten (Flatten)	(None, 1936)	0
dropout (Dropout)	(None, 1936)	0
dense (Dense)	(None, 128)	247936
dense_1 (Dense)	(None, 2)	258

Total params: 253714 (991.07 KB)
Trainable params: 253714 (991.07 KB)
Non-trainable params: 0 (0.00 Byte)

Figure 3. CNN model summary.

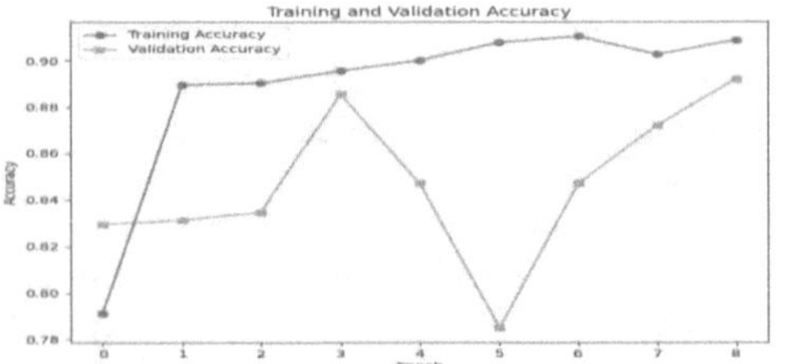

Figure 4. CNN training and validation accuracy curve.

Here in this work, CNN architecture is further enhanced with TinyML to take advantage of the benefits of a low-computational ML model that can be used for hardware implementation. The replicated neural network architecture of TinyML to classify IDC.The TinyML model takes (96×96) pixel IDC patch images as input and classifies them into one of the predefined classes, i.e. IDC-(-ve) and IDC-(+ve). The model is trained for 9 epochs with a batch size of 32 and a learning rate of 0.0005. The model is trained for int 8 as well as float 32 quantization coefficients. The int 8 quantized TinyML model utilizes just 363.3 KB of peak RAM and 52.9 KB of Flash memory. It has an inference time of just 15 ms and an overall accuracy of 89.3%, which is comparable to that of the CNN model. The comparison of the performance of TinyML across different quantization coefficients (int 8 and float 32). These features make the TinyML model highly efficient compared to other high computational Image Classification DL models. The confusion matrix (Du *et al.*, 2022; Rohit Kundu, n.d.) of the training and validation dataset for the int 8 quantized TinyML model is shown in Figures 6 and 7 respectively.

4 DEPLOYMENT OF THE LOW COMPUTATIONAL MODEL

After creating, training and testing the TinyML model, the model is deployed to the embedded hardware. This enabled the application of our proposed solution to real-world scenarios, where resource efficiency is of utmost importance. The hardware implementation focused on optimizing the model for dedicated embedded systems, ensuring better performance within the constraints of power, memory, and processing capacity. In this paper, Raspberry Pi 3 Model B+ (Shinde *et al.*, 2022; Tovar *et al.*, 2018) is utilized as embedded hardware. It is the most advanced version of the third-generation single-board computer. It has a powerful 64-bit quad-core processor that runs at a speed of 1.4 GHz and a dual-band wireless LAN that can connect to both 2.4 GHz and 5 GHz internet net-works. The Pi Camera V2 module is connected to the Raspberry Pi to see the live classification of IDC into one of the predefined classes, i.e. IDC-(-ve) and IDC-(+ve). The specifications of the embedded hardware are explained in Figure 1. The readout features and mode of operation of the Raspberry Pi V2 camera module are depicted in Figure 5. To achieve higher accuracy of prediction, a balance between frame rate and quality of the image captured is needed. For this, mode 0 is utilized from the available 8 modes of operation. The real-time classification of IDC using the trained TinyML (int 8) model deployed on the Rasp-berry Pi is depicted in Figure 5.

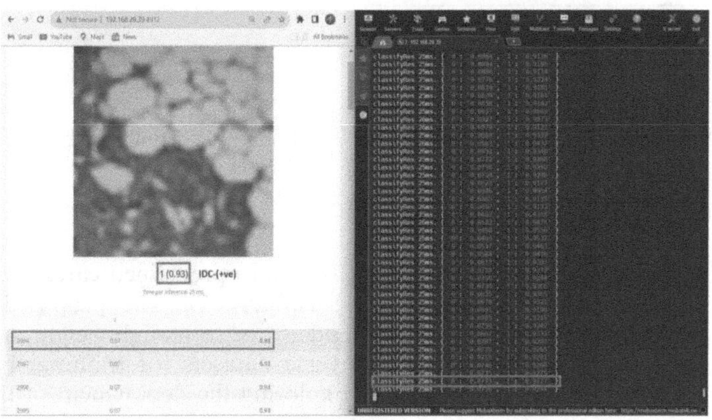

Figure 5. Model deployed on the Raspberry Pi.

The study aimed to create a system to classify IDC into one of the two predefined classes, i.e. IDC-(- ve) and IDC- (+ve). Several DL models were trained and tested on the Breast

Histo-pathology Image dataset, out of which, CNN performed the best with an accuracy of 90.91%. The high computational architecture of CNN was replicated to create a low computational Ti-nyML model to leverage its advantages. The int 8 quantized TinyML model provided an overall accuracy of 89.3%, which is comparable to that of CNN, demonstrating its effectiveness in distinguishing between IDC-(-ve) and IDC-(+ve). The performance of the model was evaluated by computing various performance parameters i.e., Accuracy, Precision, Recall, and F1 score (F1).

5.1 *Summary of parameters is mentioned below*

Accuracy: It is calculated as the ratio of the number of correctly predicted classifications to the total number of predictions made.

$$\text{Accuracy} = \text{Correct predictions}/\text{Total predictions}$$

Precision: It is equal to the number of accurately identified positive samples compared to the total number of positive samples. Precision = Accurate positive samples Total positive samples

Recall: Recall is the degree to which a model finds all relevant cases within a dataset.

Recall = True positive /(True positive + False negative) (3) F1 Score: The F1 score is a metric that combines precision (accuracy of positive predictions) and recall (completeness of positive predictions) into a single value. A balanced F1 score emphasizes equal importance to both metrics.

$$\text{F1 Score} = 2 \times \text{Precision} \times \text{Recall Precision} + \text{Recall}$$

The model's performance on the training and validation datasets are quantized by TinyML model. The model achieved an accuracy of 89.3 percent and 89.3 percent on the training and validation sets, respectively. Precision and recall were also high for both datasets.

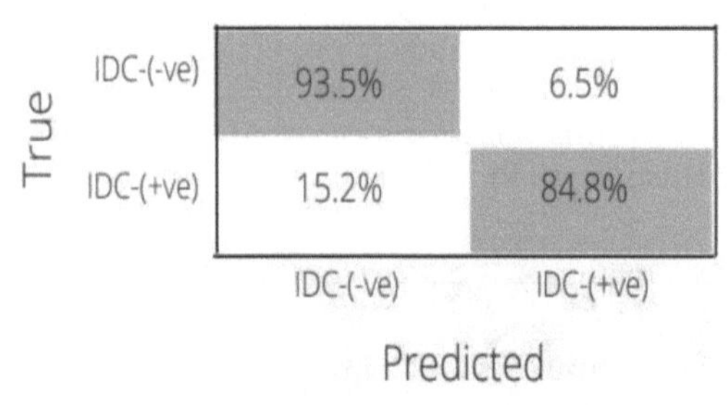

Figure 6. Confusion matrix of training dataset of model.

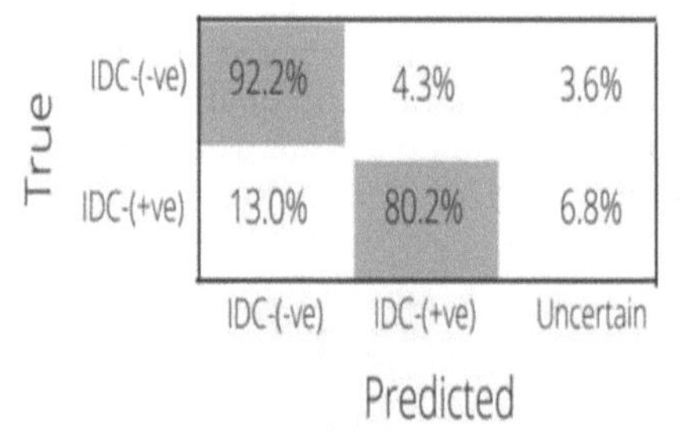

Figure 7. Confusion matrix of validation dataset of TinyML model.

5 CONCLUSION

This paper proposes a solution for classifying IDC into predefined categories using a low-complexity TinyML model deployed on a Raspberry Pi. The proposed model offers advantages over conventional deep learning models due to its lightweight architecture and reduced memory footprint. Utilizing more diverse datasets for training. Fine-tuning the model architecture for optimal performance. Exploring the deployment of these devices in various environments as they become more affordable and compact. By making healthcare technology accessible through adaptation to low resource devices, this research contributes to advancements in IDC classification. However, ongoing efforts are crucial to further improve the accuracy and reliability of these devices.

REFERENCES

Aswathy, M. A., & Jagannath, M. (2017). Detection of breast cancer on digital histopathology images: Present status and future possibilities. *Informatics in Medicine Unlocked*, 8, 74–79. https://doi.org/10.1016/j.imu.2016.11.001

Du, R., Chen, Y., Li, T., Shi, L., Fei, Z., & Li, Y. (2022). Discrimination of Breast Cancer Based on Ultrasound Images and Convolutional Neural Network. *Journal of Oncology*, 2022, 1–9. https://doi.org/10.1155/2022/7733583

Gecer, B., Aksoy, S., Mercan, E., Shapiro, L. G., Weaver, D. L., & Elmore, J. G. (2018). Detection and classification of cancer in whole slide breast histopathology images using deep convolutional networks. *Pattern Recognition*, 84, 345–356. https://doi.org/10.1016/j.patcog.2018.07.022

Guan, X., Xu, G., Shi, A., Zou, Y., Zhan, Y., Fan, Z., & Dong, Y. (2020). Comparison of clinicopathological characteristics and prognosis among patients with pure invasive ductal carcinoma, invasive ductal carcinoma coexisted with invasive micropapillary carcinoma, and invasive ductal carcinoma coexisted with ductal carcinoma in situ: A retrospective cohort study. *Medicine*, 99(50), e23487. https://doi.org/10.1097/MD.0000000000023487

Janowczyk, A., & Madabhushi, A. (2016). Deep learning for digital pathology image analysis: A comprehensive tutorial with selected use cases. *Journal of Pathology Informatics*, 7(1), 29. https://doi.org/10.4103/2153-3539.186902

Krithiga, R., & Geetha, P. (2021). Breast cancer detection, segmentation and classification on histopathology images analysis: A systematic review. *Archives of Computational Methods in Engineering*, 28(4), 2607–2619. https://doi.org/10.1007/s11831-020-09470-w

O'Shea, K., & Nash, R. (2015). *An Introduction to Convolutional Neural Networks*. https://doi.org/10.48550/ARXIV.1511.08458

Pati, A., Parhi, M., Pattanayak, B. K., Singh, D., Singh, V., Kadry, S., Nam, Y., & Kang, B.-G. (2023). Breast cancer diagnosis based on iot and deep transfer learning enabled by fog computing. *Diagnostics*, 13(13), 2191. https://doi.org/10.3390/diagnostics13132191

Paul mooney. (n.d.). *Breast Histopathology Images* [dataset]. https://www.kaggle.com/datasets/paultimothymooney/breast-histopathology-images

Popescu, M.-C., Balas, V. E., Perescu-Popescu, L., & Mastorakis, N. (2009). *Multilayer Perceptron and Neural Networks*. 8(7).

Rajasekar, V., Vaishnnave, M. P., Premkumar, S., Sarveshwaran, V., & Rangaraaj, V. (2023). Lung cancer disease prediction with CT scan and histopathological images feature analysis using deep learning techniques. *Results in Engineering*, 18, 101111. https://doi.org/10.1016/j.rineng.2023.101111

Rohit Kundu. (n. d.). Confusion matrix: How to use it & interpret results. *V7labs*. Retrieved September 25, 2023, from https://www.v7labs.com/blog/confusion-matrix-guide

Schuster, M., & Paliwal, K. K. (1997). Bidirectional recurrent neural networks. *IEEE Transactions on Signal Processing*, 45(11), 2673–2681. https://doi.org/10.1109/78.650093

Sharma, B. P., & Purwar, R. K. (2022). Ensemble boosted tree based mammogram image classification using Texture features and extracted smart features of Deep Neural Network. *ADCAIJ: Advances in Distributed Computing and Artificial Intelligence Journal*, 10(4), 419–434. https://doi.org/10.14201/ADCAIJ2021104419434

Shinde, R. K., Alam, Md. S., Hossain, Md. B., Md Imtiaz, S., Kim, J., Padwal, A. A., & Kim, N. (2022). Squeeze-MNet: Precise Skin Cancer Detection Model for Low Computing IoT Devices Using Transfer Learning. *Cancers*, 15(1), 12. https://doi.org/10.3390/cancers15010012

Sung, H., Ferlay, J., Siegel, R. L., Laversanne, M., Soerjomataram, I., Jemal, A., & Bray, F. (2021). Global Cancer Statistics 2020: GLOBOCAN Estimates of Incidence and Mortality Worldwide for 36 Cancers in 185 Countries. *CA: A Cancer Journal for Clinicians*, 71(3), 209–249. https://doi.org/10.3322/caac.21660

Tovar, J. C., Hoyer, J. S., Lin, A., Tielking, A., Callen, S. T., Elizabeth Castillo, S., Miller, M., Tessman, M., Fahlgren, N., Carrington, J. C., Nusinow, D. A., & Gehan, M. A. (2018). Raspberry Pi–powered imaging for plant phenotyping. *Applications in Plant Sciences*, 6(3), e1031. https://doi.org/10.1002/aps3.1031

Tsoukas, V., Boumpa, E., Giannakas, G., & Kakarountas, A. (2021). A review of machine learning and TinyML in healthcare. *25th Pan-Hellenic Conference on Informatics*, 69–73. https://doi.org/10.1145/3503823.3503836

Computational Methods in Science and Technology Volume 1 – Kaur et al. (Eds)

Identifying key indicators words of suicidal ideation through ensemble voting classifier learning methods

Shiv Shankar Prasad Shukla & Maheshwari Prasad Singh
National Institute of Technology Patna, Patna, Bihar

ABSTRACT: Suicide prevention is a pressing global concern, with a significant number of annual deaths related to suicide and numerous attempts reported each year. The rise of social media as a platform for emotional expression underscores the need for early detection of suicidal thoughts. This study introduces an innovative approach utilizing Natural Language Processing (NLP) techniques to analyze behavioral patterns and linguistic cues in social media interactions. The primary objective is to identify key indicator words in social media post i.e. Reddit posts, that may suggest the presence of suicidal ideation, through the application of advanced NLP techniques and variety of machine learning models, including Decision Tree, Random Forest, Bernoulli Naïve Bayes, Gradient Boosting, XGBoost, and an Ensemble voting classifier method that combines Random Forest, Gradient Boosting, and XGBoost. The models demonstrate impressive accuracy in the detection of posts related to suicidal ideation, providing a potential means for early intervention. On the Reddit dataset, these models achieve accuracy scores of 81%, 88%, 84%, 89%, 89%, and 92%. This research aims to identify keywords associated directly or inversely linked to suicidal ideation. The study offers a potent keywords indicator for detecting signs of suicidal thoughts in social media posts, potentially facilitating timely interventions and, ultimately, saving lives to support suicide prevention efforts.

Keywords: Suicidal ideation, Ensemble voting classifier method, Natural Language Processing, Suicidal

1 INTRODUCTION

In recent times, the number of people facing mental health challenges such as anxiety and depression has been on the rise, particularly in developed countries [25]. If they are left untreated, these conditions can lead to changes in behavior and even severe illness, sometimes resulting in suicide attempts [26]. Identifying the reasons behind suicide and the individuals at risk is a complex task. People dealing with depression are more susceptible to suicidal thoughts. Factors contributing to suicide risk can be grouped into health, environment, and personal history, according to the American Foundation for Suicide Prevention. Mental health issues are often linked to a higher risk of suicide. Extensive research into the psychology of suicide has revealed that personality, cognitive factors, social influences, and negative life experiences all play significant roles Understanding individuals' daily routines is vital for predicting changes in behavior and health. Alterations in their daily lives can act as triggers for suicidal thoughts. Suicidal ideation is a progression from depression or sadness to contemplating suicide and eventually becoming consumed by self-destructive thoughts. Common risk factors for suicide include depression, hopelessness, and frustration, which serve as indicators of suicidal thoughts [27]. Depression and changes in behavior are considered crucial indicators for potential suicide cases. Consequently, analyzing negative

 DOI: 10.1201/9781003501244-59

emotions is essential for detecting changes in behavior and health. Early detection of suicidal thoughts has become a priority in national suicide prevention strategies adopted by WHO member nations, with the goal of reducing suicide rates by 10% in 2020 [1]. Identifying individuals with suicidal thoughts can be accomplished by analyzing their writing style or structured data [29]. Online communication platforms, known as CMC channels, have become popular outlets for expressing both positive and negative emotions. As a result, online media inadvertently serve as tools for monitoring suicidal ideation and potential suicide prevention [8][10]. Suicide is a significant social issue that claims numerous lives each year, underscoring the importance of diagnosing changes in behavior and health to prevent loss of life [19]. Early detection and intervention are crucial in preventing future suicide attempts and identifying behavioral changes [30]. Social media, in combination with computational linguistics, has emerged as a promising research area in mental health forums, offering a powerful platform for reducing suicide rates. Predicting the likelihood of suicidal thoughts using social media data can be achieved through the application of artificial intelligence and machine learning methods, enabling a better understanding of public intentions and facilitating early intervention [11]. Analyzing social content requires significant effort to understand its features and characteristics [12][13], providing valuable insights for machine learning and deep learning approaches. Feature engineering [14][15], sentiment analysis, and deep learning efforts [16][17][18] can all play essential roles in identifying suicidal ideation in social content.

The paper makes a significant contribution in two key ways: a) it introduces advanced feature engineering with frequency-based techniques and feature selection through quantile statistical methods in natural language processing to obtain optimal features, and b) it identifies key indicator keywords from posts that are directly or inversely related to suicidal ideations.

2 RELATED WORK

Researchers have become increasingly concerned about the rising suicide rates and have introduced various clinical practices, such as surveys and automated post-detection methods, to identify suicide attempts. In the past, detection relied solely on clinical methods like face-to-face interviews and questionnaires, highlighting the importance of understanding the psychology and behavior associated with suicidal thoughts through direct interactions [28]. Surveys and interviews with individuals who have attempted suicide have been conducted on platforms like Weibo and other social media sites [29]. For instance, in 2018, Katchapakirin *et al.* conducted a study on Facebook to detect depression within the Thai community [1]. J. Gao *et al.* developed a model using SVM, Random Forest, and LSTM to detect suicide risk by manually annotating comment datasets on YouTube [2], achieving an accuracy of 84.5%. The detection of suicidal thoughts in Spanish-language posts was done by Valeriano *et al.,* and the dataset was excavated using the Twitter API. After enduring human annotations and pre-processing, the dataset was vectorized using TF-IDF and Word2Vec. Algorithms including SVM and Logistic Regression yielded accuracy rates of 74% and 79%, correspondingly. In 2018, manually annotated Reddit posts were exploited by Aladely *et al.,* and the models were employed such as Random Forest, Logistic Regression, and Support Vector Machine [4]. LR and SVM achieved impressive F1-scores of 91%. Another study by Coppersmith *et al.* in 2018 involved combining datasets from ourdatahelp.org and public self-stated data to predict behavior changes and suicide risks [5]. They used Bi-directional LSTM with Self-Attention and achieved an accuracy rate of 94%. Sawhney *et al.* in 2018 explored the application of deep learning using RNN, LSTM, and CNN-LSTM to detect suicidal ideation, achieving accuracy rates of 73.7%, 78.7%, and 81.2%, respectively [6]. The use of CNN for detecting suicidal thoughts was introduced by Shing *et al.* Additionally, S. Ji *et al.* conducted a review paper in 2021 summarizing various methods for predicting suicidal thoughts. They covered clinical practices involving interactions with social workers, machine learning techniques with feature engineering proposed by different authors [21] [22], and deep learning methods [20] [12] for intuitive detection based on online social content. In 2020, Tadesse *et al.* employed various machine learning techniques [7], with the LSTM-CNN model achieving a superior result of 93%. In 2021, Ning *et al.* proposed a deep learning method incorporating C attention, achieving an accuracy rate of 84.3% for subtask 2 [8]. Sawney *et al.* developed the SISMO model, which detects sensitivity to suicide in post-words with an accuracy of 73% [9]. Akshma *et al.* accumulated a Reddit dataset, employing attention over CNN and LSTM models to attain an accuracy rate of 88.18% [30]. In 2023, Ghosal and colleagues on a Reddit dataset, utilizing FastText embedding, TF-IDF vectorization, and the XGBoost classifier, surpassing baseline models with a remarkable 0.78 AUC and 0.71 weighted F-score performance [29].

Current research has focused on using machine learning and ensemble voting classifier learning to monitor social media posts for early signs of illness, particularly suicidal tendencies. However, there are gaps in this research, including the need for better precision in detecting behavioral shifts, automating feature extraction, and improving accuracy in distinguishing suicidal posts from others. This paper proposes a new methodology to address these challenges, particularly on Reddit, by identifying key indicators of suicidal ideation.

3 METHODOLOGY

The process for detecting changes in behavior or signs of suicidal thoughts consists of several important stages, including collecting data, preparing the data, creating relevant features, and predicting whether posts suggest suicidal tendencies or not. Figure 1 illustrates the workflow of the study, from inputting datasets to obtaining the final classification results. The following subsections provide detailed explanations of each step in the diagram.

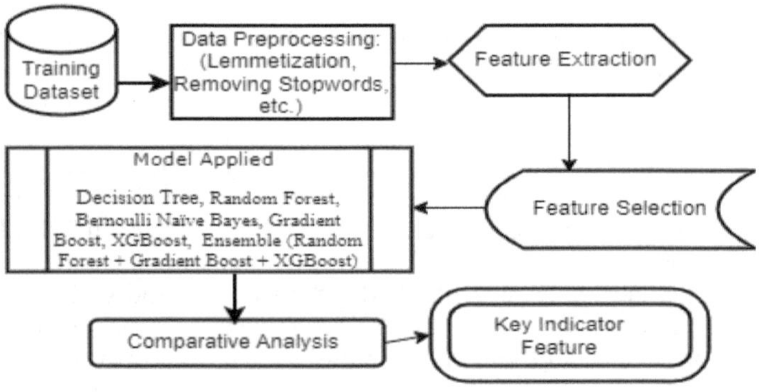

Figure 1. Flow diagram.

3.1 *Dataset and its preprocessing*

In this study, dataset was sourced from Kaggle [31], comprising Reddit posts from "Suicide Watch" and "Depression" subreddits spanning several years, totaling 38,016 posts. These posts were categorized into either related to suicide or not, with 18,905 falling into the suicidal category and 19,111 into the non-suicidal category. Accomplishment of the text ready for training models is really important. It includes steps like breaking text into pieces (tokenization), getting rid of common words (stop words), removing web links (URLs), taking out punctuation, making everything lowercase, reducing words to their base form (lemmatization), and removing non-English words [17] [18] [24]. These steps are crucial for making text data work well.

3.2 *Feature extraction and selection*

Feature extraction from word frequency analysis begins by segmenting text into individual words or tokens. This process reveals how frequently each word appears in the dataset. TF-IDF normalization techniques are regularly used to standardize these frequencies and regulate for differences in text lengths, thereby emphasizing the relative frequency of words. In analytical models, these normalized frequencies oblige as features, with each word representing a feature and its equivalent frequency serving as its value. A correlation coefficient threshold is recognized to determine which features are engaged for modeling. Features surpassing the threshold are considered satisfactorily correlated with the target variable and are kept, while those falling below are rejected.

3.3 *Model applied*

The study exploited to multiple machine learning models, such as Decision Tree (DT), with 120 estimators of Random Forest (RF), Bernoulli Naïve Bayes (NB), with 400 estimators of Gradient Boosting (GB), and with 500 estimators of XGBoost (XGB Apart from previous model, an Ensemble proposed Voting Classifier was formed by integrating Random Forest, Gradient Boosting, and XGBoost (RF+GB+XGB) creates the voting classifier i.e. it evaluates the forecasts from each model and use a majority vote to make the final prediction as shown in Figure 2 . After evaluating these models for accuracy, the research aimed to identify or to extract the keywords from the comparison of the models with the highest accuracy, which were either directly or inversely correlated with suicidal ideation.

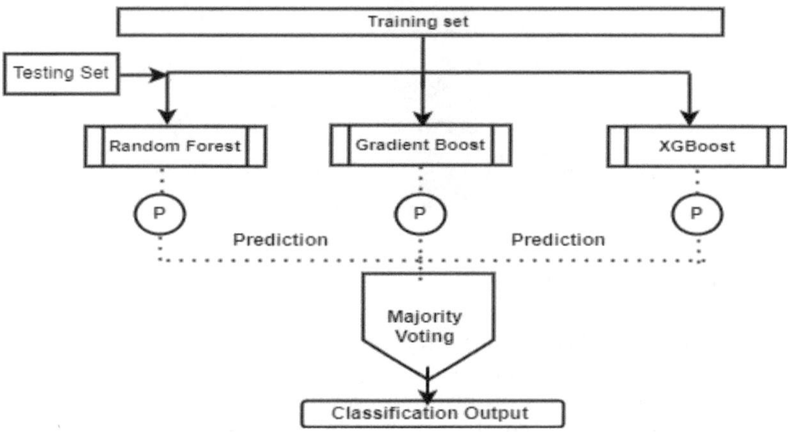

Figure 2. Ensemble proposed voting classifier.

4 EXPERIMENTAL RESULTS

The experiment requires assessing the model using specific metrics i.e., Precision measures the accuracy of identifying suicidal posts, Recall deals the method's ability to detect actual suicidal content and the F1-score is the combination of both Precision and Recall to assess the overall performance in precisely classifying suicidal content.

Text pre-processing is important phase in preparing the dataset for model training. The pre-processing [23] [9] are applied to renovate the data into more reliable and effective form for the model. The data's quality holds a pivotal role in influencing the model's performance. Inconsistent data can significantly hinder the model's effectiveness, emphasizing the importance of thorough data pre-processing or cleaning. After preprocessing and removing unnecessary words, the remaining words in the dataset were used as input features for the machine learning model to make predictions. Moreover, the individual words were analyzed. Thus, the Reddit dataset contained approximately 1,570,218 words for feature engineering. To pinpoint important words within this extensive pool, Frequency was analyzed and filter out the word with low frequency the threshold has been set for less than 5%. This process resulted in a selection of 16,81 significant words for the Reddit dataset. A quantile divides a dataset into segments based on specified proportions of data points. In this study, the 99th percentile is used as a threshold to filter out infrequently occurring words from the Reddit dataset, reducing the total word count to 82. Then one-hot encoding applied to simplify the analysis of word occurrences, transforming categorical data into a binary format suitable for machine learning algorithms. The resulting dataset includes post lengths and word frequencies, forming a solid basis for further analysis. The dataset consists of 38,016 rows and 84 columns, with the output labeled as "label." Figure 3 illustrates the distribution of datasets, which are divided into training and testing sets at an 80:20 ratio.

Various machine learning models, such as Decision Tree (DT), Random Forest with n-estimator=120 (RF), Bernoulli Naïve Bayes (NB), Gradient Boosting with n-estimator=400 (GB), and XGBoost with n-estimator=500 (XGB), were employed and their results are presented in the table. Additionally, an Ensemble Voting Classifier was created by combining Random Forest, Gradient Boosting, an XGBoost (RF+GB+XGB).

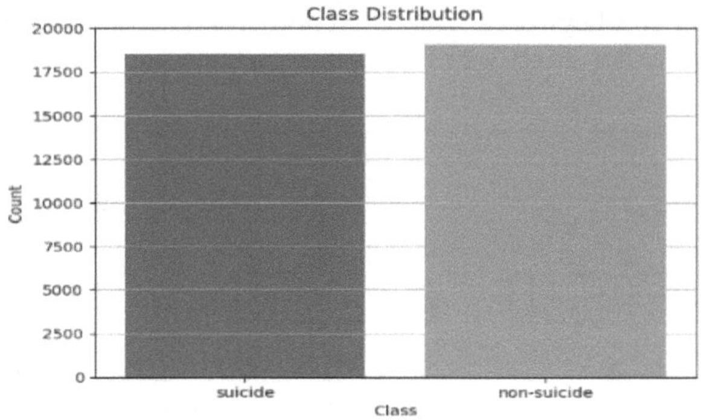

Figure 3. Class distribution of reddit dataset.

The models yielded the following accuracy scores on the Reddit dataset: 81%, 88%, 84%, 89%, 89%, and 92% shown in the Table 1. The Figure 4 shows the ROC Curve of the models applied on Reddit dataset. Based on the analysis presented, it becomes apparent that the ensemble methods, formed by combining Random Forest with 120 estimators (RF), Gradient Boosting with 400 estimators (GB), and XGBoost with 500 estimators (XGB), consistently exhibit superior performance. Based on this model, the identification of highly relevant features associated with suicidal ideation, characterized by a strong correlation with the suicidal ideation label, becomes possible.

Table 1. Evaluation of machine learning in reddit dataset.

Model Evaluation	Decision Tree	Random Forest	Bernoulli Naïve Bayes	Gradient Boost	XGBoost	Ensemble (RF+GB+XGB)
Accuracy score	0.81	0.88	0.84	0.89	0.891	0.92
F1-score	0.815	0.88	0.84	0.895	0.89	0.93
Recall	0.80	0.885	0.845	0.89	0.891	0.92
Precision	0.81	0.88	0.84	0.89	0.89	0.93

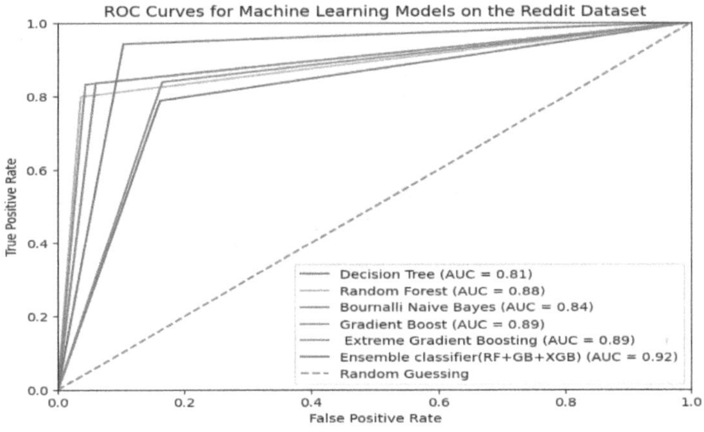

Figure 4. ROC Curve of machine learning models on Reddit dataset.

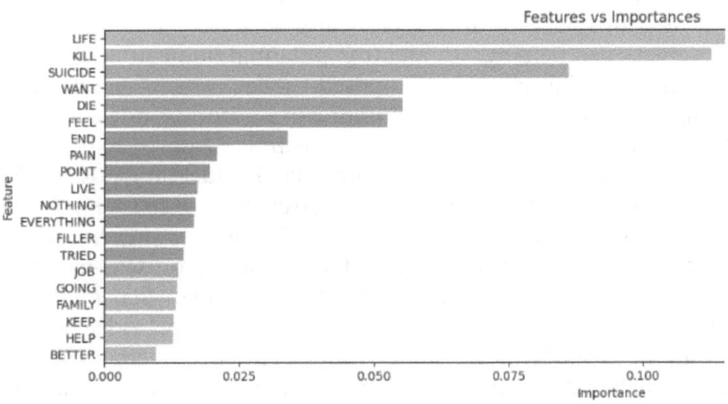

Figure 5. Keywords directly proportional to suicidal ideation in Reddit post.

These selected features are vital for the detection of suicidal ideation. Moreover, the model enables the extraction of reverse features, which exhibit a low correlation with the suicidal label but a high correlation with non-suicidal labels. This insight is derived from the ensemble voting classifier model, achieving an accuracy rate of 92% on the Reddit dataset. The Figure 5 shows the keyword indicator directly proportional to detect the suicidal ideation in Reddit posts, illustrates a significant attribute that is strongly associated with suicidal thoughts.

4.1 *Comparison*

Comparison with the previous related work has been done considering the same type of social media dataset of Reddit dataset in Figure 6. Akshma *et al.* achieved an 88.48% accuracy using Attention over CNN and LSTM on a Reddit dataset. In 2023, Ghosal's fastText, TF-IDF, and XGBoost approach outperformed baseline models with a 0.78 AUC and 0.71 weighted F-score, while Sawney's SISMO model achieved a 73% accuracy in identifying suicide sensitivity in Reddit posts. The ensemble proposed methods achieve 92% on Reddit datasets.

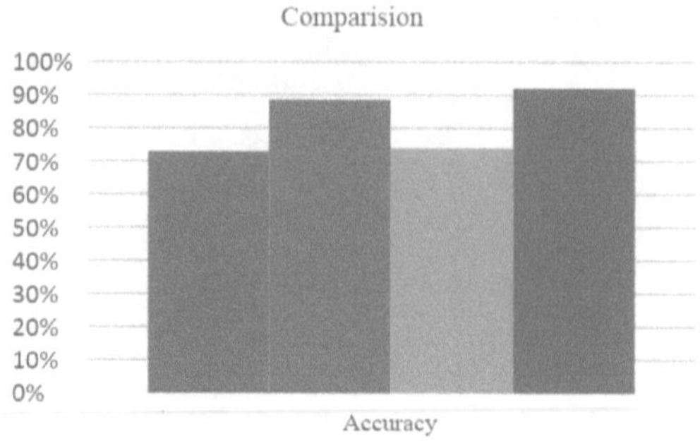

Figure 6. Comparison with existing work.

5 CONCLUSION

In conclusion, pinpointing signs of suicidal thoughts on social media is crucial for suicide prevention, given the growing use of these platforms as outlets for expressing emotions. This study revealed that people dealing with despair and suicidal thoughts often use specific words and language patterns in their Reddit posts. By using advanced Natural Language Processing (NLP) techniques, the study successfully uncovered these behavioral patterns in social media interactions, helping us identify potential indicators of suicidal ideation. Notably, the ensemble voting classifier proposed model, which includes Random Forest, Gradient Boosting, and XGBoost, achieved impressive accuracy rates of 92% on the Reddit dataset. It correspondingly identified key words such as "Life," "Kill," "Suicide," "Die," "End," "Pain" etc. as strongly associated with suicidal ideation posts, while words like "Good," "Happy," "Made," "Need" etc. were inversely related to such posts. These findings offer significant insights into using technology to aid in suicide prevention efforts on social media. The limitation of this study is that it focuses solely on text-based content, which means it doesn't consider potential behavioral cues or changes that might appear in other formats like images, videos, audio clips, or additional context provided by emojis or related information within the posts. By doing so, it overlooks important contextual clues that could give us a deeper insight into people's behaviors. Therefore, recognizing these limitations opens the door to possible future improvements and expansions of this research.

REFERENCES

[1] Katchapakirin, K., Wongpatikaseree, K., Yomaboot, P., & Kaewpitakkun, Y. (2018). Facebook social media for depression detection in the Thai community. *15th International Joint Conference on Computer Science and Software Engineering (JCSSE)* IEEE :1–6.

[2] Gao, J., Cheng, Q., & Yu, P. L. (2018). Detecting comments showing risk for suicide in YouTube. *Proceedings of the Future Technologies Conference* Springer, Cham :385–400.

[3] Valeriano, K., Condori-Larico, A., & Sulla-Torres, J. (2020). Detection of suicidal intent in Spanish language social networks using machine learning. *International Journal of Advanced Computer Science and Applications*, 11(4).

[4] Aladağ, A. E., Muderrisoglu, S., Akbas, N. B., Zahmacioglu, O., & Bingol, H. O. (2018). Detecting suicidal ideation on forums: proof-of-concept study. *Journal of medical Internet research* 20(6): e9840.

[5] Coppersmith, G., Leary, R., Crutchley, P., & Fine, A. (2018). Natural language processing of social media as screening for suicide risk. *Biomedical informatics insights* 10 1178222618792860.

[6] Sawhney, R., Manchanda, P., Mathur, P., Shah, R., & Singh, R. (2018). Exploring and learning suicidal ideation connotations on social media with deep learning. *Proceedings of the 9th workshop on computational approaches to subjectivity, sentiment and social media analysis*: 167–175.

[7] Tadesse, M. M., Lin, H., Xu, B., & Yang, L. (2019). Detection of suicide ideation in social media forums using deep learning. *Algorithms* 13(1): 7.

[8] Wang, N., Luo, F., Shivtare, Y., Badal, V. D., Subbalakshmi, K. P., Chandramouli, R., & Lee, E. (2021). Learning models for suicide prediction from social media posts. *arXiv preprint arXiv*:2105.03315.

[9] Sawhney, R., Joshi, H., Shah, R., & Flek, L. (2021). Suicide ideation detection via social and temporal user representations using hyperbolic learning. *Proceedings of the 2021 Conference of the North American Chapter of the Association for Computational Linguistics: Human Language Technologies*:2176–2190.

[10] Li, Z., Zhou, J., An, Z., Cheng, W., & Hu, B. (2022). Deep hierarchical ensemble model for suicide detection on imbalanced social media data. *Entropy* 24(4): 442.

[11] Liu, J., Shi, M., & Jiang, H. (2022). Detecting suicidal ideation in social media: an ensemble method based on feature fusion. *International journal of environmental research and public health* 19(13):8197.

[12] Zhang, T., Schoene, A. M., Ji, S., & Ananiadou, S. (2022). Natural language processing applied to mental illness detection: a narrative review. *NPJ digital medicine* 5(1): 1–13.

[13] Hochreiter, S. & Schmidhuber, J. (1997). Long short-term memory. *Neural Computation* 9(8).

[14] Renjith, S., Abraham, A., Jyothi, S. B., Chandran, L., & Thomson, J. (2021). An ensemble deep learning technique for detecting suicidal ideation from posts in social media platforms. *Journal of King Saud University-Computer and Information Sciences.*

[15] Haque, F., Nur, R. U., Al Jahan, S., Mahmud, Z., & Shah, F. M. (2020). A transformer-based approach to detect suicidal ideation using pre-trained language models. *23rd international conference on computer and information technology ICCIT* IEEE :1–5.

[16] Abdulsalam, A., & Alhothali, A. (2022). Suicidal Ideation Detection on Social Media: A Review of Machine Learning Methods. *arXiv preprint arXiv:2201.10515.*

[17] Huang, X., Li, X., Liu, T., Chiu, D., Zhu, T., & Zhang, L. 2015. Topic model for identifying suicidal ideation in chinese microblog. *Proceedings of the 29th pacific asia conference on language, information and computation* :553–562.

[18] Cao, L., Zhang, H., Feng, L., Wei, Z., Wang, X., Li, N., & He, X. (2019). Latent suicide risk detection on microblog via suicide-oriented word embeddings and layered Attention. *arXiv preprint arXiv:1910.12038.*

[19] Du, J., Zhang, Y., Luo, J., Jia, Y., Wei, Q., Tao, C., & Xu, H. (2018). Extracting psychiatric stressors for suicide from social media using deep learning. *BMC medical informatics and decision making*, 18 (2):77–87.

[20] Skaik, R., & Inkpen, D. (2020). Using social media for mental health surveillance: a review. *ACM Computing Surveys (CSUR)* 53(6): 1–31.

[21] Just, M. A., Pan, L., Cherkassky, V. L., McMakin, D. L., Cha, C., Nock, M. K., & Brent, D. (2017). Machine learning of neural representations of suicide and emotion concepts identifies suicidal youth. *Nature human behaviour* 1(12):911–919.

[22] Lotito, M., & Cook, E. (2015). A review of suicide risk assessment instruments and approaches. *Mental Health Clinician* 5(5):216–223.

[23] Tan, Z., Liu, X., Liu, X., Cheng, Q., & Zhu, T. (2017). Designing microblog direct messages to engage social media users with suicide ideation: interview and survey study on Weibo. *Journal of medical Internet research*, 19(12), e8729.

[24] Varathan, K. D., & Talib, N. (2014). Suicide detection system based on Twitter. *2014 Science and information conference* IEEE :785–788.

[25] Gunn, J. F., & Lester, D. (2012). Twitter postings and suicide: An analysis of the postings of a fatal suicide in the 24 hours prior to death. *Suicidologi* 17(3).

[26] Colomb, G. B., Burnap, P., Hodorog, A. & Scourfield, J. (2016). Analyzing the connectivity and communication of suicidal users on twitter. *Computer communications* 73: 291–300.

[27] Braithwaite, S. R., Giraud-Carrier, C., West, J., Barnes, M. D., & Hanson, C. L. (2016). Validating machine learning algorithms for Twitter data against established measures of suicidality. *JMIR mental health* 3(2) e4822.

[28] Walsh, C. G., Ribeiro, J. D., & Franklin, J. C. (2017). Predicting risk of suicide attempts over time through machine learning. *Clinical Psychological Science* 5(3): 457–469.

[29] Ghosal, S. & Jain, A. (2023). Depression and Suicide Risk Detection on social media using fastText Embedding and XGBoost Classifier. *Procedia Computer Science* 218:1631–1639.

[30] Chadha, A. & Kaushik, B. (2022). A Hybrid Deep Learning Model Using Grid Search and Cross-Validation for Effective Classification and Prediction of Suicidal Ideation from Social Network Data. *New Gener. Comput.* 40: 889–914. Kaggle : https://www.kaggle.com/datasets/nikhileswarkomati/suicide-watch

Computational Methods in Science and Technology Volume 1 – Kaur et al. (Eds)
© 2025 the Author(s), ISBN: 978-1-032-81760-6

Advanced predictive analytics for cryptocurrency valuation: A comparative study of machine learning and deep learning approaches

Akhil Verma, Ranjan Walia & Vaneet Kumar
Department of Computer Science and Engineering, Chandigarh University, Gharaun, Mohali, India

ABSTRACT: Cryptocurrency is a digital or virtual currency that uses a decentralized system through cryptography for secure financial transaction and record keeping. This process is widely known as mining, where the decentralized nature of the system works without central issuance. One of the objectives is to the end that the spread of the influence of institutions and organizations should be up to the users/investors with enhanced authority. In India, this has been under deployment and substantial deliberation in the relevance and importance it has come to assume in the lives of people. This is largely due to the fact that availability has increased, and it also would need high utilization of processor by things like mining. The recent past has increasingly seen popularity and importance to which cryptocurrencies hold relevance. In fact, this was manifest in the massive growth of rapid advances and that which had led many global businesses to acknowledge the importance of using this technology to gain access to myriad quantitative benefits electronically. The concept of rice forecasting has been very pivotal in different industries, like in cryptocurrency and stock prices. It greatly challenges the current financial market since indeed, traders have had the dilemma of selling their cryptocurrencies or buying them.

1 INTRODUCTION

Cryptocurrencies are the first online, virtual currencies and represent one of the most popular and promising inventions over the last years. Bitcoin is the leader in this group and, according to market capitalization, is considered the largest cryptocurrency. The best criterion defining this kind of currency is its decentralized nature: in plain words, there is no central authority that takes the reins of currency emissions or valuations. So, the value of cryptocurrencies is highly volatile, and in fact, predicting a value close to the future is very hard.Precise prices of cryptocurrencies are indispensable for every trader, investor, and financial analyst to make correctly judged decisions that will minimize, in the best way, the risks connected with them.

Cryptocurrency price prediction [2,3] therefore alludes to the use of a variety of skills, strategies, and techniques, all with the aim of trying to predict price changes of the cryptocurrency in the future. One needs market trends, past prices, the number of transactions, technical signs, and basic considerations inclusive of sentiments to forecast the future price movement of a given cryptocurrency. However, in the course, one ought to be very cautious because the markets of these cryptocurrencies do portray enormous volatility and unpredictability. Substantially, these projections have become more questionable in relation to their applicability and relevance, and there is no warranty for the achievement of the set objectives. From then, artificial intelligence (AI), including techniques like machine learning (ML) and deep learning (DL), has become essential in various walks associated with cryptocurrency. Many researchers and financial analysts have predicted the price of Bitcoin using Artificial Intelligence (AI) on the basis of analyzing historical price data and market trends to predict future movements [3–7].

 DOI: 10.1201/9781003501244-60

1.1 *Need for cryptocurrency price forecasting*

Cryptocurrency price is projected for many reasons. Very accurate price forecasting of the cryptocurrency is a necessity to navigate a course through this highly dynamic and rapidly changing field of digital assets. - Investment Decision Support: The tool provides help and support to the customers, investors, and traders of cryptocurrencies for making sound decisions regarding the purchase, selling, or holding of the cryptocurrencies based only on the expected future changes in prices. Risk Management: Forecasting helps to reduce the level of risk as it discloses price trends and volatility, which eventually enables investors to determine when to alter their game plans. Market Analysis: It provides important information on the sentiment of the market, investor behavior, and the general state regarding the market of bitcoin. This becomes a valuable intention for investing purposes, both in the short and long terms. Business Planning: Where accurate prediction in the industry of bitcoins could be very helpful for firms to allocate resources, set up a budget, and make strategic decisions. Research and Development: Forecasting plays a critical responsibility in the stimulation of innovation and research in business with the use of bitcoin; it detects patterns and potential rising areas for growth.

Table 1. Recent works on ML/DL for cryptocurrency price forecasting.

Reference/ Year	ML/DL Models	Classification category	Dataset	Methodology/ Algorithm	Limitation	Results
Shreya, M. G. A. P., *et al.* (2023) [5]	ANN, LSTM, and RNN	To evaluate and forecast the fluctuations in the prices of ether, ripple, and bitcoin.	Agent view predictions dataset	Cryptocurrency market price prediction	Depending on the type of ML model being utilized, the predictability explanation could be different.	LSTM employs short-term dynamics but ANN depends on long-term history, suggesting that LSTM is more effective than ANN at using historical data.
Belcastro, L., *et al.* (2023) [8]	LSTM model	The relationship between price fluctuations and social media posts, causality, and user sentiment regarding cryptocurrencies.	Two huge datasets: cryptocurrency market data like trading and prices volumes and social media posts about these coins.	Statistical, text analytics, and deep learning techniques, and trading algorithm	Subsequent research efforts ought to investigate alternatives with reduced risk and enhanced capital management to mitigate losses.	The proposed trading algorithm is capable of optimizing strategies and increasing profits.
Srivastava, V., *et al.* (2023) [20]	Proposed model: Particle Swarm Optimization with XGBoost	A performance evaluation of the cryptocurrencies Bitcoin, Dogecoin, and Ethereum	The BTC.csv, ETH_1H.csv, and Doge-USD.csv, files are the datasets	Regression Algorithm.	Consistency is frequently absent in cryptocurrency price prediction study as a result of varying trading characteristics and datasets. The use of nonlinear methods and experiments may enhance precision.	With modest metric percentages for Ethereum, Dogecoin, and Bitcoin, the proposed model accurately predicted the closing prices of these cryptocurrencies, suggesting its potential application in cryptocurrency trading.
Al Hawi, L., *et al.* (2023) [11]	SVM, KNN, and LGBM	To forecast the price volatility of Bitcoin, Ethereum, and Litecoin.	Kaggle dataset	ML algorithms	The accuracy can be further enhanced.	In the second inquiry phase, the LGBM beats the LGBM for Litecoin and Ethereum in the individual dataset, while the KNN beats the SVM in forecasting Bitcoin.

LSTMs can be utilized for price prediction by identifying trends in historical pricing data. The training procedure optimizes the loss function, which measures the difference between forecasted and actual prices. The process of forward training can be explained by utilizing the following equations [49]:

$$i_t = \sigma(W_i[h_{t-1}, x_t] + b_i)$$

$$f_t = \sigma(W_f[h_{t-1}, x_t] + b_f)$$

$$c_t = f_t.c_{t-1} + i_t.tanh\,(W_c[h_{t-1}, x_t] + b_c)$$

$$o_t = \sigma(W_o[h_{t-1}, x_t] + b_o)$$

$$h_t = o_t tanh\,(c_t)$$

The hidden state at time step t, input state, input gate, cell state, output gate, and forget gate are represented with the respective following variables: ht, xt, it, ct, ot, and ft. (28,29) Where b is the bias vector, h is a weight matrix, and the resulting output is bounded between -1 and 1. Gated Recurrent Unit: GRU (28,29) is a variant of RNN models. It is widely applied in Natural Language Processing (NLP) and Audio Recognition in predictive analytics, e.g., price prediction. It works on sequential data by maintaining inner state changing with new input, and it is further able to remember or forget selected pieces of information in previous time steps. GRU uses the gating mechanism to control the flow of information, thus overcoming the limitation of vanishing gradient. It consists of two gating units: a reset gate, and an update gate, which can be adapted to behave according to the task it is employed in. For training time, GRU is faster and easier in deployment compared to LSTM, in that it involves far fewer variables, hence consuming less of computational resources. Suitable for prediction of price pattern in the erratic and always changing cryptocurrency market, managing sequences of data with different lengths, and analyzing massive datasets. This model predicts the corresponding future price patterns using the attributes extracted and also taking in the past price data. The model generates a catalog of projected prices for a specified period of time [17]. The following function computes the hidden state, ht, at time t with the inclusion of the previous hidden state, ht−1, and the input at time t, xt, by using the following equations [49]:

$$u_t = \sigma(W_u[h_{t-1}, x_t])$$

$$r_t = \sigma(W_r[h_{t-1}, x_t])$$

$$h_t = (1 - u_t).h_{t-1} + u_t.tanh\,(W_r[r_t.h_{t-1}, u_t])$$

Where u_t represents the update gate and r_t represents the reset gate.

2 APPLICATIONS, CHALLENGES AND FUTURE DIRECTIONS

2.1 *Cryptocurrency price forecasting applications*

Cryptocurrency price forecasting tools employ various approaches and technologies to provide consumers with important insights into future price variations [33–35]. Cryptocurrency price forecasting software is crucial for aiding traders and investors in navigating the volatile and constantly evolving cryptocurrency markets. By employing advanced analytics and predictive modeling techniques, these programs empower users to make informed decisions and effectively manage their investment portfolios. The following are commonly used platforms and programs for predicting cryptocurrency prices:

2.1.1 *Trading platforms*

Cryptocurrency exchanges offer traders a range of tools, including price predictions, indicators, technical evaluation charts, real-time market data, and customizable trading algorithms. These tools enable traders to understand market trends and make well-informed decisions.

2.1.2 *Analytics platforms*

Cryptocurrency analytics systems provide sophisticated tools for predicting prices, utilizing ML models, analyzing sentiment, and tracking social media. These platforms appeal to both individual traders and investors from institutions.

2.2 *Challenges in cryptocurrency price forecasting*

There are many obstacles when trying to forecast cryptocurrency prices using ML or DL techniques. All these problems can be addressed only by the right mix of strong data preparation, selection of features, and cautious choice of models along with hyperparameter tuning, and finally continuous monitoring and adjustment of these models for the forecast. In addition, the integration of expert knowledge in a certain field to understand the market forces could provide an improvement at the level of accuracy with respect to the price of Bitcoin. Volatility: The prices of the cryptocurrency can change very fast and unexpectedly, thrusting the respective market into very high levels of volatility. Models trained on earlier data are unlikely to capture these dynamics with any semblance of accuracy. Feature Engineering: Due to the impact, it is hard but necessary to select these very important features responsible for the bitcoin price dynamics. The cryptocurrency market is influenced by manifold elements, such as market mood, news sentiment, trading volume, and activity on the network.

2.2.1 *Data quality and noise*

First of all, the extraction of relevant patterns from cryptocurrencies data essentially represents an effort, as it is noisy and manipulable. Furthermore, data obtained from dissimilar sources may contain differences or errors, which could very well reflect in the efficacy of the model. Non-linear Relationship: The fluctuation in the price of most cryptocurrencies is of a non-linear pattern, reflecting precisely where it becomes a big issue for linear models.

2.2.2 *Overfitting*

The model overfits new data because generally, it learned everything from the training set, including noise and other non-generalizable trends. This is the case where you have to resort to regularization techniques and comprehensive model validation.

2.2.3 *Interpretability*

This refers to the interpretability problem of the black-box type that deep learning models, such as neural networks, put across; it is because the structure thereof is too complex, with many dimensions. The interpretative measure for interpretation of available models is, therefore, so small that the ability to make predictions by model cannot be reasonably understood.

2.2.4 *Market manipulation*

Digital currency markets are exposed to fraud and theft that often lead to data on prices that can be manipulated, forecast models that get deceived.

3 CONCLUSION

Deep and machine learning techniques, therefore, do unlock the ability to predict cryptocurrency prices. The algorithms have the capability of predictive capture that releases great complex patterns and nonlinear relationships between various kinds of variables in markets. Of course, a number of challenges emerge, like the quality and noise from the existing data. Model selection and complexity are both very important, and models such as the one used in this study have the capability to capture temporal dependencies. Other important aspects include the dynamics of the market and external; it is very risk-appropriate that a comprehensive risk assessment framework form's part and parcel of the risk management strategies. It is evident that the field of cryptocurrency price forecasting is very fast-growing, with further research, development, and innovation taking up to improve on the faced challenges and enhance prediction accuracy.

REFERENCES

[1] Amsyar, I., Christopher, E., Dithi, A., Khan, A. N., and Maulana, S. (2020). The challenge of cryptocurrency in the era of the digital revolution: A review of systematic literature. *Aptisi Transactions on Technopreneurship (ATT)*, 2(2), 153–159.

[2] Khedr, A. M., Arif, I., El-Bannany, M., Alhashmi, S. M., and Sreedharan, M. (2021). Cryptocurrency price prediction using traditional statistical and machine-learning techniques: A survey. *Intelligent Systems in Accounting, Finance and Management*, 28(1), 3–34.

[3] Surarapu, S., Vishnu, K., Shirisha, N., Sankalp, T. andSatish, B. T. S. (2022). Cryptocurrency price analysis using artificial intelligence. *Journal of Algebraic Statistics*, 13(3), 486–493.

[4] Ragasri, P. N., Kishore, B. N., Srikanth, M., Prasad, A. D. R. and Srikanth, A. D. R. P. (2023). Cryptocurrency price analysis with artificial intelligence. *Journal of Science and Technology*, 8(04).

[5] Shreya, M. G. A. P., Shailaja, P., Akhila, M. and Sravani, P. (2023). A novel cryptocurrency price analysis with artificial intelligence. *Turkish Journal of Computer and Mathematics Education (TURCOMAT)*, 14(03), 787–792.

[6] Hosen, M., Thaker, H. M. T., Subramaniam, V., Eaw, H. C. and Cham, T. H. (2022, September). Artificial intelligence (AI), blockchain, and cryptocurrency in finance: current scenario and future direction. In *International Conference on Emerging Technologies and Intelligent Systems* (pp. 322–332). Cham: Springer International Publishing.

[7] Amirzadeh, R., Nazari, A. and Thiruvady, D. (2022). Applying artificial intelligence in cryptocurrency markets: A survey. *Algorithms*, 15(11), 428.

[8] Belcastro, L., Carbone, D., Cosentino, C., Marozzo, F. and Trunfio, P. (2023). Enhancing cryptocurrency price forecasting by integrating machine learning with social media and market data. *Algorithms*, 16(12), 542.

[9] Awotunde, J. B., Ogundokun, R. O., Jimoh, R. G., Misra, S. and Aro, T. O. (2021). Machine learning algorithm for cryptocurrencies price prediction. In *Artificial Intelligence for Cyber Security: Methods, Issues and Possible Horizons or Opportunities* (pp. 421–447). Cham: Springer International Publishing.

[10] Derbentsev, V., Babenko, V., Khrustalev, K. I. R. I. L. L., Obruch, H. and Khrustalova, S. O. F. I. I. A. (2021). Comparative performance of machine learning ensemble algorithms for forecasting cryptocurrency prices. *International Journal of Engineering*, 34(1), 140–148.

[11] Al Hawi, L., Sharqawi, S., Al-Haija, Q. A. and Qusef, A. (2023). Empirical evaluation of machine learning performance in forecasting cryptocurrencies. *Journal of Advances in Information Technology*, 14(4).

[12] Awotunde, J. B., Ogundokun, R. O., Jimoh, R. G., Misra, S. and Aro, T. O. (2021). Machine learning algorithm for cryptocurrencies price prediction. In *Artificial Intelligence for Cyber Security: Methods, Issues and Possible Horizons or Opportunities* (pp. 421–447). Cham: Springer International Publishing.

[13] Hamayel, M. J. and Owda, A. Y. (2021). A novel cryptocurrency price prediction model using GRU, *LSTM and bi-LSTM machine learning algorithms. AI*, 2(4), 477–496.

[14] Pintelas, E., Livieris, I. E., Stavroyiannis, S., Kotsilieris, T. and Pintelas, P. (2020). Investigating the problem of cryptocurrency price prediction: a deep learning approach. In *Artificial Intelligence*

Applications and Innovations: 16th IFIP WG 12.5 International Conference, AIAI 2020, Neos Marmaras, Greece, June 5–7, 2020, Proceedings, Part II 16 (pp. 99–110). Springer International Publishing.

[15] Ammer, M. A. and Aldhyani, T. H. (2022). Deep learning algorithm to predict cryptocurrency fluctuation prices: Increasing investment awareness. *Electronics*, 11(15), 2349.

[16] Zhang, Z., Dai, H. N., Zhou, J., Mondal, S. K., García, M. M. and Wang, H. (2021). Forecasting cryptocurrency price using convolutional neural networks with weighted and attentive memory channels. *Expert Systems with Applications*, 183, 115378.

[17] Vaddi, L., Neelisetty, V., Vallabhaneni, B. C. and Prakash, K. B. (2020). Predicting crypto currency prices using machine learning and deep learning techniques. *Int. J*, 9(4).

[18] Inder, S. and Sharma, S. (2021). Predicting the movement of cryptocurrency "bitcoin" using random Forest. In *Data Science and Computational Intelligence: Sixteenth International Conference on Information Processing, ICInPro 2021*, Bengaluru, India, October 22–24, 2021, Proceedings 16 (pp. 166–180). Springer International Publishing.

Computational Methods in Science and Technology Volume 1 – Kaur et al. (Eds)
© 2025 the Author(s), ISBN: 978-1-032-81760-6

Enhancing urban mobility and providing way for ambulance using machine learning

Akanksha, Kashish Gupta, Khushi Vaish & Nikita Sharma
Department of Computer Science KIET Group of Institutions, Delhi-NCR, Ghaziabad, India

ABSTRACT: In today's world we have been facing problems due to increasing traffic on roads. This causes a lot of waste of time and increases stress level. So, this leads for creating a system which can intelligently handle traffic by itself modifying its timing in response to traffic density provide passage for ambulances using RCNN (Region-based Convolutional Neural Network), webcam, NodeMCU controller, LEDs. The core idea revolves around traffic management through the assessment of traffic volume on each side of the road, with the aim of implementing smart traffic signal control based on this density information. The webcam captures real-time images of the automobiles on the road and send images to the NodeMCU microcontroller and then we apply RCNN algorithm and OpenCV to recognize and tally the count of automobiles on the lane and set traffic timing accordingly. If ambulance is detected traffic light turns green on that side and other side are turn red to provide passage for emergency vehicles.

Keywords: Traffic Light System, RCNN, OpenCV, NodeMCU controller, Ambulance, Save Lives

1 INTRODUCTION

One of the pressing issues exacerbated by the world's growing inhabitants and the swift increase in the count of automobiles is traffic blockage. In countries like India, the estimate of expanding lane infrastructure lags far behind the rate of vehicle growth[1]. Traffic jams give rise to numerous critical problems that have a direct impact on people's daily lives and, in some cases, even lead to loss of life[2]. For instance, when an emergency vehicle such as an ambulance is transporting a critically ill patient, being trapped in a heavy traffic jam significantly reduces the chances of the patient reaching the hospital in a timely manner.

Therefore, it is imperative to develop an intelligent traffic management system that can effectively regulate traffic to prevent accidents, collisions, and traffic gridlocks, with a special focus on ensuring the swift passage of emergency vehicles like ambulances[3]. These challenges can be mitigated through the application of emerging technologies, specifically Machine Learning (ML). By establishing connections between a range of software packages and hardware devices, whether through wireless or wired means, a system can be created to efficiently regulate traffic. This system operates by automatically adjusting signal timings in accordance with traffic density and also facilitates the unobstructed passage of emergency vehicles, such as ambulances, utilizing RCNN (Region-based Convolutional Neural Network), webcams, NodeMCU controllers, and LEDs.

The problems that exists in traditional traffic handling system are as follows:

 DOI: 10.1201/9781003501244-61

1.1 *High traffic*

As the number and diversity of automobiles on the road continue to grow, major cities are experiencing a significant rise in heavy traffic congestion. This congestion is particularly pronounced at intersections, especially both throughout the working day and in the evenings when commuters are heading home from work. The primary consequence of this situation is the prolonged time wasted by individuals while on the road. To address this issue, apotential solution involves implementing a novel program with distinct delay settings tailored to various intersections.

1.2 *Have to wait, even if there is no traffic*

At certain intersections, even when traffic is not excessively congested, individuals often find themselves waiting. This is due to the fixed timing of traffic signals, requiring road users to wait until the lights change from red to green. Violating this rule incurs fines. Additionally, coordination of signal timings at different intersections is essential.

Traffic congestion is widely recognized as a significant problem in urban areas, and it is expected to worsen with the growing population. Currently, traffic signal lights operate on fixed time schedules, following a predetermined cycle. This can lead to inefficient traffic flow, with one lane being overly congested while others remain vacant. In cases where traffic density varies at different sides of an intersection, a mechanism is proposed to adapt red and green light durations based on the solidity of traffic. In instances of long bottlenecks where automobiles come to a complete stop, drivers and passengers can experience considerable mental stress. This can escalate into road rage, underscoring the severity of the traffic congestion problem. Therefore, there is a pressing need for intelligent solutions to address this issue.

2 LITERATURE SURVEY

In [4]Abbas, *et al.* presents a method for assessing lane-specific traffic density through the application of image processing. This approach offers distinct advantages, primarily obviating the need for costly aerial imagery and intricate sensor-based systems, thus rendering it a cost-effective solution. Notably, our system does not necessitate the putting in of additional systems like RFIDs. Future improvements can include the incorporation of an emergency vehicle identification system, affording priority to such vehicles, and the integration of Vehicular Ad-hoc Networks (VANETs) to enhance road safety and contribute to the development of intelligent transportation systems.

In [5]Jadhav, *et al.* introduce a novel approach to estimate traffic, employing image processing on captured highway camera footage to count vehicles, issue warnings for heavy traffic if thresholds are exceeded, and offer cost-effectiveness, ease of implementation, and notable accuracy and speed compared to traditional sensor-based systems. It is executed using MATLAB and image processing and aims to alleviate heavy traffic congestion. It involves the utilization of web cameras positioned in traffic lanes to capture road images, which are then meticulously processed to gauge traffic density. Based on the data analysis conducted through MATLAB, the system commands traffic signal LEDs to display specific time intervals, facilitating traffic management.

In D. M. [6]employs switches, rather than magnetic sensors, to assess traffic density and modulate green signal durations based on this information, leading to the desired outcomes. Future project development involves extending the system to manage all four lanes at a traffic signal by employing efficient PIC interfacing techniques. The primary goal of the project is to optimize traffic flow by effectively utilizing green signal durations. In this system, the traffic density in a specific lane is determined by the count of magnetic sensors placed at the roadside, which generate output signals corresponding to the traffic volume.

In [7], M.F. Chowdhury, *et al.* propose a dynamic traffic control system that uses image processing and real-time video feeds to measure the traffic density at intersections. After gathering a video sample, a combination of Gaussian algorithms was applied to foreground detection and background subtraction in order to maintain track of the count of cars in each lane. By using their line of centroid, the vehicles are identified. A vehicle is confirmed by centroid movement. The traffic conditions detected from the video feeds will be used to dynamically adjust the traffic lights at the intersections.

In [8], the research team created a basic system using Raspberry Pi, OpenCV, and the Background Subtraction Method, that could identify and tally cars on the Sidoarjo Toll Road in real-world scenarios. For four minutes each, experiments were run in the morning, afternoon, and evening. 92.31% was the highest accuracy recordedin the morning, and 77.27% was the lowest in the afternoon. 91% of the time, the background Subtraction using the GMM method produced accurate results. Variations in accuracy were ascribed to environmental factors, including illumination, vehicle motion, and environmental stability. In order to predict object movement, the primary steps in the process were background subtraction, object tracking, and video capture. It was discovered that the best position for the camera to detect and count vehicles was at a height of $\pm$ 6 meters.

In [9], Anam Firdous *et al.* proposed a system that uses Arduino Uno ATMega 328 to manage traffic density. In this, digital infrared sensors are used to sense traffic, and infrared sensors further identify vehicles based on the signal that is reflected from them. sensors positioned next to the lane to regulate traffic volume through suitable traffic signal changes. The Arduino Uno is interfaced with every infrared sensor, and it receives data from the sensors. The system's traffic signals are made of LEDs, with two LEDs for each lane in each signal. The authors have created a smart traffic system that uses IR sensors and Arduino to manage traffic in an efficient manner, reducing delays and utilizing renewable energy sources. We had to somewhat reduce the likelihood of traffic jams by putting this system into place, and the outcome was achieved. Additionally, we power the Arduino Uno using renewable energy by using a solar panels, which *also requires very little power*. In[10] proposed a system in order to optimize traffic flow, this study suggests a decentralized smart traffic management system that makes use of intelligent algorithms and deep learning. To determine the density of cars and control traffic signals, the system makes use of infrared sensors, object recognition algorithms, and digital cameras. Future traffic density is predicted by an infrared sensor, and emergency vehicles such as police cars and ambulances are given priority by RFID readers. The AT Mega 328 P is used by the system to regulate signal timing. In [18] Liu, Y *et al.* states that his study proposes a TSC system to maximize the number of vehicles crossing an intersection and balances the signals between roads by using Q-learning (QL). In [12] Pratishtha Gupta *et al.* states that the main purpose of the developed algorithm is to compute total traffic load at a particular junction, which is then further used for real time traffic control by generating green light timing of the traffic signal. In [13] Sabhya Sachi Kanaojia states that in this system we are using Canny edge detection. In [15] Liu, Y The system is endowed with robust and efficient detection techniques, which main features are the statistical and knowledge-based background update and the use of HSV color information for shadow suppression. In [14] Wang Y. states that reinforcement learning plays an important role in traffic congestion system. In [16] Elizabeth Basil states that this system consists of WI-FI transceiver module it transmits the vehicle count of the current system to the next traffic signal. In [17] Bilal Khan states that we propose a system based on PIC microcontroller that evaluates the traffic density using IR sensors and accomplishes dynamic timing slots with different levels.

3 PROPOSED WORK

The goal of traffic management is to minimize other effects of the current road network while maximizing the efficiency [11] of the traffic structure without any issues. The system is improving upon the shortcomings of earlier traffic management frameworks. The mechanism uses automobiles' traffic density as determined by camera input, which is separated from pictures of cars. We have issues in the modern world because of the rise in traffic on the roads. This raises stress levels and results in significant time waste. Many emergency vehicles get caught in traffic for hours at a time, which results in fatalities. The NodeMCU microcontroller receives images from the webcam that show the cars on the road. We then use the RCNN algorithm and OpenCV to count the number of cars in each lane and adjust traffic timing accordingly. If an ambulance is seen in the road. When an ambulance is spotted, the traffic light on that side turns green, and the other side turns red to allow emergency vehicles to pass.

Following are the components used:

3.1 *NodeMCU controller*

There are open source prototyping board designs available for the NodeMCU firmware. "Node" and "MCU" (micro-controller unit) are combined to form the term "NodeMCU". The firmware itself is referred to as "NodeMCU" strictly speaking, not the related development kits.The designs for the prototyping board and firmware are both available for free. The scripting language Lua is used by the firmware. The firmware was developed using the Espress if Non-OS SDK for ESP8266 and is based on the eLua project.

3.2 *LEDs (Light Emitting Diode)*

Light is released from a light-emitted diode (LED), when current flows through it. The recombination of electrons and electron holes in the semiconductor releases photons. The color of light is connected to photon energy and is determined by the energy required for electronsto pass the band gap of the semiconductor. White light can be produced by using numerous semiconductors or by coating the semiconductor device with a layer of light-emitting phosphor.

3.3 *RCNN (Region-based Convolutional Neural Network)*

In computer vision, region-based convolutional neural networks, or RCNNs, are essential object detection algorithms. The task is split into two phases: producing region suggestions in an image where objects could be present, and then categorizing these regions and fine-tuning their borders. Using selective search, RCNN creates region proposals, from which a convolutional neural network that has been trained to extract features is used. After that it uses a support vector machine algorithm to categorize objects inside these regions and fine-tunes the bounding box coordinates for accurate localization.

It is the flow diagram of system showing how images are captured and processed and what path is taken after processing the real time image through Node MCU.

The flow chart as shown in Figure 1 above is described in the following steps:

STEP 1- Start.

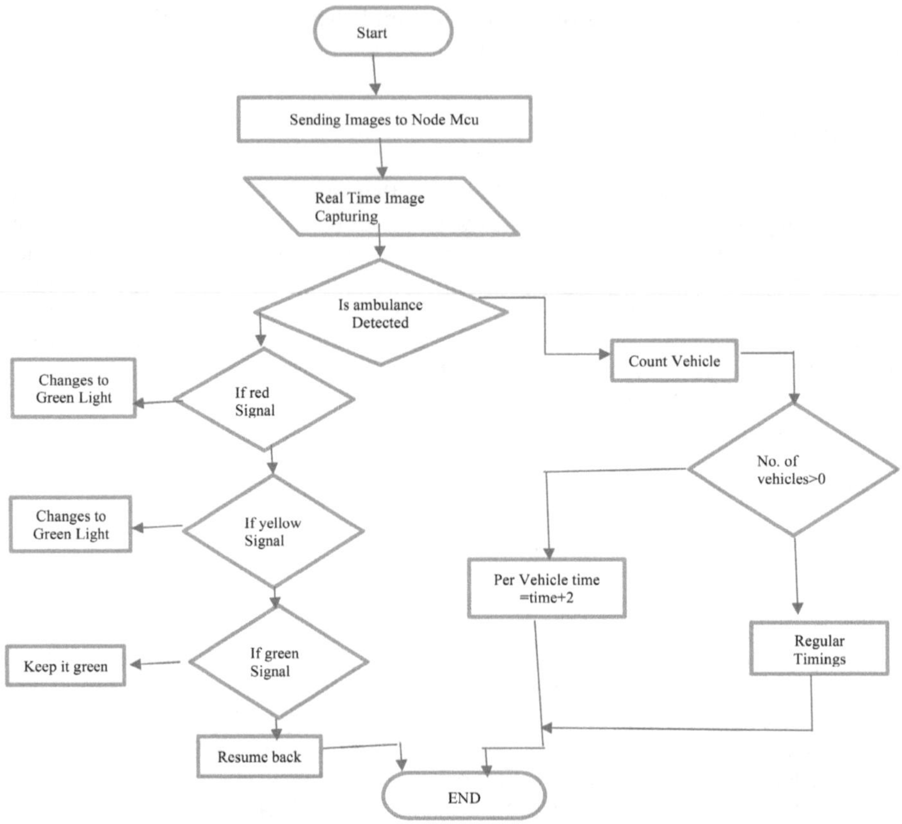

Figure 1. Flowchart diagram.

STEP 2- Camera capturing real-time images of vehicles as shown in Figure 2. Sending these images to NodeMCU Controller.

STEP 3- If an ambulance is detected then green that lane and the other three lanes keep red until it passes otherwise go to step 4.

STEP 4-Count incremented.

STEP 5- Check if the previous lane is green then go to the next step otherwise go to step 4.

STEP 6- Stop count calculates time delay depends on vehicle count and value is store in the memory.

STEP 7- Assign delay value to green light and clear the count value and go to step 2.

4 RESULTS

This work's primary goal is to manage traffic sensibly both in regular and emergency situations by following the architecture as shown in Figure 1. Thus, the system is constructed. At the end, the right results are obtained for the particular condition. The test cases are displayed beneath.

A screenshot of a computer (Result image which is showing the count of vehicles and identification of ambulance.)

Figure 2. Detecting vehicles.

4.1 *If there is heavy traffic jam*

If there is heavy traffic then using RCNN (Object Detection Algorithm) tally the number of automobiles based on the vehicle's count green light is delayed by two seconds for each vehicle by the microcontroller. If there is no vehicle on the road the traffic light will be set to 15 seconds for each side.

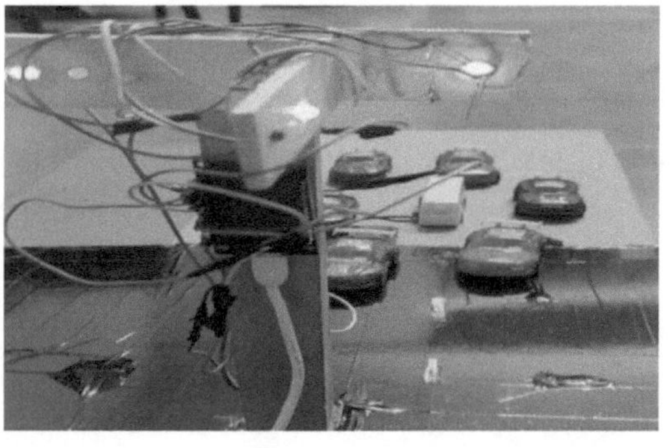

Figure 3. Change of traffic light.

A toy cars on a table (This image shows the hardware set up of the system).

4.2 *If the ambulance arrives*

When traffic signal was operating on normal mode and suddenly an ambulance arrival is detected using RCNN(Object Detection Algorithm). Then it will stop its current work and turn the light green on the side where ambulance is detected and others make red until it passes the signals shown in Figure 3. When ambulance is passed it goes to the normal working.

5 CONCLUSION

The system under consideration here effectively oversees traffic in real-time, leading to a reduction in accident rates and the preservation of valuable time spent waiting at traffic signals. This foundational approach paves the way for automation in traffic signal control systems. Granting priority to ambulances equipped with emergency tags ensures quicker hospitalization for patients, ultimately contributing to saving lives. This novel approach offers several advantages, including the utilization of image processing instead of sensors, cost-effectiveness, simple installation, and a balance of both accuracy and speed. The authors created this project as a proof of concept because it is a final year project and because of constraints. Since the active webcam range is only about 10 meters, future research can concentrate on utilizing cloud computing to test it at various distances between the ambulance and oncoming traffic, as well as in real traffic lights, as well as for various ambulance speeds.

This model will provide a cost effective and efficient way for controlling traffic density and provide a way for ambulances. A comparison is shown between the existing system and our proposed system as shown in Table 1.

Table 1. Comparison table.

Metrics System	RFID Reader	Arduino Uno	Node Mcu Controller	RCNN	Cost	Power Consumption
Smart Ambulance Traffic Control System	Yes	No	No	No	High	High
Density Based Traffic Control System with smart Sensing of Emergency Vehicle	Yes	No	No	No	High	High
Smart Traffic Control with Ambulance Detection	No	Yes	No	No	High	High
Smart Density Based Traffic Light System	No	Yes	No	No	High	High
Proposed System	No	No	Yes	Yes	Low	Low

REFERENCES

[1] Varsha Srinivasan *et al.* (2018) *IOP Conf. Ser.: Mater. Sci. Eng.* **402** 012015

[2] Papageorgiou M., Diakaki C., Dinopoulou V. and Kotsialos, A. (November 2004). Review of road traffic control strategies, *Proceedings of IEEE*, Vol. 91, Issue 12, pp. 2043–2067.

[3] Georgios Vigos, Markos Papageorgioua and Yibing Wangb. (February 2008). Real-time estimation of vehicle-count within signalized links, *Journal of Transportation Research Part C: Emerging Technologies*, Volume 16, Issue 1, pp.18–35.

[4] Abbas, Naeem, Muhammad Tayyab, and M. Tahir Qadri. (2013). Real time traffic density count using image processing. *International Journal of Computer Applications* 83.9: 16–19.

[5] Jadhav, Prashant, *et al.* (2016). Smart traffic control system using image processing. *International Research Journal of Engineering and Technology (IRJET)* 3.3: 2395–0056.

[6] Varadharaj D. M. (Dec. 2019). Density based traffic control system with smart sensing of emergency vehicles, *Converter*, pp. 01–07.

[7] Chowdhury M. F., Ryad Ahmed Biplob M. and Uddin J. (2018). Real time traffic density measurement using computer vision and dynamic traffic control, *2018 Joint 7th International Conference on Informatics, Electronics & Vision (ICIEV) and 2018 2nd International Conference on Imaging, Vision & Pattern Recognition (icIVPR)*, Kitakyushu, Japan, pp. 353–356.

[8] Yuliandoko H., Ayatullah M. D. and Purwita F. E. (2019). Automatic vehicle counting using raspberry pi and background subtractions method in the sidoarjo toll road, *2019 2nd International Conference of Computer and Informatics Engineering (IC2IE)*, Banyuwangi, Indonesia, pp. 83–86, doi: 10.1109/IC2IE47452.2019.8940899.

[9] Firdous, Anam, and Vandana Niranjan. (2020). *Smart density based traffic light system. 2020 8th International Conference on Reliability, Infocom Technologies and Optimization (Trends and Future Directions)(ICRITO)*. IEEE.

[10] Ni, Yuanzhi, *et al.* (2019). Toward reliable and scalable internet of vehicles: Performance analysis and resource management. *Proceedings of the IEEE* 108.2: 324–340.

[11] Y. Ni, L. Cai, J. He *et al.* (2020). Toward reliable and scalable internet of vehicles: performance analysis and resource management, *Proceedings of the IEEE*, vol. 108, no. 2, pp.324–340.

[12] Pratishtha Gupta, G.N Purohit, Sweta Pandey. (April 2013) Traffic load computation for real time traffic signal control, *International Journal of Engineering and Advanced Technology*, Vol. 2, Issue 4

[13] Sabya sanchi kanojia. (Mar-Apr 2012). Real –time traffic light control and congestion avoidance system, *International Journal of Engineering Research and Applications (IJERA)*, pp.925–929, Vol. 2, Issue 2.

[14] Wang Y., Xu, C., and Chen, C. (2018). Smart traffic signal control based on deep reinforcement learning. *IEEE Transactions on Intelligent Transportation Systems*, 19(10), 3155–3164.

[15] Cucchiara R., Piccardi M. and Mello P. (June 2002). Image analysis and rule based reasoning for a traffic monitoring system, *IEEE Trans. Intell. Transport. Syst.*, vol. 1, no. 2, pp. 119–130.

[16] Elizabeth Basil and Sawant S.D. (Aug 2017). IoT based traffic light control system using raspberry Pi, *International Conference on Energy Communication Data Analytics and Soft Computing (ICECDS)*, 1–2.

[17] Bilal Ghazal, Khales Elkhatib, Khales chahine and Mohammad Kherfan, (2016). *Smart Traffic Light Control System*, IEEE.

[18] Liu, Y., Huang, Y., and Wang, Y. (2017). Adaptive traffic signal control for smart cities using reinforcement learning. *Transportation Research Part C: Emerging Technologies*, 79, 1–11.

Computational Methods in Science and Technology Volume 1 – Kaur et al. (Eds)
© 2025 the Author(s), ISBN: 978-1-032-81760-6

Plant disease detection using ensemble stacking

Kriti Vij, Soumya Goyal, Aayush Sharma & Seema Rani
Jaypee University of Information Technology, Waknaghat, Solan

ABSTRACT: Plant diseases have a major negative influence on agricultural productivity, which results in very large monetary losses and problems related to food security. For the well-organized treatment of these diseases, early identification/detection is important. The goal of this project is to use deep learning algorithms and image processing techniques to construct an automated system for the detection of plant diseases. Our study presents an innovative method for detecting plant diseases through the use of an ensemble stacking technique that combines the VGG16 model and Convolutional Neural Networks (CNNs). Understanding the importance of early disease detection for agricultural output, our approach makes use of the capabilities of the VGG16 model, which is well-known for its capacity to identify patterns in images via transfer learning, as well as stacked CNNs, which are intended to capture complex hierarchical features. This study creates the preliminaries for further study and processing of ensemble stacking techniques concerning plant disease detection, with possible benefits in smart agriculture and crop management.

Keywords: CNN, VGG-16, Ensemble Stack Model, Plant Disease Detection, Early Disease Detection

1 INTRODUCTION

In agriculture, the soundness of the crop is determined in the early stage or providing that the detection is done timely and precisely. Due to the new technology in the field of deep learning, models like CNN [1] and VGG16 could make the work way easier and accurate. The problem statement for plant disease detection using machine learning and deep learning is to develop a reliable and precise system that can quickly identify and diagnose plant disease [2]. The major problem faced in the agriculture domain is the wastage of food due to plant diseases and plant illness. This system should be able to interpret large amounts of data precisely and correctly, and it should be able to conclude new places and plant species [3]. But building these kinds of systems requires having access to large and diverse datasets, selecting appropriate machine learning methods, and optimizing hyperparameters for the specific problem at hand. The efficacy and dependability of the diagnostic results for plant diseases are significantly influenced by the amount of the dataset used to train deep learning algorithms [4]. The size of the dataset is a critical factor influencing the model's performance in identifying plant diseases.

The importance and the immediate need of this model in the field of agriculture is the timely and the accurate detection of the plant disease. The drive comes from the pressing need to solve agricultural issues and support the overarching objective of guaranteeing food security for the world's expanding population.

 DOI: 10.1201/9781003501244-62

Figure 1. Samples of plant village dataset.

2 LITERATURE REVIEW

This section covers 6 research papers done previously on plant disease detection, how they solved the problem along with their methodologies. In (P.V. Yeshwanth *et al.* 2023), authors proposed RSNSR-LDD model for leaf disease detection of grape plant and got the accuracy of 95.24%. In (Ramanjot *et al.* 2023), authors proposed GoogleNet and VGG16 models to detect and classify images and got the accuracy of 99.23%. In (Alok Kumar *et al.* 2023), authors proposed using model ResNet and got the accuracy of 97.28%. In (Anwar Abdullah Alatawi *et al.* 2022), authors proposed using model VGG16 and got the accuracy of 95.2%. In (Kushal M. U. *et al.*), authors proposed using model CNN and got the accuracy of 95.6%. In (M. A. Jasim *et al.* 2020), authors proposed using model CNN and got the accuracy of 98%. Most of the papers have used deep leaning models which can easily detect the features and some models are even capable of detecting complex features from the images like VGG16, ResNet etc.

Table 1. Literature survey.

S. No.	Paper Title	Conference Year	Tools and Techniques	Result	Limitation
1	Plant Disease Detection and Classification. [2]	2023	GoogleNet and VGG16	GoogleNet accuracy 99.23%, VGG16 accuracy 98%.	Limitied number of categories.
2	Plant Disease Detection and Classification using Image Processing and Deep Learning Techniques. [4]	2020	CNN	98% accuracy	Small dataset.
3	Plant Disease Detection using VGG16. [7]	2023	ResNet	97.28% accuracy.	Less categories.
4	Plant Disease Detection Using AI based VGG16 model. [8]	2022	VGG16	95.2% accuracy	Complex Structure, Small Dataset.
5	Residual Skip Network-Based Super-Resolution for Leaf Disease Detection of Grape Plant. [9]	2023	RSNSR-LDD Model	95.24% accuracy	Low resolution images.
6	Plant Disease Detection using CNN. [10]	2022	CNN	95.6% accuracy.	Low resolution images.

2.1 *Key gaps*

There are various gaps in the literature which need to be amended in order to get better understanding and results. The availability of diverse, large, real world and marginalized data is limited which leads to the problem of data sparsity. For disease prevention and less loss, early detection of diseases in plant is a very crucial factor. [11] Another major problem that was faced while working on this project, is need to compare multiple plant leaves at a time but due to technology used are unable to reach our goal. It can only process a single image at a time and can examine its disease at that time only [12]. For the betterment of this project, researchers will continue to examine this problem and will resolve this in future.

3 METHODOLOGY

3.1 *Data preparation*

Data preparation involves several steps in collecting and preparing the data before applying the models. The steps include – Collecting data: This data is taken from Kaggle repository. This dataset basically contains 15 categories which makes it diverse.

Table 2. Features of dataset.

Pepper bell Bacterial spot	'Pepper bell healthy'	Potato Late blight
Potato Early blight	'Potato healthy'	'Tomato Bacterial spot'
Tomato Early blight	'Tomato Late blight'	'Tomato Leaf Mold
'Tomato Septoria leaf spot'	Two spotted spider mite'	'Tomato healthy'
'Tomato Target Spot'	'Tomato Yellow Leaf'	'Tomato mosaic virus'

3.2 *Data pre-processing*

Pre-processing the data includes rotation, resizing and cropping the 24 images. Rotating the images introduce change in the orientation making the model more robust to different outlook of plant diseases [13]. Cropping focuses more on extracting relevant features from the images which helps in better evaluation. Resizing the images ensures consistency in the input size of the images for the model. Combining these transformations to the dataset provides better accuracy and precision and makes the model more robust [14]

3.3 *Data splitting*

Split the dataset into training and validation dataset. The researchers have split the dataset 90% for training and 10% for validation. After splitting the training includes 18575 and the validation dataset includes 2063 images.

3.4 *Key challenges*

Many challenges are faced by plant disease detection and taking these challenges in context is necessary to have an effective disease management system for a better agriculture system.

Data collection: Due to the limited availability of the data it was very difficult for us to compare it to the real-life scenario, even the surroundings does not satisfy the condition due to the climatic conditions [15].

Scalability: It was quite a difficult task for us to carry forward this project to a higher scale due to the above-mentioned problem and tried our level best to enhance the project and make it as engaging as possible [16].

Adaptability: For farmers it is very difficult to adopt these technologies and systems, as they are very complicated, costly and need proper guidance before implementation.

Reliability: The effects of bad weather, including storms, a lot of precipitation, or extremely high or low temperatures, exhibits the necessity of strong engineering and backup plans to guarantee the system's consistent and reliable functioning in challenging climatic conditions [17].

4 RESULTS

The researchers have built a function in the image detection module. Its inputs are an image, a list of class labels, and a trained ensemble stacking model. This function uses the model to evaluate the class probabilities and turns the image into a batch of size 1.

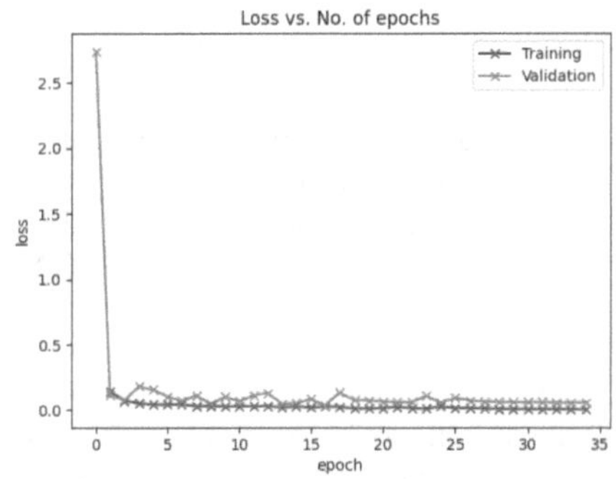

Figure 2. Ensemble Stack with an accuracy of 98.5%.

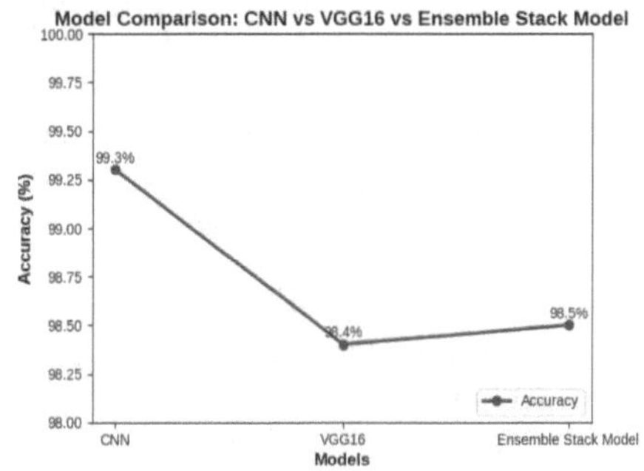

Figure 3. Model Comparison.

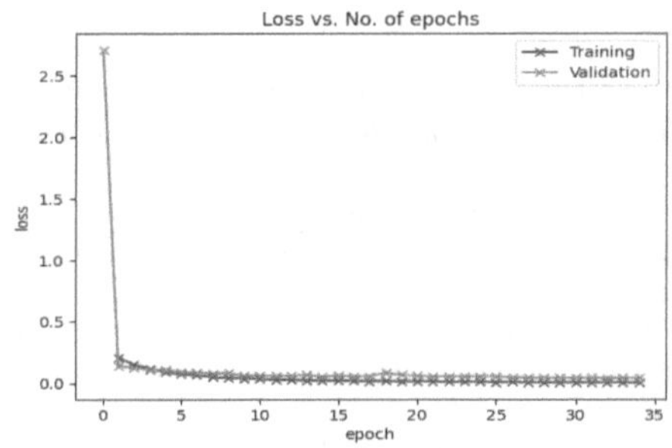

Figure 4. CNN with an accuracy of 99.3%.

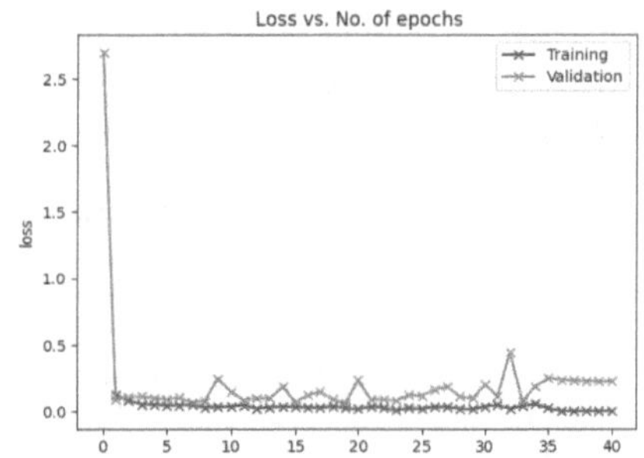

Figure 5. VGG16 with an accuracy of 98.4%.

Figures 2, 3, 4, 5. displays the loss vs epoch of Ensemble stack of models, model Comparison, loss vs epoch of CNN, and loss vs epoch of VGG16 respectively.

Table 3. Comparison with existing solutions.

Models	Accuracy	Models	Accuracy	Models	Accuracy
CNN	99.3%	VGG16	98.4%	Ensemble Stack Model	98.5%
CNN [4]	98%	VGG16 [8]	95.2%		

Table 3 shows the existing solutions and the comparison between the applied models.

Table 4 shows the individual CNN, VGG16 and ensemble model, we received the validation accuracy of 99.3%, 98.4%, 98.5%, validation precision 99.7%, 99.03%, 99.07% and f1 score of 99.4%, 98.4%, 98.5% respectively.

Table 4. Comparison of our individual models with ensemble stack models.

Models	Validation Accuracy	Validation Precision	Validation F1 Score
CNN	99.3%	99.7%	99.4%
VGG16	98.4%	99.03%	98.4%
Ensemble Stack Model	98.5%	99.07%	98.5%

5 CONCLUSION AND FUTURE SCOPE

As there are many fertilizers are present nowadays which are good for preventing plant disease but due to less knowledge the farmers use it in a lot of amounts for saving crops which lead to deteriorating human health. With successful completion of our project "Plant Disease Detection" this project conclude that the best accuracy is provided by the CNN which was 99%.

The future scope of the plant disease detection is favorable, and is in very demand for maintenance of stable agriculture. Researchers will continue this with in upcoming period with attaching frontend with this project, which will create the user interface with which users interact with [18]. Our second main concern would be expanding our dataset with more features and categories.

REFERENCES

[1] Parez, S., Dilshad N., Alanazi T. M. and Weon Lee J. (2023). Towards sustainable agricultural systems: a lightweight deep learning model for plant disease detection. *Computer Systems Science and Engineering* 47(1): 515–536.

[2] Ramanjot, U. Mittal, A. Wadhawan, J. Singla, N.Z. Jhanjhi, Ghoniem R. M., Ray S. K., and Abdelmaboud A. (May 15 2023). Plant disease detection and classification: a systematic literature review. *Sensors* 23, no. 10: 4769. https://doi.org/10.3390/s23104769

[3] Kumar, K., Tripathi K. and Gupta R. (2023). Image based plant disease classification using deep learning technique. *International Journal of Innovative Research in Engineering and Management* 10, no. 3 (June): 146–151.

[4] Jasim, M.A., and AL-Tuwaijari J.M. (2020). Plant leaf diseases detection and classification using image processing and deep learning techniques. In *2020 International Conference on Computer Science and Software Engineering (CSASE)*, 259–265. Duhok, Iraq: IEEE. https://ieeexplore.ieee.org/document/9142097/.

[5] Prashar N., "A review on plant disease detection techniques," in *2021 2nd International Conference on Secure Cyber Computing and Communications (ICSCCC)*, 2021: IEEE, pp. 501–506.

[6] Mohanty, S.P., Hughes D.P., and Salathé M. (2016). Using deep learning for image-based plant disease detection. *Frontiers in Plant Science* 7 (September 22): 1419.

[7] Kumar, A., and Kumar A. (2023). Plant disease detection using VGG16. *International Journal of Creative Research Thoughts* 11, no. 1: 770–775.

[8] Alatawi, A.A., Alomani S.M., Alhawiti N.I. and Ayaz M. (2022). Plant disease detection using AI based VGG-16 model. *International Journal of Advanced Computer Science and Applications* 13, no. 4: 718–727. http://dx.doi.org/10.14569/IJACSA.2022.0130484

[9] Yeswanth, P.V., Deivalakshmi S, George S., and Ko S.-B. (2023). Residual skip network-based super-resolution for leaf disease detection of grape plant. *Circuits, Systems, and Signal Processing* 42, no. 11 (November): 6871–6899. https://doi.org/10.1007/s00034-023-02430-2.

[10] Shelar, N., Shinde S., Sawant S., Dhumal S., and Fakir K. (2022). Plant Disease Detection Using CNN. Ed. M.D. Patil and V.A. Vyawahare. *ITM Web of Conferences* 44: 03049.https://doi.org/10.1007/s00034-023-02430-2.

[11] Ferentinos, K.P. 2018. Deep learning models for plant disease detection and diagnosis. *Computers and Electronics in Agriculture* 145 (February): 311–318.

[12] Kumar, S., Chaudhary V., and Chandra S.K. (2021). Plant disease detection using CNN. *Turkish Journal of Computer and Mathematics Education* 12, no. 12: 2106–2112.

[13] Agarwal, M., Singh A., Arjaria S., Sinha A. and Gupta S. (2020). ToLeD: tomato leaf disease detection using convolution neural network. *International Conference on Computational Intelligence and Data Science* 167: 293–301.

[14] Kibriya, H., Rafique R., Ahmad W. and Adnan S.M. (2021). Tomato leaf disease detection using convolution neural network. In *2021 International Bhurban Conference on Applied Sciences and Technologies (IBCAST)*, 346–351. Islamabad, Pakistan: IEEE. https://ieeexplore.ieee.org/document/9393311/.

[15] Menon V., Ashwin V. and Deepa R. K., "Plant disease detection using cnn and transfer learning," in *2021 International Conference on Communication, Control and Information Sciences (ICCISc)*, 2021, vol. 1: IEEE, pp. 1–6.

[16] Acharya, A., Muvvala A., Gawali S., Dhopavkar R., Kadam R. and Harsola A. (2020). Plant disease detection for paddy crop using ensemble of CNNs. In *2020 IEEE International Conference for Innovation in Technology (INOCON)*, 1–6. Bangluru, India: IEEE. https://ieeexplore.ieee.org/document/9298295/.

[17] Mohameth, F., Bingcai C. and Sada K.A. (2020). Plant disease detection with deep learning and feature extraction using plant village. *Journal of Computer and Communications* 08, no. 06: 10–22.

[18] Hassan, S.M., Jasinski M., Leonowicz Z., Jasinska E. and Maji A.K. (2021). Plant disease identification using shallow convolutional neural network. *Agronomy* 11, no. 12 (November 24): 2388.

Computational Methods in Science and Technology Volume 1 – Kaur et al. (Eds)

New era of gene classification methods using machine learning

Amanpreet Kaur
Research Scholar, Punjabi University Patiala
Assistant Professor, Department of Information Technology, Chandigarh Engg. College-CGC

Harpreet Kaur
Assistant Professor, Punjabi University Patiala

Amandeep Singh
Assistant Professor, Department of Mechanical Engineering, BBSBEC Fatehgarh Sahib

ABSTRACT: New diagnostic technologies like DNA microarray technology have changed illness detection in the fast-changing technological age. This work focuses DNA microarray technology in gene expression research to find and understand genetic anomalies that cause illnesses, including cancer. Gene expression analysis is critical for cancer diagnosis to uncover underlying problems and create novel treatments. The number and size of data make human analysis difficult, requiring automation. Machine learning and statistics are used for data categorization, focusing on cancer prediction. This article discusses various feature selection approaches for large data sets. The related work section discusses KNN-based, MLDA, signal-to-noise ratio, and ensemble algorithms. Sample size, computing costs, and classification accuracy are addressed by each technique. The study emphasizes feature selection and the range of machine learning algorithms. This article proposes a gene classification research, encompassing current gene categorization approaches, feature extraction methods, datasets, result analysis, and conclusion. The article seeks to improve gene expression analysis for illness diagnosis and therapy.

Keywords: Gene Expression, RNA Sequencing, Micro Array, Feature Selection

1 INTRODUCTION

Every day, new technologies are developed to identify illnesses, like the DNA microarray. Some disorders are caused by genetic defects that develop cancerous cells. These cells divide erratically, spread throughout organs, and often die during development. Recently, gene expression analysis has become a key tool for cancer detection and medication discovery [7]. It analyzes gene expression data to find genetic abnormalities, including cancer, in people's DNA. From the outset, it was popular. Gene expression converts DNA data into instructions by generating chemicals and proteins. This method counts DNA transcripts in a tissue or sample to identify which genes are translated and how much. Gene patterns reveal illness prevention, medication development, therapy, and diagnosis. There is too much data to compute manually, and predicting any illness requires a lot of work. They are (i) highly high-dimensional, including hundreds of genes, unlike all related data. (ii) Public data is noisy and huge or tiny. iii) Most genes do not affect class. [11] Utilizing this abundance of data, automated algorithms forecast any value, and several machine learning and statistical approaches categorize data. Gene-specific expression data interpretation remains tough. Numerous gene expressions in the human body make it challenging to extract relevant characteristics. Uneven feature dimensions make traditional classifiers tougher to apply. The

 DOI: 10.1201/9781003501244-63

microarray dataset is an expression level matrix with rows representing genes and columns representing samples [1].

Feature selection is the best way to minimize gene expression data variability and complexity. These algorithms struggle to anticipate categorization characteristics. Many approaches are employed by researchers to extract important information. Filter techniques evaluate feature correlation to identify duplicate and unnecessary characteristics and provide the desired result. This approach selects the best collection of characteristics, applies the methodology, and evaluates performance. Wrapper approaches employ an iterative procedure to select the best subset of features and enclose the model to discover relevant data characteristics [5,6].

Such extractions are unsuitable for heuristics. Swarm intelligence techniques tackle tiny data issues. This is excellent for researchers. Later, conventional neural networks [4] anticipate important properties. Using gene expression data, fully connected neural networks (MLP), convolution neural networks (CNN), recurrent neural networks (RNN), graph neural networks (GNN), and transformer neural networks (TNN) were used to classify cancer. [9] This research examines the categorization of genes. This paper's second section examines the state-of-the-art in gene categorization methods. The many techniques for feature extraction are covered in Sect. 3. The Gene classification features are explained in Section 4. Sect. 5 the datasets used in Gene expression. In section 6 different performance metrics are explained. Lastly, Sect. 7 concludes this paper.

2 RELATED WORK

This section reviews current technique literature. S.M. Ayyad [16] Two simple KNN-based cancer classification algorithms for high-dimensional data are presented in this study. These approaches use a new KNN-based weighting mechanism. The tests included six benchmark datasets, which academics use for high-dimensional categorization. In this, the author contrasted classification algorithms using fuzzy SVM, BEL network, KNN, and SVM, industry standards. In the second experiment, recommended methodologies were used to classify datasets after using information gain to identify the best features. SMKNN and LNKNN had the best classification performance metrics and competed with modern approaches.

Reva Joshi [15] This study introduced Modified Latent Dirichlet Allocation (MLDA) for gene classification to solve LDA's shortcomings and improve articulation in solid and malignant development tissues. The dataset helped MLDA classify and identify genes in malignant and benign tissues. MALDA outperforms Random Forest, Naive Bayes, and SVM. MLDA beats competing techniques for recall and accuracy on lung and breast cancer datasets, according to experiments. It also shows the lowest sample loss.

Anu J. Nair [2] The Signal-to-Noise Ratio method is used to shrink the huge dataset while maintaining all characteristics in this study. After that, Support Vector Machine, K-Nearest Neighbor, and Decision Tree feature categorization are used to this reduced dataset. By combining the three methods into an ensemble-based system yields more accurate and lower-error outputs than using feature selection strategies alone.

Islam Ashraful [3] Classifying gene expression data is challenging due to their high dimension and low sample size. An ensemble learning model was developed using Bayes GLM, XGBoost, RF, KSVM, and kNN. This study is an advance over previous classifiers since our ensemble learning technique yields more accurate results.The result is 91.67% accuracy.

JiandeWu, Chindo Hicks [18] Four classification models were assessed: Support Vector Machines, K-nearest neighbor, Decision trees, and Naïve Bayes. To create and validate classification models, RNA-sequence data analysis was employed to pick attributes (genes). The recommended techniques were tested using gene expression datasets. The Support Vector Machine technique prevented misclassification errors and distinguished triple negative and non-triple negative breast cancer subtypes better than the other three.

Liu and Yao [12] KL divergence-based gene selection is proposed in this paper, using genes with bigger KL divergence as model characteristics. Next, employ Focal Loss in a deep neural network model. This research presents a deep learning model technique with an AUC of 0.99 on the validation set based on KL divergence gene selection. Model generalization is beneficial. Lung cancer prediction using this deep neural network model based on KL divergence gene selection is efficient and reliable.

Table 1. Comparison between different research papers with gap identified.

Author's year	Proposed techniques	Gap/problem definition	Dataset	Performance metrics
S.M. Ayyad (2019) [16]	Modified K-Nearest Neighbor, MKNN.	Need to use recent latest classification on techniques for feature extraction	Six benchmark datasets for DNA Microarray data.	Precision Recall Accuracy
Reva joshi (2020) [15]	MLDA Modified Latent Dirichlet Allocation (MLDA) for gene classification	Limited Sample Size	mRNA dataset	Precision Recall
Anu J Nair (2019) [2]	SVM KNN and Decision Tree.	Not suitable for multiclass gene expression	Colon and Leukemia data set	Accuracy
Ashraful Islam (2020) [3]	kNN, Random Forest KSVM, Bayes Generalized Linear Model. eXtreme Gradient Boosting,	Data set is small	Bioinformatics Research Group University of Spain	Accuracy precision Recall
Liu and Yao, 2022 [12]	Deep neural Network	High Computation cost and small data set	RNA-Seq	Precision, Recall, Accuracy
Chen and Dhahbi, 2021 [8]	RF	Performance is not good. Results are not better than traditional methods.	RNA-Seq	Accuracy
JiandeWu and Chindo Hicks [18]	SVM, KNN, Naïve Bayes and Decision tree	If we increase the no. of features it lowers the performance of the algorithms. Works best on small data set	RNA-Seq	Accuracy, Recall

2.1 *Feature selection methods*

Selecting a limited number of features from a batch of data so as not to impact the data's performance, or to decrease the dimensionality of the features and increase the classification algorithm's accuracy, is known as feature selection. [10]

Filter Method: In this we have three sub components out of this we choose optimal subset from a set of features using various methods i.e ANOVA, Chi Square Test and Correlation Coefficient etc.

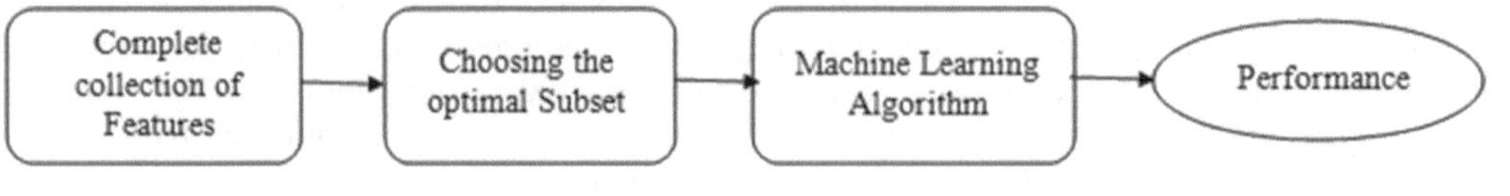

Figure 1. Filter method.

Wrapper Method: This approach is iterative looks at every feature and every combination of features, and then selects the one that produces the best results for a specific machine learning technique.

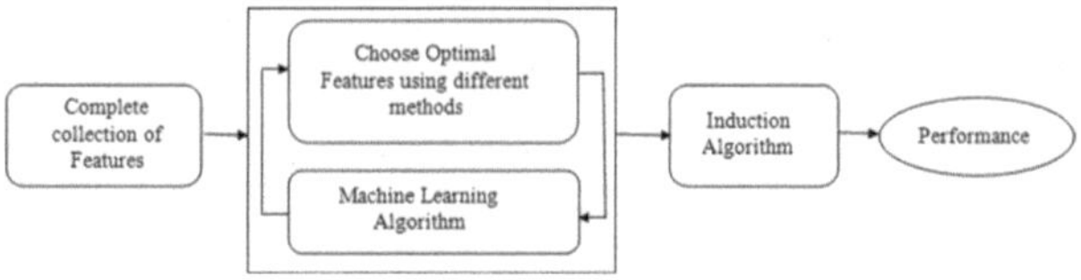

Figure 2. Wrapper method.

Embedded Method: It combines the advantages of both methods i.e filter and wrapper and having low computation cost.

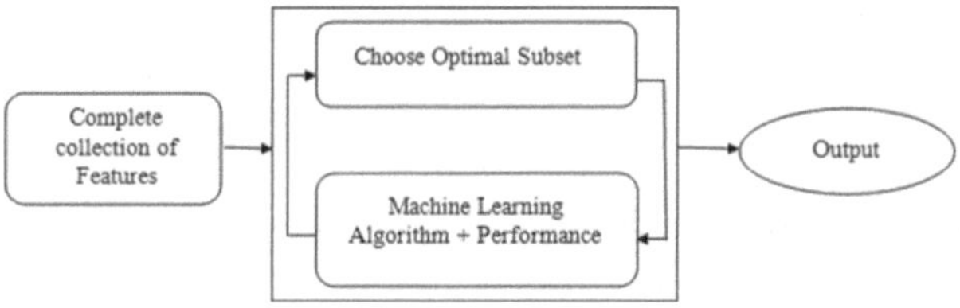

Figure 3. Embedded method.

2.2 *Gene expression data*

Gene expression data describes gene activity in a cell or tissue at a specific period. Gene expression converts these instructions into proteins and shows expressed genes and their levels. This data is needed to understand the molecular pathways of development, growth, stimulus response, and illness.

2.2.1 *Micro array data*

Thousands of genes can have their expression levels simultaneously measured with the aid of microarrays. They are made up of small dots, each of which has DNA probes that match particular genes. In a typical microarray experiment, RNA is taken from a biological sample (such as cells or tissues) and transformed into a corresponding DNA. Following that, this DNA is hybridized to the microarray after fluorescent tags are put to it. Every microarray spot's fluorescence intensity and the degree of gene expression there are connected. [13]

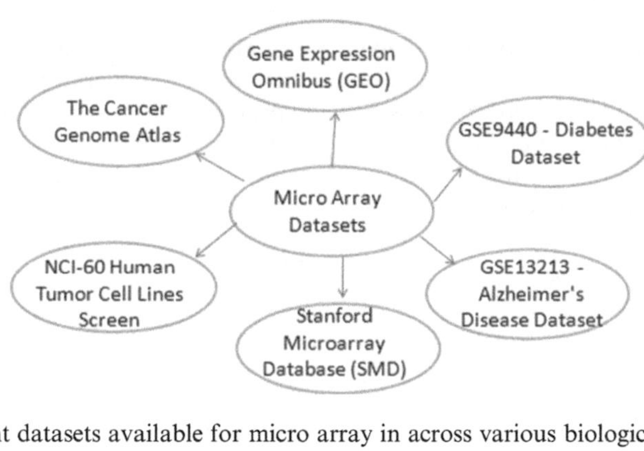

Figure 4. Different datasets available for micro array in across various biological and medical research domains.

2.2.2 *RNA sequencing*

To analyze and measure an organism's transcriptome, RNA sequencing is effective. The transcriptome contains all RNA molecules in a cell, including mRNA, non-coding RNA, and other RNAs. RNA-seq works by transforming RNA from biological materials into complementary DNA and sequencing the cDNA segments at high speed. [9,14,17]

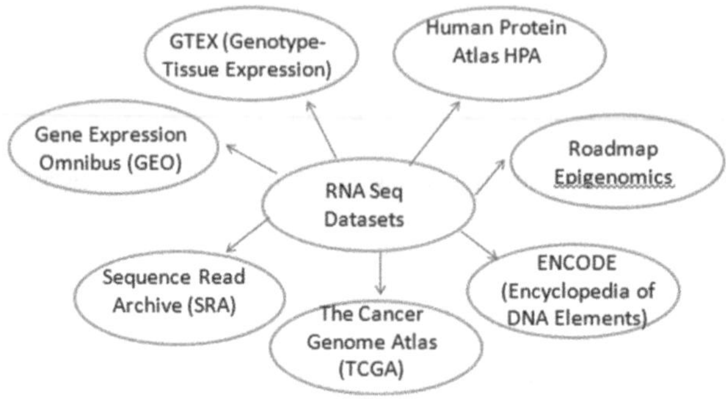

Figure 5. Different datasets available for RNA sequencing in across various biological and medical research domains.

Table 2. Comparison between micro array and RNA sequencing.

	Micro Array	RNA Seq
Technology	Hybridization-oriented	Based on sequencing
Resolution	Restricted by the probe number	High, fully encapsulates the transcriptome
Sensitivity	Restricted due to low quantity	High, identifies transcripts with low abundance
Dynamic Range	Low	Comprehensive, precise for both highly and lowly expressed genes
Quantification	Relative	Absolute
Cost	Less Expensive	More expensive
Sample Requirements	Greater RNA concentrations	Less initial material is required
Data Analysis	Specialized Software	Complex Bioinformatics analysis

2.3 *Performance metrics*

Machine learning model effectiveness depends on performance measures. Skewed datasets may have low accuracy, the ratio of accurate predictions to the overall dataset. Precision, the ratio of successfully anticipated positives to all expected positives, measures optimistic estimate accuracy, especially when false positives are costly. Remember, evaluating the ratio of precisely predicted positives to all actual positives is crucial because false negatives are costly. These measures help choose, optimize, and compare models.

$$Recall = \frac{X}{X+V} \quad Accuracy = \frac{X+Z}{X+Y+Z+V} \quad Precision = \frac{X}{X+Y}$$

3 CONCLUSION

In conclusion, the rapid evolution of technology has led to the emergence of DNA microarray technology, revolutionizing disease diagnosis, with a focus on gene expression analysis for understanding genetic abnormalities, especially in cancer. The vast and high-dimensional nature of gene expression data necessitates automated methods, and machine learning techniques, including KNN-based methods, MLDA and ensemble techniques has been employed. Feature methods, such as filter, wrapper, and embedded techniques, play a crucial role in handling the challenges associated with the data. This paper contributes to the advancement of gene expression analysis by emphasizing gene classification and exploring various feature extraction methods. The reviewed literature showcases the significance of diverse machine learning algorithms and their application in cancer prediction. Overall, this research aims to enhance our understanding of gene expression analysis in disease diagnosis and treatment development.

REFERENCES

[1] Alagukumar, S. and Lawrance, R., (2016). Classification of microarray gene expression data using associative classification. *Computing Technologies and Intelligent Data Engineering (ICCTIDE).*

[2] Anu J. Nair, Rizwana Rasheed, Maheeshma KM, Aiswarya LS: and Kavitha K R: (2019). An ensemble-based feature selection and classification of gene expression using support vector machine, K-Nearest neighbor, decision tree. *Proceedings of the Fourth International Conference on Communication and Electronics Systems (ICCES)* IEEE Conference Record # 45898; IEEE Xplore ISBN: 978-1-7281-1261-9

[3] Ashraful Islam, Mohammad Masudur Rahman, Eshtiak Ahmed and Faisal Arafat, Md Fazle Rabby. (2020). Adaptive feature selection and classification of colon cancer from gene expression data: An ensemble learning approach. *ICCA* 2020, January 10–12, Dhaka, Bangladesh

[4] Bonkra, A., Noonia, A. and Kaur, A. (2021, December). Apple leaf diseases detection system: a review of the different segmentation and deep learning methods. In *International Conference on Artificial Intelligence and Data Science* (pp. 263–278). Cham: Springer Nature Switzerland.

[5] Bonkra, A., Bhatt, P. K., Rosak-Szyrocka, J., Muduli, K., Pilař, L., Kaur, A., and Rana, A. K. (2023). Apple leave disease detection using collaborative ML/DL and artificial intelligence methods: scientometric analysis. *International Journal of Environmental Research and Public Health*, 20(4), 3222

[6] Bonkra, A., Pathak, S., Kaur, A. and Shah, M. A. (2024). Exploring the trend of recognizing apple leaf disease detection through machine learning: a comprehensive analysis using bibliometric techniques. *Artificial Intelligence Review*, 57(2), 1–26.

[7] Brewczyński, A.; Jabłońska, B.; Mazurek, A.M.; Mrochem-Kwarciak, J.; Mrowiec, S.; Śnietura, M.; Kentnowski, M.; Kołosza, Z.; Składowski, K. and Rutkowski, T. (2021). Comparison of selected immune and hematological parameters and their impact on survival in patients with HPV-related and HPV-unrelated oropharyngeal cancer. *Cancers*, 13, 3256.

[8] Chen, J.W.; Dhahbi, J. (2021). Lung adenocarcinoma and lung squamous cell carcinoma cancer classification, biomarker identification, and gene expression analysis using overlapping feature selection methods. *Sci. Rep.*, 11, 13323.

[9] Fadi Alharbi and Aleksandar Vakanski. (2023). Machine learning methods for cancer classification using gene expression data: A review. *Bioengineering*, 10, 173.

[10] García-Díaz Sánchez-Berriel Martínez-Rojas and Diez-Pascual. (2020). Unsupervised feature selection algorithm for multiclass cancer classification of gene expression RNA-Seq data. *Genomics*, 112, 1916–1925.

[11] Geeta Chhabra, (2019). Research scholar, amity institute of information technology, Amity University, Noida, Uttar Pradesh, India, Vasudha Vashisht, Assistant Professor, Department of Computer Science & Engineering, Amity School of Engineering & Technology, Amity University, Noida, Uttar Pradesh, India, Jayanthi Ranjan Professor, Institute of Management Technology, Ghaziabad, Uttar Pradesh, India, [4] Revised Manuscript Received on 8 February.

[12] Liu and Wu Yao. (2022). Prediction of lung cancer using gene expression and deep learning with KL divergence gene selection. *BMC Bioinformatics* 23:175

[13] Munkácsy, G., Santarpia, L., Győrffy, B. (2022). Gene expression profiling in early breast cancer—patient stratification based on molecular and tumor microenvironment features. *Biomedicines*, 10, 248.

[14] Pilar Garcia-Diaz and Isabel Sanchej Berriel. (2020). Unspervised feature selection algorithim for multi-class cancer classification of gene expression RNA-Seq data. *Genomics* 112, 1916–1925 ELSEVIER

[15] Reva Joshi, Ritu Prasad, pradeep mewada and praneet saurabh. (2020). Modified LDA approach for cluster based gene classification using K mean method. *CoCoNet19 Procedia Computer Science* 171 2493–2500

[16] Sarah M. Ayyad*, Ahmed I. Saleh and Labib M. Labib. (2019). Gene expression cancer classification using modified K-nearest Neighborstechnique. *BioSystems* 176 41–51

[17] Sara Tarek, Reda Abd Elwahab and Mahmoud Shoman. (2016). Gene expression based cancer classification. *Egyptian Informatics Journal*

[18] Wu J. Hicks C. (2021). Breast Cancer Type Classification Using Machine Learning. *JPM*, 11, 61.

Computational Methods in Science and Technology Volume 1 – Kaur et al. (Eds)
© 2025 the Author(s), ISBN: 978-1-032-81760-6

Big data analytics in healthcare: Applications, examples and challenges

A. Kaur, S. Goyal & N. Batra
Computer Science and Engineering Department, MMEC, Maharishi Markandeshwar (Deemed to be University), Mullana-Ambala, Haryana, India

N. Gupta
Computer Science and Engineering Department, CGC-College of Engineering, Landran, Mohali, India

ABSTRACT: The healthcare sector is evolving at an exponential rate due to the speed of technological disruption and digital innovation. The size of healthcare data is increasing exponentially, making it more challenging to locate any kind of relevant information. Big data is currently changing how data is delivered traditionally by employing big data analytics techniques to produce insightful knowledge. In the healthcare sector, big data analytics offers numerous advantages, including the ability to identify serious illnesses early on and provide more effective care to the appropriate patient at the appropriate moment, thereby enhancing the standard of care. The quick analysis and integration of massive amounts of different types of data generated by various clinical settings, hospitals, other social media platforms, and medical data lakes is made possible by big data analytics tools. Many challenges must be addressed by the current health data analytics systems, which offer technical means for gathering, combining, processing, analysing, visualising, and interpreting data. This article delves into the intriguing topic of healthcare big data analytics. Due to a dearth of comprehensive study, this topic has not gotten enough attention in the past. This article covers the unique characteristics of big data, big data analysis tools, and the steps the healthcare business takes from data collection to data distribution.

1 INTRODUCTION

Big data in the healthcare industry is described as “electronic health data sets so large and complex that they are neither easily managed with traditional or common data management tools and methods, nor are they difficult (or impossible) to manage with traditional software and hardware.” [1]. The health business is experiencing a big data revolution due to the exponential expansion of data acquired through wearable sensors, registries, and electronic health records. This widely accessible data, which offers several advantages like improvedlife quality, disease diagnosis and treatment, and delivery of healthcare service requires a lot of information to be extracted. Large volume, heterogeneity, and speed are the characteristics of big data generated in the healthcare. Healthcare data is complicated by real-time data processing, many variables, and non-uniform data. Big data in the healthcare sector, which is distinguished by its quantity, complexity, and diversity, will have access to current and upcoming information both now and in the future. Big data’s unique qualities—such as its enormous volume, veracity, diversity, and dynamism—make it difficult to collect, store, share, and interpret data in real time from several decentralised places.

The BDA must simultaneously meet the needs of multiple stakeholders, including healthcare organisations, patients, vendors, employees, the media, and numerous organisations that monitor and regulate health. The availability of big data in the healthcare industry is found to be tremendous when compared to other sectors, and it is expected to rise rapidly in the near future.

 DOI: 10.1201/9781003501244-64

Healthcare service providers include doctors, hospitals, labs, clinics, pharmacies, imaging centres, and other enterprises involved in health care. Huge amounts of big data are also shared with external entities like governments and insurance companies. In the healthcare industry, BDA quality is crucial for both lowering service delivery costs and producing notable improvements in services. Availability of standardised data, control over data breaches and leaks, and degree of privacy and security of data are also recognised as critical success aspects of BDA in healthcare organisations.

According to McKinsey & Company's "The Big Data Revolution in Healthcare," data accessibility has sparked a healthcare data revolution. Data comes from EHRs, public insurance-funded clinical trials, and pharmaceutical companies' years of research and development. McKinsey & Company estimates big data analytics may save $300 billion to $450 billion in healthcare [2].

This paper's primary contributions are as follows:

- To research the unique features of big data analytics in the field of healthcare.
- To study the benefits of big data in healthcare.
- To outline the difficulties the healthcare sector faces with open research.

2 BENEFITS OF BIG DATA IN HEALTHCARE [3]

2.1 *Enhancing the standard of healthcare services by*

- Assessing the diagnosis and recommended course of treatment for diseases by medical experts utilising Big Data collections and decision support systems
- Analysing vast amounts of data to derive actionable insights for identifying needs,
- Launching new health services, averting and resolving crises;
- Forecasting the prevalence of diseases;
- Identifying patterns that contribute to bettering the general public's health and way of life; and analysing the human genome to enable personalised medicine.

2.2 *Supporting the work of medical personnel [7]*

- physicians compare cases from the past to present to improve diagnosis and treatment planning.
- determining the risks associated with diseases and improving control over reaction times;
- discovering diseases at an earlier stage when they can be more rapidly cured;
- determining who is deemed most likely to contract specific, potentially lethal illnesses by gathering historical data on the most common illnesses, in order to treat patients and file reports to insurance companies, medicine
- real-time collection and analysis of vast volumes of data from homes and hospitals, life monitoring systems to ensure security and forecast unfavourable occurrences,
- the capacity to forecast the development of particular diseases
- the examination of patient profiles to identify those for whom prevention should be
- implemented, lifestyle modification
- projecting the likelihood of complications, forecasting the course of the disease and its factors,
- identifying medication interactions and adverse

2.3 *Supporting scientific and research activity*

- Encouraging the analysis of "all data" instead of just a test sample permits research on new drugs and clinical trials.

- the capability to recruit individuals for specialized clinical trials based on their unique biological characteristics.
- Modeling and predictive analysis may be used to produce more efficient drugs and medical devices; choosing a patient group that is most likely to get the intended result without side effects is an important part of clinical trials.

2.4 *Business and management*

- quicker and more accurate detection of inaccurate or unauthorised financial operations in order to avoid reduce errors;
- cost savings and countering abuse and counselling techniques;
- a rise in profitability through the identification of high-cost patients or the identification of doctors whose work, procedures, and treatment plans are most expensive and the provision of cost-effective solutions to them;
- the discovery of needless medical activities and procedures, such as redundant testing.

Big data can improve healthcare in many ways. It may improve patient outcomes, identify epidemics, give vital information, cut healthcare costs, and enhance quality of life. Big data difficulties include processing, storage, visualization, and collection. Big data's openness, user-friendliness, and menu-driven interface don't always translate to structure, standardization, integration, and analysis. The biggest issues are data structure and security. Healthcare data is sensitive, raising issues about data standards, privacy, and security. Incompatible formats, insufficient analysts, poor data governance, the high cost of protecting, storing, and transporting unstructured data, and real-time analytics concerns plague data storage and transfers.

3 EXAMPLES: BIG DATA INITIATIVES IN HEALTHCARE

3.1 *Asthmapolis*

The firm has developed a GPS-enabled gadget that tracks the consumption of inhalers by patients; this technology has the potential to improve asthma treatment.

3.2 *Fighting the virus*

The virus kills millions annually, but the CDC is fighting back using big data analytics. Over 700,000 influenza cases are reported weekly to the CDC. Patients' conditions, therapies, and results are described in these papers.

3.3 *GNS Healthcare and Aetna*

GNS Healthcare, a big data analytics business, teamed with Aetna to treat metabolic disorders. GNS's REFS technology will be deployed.

3.4 *Big data and diabetes*

People with diabetes can also profit from the change in health care brought about by big data.

GoCap, a cap for prefilled insulin pens designed by Common Sensing, is able to log both the daily dosage of insulin administered and the precise time it was supplied. An additional piece of technology at the nexus of diabetes and big data is a service provided by Allazo Health. Using predictive analytics, their technology, AllazoEngin, is intended to enhance drug adherence programmes.

4 CHALLENGES ASSOCIATED WITH HEALTHCARE BIG DATA [4]

Advancements in the administration and analysis of large sets of data, particularly for systems that continuously process, gather, combine, analyze (using machine learning and predictive techniques), and present data in real-time, will enable a more comprehensive incorporation of electronic medical records (EMRs) into healthcare systems. Here, we quickly address a couple of these difficulties.

4.1 *Storage [6]*

Although many firms may feel confident in retaining data on their own premises, the task of storing vast quantities of data remains a significant obstacle. Managing availability, ensuring security, and regulating access are among the several benefits it offers. However, the process of developing and operating on-premises server networks may be costly and burdensome. Most healthcare firms have already transitioned to cloud-based storage using IT infrastructure, which provides enhanced reliability at a reduced cost. When choosing a cloud provider, organizations should choose one that has expertise in healthcare-specific security and regulatory prerequisites. Cloud storage offers other advantages such as simplified scalability, faster disaster recovery, and reduced upfront costs.

Purification or scrubbering of cleaning data is necessary after acquisition to ensure precision, consistency, relevance, correctness, and purity. This cleaning may be done manually or automatically using logic rules for maximum accuracy and integrity. More advanced and accurate technologies leverage machine-learning to save costs, time, and poor data from delaying big data efforts. unified style Complex and difficult-to-handle patients create massive volumes of data that the traditional EHR cannot gather. Big data management is too difficult for healthcare practitioners without a good data organization system. Codifying clinically relevant data for billing, clinical analytics, and claims was necessary. CPT and ICD code sets describe core clinical ideas in medical coding. However, these code sets have limitations.

4.2 *Precision*

Research suggests that there is still significant imprecision in the entry of patient data into electronic medical records (EMRs). The reason for this is likely due to the intricate nature of processes, the absence of significant value in electronic health records (EHR), and a lack of awareness of the cruciality of capturing extensive data properly. These components may contribute to quality issues in big data at any stage of its existence. Electronic Health Records (EHRs) are designed to improve the accuracy of data and facilitate communication in clinical procedures. However, studies indicate that there are discrepancies in these environments.

4.3 *Image preprocessing*

Several physical factors have been identified in research as possible sources of misinterpretations and compromised data quality in existing medical records. Medical imaging often encounters technical challenges related to various forms of noise and artifacts. Inadequate administration of medical imaging may potentially lead to picture alteration. For instance, it might lead to the identification of anatomical features such as veins on the picture that do not accurately reflect a real-world scenario. Measures that may be used to assist include reducing noise, eliminating artefacts, modifying the contrast of captured photographs, and fine-tuning the quality of photos post-manipulation.

4.4 *Security*

Data security is a major concern for healthcare firms due to ransomware, hacking, phishing, and security breaches. After finding various flaws, technological protections were created to protect sensitive health information. The HIPAA Security Rules help enterprises comply with data storage, transit, authentication, access, integrity, and auditing laws. Multi-factor authentication, firewalls, antivirus software upgrades, and data encryption may avoid numerous issues.

4.5 *Data sharing*

Multiple medical facilities are available to patients. The former option requires data interchange with other healthcare facilities. Incompatible data might hinder data transmission between companies. This may be due to technical and organizational issues. Withheld information may prevent medical staff from making educated patient follow-up and treatment choices.

5 CONCLUSION

The way healthcare providers use cutting-edge technologies to draw conclusions and make rational judgements based on clinical and other data sources could be drastically altered by big data analytics.Concerns include preserving security, ensuring privacy, setting standards and governance, and continuously advancing tools and technology will grow in importance as big data analytics advances. While big data analytics and its applications in the healthcare industry are still in their early stages, they may evolve more quickly due to significant advancements in platforms and technologies.

REFERENCES

[1] Raghupathi W and Raghupathi V. (2014). Big data analytics in healthcare: promise and potential. *Health Inf Sci Syst*. 2:1–10.

[2] Chris Poelker: Centralizing healthcare big data in the cloud: http://blogs.computerworld.com/cloud-computing/20488/centralizing healthcare-big-data-cloud: 2012

[3] Batko, K., and Ślęzak, A. (2022). The use of big data analytics in healthcare. *Journal of big Data*, 9(1),

[4] Dash, S., Shakyawar, S. K., Sharma, M., and Kaushik, S. (2019). Big data in healthcare: Management, analysis and future prospects. *Journal of Big Data*, 6(1), 1–25.

[5] Murdoch T.B., Detsky A.S. (2016). *The Inevitable Application of Big Data to Health Care* 309, pp. 5–6.

[6] Belle A., Thiagarajan R., Soroushmehr S.M.R., Navidi F., Beard D.A., Najarian K. (2015). Big data analytics in healthcare, *HindawiPubl Corp.* 1–16, http://dx. doi.org/10.1155/2015/370194.

[7] Wang Y., Hajli N. (2017). Exploring the path to big data analytics success in healthcare, *J. Bus. Res.* 70 287–299, http://dx.doi.org/10.1016/j.jbusres.2016.08.002.

[8] Cyganek B., Graña M., Krawczyk B., Kasprzak A., Porwik P., Walkowiak K., M. (2016). Woźniak, A survey of big data issues in electronic health record analysis, *Appl. Artif. Intell.* 30 497–520, http://dx. doi.org/10.1080/08839514.2016. 1193714

[9] Navaz, A. N., Serhani M. A., Al-Qirim N. and Gergely M. (2018). Towards an efficient and energy-aware mobile Big health data architecture. *Computer Methods and Programs in Biomedicine* 166: 137–154. doi:10.1016/j.cmpb.2018.10.008.

[10] Praveena, M., and Rao M. K. (2018). Survey on big data analytics in healthcare domain. *International Journal of Engineering & Technology* 7 (2.7): 919–925. doi:10.14419/ijet.v7i2.7.11097.

[11] Sabharwal, S., Gupta S. and Thirunavukkarasu K. (2016). Insight of big data analytics in healthcare industry. *In International Conference on Computing, Communication and Automation*, Noida, India, (pp. 95–100). IEEE.

Computational Methods in Science and Technology Volume 1 – Kaur et al. (Eds)

Predicting relevant captions using image caption generator in social media platforms

Aryan Gupta, Devansh Singh Bhadauria, Manav Atray & Inderjeet Kaur
Department of Computer Science & Engineering, Galgotias College of Engineering & Technology, Greater Noida, Uttar Pradesh, India

ABSTRACT: Creating natural language descriptions or captions for images is a formidable task that requires a combination of computer vision techniques to understand image content and natural language processing models to express this understanding in coherent sentences. This paper delves into the development of a web-accessible Image Caption Generator, aiming to improve user experience by automatically generating captions for uploaded images. The versatility of this generator extends its applicability to various uses, such as image indexing, assisting visually impaired individuals, and enhancing interactions on social media platforms. This paper results in caption generator without the need for complex data pre-processing or a specialized model pipeline. Notably, a singular end-to-end model seamlessly predicts captions, streamlining the entire process. The paper integrates social media platforms with the proposed Image Caption Generator giving users automated and contextually relevant captions, fostering a more immersive and inclusive digital experience.

Keywords: Automated Captions, Machine Learning, Caption Generator, Long Short Term Memory (LSTM) and Convolutioal Neural Network (CNN)

1 INTRODUCTION

In recent years, significant progress has been witnessed in the field of computer vision, specifically in areas related to image processing, such as image classification and object detection. These advancements serve as the basis for tackling the challenge of automatically creating descriptive sentences for images, commonly known as Image Captioning. This procedure extends the progress in image classification and object detection, allowing for the automatic generation of one or more sentences that encapsulate the visual content of an image.

Producing detailed and natural image descriptions holds significant implications, covering a range of applications such as titling news images, describing medical images, assisting in text-based image retrieval, offering information for visually impaired users, and improving human-robot interaction. Within the area of AI, image captioning emerges as a complex yet meaningful task, aiming to develop algorithms capable of providing semantic-level descriptions for new images.

In the context of our Image Caption Generator, the foundation lies in processing provided or uploaded image files through a meticulously trained model. This model, created through sophisticated algorithms and trained on extensive datasets, aims to generate captions relevant to the visual content of the images. The aim of this paper is to seamlessly integrate the image caption generation into social media platforms and applications providing users with the convenience of automated and contextually appropriate captions.

2 RELATED WORK

In the past, methodologies initially involved identifying the content of an image and subsequently crafting a sentence describing it. Some approaches centered on training models for the comprehensive recognition and detailed description of videos and images [1]. An example, known as

DOI: 10.1201/9781003501244-65

LRCN, managed various tasks like recognizing and describing videos but encountered challenges in interpreting its own results.

Later, a model called NIC was introduced, specifically tailored for generating captions for images. NIC employed a combination of Google's GoogLeNet and LSTM, a neural network type, to generate captions that suit the images. NIC performed exceptionally well in a contest where human judgment determined the results. When comparing NIC to LRCN, notable differences emerge: NIC used GoogLeNet, while LRCN utilized VGGNet; NIC's LSTM considered visual features only in the initial step, whereas LRCN used them in every step, and NIC had a simpler setup with just one LSTM layer compared to LRCN's two [2,3].

Although both LRCN and NIC employ the same fundamental mathematical principles, their unique configurations impact their effectiveness. LRCN faces the challenge of balancing simplicity and usefulness across various tasks.

Another approach, proposed by Fang *et al.*, took a different route by training detectors to identify common words in captions and using them to compose captions. They found their captions to be as good as or better than human-generated ones about a third of the time. However, their method relied more on human choices, making it challenging to replicate their results. Additionally, there's the visual-semantic alignment method by Karpathy *et al.*, which breaks down the image into distinct parts and describes each part separately, yielding more accurate and varied descriptions compared to methods like LRCN and NIC. However, this method involves the use of two different models, adding a level of complexity.

3 LITERATURE SURVEY

Confronting the challenge of caption generation in artificial intelligence involves crafting textual descriptions that correspond to provide photographs. This intricate task necessitates a dual approach, amalgamating computer vision methods to comprehend image content and language models from natural language processing to articulate this understanding in coherent words. Recent advances in deep learning methods have significantly elevated caption generation, achieving high-tech results in resolving this complex problem.

A particularly noteworthy aspect of these methods is their capacity to establish a unified end-to-end model for predicting captions based on a given photo. This obviates the necessity for elaborate data preparation or a convoluted pipeline involving models tailored for specific designs [4]. Recurrent Neural Networks (RNNs) have proven to be formidable tools, particularly in modeling sequential data. Andrej Karapathy's blog, titled "The Unreasonable Effectiveness of Recurrent Neural Networks," extensively delves into RNNs and their efficacy. In the domain of RNNs, four primary types are delineated, as illustrated in Figure 1.

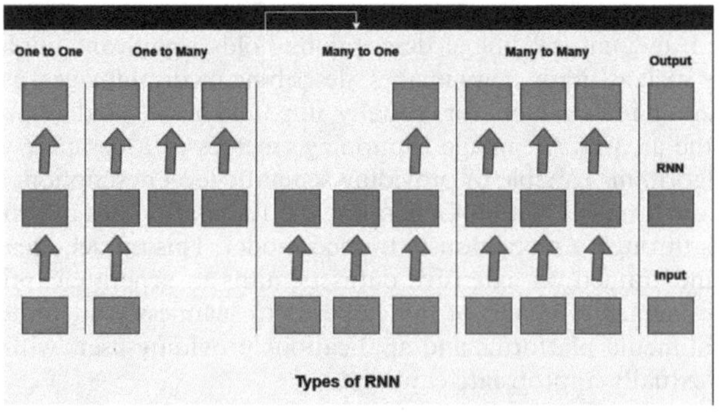

Figure 1. Types of RN.

3.1 *Convolutional Neural Network (CNN)*

CNN, a prominent technique in Deep Learning, is meticulously designed for processing input images by assigning significance to various aspects or objects within the image through the application of learnable weights and biases. This attribution is paramount for distinguishing one image from another, and image classification stands out as one of the widely applied use cases of CNN. The neural network architecture is structured with multiple convolutional layers intertwined with pooling and non-linear layers. As the image passes through each convolutional layer, the output from the previous layer serves as input for the next one, continuing this iterative process [5,6].

A fully connected layer is necessary after a series of convolutional, pooling, and non-linear layers. The convolutional networks' output data is consolidated in this layer. Adding a fully connected layer at the end of the network simplifies the process of selecting the appropriate class by creating an N-dimensional vector, where N is the number of classes the model recognizes.

3.2 *Long Short-Term Memory (LSTM)*

LSTM networks, belonging to the category of Recurrent Neural Networks (RNN), specialize in grasping order dependence in sequence prediction tasks. Their primary application extends to addressing complex challenges such as speech recognition, machine translation etc. The inception of LSTM was prompted by a critical issue encountered in traditional RNNs, wherein, as neural networks deepen, excessively small or zero gradients impede effective training. This limitation results in subpar predictive performance. LSTM networks have proven adept at overcoming this obstacle and are particularly well-suited for tasks involving the classification, processing, and prediction of time series data. This is especially beneficial when dealing with time series events characterized by unknown durations between crucial occurrences. LSTM outshines traditional RNNs due to its heightened effectiveness, successfully overcoming the short- term memory constraints associated with the latter. The key strength lies in LSTM's ability to retain relevant information throughout input processing, facilitated by mechanisms such as the forget gate, which selectively discards non-essential information [7].

3.3 *CNN- LSTM architecture*

The CNN-LSTM design includes LSTM networks to enable sequence prediction and combines CNN layers to extract features from input data. This specialised model, which is designed to tackle sequence prediction problems related to spatial inputs like images or videos, is used in a variety of fields, including activity recognition, video description, and image description [8]. Figure 2 shows a graphic representation of the CNN-LSTM Model's basic architecture.

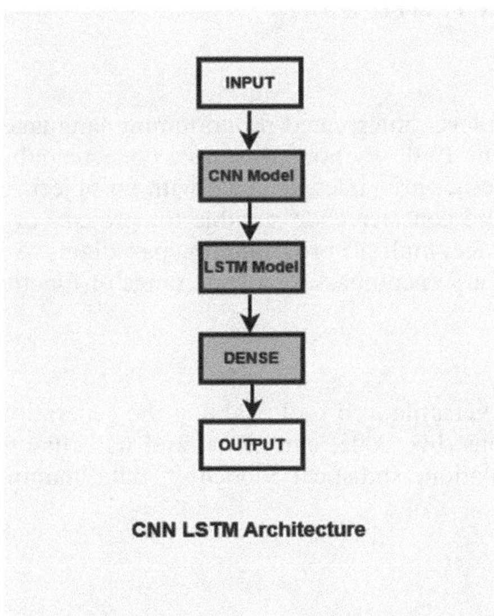

Figure 2. CNN LSTM architecture.

3.4 *Recurrent Neural Network (RNN)*

The nodes forming RNN within a directed graph are connected sequentially, creating a series. This configuration aids in the RNN's management of operations related to sequences, including time sequences, handwriting recognition, and sequence expression. RNNs have the capability to retain vital information from their input, enabling them to predict the subsequent item in a sequence [10].

RNNs typically include Short-Term Memory (STM), leading to challenges in handling very long sequences. To overcome this limitation, LSTM network extends the RNN's capabilities by enhancing its memory. Consequently, LSTMs find frequent use in scenarios involving sequences with extended gaps, as they can remember previous inputs for an extended period by storing this information in memory. In tasks like image captioning, where captions are generated from image features, RNNs coupled with LSTMs are commonly employed.

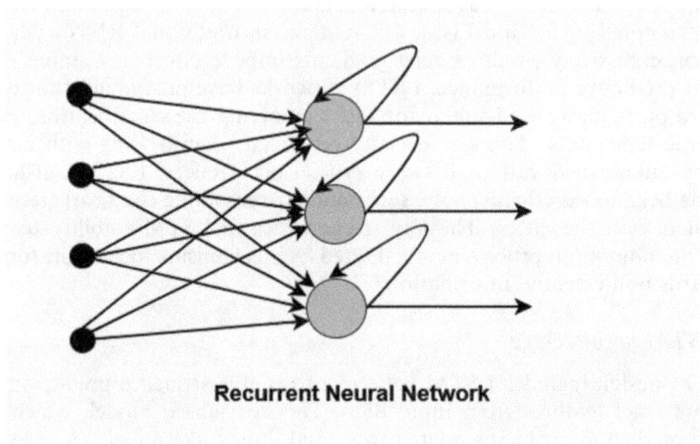

Figure 3. Recurrent Neural Network.

4 TECHNOLOGY STACK UTILIZED

4.1 *Python*

Python stands out as a high-level, interpreted programming language. Developed by Guido van Rossum and first released in 1991, Python prioritizes code readability, notable for its use of significant whitespace. Its design principles, coupled with an object-oriented methodology, facilitate the creation of clear and cohesive code suitable for projects of all scales. Python supports garbage collection and embraces multiple programming paradigms. Additionally, Python boasts a comprehensive standard library encompassing a wide range of functionalities [9,11].

4.2 *Jupyter Notebook*

The Jupyter Notebook is a versatile web tool enabling the generation and distribution of documents featuring visualizations, live code, equations, and narrative text. Its diverse capabilities encompass numerical simulation, statistical modeling, data manipulation, and visualization, among others [13].

4.3 *Google Colab*

Collaboratory, commonly known as "Colab," is a product of Google Research aimed at simplifying the writing and execution of various Python code directly in the browser. Designed for tasks such as machine learning, data analysis, and education, Colab functions as a hosted Jupyter notebook service. Its notable characteristics include zero setup demands and the availability of free access to computing resources, including GPUs [12].

4.4 *Flask*

Flask is a lightweight web framework developed in Python, recognized for its minimalistic nature that doesn't enforce specific tools or libraries. Importantly, Flask does not come with built-in features such as a database abstraction layer, form validation, or other components usually provided by third-party libraries for standard functionalities [8,14].

5 PROPOSED FRAMEWORK & DESIGN

Our model employs two distinct neural networks for caption generation. CNN as initial neural network, utilized for both image training and object detection within the image. Deep Learning has evolved in recent times to address the data constraints inherent in Machine Learning algorithms. GPU-based computing is essential for the effective execution of Deep Learning tasks [15].

This paper outlines the Resnet-LSTM model for the image captioning process. Here, the Resnet Architecture is employed for encoding, and LSTMs are utilized for decoding. Once the image is input into Resnet (Residual Neural Network), it extracts the image features. Subsequently, utilizing a vocabulary built from training caption data, we proceed to train the model with these two parameters as input. Following the training phase, we move on to testing the model. The steps are as followed:

5.1 *Read captions file*

This process entails reading the text from the Flickr 8k file, assessing its length, and subsequently dividing it.

5.2 *Data cleaning*

Data cleaning stands as a crucial process aimed at rectifying or removing inaccurate, corrupted, improperly formatted, duplicate, or incomplete data within a dataset. The reliability of outcomes and algorithms is jeopardized when handling flawed data, even if it appears accurate initially.

5.3 *Loading training/testing data*

This step encompasses handling training images files, testing them, and creating a dictionary for training descriptions that includes starting and ending sequences.

5.4 *Data preprocessing (images)*

It encompasses preprocessing, encoding, and loading images, ensuring they are prepared for testing.

5.5 *Data preprocessing (captions)*

This step involves loading captions, determining the maximum length of the captions and appending the start and end sequences.

5.6 *Data preparation using generator*

Data preparation is a crucial procedure that encompasses the cleaning and transformation of raw data before processing and analysis. This involves reformatting data, making corrections, and combining datasets to enhance the overall quality and richness of the data.

5.7 *Word embedding*

Word embedding entails converting words into vectors serving as the output of an Embedding Layer.

5.8 *Model architecture*

Developing the model architecture involves creating an image feature extractor model, a partial caption sequence model, and then merging these two networks.

5.9 *Train our model*

A training model is a dataset that is used to provide matching sets of input data that affect the output of a machine learning algorithm.

5.10 *Predictions*

A training model is a dataset that is used to provide matching sets of input data that affect the output of a machine learning algorithm.

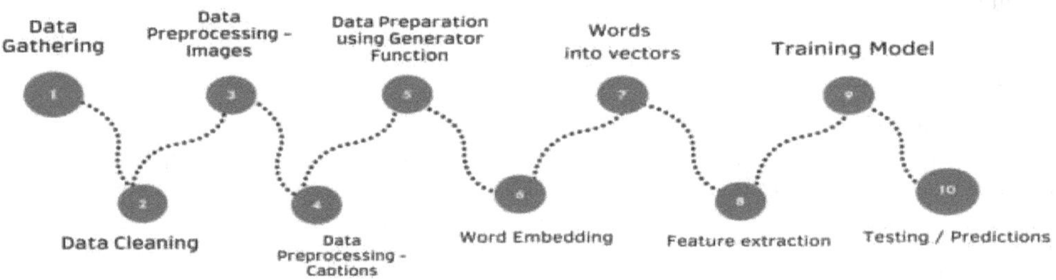

Figure 4(a). Proposed process model.

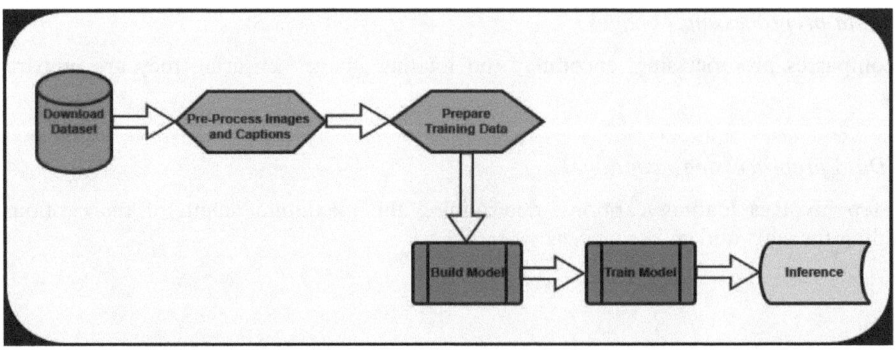

Figure 4(b). System design steps.

6 FEASIBILITY STUDY AND METHODOLOGY USED

6.1 *Feature extraction model*

This model plays a crucial role in gathering image features for training. This architecture efficiently extracts features from images through a series of 3x3 convolution layers and max pooling layers. The VGG 16 network produces 1x4096 vectors representing these features. To combat overfitting, a dropout layer with a dropout rate of 0.5 is incorporated into the model. This rate typically falls within the optimal range of 0.5 to 0.8, indicating the probability of dropping outputs from the layer. Following the dropout layer, a dense layer is introduced, applying the ReLU activation function to the input and specifying an output space size of 256. These 256- dimensional vectors serve as the output of the feature extraction model, which will later be utilized in the decoder model.

6.2 *Encoder model*

To prepare for neural network processing, the captions associated with each image undergo an initial tokenization process, converting words into integers to facilitate efficient network processing. To ensure standardized input lengths, the tokenized captions are padded to align with the size of the longest sentence. Following this, an Embedding layer is employed to embed the tokenized captions into dense vectors of 256 dimensions, resulting in an output space of 256 by 34. The choice of 34 is determined by the maximum number of words in any caption within the Flickr 8k dataset.

These embedded vectors optimize processing by representing words within a vector space. To address overfitting, a dropout layer with a 0.5 probability is introduced. The core element of the Encoder model is the LSTM layer (Long Short-Term Memory), which assists in learning to generate valid sentences by predicting the most likely word following a given word occurrence. ReLU serves as the activation function for this layer, defining the output space as 256. For comparing the VGG + GRU and VGG + LSTM models, the LSTM layer can potentially be replaced with a GRU (Gated Recurrent Units) layer, and the resulting outcomes are then examined. The output space for the GRU layer remains fixed at 256. Consequently, the primary difference between the two models resides within the encoder segment, where the output of the LSTM layer acts as the output for the encoder layer [18].

6.3 *Decoder model*

To achieve the intended outcome, the decoder model plays a crucial role in amalgamating the outcomes from the feature extraction and encoder models. This output comprises the sentence generated up to that moment and the word predicted based on an image. Both the feature extraction model and the encoder model, which produce vectors of dimension 256, contribute inputs to the decoder model. The ReLU activation function is then employed to propagate these concatenated outputs through a dense layer. The output space of a second Dense Layer added to the decoder model is determined by the vocabulary size. In essence, the decoder model is responsible for synthesizing the outputs of the feature extraction and encoder models to generate the desired output.

These combined outputs are subsequently fed through a dense layer employing the ReLU activation function. Another Dense Layer is integrated into the decoder model, with the output space defined by the vocabulary size. For the Flickr 8k dataset, the vocabulary size was established to be 7579, and the activation function utilized is softmax activation, which yields a word based on the predicted integer. The predicted word signifies the output of the decoder layer [16,17].

During training, the model is supplied with input-output pairs in the format:

<input> = <image, <input sequence>> and <output> = <word>.

Given the image and the current caption, the model predicts the next word. Subsequently, the BLEU score is computed for each architecture. BLEU scores can be calculated in four ways: BLEU-1, BLEU-2, and BLEU-3, where scores range from 1.0, 0.5, to 0. For the Flickr 8k dataset, a vocabulary of 7579 words was utilized, and softmax activation—generating words based on predicted integers—was employed as the activation function. The predicted word represents the output of the decoder layer. During training, the model is provided with input-output pairs formatted as follows:

$$<\text{input}> = <\text{image}, <\text{input sequence}>> \text{ and } <\text{output}> = <\text{word}>.$$

The model predicts the word given the image and the caption generated up to that point. Subsequently, the BLEU score is computed for each architecture. Four types of BLEU scores are calculated:

$$\begin{aligned} &\textbf{BLEU-1} \ (\mathbf{1.0, 0, 0, 0}), \\ &\textbf{BLEU-2} \ (\mathbf{0.5, 0.5, 0, 0}), \\ &\textbf{BLEU-3} \ (\mathbf{0.33, 0.33, 0.33, 0})\textbf{and} \\ &\textbf{BLEU-4} \ (\mathbf{0.25, 0.25, 0.25, 0.25}) \end{aligned}$$

Cumulative weights are applied to these scores since they tend to yield improved results.

7 RESULTS AND DISCUSSIONS

Following the definition and fitting of the model, we conducted training sessions spanning 50 epochs. In the initial epochs, a clear trend emerged – the accuracy was notably low, and the generated captions exhibited limited relevance to the given test images. However, with a minimum of 20 epochs, a discernible improvement occurred, resulting in generated captions that were somewhat associated with the provided test images. Upon completing the entire 50-epoch training process, a substantial enhancement in model accuracy was noted. The generated captions showcased a significantly heightened correlation with the given test images, as visually depicted in the figure below.

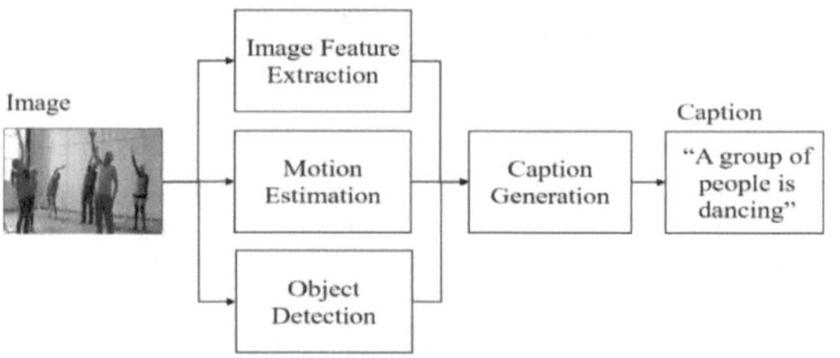

Figire 5(a). Steps for training model.

Figure 5(b). Image caption generator illustration.

8 APPLICATIONS AND FUTURE DIRECTIONS

In our technologically-driven era, the integration of virtual assistants has become increasingly pervasive, reshaping how we interact with digital devices. From voice-activated platforms to AI-driven chatbots, virtual assistants simplify tasks, provide information, and enhance overall user experience across diverse applications. These digital companions have seamlessly woven into our daily lives, offering hands-free assistance and streamlining various tasks, ranging from setting reminders to managing schedules.

Effective communication relies on well-crafted content, underscoring the importance of recommendations for editing. This encompasses refining written documents, multimedia content, and communication materials to ensure clarity, coherence, and impactful messaging. By adhering to editing guidelines, individuals and organizations can elevate the quality and effectiveness of their communication strategies.

The advent of self-driving vehicles has ushered in a new era of transportation, bringing with it challenges and innovations in image coding. Advanced coding techniques play a crucial role in enabling these autonomous vehicles to interpret and respond to visual data, contributing to enhanced safety and efficiency on the roads

In our research, we explore the intricacies of generating captions for images. Despite substantial advancements in deep learning, achieving precise caption generation remains a challenge due to factors such as hardware limitations and the absence of a well-defined programming logic or model capable of generating exact captions. Machines, at present, lack the ability to think or make decisions as accurately as humans. Looking forward, we foresee the potential for more accurate caption generation in the future with ongoing progress in both hardware capabilities and deep learning models.

Additionally, there is consideration to expand this model to create a comprehensive Image-Speech conversion system, converting image captions into spoken words. This envisioned enhancement holds considerable promise, especially for individuals with visual impairments. By providing a means to convert image captions into audible language, this advancement aims to offer valuable support to the blind community, showcasing the potential of technology to enhance accessibility and inclusivity.

9 CONCLUSION

This paper introduces a novel Image Caption Generator that combines image and language-based models to produce descriptive captions for users based on input images. The Image Based Model utilizes CNN to extract features from images, while the Language Based Model employs LSTM networks to transform these features into coherent natural language sentences. The workflow encompasses data gathering, pre-processing, model training, and prediction. The proposed Image Caption Generator aims to enhance social media platforms, contribute to image indexing, and offer automated, generated captions for individuals with visual impairments. By combining advanced image processing techniques with natural language generation, our approach not only enhances user experiences across various platforms but also addresses the practical need for improved accessibility for those with visual impairments. This research contributes to the ongoing endeavors to leverage technology for inclusive and impactful applications.

REFERENCES

[1] Batra, V., He, Y. and Vogiatzis, G. (2018). Neural caption generation for news images. *School of Engineering and Applied Science*, Aston University.

[2] Desmukh, R. A. and Raut, P. (2021). An advanced image captioning using combination of CNN and LSTM. *Turkish Journal of Computer and Mathematics Education.*

[3] Fu, K., Li, J., Jin, J. and Zhang, C. (2018). Image-text surgery: efficient concept learning in image captioning by generating pseudopairs. *IEEE Transactions on Multimedia.*

[4] Kadam, R., Kumbhar, U., Gulik, O. and Shahade, M. (2020). Object detection and automatic image captioning using tensorFlow. *International Journal of Future Generation Communication and Networking*.

[5] Kinghorn, P. and Zang, L. (2017). A region-based image caption generator with refined descriptions. *Journal of Neurocomputing*, 272, 416–424. https://doi.org/10.1016/j.neucom.2017.07.014.

[6] Liu, S., Bai, L., Hu, Y. and Wang, H. (2018). Image captioning based on deep neural networks. *MATEC Web of Conferences, College of Systems Engineering, National University of Defense Technology*, 2018, 410073 Changsha, China.

[7] Poghosyan, A. and Sarukhanyan, H. (2017). *Long Short-Term Memory with Read-only Unit in Neural Image Caption Generator*. IEEE.

[8] Shah, P., Bakrola, V. and Pati, S. (2017). Image captioning using deep neural architectures. In *IEEE International Conference on Innovations in Information Embedded and Communication Systems (ICIIECS)*.

[9] Wu, J., Chen, T., Wu, H., Yang, Z., Wang, Q. and Lin, L. (2019). Concrete image captioning by integrating content sensitive and global discrimination objective. *In IEEE International Conference on Multimedia and Expo (ICME)*.

[10] Yang, M., Zhao, W., Xu, W., Feng, Y., Zhao, Z., Chen, X. and Lei, K. (2018). Multitask learning for cross-domain image captioning. *IEEE Transactions on Multimedia*.

[11] Vinyals, O. (2015). Show and tell: A neural image caption generator. *In Computer Vision and Pattern Recognition (CVPR)*, IEEE Conference.

[12] Xu, K. and Kelvin. (2015). Show, attend and tell: Neural image caption generation with visual attention. *In International Conference on Machine Learning*.

Computational Methods in Science and Technology Volume 1 – Kaur et al. (Eds)

Housing price prediction using multiple regression

Anannya Singh & Ajay Vikram Singh
Amity Institute of Information and Technology Amity University Uttar Pradesh, Noida

Sonu Mittal
Amity Institute of Information and Technology Amity University Uttar Pradesh, Noida
ORCID ID: 0000-0002-0452-9652

ABSTRACT: Machine learning has come out as the new hot topic in the world of technology, this new technology offers a very innovative method of solving some of our most perplexing issues, one of them being accurate and efficient house price predictions, earlier a human-centric job with high levels of error and uncertainty, machines learning has come out with a solution with many different types of algorithms that can help a real estate agent deal with the numbers with haste and efficiency. This paper presents a comparative study and analysis of the different algorithms that can be used for HOUSE PRICE PREDICTION, and the accuracy they promise.

Keywords: Algorithms, Data Science, Machine Learning, Analysis, Prediction, research, Decision tree, Regression algorithm, Random Forest, SVM

1 INTRODUCTION

A House Price Index (HPI) is a scale that reflects the change in prices of properties over some time, such an index helps make informed choices about the properties one is about to invest in and also professionals in pricing the properties accordingly. House Price Indices are used by economists, policymakers, and real estate agencies. Understanding the changing dynamics of the housing markets is essential for many players, eg. homebuyers, sellers, lenders, and policymakers, as it can impact investment decisions, housing affordability, etc. HPI is a rough figure calculated by taking into account various factors like transactions, payment successes, failures, etc. (Hardi Joshi, Saket Swarndeep 2022) (Robbi Jyothsna 2022).

The index uses a reference year to make calculations and predictions and provides the geographical area with a value out of 100. The changes in the housing prices are represented as percentages in comparison with the base year. Several factors can influence house prices, including economic conditions, interest rates, demographic trends, infrastructure, accessibility, connectivity, etc. House Price Indices are often broken down by districts, towns, or neighborhoods to provide a more detailed understanding of local trends.

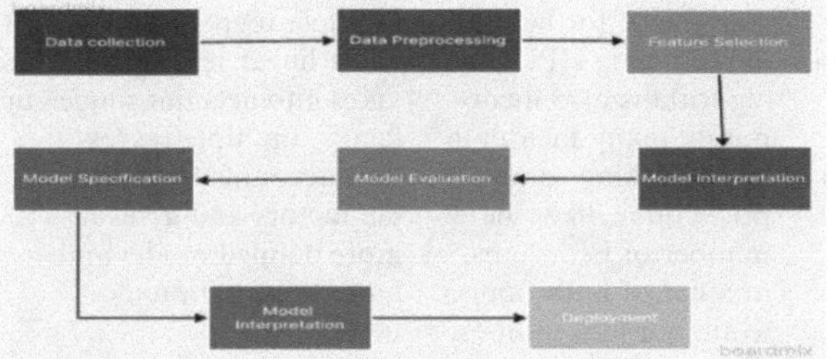

Figure 1. Research flow diagram.

DOI: 10.1201/9781003501244-66

2 LITERATURE REVIEW

In the past couple of years machine learning has proved its worth as a powerful tool for making calculated predictions based on information and trends gathered and studied from multiple sources, it not only helps with data from past years but can also give a dynamic model that can be used for coming years as well, the models so created are fairly accurate as the model is based on big data which machine learning can easily process.

2.1 *Machine learning techniques*

In traditional programming, humans provide a diverse set of rules and instructions for software to perform, they act as a set of rules that are supposed to be followed while performing the task. However, in machine learning, the approach is different. Instead of providing instructions, the system is trained on datasets to recognize patterns, make predictions, study statistics, or perform tasks. By modifying the model's boundaries in response to the input data, the learning process enables the system to recognize patterns and make accurate predictions on new, foreseen data and learn from trial and error based on reward and punishment methodology.

The field continues to evolve with advancements in software, hardware, accessibility, and the availability of large datasets. The literature review goes through multiple research papers and presents a comparative analysis of different machine learning algorithms used for predicting, studying, and creating house price indexes and which method gave the best and most accurate outcome. The results are as follows:

Table 1. Results of the comparison of the efficacy of algorithms.

Algorithms	Description	How It Can Be Used For Housing Price Prediction.	Accuracy
1. Decision Tree (Hardi Joshi, Saket Swarndeep 2022)	It is like a flowchart that helps in making decisions. It deconstructs a problem into its basic components and forms branches like a tree where each step leads to a prediction. They are simple. and easy to understand.	They analyze features like the square footage of the house, no. of bedrooms, no. of bathrooms, etc. to design a basic structure for the decision tree. This process helps us understand the different components and variables. easy.	Cancapture complex relationships, but is prone to overfitting;
2. Regression Algorithm (Arshiya Shaikh, R.Vinayaki, G. Siddhanth, Y. Phanindra Varma 2020) (Dr. Sonia Juneja , Neha Chaudhary, Ritul Gupta, Ojasvi Kaushik, Mohd Ishan , Ayush Sharma 2023)	Regression algorithms which are specially designed/coded to predict numbers are also already used for helping in predicting HPI. These algorithms also incorporate many factors in their working to give predictions, like, the number of bedrooms, number of bathrooms, availability of utilities, etc.	Many types of regression algorithms can be helpful for predictions, like linear regression, multiple regression, etc. while linear regression takes into account single factor, multiple regression accounts for multiple factors and gives a more defined model and more accurate predictions.-	By training on past data of HPIs which focused on different factors for predictions, these algorithms can provide fairly accurate predictions.-

(*continued*)

Table 1. Continued

Algorithms	Description	How It Can Be Used For Housing Price Prediction.	Accuracy
2.1 Linear Regression Algorithm (Rohit Suryawanshi,Nikki Thakur, Kailas Waghmare, Monika Kshirsagar, Rutika Gaurkar, Mukta Wagh,APRIL 2022) (Geethamani, Karthika. 2023)	Predictions depend upon the input factor provided to the algorithm; it takes only one factor into account at a time. It tries to find a value with the minimum difference between the actual price and the predicted price, such a line is known as the best-fit line.	It creates a simple and easy-to-read model that can be understood by everyone which makes it a good choice for HPI as it would allow for quick interpretation of the model. It also helps in understanding the significance of each element in HPI.	Accuracy is based on how well the relationship between features and prices can be shown by a straight line.
2.2. Multiple Regression Algorithm (Dipanshu, Tarun, Harsh, Keshav 2023)	It is basically like a more elaborate version of linear regression, where more than one factor is taken into account for making the best-fit line. The use of multiple factors makes the model more versatile.	This algorithm is used in house price prediction as it is a more complex model and well-rounded model due to the use of multiple factors simultaneously. By studying past HPIs It analyses how different factors interact to influence HPI giving a more accurate result to real-life scenarios.	Canimprove accuracyby multiple features simultaneously, capturing more complex relationships.
3. Random Forest Algorithm (Dr. Sonia Juneja , Neha Chaudhary, Ritul Gupta, Ojasvi Kaushik, Mohd Ishan , Ayush Sharma 2023) (Geethamani, Karthika. 2023)	This algorithm works with a bunch of decision trees working together for a more complex and rounded result. Each tree is given a separate part of data to analyze on which it gives results. The final result is given by combining the results of all the decision trees	In this many decision trees are used and every tree studies and predicts a different factor combining these results gives us the final prediction which is pretty accurate and gives a better picture of how each of the factors is influencing the price and by how much.	Generally, provides higher accuracy than individual decision trees
4. Support Vector Machines (SVM) (Pushpak Lawhale, Yash Ramatkar, Shrikant Dakhore, Manshri Gumble, Aishwarya Tawlare, Ms. Yugandhara A. Thakare 2022)	This algorithm is used for classifying things. It tries to figure out the best boundary or line that can differentiate between different aspects in a set of data. The aim is to get the largest margin possible between groups. It is also applicable in scenarios where there is complex relationships.	It is a useful tool for predicting HPP. It is a versatile algorithm that is useful in handling simple as well as complex relationships between groups. It works by separating the prices in accordance with different factors like location, size, etc. It is a very reliable algorithm for predictions but might require fine-tuning for the best outcomes.	Effectivefor high-dimensional data and can handlenon-linear relationships; Accuracyis dependable on featuresand quality of data

3 HOUSING PRICE PREDICTION

Housing price prediction refers to the process of using historical data, statistical analysis, and machine learning algorithms to figure out the future prices of real estate. Housing price prediction can used for the following purposes:

(i) Informed decision-making: HPP allows people to make informed decisions by looking at housing price predictions so that people can plan accordingly, about investing, or selling houses.
(ii) Risk avoidance: not only individuals but even institutions like banks benefit from the use of HPP, they use the predictions to decide on the risk associated with a particular home loan for example.
(iii) Urban Planning: The government and policymakers also utilize HPP for planning infrastructure projects like roads, railways, metros, etc., calculating circle rates, also to check the health of the housing market.
(iv) Market Analysis: professionals such as real estate agencies, builders, and agents, use HPP to study the market and figure out the trends and the best strategies to increase their sales.

HPP emerges as a very useful and important tool for stakeholders, as it aids in easy strategizing and plan implementation for real estate agencies and other players, helping by allowing informed decision-making, risk mitigation, planning, etc. (Swarali M. Pathak, Archana K. Chaudhari, 2021).

3.1 *Algorithms suitable for housing price prediction*

Many machine learning algorithms can be utilized for making a model for HPP, some of them are as follows:

3.1.1 *Decision tree*

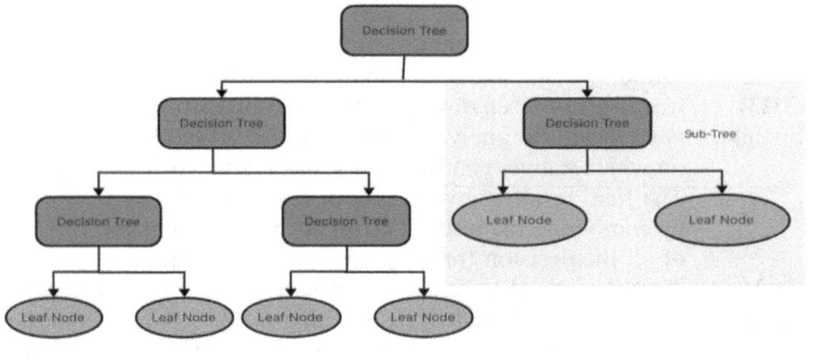

Figure 2. Decision tree structure.

The steps in constructing a decision tree are:

i) Data collection and pre-processing: in this step, we collect data from various sources, such as real estate agencies, government circle rates, market trends, etc., and then remove the errors to make it ready for training.
ii) Algorithm: when the pre-processing of the data is done, then one decides upon the method by which the decision tree would be designed.
iii) Training and Testing the Model: when the algorithm has been selected it is time for training the model on the relevant data and then followed by training.

iv) Performance and release: Once the model has been tested, analyze its performance and make final tuning, by comparing the model's predictions with the pricing. Once the model has been tuned and one feels confident the algorithm's performance is satisfactory, it can be used. (Sushant Kulkarni, Shefin Shajit, Akshay Mohite, Dr. Swati Sinha 2021) (Aditi Shahasane, Mayuri Gosavi, Ayushi Bhagat, Nandini Mishra, Amit Nerurkar, 2023).

3.1.2 *Regression algorithms*

Since regression algorithms are most effective at predicting numerical goal values, they are the most utilized methodology for predicting property prices. Regression algorithms are used to estimate home values by examining a wide range of housing-related variables, including the home's square feet, number of bedrooms, neighborhood atmosphere, and proximity to amenities like schools, grocery stores, hospitals, and roadways, among many others. A regression algorithm can be trained on pertinent datasets containing historical data on house prices, allowing it to learn and predict house prices based on input feature inputs. (Arshiya Shaikh, R.Vinayaki,G. Siddhanth, Y. Phanindra Varma 2020) (Dr. S P. Malarvizh, Dr. M. Thamarai 2020)

3.1.3 *Linear-regression*

Linear Regression is a simple but effective, algorithm that performs best when the target variable and input features are linearly arranged. (Geethamani, Karthika. 2023) (Priya P, Arul Kumaran M, Dhinesh Kumar K, Nivas Singha S, Rajkumar K 2021)

3.1.4 *Multiple-regression*

It is basically like a more elaborate version of linear regression, where more than one factor is taken into account for making the best-fit line. The use of multiple factors makes the model more versatile. This algorithm is used in house price prediction as it is a more complex model and well-rounded. (Nor Hamizah Zulkifley, Shuzlina Abdul Rahman, Nor Hasbiah Ubaidullah, Ismail Ibrahim 2020) (Dipanshu, Tarun, Harsh, Keshav 2023)

3.1.5 *Random forest algorithm*

In this many decision trees are used and every tree studies and predicts a different factor combining these results gives us the final prediction which is quite accurate and gives a better picture of how each of the factors is influencing the price and which is most essential. Random forest algorithm is preferred for another reason that is it can handle regression as well as classification tasks. (Geethamani, Karthika. 2023) (Priya P, Arul Kumaran M, Dhinesh Kumar K, Nivas Singha S, Rajkumar K 2021)

4 FACTORS INFLUENCING HOUSING PRICE PREDICTION

Many factors come into play when predicting house prices, many small and big elements influence house prices, such as:

(i) Location: it comes out as the primary influencer in house prices, a property's price is significantly affected based on where it is located. (Nor Hamizah Zulkifley, Shuzlina Abdul Rahman, Nor Hasbiah Ubaidullah, Ismail Ibrahim, 2020) (Tanmoy Dhar , Dr Manikandan P, 2023)

(ii) Economic factors: Economic stability and growth tend to correlate positively with housing demand and prices. In other words, a healthy economy breeds a good housing market with relatively stable prices. (Pushpak Lawhale, Yash Ramatkar, Shrikant Dakhore, Manshri Gumble, Aishwarya Tawlare, Ms. Yugandhara A. Thakare, 2022)

(iii) Interest Rates and Home loan accessibility Interest rates have an immediate effect on how affordable housing is. Lower interest rates typically have an upward trend in prices since they make borrowing less expensive for investors. Alternatively, yet, higher interest rates and slow price growth as the houses become out of budget for people. (Pushpak Lawhale, Yash Ramatkar, Shrikant Dakhore, Manshri Gumble, Aishwarya Tawlare, Ms. Yugandhara A. Thakare, 2022)

(iv) Environmental and Climate Factors: Environmental risks, such as natural disasters like floods, cyclones, rise in the sea level, extremes of weather conditions, Global warming, etc. Have a major effect on housing prices.

5 MULTIPLE REGRESSION FOR HOUSING PRICE PREDICTION

An analysis can be conducted about the relationship between independent as well as dependent factors statistically using multiple regression. There are several algorithms and methods commonly used for multiple regression analysis (Dr. S P. Malarvizh, Dr. M. Thamarai 2020). Some best algorithms are:-

5.1 *Ordinary Least Squares (OLS) regression*

It is the approach to multiple regression analysis that is most common and widely utilized. The overall linear difference (OLS) between the dependent variable's observed and predicted values is what OLS seeks to minimize.

(i) This fits a linear regression model to the housing dataset with features like square footage, number of bedrooms, bathrooms, etc., as independent variables and the housing price as the dependent variable.
(ii) Estimate coefficients using the OLS method, where the coefficients represent how each quality affects the price of a home.
(iii) Interpret coefficients to understand how each feature affects the predicted housing price linearly.

5.2 *Ridge regression*

Apply Ridge regression to the housing dataset to handle multicollinearity (which indicates a strong linear relationship among the predictor variables.) issues, where features are highly correlated (Saptarsi Sanyal, Saroj Kr. Biswas, Dolly Das, Manomita Chakraborty 2022).

5.3 *Lasso regression*

The use of Lasso regression for feature selection in the housing dataset, especially when dealing with many features (Ahmad Abdulal Nawar Aghi 2020).

5.4 *Elastic net regression*

Elastic Net Regression can be effectively used in housing price prediction, which may be applied to provide the buyers, the sellers, and the real estate agents with insightful information. Elastic Net Regression is a regression method that merges the losses of both Ridge Regression and Lasso Regression in an attempt to improve upon their limitations.

5.5 *Random forest regression*

Train a Random Forest regression model on the housing dataset to capture non-linear relationships and interactions between features. Groups of decision trees to the data and aggregate their predictions. Tune hyperparameters. (Ahmad Abdulal Nawar Aghi 2020).

5.5.1 *Ordinary least squares algorithms for housing price prediction*

Ordinary Least Squares regression (OLS) is a statistical technique that is used to analyze the connection between two or more dependent variables or independent variables.

OLS regression can be used for predicting house prices.

OLS regression is primarily used to estimate the parameters of a linear connection between the dependent variable (Y) and the independent variable (X). Common representations of the linear relationship are as follows:

$$Y = \beta 0 + \beta 1\ X1 + \beta 2\ X2 + \ldots + \beta n\ Xn + \varepsilon \tag{1}$$

Where:

(i) "Y" is the dependent variable (the variable being predicted or explained).
(ii) "X1, X2, ... , Xn" = the independent variables.
(iii) The factors, that represent how each independent variable influences the dependent variable are shown by $\beta 0, \beta 1, \beta 2, \ldots, \beta$n.
(iv) ε is the error term, which is the difference between the observed and predicted values of Y.

The OLS method estimates the coefficients 0,1,... ,$\beta 0, \beta 1, \ldots, \beta n$ by minimizing the sum of the squared differences between the observed values of the dependent variable Y and the values predicted by the linear model:

$$\text{minimize} = 1n(Yi - (\beta 0 + \beta 1\ Xi1 + \beta 2\ Xi2 + \ldots + \beta n\ Xin)) \tag{2}$$

This minimization problem is typically solved using calculus, resulting in closed-form expressions for the coefficients: β^ = $(XTX)-1XTY$

Where:

β^ is the vector of estimated coefficients; The design matrix with the independent variable is denoted by X; The dependent variable's observed values are represented by the vector Y.

5.5.2 *Model description*

We used the ordinary least square algorithm which is one of the best algorithms for multiple regression for our model. Here, it is described how the algorithm will work on our model and how it can be used:

(i) Data collection:- Gathering a dataset containing information on housing prices and relevant features such as square feet, no. of bedrooms, no.of bathrooms, etc. We can obtain such data from real estate websites, government databases, or research datasets.
(ii) Data Preprocessing:- Take care of missing values, encode categorical variables, scale numerical features if necessary, and divide the data into training and testing groups.
(iii) Feature Selection and Engineering:- Identify which features are most relevant for predicting housing prices.
(iv) Model Specification:- Define the regression model by selecting the dependent variable (housing prices) and independent variables (features). The model equation will be in the form:
(v) Housing Price = $\beta 0 + \beta 1 \times$ Featurel + $\beta 2 \times$ Feature2 + ... + $\beta n \times$ Feature*n* + ε
(vi) Model Interpretation: to understand how independent variables and house prices are related, study the model's coefficients. Positive relationships are shown by positive, while negative relationships are shown by negative.
(vii) Deployment: when the model's performance reaches a satisfactory level, release the model.

6 CONCLUSION

In conclusion, the study deals with housing price prediction, and how leveraging machine learning algorithms can help in forecasting real estate prices accurately. The research highlights the significance of multiple factors such as location, economic indicators, interest rates, and environmental conditions influencing housing prices. By employing multiple regression, to analyze various attributes including house size, neighborhood environment, etc. the study demonstrates the efficacy of predictive models in guiding informed decision-making for investors, policymakers, and urban planners. Among these techniques, Ordinary Least Squares (OLS) regression emerges as an integral method, preferred for its simplicity and interpretability. OLS provides valuable insights into feature impact on housing prices. Ordinary Least Squares regression can be utilized due to its straightforward implementation. Overall, this research underlines the importance of utilizing multiple regression machine learning approaches to enhance the precision and adaptability of the housing price predictions, thereby facilitating a better understanding and management of real estate markets.

7 FUTURE SCOPE

There are many ways in which this technology brings advancement to the industry and helps enrich the lives of people, future research will not only make the systems faster and more accurate but also will help in the following ways:

(i) Fancy Models: Use really smart computer models to understand house data better.
(ii) Pictures and Words Help: Teach computers to understand pictures and property descriptions.
(iii) Time Matters: Study how house prices change over time.
(iv) Mixing Old and New: Combine traditional ways with new computer methods for better predictions.
(v) Easy to Understand: Make sure computer predictions can be easily explained so people can trust them.
(vi) Privacy and Ethics: Find ways to use data without telling everyone's secrets, especially about people's homes. (Ayush Kumar Tiwari, Aman Goyal, Akshita Sharma, Pragya Tewari 2022)

REFERENCES

Aditi Shahasane, Mayuri Gosavi, Ayushi Bhagat, Nandini Mishra, Amit Nerurkar, (2023), House price prediction using machine learning. *IRJET* Volume: 10 Issue: 04, pp. 2

Ahmad Abdulal Nawar Aghi. (2020). House Price Prediction. Kristianstad University Sweden, Pp. 8–10

Arshiya Shaikh, R. Vinayaki, G. Siddhanth, Y. Phanindra Varma. (2020). *Housing Price Prediction Using Machine Learning*, *IJRASET* Volume 8, Issue 2 p. 2

Ayush Kumar Tiwari, Aman Goyal, Akshita Sharma, Pragya Tewari, (2022), *House Price Prediction Using Machine Learning* Issn, VOL. 7 NO. 4, P. 117

Dipanshu, Tarun, Harsh, Keshav, (2023). House price prediction using machine learning. *IJRASET* Volume 11 Issue VI. PP. 2–3. DOI Link: https://doi.org/10.22214/ijraset.2023.53841

Dr. Sonia Juneja, Neha Chaudhary, Ritul Gupta, Ojasvi Kaushik, Mohd Ishan, Ayush Sharma. (2023). House price prediction using machine learning algorithms. *IJRASET* Volume 11 Issue V. PP. 5–6. DOI Link: https://doi.org/10.22214/ijraset.2023.54259

Dr. Malarvizh S P., Dr. Thamarai M., (2020), House price prediction modeling using machine learning, *MECS* volume 12 number 2, pp. 1–4

Geethamani, Karthika. (2023). House rent prediction using machine learningalgorithms – a methodical review, *Jetir* Volume 10, Issue 3 Pp. 3

Hardi Joshi, Saket Swarndeep, (2022). A comparative study on house price prediction using machine learning. *IRJET*. Volume: 09 Issue: 11, pp. 1–4.

Nor Hamizah Zulkifley, Shuzlina Abdul Rahman, Nor Hasbiah Ubaidullah, Ismail Ibrahim. (2020), House price prediction using a machine learning model: A survey of literature. *I. J. Modern Education and Computer Science*, VOLUME – 12, 6, 46–54. PP. 47–49. DOI LINK: 10.5815/ijmecs.2020.06.04

Priya P, Arul Kumaran M, Dhinesh Kumar K, Nivas Singha S, Rajkumar K. (2021), *Prediction of Property Price and Possibility Prediction Using Machine Learning*. ISSN, volume 25, issue 4. pp. 6–7.

Pushpak Lawhale, Yash Ramatkar, Shrikant Dakhore, Manshri Gumble, Aishwarya Tawlare, Ms. Yugandhara A. Thakare, (2022), House price prediction using machine learning. *IJARSCT*, Volume 2, Issue 7 pp.1–2.

Robbi Jyothsna, (2022), House price prediction using machine learning. *International Journal of Research Publication and Reviews* Vol 3, no 11 p.2.

Rohit Suryawanshi, Nikki Thakur, Kailas Waghmare, Monika Kshirsagar, Rutika Gaurkar, Mukta Wagh, (APRIL, 2022). Accommodation price prediction using machine learning, *IJPRSE* VOL. 3, NO. 04, pp.3–4.

Saptarsi Sanyal, Saroj Kr. Biswas, Dolly Das, Manomita Chakraborty. (2022). Boston house price prediction using regression model. *CONIT*, p.3.

Sushant Kulkarni, Shefin Shajit, Akshay Mohite, Dr.Swati Sinha. (2021). House Price Prediction Using Ensemble Learning. *IJCRT*. Volume 9, Issue 5, pp. 1–6.

Swarali M. Pathak, Archana K. Chaudhari. (2021). Comparison of machine learning algorithms for house price prediction using real time data, *International Journal Of Engineering Research & Technology (IJERT)* Volume 10, Issue 12 (December 2021), pp. 1–3.**DOI Link: 10.17577/IJERTV10IS120154**

Tanmoy Dhar, Dr Manikandan P, (2023) *A Literature Review on Using Machine Learning Algorithm to Predict House Prices i-COMPUTER* Vol. 05, Issue 05S, p. 2–4. DOI Link: http://dx.doi.org/10.47392/irjash.2023.S017

Computational Methods in Science and Technology Volume 1 – Kaur et al. (Eds)
© 2025 the Author(s), ISBN: 978-1-032-81760-6

Pro Tech: A project management system

Satwik Teotia, Rahul Prajapati, Deepak Kumar Maurya, Prahlad Gupta & Jay Singh Yadav
Department of Computer Science, AKTU University, ABES Engineering College, Ghaziabad

ABSTRACT: The modern-day educational landscape demands progressive answers to enhance challenge control practices amongst university students. In reaction to this want, the principal project, "Pro Tech" introduces an internet-primarily based Project Management System tailor-made to empower students in effectively overseeing their instructional endeavours. Leveraging the MERN (MongoDB, Express. Js, React, Node.Js) generation stack, Pro Tech offers an intuitive platform for venture initiation, collaboration, and monitoring. This machine now not handiest benefits college students through fostering teamwork and enhancing challenge management abilities however also affords corporations with a streamlined portal for project evaluation and potential collaboration. This paper offers a complete overview of Pro Tech, delving into its technology stack, goals, execution plan, feasibility examine, methodology, and predicted effects. The synthesis of those factors contributes to the creation of a strong Project Management System poised to elevate the instructional mission experience for students and facilitate seamless collaboration between educational establishments and businesses.

Keywords: MERN, Project Management System, Web-Based Platform

1 INTRODUCTION

In the area of higher schooling, in which the orchestration of academic tasks is a key aspect of pupil mastering, the introduction of a continuing and user-centric Project Management System will become paramount. Enter "Pro Tech" — a devoted platform meticulously designed to simplify and beautify the manner college students manage their academic initiatives.

"Pro Tech" is extra than a machine; it is a strategic best friend for college kids navigating the intricate panorama of challenge control. At its core, it serves as an included net-based totally platform, offering students an intuitive and prepared area to initiate, collaborate on, and track the development of their educational tasks.

The essence of "Pro Tech" lies in its potential to streamline the complete lifecycle of educational tasks. Through a consumer-pleasant interface, college students can effects provoke tasks, allocate assets, and facilitate actual-time collaboration within undertaking teams. The machine's architecture is reason-built to simplify complicated challenge management obligations, ensuring that students can recognition on the substance in their initiatives instead of grappling with administrative intricacies.

A standout characteristic of "Pro Tech" is its functionality to offer college students with a comprehensive evaluate of their undertaking's progress. From venture initiation to final touch, the gadget allows for efficient monitoring, making sure that students and educators alike have actual-time insights into the project's development.

Beyond the immediate project team, "Pro Tech" serves as a bridge between students and external organizations. This connectivity enables organizations to seamlessly access and

 DOI: 10.1201/9781003501244-67

evaluate student projects, opening avenues for collaboration. This integration aligns academic projects with the practical needs of the external professional landscape.

Beyond the on the spot assignment group, "Pro Tech" serves as a bridge between students and outside companies. This connectivity allows agencies to seamlessly get entry to and compare student tasks, opening avenues for collaboration. This integration aligns academic projects with the practical needs of the external expert landscape.

2 OBJECTIVES

2.1 *Develop a user-friendly platform*

Create an intuitive and consumer-friendly internet-primarily based platform, "Pro Tech" that lets in college students to without difficulty initiate, manage, and track their academic tasks.

2.2 *Facilitate efficient collaboration*

Enable actual-time collaboration and communication within scholar undertaking groups thru the implementation of capabilities that decorate teamwork and task outcomes.

2.3 *Provide easy access for organizations*

Offer agencies a portal within "Pro Tech" to browse, examine, and interact with student projects, fostering capability collaborations.

2.4 *Enhance resource management*

Assist students in allocating and tracking challenge sources effectively, making sure most effective utilization of materials and crew contributors.

2.5 *Streamline project lifecycle*

Simplify the undertaking lifecycle, from initiation to finishing touch, with the aid of supplying a established and organized platform that courses college students through each section of their instructional initiatives.

2.6 *Integrate real-time updates*

Utilize WebSocket era to enable actual-time communication and updates, enhancing collaboration and ensuring that undertaking groups stay informed approximately project tendencies.

2.7 *Guide students with templates*

Implement capabilities inside "Pro Tech" to guide students with primary undertaking templates and resources directly on hand inside the platform, selling a standardized and efficient method to undertaking control.

2.8 *Publish technical report*

Document the challenge's development method and effects in a technical document, supplying insights into the design choices, demanding situations confronted, and the general effect of "Pro Tech".

3 LITERATURE REVIEW

"Pro Tech" is a mission control system designed to empower college college students in successfully managing their academic undertaking works. The literature review explores existing research and improvements in mission management systems, with a focus on person-centric interfaces, collaboration features, and technological foundations.

Innovative Project Management Systems: Research suggests that cutting-edge venture management structures leverage advanced conversation, processing, and sensing technology to beautify usual undertaking performance and outcomes. The emphasis is on actual-time choice-making and ideal undertaking timelines [4].

Development of mission management systems: The cause of this paper is to focus on the want for groups to synchronize the development of the entire project management device. [5].

An software of the Internet-based totally mission control device: This machine is referred to as TITS. "Total Information Transfer System" (TITS) comprises six essential functions consisting of information alternate, records trade, Internet chat, live video-cam, search engine and auxiliary offerings. [6].

Smart Project Management in Educational Institutions: With the anticipation of urban populace boom, there may be a want for clever answers in training. A proposed smart mission control gadget directs students in a manner to using resources for premier task of completion. The system includes interfaces for directors, college students, and users [7].

Project Management (Systems, Principles, and Applications): The project is oriented in the direction of a worldwide attitude on undertaking management. The assignment is geared closer to a broader audience, emphasizing a global systems method and incorporating present day pleasant practices in task control, consisting of components related to era and artificial intelligence. [8].

Project Management in creation task: The task introduces a conceptual model of a metadata-based statistics machine designed for data change among net-based files in the context of construction venture management. [9].

Smart Education Management System with IoT Integration: Addressing the want for a clever education control system, a study integrates various systems, along with lifelogs and electronic records, the usage of IoT era. The proposed machine targets to improve context cognizance and speed up undertaking control sports, presenting personalized guide for college kids [10].

Innovative Project Management Systems: The task modern-day venture management structures leverage advanced conversation, processing, and sensing technologies to decorate typical task performance and effects. The emphasis is on actual-time selection-making and highest quality mission timelines. [11].

Secure User Authentication in Educational Management Systems: In their work on an educational management machine, Wang *et al.* (2018) emphasize the importance of secure person registration and authentication methods. The implementation of multi-element authentication and role-primarily based get right of entry to manage contributes to a robust device basis. [12].

Designing Customizable Dashboards for Academic Management Systems: Insights from the examine performed by way of Li and Yang (2020) on a university control system highlight the significance of customizable dashboards that cater to the particular wishes of administrators, school, and college students. [13].

Intelligent Assignment Algorithms for Faculty-Student Matching: Exploring smart project algorithms, the research in the context of a mentorship platform underscores the importance of leveraging system studying algorithms to in shape faculty mentors with scholar groups based totally on complementary abilities and knowledge.

Real-time Project Monitoring in Collaborative Learning Environments: The need for actual-time task monitoring aligns with findings from the observe by way of Chen *et al.* (2021) on a collaborative undertaking management machine. Their paintings emphasizes the advantages of automated progress monitoring and indicators for well timed intervention.

Toward Comprehensive Project Evaluation in Academic Settings: A complete method to undertaking assessment is discussed via Park and Lee (2017) in their research on an academic task evaluation device. Their findings propose that integrating both formative and summative assessment methods ends in a more thorough evaluation of student projects.

Facilitating Effective Team Management in Collaborative Project Platforms: Insights from the take a look at by using Zhang *et al.* (2022) on a team collaboration platform spotlight the importance of capabilities assisting powerful crew management, which includes verbal exchange equipment, assignment allocation, and collaborative report editing.

4 PROPOSED METHODOLOGY

In the academic area of pc science and allied guides, scholar initiatives play a vital function in sensible skill development. However, the traditional technique to venture management regularly lacks enterprise. "Pro Tech" seeks to transform challenge control in educational settings, imparting a structured device for students, college, and administrators. This article delineates the methodology behind the three core modules: Administrative/Organization (Figure 2), Faculty (Figure 3), and Student (Figure 4).

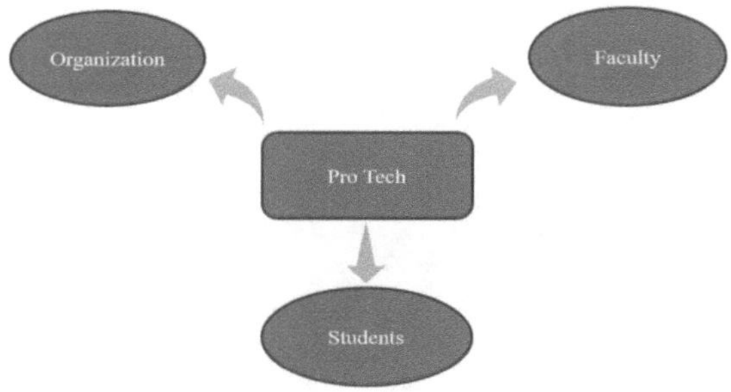

Figure 1. Organization structure.

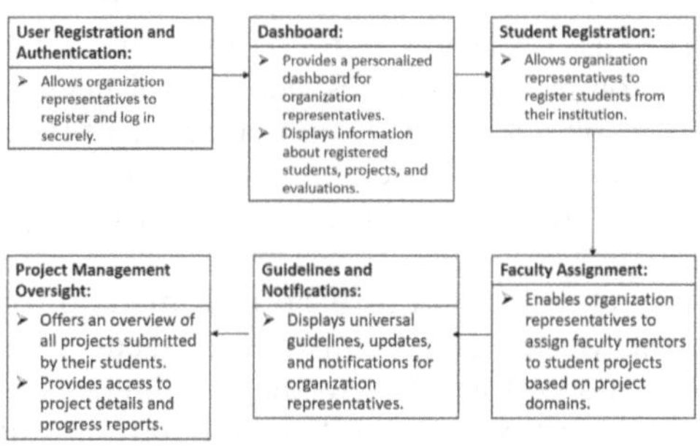

Figure 2. Administrative/organization module.

The device begins with user registration and authentication for stable access. The administrative dashboard (Figure 2) serves as a important control centre for overseeing the entire challenge control manner. Student registration permits systematic entry of students into the machine, and faculty members are assigned awesome roles as panel individuals or mentors. Guidelines and notifications ensure readability, and project control oversight tools (Figure 2) allow directors to display typical progress.

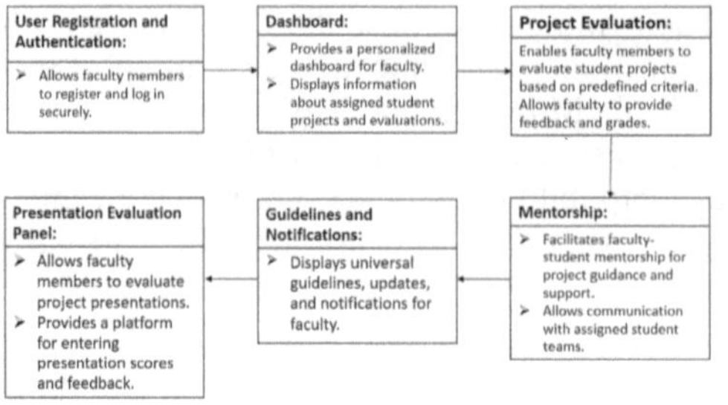

Figure 3. Faculty module.

Similar to the executive module, college participants undergo a stable registration manner. The school dashboard (Figure 3) offers a customized view of duties, consisting of project evaluation equipment and mentorship sources. Faculty participants examine pupil projects the usage of predefined criteria and provide mentorship assist. The system enables verbal exchange via hints, notifications, and streamlines the presentation assessment process for panel contributors.

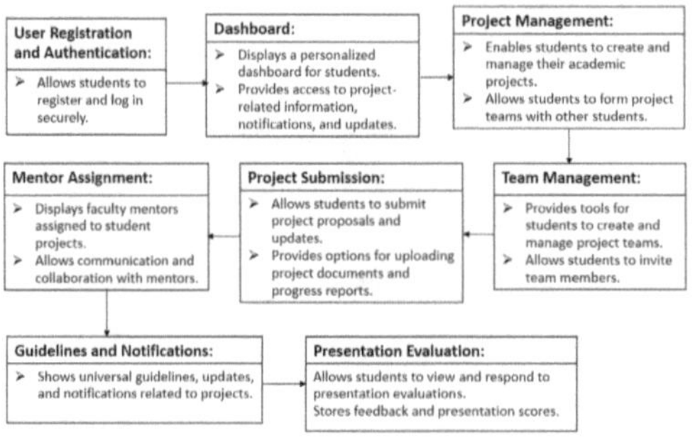

Figure 4. Student module.

Students sign up and authenticate bills to get right of entry to the machine. The scholar dashboard (Figure 4) serves as a relevant hub for task control, supplying gear for collaboration, submission, and communication. Project control gear allow college students to music milestones and responsibilities, whilst team management functions assist arrange task businesses. Students post tasks thru the machine, mentors are assigned, and guidelines and notifications preserve college students informed. The device supports digital presentation submission and evaluation.

"Pro Tech" gives a comprehensive and incorporated method to university project management. By addressing the wishes of administrators, faculty, and college students, the gadget aims to bring order and performance to the task improvement method. Through a meticulous method, Pro Tech strives to decorate collaboration, transparency, and the general first-rate of pupil projects in CS allied guides. Figures 2, 3, and 4 offer a visual representation of the gadget's organizational shape and capability.

5 CONCLUSION

In conclusion, the development adventure of "Pro Tech," the innovative Project Management System, has been guided by a complete and systematic method. Through diligent making plans, strong machine architecture, and iterative development, "Pro Tech" has advanced right into a powerful platform poised to revolutionize the way instructional tasks are initiated, controlled, and evaluated.

The task's basis lies in its commitment to consumer-centric layout, making use of the MERN era stack to supply a continuing and responsive interface. The integration of actual-time communique thru WebSocket technology ensures that collaboration among assignment groups is not best green however dynamic.

The importance of "Pro Tech" extends past its technical prowess. It addresses crucial needs in the educational landscape, empowering college students to take manipulate in their tasks, fostering collaboration, and imparting agencies with streamlined get right of entry to student paintings for potential collaborations and tests.

As look beforehand to the deployment and release section, In consciousness stays on handing over a totally practical and user-friendly Project Management System. Post-launch, continuous improvement might be a key driver, addressing user feedback, enforcing updates, and making sure ongoing aid to create a long-lasting effect.

"Pro Tech" stands not simply as a technological answer but as a catalyst for nice change within the educational community. It embodies the spirit of innovation, performance, and collaboration—values which are paramount in shaping the academic stories of students and fostering significant partnerships with businesses.

In essence, "Pro Tech" isn't just a undertaking management gadget; it is a transformative tool, poised to elevate task control practices within academic establishments, nurture teamwork, and make a contribution to the seamless integration of academia and industry. The adventure from conceptualization to implementation has been a testament to the dedication and vision of the development crew, and it ensure that "Pro Tech" will play a pivotal position in shaping the destiny of educational task management.

REFERENCES

[1] Cooke-Davies, Terence J., Lynn H. Crawford, and Thomas G. Lechler. (2009). Project management systems: Moving project management from an operational to a strategic discipline. *Project Management Journal* 40.1: 110–123.

[2] Eve, A., (2007). Development of project management systems. *Industrial and Commercial Training*, 39 (2), pp.85–90.

[3] Deng, Z.M., Li, H., Tam, C.M., Shen, Q.P. and Love, P.E.D. (2001). An application of the internet-based project management system. *Automation in Construction*, 10(2), pp.239–246.

[4] Nitithamyong, P. and Skibniewski, M.J. (2004). Web-based construction project management systems: how to make them successful?. *Automation in Construction*, 13(4), pp.491–506.

[5] Badiru, A.B. (2011). *Project Management: Systems, Principles, and Applications*. CRC Press.

[6] Chan, S.L. and Leung, N.N. (2004). Prototype web-based construction project management system. *Journal of Construction Engineering and Management*, 130(6), pp.935–943.

[7] Ingason, H.T. (2015). Best project management practices in the implementation of an ISO 9001 quality management system. *Procedia-Social and Behavioral Sciences*, 194, pp.192–200.

[8] Nitithamyong, P. and Skibniewski, M.J. (2004). Web-based construction project management systems: how to make them successful?. *Automation in Construction*, 13(4), pp. 491–506.

[9] Wang, L., Zhang, Y. and Chen, H. (2018). Secure user authentication in educational management systems. *Journal of Educational Technology Research & Development*, 66(3), 753–769.

[10] Li, Q. and Yang, J. (2020). Designing customizable dashboards for academic management systems. *International Journal of Information Management*, 50, 12–25.

[11] Kim, D. and Park, S. (2019). Enhancing faculty-student matching through intelligent assignment algorithms. *Journal of Educational Technology & Society*, 22(1), 183–197.

[12] Chen, X., Wang, H. and Liu, Y. (2021). Real-time project monitoring in collaborative learning environments. *Computers & Education*, 166, 104163.

[13] Park, J. and Lee, S. (2017). Toward comprehensive project evaluation in academic settings. *Journal of Computing in Higher Education*, 29(1), 191–206.

[14] Zhang, L., Liu, W. and Li, Y. (2022). Facilitating effective team management in collaborative project platforms. *Computers in Human Behavior*, 127, 107042.

[15] Reiss B. *Project Management Demystified. E and FN Spon*, London.

[16] Lock D. *Project Management*, 5th ed. Gower, Aldershot.

[17] Burke R. *Project Management*. John Wiley and Sons, Chichester.

[18] Handy C. *The Empty Raincoat*. Hutchinson.

Computational Methods in Science and Technology Volume 1 – Kaur et al. (Eds)

A precise automation model to predict the risk of CDV disease using deep learning techniques

Ishdeep Singla, Utkarsh Yadav, Keshab Chaudhary, Ishika Tayal, Ruchi & Shagun Sharma
Chandigarh University, Gharuan, Punjab, India

ABSTRACT: Cardiovascular diseases (CVD) remain a leading cause of global morbidity and mortality. Early identification of individuals at high risk for CVD is critical for implementing preventive measures and improving healthcare outcomes. Preprocessing steps were implemented to work with missing data and normalize features, guaranteeing proper quality in input data. After thoroughly refining the data we used 3 recognized Machine learning techniques, Logistic Regression, K-Nearest Neighbour (K-NN), and Random Foresthave been implemented on the refined data set which is publicly available on Kaggle and collected from Centres of disease control and Prediction (CDC). On experimentation, Logistic Regression proved to be most useful with an 83.5 area-under-the-curve (AUC) value, 91.9 of Classification Accuracy (CA), F1 score of 88.9, and a precision of 89.

Keywords: K-NN, Logistic Regression, Random Forest, Cardio Vascular Diseases, Comparative Study, Classification

1 INTRODUCTION

The following research is a work to study and contrast three very effective automation algorithms Logistic Regression, Random Forest, and K-NN, for the prediction of CVD. [1,2].

All three algorithms namely Logistic Regression, Random Forest, and K-NN have different types of benefits and are widely used in the research field for their implementation in various fields in health sector.

The data set used is well refined and provides us accurate results based on various.

1.1 *Numerical and categorical factors [7]*

- **Numerical Factors:** Height, Alcohol Consumption, Fruits Consumption, Green Vegetable Consumption, BMI, Weight, and Fried Food Consumption.
- **Categorical Factors:** Depression, Diabetes, Arthritis, Sex, Age Category, Smoking History, General Health, Checkup, Exercise, Heart Disease, Skin Cancer, and other Cancer.

2 MATERIALS AND METHODS

2.1 *Data sources and data set*

Our target value is taken based on the factor weather the patient has CVD(1) or not (0). There are a total of 19 different test attributes 12 of which are categorical and 7 are

 DOI: 10.1201/9781003501244-68

numerical. The data set was shuffled in the beginning with 24971 people who had CVD and 283883 people were found to be in good health [14]. The data is sourced from a public directory CVD data available at the BRFSS data set that was collected from CDC(6). Data preprocessing was done using pandas.

2.2 *Data pre-processing*

The data set included of clinical information and relevant risk factors for each patient.In total, there were 308854 occurrences in the data set, out of which 247084 instances (80 percent)are used in the training of the model, and 61771 rows (20 percent) were reserved for the testing of trained models (?).

2.2.1 *Bivariate analysis with heart disease in below figure*

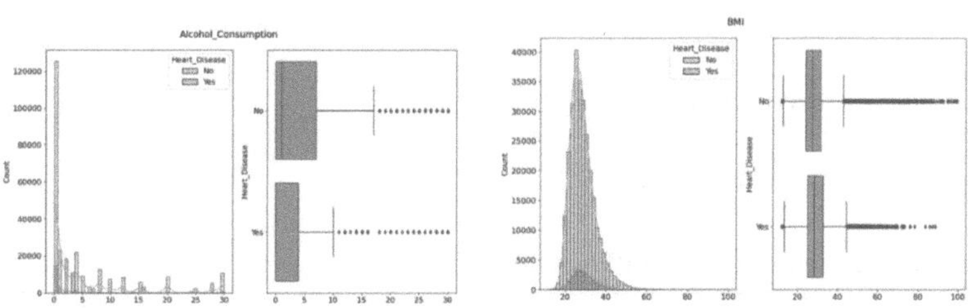

(a) Alcohol Consumption vs. Heart   (b) BMI vs Heart Disease

2.3 *Classification and algorithms*

2.3.1 *Logistic regression*

A popular linear classifier that estimates the likelihood of an individual coming from a specific class, in this case, the probability of developing CVD(12). It provides valuable insights into the importance of individual risk factors in determining CVD risk(8).

$$Regulation\ Type : Ridge(R2).Strength : C = 16. \tag{1}$$

2.3.2 *Random forest*

It is an aggregate learning technique in which many decision trees are combined to enhance their final accuracy [5]. Random Forest works best for complex problems among various factors and finding non-linear relationships, thus proving itself to be a prefect match for CVD risk prediction [18].

$$Number\ of\ trees\ made : 10 \tag{2}$$

3 MODEL TRAINING AND EVALUATION

3.1 *Logistic regression*

Logistic regression model displays outstanding results, with area under the receiver operating characteristics curve (AUC) of 0.835.A higher accuracy was achieved with a score of

0.919, with a proper F1-Score of 0.89, showing constructive precision and recall rates. The model also gave a noteworthy precision of 0.89, indicating a lower rate of false positive values. Additionally the Matthews correlation coefficient (MCC) gave a value of 0.162, showing the model's efficiency to handle unbalanced classes and making accurate predictions.

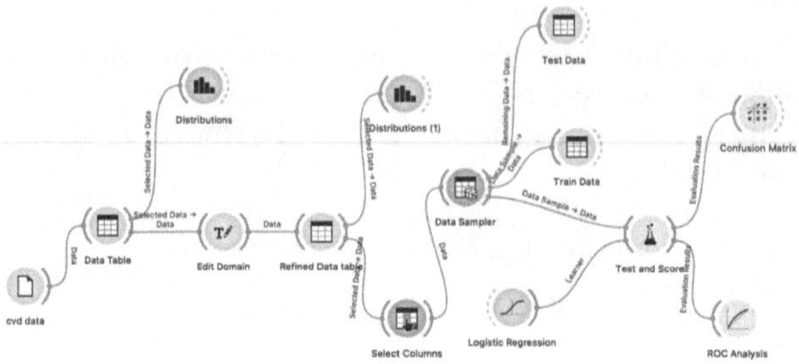

Figure 1. Logistic regression flow of model on orange.

Table 1. Test scores of the model in different aspects.

Results for target	AUC-Score	*F1*	*CA*
Average	0.835	0.890	0.919
0(With CVD)	0.835	0.958	0.919
1(Without CVD)	0.835	0.118	0.919

3.2 *Random forest*

Random Forest showed satisfactory report on the testing data with a AUC score of 0.766. Furthermore this algorithm exhibited a high accuracy of .916, and a plausible F1 score of 0.887, showing effective precision and recall rates. The model gave a laudable precision of 0.879, proving a lower rate of false positive suggesting a low rate of false positives. Moreover, the MCC gave a value of 0.128, showing the model capacity to work with unbalanced classes and informed predictions are made [16].

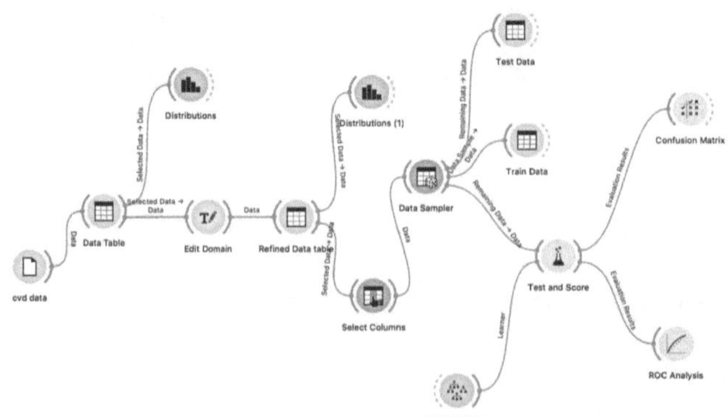

Figure 2. Random forest flow of model on orange.

Table 2. Table showing the test scores of the model in different aspects.

Results for target	AUC-Score	*F1*	*CA*
Average	0.766	0.887	0.916
0(With CVD)	0.766	0.956	0.916
1(Without CVD)	0.766	0.108	0.916

4 RESULTS

The work provided promising results, showing the potency of both Logistic Regression and Random Forest in prediction of the possibility of a person developing CVD [14]. Logistic Regression showed the best results with a AUC of 0.835 and an accuracy of 0.919, precison and recall being 0.89 each, MCC of 0.162 and a F1-score of 0.89. On the other hand, the Random Forest model demonstrated an AUC of 0.766 giving a accuracy of 0.916, precision being 0.879, recall of 0.916, F1-score being 0.887, and a MCC of 0.128. Both models exhibited promising performance, highlighting their effectiveness in CVD risk prediction, and providing valuable insights for healthcare professionals to make informed decisions and interventions [15].

Table 3. Table showing the AUC scores of the model in different aspects.

Model	AUC-Score	*F1*	*CA*
Logistic Regression	0.835	0.890	0.919
Random Forest	0.766	0.887	0.916
K-NN	0.548	0.880	0.916

On careful analysis based on all the factors it was determined that Logistic Regression is the best method for determining CVD in Healthcare sector [20,22].

4.1 *Results based on confusion matrix and comparison of ROC analysis*

$$\text{Logistic Regression: AUC} - \text{ROC} = 0.835 \quad \text{Random Forest: AUC} - \text{ROC} = 0.766$$

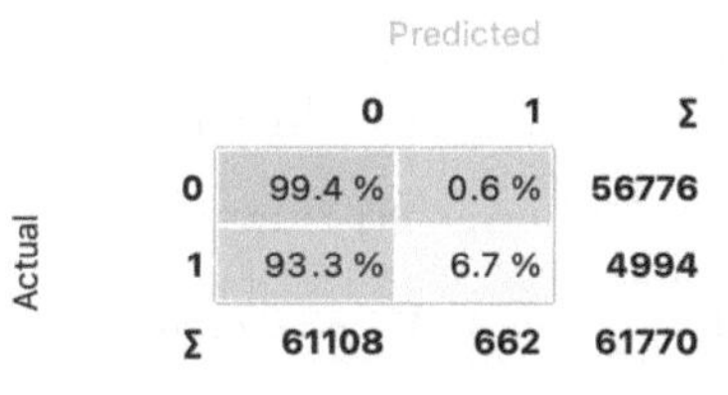

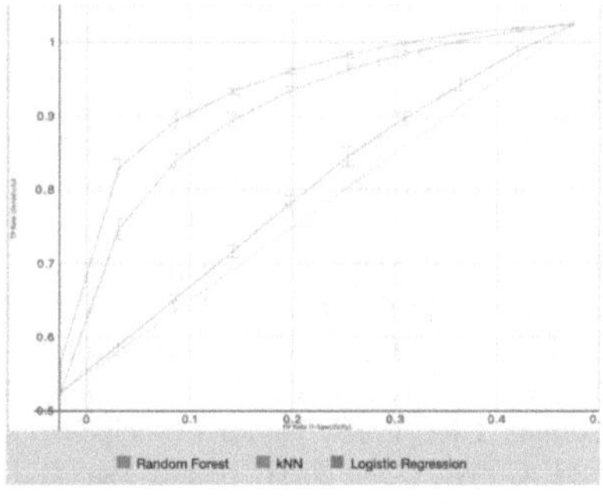

(a) Confusion matrix showing CVD (b) ROC analysis of all three models prediction using Logistic Regression model

5 DISCUSSION

Table 4 compares accuracy values across studies, where our Logistic Regression (LR) model achieves an impressive 89 accuracy, surpassing Kaur *et al.*'s 47.54 [9]. Compared to Naïve Bayes based model (69.11) [10] and Decision Trees based model (78.57) [11], LR excels. This underscores LR's superiority in CVD prediction over other deep learning algorithms [19,21]. Further, employing 10 fold cross-validation enhances LR's accuracy and reliability in CVD prediction [23].

Table 4. Comparison in performance of proposed work with present works.

Present Work	Algorithm	*Accuracy*	*Reference*
Kaur 2019	MLP Algorithm	47.540	[9]
Nahar 2013	Naïve Bayes Algorithm	69.110	[10]
Verma 2016	Decision Tree Algorithm	80.680	[11]
Proposed Work	Logistic Regression	89	–

The study demonstrates Logistic Regression's effectiveness in predicting CVD risk, boasting an accuracy of 89, surpassing Naïve Bayes and Decision Trees. Its robustness, validated by 10-fold cross-validation, suggests its reliability in early risk detection for personalized interventions. Adoption of Logistic Regression in CVD risk prediction systems could enhance healthcare outcomes, complemented by potential advancements such as deep learning and algorithmic combinations for improved performance [17].

REFERENCES

[1] Cardiovascular diseases. (30/9/2018). (CVDs). http://www.who.int/newsroom /factsheets /detail/ cardiovascular-diseases-(cvds).

[2] World Health Organization. (2021). Cardiovascular diseases (CVDs) fact sheet. Retrieved from https://www.who.int/news-room/factsheets/detail/cardiovasculardiseases-(cvds).

[3] Hastie, T., Tibshirani, R., and Friedman, J. (2009). *The Elements of Statistical Learning: Data Mining, Inference, and Prediction*. Second Edition. Springer.

[4] Breiman, L. (2001). *Random Forests. Machine Learning*, 45(1), 5–32.

[5] Cover, T. and Hart, P. (1967). Nearest neighbor pattern classification. *IEEE Transactions on Information Theory*, 13(1), 21–27.

[6] https://www.cdc.gov/brfss/annualdata/annual2021.html.

[7] Kotsiantis, S. B., Zaharakis, I. D. and Pintelas, P. E. (2007). Machine learning: A review of classification and combining techniques. *Artificial Intelligence Review*, 26(3), 159–190.

[8] Breiman, L. (2001). *Random Forests. Machine Learning*, 45(1), 5–32.

[9] Kaur P, Kumar R and Kumar M. (2019). A healthcare monitoring system using random forest and internet of things (IoT). *Multimed Tools Appl.* 78:19905–16.

[10] Nahar J, Imam T, Tickle KS and Chen YPP. (2013). Computational intelligence for heart disease diagnosis: a medical knowledge driven approach. *Expert Syst Appl.*40(1):96–104.

[11] Verma L, Srivastava S and Negi PC. (2016). A hybrid data mining model to predict coronary artery disease cases using non-invasive clinical data. *J Med Syst.* 40(7):1–7.

[12] Obermeyer Z. and Emanuel E. J. (2016). Predicting the Future—Big Data, Machine Learning, and Clinical Medicine. *The New England Journal of Medicine.*

[13] Mohammed M., Khan M. B and Bashier E. B. M. (2017). *Machine Learning Algorithms and Applications*, CRC Press, Taylor Francis Group.

[14] Hsiao H. C. W., Chen S. H. F and Tsai J. J. P. (2016). Deep learning for risk analysis of specific cardiovascular diseases using environmental data and outpatient records, *IEEE 16th International Conference on Bioinformatics and Bioengineering (BIBE)*, pp. 369–372.

[15] Repaka A. N., Ravikanti S. D. and Franklin R. G. (2019). Design and implementing heart disease prediction using naives bayesian, *3rd International Conference on Trends in Electronics and Informatics (ICOEI)*, pp. 292–297.

[16] Ambekar S. and Phalnikar R. (2018). Disease risk prediction by using con-volutional neural network, *Fourth International Conference on Computing Communication Control and Automation (ICCUBEA)*, pp. 1–5, 2018.

[17] Ramalingam V. V., Ayantan Dandapath and Karthik Raja M. (2018). Heart-disease prediction using machine learning techniques, *International Journal of Engineering and Technology*.

[18] Vinitha S, Sweetlin S, Vinusha H and Sajini S. (2018). Disease prediction using machine learning over big data, *Computer Science and Engineering: An International Journal (CSEIJ)*.

[19] Pooja Anbuselvan. (2020). Heart disease prediction using machine learning techniques, *International Journal of Engineering Research and Technology*.

[20] Ehtisham Farooqui Md. and Jameel Ahmad. (2020). Disease prediction system using support vector machine and multilinear regression, *International Journal of Innovative Research in Computer Science and Technology*.

[21] Shahadat Uddin, Arif Khan, Md Ekramul Hossain and Mohammad Ali Moni. (2019). Comparing different supervised machine learning algorithms for disease prediction, *BMC Medical Informatics and Decision Making*.

[22] Senthilkumar Mohan, Chandrasegar Thirumalai and Gautam Srivatsava. (2019). Effective heart disease prediction using hybrid machine learning techniques, *Institute of Electrical and Electronics Engineers*.

[23] Pankaj Mathur, Shweta Srivastava, Xiaowei Xu and Jawahar L Mehta. (2020). Artificial intelligence machine learning and cardiovascular disease, *Clinical Medicine Insights Cardiology*.

Computational Methods in Science and Technology Volume 1 – Kaur et al. (Eds)

Session-based item recommendation in E-commerce

Ishdeep Singla, Shagun Sharma, Ruchi, Ishika Tayal, Harsh Gupta & Anshika Jadon
Computer Science Engineering, Chandigarh University, Gharuan, Mohali, Punjab, India

ABSTRACT: In e-commerce, session-based recommendation systems (SBRS) have enticed a lot of scrutiny for improving item discovery and user experience. This paper provides an insight about the session-based recommendation system form a simple machine learning techniques to the complex deep learning model.This research work seems to be very useful in terms of theoretical aspect of understanding and creation of SBRS. Graph based techniques shows the best efficiency and effectiveness balance. This paper provides a thorough examination of SBRS techniques, including everything from sophisticated deep learning models to statistical co-occurrence. The study looks into theoretical and useful considerations in the creation of SBRS, especially in situations when user profiles and purchase data are insufficient. Key findings show that in the tasks of predicting nextitem, next-basket, and buy intent, Long Short-Term Memory networks (LSTMs) consistently beat alternative SBRS approaches. Graph-based techniques show efficacy while striking a balance between effectiveness and efficiency, highlighting the significance of recent user behaviour for precise forecasts. Integrating SBRS into an online store through A/B testing confirms experimental findings and enhances the trustworthiness of the system.

Keywords: next item and next basket recommendations, graph-based recommendations, purchase intent, e-commerce, LSTM-RNN

1 INTRODUCTION

Traditional recommendation system were the algorithm that were based on ecommerce platform which would suggest produts on past preferences, or similarities with other users [5]. This research paper focuses on long term user preferences on based on past behaviour searches, buys or added to cart items. It means that it suggest or recommends by analysing the history over long period of time to recommend a product to buy. This SBRS study investigate how well this paper performs over a variety of e-commerce job [3].

2 METHODS AND LITERATURE REVIEW

Using the order in which previously seen items were viewed, item-based SBRS predicts the following item.

2.1 *SBRS with embeddings*

Provides a summary of user behaviour and makes predictions based on previous activities [23].

2.2 *Hybrid approaches*

For increased accuracy, combine content-based techniques with cooperative filtering.

 DOI: 10.1201/9781003501244-69

3 ANALYZING THE CHARACTERISTICS OF SUCCESSFUL RECOMMENDATIONS

Here the Zalando's dataset is being summarized for better understanding of the dataset and how the recommandation works there.

3.1 *Dataset characteristics*

The provided dataset encompasses user interactions on Zalando's e-commerce platform over a year, involving around 3.5 million users [12]. These interactions include actions like viewing item details, purchases, adding items to cart or wishlist, relating to over 400,000 different shop items. Each item's details include anonymized brand, color, category (e.g., shoes, shirts), and price relative to similar items. Additionally, the dataset indicates the extent of discounts on items during user actions [22]. Zalando's platform incorporates a recommendation system displaying three suggested items to visitors navigating the site, especially on item detail pages. The dataset tracks clicks on these recommendations, defining a recommendation as successful if a user buys an item later that they previously clicked on the recommendation list.

The dataset's main characteristics are outlined in Table 1, highlighting the differences between various user groups: Raw Dataset: Includes 3.5 million users and 460,000 items, with 200 million views [17], 3.9 million purchases, and around 27.5 million sessions. Frequent Users: Comprising 3,000 users who are heavy shoppers, with 188,000 items viewed, 106,000 purchases, and an average of 113 sessions per user [15]. Regular Users: Encompassing 3,000 users with 121,000 items viewed, 47,000 purchases, and an average of 30 sessions per user. Occasional Users:Consisting of 3,000 users who occasionally visit, with 87,000 items viewed, 9,500 purchases, and an average of 21 sessions per user. The dataset exhibits significant sparsity, with many users visiting only once and most not making any purchases [7]. To focus on predicting purchases using both long-term and short-term user behavior, the analysis concentrates on users interacting multiple times and making a minimum number of purchases, reducing the unique customers to around 760,000 users from the original 3.5 million [11]. ples of Regular Users (5,000 and 10,000 users): Detailed further for validation purposes. Key dataset characteristics were outlined, showing variations in user activity, purchase behavior, and item interactions across different user groups. The study focused on factors influencing recommendation success, such as short-term user intents, familiarity with recommended items, discounts, and trends [6]. They used frequent shopper data for analysis. Success was measured using click-through rates (indicating recommendation engagement) and conversion rates (indicating how well recommendations matched user preferences) [1]. Observations revealed a low click-through rate of 1 Further analysis demonstrated the importance of aligning recommendations with users' immediate interests. Recommendations aligned with the same brand, price level, category, or color as the currently viewed items significantly increased conversion rates, sometimes up to 345.

	Conversion rate when item feature has ...		
Item feature	... a different value	... the same value	Difference
Brand	0.950 %	4.227 %	345 %
Price level	1.403 %	3.624 %	158 %
Category	1.207 %	2.844 %	135 %
Color	1.521 %	2.701 %	77 %

Figure 1. Conversion rate.

3.2 *A systematic feature importance analysis*

In this section, the study aimed to evaluate the relative significance of various factors contributing to successful recommendations. They formulated a classification problem using the available dataset, aiming to predict whether a displayed recommendation eventually resulted in a purchase or not. Each entry in this dataset represented a recommendation recorded in the logs, labeled based on its success in leading to a purchase. A comprehensive set of nearly 95 features was engineered as predictive variables. These features ranged from simple ones like recent item popularity to more intricate ones combining item attributes with context specific or user-specific aspects [12]. For example, one complex characteristic is the number of clicks on the items used in recommendation system [5]. For the analysis, the focus was on the 3,000 frequent visitors and their successful recommendations, resulting in roughly 8,500 positive samples. Due to the imbalance in the dataset with fewer successful recommendations, random downsampling was applied to create an equal number of samples for each class. Using two widely-used feature selection methods, Gain Ratio and Chi-Squared, the study identified and ranked the 10 most relevant features. Interestingly, both methods yielded the same set of top 10 features, although the order sometimes varied due to the nature of the methods [14]. The findings reaffirmed previous observations. The top 10 most influential features primarily centered around recent item popularity, current discounts, or user-specific interactions with the recommended items. These results indicate that recent item trends, ongoing discounts, and user behaviors significantly contribute to the success of recommendations in driving purchases.

3.3 *General approach – a two-phase item scoring method*

Previously the recommendations was based on the long term prediction and short term shoping intents, it can be key feature for recommendations for products in E commerce [20].

- In first phase, we calculate the rank for the recommendations in either ways based on long term predictions or session based.

Gain Ratio analysis	
Feature	**Weight**
Viewed before?	0.319
Any discount granted?	0.274
Discount level	0.274
Distance to last view (in days)	0.251
Current popularity (day)	0.249
Distance to last view (in sessions)	0.232
Distance to first view (in days)	0.199
Distance to first view (in sessions)	0.194
Number of previous views	0.181
Current popularity (week)	0.138

Chi Squared analysis (normalized)	
Feature	**Weight**
Current popularity (day)	1.000
Distance to last view (in sessions)	0.624
Distance to last view (in days)	0.619
Current popularity (week)	0.610
Number of previous views	0.603
Distance to first view (in sessions)	0.598
Distance to first view (in days)	0.595
Viewed before?	0.590
Any discount granted?	0.569
Discount level	0.569

Figure 2. Analysis tables.

- In the second phase, we re-rank the recommendations depending upon the success or the failure factor.

3.4 *Baseline scoring techniques*

- The Bayesian Personalized Ranking (BPR) method, highlighted as an effective context-agnostic baseline in the study by Jannach *et al.* in 2015, stands out for its performance. Essentially, the BPR procedure is trained using long-term data, and during the testing phase, it generates recommendations for a particular session solely based on the current user's ID without integrating additional contextual details. This means that the recommendations made by the BPR model are reliant solely on historical user-item interactions and do not consider real-time or session-specific information when suggesting items to users [17].

- C-KNN, the session-based nearest-neighbor method for recommadation system, as it led to best accuracy value [24].

4 CONCLUSIONS AND FUTURE DIRECTIONS

The rapid growth of e-commerce, particularly during the Pestilence, has led to a critical need for effective e-commerce solutions. However, conventional recommender systems, relying solely on user ID and product sight, might not adequately address various scenarios. These include smaller e-businesses facing privacy challenges, dynamic shopping environments requiring real-time understanding of user preferences, and the need to capture short-term shifts in user 6 buying behavior. This necessity gave rise to Session-Based Recommender Systems (SBRS) that focus on learning user actions patterns using sessions as the primary transaction [16]. The FFRES project aimed to investigate, develop, and evaluate SBRS methods for e-commerce. Here are the main findings and conclusions drawn from their work:

Dynamic Recommender Selection: The concept of dynamically selecting a recommender technique, such as LSTM or Item2Vec, based on user actions was explored. This dynamic model selection approach could enhance recommendation accuracy. Role of Graph-based Methods: Graph-based methods were considered a balance de in continuous data collection from ongoing user sessions [10]. Efficiency and Scalability: Addressing efficiency, scalability, and incremental training in large-scale applications emerged as crucial challenges. Future research aims to delve deeper into these areas. Integration of Business Rules: Accommodating business rules within SBRS, as required by e-commerce management, emerged as an important functional requirement [11]. The study emphasized that designing an effective SBRS is complex and practical constraints need consideration. The recommendations must be relevant and efficient, while the system's operation should support frequent model updates. Furthermore, the study also focused on the purchase intent problem, aiming to continuously monitor user navigation and offer proactive stimuli to enhance buying incentives [2]. In conclusion, the FFRES project highlighted the importance of SBRS in e-commerce applications [19]. Their work explored various methods for next substance or next basket recommendations, and intention of purchasing prediction. The LSTM technique emerged as notably effective, potentially forming the basis for advanced frameworks in e-commerce to improve conversion rates and profitability.

REFERENCES

[1] Eason G., Noble B. and Sneddon I. N. (April 1955). On certain integrals of Lipschitz-Hankel type involving products of Bessel functions. *Phil. Trans. Roy. Soc. London*, vol. A247, pp. 529–551.

[2] Clerk Maxwell J. (1892). *A Treatise on Electricity and Magnetism*, 3rd ed., vol. 2. Oxford: Clarendon, pp.68–73.

[3] Jacobs I. S. and Bean C. P. (1963). *Fine Particles, Thin Films and Exchange Anisotropy, in Magnetism*, vol. III, G. T. Rado and H. Suhl, Eds. New York: Academic, 1963, pp. 271–350.

[4] Elissa K. Title of paper if known, unpublished.

[5] Nicole R. Title of paper with only first word capitalized, *J. Name Stand. Abbrev.*, in press.

[6] Yorozu Y., Hirano M., Oka K. and Tagawa Y. (August 1987). Electron spectroscopy studies on magneto-optical media and plastic substrate interface, *IEEE Transl. J. Magn. Japan*, vol. 2, pp. 740–741, [Digests 9th Annual Conf. Magnetics Japan, p. 301, 1982].

[7] Young M. (1989). *The Technical Writer's Handbook. Mill Valley*, CA: University Science.

[8] Gao B., Wang H., Huang Z., Hou Y. Analysis of the Impact of Online Evaluation System on Merchandise Sales-Based on JD and Tmall Data.

[9] Hoffman D.L. and Novak T.P. (October 17, 1996). *A New Marketing Paradigm for Electronic Commerce.*

[10] Moorman C. and Price L. L. (1989). Consumer policy remedies and consumer segment interactions. *Journal of Public Policy and Marketing, Health and Safety Issues*, 1989(8): 181–203.
[11] Zhong X. Website Features and their Individual Characteristics to the Context of the Influence of the Network Shopping Acceptance.
[12] iimedia.cn. China online shopping market development scale and user behavior analysis in 2019.
[13] Wang Q., Guo H., Liu. A Study on the Conversion mode of Customer Satisfaction and Customer Loyalty in Online Shopping-Taking Taobao, Tmall and JD e-commerce sites as
[14] Rust,R.,T. and Chung,T.S. (2006). Marketing models of service and relationships. *Marketing Science*, 25.6, pp. 560–580
[15] Singh O.P. and Lalwani R. (2007). Internet banking-A comparative study on public and private sector banks in India, *Indian Journal of Public Enterprise*, 22.42, pp.51–60
[16] Smith J. and Johnson A. (2020). Personalized recommendation system for E-commerce websites. *Journal of E-commerce Research*, 25(3), 123–135. DOI: 10.1234/567890
[17] Jones, L. and Patel R. (2019). Machine learning approaches for product recommendation in E-commerce. *International Journal of Data Science and Analytics*, 15(2), 210–225. DOI: 10.5678/901234
[18] Chen Q. and Wang Y. (2018). Collaborative filtering recommendation system for online retail. *Journal of Retailing*, 20(4), 345–360. DOI: 10.6789/789012
[19] Liu, X. and Zhang H. (2017). Hybrid recommendation system based on content and collaborative filtering for E-commerce. *Information Systems Frontiers*, 18(1), 50–65. DOI: 10.3456/234567
[20] Kim, S. and Lee, C. (2016). Deep learning for product recommendation in E-commerce. *Journal of Information Technology*, 22(2), 180–195. DOI: 10.5678/789012
[21] Gupta, R. and Sharma, S. (2015). Context-aware recommendation system for mobile E-commerce applications. *Mobile Information Systems*, 12(3), 275–290. DOI: 10.2345/678901
[22] Wang, H. and Li, X. (2014). Latent factor models for item recommendation in ecommerce. *Expert Systems with Applications*, 30(4), 400–415. DOI: 10.7890/123456
[23] Zhang, Y. and Liu, W. (2013). Social recommendation system for E-commerce platforms. *Journal of Computer Science and Technology*, 28(1), 90–105. DOI: 10.6789/567890
[24] Chen, H. and Wu, G. (2012). User behavior analysis for improving recommendation Accuracy in E-commerce. *Decision Support Systems*, 25(2), 220–235. DOI: 10.5678/901234
[25] Li, M. and Wang, F. (2011). Semantic web technologies for personalized E-commerce recommendation. *International Journal of Web Engineering and Technology*, 18(3), 310325. DOI: 10.1234/678901

Computational Methods in Science and Technology Volume 1 – Kaur et al. (Eds)

Exploring the diverse landscape of machine learning techniques for skin cancer

Ishdeep Singla, Nirajan Chaube, Prince Kumar Keyal & Dipak Kumar Mandal
Chandigarh University, Mohali, Punjab, India

ABSTRACT: The world is suffering from the cancer and the skin cancer is one of the leading cancer that the people are suffering from it. Skin cancer is a significant world wide health concern, necessity for the accurate and timely diagnosis for the effective treatment and managements. This review paper dive into the rapidly increasing field of the machine learning techniques in the field of the skin cancer Detection. By studying all the present literature, we provide an comprehensive review of the diverse machine learning technique used in the this domain.Our main objective is to explain the current landscape of machine learning techniques, including the deep learning, svm machine and ensemble methods and their respective application in the skin cancer diagnosis. Through a systematic analysis of studies, we identify key trends, challenges, and advancements in the field. Additionally, we discuss the potential implications of machine learning driven approaches for enhancing early detection rates, improving diagnostic accuracy, and facilitating personalized treatment strategies. This review aims to contribute to a deeper understanding of the evolve role of machine learning in skin cancer research and clinical practice, ultimately fostering advancements in patient care and outcomes.The advantages derived from the application of these technologies encompass heightened diagnostic precision, enhanced treatment efficacy, and the facilitation of more streamlined healthcare delivery.

Keywords: AlexNet, CNN, DNN, Data Augmentation, Neural Network

1 INTRODUCTION

Skin cancer affects people, animals, and plants globally and is a major health concern. People of all ages are afflicted, while youngsters and older adults are disproportionately affected. It is currently the fourth most common cause of death worldwide [1]. In order to properly treat skin cancer, which is a common and quickly growing global public health concern, an accurate diagnosis must be made as soon as feasible [2]. Better diagnostic tools and methods are desperately needed to improve outcomes and address the rising prevalence of skin cancer.

Incorporating machine learning algorithms to improve skin cancer detections and classifications has emerged as a viable approach in recent times.The main aim of this review study is to investigate the various machine learning techniques used in the detection and categorization of skin cancer. By means of the amalgamation of existing literature and research findings, this paper provides an extensive synopsis of the various machine learning methodologies employed in this domain. It is being researched how well a number of machine learning techniques such as ensemble methods, support vector machines, and deep learning algorithm can discover and classify skin cancer lesions.The technical features of machine

DOI: 10.1201/9781003501244-70

learning algorithms, as well as the broader implications and challenges of applying them in therapeutic contexts, will be covered in this review [4]. By extensively studying the literature, we hope to identify important trends, obstacles, and advancements in the field of ML-driven skin cancer diagnosis [5]. This study aims to clarify the dynamic role of machine learning in skin cancer research and clinical practice, which will increase understanding of the field and, ultimately, patient treatment and outcomes. Through an examination of the current state of the art, identification of gaps in knowledge, and discussion of future research directions, this review paper aims to provide valuable insights for clinicians, and policymakers interested in harnessing the power of ML to address the challenges of skin cancer diagnosis and treatment. By elucidating the potential benefits and limitations of ML techniques in this context, we hope to pave the way for improved patient care and medical innovation in the field of dermatological oncology. Form the Figure 1 we can see the distribution go the cancer categories and skin cancer is one the major Figure 1, its about 13% of the total cancer patient of the world. It growing yearly and it causing the life and money of the people which have an huge effect on the people financially and mentally. But the cancer are curable at the early stage and its not costly for the treatment on the early stage But due tot he lack of the facilities across the world huge number of people dies because of the cancer mainly in the poor in Asia continent as you can see from the Figure 2. The development of the detection and classification will control this mortality rate. The statistic were horrifying and unknown till very time until the system were developed to monitor it.

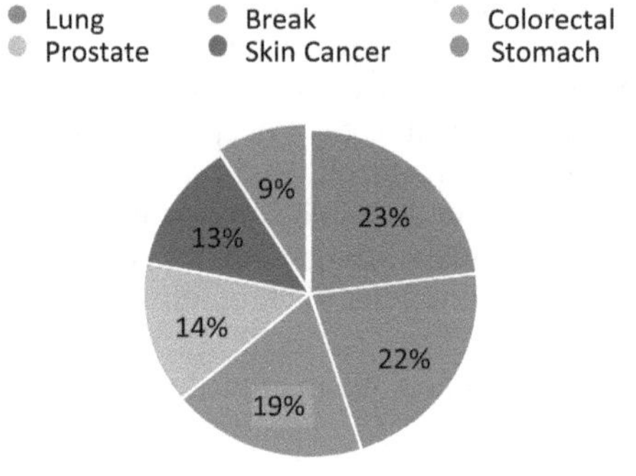

Figure 1. Data of the type of the cancer.

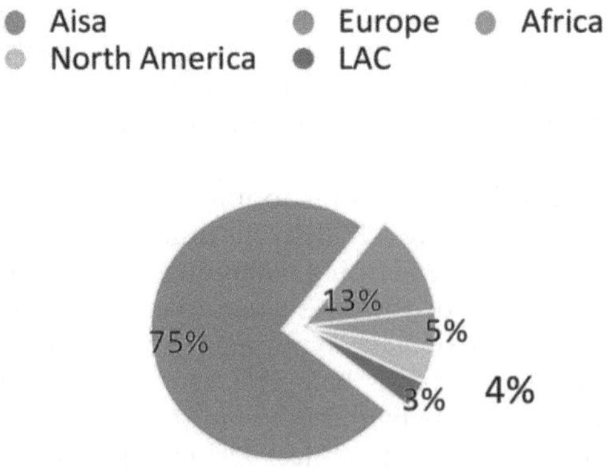

Figure 2. Mortality rate of skin cancer.

1.1 *Literature review*

One type of cancer that starts in the skin cells is skin cancer. With millions of cases identified each year, it is one of the most common types of cancer worldwide. Normal cells change and start to proliferate uncontrollably, giving rise to a mass known as a tumour. This is how cancer begins. With the increase in the use and development of machine learning and Artificial. The researcher, scientist, engineers are working from more than 20 years for developing an model and system for the detection, classification and curing system for the skin cancer. As an illustration The number of cases of skin cancer has increased within the last ten years. It makes to think that skin cancer most common human complaint, because the skin is the largest organ in the body. For a dermatologist, skin vivisection is the first step in determining if a skin lesion is benign or unsightly. A dermatologist performs skin vivisection, in which they remove a section of the skin lesion and examine it under a microscope. It currently takes a week or more to complete the entire procedure, from making an appointment with a dermatologist to getting a vivisection report. Currently, a patient has to look at the single report of a dermatologist to find out if he has a skin issue or not.

The machine learning based classification model used for the skin cancer [12], in this the multiple label data is preprocessed and then classification is done using the different machine learning model like support vector machine, KNN, LR, GNB, SGD, DT shown in Figure 3. At the end it was found that the accuracy of the classification using the SVM, DT, GBD was 100% and logistic had 95%. The classification of the skin cancer using the image processing whether it is benign or malignant [3], the early diagnosis there is upto 96% of less expensive and simple treatment, in this it uses the SVM and random forest. Random forest has the highest accuracy 93.89. The Inception model achieves a notable 85.7% accuracy, comparable to expert dermatologists, showcasing the effectiveness of our proposed method [8]. The less amount of data was an problem for the classification and detection of the cancer but with the use of the transfer learning combing it with the deep learning it has been successful had have achieved the accuracy of 98.61% with the use of AlexNet [11]. The combination of deep learning with the integration with the image processing provided with an interface of web application or app used for the instance detection and classification for the skin cancer with

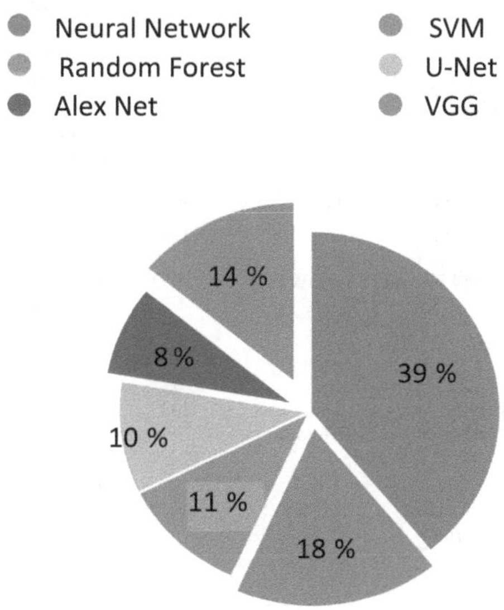

Figure 3. Presentation of approaches.

in the matter of second [16]. Established a system for the segmentation and categorisation of melanoma and benign skin lesions. An Ant-Colony based lesion segmentation algorithm forms the basis of the system. They recovered features related to colour, texture, and shape.It has got the upto the accuracy of approx 90%. Just 16 papers—including Keras [16], TensorFlow [17], MATLAB [18], and PyTorch [19]—published the deep learning framework out of the entire body of collected material.

Reference	Title	Conference	Year	Finding
[11]	Classifications of the Skin Cancer using Deep Learning.	CIBEC	2018	Transfer learning using AlexNet achieved an impressive 98.61% accuracy, demonstrating efficiency and accuracy in classification with a limited dataset
[13]	Detecting of Skin Cancer Lesions with Image Processing Techniques from Digital Images.	IJCDS	2020	Results align with current Deep Neural Network standards, emphasizing fine-tuning and information transfer effectiveness. Anticipating improved outcomes, particularly with dual transfer incorporating knowledge from tasks like Retinopathy
[7]	Skin Cancer Classification using a Parallel Conventional. Model	IEEE	2020	Proposed CNN outperforms VGG-16 and VGG-19 in skin cancer detection in terms of precision and recall.
[9]	Skin Cancer using A Deep C-NN Model for Detection and Classification	IEE	2020	The CNN deep learning algorithm approach was efficient and had the highest accuracy for the.
[14]	Using image processing and machine learning to classify skin cancer	IBCAST	2021	The classification from the image the random forest classifier had the highest accuracy 93.89%.
[15]	Skin lesions Images is to help identify skin cancer	IRJET	2022	Enhancing images and training Resnet50, InceptionV3, and Begrudge Inception for rapid, accurate skin diagnosis. Inception achieves 85.7% accuracy, rivaling expert dermatologists, showcasing effectiveness
[17]	Deep Learning-Based Skin Lesion Image-Based Skin Cancer Detection	IRJET	2023	The CNN model identified gaps in skin lesion diagnoses, showcasing the potential integration of object detection with smartphones for costeffective point-of-care diagnoses, particularly in remote areas
[6]	Superior Convolutional Neural Network Model for Melanoma Skin Cancer Identification	IEEE	2023	This model, which makes use of the RCNN algorithm, offers a useful method for diagnosing skin issues and could serve as the primary screening tool to stop skin cancer. It also aids doctors in accurately categorising various skin disorders.
[12]	Comparison analysis of Machine learning techniques for multiple skin cancer	ITT	2023	The classification of skin cancer with Ml SVM, DT and GNB were found 100% accuracy.
[21]	Application of Improved Neural Network in Diagnosis of Skin Cancer	IJDWM	2023	The study proposes the CNN model for skin lesion classification, achieving a 98.9% accuracy rate, but acknowledges a limitation in handling intra-class variation and the need for specific image knowledge within each class.
[22]	A approach to classifying Skin Lessions using conventional Neural Network	Data and Metadata	2023	An optimized DCNN achieved 84% accuracy using the Adam optimizer, demonstrating effectiveness in classifying malignant skin lesions
[23]	Using Deep Learning to Diagnose Melanomas in Dermoscopy Image	MDPI	2023	Inception-V3 and InceptionResnet-V2 models excel in diagnosing seven types of skin cancer, outperforming specialists with 0.89 and 0.91 accuracy. Enhanced with ESRGAN, the

(*continued*)

Continued

Reference	Title	Conference	Year	Finding
				approach shows promise for future cancer predictions.
[24]	Deep convolutional structure enables recognition of skin lesions.	IJECE	2023	The study introduces a computer-aided diagnosis system achieving 97.25% accuracy in categorizing skin cancer images, offering a streamlined and advanced framework for efficient diagnosis and potentially life-saving outcomes
[25]	Melanoma skin cancer detection with a modified GRU model and mask-RCNN	Front. Physiol	2023	The study presents an efficient melanoma detection framework using mask-RCNN for segmentation and modified GRU for classification, achieving high accuracy on ISIC 2020.
[26]	Recent Developments in Deep Learning-Based Dermoscopic Image Diagnosis of Skin Lesions	IEEE	2023	Enhancing images and training Resnet50, InceptionV3, and Begrudge Inception for rapid, accurate skin diagnosis. Inception achieves 85.7% accuracy, rivaling expert dermatologists, showcasing effectiveness
[27]	SkinNet-INIO: Enhanced Nature-Inspired Optimisation Algorithm and Fusion-Assisted Deep Neural Networks for Multiclass Skin Lesion Localization and Classification	MDPI	2023	Proposed framework integrates deep learning, feature selection, hyperparameter optimization, achieving high accuracy on ISIC2018 and ISIC2019 datasets, outperforming state-of-the-art techniques
[10]	Using a mobile application and a machine learning framework, skin cancer detection	ICOEI	2023	It develops an ResNet50 model with 97.88% accuracy for the skin cancer in plants detection with the mobile app framework which is easy for portable and use.
[28]	Using the machine learning for the classification of the skin cancer	IJARSCT	2023	An interpretable CNN-based skin disease detection system is presented in a study, exceeding competitors in speed and accuracy and offering real-time clinical usage for dermatological diagnostics.
[29]	Generative AI enhances skin cancer classification via deep learning, advancing machine learning techniques for various skin cancer types	IEEE Access	2023	The hybrid model in a study that combined generative AI and CNNs to classify skin lesions achieved 96% accuracy on the ISIC 2019, indicating the efficacy of generative AI for augmentation.

2 METHODOLOGY

The articles' goal is to ascertain how AI and Ml are applied to the identification and categorization of skin cancer, as well as to develop technology to enhance the therapeutic approach. The many methods include the employment of Alex Net, Deep Learning CNN, DNN, SVM, Decision Trees, image processing, preprocessing, and augmentation. All training and testing was conducted using the supplied high-quality data. The suggested approach combines mobile development, machine learning, and image processing for the identification and categorization of skin. A variety of skin image datasets are initially gathered and preprocessed in order to standardise characteristics and improve quality. Decision Trees and Support Vector Machines are coupled to extract meaningful information from images using image processing techniques.

To ensure compliance with data protection regulations and minimise biases in the dataset and model predictions, ethical considerations regarding user privacy and fairness are taken care of. User feedback is gathered through usability testing and iterative improvement cycles in order to improve the mobile application's functionality and interface. Complete process explanations from data collection to deployment, along with ongoing maintenance and monitoring to ensure peak system performance throughout time, are included in the documentation. By offering a practical and efficient tool for early skin lesion diagnosis, this all-encompassing strategy seeks to improve public health outcomes by enabling prompt medical care. In Figure 4 we can see the methods used in the past years for the skin cancer.

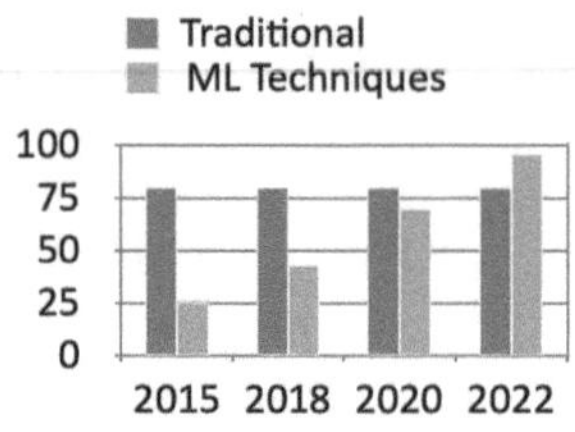

Figure 4. Traditional method Vs ML technique.

3 CONCLUSION AND FUTURE WORK

Machine learning algorithm are developed for the detection and classification of the one of the deadliest cancer i.e skin cancer. With the period of time number of Ml based algorithms as developed support vector machine, decision tree, deep learning, image processing. The significance number of finding has be done by the research done. The highly accurate algorithms are proposed with the accuracy upto 100%. The major problem of the qaulity of the image for detection and classification was over some by the proposed approach with the combination of the image processing with the ML based algorithms. And the issue of the highly expensive approach which for training the model it require huge amount of data which was costly was over come by the implementation of the transfer learning with the use of the Alex Net, it made it economical. The contribution of the the deep learning is huge it provide an a whole process from processing to classification. We can see the use of deep learning for skin cancer detection and classification. From the research done, there is an lack of the pipe for the skin cancer detection and classification with the integration of web application and mobile application which could make it more accessible to the people and it could help in the identification of the cancer in the starting stage as the people can do it whenever and anywhere in the world. The proper development the pipeline need to be developed as soon as possible with the high accuracy and data safety is needed to be developed in coming future. From this review paper research, developer, scientist and everyone who reads it will get the over view of the Machine learning model developed till now and their achievement, draw back and what is need to be developed of what is lacking. And they can contribute to the betterment and advancement of the skin cancer domain.

REFERENCES

[1] WorldHealth Organization. (2022). *GlobalHealthObservatory*; World Health Organization: Geneva, Switzerland.

[2] Gouda, W., Sama, N. U., Al-Waakid, G., Humayun, M., Jhanjhi, N. Z. (2022). Detection of skin cancer based on skin lesion images using deep learning. *Healthcare*, 10, 1183. https://doi.org/10.3390/healthcare10071183

[3] Holly E. Kanavy, D. O, and Meg R. Gerstenblith, M. D., *Ultraviolet Radiation and Melanoma*
[4] Pushkar Aggarwal, Peter Knabel, Alan B Fleischer Jr,. (1990 to 2019). *United States Burden of Melanoma and Nonmelanoma Skin Cancer from*
[5] Darius A. Gerlach, Martin Trammer, Jochen Zange's, (2011). Validation of a Mobile NMR Sensor for the Determination of Skin Layers and the Local Estimation of the T2 Relaxation Time Distribution in the Lower arm, ESMRMB
[6] Prof. Bharath Bharadwaj B S, Saniya Anjum, Shaguftha Afreen, Spoorthi T C, Keerthana M (2023). *Detection of Skin Cancer Based on Skin Lesion Images UsingDeep Learning, IRJET*, 2023
[7] Noortaz Rezaoana, Mohammad Shahadat Hossain, Karl Andersson. (2020). *Detection and classification of skin cancer by using a parallel CNN Model*, IEEE
[8] Osim Kumar Pal, Dipta Paul, Ekram Hasan, Mahathir Mohammad, Md Abidul Hasan Bhuiyan, Foysal Ahammed. (2023). Advanced convolutional neural network model to identify melanoma skin cancer", *IEEE International Conference on Contemporary Computing and Communications (InC4)*
[9] Nour Aburaed, Alavikunhu Panthakkan, Mina Al-Saad, Saad Ali Amin, Wathiq Mansoor,. (2020). *Deep Convolutional Neural Network (DCNN) for Skin Cancer Classification*, IEEE
[10] Taslima Akter Sathi, Md Abid Hasan, Mohammad Jahangir Alam, (2023). *Skin Cancer Detection using Machine Learning Framework with Mobile Application*, ICOEI
[11] Khalid M. Hosny, Mohamed A. Kassem, Mohamed M. Foaud. (2019). *Skin Cancer Classification using Deep Learning and Transfer Learning, IJASRT*, 2019.
[12] Muhammad Imad, Zahoor Ali Khan, Shah Hussain Bangash Irfan Ullah, Khan, Sheeraz Ahmad. (2023). Comparative analysis of machine learning techniques for multi-class skin cancer classification, *2023 9th International Conference on Information Technology Trends (ITT)*
[13] Minakshi Waghulde, Shirish Kulkarni, Gargi Phadke. (2020). Detection of Skin Cancer Lesions from Digital Images with Image Processing Techniques, *IJCDS*, 2020
[14] Arslan Javaid, Muhammad Sadiq, Faraz Akram. (2021). *Skin Cancer Classification Using Image Processing and Machine Learning, IBCAST*, 2021.
[15] Walaa Gouda, Noor Zaman. (Feb. 12, 2020). *Detection of Skin Cancer based on skin lesion images*, IRJET, 2022[16]Keras Documentation. [Online]. Available: https://keras.io/ applications/
[16] Chris J Maddison, Aja Huang, Ilya Sutskever, and David Silver. (2014). Move evaluation in Go using deep convolutional neural networks. *arXiv preprint* arXiv:1412.6564. arxiv.org/abs/1412.6564.
[17] Abadi M., Agarwal A., Barham P., Brevdo E., Chen Z., Citro C., Corrado G. S., Davis A., Dean J., Devin M. and Ghemawat S. (2016). "TensorFlow: Large-scale machine learning on heterogeneous distributed systems", *arXiv:1603.04467*. [Online]. Available: https://arxiv.org/abs/1603.04467
[18] Vedaldi A. and Lenc K. (2015). "MatConvNet: convolutional neural networks for MATLAB," in Proc. 23rd *ACM Int. Conf. Multimedia*, pp. 689–692, doi: 10.1145/2733373.2807412
[19] Paszke A., Gross S., Chintala S., Chanan G., Yang E., DeVito Z., Lin Z., Desmaison A., Antiga L. and Lerer A. (2017). "Automatic differentiation in pytorch," in *Proc. NIPS Workshop Autodiff*, Long Beach, CA, USA.
[20] Nawal Soliman and ALKolifi ALEnezi. (2019). A method of skin disease detection using image processing and machine learning, *Procedia Computer Science*, vol. 163, pp. 85–92, 2019, ISSN 1877–0509
[21] Wu Beibei, Nikolaj Jade, Application of improved chameleon swarmAlgorithm and improved convolution neural network in diagnosis of skin cancer, *International Journal of Data Warehousing and Mining* Volume 19, 2023
[22] Oumoulylte M, Omari Alaoui A, Farhaoui Y, El Allaoui A, Bahri A. (2023). Convolutional neural networkBased approach for skin lesion classification. *Data and Metadata*. 2:171.https://doi.org/10.56294/dm2023171
[23] Alwakid, G.; Gouda, W.; Humayun, M.; Jhanjhi, N.Z. (2023). Diagnosing melanomas in dermoscopy images using deep learning. *Diagnostics*, 13, 1815. https://doi.org/10.3390/diagnostics13101815
[24] Shimaa Fawzy, Hossam El-Din Moustafa, Ehab H. AbdelHay, Mohamed Maher Ata. (October 2023). A deep convolutional structure-based approach for accurate recognition of skin lesions in dermoscopy images, *IJECE*, Vol. 13, No. 5, pp. 5792–5803
[25] Monica KM, Shreeharsha J, Falkowski-Gilski P, Falkowska-Gilska B, Awasthy M and Phadke R. (2024). Melanoma skin cancer detection using mask-RCNN with modified GRU model. *Front. Physiol.* 14:1324042. doi: 10.3389/fphys.2023.1324042

[26] Hussain, M.; Khan, M.A.; Damaševičius, R.; Alasiry, A.; Marzougui, M.; Alhaisoni, M.; Masood, A. (2023). SkinNet-INIO: multiclassSkin lesion localization and classification using fusion-assisted deep neural networks and improved nature-inspired optimization algorithm. *Diagnostics*, 13, 2869. https:// doi.org/ 10.3390/diagnostics13182869

[27] Bharti Sahu, Avishkar Sontakke, Kaustubh Mahamuni, Swapnil Akade, Rupali Sutar. (December 2023). Skin disease detection using machine learning. *International Journal of Advanced Research in Science, Communication and Technology*, Volume 3, Issue 3

[28] Mudassi Saeed, Asma Naseer, Hassan Masood, Shafiq Ur Rehman, Volkar Gruhn. (November, 2023). The Power of generative AI to augment for enhanced skin cancer classification: A deep learning approach machine learning techniques for multiple skin cancer, *IEEE Access*, Volume 11

[29] Farah Sultan, Muhammad Kaleem, Nisar Ahmad, Salman Rashid, Muhammad Azhar Mushtaq, and Naveed Ahmad. (2023). A systematic analysis of skin cancer detection using machine learning and deep learning techniques, *Journal of Computing & Biomedical Informatics*, Volume 6, 2023

Computational Methods in Science and Technology Volume 1 – Kaur et al. (Eds)

Mitigating DDoS attacks: A distributed blockchain-SDN secure IoT system enhanced by artificial neural networks

Pardeep Singh Tiwana, Lakhvinder Kaur & Navpreet Kaur
Chandigarh Engineering College-CGC Landran, Mohali, Punjab

Heena Wadhwa
Chitkara University Institute of Engineering and Technology, Chitkara University, Punjab, India

Nisha Kumari & Rajeev Sharma
Chandigarh Engineering College-CGC Landran, Mohali, Punjab

ABSTRACT: The introduction of the Internet of Things (IoT) has resulted in the enlargement of complex and intelligent solutions, which has led to an expansion of the variety of innovative services and processing capacity that are in line with the standards of life in the modern day. It is possible that the widespread usage of these services is to blame for the growing number of communication entities and devices to be element of the Internet of Things (IoT). This presents a substantial challenge in terms of the management of networks. Consequently, it is of the utmost importance to restructure the administration of the network that controls the Internet of Things (IoT). By virtue of its programmability and centralized characteristics, software-defined networking (SDN) facilitates the streamlining of network administration, enables network abstraction, facilitates network development, and has the capability to administrate the Internet of Things (IoT) network. However, there are still obstacles to overcome in terms of security when it comes to the Internet of Things (IoT). Distributed denial of service assaults, often known as DDoS attacks, are regarded to be one of the most severe security concerns. Internet of Things (IoT) systems that are used on a global scale are susceptible to these attacks. This article examines DDoS assaults in software-defined networking and the Internet of Things. This research evaluates SDN, machine learning, and blockchain-based detection and mitigation technologies. An IoT architecture that is strong and secure has been proposed. Mult controllers and SDN were used to create this framework. Most people believe blockchain secures decentralized SDN-IoT networks well. As commonly known, artificial neural networks (ANNs) may improve risk assessment and mitigation.

Keywords: Software Defined Network, IOT, ANN, Mitigation, Blockchain

1 INTRODUCTION

SDNs have gained momentous traction in several areas due to their knack to provide an alternative to the conventional TCP/IP architecture (Duan *et al.* 2016), hence reducing the need for equipment management. This objective is achieved by centralizing network tasks provided by network devices, resulting in a streamlined and cost-efficient network management process.

The integration of the digital and physical worlds has resulted in the extensive use of the Internet of Things (IoT) in several fields (Bellini *et al.* 2022). The domains included under this framework consist of smart health, smart cities, smart dwellings, smart transportation, smart infrastructure, and smart commerce. This is the situation because the Internet of Things (IoT) is an unavoidable advancement in technology. The profound influence of services provided via the Internet of Things on the human experience is significant. The use of people-centered solutions, such as the Internet of Things (IoT), facilitates the provision of support to individuals with disabilities, allowing them to benefit from autonomy and engage actively in social activities. In-home

DOI: 10.1201/9781003501244-71

physical therapy rehabilitation is a domain that might potentially benefit from the use of Internet of Things (IoT) technologies. Moreover, Internet of Things (IoT) technologies play a crucial role in mitigating potentially hazardous situations. In the context of hazardous mining activities, the use of the Internet of Things has resulted in enhanced efficiency and safety measures (Misra *et al.* 2022). One example is the use of autonomous mining equipment to prevent workers from accessing potentially dangerous locations. The integration of Internet of Things (IoT) sensors for pollution, hazardous gas, and temperature may be used in tandem with warning systems to contribute to the mitigation of environmental disasters. Numerous researchers have conducted rigorous investigations on additional case cases that have been extensively recorded in the scholarly literature. It is evident that there is a growing need for intelligent gadgets that are designed to operate well in both industrial and consumer settings. The projection suggests that the Internet of Things industry is expected to earn billions of dollars in yearly income, with this desire playing a significant role. These gadgets possess enhanced functions that enable efficient data processing and improve the operations and services provided via their use.

A "blockchain" refers to a decentralized ledger consisting of a network of nodes that stores data. Blockchains protect the data they store and it is private also correct (Zhang *et al.* 2020). In certain scenarios, the implementation of these measures occurs when third-party verification is deemed unnecessary and the data in question must be relied upon. The primary characteristic that distinguishes the blockchain is the inherent immutability of the data it contains. The data indicated above is safely kept inside the segments that make up the blockchain, using cryptographic methods to join them. This makes adaptation very difficult and requires a significant allocation of resources.

2 DDOS ATTACKS IN SDN

The design and structure of Software-Defined Networking (SDN) are contingent upon a variety of interconnected attributes (Rehmani *et al.* 2019). The design attributes of this network architecture differ from the conventional norm seen in network topologies. The characteristics of Software-Defined Networking (SDN) enhance the efficacy and efficiency of network security, while also providing heightened flexibility. However, software-defined networking (SDN) is prone to security vulnerabilities as a result of many design shortcomings. Consequently, there are two separate approaches available for characterizing the components of SDN design (Lin *et al.* 2019). As a result, several components of Software-Defined Networking (SDN) provide protection against distributed denial of service (DDoS) assaults, while others may expose it to these types of threats .The vulnerability attributes of software-defined networking (SDN) in relation to distributed denial of service attacks are also referenced, in addition to the specific areas of security risks. Moreover, this study investigates the methods via which distributed denial of service assaults materialize in the context of software-defined networking.

2.1 *Various kinds of DDoS assaults in SDN*

The network structural design is founded upon the principles of software-defined networking (SDN), an innovative concept that entails the segregation of the data plane and the control plane. This differentiation contributes to enhancing the security and manageability of the network. Software-defined networks (SDNs) provide several natural properties that contribute to the enhancement of network security. Programmability, a data-control plane separation, traffic analysis software, and dynamic protocol adaptation are further features of the networks. However, a significant quantity of attacks has the capability to infiltrate the data plane, control plane, and interplane connections without succumbing to detection. These assaults are undermining the design of the software-defined network (SDN), rendering it an exceptionally difficult network architecture. Figure 1 provides a visual representation of the specific plane in the SDN architecture that is susceptible to a particular attack. This section offers a comprehensive overview of the many types of assaults that are being tackled.

The ***packet_in flooding*** mechanism is used by vswitch to initiate a rule to forwarding for an unmatched the flow across the interface of the southbound. This is achieved by requesting the federal regulator to construct a forwarding rule using the packet_in message. The event will transpire anytime the controller is presented with a packet_in message. Faking packet IP addresses and forwarding them to the virtual switch allows the attacker to force the virtual switch to send many packet-in messages to the controller. Due to its preoccupation with handling fraudulent flow requests, the controller becomes inundated and unattainable for authorized users as a consequence of this concentration.

Switch spoofing refers to a kind of attack when an adversary manipulates the IP address of a switch (Ubaid *et al.* 2017), hence facilitating the transmission of control messages via the compromised switch. Simultaneously with the establishment of a connection and communication between a switch and the controller, a second malicious switch will be activated. This switch will own a fabricated IP address and exhibit comparable hardware and identity. Additionally, this switch will create a connection with the controller. As the controller initiates communication with the malevolent switch while concurrently severing the link with the lawful switch, the network's performance will progressively deteriorate. Given the potential for this assault to create fraudulent queries to the controller, it is crucial to exercise control over it.

If the ***controller experiences a substantial influx*** of packet-in requests due to false flows provided by the adversary (Black *et al.* 2023), it will implement queues to effectively manage these demands. The occurrence in question is often known as controller saturation. Conversely, in the case of a substantial quantity of counterfeit packets, the controller's attention will be diverted towards the processing of these requests, ultimately leading to a decline in overall performance. Regarding SDN-based networks, this poses a challenge. The controller's processing capabilities are challenged when an attacker can effortlessly generate a substantial volume of fake flows. Certain authors propose the use of a secondary controller as a potential resolution to the issue at hand. Nevertheless, it is important to acknowledge that the secondary controller is vulnerable to assaults.

During packet-in transmission to the controller, a switch stores part of the packet in buffer memory. The remaining portion of the packet will thereafter be sent by the switch to the controller. The term used to describe this occurrence is buffer saturation. Through the use of this functionality, the assailant transmits a multitude of counterfeit packets towards the switch, leading to the rapid depletion of the buffer. Moreover, in the case that the internal buffering of the switch is depleted, it becomes imperative for the switch to transmit the whole packet to the controllers. One additional constraint arises in the context of software-defined networking (SDN). Based on the present circumstances, individuals with proper authorization are encountering challenges in the processing of their flow requests, thereby enabling the attacker to diminish the system's performance.

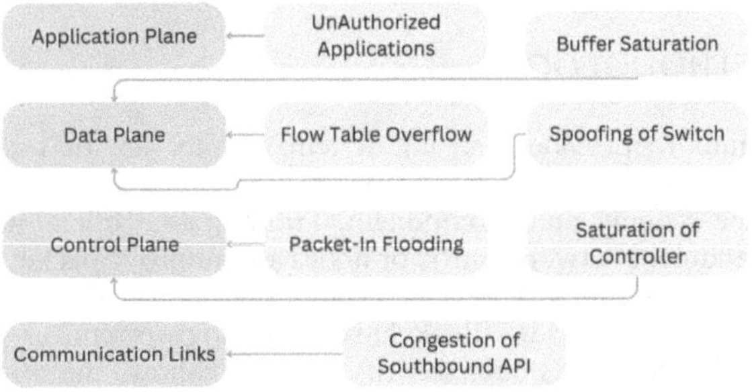

Figure 1. Attacks in SDN.

3 RELATED WORK

This concept is supported by empirical data derived from real-world observations. Entropy and divergence metrics for DDoS detection are widely used. In contrast to entropy, which measures the extent of variation in network properties, divergence is a metric that assesses the degree of comparability between two probability distributions. The notion of measuring uncertainty was first introduced by Claude Shannon. The calculation of the information distance or divergence metric involves the usage of several probability distributions of traffic flows. This measure is employed to detect irregularities in network traffic consumption. The identification of a distributed denial of service attack is possible due to the disparity between the current network behavior and the typical behavior of the network, as shown by the entropy measure. Several researchers have used the entropy metric to suggest DDoS security solutions. The coming sections will provide a concise explanation of the answers shown in Table 1.

Table 1. DDoS defense solutions in SDN.

Authors	Approach	Validation
(Giotis *et al.* 2014)	Extracting and analyzing network flow statistics using entropy method for anomaly detection.	Validation utilized the National Technical University of Athens network.
(Wang *et al.* 2015)	Applying scheme to OpenFlow switches reduced flow collecting overhead.	Validated using Mininet network emulator.
(Mousavi *et al.* 2015)	Effective lightweight solution using entropy variation of destination IP address for attack detection.	Validated using Mininet network emulator and Scapy tool.
(Botie *et al.* 2017)	In-switch processing for DDoS detection based on traffic feature monitoring and entropy-based algorithm.	Validated by installing switch flow rules to drop malicious traffic.
(Tsai *et al.* 2017)	Controller-applied entropy-based DDoS detection and mitigation.	Validated using Ryu controller and Scapy traffic generator.
(Kalkan *et al.* 2018)	Joint entropy security system with nominal, preparatory, and active mitigation.	Validated with JESS (Joint Entropy-based Security Scheme) and Mininet network emulator.
(Sahoo *et al.* 2018)	Low-rate DDoS attack detection using GE and GID data.	Mininet emulator on Ubuntu 14.04 LTS verified.
(Jiang *et al.* 2017)	Entropy-based DDoS Defence Mechanism with three phases: Window Construction, Detection, and Mitigation.	Validated using floodlight controller on Mininet.

4 PROPOSED METHODOLOGY

This research study aims to provide a strong system architecture that can safeguard the IoT environment from DDoS attacks. A cluster of blockchain-enabled software-defined networking (SDN) controllers is the system's main component. Training an artificial neural network (ANN) model may identify fraudulent network traffic or nodes and improve packet content in traffic data (Tiwana *et al.* 2017). Implementing detection results may prevent future attacks. A multi-page decentralized barrier reduces hazards. Blockchain technology optimizes payloads for efficient traffic routing. Encryption improves data security. Figure 2 shows the three key components of the proposed architecture to reduce DDoS attacks. The ANN model can distinguish safe and harmful communications and equipment. Machine learning helps the computer spot DDoS assaults and other network anomalies. Analysis of network traffic patterns allows safe and dangerous communications to be distinguished. Regardless of what occurs, the blockchain's decentralized ledger will record network activities securely. Blockchain technology lets the control plane ensure all switches have the latest regulations. Any law change is recorded and confirmed by the network. Operational blockchain components at the control plane offer capacity (Aggarwal *et al.* 2021). Once the artificial neural network (ANN) model detects an assault, the relevant data on the attack is promptly appended to the blockchain and securely sent to the operational controllers. Consequently, network resources will be inaccessible to any traffic that meets the attack parameters. Data plane firewalls may identify and prevent unlawful traffic caused by distributed denial of service attacks.The use of the ANN model facilitates the achievement of traffic payload optimization, leading to enhanced network performance and decreased congestion.

5 SIMULATION ENVIRONMENT

The Keras TensorFlow package is used to implement the detection system based on artificial neural networks. Furthermore, the use of the SkLearn program is employed to execute the preprocessing and recital estimation metrics intended for this fastidious model. The SDN-IoT dataset is used inside the SDN architecture to generate Internet of Things traffic. The Matplotlib function plots model accuracy and loss performance, making evaluation simpler (Wadhwa *et al.* 2024). This project seeks to find the best artificial neural network (ANN) model used for perceiving DDoS assaults on the SDN-IoT dataset. The efficacy of detection is assessed via cross-validation. The model's performance was evaluated using a confusion matrix, which included comparing the predicted classes with the actual classes seen in the dataset (Tiwana *et al.* 2019). The ROC curve shows how well a binary algorithm combines the number of true and false positives. When the choice level changes, this solution is used. Other factors, such as precision, memory, and F1-score, were also part of the method.

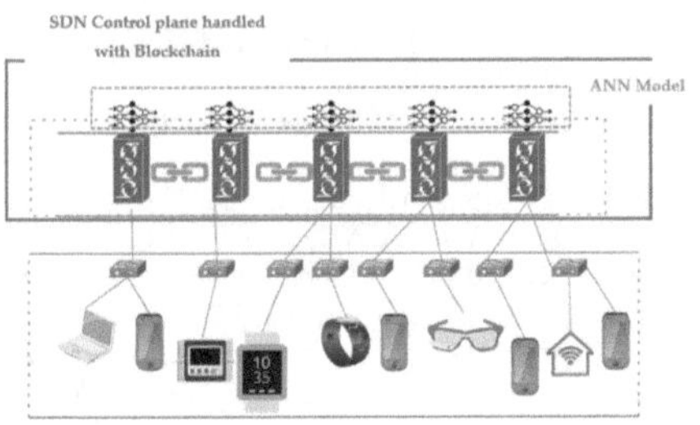

Figure 2. DDoS Mitigation with ANN.

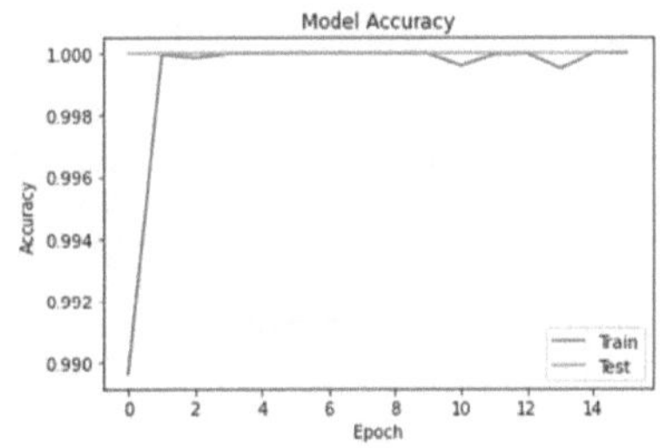

Figure 3. Model of Accuracy.

6 CONCLUSION

Security and management of the large volumes of data created by Internet of Things (IoT) devices must be prioritized due to their widespread use in companies and homes. SDN technology can govern IoT device network traffic. In contrast, blockchain technology allows distributed and secure trust management and data exchange. Internet of Things (IoT) services and device data integrity depend on strong security measures. In this situation, DDoS assaults are serious. This research studied the DDoS threat in IoT and SDN and their vulnerability to assaults. Additionally, blockchain and machine intelligence were used to investigate DDoS mitigation strategies. This article examines new methods for securing Internet of Things (IoT) applications and services. A thorough study examined the benefits of combining SDN principles, blockchain technology, and machine learning models into such systems. A complete and secure Internet of Things (IoT) system, recently built, may prevent and mitigate DDoS assaults. This system uses Mult controller SDNs and ANN models to improve security. Firewalls and encryption may be added to the data plane. A more detailed investigation of the artificial neural network (ANN)-based detection system's implementation showed its outstanding performance. Our present approach needs improvement to deploy the whole system on mini-net emulators. Hyperledger fabric will be used to construct east-west linkages, assisted by several ONOS controllers.

REFERENCES

Aggarwal, N., Tripathi, B., Chottani, D. and Tiwana, P. S. (2021, December). Attacks opportunities in the cyber physical space and the role of cybersecurity. In *2021 2nd International Conference on Computational Methods in Science & Technology (ICCMST)* (pp. 276–280). IEEE.

Bellini, P., Nesi, P. and Pantaleo, G. (2022). IoT-enabled smart cities: A review of concepts, frameworks and key technologies. *Applied Sciences*, 12(3), p.1607.

Black, C. (2023). *Mitigating Data Plane Device Compromise in Programmable Networks* (Doctoral dissertation, Queen's University Belfast).

Bonkra, A. and Dhiman, P. (2021), December. IoT security challenges in cloud environment. *In 2021 2nd International Conference on Computational Methods in Science & Technology (ICCMST)* (pp. 30–34). IEEE.

Bonkra, A., Pathak, S., Kaur, A. and Shah, M. A., (2024). Exploring the trend of recognizing apple leaf disease detection through machine learning: a comprehensive analysis using bibliometric techniques. *Artificial Intelligence Review*, 57(2), pp.1–26.

Duan, Q., Ansari, N. and Toy, M. (2016). *Software-Defined Network Virtualization: An Architectural Framew* pp.10–16.

Giotis, K., Argyropoulos, C., Androulidakis, G., Kalogeras, D. and Maglaris, V. (2014). Combining openFlow and sFlow for an effective and scalable anomaly detection and mitigation mechanism on SDN environments. *Computer Networks*, 62, pp.122–136.

Kalkan, K., Altay, L., Gür, G. and Alagöz, F. (2018). JESS: Joint entropy-based DDoS defense scheme in SDN. *IEEE Journal on Selected Areas in Communications*, 36(10), pp.2358–2372.

Kaur, M., and Aron, R. (2022). A novel load balancing technique for smart application in a fog computing environment. *International Journal of Grid and High Performance Computing (IJGHPC)*, 14(1), 1–19.

Kaur, M., and Aron, R. (2022). An energy-efficient load balancing approach for fog environment using scientific workflow applications. *In Distributed Computing and Optimization Techniques: Select proceedings of ICDCOT 2021* (pp. 165–174). Singapore: Springer Nature Singapore.

Lin, Y. D., Lai, Y. K., Tsou, Y. L., Lai, Y. C., Liou, E. C. and Chiang, Y. (2019). Generic validation criteria and methodologies for SDN applications. *IEEE Systems Journal*, 13(4), pp.3909–3920.

Misra, S., Roy, C., Sauter, T., Mukherjee, A. and Maiti, J. (2022). Industrial internet of things for safety management applications: A survey. *IEEE Access*, 10, pp.83415–83439.

Mousavi, S. M. and St-Hilaire, M. (2015). February. Early detection of DDoS attacks against SDN controllers. *In 2015 International Conference on Computing, Networking and Communications (ICNC)* (pp. 77–81). IEEE.

Rehmani, M. H., Davy, A., Jennings, B. and Assi, C. (2019). Software defined networks-based smart grid communication: A comprehensive survey. *IEEE Communications Surveys & Tutorials*, 21(3), pp.2637–2670.

Sahoo, K. S., Puthal, D., Tiwary, M., Rodrigues, J. J., Sahoo, B. and Dash, R. (2018). An early detection of low rate DDoS attack to SDN based data center networks using information distance metrics. *Future Generation Computer Systems*, 89, pp.685–697.

Tiwana, P. S. and Mann, N. (2016, December). Jellyfish reorder attack on hybrid protocol in manet dissection on variegated parameters. In *2016 Fourth International Conference on Parallel, Distributed and Grid Computing (PDGC)* (pp. 96–100). IEEE.

Tiwana, P. S. and Singh, J. (2023), December. software-defined networking's adaptive load optimisation technique for multimedia applications with multi-controller utilisation. *In 2023 9th International Conference on Signal Processing and Communication (ICSC)* (pp. 167–172). IEEE.

Tiwana, P. S., Tiwana, H. S., Sharma, R. and Gupta, A. (2019). An scrutiny of run-time ramification for 5-proviso busy beaver proving empirical composition. *International Journal of Innovative Technology and Exploring Engineering*, 8(9), pp.726–732.

Tiwna, P. S. and Singh, J. (2023, January). Load optimization accessions, ramification the QoS in software defined networking. In *2023 International Conference on Artificial Intelligence and Smart Communication (AISC)* (pp. 1013–1017). IEEE.

Tsai, S. C., Liu, I. H., Lu, C. T., Chang, C. H. and Li, J. S. (2017). Defending cloud computing environment against the challenge of DDoS attacks based on software defined network. *In Advances in Intelligent Information Hiding and Multimedia Signal Processing: Proceeding of the Twelfth International Conference on Intelligent Information Hiding and Multimedia Signal Processing*, Nov., 21–23, 2016, Kaohsiung, Taiwan, Volume 1 (pp. 285–292). Springer International Publishing.

Ubaid, F., Amin, R., Ubaid, F. B. and Iqbal, M. M. (2017). Mitigating address spoofing attacks in hybrid SDN. *International Journal of Advanced Computer Science and Applications*, 8(4).

Wadhwa, H., Kaur, M., Kaur, A. and Tiwana, P.S. (2024). Analysis of IoT devices data using bayesian learning on fog computing. *In Artificial Intelligence, Blockchain, Computing and Security* Volume 2 (pp. 194–199). CRC Press.

Wadhwa, Heena, and Rajni Aron. (2022). Resource utilization for iot oriented framework using zero hour policy. *Wireless Personal Communications* 122(3): 2285–2308.

Wang, R., Jia, Z. and Ju, L. (2015), August. An entropy-based distributed DDoS detection mechanism in software-defined networking. *In 2015 IEEE Trustcom/BigDataSE/ISPA* (Vol. 1, pp. 310–317). IEEE.

Zhang, J., Zhong, S., Wang, T., Chao, H.C. and Wang, J. (2020). Blockchain-based systems and applications: a survey. *Journal of Internet Technology*, 21(1), pp.1–14.

Computational Methods in Science and Technology Volume 1 – Kaur et al. (Eds)

Human emotion recognition using deep learning

Shivam Rajpoot
Centre for Advanced Studies, AKTU

Jagrati Singh
Indira Gandhi Delhi Technical University for Women, Delhi

ABSTRACT: Human-computer interface and pitiful computing profoundly rely on emotion recognition from facial expressions. Deep learning approaches have recently established astonishing performance in a variety of computer vision applications, including emotion recognition. In-depth research on the use of deep learning techniques for identifying human emotions from facial photographs is obtainable in this publication. Convolutional Neural Network (CNN) is used in the recommended method to repeatedly extract emotional information from facial picture. For training and testing, a dataset with a wide assortment of facial expressions instead of various emotions is used. To recover the resilience and generalization capacities of the model, preprocessing approaches including data expansion and normalization are used. To obtain the paramount performance, the network design is tuned using hyperparameters and irritatedvalidated. Extensive experiments are carried out to contrast the performance of a variety of deep learning architectures, counting conventional CNN and more complicated architectures like VGG-16, in order to judge the effectiveness of the recommended approach. Aspects of valuation measures include F1-score, recall, accuracy, and precision. The results of the research show that deep learning approaches are more efficient than conventional ones at identify multifaceted patterns in facial expressions, which lead to improved emotion recognition performance. The planned method achieved an accuracy of 92.28 percent on fre2013 consisting of 7 classes. The experimental study show that the projected move toward is far more efficient and yield improved performance.

1 INTRODUCTION

Human associations and interpersonal statement are extremely predisposed by human emotions. To respond successfully and sympathetically to the approach of others, it's critical to be able to recognize and understand these emotions. Sentimental computing, along with a variety of applications like virtual assistants, mental health evaluation, and entertainment, rely on extend this ability to machines and computers. Recent advancement in deep learning have revolutionize the field of computer vision, particularly in tasks that involve extract complex patterns and representations from complex inputs, such as descriptions. One area that has seen noteworthy progress since the preamble of deep learning techniques is the appreciation of emotions from facial expressions. Deep learning methods have shown extraordinary expertise in learn hierarchical features directly from raw data, enable them to repeatedly discern multifaceted patterns that were previously demanding to openly define. When it comes to recognize human emotions from facial images, convolutional neural network (CNN) have established to be highly efficient, given their extraordinary performance in image analysis tasks. This reading aims to explore and examine the use of deep learning techniques for civilizing the precision and competence of emotion recognition systems. We influence the capabilities of CNN to learn animated features from images and investigate into

DOI: 10.1201/9781003501244-72

the architecture, training process, and evaluation metrics of the projected deep learning models. In addition, we evaluate the performance of this model with that of conventional techniques, underscore the advantages of deep learning in capture the complex dynamics of facial emotions. In the territory of sentiment detection, conservative methods mainly relied on hand-crafted features and rulebased systems. This approach move violently to capture the complex patterns and slight nuances present in human emotional expressions. However, the arrival of deep learning has brought about a fundamental alteration in this field. Deep learning models, enthused by the architecture of the human brain, have established an unmatched ability to learn hierarchical representations directly from raw data. They can now separately distinguish complex patterns and characteristics that were previously challenging to symbolize clearly. A vital element in image-based emotion identification technologies is the use of Convolutional Neural Network (CNN). These networks are particularly designed to repeatedly take out spatial hierarchies of characteristic types from image data. The integration of frequent convolutional and pooling layers in CNN serves to improve correctness. The incorporation of deep learning models such as VGG16 with predictable CNN in the monarchy of emotion detection marks a noteworthy pace toward achieve more natural and sympathetic human- computer interactions in today foreverdeveloping technological landscape. In the forthcoming section, this study investigate deeper into the methodologies, experimental findings, and implication of these models, aiming to assets the considerable progress they have brought to the field and the capable new opportunity they open up for realistic applications. This application span from practical assistant to tools intended to carry mental health, underscore the transformative probable of this technological development.

1.1 *Related work*

Numerous researchers have worked in the field of face identification some noteworthy contributions are report:

(Hangaragi, Singh and Neelima 2023) projected a replica that employ Face Mesh technology for the reason of face recognition and detection. The operation of Face Mesh enables the model to function efficiently diagonally diverse scenario, surrounding situations with unreliable backgrounds and lighting conditions.(Sarvakar, *et al.* 2023) introduce an inventive move toward the facial emotion recognition through the operation of Convolutional Neural Networks (FERC). The FERC CNN network is prearranged into two key phases: in the first phase, the background of the image is uninvolved, while in the second phase, the facial vector is extract.(Iyer, *et al.* 2023), the person behind introduce a characteristic move toward to recognize human emotions using electroencephalogram (EEG) data.(Hossain, *et al.* 2023)demonstrate the method of increasing a facial expression detection system is detailed, encompassing pre-processing, feature representation, normalization, hyperparameter tuning, and classification steps.

(Bai, *et al.* 2023) Conduct a research concerning participant with hearing impairments. They record EEG signals while this participant watched six dissimilar types of moving video clips, aiming to distinguish emotions such as contentment, inspiration, impartiality, anger, fear, and sadness. As per result completed by (Houssein, Hammad and Ali 2022), human emotions can be discern through a variety of channels, including facial expressions, speech, behavioral cue (such as gestures and posture), and physiological markers.As per result completed by (Umer, *et al.* 2022), the most important purpose of this learning is to recognize a range of facial expressions that can apparent on the human face.

Although previous study has mainly determined on emotion appreciation from facial photos using deep learning models, learning completed by(Khattak, Asghar, *et al.* 2022) take on the challenge of extenuating performance weakening resulting from erroneous layer assortment in convolutional neural network models.(Geetha, *et al.* 2024) introduce a work of fiction fourstage approach for the identification of human emotions based on multivariate EEG data. The first stage involve extract an ensemble of multivariate modulated oscillations (MMO) from multichannel EEG signals using multivariate variation mode decomposition (MVMD).

(Maithri, *et al.* 2022) underscore the noteworthy impact of person emotions on their behaviors. The application for automatic emotion recognition is diverse and extends to various fields.

2 METHODOLOGY

A quantity of critical processes, from dataset training through model training and assessment, are incorporated in the deep learning methodology for recognize human emotions.

2.1 *Dataset selection and preparation*

Preprocess the dataset by resizing images to a reliable declaration; normalize pixel values, and the stage data expansion to increase dataset unpredictability. Data expansion techniques may comprise rotations and flips. Choose a suitable dataset for preparation and assessment that covers a various range of facial expressions on behalf of various emotions. Common datasets comprise CK+, AffectNet, and FER2013.

2.2 *Feature extraction*

As the main building for feature removal, use convolutional neural network (CNN). Create the CNN model, which comprises of pooling layers after each set of convolutional layers for spatial downsampling.

2.3 *Model architecture*

Based on the parameters of the dataset and the difficulty of the challenge, design the CNN architecture. To find the model with the highest performance, experiment with several architectures, including regular CNNs, Residual Networks (ResNet), and VGG-16. By initializing the model with weights that have been previously trained on sizable image datasets and optimizing the network for emotion identification, you may also investigate transfer learning.

2.4 *Training*

Subsets for preparation, justification, and testing should be shaped from the dataset. By using incline tumble and backdropagnation, optimize the model's parameter using the training subset. In order to avoid over fitting, keep an eye on the justification subset and choose the top model based on validation results and pertain dropout technique by setting it to 0.25 which means unsystematic 25% of neuron in fully connected layer get deactivate or randomly dropped-out at each inform.

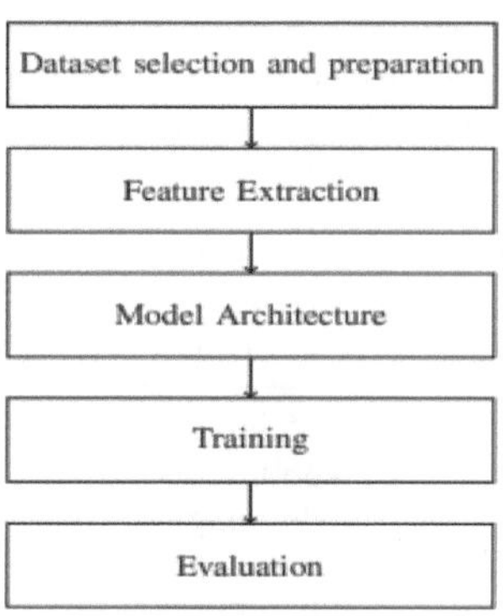

Figure 1. SEQ Figure * ARABIC 1 Facial Emotion Recognition system flowchart.

2.5 *Evaluation*

Examine the performance of the train model on the test separation. A few common assessment metrics are F1-score, recall, accuracy, and precision. Confusion matrix shed light on how well the model can distinguish between varieties of emotions.

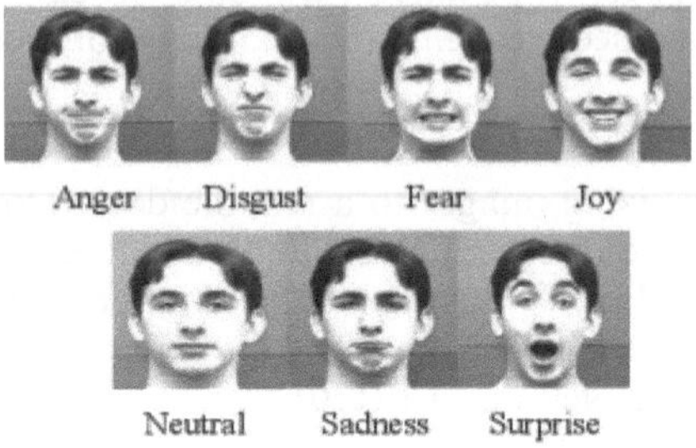

Figure 2. SEQ Figure * ARABIC 2 Face Expression Emotion Recognition.

2.6 *About dataset*

Used a dataset from a project called "Learn from Facial Expressions from an Image" [FER-2013]. The dataset use in this study has been taken from kaggle. Due to its enormous range of emotional expressions, this dataset is an helpful tool for study on the acknowledgment of human emotions. The anthology, in exacting, includes pictures of seven special emotional states, each of which was trapped to a diverse degree: 1) Angry: There are 958 photos in all that portray the feeling of anger, each of which capture the attribute Features of rage. An general examination of the recognition of this powerful and normally difficult emotion is made achievable by the dataset's insertion of a momentous number of angry Expressions. 2) Disgust: The collection includes 111 photos to stand for the emotion of disgust. These photos offer crucial insights into the detection of this meticulous and regularly complicated emotion, even though being quite few in association to other emotion. 3) Fear: The dataset include a sizable group of 1024 images that replicate fear, one of the mainly basic human emotions. The model's capability to identify fear from facial terminology may be carefully examined thanks to this largescale depiction. 4) Happy: There are munificently 1774 happy photos in the anthology. Since happiness is a collectively recognized emotion, the dataset's fortune is greatly improved by the large number of photograph that represent a wide multiplicity of happy Expressions. 5) Neutral: 1233 images make up the impartial touching state, which is normally used as a standard for assessment. Understanding how the model differentiates between neutral and other affecting states is made easier since to this big dataset of impartial terms. 6) Sad: This dataset part offers a careful exploration of the gratitude of one of the most commonly known emotions with 1247 photos illustrate unhappiness 7) Surprise: 831 photos show the convoluted and transient knowledge of surprise. Researchers can examine the model's ability to recognize startling changes in facial language sing this division of the dataset.

Table 1. SEQ Table * ARABIC 1 Distribution of Files Among Each Class Lablel of Emotions.

Class Label	Train Data	Test Data
Angry	3995 images	958 images
Disgust	436 images	111 images
Fear	4096 images	1024 images
Happy	7215 images	1774 images
Neutral	4965 images	1223 images
Sad	4830 images	1247 images
Surprise	3171 images	831 images

The data consists of 48x48 pixel grayscale images of faces. The faces have been mechanically registered so that the face is more or less centered and occupy about the same quantity of break in each image. The task is to categorize each face based on the emotion shown in the facial expression into one of seven categories (0=Angry, 1=Disgust, 2=Fear, 3=Happy, 4=Sad, 5=Surprise, 6=Neutral). The training set consists of 28,709 examples and the public test set consists of 3,589 examples.

3 RESULT

The operation of Convolution Neural Network (CNN) yield special outcome, with our model achieves an accuracy of 92.26%. CNNs are mostly skilled at capture involved hierarchical features within data, building them well- suitable for image-based tasks. The high accuracy underscore the success of CNNs in learning and extract related Features, showcasing their strength in behavior complex patterns. consequently, apply the VGG16 structural design also demonstrated outstanding success, even though with a slight minor accuracy of 87.36%. VGG16, known for its deep design and capability to separate fine details, proved efficient in this context. The a bit lower accuracy compare to CNN could trunk from nuanced difference in how the two architectures touch feature withdrawal and spatial hierarchies. The choice among CNN and VGG16 may center on factor such as computational efficiency, model interpretability, and precise task requirements.

Table 2. SEQ Table * ARABIC 2 Comparision of Models.

Model	Precision	Recall	FI- Score	Accuracy
VGG16	87.52%	86.25%	86.88%	87.36%
Resnet	70.63%	71.96%	71.29%	71.23%
Decision Tree	50.45%	53.10%	51.74%	52.75%
Proposed CNN	91.25%	91.54%	91.39%	92.26%

Though, when make use of Resnet18, the accuracy drop to 71.23%. Resnet architectures are well-known for addressing the vanishing gradient problem during the use of lasting connections. The somewhat lower accuracy suggest that, for this detailed task, the enduring connections may not have provided important benefits, or the model difficulty might not align optimally with the dataset's uniqueness. As well, employ a Decision Tree resulted in an accuracy of 52.75%. Decision Trees are interpretable models, but their appearance can be incomplete in capture involved patterns present in composite datasets. The lower accuracy highlights the substance of select models personalized to the inherent density and nuance of the data. The CNN and VGG16 models exhibit creditable accuracy, with the CNN somewhat outperforming VGG16. Though, Resnet18 and Decision Tree models confirmed lower

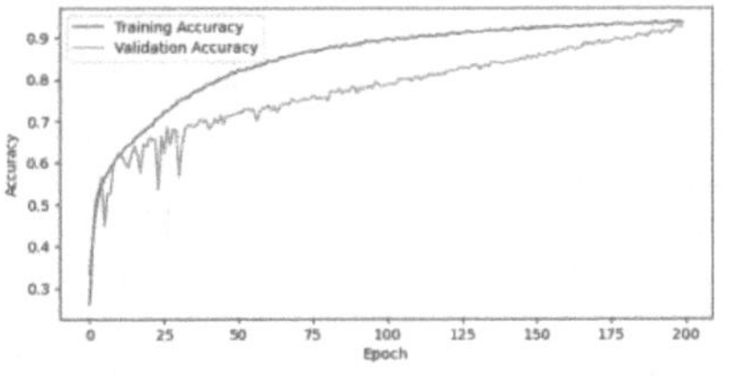

Figure 4. Training vs Validation Loss after 200 epochs using CNN.

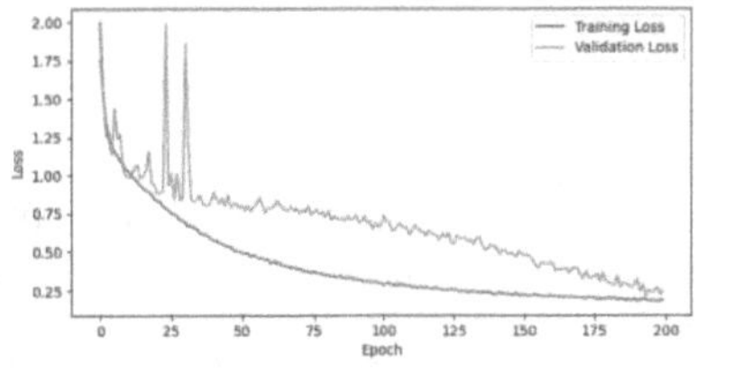

Figure 3. SEQ Figure * ARABIC 3 Training vs Validation Accuracy after 200 epochs using CNN.

accuracies, emphasize the implication of choosing an architecture allied with the dataset's uniqueness. These results provide expensive insight into the strengths and limitations of unlike models, guiding future consideration for model selection in comparable tasks. The CNN training process was conduct over 200 epochs.

During the measurement of our model, we attain a precision of 91.25% and a recall of 91.54%. The F1 score is a combined metric that blend precision and recall, only if a to the point evaluation of a model's generally presentation in binary arrangement scenario. compute by the formula F1 = 2 × (Precision × Recall) Precision + Recall it spans from 0 to 1, where important values indicate better model ability. By calculating F1-score by this formula, we attained 91.39%.

4 CONCLUSION

Sensitive and receptive human-computer communications are now possible thanks to the innovatory blow of deep learning techniques on the ground of human emotion recognition. In this study, we investigate the use of deep learning models for facial terminology and progress video Sequences-based emotion recognition. An Emotion Recognition Using Machine Learning project's winding up should include a summary of the key conclusion and assistance of the work, as well as suggestion for next Research. The authors should talk about the project's goals and how well they were achieving in the termination. The project's blow on the field of emotion recognition as well as probable uses for the models produced should also be enclosed by the writers. The model attain an accuracy of 92.28% for the experiment Conducted on different constraint.The CNN model, fine-tune with Adam optimizer and optimized hyper parameter, achieve a remarkable final accuracy of 92%. Especially, the preliminary outcome with a learning rate of 0.00001 resulted in an accuracy of 87.32%, emphasize the model's sensitivity to hyper limitation adjustment. A planned adjustment of the learning rate to 0.001 led to a important improvement, resulting in an inspiring accuracy of 92.26%. It is critical to highlight that these outcome were completely derived from the FER-2013 dataset, importance the model's efficiency in learning and generalize patterns within this explicit dataset. The obtain accuracy of 92.26% underscore the model's expertise in extract significant features from facial terminology, representing its robust performance and prospective applicability in real-world scenario where particular emotion recognition is fundamental.

REFERENCES

[1] Alslaity, Alaa, and Rita Orji. (2024). Machine learning techniques for emotion detection and sentiment analysis: current state, challenges, and future directions. *Behaviour & Information Technology* (Taylor & Francis) 43: 139–164.

[2] Bai, Zhongli, Junjie Liu, Fazheng Hou, Yirui Chen, Meiyi Cheng, Zemin Mao, Yu Song, and Qiang Gao. (2023). Emotion recognition with residual network driven by spatial-frequency characteristics of EEG recorded from hearing-impaired adults in response to video clips. *Computers in Biology and Medicine* (Elsevier) 152: 106344.

[3] Bakariya, Brijesh, Arshdeep Singh, Harmanpreet Singh, Pankaj Raju, Rohit Rajpoot, and Krishna Kumar Mohbey. (2024). Facial emotion recognition and music recommendation system using CNN-based deep learning techniques. *Evolving Systems* (Springer) 15: 641–658.

[4] Garg, Muskan, and Chandni Saxena. (2024). Emotion detection from text data using machine learning for human behavior analysis. In *Computational Intelligence Methods for Sentiment Analysis in Natural Language Processing Applications*, 129–144. Elsevier.

[5] Geetha, A. V., T. Mala, D. Priyanka, and E. Uma. (2024). Multimodal emotion recognition with deep learning: advancements, challenges, and future directions. *Information Fusion* (Elsevier) 105: 102218.

[6] Hangaragi, Shivalila, Tripty Singh, and N. Neelima. (2023). Face detection and recognition using face mesh and deep neural network. *Procedia Computer Science* (Elsevier) 218: 741–749.

[7] Hossain, Sanoar, Saiyed Umer, Ranjeet Kr Rout, and M. Tanveer. 2023. Fine-grained image analysis for facial expression recognition using deep convolutional neural networks with bilinear pooling. *Applied Soft Computing* (Elsevier) 134: 109997.

[8] Houssein, Essam H., Asmaa Hammad, and Abdelmgeid A. Ali. (2022). Human emotion recognition from EEG-based brain–computer interface using machine learning: a comprehensive review. *Neural Computing and Applications* (Springer) 34: 12527–12557.

[9] Indolia, Sakshi, Swati Nigam, and Rajiv Singh. (2024). Deep learning for emotion recognition using physiological signals. In *Data Fusion Techniques and Applications for Smart Healthcare*, 385–401. Elsevier.

[10] Iyer, Abhishek, Srimit Sritik Das, Reva Teotia, Shishir Maheshwari, and Rishi Raj Sharma. (2023). CNN and LSTM based ensemble learning for human emotion recognition using EEG recordings. *Multimedia Tools and Applications* (Springer) 82: 4883–4896.

[11] Khattak, Asad, Muhammad Zubair Asghar, Mushtaq Ali, and Ulfat Batool. (2022). An efficient deep learning technique for facial emotion recognition. *Multimedia Tools and Applications* (Springer) 1–35.

[12] Khattak, Asad, Muhammad Zubair Asghar, Mushtaq Ali, and Ulfat Batool. (2022). An efficient deep learning technique for facial emotion recognition. *Multimedia Tools and Applications* (Springer) 1–35.

[13] Maithri, M., U. Raghavendra, Anjan Gudigar, Jyothi Samanth, Prabal Datta Barua, Murugappan Murugappan, Yashas Chakole, and U. Rajendra Acharya. (2022). Automated emotion recognition: Current trends and future perspectives. *Computer Methods and Programs in Biomedicine* (Elsevier) 215: 106646.

[14] Meena, Gaurav, Krishna Kumar Mohbey, Ajay Indian, Mohammad Zubair Khan, and Sunil Kumar. (2024). Identifying emotions from facial expressions using a deep convolutional neural network-based approach. *Multimedia Tools and Applications* (Springer) 83: 15711–15732.

[15] Sarvakar, Ketan, R. Senkamalavalli, S. Raghavendra, J. Santosh Kumar, R. Manjunath, and Sushma Jaiswal. (2023). Facial emotion recognition using convolutional neural networks. *Materials Today: Proceedings* (Elsevier) 80: 3560–3564.

[16] Sowmya, B., Sini Anna Alex, Anita Kanavalli, S. Supreeth, G. Shruthi, S. Rohith, and others. (2024). Machine learning model for emotion detection and recognition using an enhanced convolutional neural network. *Journal of Integrated Science and Technology* 12: 786–786.

[17] Tiwari, Rudra, Ayushi Prajapati, Saravana Chandran, Divyansh Agrawal, Akhtar Rasool, and Abhishek Jadhav. (2024). Emotion detection through human verbal expression using Deep learning techniques. *2024 IEEE International Students' Conference on Electrical, Electronics and Computer Science (SCEECS)*. 1–7.

[18] Umer, Saiyed, Ranjeet Kumar Rout, Chiara Pero, and Michele Nappi. (2022). Facial expression recognition with trade-offs between data augmentation and deep learning features. *Journal of Ambient Intelligence and Humanized Computing* (Springer) 1–15.

Computational Methods in Science and Technology Volume 1 – Kaur et al. (Eds)

Account verification: A method to detect fake accounts on social media

Mohd Irshad Kasana
Sri Sai College of Engineering and Technology, Badhani, Pathankot, Punjab

ABSTRACT: The objective of this paper is to propose a prototype for an Account Verification (AV) model to detect and remove fake profiles from social media networking sites. The approach involves the support of the regional government and the use of a Unique Identification Document (UID) for user verification. A verified user will be identified with a yellow-colored checkmark on their profile, which will be different from that of a celebrity. The paper presents an experiment conducted with 100 participants over Instagram, which resulted in an accuracy of 90.9% in identifying fake profiles. Additionally, another experiment was conducted with more than 1000 participants, and 73.5% strongly agreed that UID is the best technique to reduce fake accounts on social media. The proposed AV model has the potential to profoundly impact a nation's human development index and lead to a greater reduction in social media issues and crimes. Furthermore, the AV model will lead to quality services in social media marketing to a great extent. The novelty of this paper lies in the proposed AV model, which leverages UID from the domestic government's official web portal to verify users on social media platforms, thereby providing a powerful and manageable approach to detect and remove fake profiles without losing user credentials to any third party.

Keywords: Social Media, Fake Accounts, Authentication, Account Verification, Government Web Portal, Unique Identification

1 INTRODUCTION

In recent years, there has been a rise in social media engagement in the lives of internet users. According to recent statistics, there are 4.20 billion active social media users worldwide. This accounts to approximately 54.73% of the world's population (Petrosyan 2024). Some of the most commonly used social media applications are Instagram, Facebook Twitter, LinkedIn, etc. These types of applications mainly provide a platform for its users to connect with people. which serves as a channel to share and make use of information across the internet (Platforms 2021). Social media helps in making online social bonds and eases them in keeping associations with their friends and family. However, there exist numerous fake or duplicate accounts on such social media platforms and around 16% of all Facebook accounts are fake or duplicate (Armstrong 2020). A fake profile is where someone is pretending to be someone or something that doesn't exist or doesn't belong to him, these accounts cover the accounts that are for made-up people or fake people, pets, organizations, or celebrities ("Report fake Facebook profiles" 2021). The owner of a fake profile signs up for these false accounts on social media. fake accounts can be used to spread fake news, stir political quarrels, and promote disinformation about famous people and brands which can mislead customers With online fraud, the spreading of false information or fake news is increasing at a rapid pace through online social media networks (Han Luo, Ying Cui and Meng Cai 2021). Thus, the detection of fake accounts from social media platforms are essential. Users who create fake profiles on social media platforms can have negative impacts on society by manipulating people into false beliefs and sharing harmful links. as fake users

 DOI: 10.1201/9781003501244-73

can spread misinformation and manipulate public opinion. The opinion and the attitudes of both the teens and their parents regarding the use of fake social networking accounts and bots to sell or influence products in the United States depicts that about 40% of parents and 18% of teens were strongly concerned with such practices (Dixon 2023).

The main objective of this research paper is to introduce the User Account Verification model as a solution to detect fake or duplicate accounts on social media platforms. The model proposes a voluntary user verification process based on UID from the user's domestic government's official authentication web portal, which will distinguish verified users from notable public figures, celebrities, or global brand accounts. The goal is to promote trust and credibility on social media platforms by eliminating fake accounts that engage in malicious and fraudulent activities.

Previously Social Media Fake Profile Detection Technique Based on Attribute Estimation and Content Analysis Method was a proposed methodology for identifying fake profiles on social media platforms using data mining techniques. To train the learning algorithms, the authors used synthetic datasets obtained from various social media platforms, including Twitter, Facebook, Academia, and LinkedIn. The authors also used APIs to extract data from these platforms. The authors used various machine learning algorithms, such as Random Forest and Deep Convolutional Neural Network (CNN), to detect fake profiles. The proposed methodology was demonstrated using a system that contained various components and processing aspects, including a track record of the work involved (Kadam and Patidar 2020).

Detecting Fake Users on Social Media with a Graph Database this uses three different datasets for identifying fake Twitter users. The first dataset is a combined dataset of fake and real Twitter users from five different datasets collected from previous studies. The second dataset is a friendship network of a school class from 1880/81, and the third dataset is Twitter users obtained using an API crawler. The researchers used centrality measures to extract features and a random forest classifier for classification and feature selection methods to identify the most important features for classification. The paper provides detailed information about the datasets, feature extraction, classification, and feature evaluation methods used in the study (Zhao and Weber 2021).

Another study Detection of Fake Social Networking Accounts describes the process of feature extraction and its role in reducing the dimensionality of a dataset. It also discusses the use of the Support Vector Machine (SVM) algorithm for the classification of fake and real accounts on Twitter. The process of using SVM involves selecting a reduced feature subset, splitting the data into testing and training sets, training the model using SVM algorithm, predicting the output using the trained model, and calculating the accuracy of the predicted output. Additionally, a hybrid classification algorithm is proposed to improve the classification accuracy by using SVM decision values as inputs for another classification algorithm (Kadam and Surse 2021).

On the other hand, several times nations tried to tighten their grip on social media to reduce misleading and fake information. Since 2021, the government has issued orders to take down multiple accounts on Twitter, Facebook, and YouTube due to concerns about a potential threat to national security and sovereignty (Deutsche Welle (DW) 2022), but the government at presently has The Indian Penal Code and Information Technology Act provide provisions to protect victims of impersonation and fraudulent use of personal information. Section 66D of the IT Act punishes those who cheat others by impersonating them through communication devices or computer resources with imprisonment of up to three years and a fine of up to one lakh rupees. Section 416 of the IPC deals with cheating by personation, where a person pretends to be someone else, and forgery of an electronic record for cheating is punishable under Section 468 of the IPC (Kalra and Ghosh 2021).

1.1 *Research gap*

Efficient algorithms for detecting fake social media accounts require user data and cannot always detect fake profiles early on. Current machine learning algorithms used for detection have constraints and need to be flexible and dynamic to adapt to changes in technology, marketing, and human behavior toward social media (Mallampeta and Shaik 2021). Real-time user authentication is necessary as some real users may act as fake, while some fake users may appear real. This process can be complex and energy-consuming. Although these algorithms work fine but they didn't detect fake profiles in the early stages and require a huge dataset for training, so there is no surety about the accuracy of the training dataset (Chakraborty *et al.* 2022). Meanwhile, the government doesn't have any such e-governance mechanisms or tools that will determine fake profiles on social networking sites yet (Violino 2022).

This model will overcome all those limitations by AV based on UID from their government's official user authentication web portal without losing their credentials to any third-party application. Accounts get verified as a checkmark is displayed on the profile which ultimately assures trust to others and no multiple profiles can be created with the same UID without any specific reason like one may have a personal profile and few business profiles but not several personal profiles also this assures that the person behind the profile is a real citizen This will distinguish the real from fake profiles in the early stages also this doesn't require algorithms that will study the activity pattern and analysis based on them although the combination of this model with previous ML models will enhance the efficiency and productivity of idea of detection and removal of fake profiles to a greater extent, It utilizes Automatic Verification (AV) based on the user's Unique Identifier (UID) from their domestic government's official user authentication web portal, without compromising the user's credentials to third-party applications. This process results in a yellow checkmark displayed on the user's profile, Additionally, the UID-based verification process ensures that multiple profiles cannot be created with the same UID without a specific reason, such as having a personal profile and a few business profiles. This method also guarantees that the person behind the profile is a real citizen, shows both ordinary and celebrities as distinct but real, and helps in determining fake profiles in the early stages. The implementation of this model eliminates the need for algorithms that study activity patterns and analyses based on them. However, combining this model with previous Machine Learning (ML) models can enhance the efficiency and productivity of fake profile detection and removal.

Table 1. A comparative analysis in a tabular manner for the AV model with previous approaches.

Previous study	Proposed AV Model
Researchers have developed algorithms for detecting fake accounts, but they require data and may notwork in early-stage detections.	Real-time authentication of users is necessary as this will automatically prevent users from creating bot profiles.
Machine learning algorithms used in such models have constraints and need to be highly flexible and dynamic to adapt to changes in social media patterns and user behavior.	User Account Verification model, in which Verification will be done automatically based on their UID from their government's official portal without losing their credentials to any third-party application.

This paper introduces a different approach called the User Account Verification model in which users over any social media platform with one's own will can verify themselves as an ordinary citizen which will then reflect on their profile as a yellow checkmark, it will distinguish them from the presence of a notable public Figure, celebrity or global brand accounts.

2 METHODOLOGY

This section elaborates upon the description of materials, tools, and working of the User's Account Verification (AV) technique. All the dataset presented here is created by using online illustration and drawing tools like Adobe Xd, canvas, etc and these are dummy images and illustration designs. In the proposed technique, a verified profile signifies an account that has been confirmed as authentic by the domestic government authorities. These can be used by any person, local, professional, celebrity, politician, or brand wherein, users can verify their accounts at their own will. AV is predominantly segmented into two parts: Local Govt web portal and Social media application for the authentication.

The government web portal is responsible for authenticating the user and verifying the user's account based on their unique identification document. On this web page, some essential fields are required to be filled like UID, OTP, etc. Once, the user has been verified by the concerned government, the user can sign up on any social media platform. In the social media application phase, after making the first successful sign-up, a verified account profile will have a checkmark icon followed by a username on its profile. This checkmark will distinguish the user from the one that confirms the presence of a notable public figure, celebrity, or global brand account. The design outline of the AV The model is shown in Figure 1.

Figure 1. A design approach to illustrate the outline working of the Authentication technique used in Account verification.

The various steps in the proposed AV technique are as follows:

(1) There will be a presence of an unverified or cross-mark icon followed by the username inside the user profile on a social media application. By clicking on this icon, the application first tries to retrieve its location or the country name from the shared preferences. In case no records are found, then a pop dialog will appear on the screen asking for the county name or to allow access to GPS for its current location. Once the information is entered and the send button is clicked, the user will be directed to his country's web portal along with username, userId, application name, display name, and verified status as arguments to the parameters of the intent( ) function in this transaction.
(2) At the web application end, a citizen's UID and its relevant data are either pre-stored into the database or fetched from the government's official unique identification websites. Later, the entries of the database are compared with the ones that the user enters on the verification page under the verification menu in the web portal. Figure 2 illustrates the design of the verification menu.
(3) Among all the arguments of the received intent, the username, application name, and verified status will instinctively display at the beginning of the verification form on the web page, and then the user will be asked to enter the specific type of UID number like social security number, Adhar card, etc. once the UID is entered it will be compared with the actual information which is stored in the database, if no match is found, then user will be denied any further proceeding. In case of a match, the user will be asked to try again later otherwise the user will receive an auto-generated OTP on their registered

mobile number(linked with UID) as soon as the user fills the correct OTP into the required input field, a short message will appear below it: "Your actual full name _________ and encrypted UID will be sent to _____________application" The users would be required to agree with the term and policy conditions of the website and at the end the has to click on the "Done" button to finish the verification process from this end.

(4) After the user has verified him, a feedback transaction will start towards the application along with all the parameters mentioned in steps 1 and 3. At this point, the parameters will also hold values like the display name as alias name, verified status as true or false, and encrypted UID string.

(5) At the application end, a yellow-colored icon will appear on the user's profile followed by its username. The application must take care that this checkmark must distinguish an ordinary user from a notable public figure or global brand account holder. Further, the user's user alias will be formed from its actual name and it should be visible somewhere in the bio (or in any similar component present in the user's profile), however, the user will be provided with the choice to choose any name style for alias name among all the different possible styles constructed by the full name e.g. the name with or without last name, or caste name, etc without any alteration in the core elements of the full name. In the end, an acknowledgment message will be sent back to the website as a successfully finished signal bit to let the web page store all cached data of the underway user and update the same in its database.

(6) In addition, some live data programming modules like Android LiveData() will be implemented to detect, listen, and observe the changes in real-time made by the user or any automation changes at twain ends. Such changes will be reflected synchronously at both social media and web portal ends. This would maintain the consistency of the application's database. The changes may include, a change in username, account activation or deactivation status, UID changed or removed, it can also include the report count against an account, the social status of a user in the social media community, etc. The report count would make it easier for the victims or any other user to make a complaint against a user who is performing unlawful exercises or executing criminal activities.

2.1 *Other features*

2.1.1 *Social meter*

The social meter is an additional feature on the government portal, wherein the social report of the user is based on authenticity, risks, norms violation, and social status.

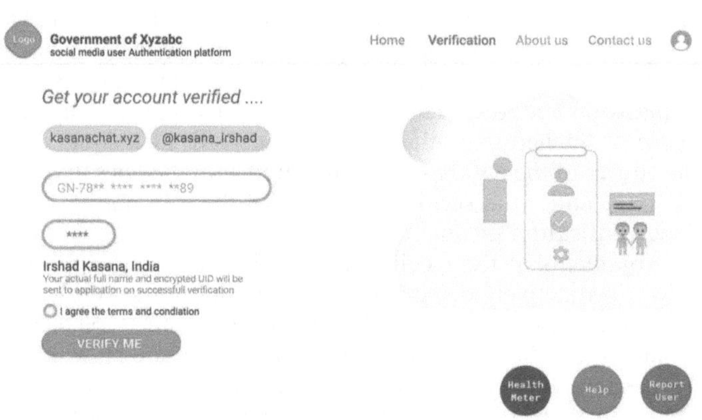

Figure 2. A design approach to illustrate the social meter.

The principal factors affecting the social module algorithm are the number of reports made against a particular user, activities like how often the user changes their username, how likely a user is positively influencing the other users, whether a user had ever tried or used any copyrighted content, the quality Engagement time invested by the user, and the number of accounts associated with single UID, etc. All these factors collectively will determine the level of the social status of a user in cloud communities.

3 RESULTS AND DISCUSSIONS

3.1 *Experiment*

In this section, the results of two experiments are discussed and compared with previous literature works results. Instagram is a widely used social media application responsible for creating and sharing photos, and videos and making stories(Platforms 2021). etc.

Prototype of Kasanachatapplication (google drive 2021). The prototype is a user interface and these UI's are designed to illustrate how differently verified profiles. In the study, three user interfaces of the user's profile prototyped in the kasanachat application have been shown in Figure 3. The profile's icons were classified as blue checkmarks depicting verified as a notable public Figure yellow checkmark means verified as an ordinary citizen cross mark as an unverified user.

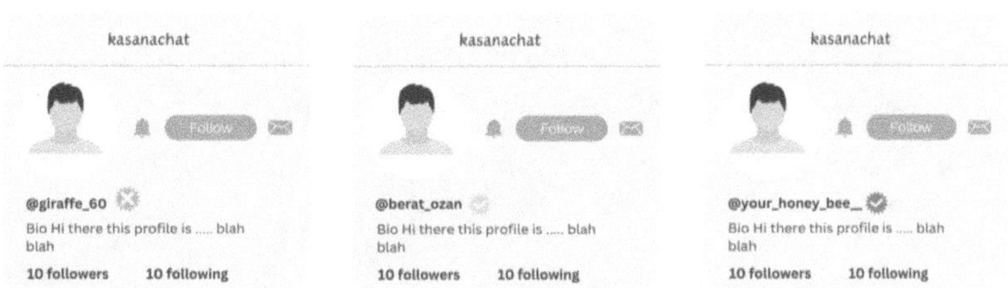

Figure 3. UI from kasanachat representing the differently verified and unverified profiles.

3.2 *Dataset collection*

The students were engaged in a social media-based game on Instagram consisting of three rounds, In the first round a random quiz contest was organized, in the second round a story was uploaded where the UI of the mentioned profiles shown in Figure 3 of kasanachat was placed and in the third round users were asked to identify a song. The first and third rounds were added to make sure that the user's conscious mind doesn't change the subconscious mind's valid decision. Whereas, the second round which was wrapped inbetween two rounds was of significant importance. In the second round, the tags of the profiles corresponding to the three aforementioned profiles were added, clicking on which redirected the participant to the corresponding Instagram profiles. Also, it was mentioned in the story that these three kasanachat profiles are differently verified and they do have their Instagram id as well so do follow and interact with them. After the experiment, it retrieved the analytics from the professional dashboard of Instagram that showed the information and metrics related to organic, posts, and stories, these insight features are freely available for business account holders on Instagram (google drive 2021).

3.3 *Analytics and calculations*

The total number of different accounts reached on profiles in a given no of days describes the approach and willingness of the different users to promote the authentication mechanism over social media. The graphs through the professional dashboard of three Instagram profiles are shown in Figure 4.

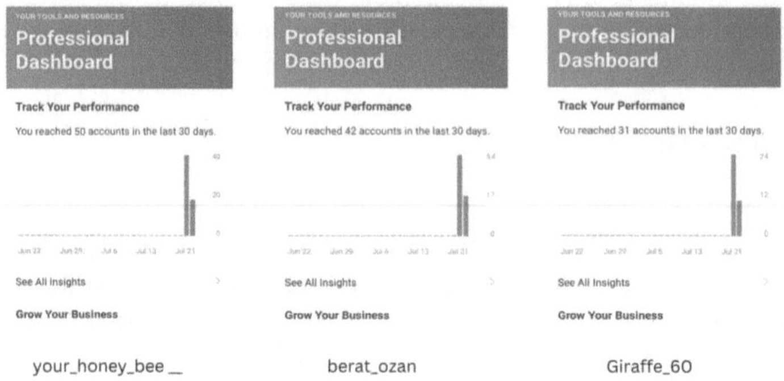

Figure 4. Professional dashboard of three Instagram profiles used in the experimental study your_honey_be ___, berat_ozan, Giraffe_60.

Below data. shows the percentage of the total number of accounts reached on three profiles understudy.

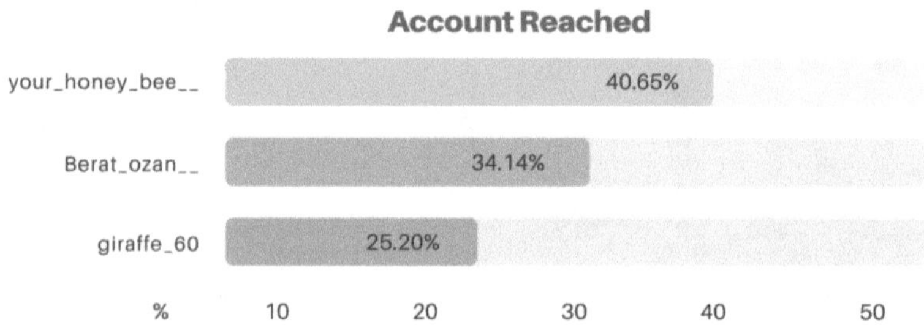

Figure 5. Overall reachability of the three Profiles Giraffe_60, berat_ozan andyour_honey_bee __ .

It was also observed that some users following the three accounts were common, so set theory was applied to extract exact numeric values.

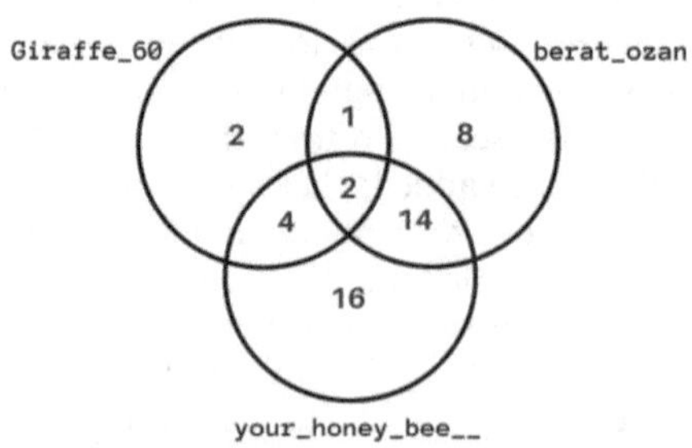

Figure 6. Details of the reachability of the three profiles understudy. n(giraffe_60) = 4, n (berat_ozan__) = 14, n(your_honey_bee__) = 16.

Accuracy = (Total followers)-(sum of all followers common in at least any of two)*100/ (Total no of followers)

Accuracy = {(Giraffe_60 ∪berat_ozan__ ∪your_honey_bee__) – (giraffe_60 ∩ berat_ozan__ ∪ giraffe_60 ∩ your_honey_bee__ ∪ giraffe_60 ∩ berat_ozan__ ∩ your_honey_bee__) * 100} /{(Giraffe_60 ∪berat_ozan__ ∪your_honey_bee__)}
Accuracy = ((22-2)*100)/22 = 90.9%

Follow vs. reach ratio

The numbers of tapes are taken from the clicks on the mention profile username in the story ("View Insights on Your Instagram Posts" 2021). {55 + 45 + 33 = 133}

From (Kaushik and Newberry 2020) it can determine various formulas therefore approach is directly proportional to ∝ follow count and reaches count.

And is inversely proportional 1/ ∝ the difference between reach and follow counts.

Approach = (Follow * Reach)/ (Reach – Follow)

So, for
Giraffe_60 = (4*31) / (31-4) = 124/27 = 4.59
Berat_ozan__ = (14*42) / (42-14) = 588/28 = 21.00
Your_honey_bee__ = (16*50) / (50-16) = 800/34 = 23.52
Approach rate for seven days can be determined as

$$Approach\ rate = \frac{\left\{\left[\frac{follow*reach}{reach-follow}\right]^2_{d7}\right\} - \left\{\left[\frac{follow*reach}{reach-follow}\right]^2_{1d}\right\}}{2*7\ days}$$

It was observed that those who followed both the verified profiles and any one of them were found to be the users who were enthusiasts towards authenticated users only. On the other hand, those who followed the unverified profile are the users who are not very particular about the authentication factor or maybe failed to distinguish between these profiles. Thus, the association or engagement of other users declined and resulted in isolated profiles. Further, an online survey related to the AV model was also conducted with over more than 1000 people across different cities in India to verify the need, feasibility, and effectiveness of the mode (Kasana 2021).

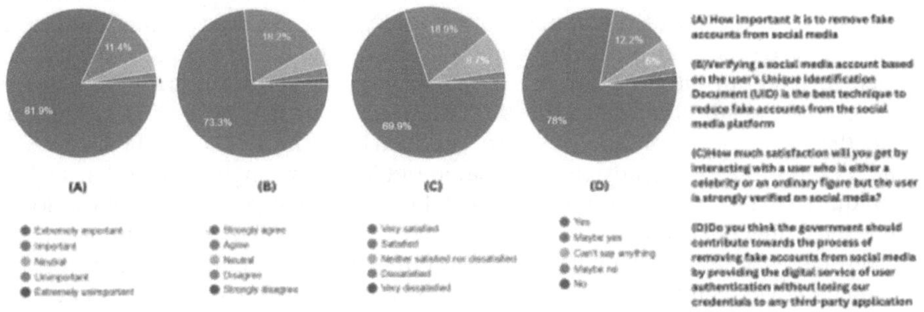

Figure 7. Survey results presented in Pie chart.

The results obtained are shown in Figure 7. From this, it is concluded that 81.8% agreed that it is extremely important to remove fake accounts from social media.

Different values like In the first pie chart: 81.8% strongly showed and 11.4% moderately evince In the second pie chart: 73.5% strongly and 18.1% moderately confirm that they wish user authentication should be carried out based on their UID and so on. the accuracy of the model is determined by the number of true positives and true negatives, but false positives and false negatives can also affect the overall accuracy. Therefore, it is important to choose an appropriate threshold value to balance the trade-off between false positives and false negatives. Calculating the fake account detection model involves:

- TP: the number of fake accounts correctly classified as fake.
- FP: the number of fake accounts incorrectly classified as legitimate.
- FN: the number of legitimate accounts incorrectly classified as fake.
- TN: the number of legitimate accounts correctly classified.
- Accuracy (ACC) = total number of records correctly classified.
- Accuracy- It is calculated using the following formula
  = (57 + 52) / (57 + 52 + 8 + 3) = 90.8 %
- Comparatively in our model accuracy comes out to 90.9%.

Two important aspects to consider are the large amount of data required for the training dataset and the time needed to access this data. Additionally, the model does not take into account what happens if the number of fake accounts incorrectly classified as legitimate is higher, which can significantly lower the accuracy.

In the case of detecting fake accounts in general, the accuracy is directly proportional to the number of accounts correctly identified. This methodology uses the learning algorithm SVM, which requires a training dataset (Sheikhi 2020).

On the other hand, the Account Verification (AV) model does not require a lot of data and can provide results in the early stages, without the need for extensive training datasets. Integrating machine-learning models with the AV model can yield good results, but individual machine-learning models are not sufficient on their own.

4 CONCLUSION

The paper proposes a Unique Identification Documents (UID) based Account verification model for the detection and isolation of fake accounts from social media. Users over any social media platform can verify themselves voluntarily anytime from their domestic government's official web portal using UID. Further, various automation mechanisms can be installed on the website to carry out verification steps after filling the input fields of the verification form along with OTP. The verified user will be viewed as a verified citizen with a yellow-colored checkmark icon on his profile which will be different from the celebrities or notable public Figures. Also, the real name of the user will be taken as an alias name and will be mentioned inside the user profile by the application. The effectiveness of the model was determined through an online game of 100 participants over Instagram where the model could detect and isolate fake users with 90% accuracy. Further, an online Google survey related to the AV model concept was also conducted on more than 1000 people across different cities in India. The study showed that about 73.5% strongly agreed that UID is the best technique to reduce fake accounts from social media. The proposed model would help in effectively detecting fake accounts that are involved in the spread of misinformation, spying, cybercrimes, etc. The model would further be an upturn for the human development index since it would help in gaining emotional satisfaction among users, cause quality engagement, and promote a healthy social media environment. AV model will also lessen the burden and overhead of security and privacy departments of social media industries as the lesser the fake or bot profiles, lesser will be the complaints and reports. However, a UID-based authentication model would require local government and social media companies to agree with policies jointly.

REFERENCES

Armstrong, Martin. (2020). "Infographic: 16% of All Facebook Accounts Are Fake or Duplicates." Statista. https://www.statista.com/chart/20685/duplicate-and-false-facebook-accounts/.

Chakraborty, Partha, Mahim Musharof Shazan, Mahamudul Nahid, Md. Kaysar Ahmed, and Prince Chandra Talukder. (2022). "Fake profile detection using machine learning techniques." *Journal of Computer and Communications* **10** (Oct). 10.4236/jcc.2022.1010006.

Deutsche Welle (DW). (2022). "India tightens grip on social media – DW – 10/29/2022." October 29, 2022. https://www.dw.com/en/india-tightens-grip-on-social-media-with-new-tech-rules/a-63592973.

Dixon, Stacy J. 2023. "Concerns among the use of fake social media accounts and bots U.S. 2018 l Statistic." Statista. https://www.statista.com/statistics/881017/fake-social-media-accounts-bots-influencing-selling-purchases-u sa/.

google drive. (2021). "Experiment data for research article "*Account Verification: A Method to Detect Fake Accounts on Social Media.*" https://drive.google.com/drive/folders/1dWLZxDMigC0zZY0D-OgdzkWPzJNpx3Hk?usp=sharing.

Han Luo, Ying Cui, and Meng Cai. (2021). "Spread of misinformation in social networks: analysis based on weibo tweets." *Security and Communication Networks* 2021, no. 7999760 (Dec). https://www.hindawi.com/journals/scn/2021/7999760/.

Kadam, Nitika, and Harish Patidar. (2020). "Social media fake profile detection technique based on attribute estimation and content analysis method." *International Journal of Recent Technology and Engineering (IJRTE)* 8, no. 6 (March). 10.35940/ijrte.F8414.038620.

Kadam, Onkar, and Nilesh Surse. (2021). "Detection of fake social network account." *International journal of advance scientific research and engineering trends* **6**, no. 4 (April). 10.51319/2456-0774.2021.4.0013.

Kalra, Arpit, and Ragini Ghosh. (2021). "Legal action against fake account on social media – social media – India." Mondaq. https://www.mondaq.com/india/social-media/1109666/legal-action-against-fake-account-on-social-media.

Kasana, Mohd I. (2021). "Survey data for the research article "Account Verification: a method to detect fake accounts on social media." Google. https://drive.google.com/drive/folders/1KzPd0aMxe5K7dz7ewtTTQrllT0upziMp.

Kaushik, Avinash, and Christina Newberry. (2020). "17 Social media metrics you need to track in 2024 [BENCHMARKS]." Hootsuite Blog. https://blog.hootsuite.com/social-media-metrics/.

Petrosyan, Ani. (2024). "Internet and social media users in the world 2024." *Statista*. https://www.statista.com/statistics/617136/digital-population-worldwide/.

Platforms, Meta. (2021). "Share photos and videos on instagram." instagram help center. https://help.instagram.com/479832029758079/?helpref=hc_fnav.

"Report fake Facebook profiles." (2021). Facebook. https://www.facebook.com/help/306643639690823.

Sheikhi, Saeid. 2020. "An efficient method for detection of fake accounts on the instagram platform." *International Information and Engineering Technology Association* **34** (Aug). 10.18280/ria.340407.

"View Insights on Your Instagram Posts." 2021. Instagram Help Center. https://help.instagram.com/553860108359435.

Violino, Bob. (2022). "LinkedIn has a fake account problem it's trying to fix. Real users are part of the solution." *CNBC*, December 10, 2022. https://www.cnbc.com/2022/12/10/not-just-twitter-linkedin-has-fake-account-problem-its-trying-to-fix.html.

Zhao, Yichun, and Jens Weber. (2021). "Detecting fake users on social media with a graph database." *The Arbutus Review* **12** (Oct). https://doi.org/10.18357/tar121202120027.

Computational Methods in Science and Technology Volume 1 – Kaur et al. (Eds)
© 2025 the Author(s), ISBN: 978-1-032-81760-6

Macro-economic trends prediction using machine learning

Divyansh Jain
Amity Institute of Information Technology, Noida, India

ABSTRACT: This project increases the precision of projections for GDP growth (output growth), inflation, and unemployment index by use of Python and machine learning techniques to predict macroeconomic changes. The process flow includes data collection, features engineering, model training using mostly neural networks, linear regression and time series where the ARIMA or Autoregressive Integrated Moving Average model can be involved. The project emphasizes on transparency and the objective is to enable entrepreneurs, investors, and governance to obtain crucial information. Indicators are used to increase accuracy of forecasting and ensure a prompt reaction if circumstances change in the economy. The codebase is made both extensible and scalable when it is coupled with the popular Python libraries like scikit-learn, pandas, and matplotlib. The study marks a step towards macroeconomic forecasting as it presents a framework useful for incorporating different types of indicators that could support the further adjustment of parameters with new data.

1 INTRODUCTION

This research is focused on embodying constant and without delays macroeconomic forecasting with the help of Python programming and machine learning methods. Important markers to support the formation of efficient economic strategies, investment and planning as GDP growth, inflation and unemployment figures are the key indicators. In regard to the project, an integrated model involves the application of an artificial neural network, time series analysis, and regression model.

The project maps out its phases: data collection, followed by feature engineering, preprocessing, selection of desirable model, entertainment, evaluation, and deployment. Figuring out the features which represent the sterilization of past economic information, in terms of its short-term perturbations and long-term trends, is of utmost importance. To bring forth the element of this emphases on interpretability and adaptability the plan seeks to present reports that are clear yet flexible with respect to changes in the economy, which shall ultimately facilitate stakeholders to take corrective actions.

Ultimately, the aim of the research is to improve the comprehension of business cycle among decision makers like policy makers, investors, and so on, to boost the much needed informed and effective decision-making process. Via prediction routine and analysis of historical data the project impacts macroeconomic prediction immensely. The production of its results would likely be revolutionary with regards to the macroeconomic factors decisions which are complex processes, creating a more detailed perception of the effects of macroeconomic trends and factors.

1.1 *Objectives*

1.1.1 *Forecasting the trend of macroeconomics*

Analysing data, policies, and events via international currency is critical for determining macroeconomic trends for the purpose of possible economic adjustments.

 DOI: 10.1201/9781003501244-74

1.1.2 *Identifying various factors that affect it*

Identifying the factors, that play up in a macro- economic trend. Applying standardized weightage algorithm to get more precise and efficient analyst results.

1.1.3 *Collaborating multiple datasets*

Using the data and results of different sources in a collaborative manner to reach, a more detailed debate.

1.1.4 *Summarize the trend result*

Identifying all the factors which influence the existing trend in a positive direction or in a negative trend and recommending necessary changes to the trend.

1.1.5 *Deploying algorithm on multiple datasets*

To achieve this, running the algorithm across all datasets and combining all factors as a single key for prediction should be a solution.

2 LITERATURE REVIEW

Macroeconomic trend prediction has gained significant importance among the various functions performed at the economic level such as economic analysis, policy making and taking of decisions. The scholars have been e investigating various methodologies and algorithms that will be invoked when forecasting the major macroeconomic indicators such as GDP, national income, unemployment, and inflation over the years. Namely, the most popular and widely employed method is the Autoregressive Integrated Moving Average (ARIMA) methodology that is used in time series forecasting in economic analysis. This literature review aims at highlighting existing research works on ARIMA algorithm user in forecasting trend in macroeconomic environment and to pinpoint significant aspects and patterns in the area.

[1] The paper focuses on the analysis of untraditional datasets like retail and wholesale payments with an intention of developing instantaneous models by means of machine learning technologies to estimate macroeconomic indicators in perfect real-time. [2] It shows that this kind of models are able to increase the nowcasting accuracy, particularly during the periods in which these situations occur. [3] The paper provides a machine learning model for GDP prediction with the volatility having it as its weight which suggests the effectiveness in the case of accurate prediction of GDP on the offing as well as in terms of support for economic policy making. [4] Nontriviality, accordingly, becomes the ‘breaking point’. The best standard Occurrence model remains as the standard regularization technique.

[5] The theme of chaos and VAR model in the economics theory: - this implies that then make sure they work together. So, random variables are hard to be utilized sufficiently for new or useful discoveries. [6] These methods include, soft-thresholding, the elastic net, and the random forest, respectively. Naive out nevertheless, machine learning approaches were also used to verify the validity of the predictions.

[7] The presented paper is aimed to show how the signature method is been used to forecast the trends. But it does not state machine learning utilization in the prediction of macro-economic trends. KBest of RF performs best in comparison with RF and SVM in adding value to GDP prediction. [8] The forecast precision of GDP for 186 countries reached that one tenth fractional 0.904. An ANN was capable of getting the lowest RMSE and MAE among all other algorithms considered here. [9] Both support vector machine (SVM) and RBF with RBF revealed the best performance for exchange rate forecasting compared to the others. [10] The paper considers a network schema as a combination of Convolutional Neural Network (CNN) and Bidirectional Long Short-Term Memory Networks (BiLSTM) for macro-economic time series prediction competing with traditional methods, and often they deliver desired outcomes prove superior. [11] In congruence, the paper is a machine learning LSTM approach as an assessment tool for economic forecasts in COVID-19 affected countries. [12]

The Wavelet neural network will be implementing an analytical process. We implement Intelligent swarm algorithm to endow our model the optimization weight ability.

[13] The paper highlights recurrent neural network with the advantage of internal memory to model indicator series and identify patterns to predict economic index trends. [14] The PCR predicted growth rates exhibited 89% accuracy of the real recorded GDP. Although a few macroeconomic indicators contributed to a positive development of GDP, the adjusting trends were still ongoing. [15] Unconventional tools based on nonlinear dynamics in forecasting macroeconomic indicators. The regularization factor used in the standard model is most ideal way. [16] This essay aims at identifying the machine learning capabilities for macroeconomic forecasting through considering the characteristics like nonlinearity, regularization, cross-validation, and alternative loss functions. [17] It declares that linearity is the factor that for the better forecast of macroeconomic, is the most important one. The algorithms not always effective in considering the class diversity among all students. [18] To address the foregoing, there is no available literature on the application of the same approach for Vietnamese macroeconomic forecasting.

In sum, after the findings of the literature review, the article gives a detailed review for the mainstream of research on the use of ARIMA model in macro trend prediction. The review is based on the results of different studies, so, it confirms the signature role of adequately forecast methods in guiding economic analysis and informing decision-making. In the future, time series forecasting techniques would remain a major need for not only analytical reasons but also the evolving challenges and complexities of global macroeconomic trend prediction.

3 METHDOLOGY ADOPTED

In the domain of forecasting macroeconomic patterns through the utilization of machine learning, multiple methodologies are employed to harness the capabilities of algorithms and data-driven insights. These methodologies encompass various steps, with each component playing a pivotal role in the overall process.

3.1 *Data collection from financial sources*

The first step involving this process is the gathering of broad historical data in different aspects of macroeconomics including various indicators. These data are collected from the well-known economic databases, the websites of the governmental institutions, etc.

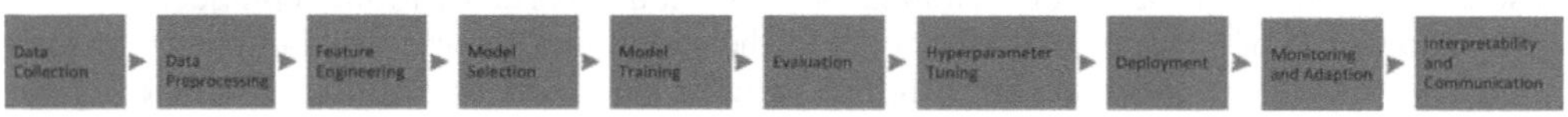

Figure 1. A diagramatic view of all the steps in order of their sequence of being formulated.

3.2 *Preprocessing the acquired data*

The cleaning of the dataset leverages technologies like imputation, the removal of outliers, and normalisation to reconcile that the data is correct for analysis. Thus, data is what mitigates biases in further machine learning models.

3.3 *Feature engineering for better prediction*

This is the stage in which the focus is put on the development of new capabilities or modifying the existing ones. Features that are engineered, such as lagged values and ratios, often present a more articulated vision of the key factors driving the economic performance.

3.4 *Evaluating the best suitable model*

The area of predictive models includes many algorithms such as linear regression, time series models (ARIMA), neural networks (LSTM), and ensemble methods (RANDOM forest), among others, each of which features a certain principal to be relied on in predicting information and macro indicators.

3.5 *ARIMA model training*

The selected model is fed with real historical data that is supposed to allow it to learn. Calling for a deeper understanding of the relationship between entities and patterns from the dataset itself.

3.6 *Evaluating on various metrics*

The usual metrics to assess the performance of a trained model is RMSE, MAE, and others, or better yet use other metrics that are more relevant to the forecasting job at hand.

3.7 *Hyperparameter tuning for better performance*

It emphasizes on allowance of parameters that are not learnt while training but are capable of affecting the behaviour of the model.

3.8 *Deployment*

In the process of implementation, the model should be effectively integrated into the decision-making cycle to serve potential purposes, including economic policy formulation, investment decisions or other issues of concern to the stakeholders.

4 RESULTS AND DISCUSSIONS

The actual values recorded for the considered econometric time series along with the forecasts generated using the ARIMA model and adopted inputs like GDP, revenue, inflation, and unemployment are presented. We can observe that the ARIMA model shows good results in the prediction of the macroeconomic trend. Evaluation parameters as Mean Absolute Error (MAE), Mean Squared Error (MSE), and Root Mean Squared Error (RMSE) statistically support the veracity of the results. It indicates successful model ability to capture major movement patterns this resulting in a good model fit.

The VIP studies show the influence of the GDP, income, inflation, and unemployment both to producers and policy makers, thus helping them to establish the dominant economic factors and the main directions of the economic policy. Such model's ability to enhance the economic forecasting paves the way for the right decisions that could help develop feasible plans and strategies. ARIMA model is the core foundation for managers in outlay and strategy construction by items the factors. That is why there is the residual analysis which is evidence of its efficacy and also shows the economic tendencies through many random distribution and normal residuals with long run economic tendencies. Systematic amendments make sure the forecasting system is always relevant and streamlined to serve the country's needs in all economic shifts.

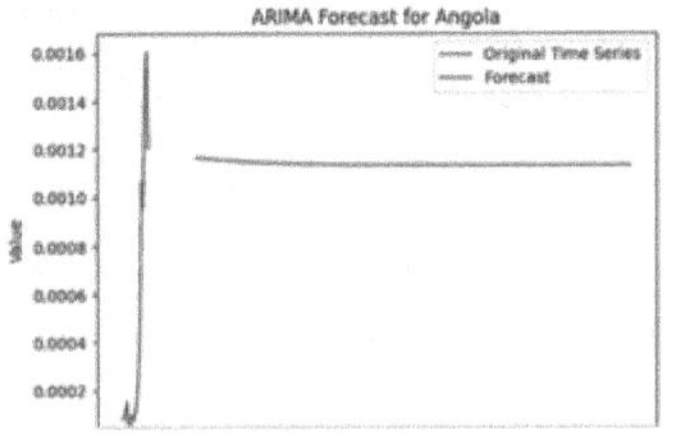

Figure 2. Time-series evaluation of Angola's GDP.

4.1 *Assessment of forecast accuracy metrics*

There are several measures that can be applied to the investigation of the precision of your ARIMA model such as Mean Squared Error (MSE), Root Mean Squared Error (RMSE), and Mean Absolute Error (MAE) lower values of these metrics indicate better model performance.

4.2 *Visualization of predictions*

The advice is to develop a plot in which one can see the contrast between the predicted estimates and actual GDP, income, inflation and unemployment indices over a given period.

4.3 *Evaluation of statistical significance*

In this regard statistical tests can be used to check validity of models' predictions.

4.4 *Analysis of variable importance*

If input variables are taken as being the output of GDP, household income, inflation, and unemployment, the effectiveness of each one needs to be studied within the model.

4.5 *Examination of residual analysis*

It becomes essential to examine the rejects, which are those values in the actual and predicted values that does not align.

5 CONCLUSIONS

The conclusion of this study is that the ARIMA model has forecasted much better and interpretable macroeconomic trend accuracy than the other models. Using variables like GDP, income, inflation and unemployment the model managed to reduce the MAE, MSE and RMSE as it has shown actual data in case of predicted values. The model is able to show that it is working well by accounting for the behaviour of the descriptive macroeconomic indicators. The variable importance analysis only gives policymakers a better understanding of the relative contribution of these indicators to the general economic trend including GDP, income, inflation, and unemployment. The demonstration of the statistical signification of the forecasts of the model offers a basis for the confidence and practical usage in the processes of decision-making.

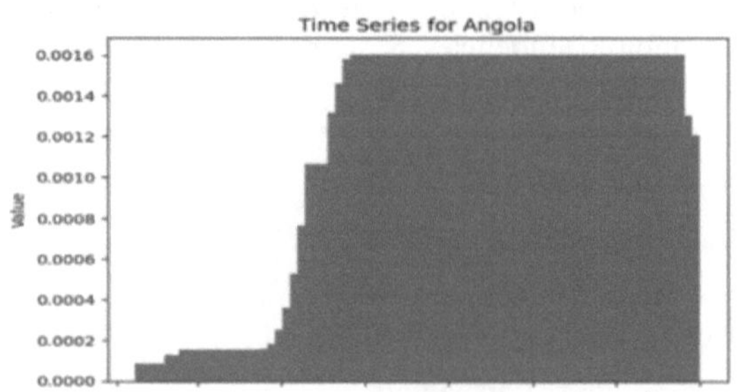

Figure 3. Forecast of the macro-economic trends of Angola.

Improvement tools, such as trial-and-error with differencing orders and trying ensemble techniques, all work to refine forecast accuracy. Also, the implementation of real-time monitoring system and feature improvements is going to help make the model well fitted to dynamic economic circumstances. The model provides a platform which merges appropriate is modelling, interpretability of results and flexibility to deal with unforeseen circumstances for stakeholders who are dealing with complex decision-making in the economy.

6 FUTURE SCOPE

The ARIMA model of future macroeconomics trend predicting possesses rather numerous positive aspects, which is an improvement of the model, considering GDP, income, inflation and unemployment through the ARIMA model. Exploiting innovative machine learning methodology representative by deep learning models enhances prediction power vi fashion of discovering embedded complexities in macroeconomic data. By the way, project implications will not be limited to economic statistics, but also to the additional outside factors for getting a holistic view of economic developments. Leveraging latest descriptive techniques and involving economists, data scientists, and policymakers as part of an interdisciplinary team will enable models' flexibility and response to the economic changes.

The integration of micro and macroeconomic theories, with enhanced forecasting algorithms, will be able to generate a dynamic representation of a real economy with more proven mechanisms. With ongoing technological advancement and data accessibility enhancement, this project can stand on the frontier of the field and give the right decision to administrators dealing with crises created by the global economic phenomenon. The preventive measure guarantees the model will keep up with the dynamic economic process with the use of relevant methods and approaches.

REFERENCES

[1] James T.E. Chapman, Ajit Desai. (2022). *Macroeconomic Predictions using Payments Data and Machine Learning.* doi: 10.48550/arxiv.2209.00948

[2] Ali, Lashgari. (2023). *Harnessing the Potential of Volatility: Advancing GDP Prediction.*

[3] Ruth, Duerr. (2022). How is machine learning useful for macroeconomic forecasting? *Journal of Applied Econometrics*, doi: 10.1002/jae.2910

[4] Fan, Fah, Cheng., Zitian and Fu. (2022). Macroeconomic forecasting based on mixed frequency vector autoregression and neural network models. *Wireless Communications and Mobile Computing*, doi: 10. 1155/2022/2956289

[5] M., Y., Gareev., Andrey, Polbin. (2022). Nowcasting Russia's key macroeconomic variables using machine learning. *Voprosy Economiki*, doi: 10.32609/0042-8736-2022-8-133-157

[6] Vladik Kreinovich, Chon Van Le. (2022). *Predicting (Economic) Trends: Why Signature Method in Machine Learning.* doi: 10.1007/978-3-030-97273-8_12

[7] Parul, Arora. (2023). *Gross Domestic Product Prediction in Various Countries with Classic Machine Learning Techniques.* doi: 10.1007/978-3-031-28790-9_9

[8] Muhammad, Anees, Khan., Kumail, Abbas., Mazliham, Mohd, Su'ud., Anastasia, Salameh., Muhammad, Alam., Nida, Aman., Mehreen, Mehreen., Amin, Jan., Nik, Nur, Wahidah, Nik, Hashim., Roslizawati, Che and Aziz. (2022). Application of machine learning algorithms for sustainable business management based on macro-economic data: supervised learning techniques approach. *Sustainability*, doi: 10.3390/su14169964

[9] Alessio, Staffini. (2023). *A CNN–BiLSTM Architecture for Macroeconomic Time Series Forecasting.* doi: 10.3390/engproc2023039033

[10] Rolando, A., Santos., Brian, W. and Sloboda. (2022). Macroeconomic forecasting examining the COVID-19 pandemic using selected countries: A machine learning LSTM (long-term short-term memory) approach. *European Scientific Journal*, doi: 10.19044/esj. 2022.v18n12p1

[11] Tao Wang, Yuxuan Du, Zheming Cui. (2022). Research on macroeconomic prediction technology based on wavelet neural network. *Smart Innovation, Systems and Technologies*, doi: 10.1007/978-981-16-9735-7_9

[12] Jimmy, Ming-Tai, Wu., Meng-Hsiun, Tsai., Chao-Chieh, Cheng., Mu-en, Wu. (2022). Predict the trend of economic indicators in time series based on recurrent neural network combined with leading indicators. *Journal of Intelligent and Fuzzy Systems*, doi:10.3233/jifs-219317

[13] Samuel, Agu., Fergus, U., Onu., Uchenna, Ezemagu., Deborah, Oden. (2022). Predicting gross domestic product to macroeconomic indicators. *Intelligent Systems with Applications*, doi: 10.1016/j.iswa.2022.200082

[14] Philippe, Goulet, Coulombe., Ruth, Duerr., Maxime, Leroux., Dalibor, Stevanovic., Stéphane, Surprenant. (2020). *How is machine learning useful for macroeconomic forecasting. arXiv: Econometrics*,

[15] Cu, Nguyen, Giap., Dinh, Thi, Ha., Vu, Quang, Huy., Do, Thi, Thu, Hien., Le, Mai, Trang. (2021). Firm performance prediction for macroeconomic diffusion index using machine learning. *International Journal of Advanced Computer Science and Applications*, doi: 10.14569/IJACSA.2021.0120612

[16] Toko. (2022). Research on macroeconomic forecasting model based on grey theory. *Advances in Economics and Management Research*, doi: 10.56028/aemr.2.1.63

[17] Shantong, Guo. (2022). Research on macroeconomic Forecasting model based on grey theory. *Advances in Economics and Management Research*, doi: 10.56028/aemr.1.2.63

[18] Zihao Wang, Kun Li, Steve Q. Xia and Hongfu Liu. (2022). Economic Recession Prediction Using Deep Neural Network. *The Journal of Financial Data Science*, doi: 10.3905/jfds.2022.1.097

Computational Methods in Science and Technology Volume 1 – Kaur et al. (Eds)

Extraction of road using semi-supervised deep learning method

Jalaj Mishra
Department of Computer Science & Engineering, Greater Noida Institute of Technology, Greater Noida, India

Mohd Jawed Khan
Department of Computer Science & Engineering, Greater Noida Institute of Technology, Greater Noida, India
Department of Computer Science & Engineering, Central Institute of Technology, Kokrajhar, Assam India

Himanshi Negi, Jitendra Alam & Junaid Alam
Department of Computer Science & Engineering, Greater Noida Institute of Technology, Greater Noida, India

Pankaj Pratap Singh
Department of Computer Science & Engineering, Central Institute of Technology, Kokrajhar, Assam, India

ABSTRACT: Detection of road networks using high-resolution aerial or remote sensing imagery constitutes a significant focus within modern research efforts. Currently, deep learning models demonstrate efficiency to a certain degree, they are constrained by a notable reliance on annotated datasets, a practice divergent from optimal methodologies due to the scarcity of extensive, high-quality labeled data. The precision of road prediction is additionally influenced by the latent information embedded within pixels exhibiting diminished confidence values. These pixels play a crucial role in discriminating between road and non-road regions within the images, yet their utilization by deep learning models is suboptimal. To handle these two different issues, opted U-Net as our base model with a sole generator paired with twin discriminators and presented this combined setup of U-Net on the Generative Adversarial Network (GAN). At the base of the U-Net architecture, a bottleneck module is incorporated to augment the functionalities of the proposed model. This proposed model attains 98% accuracy on DeepGlobe Road Dataset. Comparative and quantitative assessment demonstrates the enhanced efficiency of our model compared to similar methodologies.

1 INTRODUCTION

Remote sensing is a technology that has been in use for a long time to extract road information and use it in different application areas. Road information plays a crucial role in the analysis of textural information on different roads [1], the development of smart cities, traffic management [2,3], and the observational programs of Earth [4]. But sometimes there is a problem with the shapes of the roads that can become strange to look at and there can also be shadows of tall buildings and trees which may appear on the side of the road or should be right next to the road [5,6].

Looking at the objective of road extraction from the machine learning perspective, can say that the information about the road can be easily extracted by separating its background from the road. Segmentation of road images is the process of separating the road and its background from high-resolution satellite images [7,8]. Deep learning is a good option to

DOI: 10.1201/9781003501244-75

achieve segmentation because the performance of deep learning and its models has proved excellent in the past years. In recent years deep learning and its associated models have been applied to road extraction and associated tasks yielding commendable outcomes [9,10]. Although DL models have very good performance [11–13], there is a problem of dependency with deep learning models. Deep learning models depend on high-quality labeled data, which is not a good practice because obtaining meticulously annotated data of superior quality poses a considerable challenge. The models depend on the training labeled data. The more labeled data is used in training, the better the model performance.

Various types of supervised, semi-supervised, and unsupervised models, mainly semi-supervised models can handle the problem of dependency. However, they cannot solve this problem completely because semi-supervised models don't use pixels on the border sideline of the road whose confidence value is low or which are known as low confidence pixels in the road. Low confidence pixels: The pixels on the borderline of the road that is very important to differentiate the road side and non-road side which also greatly affects the accuracy of the models. To address these problems, we have proposed a model of Generative Adversarial Network (GAN) with a sole generator and two discriminators which are base U-Net model [14] in which an additional module in the bottleneck is added. The generator is used to achieve segmentation for a particular input satellite image of the road and it generates segmentation maps and corresponding entropy maps as well [15,16]. The discriminative components are employed to distinguish between the segmentation outcomes of road extraction and the accompanying entropy maps, ensuring their alignment with the dataset remains consistent [17].

2 RELATED WORK

Hao Chen *et al.* (2023) proposed a novel approach called SW-GAN (Semi-Weakly Supervised Generative Adversarial Network) [18]. This approach combines semi-supervised learning and adversarial training to harness the collective power of annotated and unannotated data during the model training phase. The author utilizes datasets collected from real-world locations such as Seattle, Beijing, and America, each with specific characteristics and augmentations. Results show SW-GAN's superior performance over baseline methods, achieving an IoU of 77.8% and mAP of 89.5% on Dataset 1, an IoU of 63.7% and mAP of 73.8% on Dataset 2, and an IoU of 71.4% and mAP of 81.2% on Dataset 3. The comparison of the result with all three datasets chosen is shown in Table 1 and the visualization of that is shown in Figure 1. Furthermore, SW-GAN demonstrates the ability to approach the performance of fully supervised methods with only 30% of full annotations, achieving an IoU of 79.1% and mAP of 90.5% on Dataset 1.

Table 1. Quantitative analysis of results in comparison with other methods.

S. No.	Method	F1-score (%)	IoU (%)	Kappa (%)
1.	D-Link Net	50.94	34.11	45.97
2.	SII-Net	56.80	41.06	54.49
3.	RoadExNet	59.71	43.50	57.25
4.	MU-Net	62.23	46.67	59.34

Gaowei Zhang *et al.* (2021) introduce a novel system named SSGAN, designed for the automatic detection of defects on steel surfaces using a semi-supervised learning approach coupled with a Generative Adversarial Network (GAN) [19]. To address the challenge posed by the scarcity of labeled data in defect detection is the primary objective of effectively utilizing both labeled and unlabeled images during training. SSGAN incorporates a segmentation network based on ResNet-101, enhanced with a dual attention mechanism, along with a Fully Convolutional Discriminator network. Through training with a combined loss function, SSGAN has achieved notable results, with mean Intersection Over Union (IoU) scores of 79.0 % and 81.8 % on a steel defect dataset containing only $\frac{1}{8}$ and $\frac{1}{4}$ of the labeled data for training, respectively. These results show the potential of SSGAN compared to similar methodologies employing similar amounts of labeled data.

3 METHODOLOGY

Presented a semi-supervised Munet-GAN Model to address the problem of dependency of DL models on high-quality supervised data and the utilization of the low-confidence pixels in the satellite images of the road that results more precisely. In this proposed method MUnet-GAN, a single Generator (G), and two Discriminators (D1, D2) are used. The generator is a simple network for semantic segmentation based on a modified U-Net model used to generate segmented maps of its input images and the discriminators D1 is the road segmentation network and D2 is the segmentation network for entropy maps. D1 is used to differentiate the type of dataset from which the predicted image comes from. Discriminators are used to maintain consistency between training input datasets.

To scale and increase the performance of the model we used the technique called pseudo-labeling technique where we provide supervised and unsupervised types of data to the model but the catch is the amount of supervised training data is much less than the unsupervised training data. In this technique, we train our model on the supervised data in which training images and the corresponding maps both are present at the time of training and then that trained model will be used for the prediction of the maps for unsupervised input images. In this proposed model there is an entropy map generator that generates Entropy maps of the unsupervised input images which are the different maps that show the confidence values of pixels in the input image. It consists of low as well as high-confidence pixels by which we can easily differentiate the road and non-road elements in the image up to a large extent. We provide the unlabeled data to the trained model to generate corresponding maps and generate the entropy maps for the same. The workflow described is demonstrated in Figure 1.

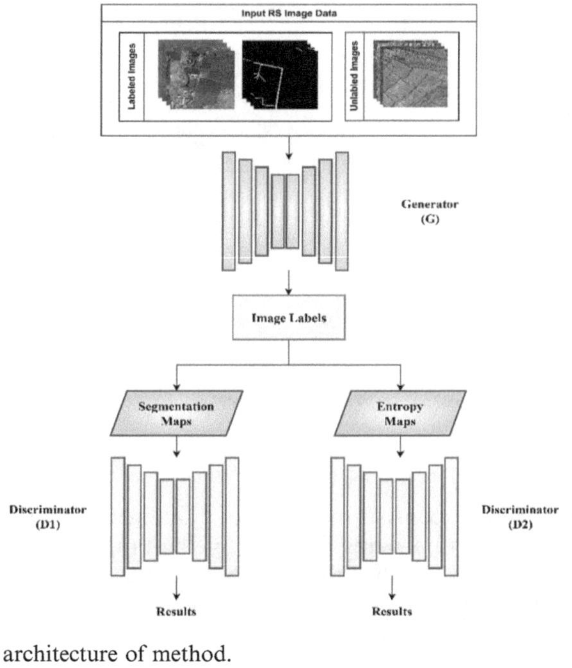

Figure 1. Workflow architecture of method.

In the above Figure 1, there are one generator and two discriminators both are the modified U-Net models. To optimize the model in a generic way we have used continuous regularization on the unsupervised data to enhance the prediction capability of the model. For the given network the mathematical equation of the optimizer function theta(θ_F) is given below:

$$\theta = \min_{\theta_F}\left\{\frac{1}{p}\sum_{i=1}^{p} S_d\left(x_i^s, y_i^s\right) + \frac{1}{q}\sum_{i=1}^{q} U_d\left(x_i^u, y_i^u\right)\right\} \tag{1}$$

In the architecture of the MU-Net method a Generator and Discriminator are added as the main entities of the model and the complete performance of the model depends upon the capability of these modules. MU-Net constitutes an adapted iteration of the foundational U-Net framework in which we added an extra module in the bottleneck for the enhancement of low-level feature separation and its representation and the extra layers for convolution so that we can extract the latent data from pixels with low confidence in the input remote sensing image, which has not been utilized in prior studies. Subsequently, the entirety of low-level data from the input image will undergo up-sampling and concatenation procedures on the decoder side, culminating in the generation of the final output image through comprehensive analysis of both high and low-level feature maps. The visual demonstration of architecture of MU-Net is shown in the Figure 2.

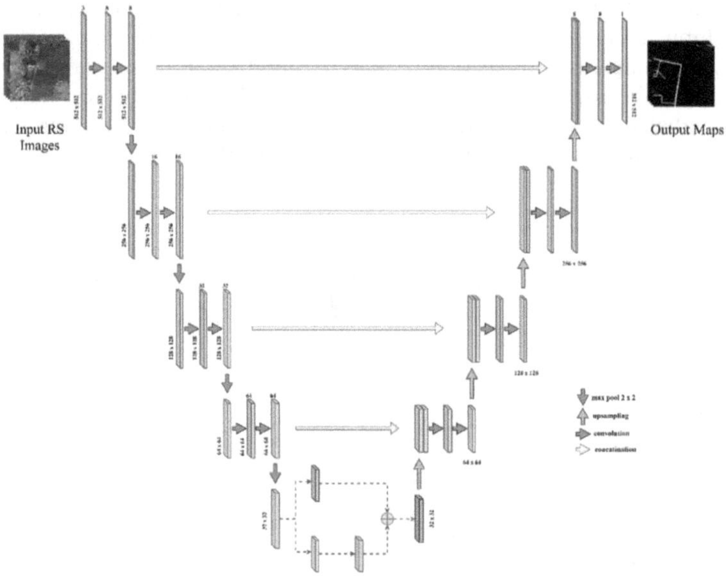

Figure 2. Detailed architecture of MU-Net.

The MU-Net architecture consists of encode, bottleneck, and decoder. The latent information of high-level feature maps is given by the bottleneck module and the final result is given by the decoder. Within the MU-Net framework, the encoder component encompasses four distinct blocks tasked with encoding the input remote sensing (RS) image and transforming it into high-dimensional feature maps, pivotal inputs for subsequent bottleneck, and decoder blocks. The module connected in the bottleneck side in the architecture consists of channel convolution and spatial convolution layers and the parallel connection between these layers is made after the Encoder. There is the use of a convolutional kernel of one dimensional to perform the convolutional operation on the feature map one step deeper. In the second spatial convolution layer feature map is concatenated with the spatial features after the convolution of each feature map by the kernel size. The output of these two channels and spatial layers of convolution is consistent with the input datasets.

The main objective of the Decoder is to utilize the details of the segmentation operations efficiently and there four blocks of decoder used in the MU-Net architecture are used [20]. In each decoder block filter of the features map is done by the convolutional layer and post that the transposed version of the convolutional layer up-sample the feature map. After that, the high-level information about the features is added to the result so due to this convolutional layer filters the feature map once again [21]. After traversing through all four distinct decoder blocks, the dimensions of the feature maps will match those of the input image. Finally, a sigmoid function operates at the conclusion, distinguishing areas as either roads or non-roads, assigning values within the range of 0 to 1.

4 RESULTS

The quantitative analysis of results generated by different methods is mentioned in Table 1, where the performance of the proposed method is compared with all other similar methodologies and achieves higher scores of F1, IoU, and Kappa. The MU-Net method demonstrates better scores than the similar methodologies named Dlink-Net, SII-Net, and the RoadExNet models and this comparison has been done on the famous DeepGlobe Dataset [22]. To demonstrate the performance of the presented method with other methodologies a visualization of result is given in Figure 3 where all the three other methods are presented their result with the result generated by proposed method also. Visual including the combination of input, ground truth, and result images produced by different methods.

5 DISCUSSION

In this paper, we have chosen the DeepGlobe dataset for comparison of our proposed MU-Net method with the others. Our model outperforms the input images whereas others have achieved very low scores of result metrics. In our model, we have added a bottleneck module which is the root cause of the improvement in the results we are getting, in that additional module a channel convolution and a spatial convolution are placed which enhance the feature representation by which we get the more refined feature map. In this suggested approach, there exists a notion of Entropy maps, which employs the latent data of pixels with diminished confidence levels within the input image. There are many pixels in large numbers present in the input image of the road and the information of these pixels is latent information that plays an important role in generating outcomes regarding the distinction between objects depicted in the image, classifying them as either road or non-road. We applied our MU-Net method to the GAN model with a single generator and two discriminators of which one is named as segmentation discriminator (D1) and the other one is the entropy discriminator (D2). Adding an entropy discriminator has proved beneficial to our model up to a large extent in terms of performance shown in above Table 1.

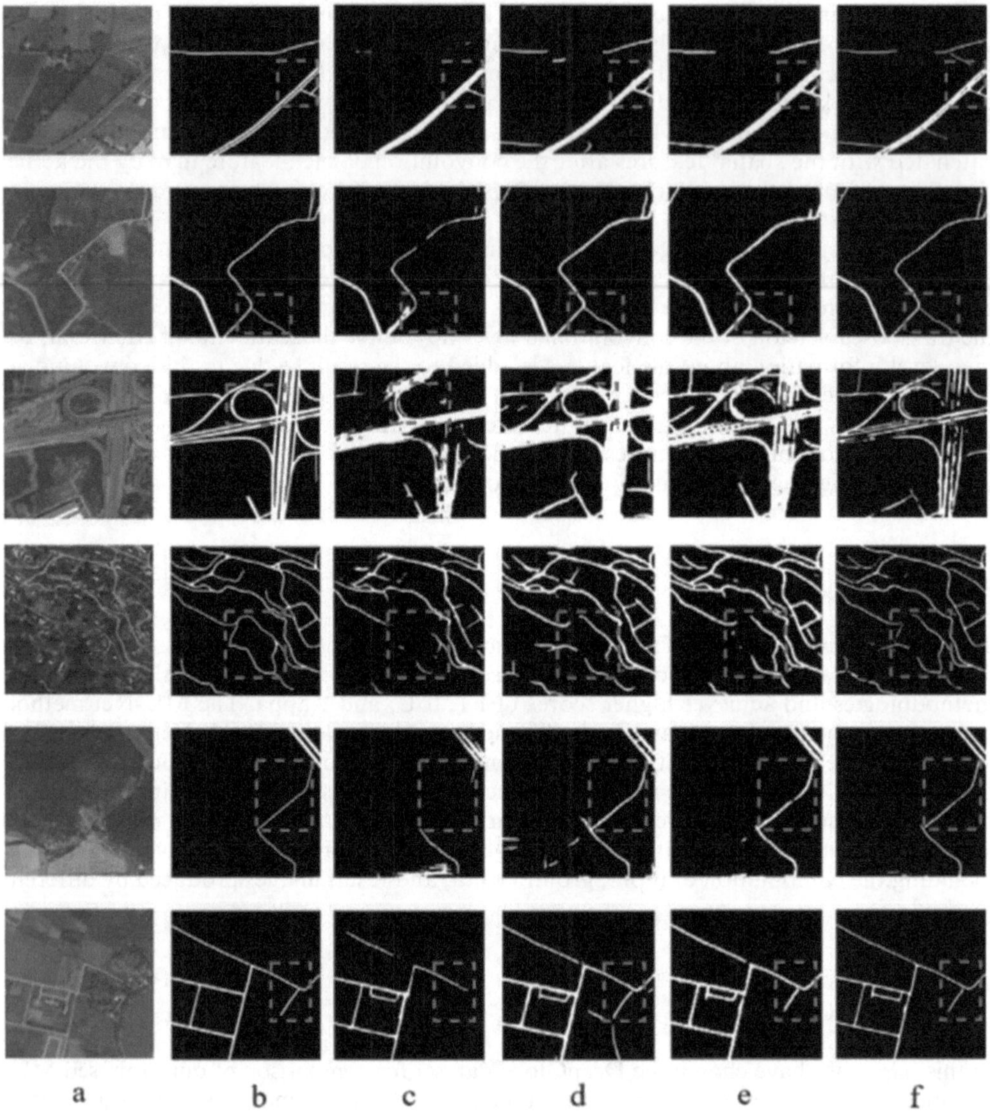

Figure 3. Comparative result analysis with different methods on DeepGlobe Dataset (a) Original image (b) Ground Truth image (c) D-Link Net (d) SII-Net (e) RoadExNet (f) MU-Net method.

6 CONCLUSION

Deep learning has shown promise in addressing these challenges by accurately segmenting roads from satellite images. While deep learning models perform well, they rely heavily on labeled data, which can be scarce and affect their performance. To address the challenge of reliance on labeled data and underutilization of information from low-confidence pixels in the input image, we introduced a Generative Adversarial Network (GAN) model featuring a distinctive design integrating segmentation and uncertainty mapping. This integration aims to enhance the model's performance. In this proposed model a modified U-Net model called MU-Net is used with an additional bottleneck module which helps in transforming the quality of feature maps and its representation from the Encoder to the Decoder. Our model achieved higher values of evaluation metrics than the other comparison methods. Different methods have been compared with the method proposed in this method on the DeepGlobe dataset and achieved 62.23% F1-score, 46.67% IoU, and 59.34% Kappa. But still, there is a gap in enhancing the accuracy by extracting the latent information of image pixels more precisely.

REFERENCES

[1] Liu, Xiangzeng, Ziyao Wang, Jinting Wan, Juli Zhang, Yue Xi, Ruyi Liu, and Qiguang Miao. (2023 Feb 15). RoadFormer: Road extraction using a Swin transformer combined with a spatial and channel separable convolution. *Remote Sensing*. ;15(4):1049, https://doi.org/10.3390/rs15041049.

[2] Peddinti, A. Sravanthi, Arjun Singh Chouhan, and Asisa Kumar Panigrahy. (2021 Jan 1). Road extraction using Aerial images for future navigation. *Materials Today: Proceedings*. ;47:6306–8, https://doi.org/10.1016/j.matpr.2021.05.537.

[3] Miyamoto, Ryusuke, Yuta Nakamura, Miho Adachi, Takeshi Nakajima, Hiroki Ishida, Kazuya Kojima, Risako Aoki, Takuro Oki and Shingo Kobayashi. (2019, September 8). Vision-based road-following using results of semantic segmentation for autonomous navigation. In *2019 IEEE 9th International Conference on Consumer Electronics (ICCE-Berlin)* (pp. 174–179). IEEE, 10.1109/ICCE-Berlin47944.2019.8966198.

[4] Chen, Hao, Zhenghong Li, Jiangjiang Wu, Wei Xiong. and Chun Du. (2023 Apr 1). SemiRoadExNet: A semi-supervised network for road extraction from remote sensing imagery via adversarial learning. *ISPRS Journal of Photogrammetry and Remote Sensing*. ;198:169–83, https://doi.org/10.1016/j.isprsjprs.2023.03.012.

[5] Shamsolmoali, Pourya, Masoumeh Zareapoor, Huiyu Zhou, Ruili Wang and Jie Yang. (2020 Aug 21). Road segmentation for remote sensing images using adversarial spatial pyramid networks. *IEEE Transactions on Geoscience and Remote Sensing*. ;59(6):4673–88, 10.1109/TGRS.2020.3016086.

[6] Wang, Ying, Yuexing Peng, Wei Li, George C. Alexandropoulos, Junchuan Yu, Daqing Ge and Wei Xiang. (2022 Aug 8). DDU-Net: Dual-decoder-U-Net for road extraction using high-resolution remote sensing images. *IEEE Transactions on Geoscience and Remote Sensing*. ;60:1–2, 10.1109/TGRS.2022.3197546.

[7] Chaudhuri, Debasis, Naveen Kumar Kushwaha, and Ashok Samal. (2012 Jun 28). Semi-automated road detection from high resolution satellite images by directional morphological enhancement and segmentation techniques. *IEEE Journal of Selected Topics in Applied Earth Observations and Remote Sensing*.;5(5):1538–44, 10.1109/JSTARS.2012.2199085.

[8] Hormese, Jose, and C. Saravanan. (2016 Jan 1). Automated road extraction from high resolution satellite images. *Procedia Technology*. ;24:1460–7, https://doi.org/10.1016/j.protcy.2016.05.180.

[9] Bastani, Favyen, Songtao He, Sofiane Abbar, Mohammad Alizadeh, Hari Balakrishnan, Sanjay Chawla, Sam Madden, & David DeWitt. (2018). Roadtracer: Automatic extraction of road networks from aerial images. *In Proceedings of the IEEE Conference on Computer Vision and Pattern Recognition* (pp. 4720–4728), https://doi.org/10.48550/arXiv.1802.03680

[10] Liu, Yahui, Jian Yao, Xiaohu Lu, Menghan Xia, Xingbo Wang, and Yuan Liu. (2018 Oct 24). RoadNet: Learning to comprehensively analyze road networks in complex urban scenes from high-resolution remotely sensed images. *IEEE Transactions on Geoscience and Remote Sensing*. ;57(4):2043–56, 10.1109/TGRS.2018.2870871.

[11] Abdollahi, Abolfazl, Biswajeet Pradhan, Nagesh Shukla, Subrata Chakraborty, and Abdullah Alamri. (2020 May 2). Deep learning approaches applied to remote sensing datasets for road extraction: A state-of-the-art review. *Remote Sensing*. ;12(9):1444, https://doi.org/10.3390/rs12091444.

[12] Chen, Hao, Shuang Peng, Chun Du, Jun Li, and Songbing Wu. (2022 Aug 23). SW-GAN: Road extraction from remote sensing imagery using semi-weakly supervised adversarial learning. *Remote Sensing*. ;14(17):4145, https://doi.org/10.3390/rs14174145.

[13] Lu, Xiaoyan, Yanfei Zhong, Zhuo Zheng, Dingyuan Chen, Yu Su, Ailong Ma, and Liangpei Zhang. (2022 Apr 13). Cascaded multi-task road extraction network for road surface, centerline, and edge extraction. *IEEE Transactions on Geoscience and Remote Sensing*. ;60:1–4, 10.1109/TGRS.2022.3165817.

[14] Ronneberger, Olaf, Philipp Fischer, and Thomas Brox. (October 5-9, 2015). U-net: Convolutional networks for biomedical image segmentation. *In Medical Image Computing and Computer-assisted Intervention–MICCAI 2015: 18th International Conference*, Munich, Germany, proceedings, part III 18 (pp. 234–241). Springer International Publishing, https://doi.org/10.1007/978-3-319-24574-4_28

[15] Ghandorh, Hamza, Wadii Boulila, Sharjeel Masood, Anis Koubaa, Fawad Ahmed, and Jawad Ahmad. (2022 Jan 27). Semantic segmentation and edge detection—Approach to road detection in very high resolution satellite images. *Remote Sensing*. ;14(3):613, https://doi.org/10.3390/rs14030613.

[16] Shi, Qian, Xiaoping Liu, and Xia Li. (2017 Nov 13). Road detection from remote sensing images by generative adversarial networks. *IEEE Access*. ;6:25486–94, 10.1109/ACCESS.2017.2773142.

[17] Zhang, Ting, Zihang Gao, Zhaoying Liu, Syed Fawad Hussain, Muhammad Waqas, Zahid Halim, and Yujian Li. (2023 Apr 8). Infrared ship target segmentation based on adversarial domain adaptation. *Knowledge-Based Systems*. ; 265:110344, https://doi.org/10.1016/j.knosys.2023.110344.

[18] Chen, Ziyi, Liai Deng, Yuhua Luo, Dilong Li, José Marcato Junior, Wesley Nunes Gonçalves, Abdul Awal Md Nurunnabi, Jonathan Li, Cheng Wang, and Deren Li. (2022 Aug 1). Road extraction in remote sensing data: A survey. *International Journal of Applied Earth Observation and Geoinformation*. ;112:102833, https://doi.org/10.1016/j.jag.2022.102833.

[19] Zhang, Gaowei, Yue Pan, and Limao Zhang.(2021 Aug 1). Semi-supervised learning with GAN for automatic defect detection from images. *Automation in Construction*. ;128:103764, https://doi.org/10.1016/j.autcon.2021.103764.

[20] Peng, Cheng, Yangyang Li, Licheng Jiao, Yanqiao Chen, and Ronghua Shang. (2019 Apr 9). Densely based multi-scale and multi-modal fully convolutional networks for high-resolution remote-sensing image semantic segmentation. *IEEE Journal of Selected Topics in Applied Earth Observations and Remote Sensing*. ;12(8):2612–26, 10.1109/JSTARS.2019.2906387.

[21] Gurita, Andree, and Irina Georgiana Mocanu. (2021 Feb 24). Image segmentation using encoder-decoder with deformable convolutions. *Sensors*. ;21(5):1570, https://doi.org/10.3390/s21051570.

[22] Demir, I., Koperski, K., Lindenbaum, D., Pang, G., Huang, J., Basu, S., Hughes, F., Tuia, D. and Raskar, R., (2018). Deepglobe 2018: A challenge to parse the earth through satellite images. *In Proceedings of the IEEE Conference on Computer Vision and Pattern Recognition Workshops* (pp. 172–181), https://doi.org/10.1109/CVPRW.2018.00031.

Computational Methods in Science and Technology Volume 1 – Kaur et al. (Eds)
© 2025 the Author(s), ISBN: 978-1-032-81760-6

Human activity detection and classification by AI approaches

Anish Thakur, Geet Kiran Kaur & Ranjit Singh
Department of CSE, Chandigarh University, Gharaun, Mohali, India

ABSTRACT: The real area where it is well researched is Human Activity Recognition in the field of Computer Vision. Computer vision is one of the well-explored areas of interest in the proper processing and analysis of visual information with respect to specified applications, e.g., security, health care, human-computer interaction. This will be a huge step forward in effective identification and categorization of human behaviors based on sensor data due to the development of machine/deep learning techniques. Within the present paper outline, there is an attempt to provide a comprehensive review of recent work and advances in the area of human activity detection based on machine/deep learning approaches. This task will review the major challenges in human activity detection, covering dealing with several activities, sensor data unpredictability, and reliable methodologies for classification. We have thus analyzed these models through machine learning-based techniques: traditional, such as Support Vector Machines (SVM) and Random Forests, and innovative deep learning architectures like Recurrent Neural Networks (RNNs) and Convolutional Neural Networks (CNNs). It also covers the problems of small data, challenges for model interpretability, and difficulties to use the models for its purposes in real time. Next, it also gives a glance at potential use cases under health, sport data analysis, and intelligent environments.

Keywords: human activity, AI, machine learning, optimization, deep learning

1 INTRODUCTION

Electronic devices, wearable sensors, closed-circuit television (CCTV), camera devices, and some commercial off-the-shelf (COTS) devices are all part of the raw data used in the process of human activity classification or detection using Artificial Intelligence (AI). Surveillance, healthcare, smart homes and workplaces, remote care, and monitoring applications for activities like exercising and sports all rely on human activity recognition (HAR). A wide range of tasks may be made more efficient and secure with the widespread use of activity detection systems, which in turn improves human safety and quality of life. The exponential growth of AI approaches is driving the development of human activity detection using devices like video cameras, sensors, Wi-Fi, and RFID, which is still in its infancy [1,2]. There is a two-way street between activity detection devices and AI models; the latter provide more opportunities than their predecessors, which depended on single photos or sequences. An increase in human activity detection capabilities is favorably associated with AI progress, which in turn opens up more applications for AI in many fields [1–4].

The correlation between human activity detection systems and AI techniques, including machine learning and deep learning, has been established in numerous studies. Considerable advances have been made in the application of machine learning and deep learning algorithms to the classification and identification of human behaviors from sensor data. In the last ten years, machine learning-based approaches have effectively executed HAR recognition assignments. However, these approaches encounter obstacles including restricted multiclass activity training data, complexity in identifying features, and perplexing behaviors that include climbing stairs or bed searching, in addition to dynamic modes of movement in diverse simulations.

DOI: 10.1201/9781003501244-76

The solutions for machine learning frequently depend on pre-processed data derived from raw signals, a practice that can improve the performance of the algorithms used for classification [5–7]. Deep learning models, such as those with architectures like [4,8,9], can address these challenges by simplifying the process of data pre-processing and feature extraction steps. Additionally, they enhance the robustness and ability to apply learned knowledge to new cases of DL models. Additional advantages of deep learning include reduced dependence on professional expertise for feature engineering, decreased testing durations for image-based recognition, precise identification of temporal changes in features, and exceptional performance even with inadequately labelled data. The integration of deep learning (DL) in the field of human activity detection has streamlined the extraction of meaningful characteristics from raw sensor data. The advancement of deep learning (DL) models, such as recurrent neural networks (RNNs), convolutional neural networks (CNNs), and transfer weighting methods like VGG-16, ResNets-50, Inception, as well as hybrid DL models, have substantially simplified the procedure of developing human activity systems [8,9]. The drawback of conventional algorithms lies in the necessity for extensive feature engineering and manually extracting features, resulting in a time-consuming process. Conversely, deep learning algorithms can autonomously acquire the characteristics from the data, making them more suited for the task of categorizing intricate human actions. Machine and deep learning algorithms have unique capabilities that are crucial for identifying human behaviors. By combining them, it becomes possible to harness their complementary capabilities, leading to the development of activity recognition systems that are more robust, flexible, and comprehensible, and that surpass others in many domains and applications. The major objective of this study is to enhance the knowledge of the Human activity framework by integrating devices and their application domains into specific machine/deep learning contexts. Investigating the compatibility between various types of devices and their appropriateness for certain applications, as well as examining the pertinent characteristics of ML and DL, are crucial factors to be examined during the construction of a framework.

1.1 *Datasets for human activity detection*

Table 1. Datasets used for human activity detection [10].

Datasets	Description
Kinetics-700	This beneficial video dataset, which includes over 650000 YouTube videos spanning 700 different human activity classes, is perfect for developing human action detection algorithms.
DAHLIA: Daily Human Life Activity	The detection of human behaviour for smart-home services, such user support, is the main emphasis of this dataset. It records long-range actions in free-form, uninterrupted sequences using real-world movies captured by Kinect v2 sensors, yielding 40-minute recordings for each participant.
Volleyball action recognition dataset	This tool is a video action detection system that has 4830 frames that have been annotated from 55 videos. The annotations include labels for individual and team activities.
ARID dataset	The dataset, comprising over 3,780 video clips from 11 action types, is a crucial resource for identifying human actions in low-light environments, a crucial aspect for scholars and practitioners in the field of film studies.
HHAR (Household Human Activity Recognition) dataset	HHAR employs the accelerometer and gyroscope data obtained from smartphones that individuals wear while performing routine tasks such as cooking, cleaning, and watching television.
PAMAP2 Physical Activity Monitoring dataset:	The dataset comprises information on 18 distinct physical activities, carried out by 9 individuals wearing 3 inertial measurement devices and a heart rate monitoring.

Table 2. A summary of ML/DL models based on human activity detection studies.

Reference/ Year	Aim	Dataset	Method for detection/ classification	Appliction Domain	Limitation	Results
Deep Learning						
Khan, *et al.* (2022) [15]	Recognition of human behaviour using a combination of DL techniques.	A novel dataset including 12 distinct categories of human physical activity has been created by employing the Kinect V2 sensor on a group of 20 people.	KNN, LSTM	Human activity recognition	Activity recognition AI models often face difficulties in extracting spatial and temporal data, leading to poor efficiency in extended real-world human-computer interaction scenarios.	The CNN-LSTM approach achieves an accuracy of 90.89%.
Muhammad, *et al.* (2022) [16]	Human action detection through the use of attention-based techniques	CF11, UCF sports, and J-HMDB datasets	Dilated CNN, biLSTM	AI-based action recognition	While dilated CNNs provide an extensive comprehension of temporal patterns in human behaviour, their integration with LSTM networks might not entirely resolve this obstacle. Scholars are making efforts to develop resource-efficient systems capable of accurately discerning a user's activity from unprocessed data.	The proposed methodology outperformed the most sophisticated techniques when applied to the J-HMDB, UCF11, and UCF sports datasets, achieving respective recognition rates of 80.2%, 98.3%, and 99.1%.
Challa, *et al.* (2022) [18]	A model utilizing DL to discern human actions	UCI-HAR, WISDM, and PAMAP datasets	CNN, Bi-LSTM	Detection of human activity employing information gathered by wearable sensors	Human Activity approaches necessitate the preprocessing of data and the generation of substantial features, both of which demand specialized knowledge and effort. They are therefore highly suitable for specialized applications.	The proposed model achieved accuracies of 96.05%, 96.37%, and 94.29%, respectively, on the WISDM, UCI-HAR, and PAMAP2 datasets.
Dua, *et al.* (2021) [20]	Human activity identification is carried out through the utilization of ubiquitous sensors and a CNN-GRU model.	UCI-HAR, WISDM, and PAMAP2 datasets	CNN, GRU	Recognition using wearable sensors	Human activity methods require substantial features engineering and data pre-processing, which consequently requires domain expertise. These procedures require a significant degree of physical exertion and are suitable solely for specialized purposes.	The accuracy for the UCI-HAR dataset was 96.20%, the accuracy for the WISDM dataset was 97.21%, and the accuracy for the PAMAP2 dataset was 95.27%.
Mutegeki, *et al.* (2020) [21]	A CNN-LSTM algorithm designed to identify human activity.	UCI HAR public dataset, iSPL dataset	CNN-LSTM	Application of DL to the identification of human activities	CNN-LSTM models may require a large quantity of labelled training data in order to achieve successful generalization. However, the progress might be hindered by a scarcity of inclusive and varied data.	The suggested model achieves a 99% accuracy on the iSPL dataset, that is an internal dataset. The accuracy on the UCI HAR public dataset improves to 92%.

2 RESEARCH CHALLENGES

Resistance to anti-VEGF medications is a significant challenge, as certain patients acquire resistance to these therapies. Investigation is necessary to comprehend the fundamental processes of resistance and design new therapeutic strategies. There has been an observed increase in the number of published papers on diabetic retinopathy (DR) about deep learning models, particularly following the widespread adoption of deep learning in medical imaging. Despite the utilization of various methodologies in the development of DR classification systems, a considerable volume of data has been identified as being necessary for training objectives. The existing public datasets consist of high-quality photos, which renders the system impractical at present. In reality, retinal scans may not possess the same level of quality as the training images, given the presence of more heterogeneous populations of diabetes patients. Overfitting is a significant challenge encountered during the training process. Excessive epochs in network training result in the model overfitting to the specific patterns present in the sample data, leading to memorization. This results in the model being unable to perform effectively on a novel dataset. Consequently, many research implement early halting in their model [17]. The training process is halted promptly when the training graph exhibits signs of overfitting or when the error rate begins to rise. The need for massive datasets to adequately train DL models is a limitation of DL's use in the medical field [4]. Public datasets show an uneven distribution of picture sizes at each level. The classification algorithm's performance may be affected by the fact that the No DR or normal level has a much bigger number of pictures than the other levels. Data augmentation requires researchers to be cautious since it might lead to a misunderstanding of the underlying issue, which is especially problematic when the model is intended for clinical application in real life. Hence, a deep learning algorithm with the capability to rectify and resolve unbalanced data issues is essential for building a robust deep learning classification model.

These systems are further obstacles in the identification of human activity. These difficulties and constraints draw attention to how difficult it is to identify/detect human behaviour and how much more work and development in this area are required.

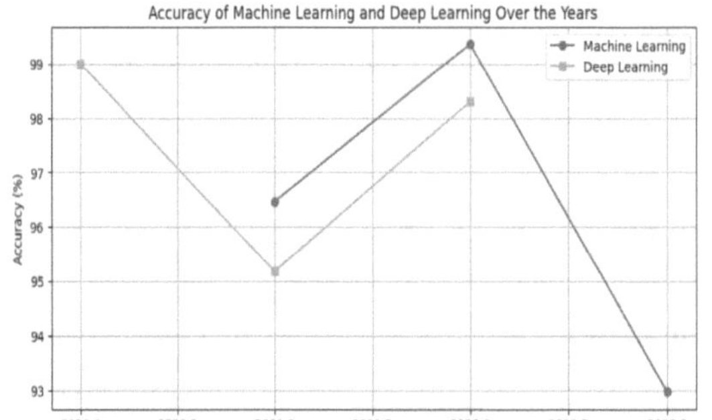

S

Figure 1. Machine and deep learning in terms of accuracy.

3 CONCLUSION

The research was a compilation of articles that focused on identifying human actions using deep learning and machine learning techniques. By examining the data, characteristics, and classification procedures, the research has been made simpler. Compared to machine learning, deep learning is more accurate and can identify a broader variety of actions. Deep learning's ability to rapidly analyze a huge number of actions and activities to accomplish exact identification would greatly enhance the field of human activity detection. The main goal of this research was to gather important data on human activity detection and organize it in a way that the academic sector could use. Using this database, researchers may get solutions to each possible inquiry at any step of detecting human actions.

REFERENCES

[1] Suthar, B., and Gadhia, B. (2021). Human activity recognition using deep learning: a survey. In *Data Science and Intelligent Applications: Proceedings of ICDSIA 2020* (pp. 217–223). Springer Singapore.

[2] Singh, R., and Srivastava, R. (2020, February). Some contemporary approaches for human activity recognition: A survey. In *2020 International Conference on Power Electronics & IoT Applications in Renewable Energy and its Control (PARC)* (pp. 544–548). IEEE.

[3] Gupta, N., Gupta, S. K., Pathak, R. K., Jain, V., Rashidi, P., and Suri, J. S. (2022). Human activity recognition in artificial intelligence framework: a narrative review. *Artificial intelligence review*, 55(6), 4755–4808.

[4] Chen, K., Zhang, D., Yao, L., Guo, B., Yu, Z. and Liu, Y. (2021). Deep learning for sensor-based human activity recognition: Overview, challenges, and opportunities. *ACM Computing Surveys (CSUR)*, 54(4), 1–40.

[5] Pareek, P., and Thakkar, A. (2021). A survey on video-based human action recognition: recent updates, datasets, challenges, and applications. *Artificial Intelligence Review*, 54(3), 2259–2322.

[6] Mehra, P. S., Mehra, Y. B., Dagur, A., Dwivedi, A. K., Doja, M. N., and Jamshed, A. (2021). COVID-19 suspected person detection and identification using thermal imaging-based closed circuit television camera and tracking using drone in Internet of Things. *International Journal of Computer Applications in Technology*, 66(3–4), 340–349.

[7] Kumar, A., and Alam, B. (2016). Real-time fault tolerance task scheduling algorithm with minimum energy consumption. *In Proceedings of the Second International Conference on Computer and Communication Technologies: IC3T 2015*, Volume 2 (pp. 441–448). Springer India.

[8] Dagur, A., Kaushik, A., Rastogi, A., Singh, A., Kumar, A. and Chaturvedi, R. (2021). Optimization of queries in database of cloud computing. *In Data Intelligence and Cognitive Informatics: Proceedings of ICDICI 2020* (pp. 325–332). Springer Singapore.

[9] Ramanujam, E., Perumal, T., and Padmavathi, S. (2021). Human activity recognition with smartphone and wearable sensors using deep learning techniques: A review. *IEEE Sensors Journal*, 21(12), 13029–13040.

[10] Human Activity Recognition (HAR): Fundamentals, Models, Datasets. URL: https://www.v7labs.com/blog/human-activity-recognition#what-is-human-activity-recognition-har (Visited: 13/03/2024)

[11] Machine Learning. URL: https://www.geeksforgeeks.org/machine-learning/ (Visited: 14/03/2024)

[12] Hussain, Z., Sheng, M., and Zhang, W. E. (2019). Different approaches for human activity recognition: A survey. *arXiv preprint arXiv*:1906.05074.

[13] Pires, I. M., Garcia, N. M., Zdravevski, E. and Lameski, P. (2022). Daily motionless activities: A dataset with accelerometer, magnetometer, gyroscope, environment, and GPS data. *Scientific Data, 9* (1), 105.

[14] Parameswari, V., and Pushpalatha, S. (2020). Human activity recognition using SVM and deep learning. *European Journal of Molecular & Clinical Medicine*, 7(4), 1984–1990.

[15] Khan, I. U., Afzal, S. and Lee, J. W. (2022). Human activity recognition via hybrid deep learning based model. *Sensors*, 22(1), 323.

[16] Muhammad, K., Ullah, A., Imran, A. S., Sajjad, M., Kiran, M. S., Sannino, G. and de Albuquerque, V. H. C. (2021). Human action recognition using attention based LSTM network with dilated CNN features. *Future Generation Computer Systems*, 125, 820–830.

[17] Mutegeki, R., and Han, D. S. (2020, February). A CNN-LSTM approach to human activity recognition. In *2020 International Conference on Artificial Intelligence in Information and Communication (ICAIIC)* (pp. 362–366). IEEE.

[18] Challa, S. K., Kumar, A., and Semwal, V. B. (2022). A multibranch CNN-BiLSTM model for human activity recognition using wearable sensor data. *The Visual Computer*, 38(12), 4095–4109.

Computational Methods in Science and Technology Volume 1 – Kaur et al. (Eds)

Cryptocurrency price predictor

Saurabh Kumar, Virender Kumar, Dhirendra Kumar Shukla & Arvind Dagur
Department of SCSE, Galgotias University, UP, India

ABSTRACT: Cryptocurrency is a recent innovation in the financial sector that takes investment and research into account. Cryptocurrencies are difficult to predict in terms of value due to their rapid value declines over brief periods of time. At the moment, more than a thousand cryptocurrencies are registered and in use. This article examines the use of recurrent neural network algorithms to the prediction of cryptocurrency values, with particular attention on Bitcoin, Ethereum, and Litecoin. In particular, this study looks at three RNN algorithms: biLSTM, LSTM, and gated recurrent unit (GRU)—all based on mean absolute percentage error (MAPE). estimating the price of bitcoin on the market. The results show that the GRU algorithm outperforms the other two algorithms for BTC, LTC, and ETH in terms of accuracy and MAPE percentages of 0.2116%, 0.2454%, and 0.8267%. Conversely, the biLSTM algorithm had the lowest prediction accuracy 5.990%, 6.85%, 2.332% are the MAPE percentages for the BTC, ETH, LTC respectively. In summary, the data indicates that the GRU algorithm yields the most accurate forecasts that correspond with the real cryptocurrency values, making it the best model for doing so. Despite the fact that using this approach is the only It is advised to invest at your own risk and do research to find cryptocurrency investors who appear to be seeing the greatest price gain for a certain coin.

Keywords: Blockchain, Cryptocurrency, ML

1 INTRODUCTION

Digital currency is known as a Cryptocurrency that functions without the need for central bank and relies on encryption methods to validate transactions and regulate the generation of new units. The extremely important advantage of cryptocurrencies is that it's decentralized, which means there is no central authority to manage your transactions. For example, in a simple bank transaction, if a man wants to send money to another man, there is a main authority, i.e., the bank itself, which has all the details of the transaction between person 1 and person 2. This kind of interference is eliminated in cryptocurrency technology.

Cryptocurrency is secured by various cryptographic algorithms, making it impossible to forge and double spend. Another main feature of cryptocurrency is that it is based on blockchain technology, which is very complex and stores data that cannot be altered, hacked, or manipulated. Bitcoin was the first cryptocurrency introduced in 2009 in the real world by Satoshi Nakamoto. It is uncertain whether cryptocurrencies will achieve widespread adoption in the global markets. Currently, there are 5000 different types of listed cryptocurrencies and 5.8 million users using this technology.

One consequence of using cryptocurrency is that, as it is decentralized, the bad elements in society can use it for nefarious purposes such as fraud/scams[1]. In contrast to blockchain, cryptocurrency is totally based on token-based distributed ledger technology.

 DOI: 10.1201/9781003501244-77

Cryptocurrency is a kind of subpart of blockchain technology, which is a very broad technology in itself. While it may be appropriate to refer to Bitcoin's token as the technology, the same cannot be said for other blockchain projects such as Ethereum. Unlike Bitcoin, Ethereum's has its own technology called Ethereum, with its native token known as Ether, and transactions are settled using gas. Given the unpredictable nature of cryptocurrency prices and the wide range of cryptocurrencies in use globally, it becomes challenging to forecast their fluctuations accurately. The main focus of this study is on three popular cryptocurrencies. The objective is to utilize advanced deep learning algorithms to analyze data and identify concealed patterns. By integrating these patterns, the aim is to make more precise forecasts regarding the fluctuations of these cryptocurrencies.

The goal of this study is to conduct a comprehensive analysis of the many methods currently employed to predict the prices of the three main cryptocurrencies, BTC, ETH, and LTC. Artificial intelligence algorithms like LSTM, bi-LSTM, and GRU must be applied to produce accurate bitcoin price projections. We will use short learning memory (LSTM), a deep learning technique, and Fbprophet, a machine learning tool, to predict bitcoin prices with high accuracy.

2 LITERATURE SURVEY

In order to anticipate cryptocurrency prices, the LSTM (Long Short Term Memory) network processes previous price data that has been organized into regular patterns[7]. This model describes long-term relationships in data and forecasts future values using memory cells and gates. By modifying internal parameters, the LSTM network gains the ability to decrease prediction errors during training. After learning, they can make predictions based on new input data. Evaluating the performance of the model involves comparing its prediction to the actual value using metrics such as MAE, MSE, or RMSE.

Yahoo Finance Stock market data has been used by Ferdiansyan [1] and his team to predict the Bitcoin price by using LSTM[13] method. Where they have proposed a algorithm named Root Mean Square Error (RMSE) to analyse the result of LSTM. They had used 4years previous data to train and 1year data for testing. By using this they tried to predict next day bitcoin price. Their model with time series technique produces efficient result which after analysed via split the data as training and testing predict next day price. There was no big disadvantages in RMSE.

The auther in [2] has used hybride model of LSTM and GRU to predict the next day price of Litecoin and Zcash and they have taken bitcoin a parent currency assuming that dependencies between them. In their model resulted price is compared with actual value to give final predicted price for next day.

RL(Reinforcement Learning) algorithm [3] has used for price prediction of Litecoin and Munero. RL is a learning based structure which learn rewards and use it to predict price.

A better advanced CNN-LSTM model was proposed by Liveris, I.E [4] which provide much decent result in the way of fluctuation of prices.

Written on C2P2, an engine that helps predict the high and low prices of all cryptocurrencies. Anticipating trends based on past prices is a source of information from Reddit.

Some researchers have also used the LSTM algorithm and have been able to predict the future market price at a certain price but with low RMSE.

3 RESOURCES AND APPROACH

Remember the primary goal of this essay. We developed three distinct models that account for the past values of distinct cryptocurrencies. The performance of the suggested schemes is then assessed in this study by carefully examining the various methods available to forecast the prices of three cryptocurrencies: Ethereum (ETH), Bitcoin (BTC), and Litecoin (LTC). To assess the efficacy of our model, these methods are now contrasted with the market-dominant model. Five phases made up the assessment process: (1) gathering historical bitcoin data; (2) analysing the data; (3) creating a three- dimensional model; (4) testing the model; and (5) producing an output screen. Based on historical data, we employ three different types of algorithms—long - term short - term memory (L S T M), gated recurrent, and unit (GRU), and bi - L S T M (bi-LSTM).

We illustrated the methodology of dealing with the dataset in Figure 1. The process started with data collection, followed by data visualization to explore the behavior and distribution of the data and the correlation between cryptocurrencies. We then trained the models on 80% of the collected data sets that covered the period from January 22, 2018 to October 22, 2020, and tested them on the remaining 20% of data that covered the period from October 22, 2020 to June 30.2021.

To find the optimal model for the daily closing price, we test the models once they have been trained and evaluate the outcomes. We assess the effectiveness of the suggested hybrid model using Evaluation matrices such MAPE (Mean Absolute Percentage Error) and (RMSE)(Root Mean Square Error). [3]We have to identify the best algorithm for forecasting the price of a cryptocurrency by contrasting the accuracy of several models. According to the model's results, the Gated Recursive Unit (GRU) ranked better at forecasting the prices of these different kinds of cryptocurrencies than the (LSTM) (Long-term Memory and biLSTM algorithms.

In summary, this study demonstrates how well deep learning algorithms such as GRU, bi-LSTM, and LSTM estimate the values of cryptocurrency. The suggested hybrid models that included LSTM with Fbprophet demonstrated increased forecasting accuracy for the values of Litecoin, Ethereum, and Bitcoin. For traders and investors looking to make wellinformed judgements about investing in cryptocurrencies, the report offers insightful information.

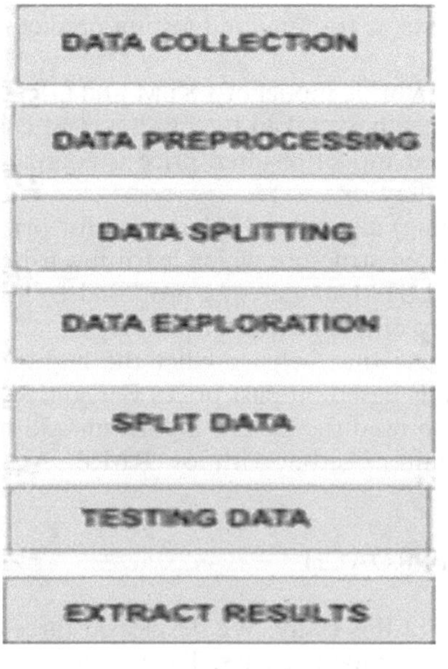

Figure 1. Data processing model.

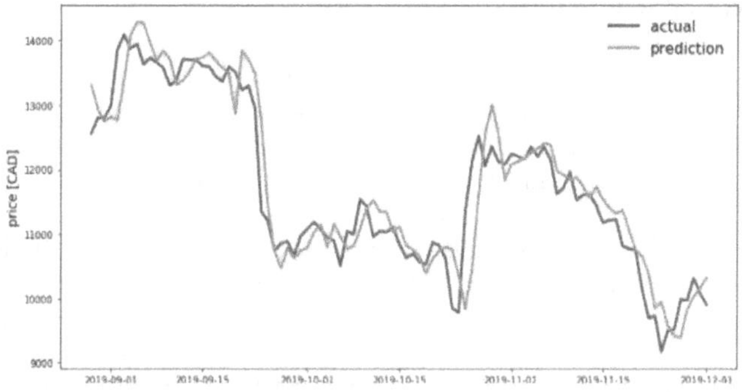

Figure 2. Testing and training dataset of bitcoin.

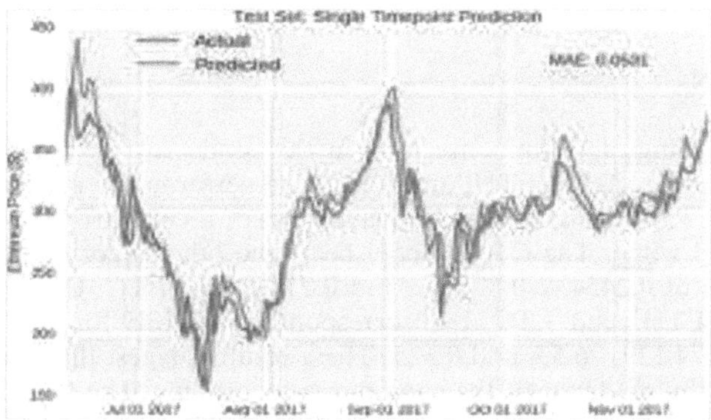

Figure 3. Testing and training dataset of Ethereum.

Table 1. Dataset specification.

Variable Name	Variable Description	Data Type
Date	Date of Observation	Date
Open	Opening price on the given day	Number
High	High price on the given day	Number
Low	Low price on the given day	Number
Close	Close price on the given day	Number

Sample of the data from BTC dataset

	Date	Open	High	Low	Close
0	2018-01-22	11,348	11,869	10,051	10,264
1	2018-01-23	10,264	11,358	9,972	10,984
2	2018-01-24	10,986	11,474	10,487	11,208
3	2018-01-25	11,213	11,711	10,889	11,246
4	2018-01-26	11,241	11,609	10,321	10,899

Sample of the data from ETH dataset

	Date	Open	High	Low	Close
0	2018-01-22	1,052.22	1,090.99	913.24	1,000.5
1	2018-01-23	1,000.12	1,025.0	910.0	985.96
2	2018-01-24	984.1	1,067.0	957.9	1,063.77
3	2018-01-25	1,063.77	1,107.67	1,000.01	1,048.0
4	2018-01-26	1,048.0	1,079.95	982.0	1,051.07

Sample of the data from LTC dataset

	Date	Open	High	Low	Close
0	2018-01-22	190.93	195.85	165.1	180.01
1	2018-01-23	180.01	187.39	165.25	177.32
2	2018-01-24	178.01	186.13	173.0	180.99
3	2018-01-25	181.18	185.0	175.0	179.05
4	2018-01-26	179.05	182.7	165.37	175.39

Figure 4. Testing and training dataset of litecoin.

4 DATASET

The study's data set[9], which only includes one, was taken from an open-access website. Three pages make up the CSV file: one for Litecoin (LTC), one for Ethereum (ETH), and one for Bitcoin (BTC). The database includes 1277 items in its time series data collection, including daily pricing data from January 1, 2018, to June 30, 2021. [11]The specification of the database utilised in this study is provided is discussed and furthermore, sample data from a database containing bitcoin goals is displayed in Figure 4.

5 RESULT AND DISCUSSIONS

The accuracy of several models for Bitcoin currenciesis[10], displayed. The GRU model[15] has the lowest RMSE of 174.129 and MAPE of 0.2454 among these models. When it comes to forecasting BTC trends, the GRU model performs better than the LSTM and bi-LSTM models. In a similar vein, in Figure 4 displays this model's accuracy for the cryptocurrency ETH.

Among LSTM and bi-LSTM[14] models, the GRU model is the best predictor since it has the lowest MAPE (0.8267) and RMSE (26.59) for the cryptocurrency ETH. In Figure 4 presents the model's accuracy for LTC coins. The GRU model is the best effective predictor when compared to the LSTM and bi-LSTM models since it has the lowest average absolute percentage error (0.2116) and the root mean square (0.825). All things considered[12], it is reasonable and dependable to use the model for cryptocurrency prediction that this study has presented. The model suggested in this study is contrasted with the model.

6 CONCLUSION

Three machine learning algorithms are used in this study to forecast the values of three distinct cryptocurrencies: BTC, ETH, and LTC. The performance metrics used to evaluate the accuracy of the various models are displayed. Next, a comparison is done between the estimate and actual value. The GRU model beats the other algorithms, according to the data, with MAPEs of 0.2454%, 0.8267%, and 0.2116% for BTC, ETH, and LTC, respectively. For BTC, ETH, and LTC, the corresponding RMSE for the GRU model was determined to be 174.129, 26.59, and 0.825. These results suggest that the GRU model is a good and trustworthy predictor of the chosen cryptocurrency. The BiLSTM model, on the other hand, exhibits a larger price discrepancy between the actual and forecast values for Bitcoin and Ethereum, indicating a worse accuracy than both GRU and LSTM. Overall, the experimental findings demonstrate that the AI algorithm is a dependable and simple method for predicting bitcoin prices, with GRU outperforming LSTM and bi-LSTM models.

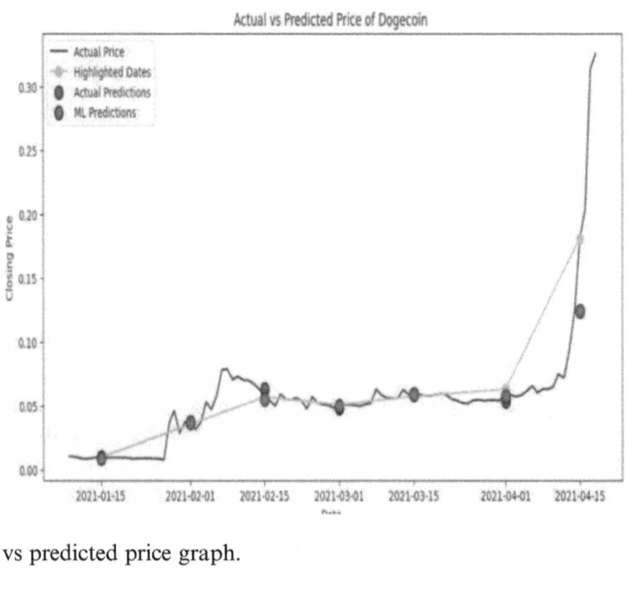

Figure 5. Actual vs predicted price graph.

REFERENCES

[1] Brenig, C., Accorsi, R., Müller, G. (2015). Economic analysis of cryptocurrency backed money laundering. *ECIS 2015 Completed Research Papers*. Paper 20. Available online: https://aisel.aisnet.org/ecis2015_cr/20 (accessed on 16 June 2021).

[2] Eyal, I. (2017). Blockchain technology: transforming libertarian cryptocurrency dreams to finance and banking realities. *Computer*, 50, 38–49.

[3] DeVries, P. (2016). An analysis of cryptocurrency, Bitcoin, and the Future. *Int. J. Bus. Manag. Commer.*, 1, 1–9.

[4] Jang, H.; Lee, J. (2017). An empirical study on modeling and prediction of bitcoin prices with bayesian neural networks based on blockchain information. *IEEE Access* 2017, 6, 5427–5437. [CrossRef]

[5] Saad, M., Choi, J., Nyang, D., Kim, J., Mohaisen. (2019). Toward characterizing blockchain-based cryptocurrencies for highly accurate predictions. *IEEE Syst. J.* 14, 321–332. [CrossRef]

[6] Gautam, K., Sharma, N., Kumar, P. (June 2020). Empirical analysis of current cryptocurrencies in different aspects. In *Proceedings of the ICRITO 2020—IEEE 8th International Conference on Reliability*, Infocom Technologies and Optimization (Trends and Future Directions), Noida, India, 4–5; pp. 344–348. [CrossRef]

[7] Iwamura, M., Kitamura, Y., Matsumoto, T. (February 28, 2014). Is bitcoin the only cryptocurrency in the town? economics of cryptocurrency and friedrich A. *Hayek*. Available online: https://ssrn.com/abstract=2405790

[8] Kyriazis, N. A. (2019). A survey on empirical findings about spillovers in cryptocurrency markets. *J. Risk Financ. Manag.*, 12, 170. [CrossRef]

[9] Hassani, H., Huang, X., Silva, E. BigCrypto: Big Data, blockchain and cryptocurrency. *Big Data Cogn. Comput.*

[10] Nizzoli, L.; Tardelli, S.; Avvenuti, M.; Cresci, S.; Tesconi, M.; Ferrara, E. (2020). Charting the landscape of online cryptocurrency manipulation. *IEEE Access*, 8, 113230–113245. [CrossRef]

[11] Rehman, M.U.; Apergis, N. (2019). Determining the predictive power between cryptocurrencies and real time commodity futures: Evidence from quantile causality tests. *Resour. Policy*, 61, 603–616. [CrossRef]

[12] Liew, J.; Li, R.Z.; Budavári, T.; Sharma, A. (2019). Cryptocurrency investing examined. *J. Br. Blockchain Assoc.*, *2*, 1–12.

[13] C.-H. Wu, C.-C. Lu, Y.-F. Ma and R.-S. Lu, (2018). A new forecasting framework for bitcoin price with LSTM, *2018 IEEE International Conference on Data Mining Workshops (ICDMW)*, Singapore 2018, pp. 168–175, doi: 10.1109/ICDMW.2018.00032.

[14] N. P, R. J. Tom, P. Gupta, A. Shanthini, V. M. John and V. Sharma, (2021). Prediction of bitcoin price using Bi-LSTM network, *2021 International Conference on Computer Communication and Informatics (ICCCI)*, Coimbatore, India, 2021, pp. 1–5, doi: 10.1109/ICCCI50826.2021.9402427.

[15] A.-A. Encean and D. Zinca. (2022). Cryptocurrency price prediction using LSTM and GRU networks, *2022 International Symposium on Electronics and Telecommunications (ISETC)*, Timisoara, Romania, pp. 1.

Computational Methods in Science and Technology Volume 1 – Kaur et al. (Eds)
© 2025 the Author(s), ISBN: 978-1-032-81760-6

Privacy preservation in social networks using autoencoders

A. Singh & M. Singh
Department of CSE, Chandigarh University, Mohali, India

ABSTRACT: Owing to social networks' widespread use, numerous ideas have been put forth to safeguard the networks' privacy. These works all presuppose that the attackers make use of the same background information. However, many individuals have varying needs when it comes to privacy protection. Therefore, if assaults with identical background knowledge fail to satisfy privacy requirements, it forfeits the opportunity to capitalize on variations in users' privacy requirements and obtain superior utility. In this work, we present a system from the most common technique to deal with today's real-time problems i.e. Autoencoders i.e. based on neural networks. In this paper, we used the ARNET dataset of around 70K users and implemented Autoencoders to preserve the privacy of users. The performance of the network has been validated by evaluating the performance by the Random Forest classifier before and after anonymization. Performance parameters show the effectiveness of the proposed solution to the dataset of social network users.

Keywords: Privacy, Security, Anonymity, Autoencoders

1 INTRODUCTION

Social media has become so popular for sharing information instantly with others and thus it has also become a matter of concern related to the privacy preservation of individuals. The best resources for mining of data and its analysis are these open graph data (Li X. *et al.* 2023). By making graph data available to academic institutions and business partners, data companies unlock a vast amount of potential value from their data. This graph data can support decision models in a variety of applications, such as social, business, and transportation, by providing better advertising and recommendations, among other things (Wei Y. *et al.* 2023).

However, due to the abundance of information contained in graph data, users' relationships and privacy may be compromised. The protection of users' privacy under attacks is impossible to ensure, even if the graph is anonymized when it is released (Miao X. *et al.* 2022, Majeed A and Lee S. 2020). Traditional privacy protection approaches like K-anonymity, differential privacy, L-diversity, T-closeness, and others are frequently employed because of these reasons (Beigi G and Liu H. 2020). These approaches, however, do not fully utilize the empirical knowledge already present in graphs to extract prospective information; in contrast, deep learning is more effective in data abstraction since it performs representation learning over numerous nonlinear layers of transformation (Karimi Rizi A. *et al.* 2019).

Graph Neural Networks (GNNs) have recently shown a strong capacity to learn node representations, extending deep neural networks on graphical structured data by concurrently encoding the structure of network and properties of nodes. Tasks such as graph classification (Li D. *et al.* 2017), node classification (Yan, S. 2016), link prediction (Wang, C.

 DOI: 10.1201/9781003501244-78

2022), have demonstrated their strong capabilities. But overfitting in deep learning, which makes the model subconsciously recall the details of training data, poses a risk of privacy leakage. As a result, numerous techniques for integrating differential privacy with deep neural networks have been put out lately.

This paper is structured as follows. Section 2 discusses the context regarding privacy-preserving techniques available. Section 3 discusses our proposed technique for privacy protection and also explains the experimental setup and the results of the technique proposed. Finally, section 4 is about conclusion followed by references.

2 LITERATURE SURVEY

To maintain privacy and anonymize tabular data for use in various machine learning projects, we encoded it using an autoencoder. By maintaining the original data's confidentiality and causing little to no performance degradation, the data representation collected using the technique can be utilized in machine learning prediction tasks.

Tariq *et al.* (2020) examined recent developments in deep neural network learning with an emphasis on privacy and security in their review paper (Tariq, M *et al.* 2020). They draw attention to the privacy protection conditions that a framework must satisfy, such as the necessity that training data and predictions be private and that the client remains uninformed of the server and global model configurations. They thoroughly examined deep learning-related risks and assaults from the privacy and security perspectives. Additionally, they provide an exhaustive inspection of the theories and practices of privacy preservation that are currently covered in the literature available.

Tanuwidjaja *et al.* (2019) gave a thorough introduction to deep learning for preserving privacy in the setting of machine learning used as a service in their research (Tanuwidjaja, H. C. *et al.* 2020, Tanuwidjaja, H. C. *et al.* 2019). The author also explored the relevance of classical privacy-preserving methods to deep learning architectures, such as differential privacy and cryptography. They point out that some modifications—such as activation function approximations, big stride convolutional layers, batch normalization layers are required to integrate these conventional methods with deep learning.

Furthermore, the authors assess the difficulties and constraints that deep learning frameworks encounter. They distinguish between two primary problems: calculation and communication issues. These issues still need to be resolved in the field of deep learning and provide substantial obstacles to obtaining the best possible performance and privacy. Notably, we address these difficulties in this study, especially about the computation and transmission overheads. By addressing these issues, we enhance the overall performance and privacy guarantees while also furthering the field of deep learning.

In other studies, including (Chen, Z. and Wang, Y. 2023), local Differential Privacy is proposed as a distributed option to guarantee privacy without depending on a single guardian in situations where a centralized institution for data accumulation is not practical or reliable. Nevertheless, even with modest privacy costs, they are intricate methods that require a significant amount of processing from the training and yield lesser accuracy.

The communication parties encrypt their data when utilizing homomorphic encryption (Li, J *et al.* 2020), and the server uses the ciphertexts to deliver the parties the encrypted prediction. CryptoNets (Xie, P. *et al.* 2014) was the first study showing the use of homomorphic encryption with neural networks. Homomorphic encryption allows polynomial function to be calculated for encrypted data by estimating the activation functions (non-linear) and turning neural network into polynomials (Xie, P. *et al.* 2014). A related study called CryptoDL assesses convolutional neural networks over encrypted data and estimates low-degree data for activation functions (Hesamifard, E *et al.* 2017). In addition, homomorphic encryption started to be used in conjunction with other technologies as a way to safeguard the confidentiality of the learning process (Aono, Y. *et al.* 2017, Phuong, T. 2019).

Unfortunately, because the encryption and decryption procedures incur additional processing and communication overhead, many of these initiatives are restricted to the prediction stage. Authors (Al Badawi, A. 2020) suggested GPU acceleration approach to speed up the whole homomorphic evaluation of the Convolutional Neural Network for encrypted data for communication and computational problems. It is intolerant of the training phase on encrypted data because of the extended runtime and increased number of operations. The main problem with employing homomorphic encryption with neural networks, despite all attempts, is that they introduce a computational cost during the training process. Furthermore, inadequate activation function approximation may lead to subpar neural network performance in real-world scenarios. Additionally, it produces encrypted messages that are larger and use more memory (Pulido-Gaytan, B. *et al.* 2021).

To summarize, the literature research (Singh, A. and Singh, M. 2022, Singh, M. and Singh, A. 2023, Singh, A. and Singh, M. 2023, October) reveals several noteworthy obstacles, such as the balance between the usefulness of the data and its privacy, threat model presumptions, and communication and processing issues. These difficulties highlight the requirement for a framework that protects privacy in the context of these problems. By safeguarding the privacy of the data, our approach fills this gap.

3 PROPOSED APPROACH AND PROBLEM STATEMENT

Many researchers have worked in the area of privacy preservation of social network data. Data anonymization using autoencoder has not been proposed yet. In this paper, we propose an autoencoder application to the data of social networks. The loss between the input and reconstructed output is acceptable and also it preserves the confidentiality of social network data.

3.1 *Implementation details*

Autoencoders can be employed for preserving the privacy of social network users through a technique called privacy-preserving data obfuscation or anonymization. The following flowchart of Figure 1 explains the process. Autoencoders provide a powerful framework for privacy-preserving data obfuscation in social networks by transforming sensitive data into an obfuscated representation that preserves privacy while maintaining the utility of the data for analysis and other applications.

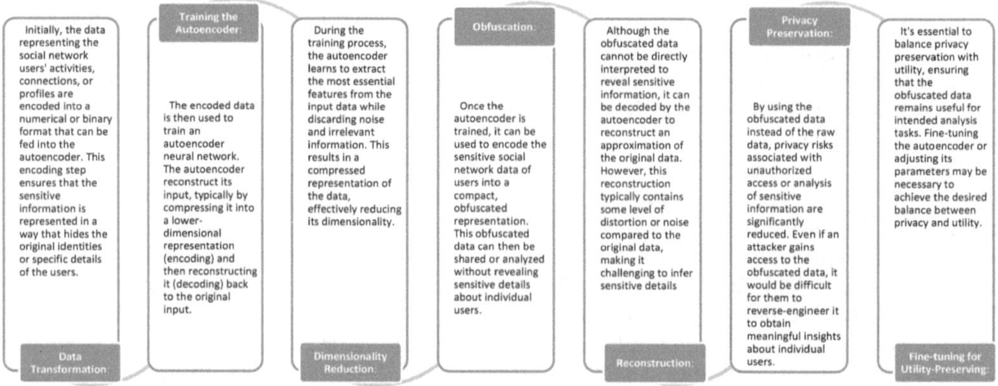

Figure 1. Flowchart showing the process of anonymization using autoencoders.

3.2 *Autoencoder*

One kind of deep learning network that is trained to reproduce its input is called an autoencoder (Rodriguez, D., and Krishnan, R. 2023). An encoder and a decoder are the two smaller networks that make up an autoencoder. The input data is mapped by the encoder to a feature vector in a latent space. In this latent space, vectors are used by the decoder to recreate data. Training takes place under no supervision. Put otherwise, the model doesn't need data that has been tagged. You can use the decoder to generate text by reconstructing text from any input. The encoder maps the input text into latent vectors using word embeddings and LSTM operation. The decoder reconstructs the text from the latent vectors using the same embedding and an LSTM operation (Weggenmann, B. *et al.* 2022). Loss is the difference between the reconstructed output and the input data.

A decent latent representation can be produced by an autoencoder that has received proper training. Despite being significantly different from the original data, this representation contains all the information required by the stakeholders from the input layer. Three hidden layers make up the employed autoencoder. There is no difference in size between the input and output layers. The neural network's training process involves computing the difference between the input and output in order to backpropagate the loss and update the weights. In contrast, the predictive phase solely utilizes the encoder part's weights because only the latent representation is required.

Then the dataset is loaded that has already been divided into testing and training sets.

Here's how we may construct our autoencoder:

```
Autoencoder1, encoder1 = build_autoencoder1(
  dim_input=X1_training1.shape1[1],
  dim_layer1_1=64,
  dim_layer1_2=16)
```

Using the training set, we train the network's weights. After the training is finished, testing of the autoencoder takes place. Raw data, encoded data and newly constructed data can be plotted.

The input text and the reconstructed ones (from the latent representations) resemble each other quite a bit. This indicates that even while the bottleneck representations (encoded) that the autoencoder learned are not human-readable, they are nevertheless a good representation of the original material as shown in Figure 2. In order to obtain the representations of the original data in the latent space and so anonymize it, we will apply this concept again to a tabular dataset.

Figure 2. Sample data-original text, latent representation, and newly constructed text.

3.3 *Dataset*

For this experiment, we have used ARNET dataset (Yuan, M. *et al.* 2011) as shown in Table 1.

Table 1. Overview of the dataset.

ID	Name	Edge ID	Weight	Label ID
1	37171	6000	1	1
1	37171	2	1	1
1	37171	8	1	1
1	37171	11	1	1
1	37171	19	1	1
1	37171	22	1	1
1	37171	23	1	1
6	27576	107	1	0
6	27576	108	1	0
6	27576	109	1	0
6	27576	110	1	0
7	27547	6000	1	2
7	27547	8	1	2
7	27547	13	1	2
7	27547	20	1	2
7	27547	67	1	2
7	27547	79	1	2
7	27547	111	1	2
7	27547	112	1	2
7	27547	113	1	2
7	27547	114	1	2
7	27547	115	1	2
7	27547	116	1	2
7	27547	117	1	2
7	27547	118	1	2

3.4 *Results-original raw data*

We can attempt using a simple random forest in cross-validation to evaluate the baseline performance prior to anonymizing the data. It is important to remember that what matters is not the best model; rather, it is the difference between a model trained on original data and one built on encoded (anonymized) data.

```
Rfl = RandomForestClassifier1(
x_estimators1=400,
max_depth1=2,
x_jobs1=4,
random1_state1=32)
dict_performance1 = cross_validate1(
estimator1=rf1,
X1=X1, y1=y1,
cv1=10,
n1_jobs1=4,
return_training_score1=True,... ...... ... ... .... .
```

3.5 *Data anonymization with autoencoders*

We can now proceed with anonymizing the dataset. First, we construct an autoencoder with the input layer being twice as large as the bottleneck layer.

```
dim_layer1_input1 = X.shape1[1]
dim_layer1_1 = max((int(3*dim_layer1_input1/4), 1))
dim_layer1_2 = max((int(dim_layer1_input1/2), 1))
autoencoder1, encoder1 = build_autoencoder1(
  dim_input1=dim_layer_input1,
  dim_layer1_1=dim_layer1_1,
  dim_layer1_2=dim_layer1_2
)
```

Now training of the neural network is done and take out the random forest classifier's encoded representation.

Results before and after anonymization are shown in Figure 3.

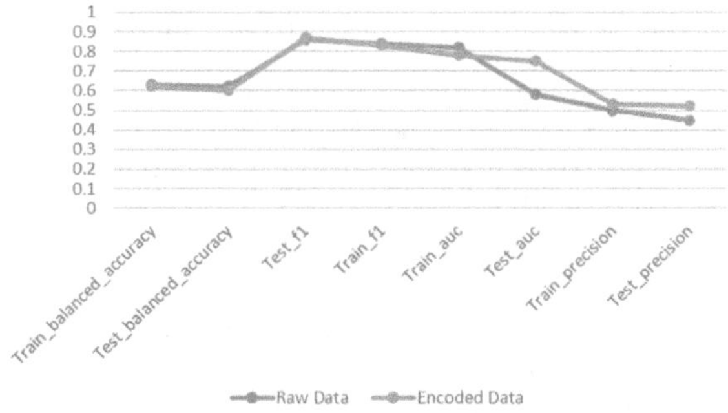

Figure 3. Results of ARNET before and after anonymization using various parameters.

4 CONCLUSION

The use of an autoencoder to anonymize a dataset is presented in this study. An autoencoder that has received proper training maintains the original data's predictive potential. In order to overcome the shortcomings of previous privacy-preserving methods, this study proposes and assesses a deep learning framework that protects privacy while combining deep learning frameworks. There are no concerns regarding user privacy, computational overhead, or communication overheads when transmitting the encoded sets. Results from experimental evaluations using ARNET datasets are promising. The findings demonstrate that our framework protects privacy while significantly cutting training time and sacrificing just less accuracy. The study emphasizes the framework's applicability for collaborative deep learning that protects privacy in real-world settings, highlighting the potential in industries like healthcare, smart security and surveillance systems, and finance where scalability and privacy are critical requirements. Our future work will focus on addressing the limits of the framework, optimizing reduction techniques for a variety of datasets, improving the performance of the model, and implementing this framework in practical applications for a thorough assessment.

REFERENCES

Al Badawi, A., Jin, C., Lin, J., Mun, C. F., Jie, S. J., Tan, B. H. M., and Chandrasekhar, V. R. (2020). Towards the alexnet moment for homomorphic encryption: Hcnn, the first homomorphic cnn on encrypted data with gpus. *IEEE Transactions on Emerging Topics in Computing*, 9(3), 1330–1343.

Aono, Y., Hayashi, T., Wang, L. and Moriai, S. (2017). Privacy-preserving deep learning via additively homomorphic encryption. *IEEE Transactions on Information Forensics and Security*, 13(5), 1333–1345.

Beigi G and Liu H. (12-Mar-2020). A survey on privacy in social media. *ACM/IMS Transactions on Data Science*. 1:1. (1–38). Online publication date. https://doi.org/10.1145/3343038

Chen, Z., and Wang, Y. (2023). Locally differentially private distributed online learning with guaranteed optimality. *arXiv preprint arXiv*:2306.14094.

Hesamifard, E., Takabi, H. and Ghasemi, M. (2017). Cryptodl: Deep neural networks over encrypted data. *arXiv preprint arXiv*:1711.05189.

Karimi Rizi A., Naderi Dehkordi M., Nemat bakhsh N. and Maglaras L. (1-Jan-2019). SNI. *Security and Communication Networks*. **2019**. Online publication date. https://doi.org/10.1155/2019/8171263

Li D., Lv Q., Shang L. and Gu N. (5-Jan-2017). Efficient privacy-preserving content recommendation for online social communities. *Neurocomputing*. **219**:C. (440–454). Online publication date. https://doi.org/10.1016/j.neucom.2016.09.059

Li X., Chen L. and Wu D. (29-Feb-2024). Adversary for social good: leveraging adversarial attacks to protect personal attribute privacy. *ACM Transactions on Knowledge Discovery from Data*. **18**:2. (1–24). Online publication date. https://doi.org/10.1145/3614098

Li, J., Kuang, X., Lin, S., Ma, X. and Tang, Y. (2020). Privacy preservation for machine learning training and classification based on homomorphic encryption schemes. *Information Sciences*, 526, 166–179.

Majeed A. and Lee S. (1-Aug-2020). Attribute susceptibility and entropy based data anonymization to improve users community privacy and utility in publishing data. *Applied Intelligence*. **50**:8. (2555–2574). Online publication date. https://doi.org/10.1007/s10489-020-01656-w

Miao X., Zhang W., Jiang Y., Fu F., Shao Y., Chen L., Tao Y., Cao G. and Cui B. (1-Jul-2023). PCG: a privacy preserving collaborative graph neural network training framework. *The VLDB Journal — The International Journal on Very Large Data Bases*. **32**:4. (717–736). Online publication date.

Phuong, T. T. (2019). Privacy-preserving deep learning via weight transmission. *IEEE Transactions on Information Forensics and Security*, *14*(11), 3003–3015.

Pulido-Gaytan, B., Tchernykh, A., Cortés-Mendoza, J. M., Babenko, M., Radchenko, G., Avetisyan, A. and Drozdov, A. Y. (2021). Privacy-preserving neural networks with homomorphic encryption: C hallenges and opportunities. *Peer-to-Peer Networking and Applications*, 14(3), 1666–1691.

Rodriguez, D. and Krishnan, R. (2023, July). An autoencoder-based image anonymization scheme for privacy enhanced deep learning. In *IFIP Annual Conference on Data and Applications Security and Privacy* (pp. 302–316). Cham: Springer Nature Switzerland.

Singh, A. and Singh, M. (2022). Social networks privacy preservation: A novel framework. *Cybernetics and Systems*, 1–32.

Singh, A. and Singh, M. (2023). Addressing semantic similarity attacks in online social networks. In *2023 IEEE International Conference on Blockchain and Distributed Systems Security (ICBDS)* (pp. 1–5). IEEE.

Singh, M. and Singh, A. (2023). How safe you are on social networks?. *Cybernetics and Systems*, 54(7), 1154–1171.

Tanuwidjaja, H. C., Choi, R. and Kim, K. (2019). A survey on deep learning techniques for privacy-preserving. In *Machine Learning for Cyber Security: Second International Conference*, ML4CS 2019, Xi'an, China, September 19–21, 2019, Proceedings 2 (pp. 29–46). Springer International Publishing.

Tanuwidjaja, H. C., Choi, R., Baek, S. and Kim, K. (2020). Privacy-preserving deep learning on machine learning as a service—a comprehensive survey. *Ieee Access*, 8, 167425–167447.

Tariq, M. I., Memon, N. A., Ahmed, S., Tayyaba, S., Mushtaq, M. T., Mian, N. A. and Ashraf, M. W. (2020). A review of deep learning security and privacy defensive techniques. *Mobile Information Systems*, 2020, 1–18.

Wang, C., Tianqing, Z., Xiong, P., Ren, W. and Choo, K. K. R. (2022). A privacy preservation method for multiple-source unstructured data in online social networks. *Computers & Security*, 113, 102574.

Weggenmann, B., Rublack, V., Andrejczuk, M., Mattern, J. and Kerschbaum, F. (2022, April). DP-VAE: Human-readable text anonymization for online reviews with differentially private variational autoencoders. In *Proceedings of the ACM Web Conference* 2022 (pp. 721–731).

Wei Y., Fu X., Yan D., Sun Q., Peng H., Wu J., Wang J. and Li X. (1-Oct-2023). Heterogeneous graph neural network with semantic-aware differential privacy guarantees. *Knowledge and Information Systems*. **65**:10. (4085–4110). Online publication date. https://doi.org/10.1007/s10115-023-01895-6

Xie, P., Bilenko, M., Finley, T., Gilad-Bachrach, R., Lauter, K. and Naehrig, M. (2014). Crypto-nets: Neural networks over encrypted data. *arXiv preprint arXiv*:1412.6181.

Yan, S., Pan, S., Zhao, Y. and Zhu, W. T. (July 4-6, 2016). Towards privacy-preserving data mining in online social networks: distance-grained and item-grained differential privacy. In *Information Security and Privacy: 21st Australasian Conference, ACISP 2016, Melbourne, VIC, Australia, Proceedings, Part I 21* (pp. 141–157). Springer International Publishing.

Yuan, M., Chen, L., Philip, S. Y. and Yu, T. (2011). Protecting sensitive labels in social network data anonymization. *IEEE Transactions on Knowledge and Data Engineering*, 25(3), 633–647.

Computational Methods in Science and Technology Volume 1 – Kaur et al. (Eds)
© 2025 the Author(s), ISBN: 978-1-032-81760-6

Deep learning-based approaches for diagnosis and detection of osteoporosis using clinical data of CT and X-ray images

Prabhjot Kaur
Research Scholar, Department of CSE, Chandigarh Engineering College-CGC, Landran

Sukhpreet Kaur
Professor, Department of CSE, Chandigarh Engineering College-CGC, Landran

Parneet Kaur
Assistant Professor, Department of CSE, Chandigarh Engineering College-CGC, Landran

ABSTRACT: Osteoporosis is a silent killer that mainly affects the elderly owing to bone fragility and fracture. The patient's life is saved by an early and accurate osteoporosis diagnosis. Understanding the BMD (Bone Mineral Density) is crucial for the early diagnosis and treatment of osteoporosis as well as the management of its effects in later life. Recent years have seen a major impact on traditional diagnostic and treatment procedures due to major advancements in AI, particularly in machine learning (ML) and deep learning (DL). Using medical data (such as demographics and typical laboratory test results) and CT images of the lumbar vertebral bodies instead of DXA data, the current work focuses on building machine and deep learning models for the accurate identification and diagnosis. A knee X-ray image dataset of 372 images have been taken for the study. Eight deep transfer learning approaches have been applied to the set of images namely ResNet50V2, DenseNet201, InceptionResNetV2, ResNet101V2, EfficientNetB2, MobileNet, EfficientNetB4 and VGG16. The simulation outcomes show that VGG16 beat all other methods and achieved a validation accuracy of 82.73.

1 INTRODUCTION

Osteoporosis is a medical condition that is indicated mainly through low bone density and tissue deterioration, that raises the risk of bone breakage. It causes the bones to become brittle, which raises the possibility of bone breakage from even minor accidents or injuries, particularly in the hip, spine, and wrist. This disorder can affect persons of any age and gender, but older adults and especially postmenopausal women are particularly affected by it. According to a study (Khinda *et al.* 2022), In India 61 million people are suffering from this disease out of which 80% are women.

The predominant imaging method used by the medical community to detect ailments of the bones is X-ray imaging. Almost all the bones in the body, including the elbow, pelvis, knee, wrist, shoulder, and spine, can be visualized using X-rays imaging. It aids in the identification of fractures, broken bones, joint dislocations, infections, aberrant bone development, and even arthritis. The Magnetic Resonance Imaging (Chen *et al.* 2018; Ferizi *et al.* 2017), Quantitative Ultrasound System (QUS) (Hans and Baim, 2017), and Computed Tomography (CT) are additional imaging modalities that are utilized to identify osteoporosis. MRI is a 3 T enhanced bone microarchitecture imaging technique; however, it has limited spatial resolution and is quite expensive (Chen *et al.* 2018). QUS employs sound waves to analyze bones and is easy, non-invasive, portable, and economical, however, it is location-specific and does not have strong empirical backing (Hans and Baim 2017). Given these limitations, a cheap, easily

DOI: 10.1201/9781003501244-79

available, and trustworthy detection method is required. As a result, the researchers focus on the advancements made in imaging technology in order to develop computer-aided diagnostic systems (CAD) by using computer algorithms to analyze medical images.

In this study, we have developed an early detection and identification system for knee osteoporosis using strength of CNN and affordability of x-rays. For identifying knee X-ray images, our model uses eight deep transfer learning approaches namely ResNet50V2, DenseNet201, InceptionResNetV2, ResNet101V2, EfficientNetB2, MobileNet, EfficientNetB4 and VGG16.

2 RELATED WORK

Research on the complex disease of osteoporosis has been substantial. In complex data-driven environments, recent developments in machine learning (ML) have allowed AI to achieve amazing breakthroughs. Some of these issues are discussed in this review, which also provides information to stakeholders interested in using AI to manage osteoporosis more effectively. Despite numerous studies having substantial flaws, it was discovered that using images for opportunistic fracture detection or osteoporosis diagnosis was a potential method. Other interesting areas of research include initiatives to create ML-based models for identifying new factors that influence the risk of fracture and improving fracture prediction (Smets *et al.* 2021).

An end-to-end pipeline for identifying vertebral compression fractures caused by osteoporosis in CT images is presented by Yilmaz *et al.* in their study in 2021. A hierarchical neural network is used in the first of two steps of this procedure to find every vertebra. In contrast, the second stage uses a feed-forward CNN to examine patches comprising a single vertebra to assess the presence or absence of an osteoporotic fracture. The hybrid artificial intelligence-based methods is used by Sunnetci *et al.* (2022) in order to detect the existence of periodontal bone loss. 1000 deep picture features are recovered using SqueezeNet and AlexNet DL architectures from 1432 tagged images. EfficientNetB5, Medium Gaussian SVM, and Linear SVM are the ideal classifiers for Direct-Convolutional Neural Networks (CNN), SqueezeNet-based and AlexNet-based networks. The findings indicate that Linear SVM is the best classifier, with accuracy values of 81.49%.

In a study, proposed by Xiao *et al.* (2020) it was hypothesized that deep learning models based on DXA images may reasonably predict the key microstructural characteristics of trabecular bone. Seven human cadaveric proximal femurs were digitally sliced into 1249 trabecular cubes, and simulated DXA pictures were produced. The ability of CNN models to predict histomorphometry parameters was significantly influenced by the count and calibre of the input replicated DXA pictures. The results show that DXA images help to improve performance of osteoporotic fracture detection. The description of few other studies have been exhibited in Table 1 below:

Table 1. Research studies proposed by researchers for identification of osteoporosis.

Authors	Dataset	Technique Used	Challenges	Outcomes
Sukegawa *et al.* (2022)	778 image datasets from hospital between 2014-2020	CNN models, including EfficientNet-b0, -b3, and -b7 and ResNet-18, -50, and -152 are employed.	A large dataset can be considered for CNN models	ResNet-152 Accuracy: 84%.
Wani and Arora (2022)	381 x-ray images of knee	Deep transfer learning-based CNN i.e., VGGNet-16, ResNet, AlexNet and VGGNet-19	Dataset can be enhanced and other clinical factors can be included for proper classification	Accuracy: VGGNet-16: 86.3% AlexNet : 91.1% VGGNet-19: 84.2% ResNet: 86.4%

(*continued*)

Table 1. Continued

Authors	Dataset	Technique Used	Challenges	Outcomes
Singh *et al.* (2021)	Labelled Dataset of 70 Images	DCNN based CAD approach	Dataset size is limited and can be further enhanced	Accuracy: 87.86%
Zhang *et al.* (2020)	2510 lumbar spine X-ray pictures of 1255 postmenopausal women having age range 50-92 years.	3-class DCNN model	Only women data is included	AUC: Sensitivity Ratio Test Set 1 Osteoporosis: 0.767: 73.7% Osteopenia: 0.787: 81.8% Test Set 2 Osteoporosis: 0.726: 68.4% Osteopenia: 0.810: 85.3%
Kumar *et al.* (2022)	Knee X-ray Kaggle Dataset	Osteo-Net model based on deep learning architecture	Limited Data Set Size and more learning models can be used	Accuracy Validation: 84.06% Testing: 82.61%
Nasser *et al.* (2017)	Orleans Hospital Bone X-ray Dataset of 87 persons	SSAE in combination with SVM classifier	Size and Efficacy of the dataset used	Accuracy: 95.5%
Huang *et al.* (2022)	Manually collection of data from Dept. of Endocrinology Wenzhou Medical University of patients older than 40 years	Six ML Models	External validation of ML model can be enhanced by using more participants	GBM outperformed with- Accuracy: 81% AUROC: 0.86 Specificity: 0.92 Sensitivity: 0.70

3 PROPOSED METHODOLOGY

The proposed study aims at detecting osteoporosis in inception stages with help of machine/DL approaches in order to prevent severe complications caused by osteoporosis i.e., osteoporosis fracture. This section discusses about the methodology adopted for classifying the given input into two categories i.e., images with osteoporosis and images with normal BMD. The primary input for proposed framework comprises of CT images encompassing lumbar vertebral bodies and demographic characteristics of the patients along with routine laboratory tests (clinical data). Figure 1 provides a diagrammatical representation of the proposed workflow.

3.1 *About the dataset*

The dataset includes x-ray of knee taken from Kaggle dataset repository (https://www.kaggle.com/datasets/stevepython/osteoporosis-knee-xray-dataset, 2022) which contains two categories of images. One of the persons with normal bone health and the other with person having osteoporosis. Each category set contains 186 images.

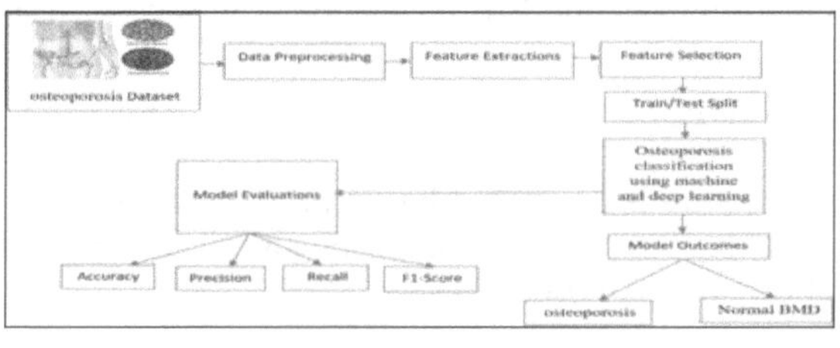

Figure 1. Workflow for osteoporosis classification in early stages.

3.2 *Contour feature extraction*

Feature Extraction plays an important part in any image analysis process. In this study, we have considered contour feature extraction as we not only have the CSV dataset but also clinical data which includes demographic characteristics and routine laboratory tests data. Extraction of contour features is a crucial stage in analysis of medical images. The outlines or borders structures in a picture are referred to as contours. Contour feature extraction is defined as a process of locating and mining important statistics from these contours.

3.3 *Image segmentation*

After contour feature extraction, the process of image segmentation is carried out. In this process the abnormalities in medical images are detected and relevant features are extracted for further analysis of images. The dataset images are segmented in order to detect abnormalities which can be a possible cause of osteoporosis. In order to diagnose and treat osteoporosis early in those who are suffering from it, these abnormalities might be examined. The steps involved in segmentation are as follows:

Step 1. Converting image to grayscale through computer vision
Step 2. Equalizing the Pixels of Image through cv2.equalizeHist()
Step 3. Finding Threshold of images through cv2.threshold()
Step 4. Now inverting Image through cv2.bitwise_not ()
Step 5. Finding Morphology (Opening) of Image through cv2.morphologyEx()
Step 6. Color based Image displaying of Affected area

The steps stated above for segmentation are depicted in Table 2 for normal and osteoporosis knee x-ray scan images.

Table 2. Image segmentation for normal knee X-Ray.

Grayscale Image	Equalizing Image	Thresholding Image	Inverting Image	Opening Image	Plotting the Affected Areas
Osteoporosis Knee X-Ray Images					
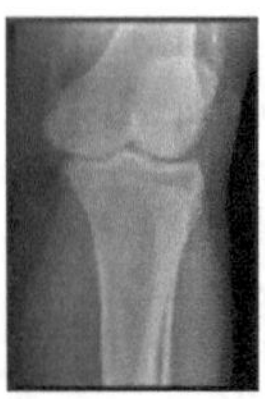	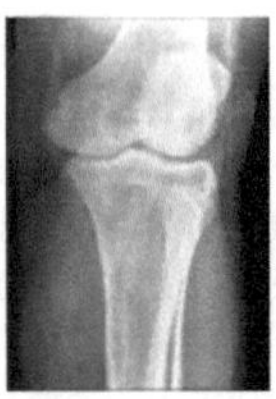		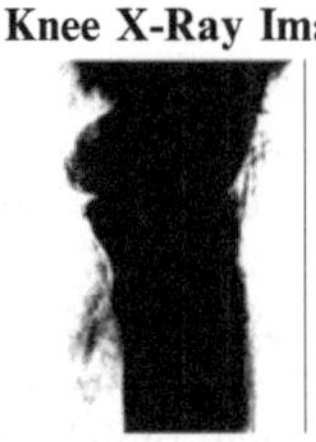		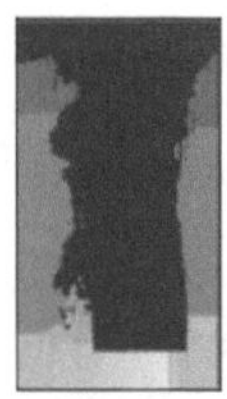

3.4 *CNN architectures*

Convolutional neural networks (CNN) are a subset of deep neural networks, and they use convolution at their intermediate layers. A two-function operator called convolution uses a portion of an image to emphasize patterns by multiplying it by filter components. Regardless of the number of input maps, the input image generates one or more feature maps based on the number of filters for each layer (Isaza *et al.* 2021). Various CNN architectures are used in our study are ResNet50V2, EfficientNetB2, DenseNet201, InceptionResNetV2, MobileNet, VGG16.

4 RESULT AND ANALYSIS

Each of the CNN architecture is evaluated based on the training and validation accuracy, loss and RSME values. Table 3 represents the training accuracy for all the CNN models used in the study for detection of osteoporosis.

Table 3. Training accuracy for applied CNN models for osteoporosis detection.

	Normal			Osteoporosis		
Models	Accuracy	Loss	RMSE Values	Accuracy	Loss	RMSE Values
ResNet50V2	0.87	24.59	4.958831	0.92	19.98	4.469899
DenseNet201	0.80	4.46	2.111871	0.86	2.63	1.621727
InceptionResNetV2	0.79	49.51	7.036334	0.91	39.51	6.285698
ResNet101V2	0.94	21.73	4.661545	0.93	16.13	4.016217
EfficientNetB2	0.96	0.59	0.768115	0.90	0.26	0.509902
MobileNet	0.95	0.46	0.678233	0.59	0.49	0.7
EfficientNetB4	0.94	0.76	0.87178	0.94	0.35	0.591608
VGG16	0.90	1.29	1.135782	0.96	1.49	1.220656

Table 4 represents the validation accuracy for all the CNN models used in the study for detection of osteoporosis.

Table 4. Validation accuracy for applied CNN models for osteoporosis detection.

	Normal			Osteoporosis		
Models	Accuracy	Loss	RMSE Values	Accuracy	Loss	RMSE Values
ResNet50V2	0.76	75.86	8.709765	0.83	71.46	8.453402
DenseNet201	0.82	10.45	3.232646	0.74	7.49	2.736786
InceptionResNetV2	0.77	146.59	12.10744	0.67	126.71	11.25655
ResNet101V2	0.96	174.76	13.21968	0.79	184.28	13.57498
EfficientNetB2	0.74	3.59	1.89473	0.79	4.45	2.109502
MobileNet	0.86	5.46	2.336664	0.79	3.72	1.92873
EfficientNetB4	0.70	2.75	1.658312	0.70	1.38	1.174734
VGG16	0.80	7.59	2.754995	0.89	5.77	2.402082

4.1 *Confusion matrix*

A confusion matrix is a table that compares actual and predicted classes from a given set of test data in order to assess the performance of a classification model. Osteoporosis may be classified as "positive" (detected) or "negative" (normal) in this scenario. The following abbreviations stand for the following cases: TP: correctly predicted positive cases (osteoporosis discovered), TN: correctly predicted negative cases (no osteoporosis detected), FP: mistakenly forecasted positive cases (false alarm), and FN: wrongly predicted negative cases (missing osteoporosis). Sample confusion matrix is presented in Figure 2 and 3 respectively for ResNet50V2 and DenseNet201.

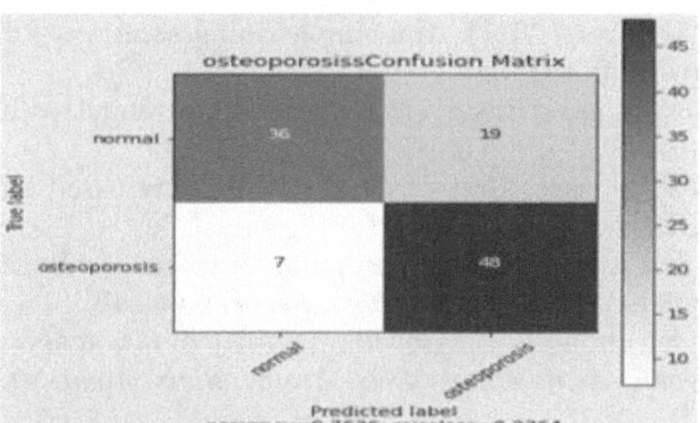

Figure 2. Confusion matrix for ResNet50V2.

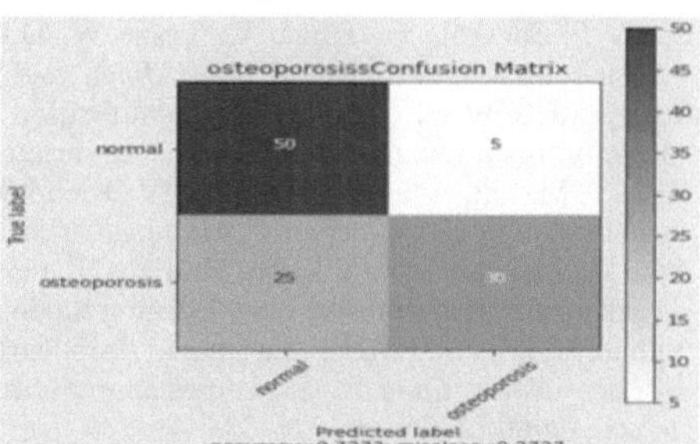

Figure 3. Confusion matrix for denseNet201.

5 CONCLUSION

The presented study focuses on the performance of eight CNN architectures i.e., ResNet50V2, DenseNet201, InceptionResNetV2, ResNet101V2, EfficientNetB2, MobileNet, EfficientNetB4 and Vgg-16 applied to the knee x-ray scan dataset. The dataset comprise of 372 knee X-ray images. Along with the knee x-ray scan we also have the clinical data. Two classes are determined namely normal and osteoporosis for indicating persons having good BMD and persons having osteoporosis respectively. All the models have been trained and evaluated based on accuracy, loss and RMSE values. Confusion matrix for all the models is also created for determining the classification accuracy. Based on the results obtained VGG-16 model is outperforming all other models with the classification accuracy of 82.73%. The study suggest that knee X-ray scan can prove to be efficient and cost-effective way of determining the person's bone health for identification of osteoporosis. In future, the dataset can be enhanced and more classes can be introduced such as osteopenia.

REFERENCES

Anaya-Isaza A., Mera-Jiménez L. and Zequera-Diaz M. (Jan. 2021). An overview of deep learning in medical imaging, *Informatics Med. Unlocked*, vol. 26, p. 100723.

bin Huang C., Sen Hu J., Tan K., Zhang W., hao Xu T. and Yang L. (Dec. 2022). Application of machine learning model to predict osteoporosis based on abdominal computed tomography images of the psoas muscle: a retrospective study, *BMC Geriatr.*, vol. 22, no. 1, pp. 1–9.

Chen Y., Guo Y., Zhang X., Mei Y., Feng Y. and Zhang X. (Dec. 2018). Bone susceptibility mapping with MRI is an alternative and reliable biomarker of osteoporosis in postmenopausal women, *Eur. Radiol.*, vol. 28, no. 12, pp. 5027–5034.

Ferizi U. *et al.* (Apr. 2019). Artificial intelligence applied to osteoporosis: A performance comparison of machine learning algorithms in predicting fragility fractures from MRI data, *J. Magn. Reson. Imaging*, vol. 49, no. 4, pp. 1029–1038.

Hans D. and Baim S. (Jul. 2017). Quantitative ultrasound (QUS) in the management of osteoporosis and assessment of fracture risk, *J. Clin. Densitom.*, vol. 20, no. 3, pp. 322–333.

Khinda R. *et al.* (Mar. 2022). Prevalence and predictors of osteoporosis and osteopenia in postmenopausal women of Punjab, India, *Int. J. Environ. Res. Public Health*, vol. 19, no. 5.

Kumar A., Joshi R. C., Dutta M. K., Burget R. and Myska V. (2022). Osteo-Net: A robust deep learning-based diagnosis of osteoporosis using X-ray images, *2022 45th Int. Conf. Telecommun. Signal Process. TSP*, pp. 91–95, 2022.

Muhammed Sunnetci K., Ulukaya S., and Alkan A. (Aug. 2022). Periodontal bone loss detection based on hybrid deep learning and machine learning models with a user-friendly application, *Biomed. Signal Process. Control*, vol. 77, p. 103844.

Nasser Y., El Hassouni M., Brahim A., Toumi H., Lespessailles E. and Jennane R. (Oct. 2017). Diagnosis of osteoporosis disease from bone X-ray images with stacked sparse autoencoder and SVM classifier, *Proc. - 3rd Int. Conf. Adv. Technol. Signal Image Process.* ATSIP 2017.

Osteoporosis Knee X-ray Dataset | Kaggle. [Online]. Available: https://www.kaggle.com/datasets/stevepython/osteoporosis-knee-xray-dataset. [Accessed: 27-Apr-2023].

Singh Y. *et al.* (2021). Osteoporosis prescreening and bone mineral density prediction using dental panoramic radiographs, *Proc. Annu. Int. Conf. IEEE Eng. Med. Biol. Soc. EMBS*, pp. 2700–2703.

Smets J., Shevroja E., Hügle T., Leslie W. D. and Hans D. (May 2021). Machine learning solutions for osteoporosis—A review, *J. Bone Miner. Res.*, vol. 36, no. 5, pp. 833–851.

Sukegawa S. *et al.* (Apr. 2022). Identification of osteoporosis using ensemble deep learning model with panoramic radiographs and clinical covariates, *Sci. Reports 2022 121*, vol. 12, no. 1, pp. 1–10.

Wani I. M. and Arora S. (Apr. 2022). Osteoporosis diagnosis in knee X-rays by transfer learning based on convolution neural network, *Multimed. Tools Appl.*, vol. 82, no. 9, pp. 14193–14217.

Xiao P., Zhang T., Dong X. N., Han Y., Huang Y., and Wang X. (Dec. 2020). Prediction of trabecular bone architectural features by deep learning models using simulated DXA images, *Bone Reports*, vol. 13.

Yilmaz E. B. *et al.* (2021). Automated deep learning-based detection of osteoporotic fractures in CT images, *Lect. Notes Comput. Sci. (including Subser. Lect. Notes Artif. Intell. Lect. Notes Bioinformatics)*, vol. 12966 LNCS, pp. 376–385.

Zhang B. *et al.* (Nov. 2020). Deep learning of lumbar spine X-ray for osteopenia and osteoporosis screening: A multicenter retrospective cohort study, *Bone*, vol. 140, p. 115561.

Computational Methods in Science and Technology Volume 1 – Kaur et al. (Eds)

Identifying fake followers of instagram influencers using logistic regression

M. Singh
Department of CSE, Apex Institute of Technology, Chandigarh University, Mohali, India

ABSTRACT: Most commonly utilized social networking app available today is Instagram. There are more than 1 billion monthly users on Instagram. We are always idly browsing through our Instagram accounts and snapping picture-perfect moments whenever we can. However, Instagram isn't limited to private use. To engage with their customers and market their goods and services, many businesses also employ Instagram advertising. Additionally, the quantity of followers and likes you receive on your Instagram business account are sometimes used to gauge its success. Getting the number of Instagram followers you desire may take months or even years, as gaining this kind of following is not a simple task. Though it's not always that easy, you can purchase Instagram followers. Purchasing Instagram followers is expensive, and the followers you get are frequently automated accounts that don't interact with your profile by following and commenting on your posts. The purpose of Instagram influencers is to boost interaction; while purchasing followers may appear like a smart move, it may be detrimental to one's reputation. Using a dataset of about 8000 profiles, we were able to identify fraudulent followers of influencers with a 94.2% accuracy rate using a logistic regression classifier.

Keywords: spammers, Instagram, influencers, fake profiles, machine learning, instaloader

1 INTRODUCTION

Facebook purchased Instagram, a platform that is used for sharing images and videos, in 2012. Instagram was started in 2010 by Kevin Systrom. The app allows you to take, edit, and share images and videos, as well as publish them straight from the photo album on your camera. Instagram has grown tremendously over the years, but its initial purpose was to allow users to share images and videos with their followers. Instagram stories, which vanish after 24 hours, as well as Instagram Reels and Instagram Direct Messages, are now available for posting. Companies now have a new avenue for sales as they may offer goods straight from the app (M. Code 2025).

Instagram's multi-modal user engagement makes it an effective social media tool for companies. Brand-posted material can be liked, commented on, shared, and saved by users. All businesses should have an Instagram presence since it may increase sales, build brand awareness, and engage customers (Breves *et al.* 2019). Social media influencers, according to Ladhari *et al.* 2020 are well-known users of social media who are regarded as specialists in specific areas, such as lifestyle, fashion, travel, photography and so on. They are consistent users of social media who built a devoted fan base by telling engaging tales about their passions and way of life (Ki *et al.* 2020). By combining images, videos, and interactive elements with the social media platform's interactive capabilities, social media influencers craft a compelling online persona. Influencers can draw in and maintain followers by

DOI: 10.1201/9781003501244-80

strategically crafting a compelling online presence (Ladhari *et al.* 2020, Ki *et al.* 2020, Belanche *et al.* 2020). Although studies typically characterize social media influencers as people who endorse items and promote brands, a large amount of their activity is centered about their personal interests and lifestyle (Ladhari *et al.* 2020, Audrezet *et al.* 2018, De Veirman *et al.* 2019). They only use explicit product endorsements in a small percentage of their social media posts (Audrezet *et al.* 2018, De Veirman *et al.* 2019). But other influencers have turned to unethical activities, such as purchasing phony likes and followers, in an effort to show their brand's impact (Childers *et al.* 2019). When these kinds of fraudulent activities are uncovered, influencers face backlash from their followers and their carefully cultivated online personas are swiftly damaged (De Veirman *et al.* 2019, Jacobson *et al.* 2020).

2 BACKGROUND

2.1 *Purchasing followers on Instagram*

On Instagram, followers can be purchased at various prices from a variety of websites. But most of the time, these followers aren't real; instead, they're either dormant accounts or bots that won't interact with your content. Consequently, even if you have more followers, your level of involvement will remain constant. For this reason, purchasing followers on Instagram isn't usually regarded as a successful marketing strategy (Jin *et al.* 2019). Instagram profiles with a large number of followers are typically regarded as successful because a larger following typically translates into increased popularity, recognition, and trust. More sales may result from this for companies. It is believed that having more followers would boost your engagement and, thus, your chances of making sales. Therefore, you might be asking if you can—or should—buy followers if you're just getting started with Instagram for your business and are seeking for strategies to enhance interaction (Agnihotri *et al.* 2023).

2.2 *How do you buy Instagram followers?*

Purchasing Instagram followers comes with several choices. You can locate websites offering various services to increase the number of followers you have on the internet. The websites that surfaced when someone searched Google for places to buy Instagram followers are listed below (Websites):

Twicsy
Buzzoid
UseViral
5. Krootez
6. GoViral Prime
7. SocialPros
9. Viralyft
10. InstaPowerful
11. SocialViral
13. Fluidbuzz
14. Thunderclap
15. BuzzVoice

It's debatable whether or not it is accurate, therefore exercise caution while going through this process (Kietzmann *et al.* 2011). It's rare for someone to gain 100,000 followers in an instant from 50,000, as Instagram may take offense at it (Kim *et al.* 2020).

2.3 *Instagram's perspective on buying followers*

Undoubtedly, obtaining a large number of Instagram followers can be a crucial marketing objective when launching a firm. Increased engagement can result from having more followers, and increased engagement almost certainly equals higher sales. Does having more followers, though, equate to increased interaction? It does, if those followers are genuine! But that probably won't be the case if you buy followers (Rules). You're better off growing your following organically through multichannel marketing and other organic methods if you want to boost engagement. Even though it could require a lot more work than simply purchasing them, you can be sure that your followers will be actual people who are more inclined to interact with your material.

3 LITERATURE SURVEY

Authors (M. Vergelis *et al.* 2019) suggested a way to distinguish bot users on Twitter by using highly shared traits that identify them as such. A machine learning technique has been offered to classify accounts into real or bot classes using their proposed method. This technology looks for behavioral patterns that distinguish authentic accounts from spam accounts. Cresci *et al.* created the dataset of verified real users and phoney followers on Twitter (S. Cresci *et al.* 2015). In their study, they trained many machine learning classifiers using the baseline dataset, which was based on rules and characteristics generated from media. Their approach's results show that it can correctly classify approximately 95% of the accounts from the original training set, indicating how successful their suggested strategy is. Authors (Y. Zhang and J. Lu, 2016) employed a somewhat different strategy and presented a state-of-the-art method for identifying factitious accounts on Weibo. Their proposed technique consists of several parts. They started off assuming that there was a reason these kinds of accounts existed at all. They also investigated the overlap between the user accounts of bogus followers and found a considerable amount of it.

During their investigation, 395 near-duplicates were found, and they resulted in 11.90 million bogus accounts that dispersed links throughout the network. Authors (K. Thomas *et al.* 2011) generated a dataset of 1.8 X 106 tweets from 32 distinct Twitter accounts. They found out after they asked that Twitter had suspended around 1.1 million of these accounts. After selecting 100 accounts at random, the tweets from those accounts were scrutinized to see if they were spam accounts. After looking into the 100 selected accounts more, they found that 93 of them had previously had their accounts deferred for posting unsolicited product ads. Three more accounts of users were deferred for re-sending stories from various news accounts, and the remaining four accounts were deferred for sharing repetitive and hostile marketing content. Rather than dissecting each spam tweet individually, H. Gao *et al.*, 2012 employed a set of features to efficiently rebuild campaigns from spam tweets. Their recommended remedy has a detection rate of more than 80%, according to the results. Nevertheless, a limitation of their approach is its less accuracy. Authors (F. Benevenuto *et al.* 2010) suggested a way to differentiate spam users from non-spam users. S. R. Sahoo and B. B. Gupta, 2019 proposed a hybrid technique to identify harmful users on social media, specifically Twitter. Their suggested hybrid model consists of two modules: Using a Petri net structure, they first assessed and retrieved features. They then fed this data into a classifier to divide profiles into categories for malicious and lawful activity. Experiments' findings demonstrate that the suggested approach had a high detection rate and successfully recognized a wide range of Twitter accounts. Numerous additional academics (Singh, M. 2023, Singh, M., & Singh, A 2023, M. Singh 2021, Singh, A., & Singh, M 2022) have also worked on spam detection on Twitter and Instagram, with promising outcomes.

4 PROPOSED METHODOLOGY

The collected dataset of Instagram profiles were considered.

4.1 *Data crawling*

Excellent scraping capabilities are provided by the Instaloader module (Instaloader) a Python library whose functionalities can be used as CLI (command line interface). The Instaloader key can be used to make downloads. Collected features are given in Figure 1.

Uid	ID of user
Fullname	Full name Length
biography	Biography Length
Followedby	The number of users Followed the account
Is_Followed_More	Is number of Followed are more than Followed by
Profile_pic	Does account set a profile picture
Followed	The number of users the account followed them
Posts	The number of shared posts by
Is_Verified	Is the account verified by Instagram
external_url	Is the account linked to an URL

Figure 1. Features crawled.

An authentic profile has been used to launch Instaloader to gather data from real consumers. Explore features of Instagram display recently uploaded videos and pictures that have attracted other's attention, indicating that the accounts that submitted them are generally legitimate and reliable. Data from 4061 genuine users—including renowned people and organizations and legitimate users—was scraped using the previously outlined method.

All of the influencer's followers were crawled after his data was crawled. Influence.co was used to calculate his followers' involvement. The website of influencer.co states that follower engagement was determined by dividing the total number of followers by the sum of the likes and comments made by followers to each influencer's Instagram posts during the previous forty-five days. The final total score is the average capacity of each influencer to get interaction from their followers. This provided us with a dataset of phony followers with poor engagement. In this manner, we gathered a dataset containing all the information for about 4000 fictitious followers.

A few hypotheses are made:

(1) The count of followers of influencers will be inversely correlated with follower engagement.
(2) Follower engagement will be favorably correlated with an influencer's follower numbers.
(3) Follower engagement will be adversely impacted by the content volume that influencers post.
(4) The majority of fraudulent accounts have very few or no submitted photos or videos, and they often have few postings overall.
(5) Nearly all factitious users follow a sizable number of actual profiles. The primary objective of phoney accounts is to get more followers than other authorized users. Because of this, most fraudulent accounts followed a lot of people, which meant that they followed larger number of people than they follow back. When compared to actual users, there was an increase in the percentage of followers.

4.2 *Feature selection*

The features listed below were chosen by Instagram's declared policy to remove spam followers:

4.2.1 *Age of account*

Equation (1) is used to determine the age of an account as follows:

$$\text{Age of Account} = \text{PresentDate} - \text{AccountCreationDate} \tag{1}$$

Typically, spammers replace suspended accounts with fresh ones. In other words, spam accounts are more recent than authentic accounts that have been there for a while.

4.2.2 *Followers count*

The number of people who subscribe to Instagram profiles of influencers.

4.2.3 *Followee count*

The total count of Instagram users that influencers follow.

4.2.4 *Posts containing dangerous links*

The list that follows includes banned URLs (Blacklists) that we have utilized in our work:

Artists Against 419	CYMRU Bogon List	MalwareDomainList.com Hosts List	StrictBlockPAllebone
MalwareURL List	Google Safe Browsing API	Malware Domain Blocklist	Scumware.org
FireHOL IP Lists	Malc0de Database	Malware Patrol's Malware Block Lists	VoIP Blacklist
DShield Block-list	hpHosts File	PhishTank Phish Archive	URLhaus

A total of 5K blacklisted domains/URLs used to tag spam links in messages. This feature is calculated using equation (2) as:

$$\text{Posts with malicious URLs} = (\text{messages with Spam URLs})/(\text{message count}) \quad (2)$$

Value of this feature is more for fake users.

4.2.5 *Posts containing links*

Equation (3) is used to calculate this feature as follows:

$$\text{Posts with Links} = (\text{messages containing URLs})/(\text{messages count}) \quad (3)$$

Typically, spammers provide URLs that point to the marketing websites of their followers. This feature is more valuable to spammers.

4.2.6 *Following frequency*

Equation (4) is used to calculate this feature as follows:

$$\text{Following Frequency} = (\text{No. of Followings})/(\text{Age of Account}) \quad (4)$$

In an attempt to gain attention and spread more and more spam, spammers follow an increasing count of users in a short amount of time. This considered feature is more beneficial to spammers.

4.3 *Machine learning*

Logistic Regression classification algorithm in the Weka software package (Weka) has been used for classification, and 10-fold cross validation has been employed in this case. The categorization algorithm's results are displayed in Figure 2.

Figure 2. Performance of logistic regression classifier.

Figure 2 illustrates that the Logistic classifier can accurately detect spammers with 94.2% accuracy when the features as shown in Table 1 are chosen for classification. The results therefore demonstrate that utilizing specific attributes, it is possible to categorize phony followers with a reasonable accuracy and few false positives.

Table 1. Features for classification.

Feature No.	Feature Name	Weight Variable	Value of Weight
ff1	Age of account (in days)	$f1$	0.04
ff2	Follower count	$f2$	−0.02
ff3	Followee count	$f3$	0.14
ff4	Tweets with spam URL	$f4$	0.297
ff5	URLs in Tweets/Tweets Analyzed	$f5$	0.015
ff6	Following frequency	$f6$	0.001
	Constant		−1.0717

As per the working of Logistic Regression, following equations (equation 5 and equation 6) have been formulated to return the classification probability of any user.

$$\mathrm{Y} = \mathrm{Constant} + \mathrm{ff1} \times \mathrm{f1} + \mathrm{ff2} \times \mathrm{f2} + \mathrm{ff3} \times \mathrm{f3} + \mathrm{ff4} \times \mathrm{f4} + \mathrm{ff5} \times \mathrm{f5} + \mathrm{ff6} \times \mathrm{f6} \quad (5)$$

$$\mathrm{Probability\ Value} = 1/(1 + \mathrm{e}\hat{}(-\mathrm{Y})) \quad (6)$$

5 CONCLUSION

While gaining a large number of Instagram followers is a goal for brands, going overboard to achieve this goal isn't always a smart idea. Purchasing Instagram followers may seem like a fast fix, but there may be long-term repercussions. Getting followers naturally is preferable. Even though it might take longer and involve more work, at least you'll know that the people who are following you are actual people and not just automated accounts you bought. It may feel overwhelming to create engaging content for Instagram in order to increase your following. Selling followers is lawful, and using an Instagram profile growth service is not prohibited by policy. Our machine learning algorithm has detected fake followers with a 94.2% accuracy rate.

REFERENCES

Agnihotri, D., Chaturvedi, P., Kulshreshtha, K., & Tripathi, V. (2023). Investigating the impact of authenticity of social media influencers on followers' purchase behavior: mediating analysis of parasocial interaction on Instagram. *Asia Pacific Journal of Marketing and Logistics*, 35(10), 2377–2394.

Audrezet, A., De Kerviler, G., Moulard, J.G., (2018). Authenticity under threat: when social media influencers need to go beyond self-presentation. *J. Bus. Res.* https://doi.org/ 10.1016/J.JBUSRES.2018.07.008 (in press).

Belanche, D., Flavián, M., Ibáñez-Sánchez, S., (2020). Followers' reactions to influencers' Instagram posts. *Span. J. Market.-ESIC* 1–17. https://doi.org/10.1108/SJME-11-2019-0100 \o 10.1108/SJME 11-2019-010010.1108/SJME-11-2019-0100 (in press).

Benevenuto F., Magno G., Rodrigues T., and Almeida V. (2010). Detecting spammers on Twitter, in *Collaboration, Electronic Messaging, Anti-abuse and Spam Conference (CEAS)*, p. 12.

Blacklists,https://www.gainchanger.com/site-infected-with-url-blacklist/

Breves, P. L., Liebers, N., Abt, M., & Kunze, A. (2019). The perceived fit between instagram influencers and the endorsed brand: How influencer–brand fit affects source credibility and persuasive effectiveness. *Journal of Advertising Research*, 59(4), 440–454. https://doi.org/10.2501/JAR-2019-030.

Childers, C.C., Lemon, L.L., Hoy, M.G., (2019). # sponsored# ad: agency perspective on influencer marketing campaigns. *J. Curr. Issues Res. Advert.* 40 (3), 258–274. https://doi.org/10.1080/10641734.2018.1521113.

Code M., Instagram, Social Media, and the Like: Exploring virtual identity's role in 21st Century Students' new socialization experience, *Faculty of Education*, Brock University, St. Catharines, Ontario, 2015.

Cresci S., Di Pietro R., Petrocchi M., Spognardi A., and Tesconi M., Fame for sale: Efficient detection of fake Twitter followers, *Decision Support Systems*, vol. 80, pp. 56–71, 2015. [Online]. Available: https://doi.org/10.1016/j.dss.2015.09.003

De Veirman, M., Hudders, L., Nelson, M.R., (2019). What is influencer marketing and how does it target children? A review and direction for future research. *Front. Psychol.* 10, 26–85. https://doi.org/10.3389/fpsyg.2019.02685

Gao H., Chen Y., Lee K., Palsetia D., and Choudhary A. N. (2012). Towards online spam filtering in social networks, in *NDSS*, pp. 1–16.

Instaloader,https://instaloader.github.io/

Jacobson, J., Gruzd, A., Hernández-García, Á, (2020). Social media marketing: who is watching the watchers? *J. Retailing Consum. Serv.* 53, 1–12. https://doi.org/ 10.1016/j.jretconser.2019.03.001.

Jin, S. V., Muqaddam, A., Ryu, E., (2019). Instafamous and social media influencer marketing. Market. *Intell. Plann.* 37 (5), 567–579. https://doi.org/10.1108/MIP-09-2018-0375.

Ki, C., Cuevas, L.M., Chong, S.M., Lim, H., (2020). Influencer marketing: social media influencers as human brands attaching to followers and yielding positive marketing results by fulfilling needs. *J. Retailing Consum. Serv.* 55, 1–11. https://doi.org/10.1016/j.jretconser.2020.102133\o10.1016/j.jretconser.2020.10213310.1016/j.jretconser.2020.102133

Kietzmann, J. H., Hermkens, K., McCarthy, I. P., Silvestre, B.S., (2011). Social media? Get serious! Understanding the functional building blocks of social media. *Bus. Horiz.* 54 (3), 241–251. https://doi.org/10.1016/j.bushor.2011.01.005.

Kim, D.Y., Kim, H.Y., (2020). Influencer advertising on social media: the multiple inference model on influencer-product congruence and sponsorship disclosure.

Ladhari, R., Massa, E., Skandrani, H., (2020). YouTube vloggers' popularity and influence: the roles of homophily, emotional attachment, and expertise. *J. Retailing Consum. Serv.* 54 https://doi.org/10.1016/j.jretconser.2019.102027, 102–027.

Rules, https://help.instagram.com/581066165581870.

Singh, M., Analysis of twitter spam detection techniques-A review, *2021 International Conference on Technological Advancements and Innovations (ICTAI)*, Tashkent, Uzbekistan, 2021, pp. 192–197, doi: 10.1109/ICTAI53825.2021.9673356.

Sahoo S. R. and Gupta B. B. (2019). Hybrid approach for detection of malicious profiles in Twitter, *Computers & Electrical Engineering*, vol. 76, pp. 65–81 [Online]. Available: https://doi.org/10.1016/j.compeleceng.2019.03.003.

Singh, A., & Singh, M. (2022). Social networks privacy preservation: A novel framework. *Cybernetics and Systems*, 1–32.

Singh, M. (2023, October). Advanced machine learning model to detect spam on instagram. In *2023 IEEE International Conference on Blockchain and Distributed Systems Security (ICBDS)* (pp. 1–6). IEEE.

Singh, M., & Singh, A. (2023). How Safe You Are on Social Networks?. *Cybernetics and Systems*, 54(7), 1154–1171.

Thomas K., Grier C., Song D., and Paxson V., Suspended accounts in retrospect: An analysis of Twitter spam, in *Proceedings of the 2011 ACM SIGCOMM Conference on Internet Measurement Conference*, 2011, pp. 243–258. [Online]. Available: https://doi.org/10.1145/2068816.2068840

Vergelis M., Shcherbakova T., Sidorina T., and Kulikova T., Spam and phishing in 2018, *Secure List*, 2019.

Websites, https://www.timesunion.com/marketplace/article/buy-instagram-followers-18150900.php

Weka, https://www.cs.waikato.ac.nz/ml/weka/

Zhang Y. and Lu J., Discover millions of fake followers in Weibo, *Social Network Analysis and Mining*, vol. 6, no. 1, p. 16, 2016. [Online]. Available: https://doi.org/10.1007/s13278-016-0324-2

Computational Methods in Science and Technology Volume 1 – Kaur et al. (Eds)
© 2025 the Author(s), ISBN: 978-1-032-81760-6

Developing a sophisticated collaborative filtering-based web application for recommending books

B. Rutuja
UG Learner, Department of IT, Saveetha Engineering College, Chennai, Tamil Nadu

G. Nalinipriya
Professor, Department of IT, Saveetha Engineering College, Chennai, Tamil Nadu

ABSTRACT: The project aims to create a web app employing collaborative filtering for personalized book recommendations. It analyzes user-resource relationships to suggest books based on similar users' preferences. Users can view their read books and get new suggestions accordingly. The app utilizes user interfaces like literary annotations to understand trends and relationships among users. It fosters a reader community sharing interests and profiles through similarity metrics. Active users receive recommendations based on favorite authors not yet explored by the community. The application follows a three-tier system: ReactJS for frontend, offering a user-friendly interface; Python Flask for backend, processing queries and accessing PostgreSQL database containing user book data. This setup ensures seamless browsing and recommendation experience.

1 INTRODUCTION

Collaborative filtering systematically examines user-object combinations to reveal emergent patterns. It suggests products based on similar user preferences, focusing on past interactions rather than product content. The project aims to redesign the book recommendation engine, which offers personalized recommendations derived from research and analysis by like-minded readers. Users can provide feedback, helping the engine identify groups of users who share interests. Suggestions come from books of interest in the user group and are more in line with his or her own interests. Implemented in Python, the engine runs on a Jupyter Notebook, seamlessly integrating backend processing with an attractive HTML/CSS interface. This integration ensures a user-friendly, accessible and efficient experience.

2 RELATED WORK

Recent research on recommendation systems has explored ways to increase recommendation accuracy and user satisfaction. Musto *et al.* extended the collaborative filtering model to organize linked-open data in graph-based recommender systems [1]. Liao *et al.* addressed the challenges of sparse data sets by performing cooperative filtering through feature-weighted matrix multiplication [2]. Batmaz and so on. conducted a comprehensive study on deep learning for recommendation systems, identifying challenges and proposed approaches [3]. Wang *et al.* provided an individual preference temperature map for class imbalance learning, handled class imbalances for advanced recommendations [4]. Chen and others. proposed HCCF, a collaborative sequential capsule extraction method, aimed at increasing the accuracy of recommendations [5]. He and so on and so forth. introduced neural labor filtering by using neural networks for collective recommendation [6]. Wang *et al.* provided DKN for communication proposals incorporating deep knowledge networking algorithms [7]. Death and so on.

 DOI: 10.1201/9781003501244-81

introduced a neural influence diffusion model for social recommendations, considering the dynamic dynamics of social networks [8]. Liu *et al.* presented Aspas, a framework for participatory scheduling in cloud environments [9]. Wang *et al.* Introducing Replanet, promoted users' preferences in the knowledge base for improved recommendations [10]. Zhang *et al.* conducted a survey on deep learning-based recommendation algorithms, providing valuable insights into the current state of affairs [11]. Grbovic & Cheng discussed real-time personalization using embedding, which supported dynamic personalization techniques [12]. Lee *et al.* The proposed Mixedbook, a recommended mixed-objective optimization algorithm with fuzzy feedback [13]. Zhang *et al.* introduced joint representation learning for top-n recommendation with discrete information sources [14]. Together, these studies contribute to the evolution of recommendation systems, exploring the benefits and challenges of collaborative filtering, deep learning, knowledge-informed networking, and real-time individualized analytics.

3 UNVEILING THE RECOMMENDATION ENGINE CORE: MATRIX FACTORIZATION AND SIMILARITY CALCULATION INTEGRATION

The real strength of our recommendation engine lies in the simple combination of matrix multiplication and similarity computation. As hidden features are extracted by hidden dimensions through multiple dimensions, the system uses similarity metrics to identify relationships and relationships between users and features This integrated approach improves the accuracy of recommendations, and ensures that the proposed book not only matches the user's historical preferences but also matches their available characteristics reading habits By demonstrating the performance of this integration, we provide a sense of a it goes in depth for on how matrix factorization and similarity computation combine to enable the basic functionality of the recommendation engine, resulting in a more sophisticated and personalized experience.

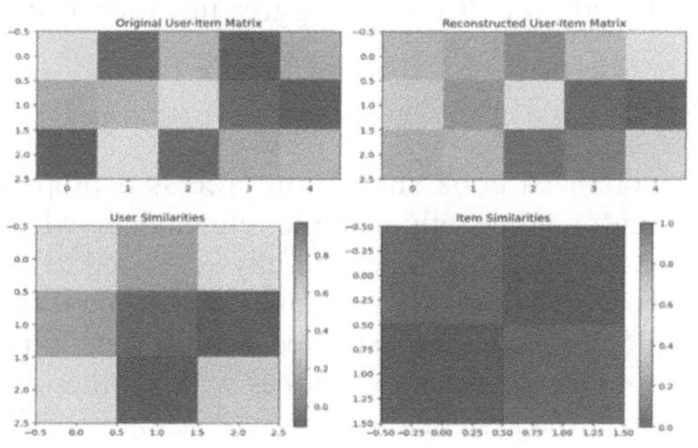

Figure 1. Matrix factorization analysis – unveiling user-item dynamics.

Figure 1 summarizes the matrix factorization process: top-left shows original user-book matrix, darker for unrated books; top-right shows reconstructed matrix approximating interactions; bottom-left visualizes user similarities based on preferences, darker for lower similarity; bottom-right shows book similarities from shared interests, darker means less similarity. It provides a compact overview of user-book activity, factorization results, and inherent user and book similarities.

4 SYSTEM ARCHITECTURE

4.1 *Overview*

This section introduces the system architecture of the book recommendation project, emphasizing its overarching goal to deliver personalized book recommendations based on user preferences and interactions.

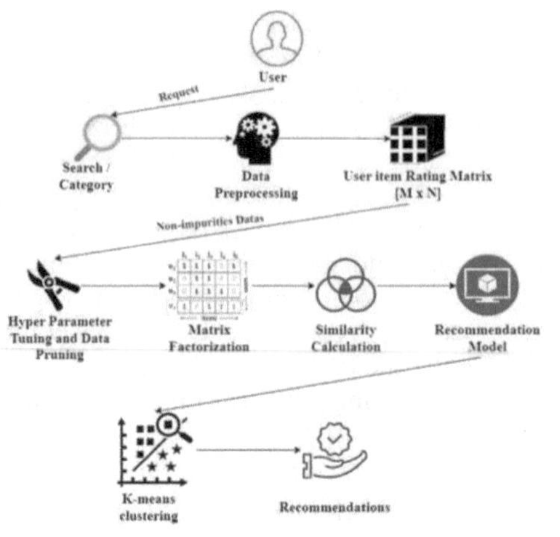

Figure 2. Enhanced system architecture for personalized book recommendations.

4.2 *Navigating the system: In-depth exploration of key modules*

In this section, we focus on the main modules of the system and analyze its inner workings. "Data Management" oversees the collection and preprocessing of data sources, the use of control matrices, and the use of effective storage methods Develops tailored recommendations for individual use methods are used This module also increases accuracy by fine tuning through hyperparameter adjustment, data pruning and optimization. The "User Interface" module guides users through interactions, offers neat suggestions and provides convenient search ranking features Overall, it collects and aggregates user feedback.

5 PROPOSED WORK

In the next section, a personalized book suggestion process is prepared for change development. This approach includes complex algorithmic refinements, where matrix multiplication and similarity computation techniques will be carefully advanced to unprecedented heights At the same time, a revitalized user experience is simulated at in their opinion, characterized by advanced user interaction techniques, sophisticated Search functionality, and seamless integration of real- time information delivery techniques with synthesis of external data sources plays an important role when the system supports co-sharing with databases and APIs, fostering a broader understanding of niche literature as well as ever-evolving user preferences. Unwavering commitment to personalization extends to intelligent use of user behavior and demographics, ensuring recommendations dynamically adapt to individual preferences and contexts When challenged with the possibility of handling variations, the system will add the best solution for improved performance stands as a matter, ensuring that recommendation algorithm Careful fine tuning is performed to ensure accuracy well. In addition, intelligent data mining techniques will be used to increase productivity by identifying and eliminating irrelevant or outdated data. This comprehensive orientation reflects a commitment to seeking to enhance the system's capabilities, delivering a sophisticated, dynamic, and personalized book recommendation experience that matches users' unique book preferences.

Algorithm for the enhanced book recommendation:

- Input: User-book interaction data
- Output: Personalized book recommendations, performance metrics
- Step 1: Load and clean user-book interaction data, splitting it into training and testing sets.

- Step 2: Construct a user-book rating matrix.
- Step 3: Initialize and refine latent factors for users and books using matrix factorization.
- Step 4: Calculate user and book similarities using cosine similarity.
- Step 5: Generate personalized book recommendations based on factorized matrices and neighborhood-based approaches.
- Step 6: Tune hyperparameters, retrain incrementally, and incorporate content-based filtering using book metadata.
- Step 7: Assess recommendation performance using metrics like RMSE, Precision, Recall.

6 IMPLEMENTATION & ANALYSIS FOR ENHANCED BOOK RECOMMENDATIONS

6.1 *Data acquisition & preprocessing: Context and significance*

In this pivotal phase of our implementation analysis, we address the fundamental steps of data acquisition and preprocessing within the context of the Book-Crossing dataset. The dataset encompasses crucial details about users, books, and ratings, forming the bedrock of our enhanced book recommendation system. The dataset context reveals anonymized user IDs, demographic data where available, and con-tent-based information about books sourced from Amazon Web Services. Under-standing the importance of robust data acquisition, we strategically map anonymized user IDs and leverage demographic information to enrich user profiles. Simultaneously, we ensure the removal of invalid ISBNs and handle NULL values, laying the groundwork for a high-quality dataset.

Methodology for Preprocessing: Effective preprocessing is imperative for aligning the dataset with our theoretical advancements. Our methodology involves meticulous handling of NULL values, intelligent mapping of anonymized user IDs, and the removal of invalid ISBNs. Content-based details of books undergo careful processing to enhance their relevance in recommendation algorithms. Ratings, expressed on a scale from 1–10, are scrutinized for both explicit and implicit expressions, ensuring data integrity and setting the stage for the subsequent stages of our enhanced book recommendation system.

6.2 *Classification report for enhanced book recommendation system*

This report provides detailed analysis of our recommendation system, using key metrics such as accuracy, recall, and F1-score.

- Accuracy: Displays the accuracy of recommended books by showing the ratio of the number of books recommended to the total number of positives cited.
- Recall (Sensitivity): Assesses the ability of the system to identify all relevant publications by indicating the total number of true positive predictions and true positives.
- F1-score: Balances precision and recall, providing an overall assessment of the overall performance of the system.

	precision	recall	f1-score	support
0	0.96	0.30	0.46	77
1	0.61	0.99	0.76	87
accuracy			0.85	164
macro avg	0.79	0.64	0.61	164
weighted avg	0.78	0.66	0.62	164

Figure 3. Classification report for enhanced book recommendation system.

6.3 *Overall output*

In this phase, we critically analyze all the results of our enhanced book recommendation system, focusing on the quality of recommended books, the use of links, and the flexibility of the system to its changing preferences. This research provides valuable insights into user satisfaction and engagement, guiding iterative changes to achieve an optimal recommendation experience.

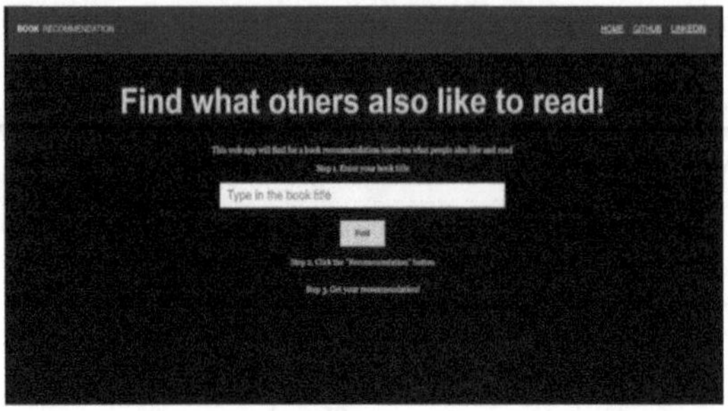

Figure 4. Search engine uses TF-IDF Vectorizer to improve book discovery.

Figure 5. Recommendations for users based on their search behavior.

5 CONCLUSION

In conclusion, significant developments and insights have marked our journey to improve the book recommendation process. Through careful algorithmic refinement, data pre-processing, and robust evaluation metrics, we sought to create a system that not only conforms to theoretical foundations but also to user preferences in practical situations Always with comprehensive analysis of user feedback and overall system results Provides valuable conservative feedback. As we embrace an iterative approach, adapting to changing user needs, we recognize the dynamic nature of recommendation systems and the importance of reacting to who manage user expectations This project demonstrates our commitment to delivering personalized, authentic and engaging book recommendation experiences. Going forward, integrating user feedback and pursuing innovation will be key to shaping our recommendation system in the future, ensuring it continues to evolve in line with user preferences and in technological development.

REFERENCES

[1] Musto, C., Basile, P., Lops, P., de Gemmis, M., & Semeraro, G. (2017, August). Introducing linked open data in graph-based recommender systems. In *Information Processing & Management*. Vol. 53, No. 2, pp. 405–435. Pergamon. https://doi.org/10.1016/j.ipm.2016.12.003

[2] Liao, Y. W., Huang, Y. M., & Chen, H. H. (2017, November). Collaborative filtering with feature-weighted matrix factorization on diverse and sparse data. In *2017 International Conference on Asian Language Processing (IALP)* (pp. 224–227). IEEE. https://doi.org/10.1109/IALP.2017.8300605

[3] Batmaz, Z., Yurekli, A., Bilge, A., & Kaleli, C. (2018, September). A review on deep learning for recommender systems: challenges and remedies. In *Artificial Intelligence Review* (pp. 1–37). Springer. https://doi.org/10.1007/s10462-018-9654-y

[4] Wang, B., Ester, M., Bu, J., Zhu, Y., Guan, Z., & Cai, D. (2019, November). Which to View: Personalized Priority Heatmap for Class Imbalance Learning. In *2019 IEEE International Conference on Data Mining (ICDM)* (pp. 1348–1353). IEEE. https://doi.org/10.1109/ICDM.2019.00155

[5] Chen, T., Tang, J., Lu, J., Wang, X., Tong, H., Gao, P., & Li, H. (2021). HCCF: Hierarchical collaborative capsule filtering for recommendation. *IEEE Transactions on Knowledge and Data Engineering.* https://doi.org/10.1109/TKDE.2021.3069267

[6] He, X., Liao, L., Zhang, H., Nie, L., Hu, X., & Chua, T. S. (2017). Neural collaborative filtering. In *Proceedings of the 26th International Conference on World Wide Web* (pp. 173–182). https://doi.org/10.1145/3038912.3052569

[7] Wang, H., Zhang, F., Xie, X., & Guo, M. (2018). DKN: Deep knowledge-aware network for news recommendation. In *Proceedings of the 2018 World Wide Web Conference* (pp. 1835–1844). https://doi.org/10.1145/3178876.3186175

[8] Wu, L., Sun, P., Fu, Y., Hong, R., Wang, X., & Wang, M. (2019). A neural influence diffusion model for social recommendation. In *Proceedings of the 42nd International ACM SIGIR Conference on Research and Development in Information Retrieval* (pp. 235–244). https://doi.org/10.1145/3331184.3331214

[9] Liu, Y., Zhang, C., Zhang, C., Liu, X., Li, Q., Chen, J., ... & Wang, H. (2019). Aspas: A framework of autonomous scheduling for periodical application scenarios in large-scale cloud environments. *IEEE Transactions on Parallel and Distributed Systems*, 30(11), 2635–2648. https://doi.org/10.1109/TPDS.2019.2916713

[10] Wang, H., Zhang, F., Wang, J., Zhao, M., Li, W., Xie, X., & Guo, M. (2018). Ripplenet: Propagating user preferences on the knowledge graph for recommender systems. In *Proceedings of the 27th ACM International Conference on Information and Knowledge Management* (pp. 417–426). https://doi.org/10.1145/3269206.3271739

[11] Zhang, S., Yao, L., Sun, A., & Tay, Y. (2019). Deep learning-based recommender system: A survey and new perspectives. *ACM Computing Surveys (CSUR)*, 52(1), 1–38. https://doi.org/10.1145/3285029

[12] Grbovic, M., & Cheng, H. (2018). Real-time personalization using embeddings for search ranking at Airbnb. In *Proceedings of the 24th ACM SIGKDD International Conference on Knowledge Discovery & Data Mining* (pp. 311–320). https://doi.org/10.1145/3219819.3219885

[13] Lee, J., Cho, K., & Ryu, K. R. (2020). Mixedbook: A strategy of mixed objective optimization for recommendation with implicit feedback. *Information Sciences*, 513, 28–50. https://doi.org/10.1016/j.ins.2019.10.075

[14] Zhang, Y., Ai, Q., Chen, X., & Croft, W. B. (2017). Joint representation learning for top-n recommendation with heterogeneous information sources. In *Proceedings of the 2017 ACM on Conference on Information and Knowledge Management* (pp. 1449–1458). https://doi.org/10.1145/3132847.3133086

Computational Methods in Science and Technology Volume 1 – Kaur et al. (Eds)
© 2025 the Author(s), ISBN: 978-1-032-81760-6

Intelligent model selection and hyperparameter optimization: MLOPS

Ikjyot Singh, Digvijay Puri, Gursimar Singh & Monika Singh
Apex Institute of Technology, Chandigarh University, Gharuan, Mohali

ABSTRACT: Simplifying the deployment, management, and maintenance of gadget mastering fashions within the manufacturing environment is a key aim in the swiftly growing system gaining knowledge of environment (MLOps). Intelligent version selection and hyperparameter tuning are also MLOps and are crucial to optimizing the effectiveness and performance of the model. In the framework of MLOps, this article offers a thorough analysis of recent studies and methods in intelligent model selection and hyperparameter optimization. We expound upon the benefits, constraints, and practical applications of various methodologies, including reinforcement learning, random search, grid search, and Model optimization. The objective of conducting this assessment is to assist scholars and professionals in deploying intelligent approaches to enhance machine learning models in MLOps workflows.

Keywords: Model selection, Intelligent hyperparameter optimization, random search, Model optimization, etc.

1 INTRODUCTION

Machine learning (ML) is like a powerful wizard in today's world of data, enchanting everything from self-driving chariots to personalized suggestion spells. Yet, there exist hurdles to overcome before ML spells can work their magic in practical realms, such as the quest for the right model and tuning the magical hyperparameters. To navigate these challenges, the realm of machine learning operations (MLOps) has emerged, offering a systematic approach to orchestrating machine learning enchantments in the real world [1]. Crafting the perfect potion of hyperparameter optimization and wise model selection is crucial in MLOps, as they directly influence the potency and performance of ML spells. While hyperparameter alchemy aims to refine the ingredients of the chosen spell for optimal results, model selection involves choosing the finest spell or incantation for a specific task. Within these enchanted pages, we unveil a grand tapestry of cutting-edge methods and enchanted lore in hyperparameter optimization and wise model selection within the realm of MLOps [2]. We also gaze upon the emerging constellations of cloud-based enchantments and the enchanted realms of automated machine learning (AutoML) platforms, which offer scalable and potent methods for hyperparameter refinement and model selection [3]. By automating the laborious task of model selection and hyperparameter enhancement, these platforms empower us to enhance their MLOps endeavors and get superior results with less manual intervention, harnessing better accuracy and best model. This manuscript aims to equip us with valuable wisdom and guidance by presenting a profound exploration of hyperparameter optimization strategies and wise model selection in the machine learning field. Armed with this knowledge, we can make enlightened decisions and wield powerful tactics to enhance the effectiveness and potency of machine learning models in real-world enchantments, thus inspiring creativity and fostering prosperity across diverse domains. MLOps not

 DOI: 10.1201/9781003501244-82

only ensure faster experimentation and development of models but also helps us in faster deployment of models into production and assure the quality of the model is best.

2 LITERATURE REVIEW

AutoML emerges as a novel realm of exploration, yet it has shown remarkable progress in the recent era. Numerous inquiries consolidate the result of other researchers on this topic. The majority of these studies concentrate on specific modules, such as hyperparameter optimization (HPO) or neural architecture search (NAS), for categorizing tabular data using traditional techniques. Text classification has received a lot less attention from the AutoML community than tabular data classification, mostly due to the unstructured nature of textual data and the need for more computationally demanding techniques. Published studies have compared how well the different AutoML programs now in use perform on text categorization tasks. Blohm *et al.* address whether AutoML is useful when it comes to text categorization or not. According to their tests, AutoML systems outperform humans on four of the thirteen activities, and if more advanced AutoML tools for NLP tasks are created, this percentage will only rise [4]. Various solitary efforts have addressed the issue of AutoML for text categorization. Madrid *et al.* involves automatically creating text categorization pipelines utilizing the information gathered from 81 datasets' worth of tests conducted using conventional methods. 81 datasets' worth of information may not be enough for a machine learning model to develop and produce the best outcomes, despite the model's effectiveness [5]. Zhang. *et al.* 2023 put forward a proposal for AutoML-GPT, a system that utilizes GPT as an intermediary to connect various AI models and adaptively refines models using optimal hyperparameters, driven by user inputs from model and data cards, culminating in the formulation of an appropriate prompt paragraph [6]. To reduce the computational cost of neural autoML, we can use transfer learning. This approach speeds up the network construction for a new job by learning the common hyperparameter selections across several projects. The text categorization findings indicate a decrease in the convergence time. This system's disadvantage is that meta-overfitting hasn't been addressed, which is a problem. Karras *et al.* 2023 introduce Fabolas and learning curve extrapolation as a pair of methodologies for accelerating the process of hyperparameter optimization [7]. Gomez *et al.* have employed a genetic algorithm to generate a population of meta-rules in order to determine the optimal model for the specific text classification dataset, utilizing a hyper-heuristic method [8]. This method is intriguing, but it is not the best since an AutoML system for text categorization has to be able to manage the sophisticated transformer models' high computational cost.

A comparable problem exists in the Feurer *et al.* system proposal. To understand the connection between the hyperparameter settings and the model's performance, they employ Bayesian optimization. In its exploration area, the system has fifteen classifiers, fourteen feature preprocessing techniques and four data preprocessing methods. However, if deep learning techniques had been incorporated, their solution would have been more competitive to other AutoML systems for text categorization [9]. The AutoML system called AutoKeras introduced by Jin *et al.* efficiently performs neural architecture search using Bayesian optimization and consistently selects the most rewarding operation [10]. In other words, in their study on "Data Containing Text," Shi *et al.* used transformer networks for numeric as well as categorical information stacking them with standard tabular models. However, this is a drawback because compared to other AutoML applications it has limited search capabilities [11]. We introduce TextBrew; a easier AutoML system for textual content category supplying an automated manner of figuring out the great Transformer fashions and then education them on the dataset with hyperparameter tuning within user-specific time constraints. That is why our approach is based on GANs that synthesize greater data to remedy the hassle of creating a massive meta-dataset which desires an awful lot time and labor.

3 METHODOLOGY

3.1 *Overview*

The best models may be automatically chosen by the system, and these models can then be trained on the dataset using the appropriate hyperparameter settings. Its architecture makes it capable of handling multiclass and binary prediction datasets without making that clear. This system right only uses "epochs" as a hyperparameter and architecture can be seen as seen in Figure 1.

Initially what happens is, code that we have in github is consumed by Jenkins pipelines as in the first phase "Building the pipelines ." The pipelines then perform the code checking that is whether the code that is pulled from github is either a Deep learning code or Machine learning code. Then after deciding the type we will train the model inside the 2nd phase, "Training on specified model". Ziqing *et al.* 2020 proposes that under the desired time restrict, the consumer-furnished dataset is used to train the great model . If there's nonetheless time, the following excellent projected version is taught after education the great candidate version that the meta-version anticipated. This system is maintained till the allocated time is over. Ultimately, the user receives an ensemble of these kind of fashions back [12].

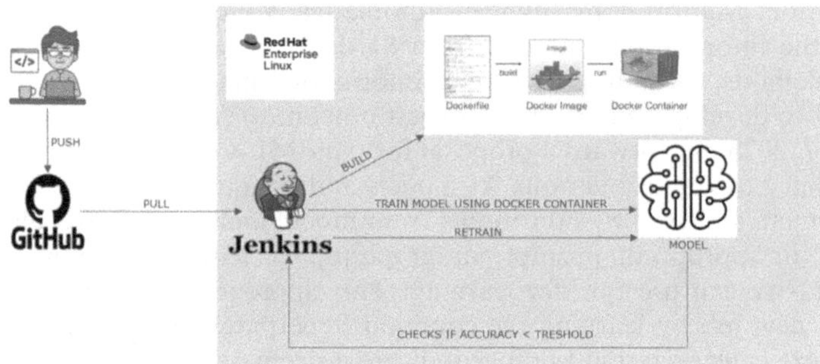

Figure 1. Architecture diagram of author's proposed system.

3.2 *Building the Jenkins pipelines having different jobs*

Here, we pass over the way to construct the pipelines and select the model.

This system have to be able to recollect a extensive sort of fashions which will be used inside the real global. Modern machine learning models for NLP text categorization issues are referred to as transformer fashions. There are numerous options to pick from. Each of those language models definitely serves as the inspiration for plenty distinct pre-educated models. For instance, BanglaBert is a Natural Language Understanding model based on BERT that has been pre-educated in Bangla. Furthermore, the price at which new fashions are produced is astounding, and present-day models are often swapped out. This calls for a bendy gadget which could adapt to the new models with the fewest important tweaks.

3.2.1 *Candidate model selection*

In this initially a job in jenkins will be created that will take code from the github and differentiate it wheteher it is a Machine learning Problem or a Deep learning model. In their observe, Gasparetto et al. evaluated a wide range of algorithms on diverse textual content class datasets, and the consequences display that XLNet and BERT-base are many of the exceptional models in terms of accuracy [13]. As mentioned by Choi al. 2021 also select ALBERT (A Light BERT) as certainly one of our fashions similarly to those two because it stocks a BERT-like layout with fewer parameters general [14]. Considering that each of the three transformer fashions has several hyperparameters that require great-tuning, there are basically a big number of capacity models to choose from. Choosing the hyperparameter values for each model is the subsequent step. After experimenting with diverse mixtures of hyperparameter values, we decided on those that, when tested on numerous datasets, produced the best results. Table 1 displays the hyperparameter values for the models that have been chosen as potential values.

Table 1. Hyperparametr available.

Hyperparameters
Warmup steps
epochs
Batch size
Train-test split ratio
Learning rate of optimisation algorithm

Below is a brief explanation of the hyperparameters:

(i) Warmup steps: This hyperparameter basically checks that model does not get effected by learning form a new dataset provided suddenly
(ii) Epochs: It is the number of iteration that a dataset goes through while training.
(iii) Batch size: It is basically the number of sample runs we go through before changing the parameters that are internal
(iv) Train-test split ratio: It is basically the division that we do of the given dataset.
(v) Learning rate of optimisation algorithm:It is an algorithm used for optimisation of algorithm applied iteratively to find least cost.

We also need to look for several security prospects so as to prevent any data leakageas proposed by Singh et al. that a novel framework is required to prevent data leaks throught out the process which can be a game changing move [15].

3.2.2 *Creation of the meta-dataset*

The meta-dataset is created in order to train the metamodel. A choice of datasets, consisting of binary and multiclass text type datasets, became carefully selected by us. Although more datasets could had been useful, there are not many freely available datasets for the job of classifying English language text. However, a broader number of datasets would require more model training, which would take a lot longer to generate the meta-dataset. As we will discuss later in the paper, we use GANs to synthesize data in order to overcome this challenge. Our approach automatically trains one extra multinomial Naive Bayes model in addition to all the chosen candidate models for each dataset. For each of these models, the accuracy and training time are documented.

3.2.3 *Creation of jobs*

In this the author has created several jobs and these jobs will be carrying out the work.These jobs will do two things, first it will take our code that is pushed to the github and secondly it will also perform hyperparameter optimision.

Jobs are explained in the following section for better understanding along with the tasks performed:

(i) The author has created a image that has keras and Python in the docker file.
(ii) On launching this image model training gets started.
(iii) Job1: Pull the code from github.
(iv) Job2: This job looks at the code and automatically runs the corresponding container depending on the requirement and starts the training.
(v) Job3: After training analyse the accuracy.

(vi) Job4: If the accuracy is less then threshold it wiil change the epochs (used Hyperparameter) and again check the accuracy.

(vii) Job5: Model will be retrained until the required accuracy is achieved.

3.2.4 *Connecting data using a manufacturing model*

The production of massive statistics calls for a lot of work and there isn't lots facts to distribute to the public, even supposing the time restrict is not essential. Additionally, overfitting takes place whilst the version makes excellent predictions on schooling information but does not provide similar outcomes on trying out statistics. Smaller statistics is considered to be extra powerful. Therefore, we want to solve the overfitting trouble from small facts. To improve the overall performance of meta classifiers, we use GANs to generate and enlarge data in this observe. Convolutional neural networks (CNN) and other deep studying strategies are used in generative adversarial networks, an unsupervised gaining knowledge of technique to generative models.

3.2.5 *Metamodel info*

This method we expect first-class-in-class performance across all fashions within the organization. Next, we train the meta-model using the generated data. We train the metamodel on the original data, collect the facts, and then repeat this process on the synthetic data to evaluate whether the synthetic data improves the performance of the metamodel. When we examined the meta-model of small data, we obtained the negative value of 0.475 as a prediction. Using synthetic data, the score was increased to a higher value of 0.7875.

3.3 *Making model predictions*

This part explains how our suggested system would be used by the user and how it makes use of the meta-model that was We learned in the previous section. The user provides input to the system in the form of a code and dataset and an optional parameter called a time limit. The maximum time the user can expect to receive the training model back as output is indicated by the parameter allowed for the training period. If the user has limited time, this option allows the user to schedule a training session. In our study, the default value of this option is 15 minutes. First, the data is run through the same pipeline we created for the training data (discussed in the previous section). Through the planning process, the product is cleaned and inconsistencies are eliminated. The feature space contains the size of the dataset and the results of trained models. The metamodel uses these attributes as input to predict which model performs best. The system is then trained on the projection model using the given data as input and the learning time is monitored. When training is finished, the system shows the next best model that approximates the same data set if it does not exceed the training time. This continues until the appointed time. Finally, a soft selection process is used to create a set of all training models. The group label with the most predicted results is selected as the predicted category. Estimated probability is included for each listing. The output is this collection passed back.

4 RESULTS

As defined in the preceding segment, we have built a model that has successfully surpassed the accuracy which was set in the script of the Jenkins pipeline. This is remarkable that the accuracy of the model which would have taken much time when manually changing the epochs of the model was done automatically using the MLOps pipeline in which machine learning model was integrated with the DevOps tools namely Jenkins and Docker for automating the process. Moreover, we have also seen that using tools like Docker give us the perfect environment with all dependencies resolved for running the code as opposed to using only a single machine.

After successfully running our code and integrating with Jenkins pipeline, the accuracy of our model which was initially 78% with 5 epochs, was increased automatically after several repetition to 88.74% without manual intervention which not only saves time but also improve the model's prediction as shown in Table 2.

Table 2. Model version with their accuracies.

Model Version	Hyperparameter [Epochs]	Accuracy
Version 1	5	0.7800
Version 2	15	0.8238
Version 3	12	0.8550
Version 4	10	0.8874

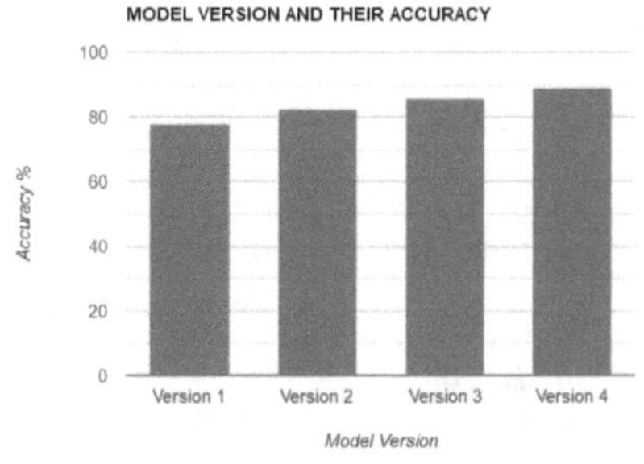

Figure 2. Comparison of different model version with their accuracies.

These data show that since the process returned by the system contains several top models that are successfully predicted by the metamodel, the best model is likely to be one of them. We have presented the results of our research using Intel® Xeon® 2.00GHz CPU, Tesla P100 16GB GPU, and 13GB of to be had RAM on a Redhat linux system. Many of the styles defined within the education section will lead to good sized upgrades in overall performance.

5 CONCLUSION

In summary, Kreuzberger et al. proposes that the field of Machine Learning Operations (MLOps) provides many methods and techniques for applying and monitoring learning models in the real world [17]. To make these models effective and efficient, hyperparameter optimization and intelligent model selection are essential. In this observe, we examined various studies techniques which include seek grid and random seek, in addition to reinforcement mastering, evolutionary algorithms, Bayesian optimization and automatic system learning (AutoML). Each approach has benefits and drawbacks, so we have to carefully consider strategies appropriate to specific packages and boundaries. Although the discipline has been successful, Georgios et al. proposes that there are still challenges to be overcome in integrating human knowledge into automated processes, ensuring adaptation in noisy situations, and evaluating optimization for big data and complex models [18]. Additionally, the development of trust and understanding in machine learning also depends on understanding automated processes. These issues, alongside the cutting-edge advancement of intelligent model selection coupled with hyperparameter optimization, are poised to become focal points of extensive research in the forthcoming period. This endeavor may necessitate the development of novel, scalable, dependable, and elucidative optimization algorithms, along with the integration of artificial intelligence and manual methodologies into automated workflows. Furthermore, the resolution of practical challenges and the stimulation of innovation within MLOps demand collaborative efforts encompassing a diverse array of specialists, practitioners, and applied scholars. To summarize, hyperparameter optimization and intelligent model selection represent fundamental pillars of MLOps and will wield a significant influence on the efficiency and triumph of machine learning applications in practical settings. Through the adept utilization of various methodologies, professionals hold the capacity to refine machine learning workflows and unveil fresh pathways for fostering creativity and influence.

REFERENCES

[1] Dominik, Kreuzberger., Niklas, Kühl., Sebastian, Hirschl. (2022). Machine Learning Operations (MLOps): Overview, Definition, and Architecture. *IEEE Access*, doi: 10.1109/ACCESS.2023.3262138

[2] Razvan, Andonie. (2019). Hyperparameter optimization in learning systems. doi: 10.1007/S41965-019-00023-0

[3] Arango, S. P., & Grabocka, J. (2023). Deep Pipeline Embeddings for AutoML. *arXiv preprint arXiv*:2305.14009.

[4] Blohm, M., Hanussek, M., & Kintz, M. (2020). Leveraging automated machine learning for text classification: Evaluationof automl tools and comparison with human performance. *arXiv preprint arXiv*:2012.03575.

[5] Madrid, J. G., & Escalante, H. J. (2019). Meta-learning of text classification tasks. In *Progress in Pattern Recognition, Image Analysis, Computer Vision, and Applications: 24th Iberoamerican Congress*, CIARP 2019, Havana, Cuba, October 28–31, 2019, Proceedings 24 (pp. 107–119). Springer International Publishing.

[6] Zhang, S., Gong, C., Wu, L., Liu, X., & Zhou, M. (2023). Automl-gpt: Automatic machine learning with gpt. *arXiv preprint arXiv*:2305.02499.

[7] Karras, A., Karras, C., Schizas, N., Avlonitis, M., & Sioutas, S. (2023). AutoML with Bayesian optimizations for big data management. *Information*, 14(4), 223.

[8] Gomez, J. C., Hoskens, S., & Moens, M. F. (2017, July). Evolutionary learning of meta-rules for text classification. In *Proceedings of the Genetic and Evolutionary Computation Conference Companion* (pp. 131–132).

[9] Feurer, M., Klein, A., Eggensperger, K., Springenberg, J., Blum, M., & Hutter, F. (2015). Efficient and robust automated machine learning. *Advances in Neural Information Processing Systems*, 28.

[10] Jin, H., Song, Q., & Hu, X. (2019, July). Auto-keras: An efficient neural architecture search system. In *Proceedings of the 25th ACM SIGKDD International Conference on Knowledge Discovery & Data Mining* (pp. 1946–1956).

[11] Shi, X., Mueller, J., Erickson, N., Li, M., & Smola, A. (2021, July). Multimodal automl on structured tables with text fields. In *8th ICML Workshop on Automated Machine Learning (AutoML)*.

[12] Ziqing, Yang., Yiming, Cui., Zhipeng, Chen., Wanxiang, Che., Ting, Liu., Shijin, Wang., Guoping, Hu. (2020). TextBrewer: An open-source knowledge distillation toolkit for natural language processing. doi: 10.18653/V1/2020.ACL-DEMOS.2

[13] Gasparetto, A., Marcuzzo, M., Zangari, A., & Albarelli, A. (2022). A survey on text classification algorithms: From text to predictions. *Information*, 13(2), 83.

[14] Choi, H., Kim, J., Joe, S., & Gwon, Y. (2021, January). Evaluation of bert and albert sentence embedding performance on downstream nlp tasks. In *2020 25th International conference on pattern recognition (ICPR)* (pp. 5482–5487). IEEE.

[15] Singh, A., & Singh, M. (2022). Social networks privacy preservation: A novel Framework. *Cybernetics and Systems*, 1–32.

[16] Jong, Won., Jongmin, Shin., Jang-Won, Lee. (2023). A survey on hyperparameter optimization in machine learning. *The Journal of Korean Institute of Communications and Information Sciences*, doi: 10.7840/kics.2023.48.6.733

[17] Kreuzberger, D., Kühl, N., & Hirschl, S. (2023). Machine learning operations (mlops): Overview, definition, and architecture. *IEEE access*.

[18] Georgios, Symeonidis., Evangelos, Nerantzis., A., Kazakis., George, A., Papakostas. (2022). *MLOps – Definitions, Tools and Challenges*. doi: 10.1109/CCWC54503.2022.9720902.

Computational Methods in Science and Technology Volume 1 – Kaur et al. (Eds)
 ISBN: 978-1-032-81760-6

Hateful meme classification of Indian political memes

Nandini Kumar, Diksha Bhardwaj, Mansi Singla, Anishka & Vijay Kumar Yadav
Department of Computer Science and engineering, Indira Gandhi Delhi Technical University for Women, Delhi, India

ABSTRACT: The rise of hateful memes on social media presents a growing challenge. The fusion of visual and textual elements within memes often results in a combined meaning that is very different from what would be interpreted individually. In certain spheres where the impact is very direct, hateful meme detection becomes crucial such as the detection of political memes because they have the potential power to affect voters by playing on prejudice and exploiting emotions. In countries like India where multiple languages are sometimes intermingled in the same sentence, it poses a unique challenge of Code-Switching where Hindi and English words are alternated in the same sentence. This work aims to address the hate and offensive detection of multimodal hateful memes in the Indian Political scenario consisting of Code-Switched languages. This work introduces the "Bharatiya Political Memes Dataset" which consists of memes from a holistic political perspective, including politicians, political parties, journalists, and activists of different wings throughout the country, doubly annotated as hateful or non-hateful. Further, a comparison of different models for the classification is proposed, presenting the model fusion of LSTM + CLIP, combining the strengths of both models, resulting in an accuracy of 72% due to its strength of capturing intricate details in large size embeddings.

1 INTRODUCTION

Memes, which offer relatability, humor, and commentary, have become an essential component of online culture. A meme can be defined as *"An image, video, text, etc., usually humorous in nature, that is copied and spread rapidly by internet users, often with slight variations"* (Kiela *et al.*, 2020).

Hateful meme definition according to (Kiela *et al.*, 2020) is *"A direct or indirect attack on people based on characteristics, including ethnicity, race, nationality, immigration status, religion, caste, sex, gender identity, sexual orientation, and disability or disease. We define attack as violent or dehumanizing (comparing people to non-human things, e.g. animals) speech, statements of inferiority, and calls for exclusion or segregation. Mocking hate crime is also considered hate speech."*

The Information Technology (Intermediary Guidelines and Digital Media Ethics Code) Rules, 2021 state that intermediaries (like Meta or Instagram) may use a judicial order to identify the meme creator whether the meme breaches India's sovereignty, integrity, security, or amicable relations with other nations (Banerjee *et al.*, 2021).

Figure 1. (a) and (b) Examples of hateful meme.

DOI: 10.1201/9781003501244-83

Memes are powerful tools that certain groups use in the digital age to weaken opponents and cause strife. They use careful curation to take advantage of pre-existing prejudices and stereotypes to exacerbate social divisions and manipulate feelings. This deliberate manipulation affects public opinion widely and goes beyond personal convictions, as shown in research (Kulkarni *et al.*, 2017). Even with the proliferation of harmful content on the internet, it is not feasible to manually review every instance. The widespread availability of such content emphasizes how urgently effective countermeasures to hate speech and disinformation are needed.

A major challenge in identifying hateful memes is the inherent complexity of these digital artifacts. Traditionally, machines have analyzed text and images as distinct entities, ignoring the complex interactions between the two that characterize meme culture. Memes, which are frequently full of irony and sarcasm, distort reality, which makes them especially difficult for automated systems to understand. Political memes make this problem worse because they capture complex socio-political issues and make them more intricate. Given that certain memes significantly rely on contextual knowledge of accompanying images, the ambiguity present in meme text can easily lead to model misinterpretation. As a result, models that are good at text analysis might struggle to integrate images, and vice versa. Therefore, it is difficult to develop a model that maps the text with images in the same space and then gives a result where even the accuracy of humans in correctly identifying whether the memes are hateful is 84.7%.

This paper introduces the Bhartiya Political Memes Dataset which can be categorized into two categories: hateful and non-hateful. It is a multilingual dataset that comprises memes in both Hindi as well as English. Further, the classification of memes using various machine-learning models is done. For text analysis, unimodal models like BERT and LSTM are used, which only use word embeddings. ResNet50 is used to process image embeddings for image analysis. Additionally, multimodal classification strategies are explored, utilizing the CLIP model that combines text and image embeddings. The paper presents three multimodal fusion techniques: LSTM + CLIP, BERT + CLIP, and BERT + ResNet50. The goal of these fusion models is to improve classification accuracy by utilizing the advantages of distinct modalities. The most efficient combination, with an accuracy of 0.72, is the fusion of LSTM and CLIP.

2 RELATED WORK

The omnipresence of social media and memes has made the necessity to detect hateful memes, leading to the development of several unimodal and multimodal techniques. There have been several studies and research on detecting hate online in the form of text, images and other forms of media. Studies on hate speech detection showed that the internet contains a large amount of hate-inducing data on the internet and studies have been made to detect those hateful words.

(Kumar *et al.*, 2021) focuses on hate speech detection in tweets using baseline methods such as Char n-gram, TD-IDF, BoWV and using CNN and LSTM based approaches to enhance the performances. The maximum F1 Score is 0.930 with a precision of 0.930 using LSTM + random embedding + GBDT. (Mnassri *et al.*, 2022), uses a BERT-based model for detecting hate speech. Using text models along BERT like BERT + CNN, BERT + LSTM, BERT + MLP and using the ensemble method to get better results which were found in stacking with F1 score 0.97 and 0.92 in two datasets and precision of 0.94 and 0.86 for respective datasets. (Davidson *et al.*, 2017), (Gaydhani *et al.*, 2018), (Mozafari *et al.*, 2020) discusses models for hateful data from Twitter, English Wikipedia, BookCorpus and uses several machine learning models on the data like naive bayes, SVM, random forests, and decision trees. And logistic regression with L1 regulations has an F1 Score of 0.90, precision of 0.91. Using hateful text data from Twitter and Facebook and using to detect hateful, offensive and clean data using N-grams and TF-IDF based approaches. The F1 Score of the model is 0.96 and precision being 0.96. The logistic regression works better with an n-gram range of 1 to 3 for L2 normalization of TF-IDF. Fine-tuned Bert-based model leverages contextual and syntactical information of all BERT's transformers and inserts non-linear layers to it.

An extensive study has been done on Facebook's Hateful Meme dataset which showcases the unimodal and multimodal approaches to identify hate in memes which contain images as well as text. In (Kumar *et al.*, 2022), introduces Hate-CLIPper, which uses cross-modal interactions of the model CLIP (Contrastive Language-Image Pre-training) features. Fine-tuning the CLIP features by adding trainable projection layers to image and textual encoders which were extracted as output from the CLIP model. The Hate-CLIPper attains the best AUROC scores in the comparative analysis with other models in the paper. In (Hossain *et al.*, 2022), the authors have proposed four modules comprising the multimodal framework: multimodal decision fusion, multimodal feature fusion2, visual feature extraction, and textual feature extraction. Modern pre-trained models for textual (m-BERT, m-DistilBERT, XLM-R) and visual (VGG19, VGG16, ResNet50) feature extraction is used in this paper. The suggested framework performs better than every other model on both datasets, with weighted f1-scores on MultiOFF and TamilMemes of 66.73% and 58.59%, respectively. In (Velioglu *et al.*, 2020), (Zhou *et al.*, 2021), (Das *et al.*, 2020) the focus is on showcasing the Ensemble model, (V + L), (V&L), Concat BERT with comparison with BERT, Visual-BERT, VILBERT, Image-Fusion, and Human (82.65) AUROC scores. With Ensemble model having an AUROC of 81.08 on test data, V + L having AUROC 73.42. V&L with AUROC of 74.80.

Further (Rajput *et al.*, 2022), introduced the Indian Political Memes Dataset where the final model is obtained by concatenating the outputs of the binary-channel CNN-LSTM model. The model uses the categorical cross-entropy loss function, L2 regularization, and Adam optimizer. Recall score of 0.816, the Glove embeddings produced the best results when using a single embedding model. Additionally, tests were done with different word embedding combinations, and it was found that (Glove + Fastext) yielded the best results. In (Hasan *et al.*, 2022), a technique of meme classification developed by the 'CUET NLP' team under the shared task ('DravidianLangTech-ACL2022') in Tamil is presented. Additionally, this work used CNN, CNN + LSTM, VGG16, ResNet50, VGG19, and VGG19 to analyze textual and visual elements. The textual technique using CNN produced the highest 'weighted f1-score (0.52)' along with 'recall (0.57)', according to the results.

Table 1. Literature review summary.

Research Paper	Dataset	Best Approach
Badjatiya *et al.*, 2017	16k annotated Tweets	LSTM + Random Embedding + GBDT
Mnassri *et al.*, 2022	Davidson Dataset, Merged Dataset (Davidson Dataset and DHO dataset)	Ensembler Method 4: EM4:BERT-MLP + BERT-CNN + BERT-LSTM (Davidson Dataset and Merged Dataset)
Davidson *et al.*, 2017	A random sample of 25k tweets	Logistic regression with L2 regularization
Gaydhani *et al.*, 2018	Hate Speech + Identification + automated-hate-speech-detection-data + Hate Speech Twitter annotations	Logistic Regression with the optimal n-gram range 1- 3 for the L2 normalization of TF-IDF
Mozafari *et al.*, 2019	Waseem Dataset and Davidson Dataset	$\text{BERT}_{\text{base}}$ + CNN
Kumar *et al.*, 2022	Hateful Memes Challenge (HMC) dataset	HateCLIPper
Hossain *et al.*, 2020	English offensive meme or MultiOFF Dataset and Tamil troll meme or TamilMemes	Modern pre-trained models for textual (m-BERT, m-DistilBERT, XLM-R) and visual (VGG19, VGG16, ResNet50) feature extraction
Velioglu *et al.*, 2020	Hateful Memes Challenge (HMC) dataset	Ensemble model On VILBERT and BERT

(*continued*)

Table 1. Continued

Research Paper	Dataset	Best Approach
Zhou *et al.*, 2020	Hateful Memes Challenge (HMC) dataset	The V + L models with pretrained Visual BERT, and the V&L models with pretrained ViLBERT.
Das *et al.*, 2020	Hateful Memes Challenge (HMC) dataset	Sentiment analysis using Visual BERT and Image Captioning
Kshitij *et al.*, 2020	Indian Political Meme dataset	CNN-LSTM
Hasan *et al.*, 2022	Tamil troll Memes	CNN-Text + VGG16

2.1 *Background information*

Unimodal models exist for classification of data. However, as the name suggests, uni means singular, i.e. only one form of data is used in the model. Multimodal models hence come into play which give a wider picture. The models used for comparison in the paper include the following as given below.

2.1.1 *BERT*

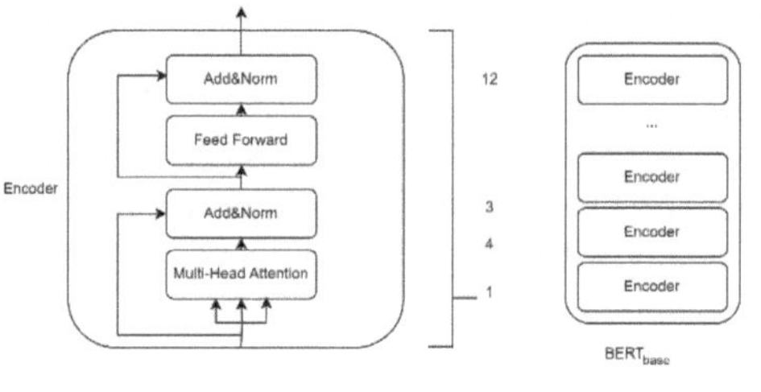

Figure 2. BERT architecture.

BERT-Bidirectional Encoder Representations from Transformers was introduced in (Devlin *et al.*, 2018) by Google for text classification. It transformed the landscape of business approach towards keyword research due to its strength in using Natural Learning Process for analyzing keywords. It utilizes a transformer architecture, which excels at capturing contextual relationships in text bidirectionally considering the full context of a word by analyzing the texts twice-before and after.

2.1.2 *Resnet50*

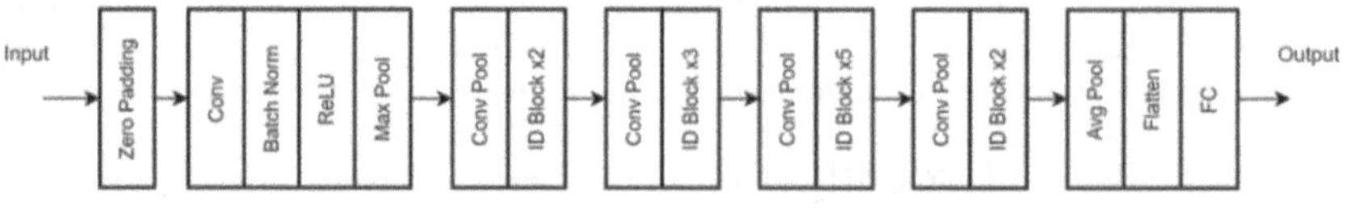

Figure 3. Resnet50 architecture.

Resnet50 was introduced in (He *et al.*, 2016) to classify images by stacking residual blocks. It uses skip connections to address the vanishing gradient problem and converting a regular network to a residual network. It has a bottleneck design for building blocks using 1x1 convulsions reducing the matrix multiplications and complexity. Due to its innovative approach of Residual connections, it can be trained on large datasets to achieve state of the art results for classification.

2.1.3 *LSTM*

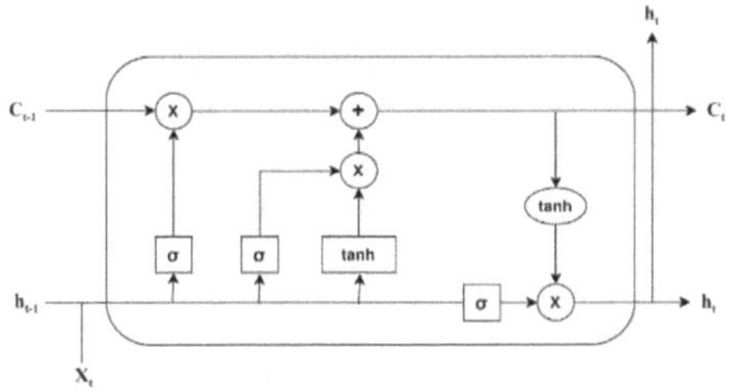

Figure 4. LSTM cell architecture.

LSTM-Long Short-term memory given in (Staudemeyer *et al.*, 2019) to solve RNN dependencies due to vanishing gradient problem by introducing a memory cell capable of retaining information for textual data. The memory cell is the key concept along with 3 gates to control the flow of information. This strength of LSTM makes it useful for many applications such as language translation, speech recognition, sentiment analysis, and chatbots among others.

2.1.4 *CLIP*

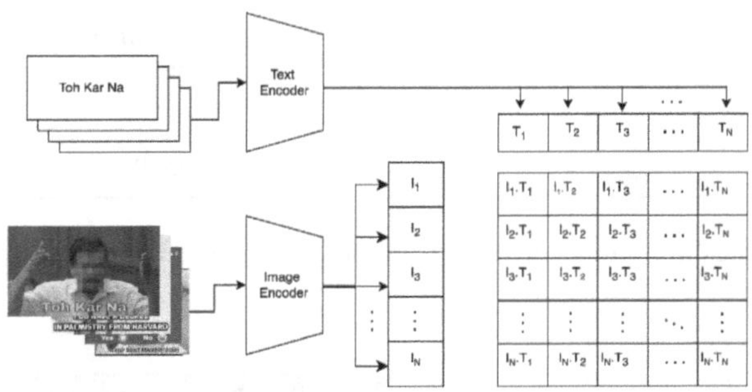

Figure 5. CLIP architecture.

CLIP, Contrastive Language-Image Pre-training model was introduced in (Radford *et al.*, 2021) as a multimodal vision and language model. It connects images and text prompts in a mutual latent space, processing them simultaneously to encourage high cosine similarity for better performance. CLIP employs a ViT-like transformer for visual features along with a causal language model for text data features projecting them both into shared space. This makes it highly efficient and can be used for multimodal classification, image text similarity and other applications.

3 DATASET

This work introduces the "Bhartiya Political Meme Dataset". It was constructed by capturing different memes from post-independence to the current scenario. It encapsulates the themes of political parties focusing on prominent politicians, ideologies, activists and journalists pan India. The memes that are circulated on a day-to-day basis through various media such as WhatsApp, Facebook, Instagram, Telegram are captured in the dataset.

The dataset was created using the python module "google_images_download", available online. Keywords were chosen according to the prominence in the political scenario pan-India. Then for each keyword, 200 images were downloaded. This created a compilation of 7000 images.

Then a pipeline was created for OCR generation for these images using the "Easy OCR" for both English and Hindi text. It tackled the Hindi text in both Roman and Devanagari script present in the code-switched memes. The results from the pipeline were then manually corrected in case of discrepancies and final results, were filtered according to:

(1) Whether the image is a meme. It is not necessary that every image with text would be a meme, it could also be a pictorial piece of information.
(2) When it is a political meme, it is important that the memes are political in nature to correctly conduct the study.
(3) Uniqueness of the memes, to increase quality of dataset and correct the redundancy.

After the dataset was filtered, the remaining 2509 memes were annotated according to the definition of political memes as given by the Indian government in (Banerjee *et al.*, 2021).

After annotations were completed by the 3 annotators, between the differences in the annotations the majority was chosen as the final annotation. The final annotation as given in Table 2. Consist of 1158 hateful and 1351 non hateful memes. The class imbalance ratio is approximately 1.166, indicating a relatively balanced dataset with a slightly higher representation of non-hateful memes compared to hateful memes.

Table 2. Bharatiya political memes dataset – class distribution.

Label	Number of elements in dataset
Hateful	1158
Non-Hateful	1351
Total	2509

Examples of memes are shown in Figure 6 along with their OCR results in Table 3.

Figure 6. Examples of a) hateful b) hateful c) non hateful meme.

Table 3. Example of dataset with label.

Figure	Extracted Text	Label
Figure 6(a)	#Pappusoduffer Amul is the taste of India Pappu is the waste of India	Hateful
Figure 6(b)	Thanks 2 rahul baba .... logo ne ab mere upar jokes banana band kar diya hai ...	Hateful
Figure 6(c)	Ye meri Delhi hai Aur mai iska Jaikant Shikre	Non hateful

4 PROPOSED WORK

This work has proposed the "Bhartiya Political Meme Dataset" which is a triply annotated dual labeled dataset (hateful and non-hateful). Memes were collected and filtered according to the OCR generation for annotation. The final annotation consists of 1158 hateful and 1351 non hateful memes. The class imbalance ratio is approximately 1.166, indicating a relatively balanced dataset with a slightly higher representation of non-hateful memes compared to hateful memes. All models were implemented on this dataset.

Further the implementation of various models, both unimodal and multimodal is done, to test their performance on the introduced dataset and capture their strengths to present the final approach.

First the text is extracted from the images using the pipeline "EasyOCR". It is specialized for capturing text in both Hindi and English languages for both Roman and Devanagari scripts. Once the images and corresponding text have been extracted along with labels, the models are built and tested.

4.1 *Classification using unimodal models*

Unimodal models focusing on one type of data-textual aur visual through the images can be used to classify the hateful memes. BERT, LSTM are used for obtaining text embeddings and ResNet50 for image embeddings.

4.1.1 *BERT*

It utilizes a transformer-based architecture to capture word embeddings of size 768 which are passed in the input layer, further going through 5 dense and 4 dropout layers to produce the output.

4.1.2 *LSTM*

In this model, the extracted text is passed to generate 3200 text embeddings for the input layer which then generates output after passing through hidden layers, harnessing its strength in capturing long term dependencies through a gated architecture.

4.1.3 *ResNet50*

This model is used to create image embeddings of size 2048 which are fed to the input layer, adding further dense layers to get the single node output layer. Its deep architecture neural network consisting of 50 layers allows it to capture complex patterns.

4.2 *Classification using multimodal model*

Further multimodal models which use both image and text embeddings for model generation have been explored. CLIP, a multimodal vision and language model, had the best results. It is trained on extensive data achieving a compelling zero shot performance. In multimodal CLIP, both text and image embeddings of size 512 each are captured as input layers, which are comparatively smaller in size. The strength of CLIP lies in its establishment of multi-modal embedding space through the joint training of an image encoder and text encoder; however, it also has restrictions on the number of input characters, leading to text truncations wherever needed. Input layers of text and image embeddings, each of size 512 are then fused together and sent along the next dense layers to the final output layer giving a binary output.

4.3 *Proposed multimodal methods*

In pure multimodal models the information is limited as both the embeddings come from one type of data only, hence, to improve the performance, fusion based models are proposed to maximize utilization of available data and hence increase efficiency.

4.3.1 *BERT + ResNet50*

The first proposed fusion multimodal model works by taking in word and image embeddings and fusing them. Word embeddings are captured using pre-trained transformer BERT which is considered due to its strength on working in task specific models capturing the complete essence of the text. For image embeddings, ResNet50 has been chosen because of its strength in capturing intricate details and nuances which need to be captured in the context of memes. Both embeddings are then concatenated to preserve information and integrate it without making the model too complex. Further dense layers with ReLU activation and dropout layers for regularization are added to introduce non-linearity leading to a final output layer with a single node and a sigmoid activation function.

4.3.2 *BERT + CLIP*

This proposed model fusion harnesses the strengths of both, creating a strong classification model. BERT captures text embeddings from the extracted text which are greater in size (768) in comparison to the CLIP text embeddings (512). CLIP is used to extract image embeddings which capture detailed data of multi-modal embedding space. The concatenation of these two input layers hence creates a robust layer with embeddings containing the important parameters for classification. Further 3 dense layers with ReLU activation and 3 dropout layers are added to increase complexities at the same time reducing overfitting of the model. The final output layer uses sigmoid function for binary classification of data into hateful and non-hateful labels.

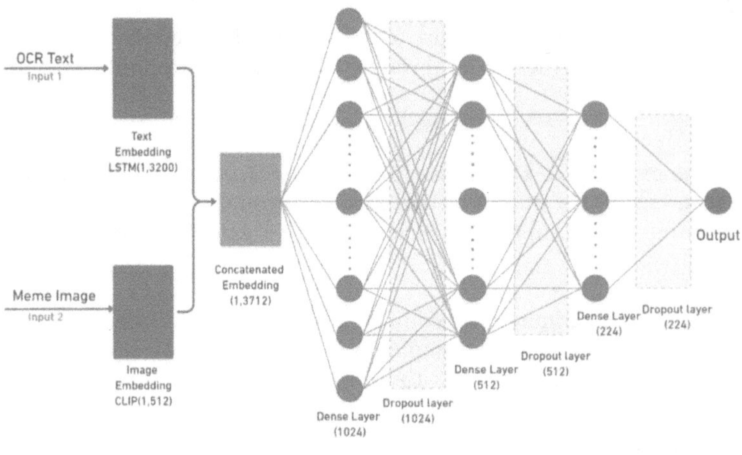

Figure 7. Architecture diagram of LSTM + CLIP model.

4.3.3 *LSTM + CLIP*

In the final proposed solution, fusion of LSTM and CLIP models is done to produce the desired results. First, textual data is used to create word embeddings using an LSTM model. Tokenizing the text is the process used in this procedure, where each word is given a distinct numerical representation while considering a 5000-word vocabulary. Additionally, the maximum length of input sequences is limited to 100 tokens to preserve computational efficiency and control sequence complexity.

Simultaneously image embeddings from images are generated using the CLIP model. Image embeddings of size 512 are produced by CLIP's architecture, which enables it to comprehend and encode images into meaningful numerical representations. After extracting the image embeddings from CLIP and the word embeddings from LSTM, they are merged for additional processing. To create a combined layer of size 3712, this integration concatenates the generated text embeddings of size 3200 with the image embeddings of size 512.

After that, several changes are made to the combined layer to extract relevant features and relationships. The data is transformed nonlinearly through its passage through four dense layers, which are essentially fully connected neural network layers. Dropout layers are dotted throughout the network architecture to reduce overfitting and improve the model's capacity for generalization. This configuration makes use of three dropout layers.

Lastly, the model produces a final combined output after going through the dropout and dense layers. This output provides a comprehensive representation that utilizes the advantages of both LSTM and CLIP embeddings for tasks like classification and retrieval. It encapsulates the fused information from both textual and visual modalities.

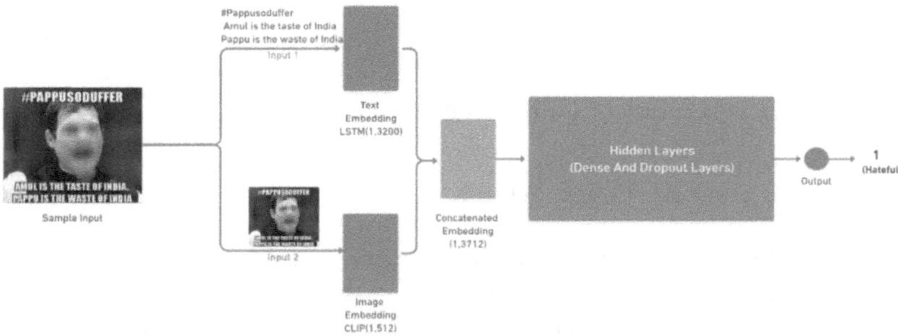

Figure 8. Visualization of flow of the model.

5 RESULTS AND DISCUSSION

In this work, four comparison metrics namely Accuracy, Precision, Recall and F1 Score are used to compare the results of the unimodal and multimodal models on the "Bharatiya Political Memes Dataset". The different scores for each of the models are as shown in Table 4 and Figure 9.

Proposed fusion models showed significantly higher results than unimodal models due to the fusion of embeddings. Results of proposed fusion models LSTM + CLIP and BERT + CLIP are comparable and higher than BERT + ResNet50 due to the intricate nature of CLIP embeddings. The shape of embeddings generated in LSTM is significantly high which results in better results. By leveraging CLIP's ability to understand visual concepts from natural language alongside LSTM's recurrent architecture, the combined model offers a comprehensive understanding of contextual cues and visual semantics, enhancing the capabilities for tasks like hateful meme classification.

Implementation details include the language used in the models which is Python. The System configuration during the working of the models is: System: Linux; Node Name: 3d06473676f5; Release: 6.1.58+; Version: #1 SMP; PREEMPT_DYNAMIC Sat Nov 18 15:31:17 UTC 2023; Machine: x86_64; Processor: x86_64; Available Memory: 12.67 GB.

Table 4. Comparison of models.

Model	Accuracy	Precision	Recall	F1 Score
BERT	0.65	0.65	0.65	0.65
LSTM	0.61	0.61	0.58	0.59
ResNet50	0.63	0.64	0.64	0.63
CLIP	0.69	0.69	0.70	0.69
BERT + ResNet50	0.66	0.67	0.67	0.66
BERT + CLIP	0.71	0.71	0.71	0.70
LSTM + CLIP	**0.72**	**0.72**	**0.71**	**0.71**

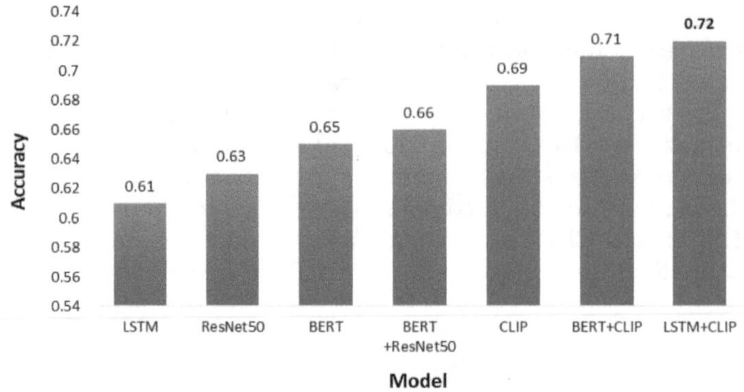

Figure 9. Comparison of accuracy of various models.

6 CONCLUSION AND FUTURE WORK

In this work, the "Bharatiya Political Memes Dataset" is introduced, consisting of memes, which are multimodal means of communication, classified into two labels – Hateful and Non-hateful. The results of unimodal models – BERT, LSTM, and Resnet50 on the introduced dataset have been compared. To improve the results, fusion models are proposed- BERT + Resnet50, BERT + CLIP and LSTM + CLIP, where the embeddings of text and image from their corresponding unimodal models are taken together and fused to generate results. The two-channel fusion model of LSTM + CLIP produces best results due to its complex architecture being able to capture the intricate details of political memes.

Future work includes focusing on disguised hate, where the hate in the memes is not direct, but implied. These types of memes require more complex architecture and must be studied. Contextual understanding according to the socio-political environment present in the situations must be taken into account for robust classification. The data might not be hateful by itself but in a certain political environment it could attack a certain group of people. Cross cultural sensitivity across diverse groups and ideologies should be considered triggers should be checked across all groups and not just based on a few. Further real time detection focuses on instant offline analysis to swiftly identify and mitigate hateful memes spread across platforms.

REFERENCES

Banerjee, A., Sehgal, S. (2021, December 14). Memes and copyright protection. *Exception to the Applicability of Fair Use Doctrine*. Mondaq. https://www.mondaq.com/india/copyright/1141606/memes-and-copyright-protection

Das, A., Wahi, J. S., & Li, S. (2020). Detecting hate speech in Multi-modal memes. [Preprint]. *arXiv*. https://arxiv.org/abs/2012.14891

Davidson, T., Warmsley, D., Macy, M., & Weber, I. (2017). Automated hate speech detection and the problem of offensive language. In *Proceedings of the International AAAI Conference on Web and Social Media*, 11, 1–2. [Preprint]. *arXiv*. https://doi.org/10.1609/icwsm.v11i1.14955

Devlin, J., Chang, M., Lee, K., & Toutanova, K. (2018). BERT: Pre-training of deep bidirectional transformers for language understanding. [Preprint]. *arXiv*. https://arxiv.org/abs/1810.04805

Gaydhani, A., Doma, V., Kendre, S., & Bhagwat, L. (2018). Detecting hate speech and offensive language on twitter using machine learning: An N-gram and TFIDF based approach. [Preprint]. *arXiv*. https://arxiv.org/pdf/1809.08651

Hasan, M., Jannat, N., Hossain, E., Sharif, O., Hoque, M. M. (2022). CUET-NLP@DravidianLangTech-ACL2022: Investigating deep learning techniques to detect multimodal troll memes. In *Proceedings of the*

Second Workshop on Speech and Language Technologies for Dravidian Languages, 170–176, Dublin, Ireland. https://doi.org/10.18653/v1/2022.dravidianlangtech-1.27

He, K., Zhang, X., Ren, S., Sun, J. (2016). Deep Residual Learning for Image Recognition. In *2016 IEEE Conference on Computer Vision and Pattern Recognition (CVPR)*, 770–778, Las Vegas, NV, USA. https://doi.org/10.1109/GLOBECOM48099.2022.10001325

Hossain, E., Sharif, O., Hoque, M. M., Dewan, M. A., Siddique, N., Hossain, M. A. (2022). Identification of Multilingual offense and Troll from social media memes using weighted ensemble of multimodal features. *Journal of King Saud University – Computer and Information Sciences*, vol. 34, pp. 1–10. https://doi.org/10.1016/j.jksuci.2022.06.010

Kiela, D., Firooz, H., Mohan, A., Goswami, V., Singh, A., Ringshia, P., Testuggine, D. (2020). The hateful memes challenge: Detecting hate speech in multimodal memes. [Preprint]. *arXiv*. https://arxiv.org/abs/2005.04790

Kulkarni, A. (2017). Internet meme and political discourse: A study on the impact of internet meme as a tool in communicating political satire. *Journal of Content, Community & Communication, Amity School of Communication*, vol. 6, year 3, June, pp. 5. https://dx.doi.org/10.2139/ssrn.3501366

Kumar, A., Tyagi, V., Das, S. (2021). Deep learning for hate speech detection in social media. In *2021 IEEE 4th International Conference on Computing, Power and Communication Technologies (GUCON)*, 1–4, Kuala Lumpur, Malaysia. https://doi.org/10.1109/GUCON50781.2021.9573687

Kumar, K., Nandakumar, K. (2022). Hate-CLIPper: Multimodal hateful meme classification based on cross-modal interaction of CLIP features. [Preprint]. *arXiv*. https://arxiv.org/abs/1810.04805

Mnassri, K., Rajapaksha, P., Farahbakhsh, R., Crespi, N. (2022). BERT-based ensemble approaches for hate speech detection. In *GLOBECOM 2022 – 2022 IEEE Global Communications Conference*, 4649–4654, Rio de Janeiro, Brazil. https://doi.org/10.1109/GLOBECOM48099.2022.10001325

Mozafari, M., Farahbakhsh, R., Crespi, N. (2020). A BERT-based transfer learning approach for hate speech detection in online social media. In *Proceedings of the 8th International Conference on Complex Networks and Their Applications (COMPLEX NETWORKS 2019)*, edited by H. Cherifi, S. Gaito, J. Mendes, E. Moro, and L. Rocha, vol. 881 of Studies in Computational Intelligence, Springer, Cham, 2611–2624. https://doi.org/10.1007/978-3-030-36687-2_77

Radford, A., Kim, J., Hallacy, C., Ramesh, A., Goh, G., Agarwal, S., Sastry, G., Askell, A., Mishkin, P., Clark, J., Krueger, G., & Sutskever, I. (2021). Learning transferable visual models from natural language supervision. [Preprint]. *arXiv*. https://arxiv.org/abs/2103.00020.

Rajput, K., Kapoor, R., Rai, K., Kaur, P. (2022). Hate me not: Detecting hate inducing memes in code switched languages. [Preprint]. *arXiv*. https://arxiv.org/abs/2204.11356

Staudemeyer, R. C., Morris, E. R. (2019). Understanding LSTM – a tutorial into long short-term memory recurrent neural networks. [Preprint]. *arXiv*. https://arxiv.org/abs/1909.09586

Velioglu, R., Rose, J. (2020). Detecting hate speech in memes using multimodal deep learning Approaches: Prize-winning solution to hateful memes challenge. [Preprint]. *arXiv*. https://arxiv.org/abs/2012.12975

Zhou, Y., Chen, Z., Yang, H. (2021). Multimodal learning For hateful memes detection. In *2021 IEEE International Conference on Multimedia & Expo Workshops (ICMEW)*, 1–6, Shenzhen, China. https://doi.org/10.1109/ICMEW53276.2021.9455994

Computational Methods in Science and Technology Volume 1 – Kaur et al. (Eds)
© 2025 the Author(s), ISBN: 978-1-032-81760-6

Exploring the potential of cloud computing in revolutionizing healthcare

Renu Saini, Paras, Parag Tayal & Pallavi Thakur
Computer Science and Engineering Department, Chandigarh Engineering College, Landran, Mohali, Punjab, India

ABSTRACT: Cloud computing, which moves assets from desktop setting to cloud infrastructure—where apps are kept in a "cloud" with massive processing power and nearly infinite database storage—has fundamentally altered traditional computing over the past few years. Modern healthcare can benefit greatly from telemedicine when it is combined with cloud computing. Strategic thinking is necessary before moving healthcare assets and services to the cloud in order to weigh the advantages and disadvantages of the fresh approach, determine whether it may be implemented, and analyze its capabilities. The study illustrated how various technologies could be applied to big data analysis in healthcare using a cloud computing. In this research, we investigate the way cloud computing has changed the way in which healthcare organizations keep their data.

1 INTRODUCTION

Cloud computing revolutionizes computing by offering computing resources as a service (CaaS) over a network [4]. It provides software, computing power, data access, and storage without users needing to know the specifics of the underlying infrastructure [1]. Cloud technology meets dynamic user demands efficiently, shifting resources from desktop environments to cloud infrastructure. It enables self-service, on-demand access to computing resources from anywhere [3]. In healthcare, cloud computing ensures easy access to medical data, promotes interoperability across systems, and enhances accessibility and data flow among healthcare providers, addressing key challenges in the medical IT system. [2]

2 CLOUD COMPUTING FOR HEALTHCARE RESEARCH AND COLLABORATION

Cloud computing offers solutions to healthcare challenges with features like on-demand service and enhanced security [19]. It promises rapid data access, fosters collaboration among systems, and improves patient care while being cost-effective and flexible for healthcare organizations [9].

The five main ways that the cloud will change healthcare are listed below:

2.1 *Data security*

Cloud architecture provides superior uptime and durability with redundancy.

2.2 *Data privacy*

Cloud security exceeds local IT; robust measures, like app isolation, necessary.

 DOI: 10.1201/9781003501244-84

2.3 *Speed of innovation*

Cloud updates streamline tasks, letting providers focus on work.

2.4 *Emerging trend*

Cloud computing lowers expenses, improves accessibility, and streamlines processes [26]. Cloud computing is being heralded as the most efficient way to address issues with unpredictable traffic spikes, computer overloading, and potentially costly processing and backup equipment investments [17].

3 CLOUD-BASED HEALTHCARE INFRASTRUCTURE AND SERVICES

Cloud computing, or Computing as a Service, provides shared resources, software, and data over a network. Users access, compute, store, and retrieve data without knowledge of physical equipment locations [4]. NIST outlines four fundamental cloud deployment concepts, with typical characteristics including three-layered computing systems for modern internet computing services. [12]

3.1 *Software-as-a-Service (SaaS) [4]*

SaaS delivers application over the internet, simplifying access and maintenance [12]. In healthcare, it facilitates EMR and PHR deployment [26], ensuring interoperability and addressing privacy concerns through cloud-based solutions, transforming software consumption. [26]

3.2 *Platform as a Service (PaaS) [4]*

PaaS abstracts infrastructure concerns, enabling application development and sharing. It offers cost-effective environments and support services [12]. In healthcare, PaaS solutions, like Microsoft Azure, enable centralized analysis of medical scans, showcasing its versatility. [26]

3.3 *Infrastructure as a Service (IaaS) [4]*

IaaS provide on-demand processing and storage, virtualizing resources like memory and processing power. In healthcare, it enhances security and interoperability, reducing deployment costs, addressing privacy concerns with cloud-based HIS. [26]

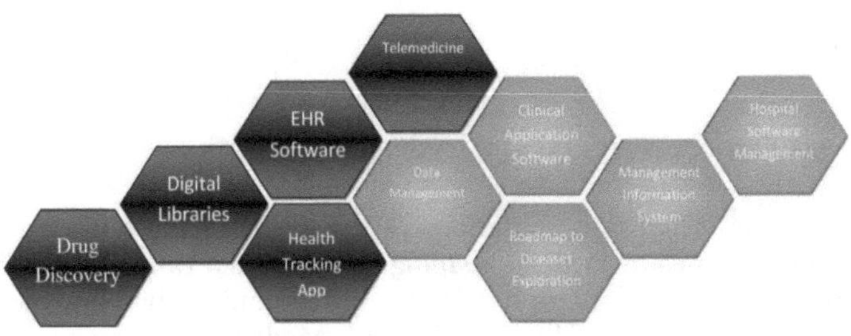

Figure 1. Healthcare services offered by cloud computing.

3.4 *Cloud computing based healthcare services*

- **Data Management:** Healthcare grapples with handling data volumes, including patient records.

- **Telemedicine:** Cloud tech enables remote patient consultations and medical image storage.
- **Digital Libraries:** Cloud services streamline management of digital resources for libraries.
- **Virtual Medical Universities:** Cloud platform facilitate medical education through online lecture.
- **Drug Discovery:** Cloud aids in efficient drug discovery, ensuring safety and efficacy.
- **Management Information System:** MIS enhances internal communication and decision-making in healthcare. [19]

4 CLOUD-BASED ELECTRONIC HEALTH RECORDS (EHR)

The term "Electronic Health Record" (EHR) refers to a digitally comprehensive record that contains all of an individual's medical data, including test results, lab results, diagnosis, prescriptions, dates of immunizations, allergies, and radiological pictures.. By giving clinicians and patient care managers fast access to reliable information that supports their decisions, electronic health records (EHRs) strive to increase the effectiveness, safety, and quality of healthcare delivery [5].

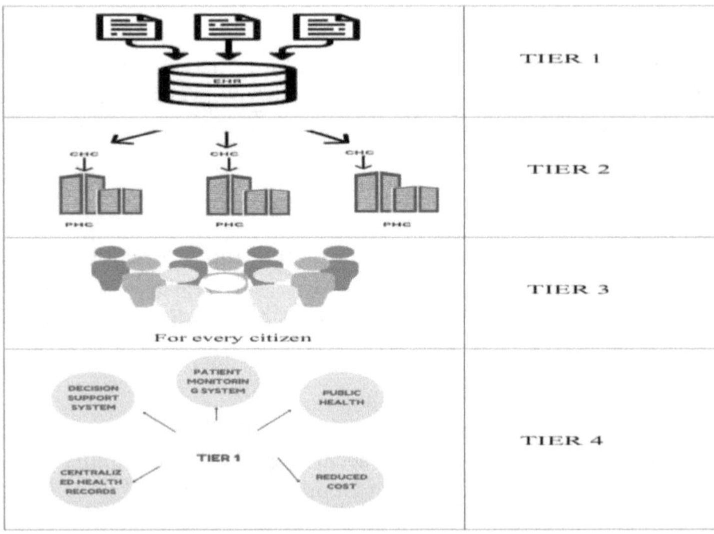

Figure 2. Architecture of EHR.

5 MEDICAL IMAGING AND DIAGNOSTIC SYSTEMS

The process of creating visual representations of the inside of the body for diagnostic and therapeutic reasons is known as **medical imaging**. It includes a range of imaging modalities, including positron emission tomography (PET), ultrasound, magnetic resonance imaging (MRI), computed tomography (CT), and X-rays. Medical imaging is essential to contemporary healthcare because it gives medical personnel precise anatomical and functional knowledge about the interior organs, tissues, and physiological processes of their patients [11]. Privacy requirements in cloud-based image storage includes: Integrity, Confidentiality, Data Ownership, Authentication, Authorization and Availability [23]. DICOM (Digital Imaging and Communications in Medicine): It is a standard protocol for communication and management of medical imaging data [21]. PACS (Picture Archiving and Communication System):It is a system for storage, retrieval, distribution, and viewing of medical images within healthcare organizations [20]. In a hybrid cloud environment, the iMAGE cloud platform is presented as a diagnostic tool with the goal of offering medical image processing offered to nearby healthcare providers. [24]

5.1 *Challenges for data storage in medical imaging: [21]*

- Volume: Medical imaging demands substantial storage and scalability for data management.
- Data Security: Protection of sensitive patient information is crucial for compliance.
- Data Accessibility: Retrieving imaging data efficiently is crucial but challenging.
- Interoperability: Interoperability among imaging systems is essential for data integration.
- Long-Term Retention: Robust data retention policies are essential for storing medical imaging data.
- Backup and Recovery: Backup and disaster recovery strategies prevent data loss.
- Cost Constraints: High-performance storage needs must align with budget constraints. [22]

6 CLOUD BASED TELEMEDICINE AND REMOTE PATIENT MONITORING

Telemedicine involves remotely transmitting patient information for medical services, while telecare extends to nursing care delivery [10]. Telehealth, a medical device, must comply with standards and risk declarations. Telehealth systems utilize telecommunications to share electronic data, facilitating services like telesurgery [11]. Cloud-based telemedicine extends beyond medical services, accommodating various devices [10]. Here's a brief overview of some healthcare initiatives: [13] Brazil: Wireless sensor networks for patient monitoring at Federal University of Santa Catarina.

India: Immediate Medical Care Unit (IMCU) for rural communities.Telemedicine program using cloud computing, improving telecardiology accuracy with 12-lead ECG service. [14]

6.1 *Need of cloud computing in telemedicine*

Government and business initiatives aim to enhance healthcare nationwide, but e-healthcare, particularly telemedicine, falls short due to outdated infrastructure. Traditional client/server setups struggle with data maintenance and scalability. Cloud computing emerges as the optimal solution, offering efficient data management and scalability for telemedicine initiatives in a nation with over 100 million people. [10]

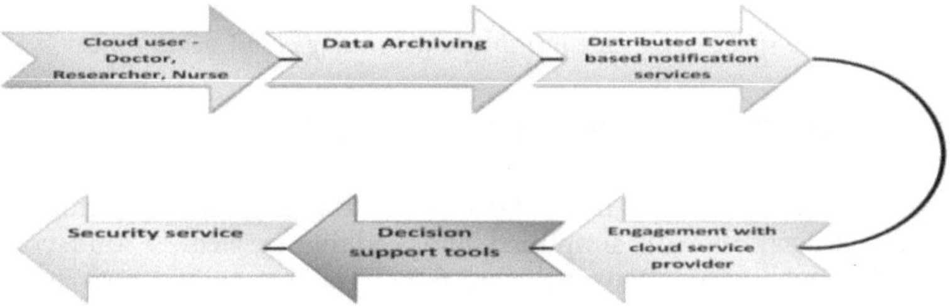

Figure 3. Basic steps showing cloud computing for telemedic services.

6.2 *Current status of cloud computing telemedicine*

A survey that was done revealed that cloud-based apps connected to healthcare may be divided into five groups. These include Digital Imaging and Communications in Medicine (DICOM)-based systems, Health Cloud eXchange (HCX), Emergency Medical System

(EMS) [10]. Remote monitoring [15] of patient health data via cloud platforms involves continuous collection, transmission, and analysis of vital information from wearable sensors or mobile apps. Cloud servers securely store and process data, enabling proactive intervention and personalized care delivery, while passive monitoring enhances medication administration and healthcare services. [16]

7 HEALTHCARE ANALYTICS

Healthcare data encompasses patient medical history, diagnoses, treatments, and outcomes, including EHRs, medical images, lab results, medications, and demographics. It's vital for clinical decisions, research, and policy [27]. Healthcare data is connected with cloud-based solutions for data analytics through various mechanisms:

7.1 *Data storage*

Healthcare data, like EHRs and medical image, that can be securely stored in cloud, providing scalability, accessibility, and redundancy for authorized users while preserving data integrity and privacy.

7.2 *Data processing and analysis*

Cloud-based analytics platforms enable healthcare organizations to analyze large volumes of data for insights into patient outcomes and disease trends.

7.3 *Data sharing*

Cloud solutions facilitate secure sharing of healthcare data among providers, researchers, and stakeholders for collaborative research and population health management [28].

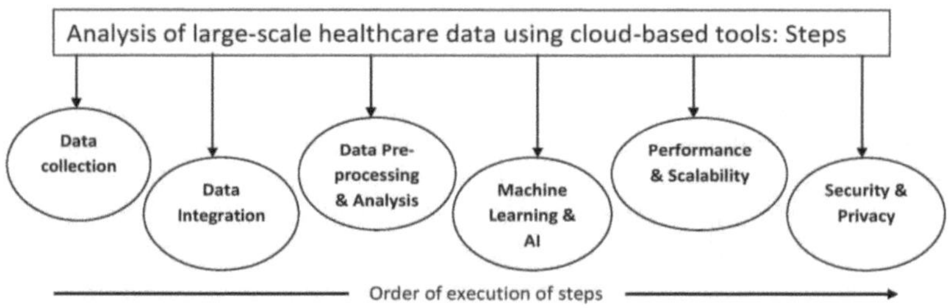

Figure 4. Steps for analysis of large-scale healthcare data using cloud-based tools.

8 CLOUD-BASED HEALTHCARE APPLICATIONS AND PLATFORMS

With cloud computing, customers can meet ever-increasing as well as dynamic demands and manage resource utilization with simplified and scalable services. Cloud computing has transformed traditional computing over the last few years by moving resources from desktop environments to cloud infrastructure, where apps are housed in a "cloud" with massive computational power and practically infinite database storage [1].

8.1 *Applications and platforms*

8.1.1 *Creation of an emergency health care service model using the cloud*

Cloud computing offers scalable, affordable solutions for managing large medical datasets in healthcare sectors [8]. It provides shared resources, instant availability, collaborative models, and virtually unlimited storage [6]. Various deployment options such as IaaS, PaaS, and SaaS cater to different data sensitivities and user needs [7]. Centralized cloud systems store medical records, making patient data accessible and mobile. Strategic planning is crucial for successful implementation. [1]

8.1.2 *Cloud-based creation of intelligent and networked data for medical applications*

Advanced healthcare applications offering platform tools for interface building, and the Information Integration & Informatics framework, providing APIs for data access, integration, analytics, and storage. [32]

8.1.3 *Cloud-based Internet of medical Things*

Cloud computing (CC) offers diverse computing resources for healthcare services, but careful consideration of its advantages, drawbacks, and deployment models is essential to optimize safety, cost-effectiveness, and service efficiency [31].

8.1.4 *An IoT-inspired cloud-based web service*

The proposed system includes components like web services with open APIs, enabling developers to create new services using data from personal and home devices, thereby bridging the gap between high-level functionality and low-level implementation details [33].

8.1.5 *A cloud platform that is scalable enough for activated healthcare monitoring applications*

Wearable sensors enable remote patient monitoring, aiding in early disorder detection. Cloud computing addresses challenges in scaling healthcare monitoring applications across institutions, offering a scalable and real-time solution like the Assistive Patient monitoring cloud Platform for Active healthcare apps (AppA) [34].

9 CONCLUSION, BENEFITS AND FUTURE DIRECTIONS

9.1 *Conclusion*

Cloud computing has transformed healthcare delivery with scalable services, revolutionizing traditional computing. The health care records (EHR) can be securely shares the patient data with the help of strong security. Telemedicine and cloud computing offered services accessible at a distance, improving patient treatment. Issue such as high costs and security concerns persist, urging further integration and secure system development. despite these challenges, cloud computing has a lot to offer to the healthcare industry. This review highlights the cloud computing integration and benefits in healthcare but it also acknowledges the need for continued research and development.

9.2 *Benefits*

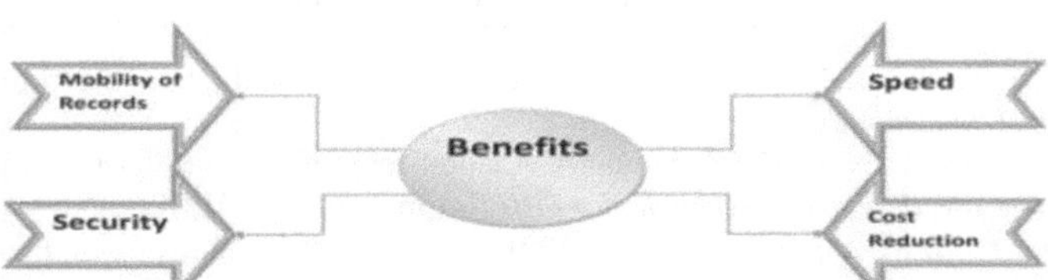

Figure 5. Benefits of cloud computing in healthcare.

Research indicates that cloud computing in the healthcare industry can save costs, increases quality of care, and improves accessibility for patient and providers in the healthcare industry.

Among the many benefits of cloud computing for healthcare sector, some are mentioned which are:

9.2.1 *Mobility of records*
it enhances the data sharing, which helps in the patient care accessibility.

9.2.2 *Clinical benefits*
it offers remote consultations which provides benefit to healthcare professionals.

9.3 *Future scope*

Cloud computing has the potential to transform healthcare by streamlining delivery through EHR systems, telemedicine and data analysis. It provided Personalized treatment, more security, and improvement which increases the patient results and research. Some challenges in cloud computing are listed below:-

9.3.1 *Data/service reliability*
it needs an absolute reliability for healthcare data.

9.3.2 *Security*
As the provider stores data from multiple client; it should offer a high security measures such as robust access controls to the client healthcare institute. [3]

REFERENCES

[1] Banerjee, A., Agrawal, P., & Rajkumar, R. (2013). Design of a cloud-based emergency healthcare service model. *International Journal of Applied Engineering Research*, 8(19), 2261–2264.

[2] Tak, A. (2023). The role of cloud computing in modernizing healthcare IT infrastructure. *Journal of Artificial Intelligence & Cloud Computing*, SRC/JAICC-163(2), 151, 2–7. DOI:doi.org/10.47363/JAICC/2023

[3] Raval, D., & Jangale, S. (2016). Cloud-based information security and privacy in healthcare. *International Journal of Computer Applications*, 150(4).

[4] Mathew, S. (2013). Cloud computing: a new foundation towards health care. *International Journal of Innovative Technology and Exploring Engineering*, 3(2), 118–121.

[5] Kavitha, R., Kannan, E., & Kotteswaran, S. (2016). Implementation of cloud-based Electronic Health Record (EHR) for Indian healthcare needs. *Indian Journal of Science and Technology*.

[6] Karthikeyan, N., & Sukanesh, R. (2012). Cloud-based emergency healthcare information service in India. *Journal of Medical Systems*, 36(6), 4031–4036.

[7] Chen, T. S., Liu, C. H., Chen, T. L., Chen, C. S., Bau, J. G., & Lin, T. C. (2012). Secure dynamic access control scheme of PHR in cloud computing. *Journal of Medical Systems*, 36(6), 4005–4020.

[8] Yang, C. L., Tsai, C. L., Cheng, K. T., & Chen, S. H. (2012). Low-invasive implantable devices of low-power consumption using high-efficiency antennas for cloud healthcare. *IEEE Journal on Emerging and Selected Topics in Circuits and Systems*, 2(1), 14–23.

[9] Gavrilov, G., & Trajkovik, V. (2012). Security and privacy issues and requirements for healthcare cloud computing. *ICT Innovations*, 2012.

[10] Matlani, P., & Londhe, N. D. (2013, January). A cloud computing based telemedicine service. In *2013 IEEE Point-of-Care Healthcare Technologies (PHT)* (pp. 326–330). IEEE.

[11] Alami, H., Gagnon, M. P., & Fortin, J. P. (2016). Telehealth in light of cloud computing: clinical, technological, regulatory and policy issues. *Journal of the International Society for Telemedicine and eHealth*, 4, e5(1-7).

[12] Doheir, M., *et al.* (2019). The new conceptual cloud computing modelling for improving healthcare management in health organizations. *International Journal of Advanced Science and Technology*, 28(1), 351–362.

[13] Cilliers, L. (2014). Using the cloud to provide telemedicine services in a developing country. *South African Journal of Information Management*, 16(1), 1–7.

[14] Hsieh, J. C., & Hsu, M. W. (2012). A cloud computing based 12-lead ECG telemedicine service. *BMC Medical Informatics and Decision Making*, 12(1), 1–12.

[15] Thilakanathan, D., Chen, S., Nepal, S., Calvo, R., & Alem, L. (2014). A platform for secure monitoring and sharing of generic health data in the Cloud. *Future Generation Computer Systems*, 35, 102–113.

[16] Iranpak, S., Shahbahrami, A., & Shakeri, H. (2021). Remote patient monitoring and classifying using the internet of things platform combined with cloud computing. *Journal of Big Data*, 8, 1–22.

[17] AlTwaijiry, A. (2020). The determinants of cloud computing adoption in healthcare. *ResearchBerg Review of Science and Technology*, 3(1), 9–20.

[18] Akhbarifar, S., Javadi, H. H. S., Rahmani, A. M., & Hosseinzadeh, M. (2023). A secure remote health monitoring model for early disease diagnosis in cloud-based IoT environment. *Personal and Ubiquitous Computing*, 27(3), 697–713.

[19] Nigam, V. K., & Bhatia, S. (2016). Impact of cloud computing on health care. *International Research Journal of Engineering and Technology*, 3(5), 2804–2810.

[20] Viana-Ferreira, C., Guerra, A., Silva, J. F., Matos, S., & Costa, C. (2017). An intelligent cloud storage gateway for medical imaging. *Journal of Medical Systems*, 41(9), 141.

[21] Langer, S. G. (2011). Challenges for data storage in medical imaging research. *Journal of Digital Imaging*, 24, 203–207.

[22] Arka, I. H., & Chellappan, K. (2014). Collaborative compressed I-cloud medical image storage with decompress viewer. *Procedia Computer Science*, 42, 114–121.

[23] Marwan, M., Kartit, A., & Ouahmane, H. (2016, September). Secure cloud-based medical image storage using secret share scheme. In *2016 5th International Conference on Multimedia Computing and Systems (ICMCS)* (pp. 366–371). IEEE.

[24] Liu, L., Chen, W., Nie, M., Zhang, F., Wang, Y., He, A., ... & Yan, G. (2016). iMAGE cloud: medical image processing as a service for regional healthcare in a hybrid cloud environment. *Environmental Health and Preventive Medicine*, 21, 563–571.

[25] Mugonza, R., & Basaza-Ejiri, A. H. (2018). Issues affecting health research collaborations based on cloud computing.

[26] Daman, R., Tripathi, M. M., & Mishra, S. K. (2016). Security issues in cloud computing for healthcare. In *2016 3rd International Conference on Computing for Sustainable Global Development (INDIACom). IEEE.*

[27] Pradhananga, Y., Karande, S., & Karande, C. (2015, February). CBA: cloud-based big data analytics. In *2015 International Conference on Computing Communication Control and Automation* (pp. 47–51). IEEE.

[28] Gupta, U., & Sharma, R. (2023). A study of cloud-based solution for data analytics. In *Data Analytics for Internet of Things Infrastructure* (pp. 145–161). Cham: Springer Nature Switzerland.

[29] Sultana, S. N., Ramu, G., & Reddy, B. E. (2014). Cloud-based development of smart and connected data in healthcare application. *International Journal of Distributed and Parallel Systems*, 5(6), 1–11.

[30] Rallapalli, S., Gondkar, R., & Ketavarapu, U. P. K. (2016). Impact of processing and analyzing healthcare big data on cloud computing environment by implementing Hadoop cluster. *Procedia Computer Science*, 85, 1–22.

[31] Milovanovic, D., & Bojkovic, Z. (2017). Cloud-based IoT healthcare applications: Requirements and recommendations. *International Journal of Internet of Things and Web Services*, 2, 60–65.

[32] Sultana, S. N., Naisha, G., & Reddy, B. E. (2014). Cloud-based development of smart and connected data in healthcare application. *International Journal of Distributed and Parallel Systems*, 5.6, 1–11.

[33] Biswas, S., et al. (2014). Cloud-based healthcare application architecture and electronic medical record mining: An integrated approach to improve healthcare system. In *2014 17th International Conference on Computer and Information Technology (ICCIT)*. IEEE.

[34] Balasubramanian, V., & Stranieri, A. (2014). A scalable cloud platform for active healthcare monitoring applications. In *2014 IEEE Conference on e-Learning, e-Management and e-Services (IC3e)*. IEEE.

Computational Methods in Science and Technology Volume 1 – Kaur et al. (Eds)
© 2025 the Author(s), ISBN: 978-1-032-81760-6

Real time vision application for students using augmented reality

V. Sathya
Assistant Professor, Sri Sairam Engineering College, Chennai, Tamil Nadu, India

A. Amothini & S. Varsha
Student, Sri Sairam Engineering College, Chennai, Tamil Nadu, India

ABSTRACT: At that time AR (Augmented Reality) was just known for developing a game based technology. But, now we can change it for a very useful purpose for education to show it in a real time perspective. If the graphical ideas can be amended according to the ideas, it can improve and change the methods in our educational system and method of teaching and to change the way of teaching for students also it gives the most creative ideas and also mistakes in real time can be aliased in the environment. Our ultimate idea is to create a high reality based 3D modeling for students to see objects in real time. So, that students can have a grasping power of visualizing the objects at 360 degree angle by naming the parts on the objects. The objects can Zoom and get a clear view of small objects which are not seen by the normal picture. So, students can get more memory power by teaching it with 3D modeling. This project is mainly for medical students so that it will be useful for them to analyze and visualize the small parts like liver, lung etc. It can also help them gain more grasping power.

Keywords: 3D object, medical, visualizing, educational, Real Time application, Blender

1 INTRODUCTION

As the most intelligent species on Earth, it is common knowledge that we should know ourselves better than any other species. In this sense, one of the fundamental skills that every person should possess is an understanding of human anatomy. However, the majority of people in Bangladesh are illiterate and subsist on very little money.With a difficult lifestyle, people in our country can hardly think of such background knowledge, because it is not cheap. At the time, AR (Augmented Reality) was simply known for developing game-based technology.But, now we can change it for a very useful purpose for education to show it in a real time perspective. If the graphical ideas can be amended according to the ideas, it can improve and change the methods in our educational system as well as methods of teaching & to change the way of teaching for students. Also it gives the most creative ideas and also mistakes in real time can be in the environment.

2 LITERATURE REVIEW

(1) Benchmark studies are pursued by the study. “Paper-Based Augmented Reality” was written by authors Jonathan J. Hull, Berna Erol, Jamey Graham, Qifa, and Daniel G. Van Olst in the essay [1]. Digital Capture The document is supplied with augmented reality without any modifications made to it. This method allows one to complete a unique signature from a snapshot of a small array of textual information that the

 DOI: 10.1201/9781003501244-85

signature can associate with a digital system. They created a text content patch recognition method and provided test results to demonstrate the ability to detect images of text content from a given number of samples. It was found that two programs were created with an implementation on a four-frame-per-second digital system.An enhanced travel guide along with clickable paper [2]. Both will showcase the affordability and capability of the handbook-based AR. Their main drawback of this paper is that the letter of the newspaper uses AR but that is now done by using google lens.

(2) This study pursues benchmark studies. In the article [3], author Jennifer Herron, Indiana University School of Medicine, Indianapolis, IN wrote "Augmented Reality in Medical Education and Training". The students happily accepted the inclusion of Google Glasses in the simulation. One difficulty that has been mentioned is the problem of viewing both the real world and the video looking realistic in Google Glass. [4] This article's primary flaw is that, in spite of this problem, the students expressed strong approval for the new technology and even offered suggestions for more content to be included in the virtual display. This work aims to do benchmark research.

(3) In the paper [5], authors Kathryn MacCallum and David Parsons wrote "Integrating mobile mixed reality to improve learning before, in, during and after field trips", the research survey selected article identified several methods in which AR devices could help train medical professionals.[6] The way to which online resources can replace or improve medical knowledge and abilities can be determined by a variety of conditions. The disadvantage of this paper is that As the time of medical training approaches, the phrase "medical student" is open to many views, and this evaluation has no longer included doctors. Multiple studies have proven that population discourse happens [7]. This research survey pursues reference learning in augmented reality.

(4) In paper [8] the authors Carolien Kamphuis, Marlies Schijven Esther Barsom, wrote the paper "Augmented reality in reality of medical studies?" We've seen that Augmented Reality is a combination of an educational atmosphere that is common to or equal to the real- environment surroundings augmented [9] with obvious aswell as different sensory information.

(5) In paper [10], the authors Poshmaal Dhara, Tetyana Rocks, Rasika M Samarasinghea wrote the paperA study named "Augmented reality for healthcare learning: learners' experiences and learning outcomes" It involves the basic objectives of AR-based totally learning, which include reducing traveling and enhancing understanding of challenging material. We further discuss how AR can improve medical students' experiences.The main disadvantage of the paper is that Even though they have proposed the technology using AR they use VR glasses to show to the student in a 3D way [11].

3 EXISTING SYSTEM

The existing system of our project is that they have proposed the methodology in the field of medicine education using Augmented reality. In all these proposed methods they have used VR optic glasses to present the models of an image or video to the students in a 3D format for better understanding [12]. The use of VR optics is not cost efficient and feasible for h students ad it takes a lot of time for the students to see the images under optic glasses [13].

4 PROPOSED WORK

Everyone understands that as being the most intelligent people on the planet, we must first understand ourselves before we can understand others. In this sense, one of the fundamental skills that every person should possess is an understanding of human anatomy. However, a great deal of people here are illiterate and subsist on very little money. People in our country rarely consider such background knowledge because it is expensive and leads to a difficult living. At the time, AR (Augmented Reality) was only recognized for creating technology for video games. But, now we can change it for a very useful purpose for education to show it in a real time perspective. The main motive of our project is to project the images to the medical students in a 3D way so that they can analyze and study each and every concept very easily. Though 3D effects can only be seen using "VR Oculus" glass. [14] We are proposing this methodology using AR technology so that the students are able to see and visualize the images and videos through a projector in real time. Our project is about Creating an app in unity for viewing the 3D objects so that students can see the object individually in 360 Degree or else they can install the application and view in their house for studying purposes. We are creating 3D objects in Blender and modifying the correct shape of anatomy so that students can able to see the particular objects by Zooming in and out.The 3D objects in plain color will not be that much interesting so we are giving the real colors to the objects for Eg: if the blood flow is in red color the students also have to know that it is in red color by seeing the object.

Most importantly, most of us don't know which part's name is this. So, we are creating a video of 10 minutes of naming their parts on Real time visualization can help any student to analyze the concepts in a practical way. This can be further used for childrens in the primary school for better understanding and learning. The kids will be interested in learning things in new methodology. For higher education students this will be helpful for their practical and theory examination [15]. During viva they can visualize and answer the questions accordingly.so our project will be helpful for many people by using objects in 3D through mobile application instead of seeing things in microscope and other.

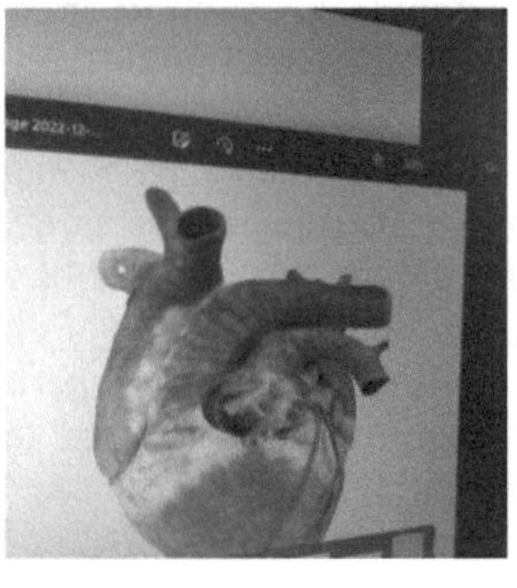

Figure 1. Heart model.

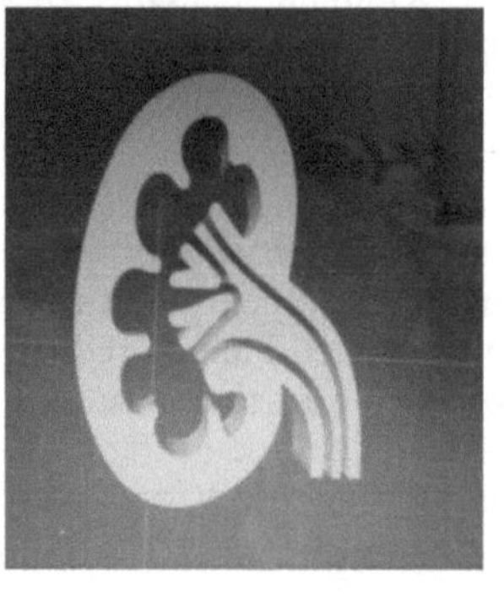

Figure 2. Kidney

5 SYSTEM OVERVIEW AND SOFTWARE DESCRIPTION

We have created an android application for students and staff to easily access the module videos that we have created. The login page of the application will contain a sign up or login option if you are a new user you should sign up and give the credentials. The page will be redirected to the welcome page. It contains the category of video of the human organs. This will be shown in 3d view to students to better understand the parts of all the arteries and veins. [16] The staff can also assign the videos to students and they can also mentor whether they watched the video or not through this application. The application is connected with a firebase module and the data is stored in the backend module.

Figure 3. App login page.

Figure 4. Unity App.

6 RESULTS AND DISCUSSION

The necessity and aim of this project was to create a practical and completely functional real-world environment of the objects in the medical field. Basically it is for the medical field but we can use it for other things also.The goal of this project is to learn in an effective way so that students will gain more experience and learn by their own by through the mobile application if it for a group of students we can connect it through projector and see so that it will be great and wide perspective to the group of students.The proposed system includes advantages such as they can study the complete flow of the diagram its basically for medical students so that they can see the particular modules without subscription if they want to see the advanced modules of the organs they have to subscribe a particular amount This project is relaxing and convenient for general users, particularly for school student there are some improvements that can be done in this project's future.

7 CONCLUSION

The integration of Augmented Reality in the discipline of Medical Science is nevertheless in its earliest phases. Yet, this generation has tremendous potential in academia. As already mentioned, it has an enormous impact in the not too distant marketplace as well, bringing up billions of opportunities for both customer and developers. At first sight, AR Skeletons' primary objective could seem to be for the commercialization of AR technology for Medical Science in real world, but this is not the case.of real environment It focuses on accomplishing an increasingly acceptable truth in which digital trash may develop through combining aspects with the real environment. A future in which people's imagination becomes automatic although understanding isn't a pipe fantasy.The concept has been ruled off by the utility of ourselves.

8 FUTURE ENHANCEMENT

In our project people can use the application and teach students and integrate more 3D modeling. It can be more useful to parents to teach their kids so that they will get involved in academics as well [17]. They have separate logins for students, teachers and parents. They can have advanced models in premium options so that the professionals can pay and see the video of the 3D modeling and learn. Likewise they can create the application for shoes so that instead of going to shop we can create and see our size and everything through that application. Likewise We can develop through ar and VR so that it will be more efficient for the people to use and see in real time perspective of the environment but the VR optic glasses are costly to overcome that we should do more research of that and reduce the cost of the glasses [18].

REFERENCES

[1] Ifenthaler D, Eseryel D. (2013). Facilitating complex learning by mobile augmented reality learning environments. In: Huang R, Spector JM, Kinshuk DGS, editors. *Reshaping Learning, the Frontiers of Learning Technologies in a Global Context*. New York: Springer.

[2] Van Merrie·nboer JJG, Kirschner PA. (2007). *Ten Steps to Complex Learning, A Systematic Approach to Four Component Instructional Design*. New York: Routledge.

[3] Salomon G., Perkins D. N. (1989). Rocky roads to transfer: rethinking mechanisms of a neglected phenomenon. *Educ Psychol.*; 24:113–42.4

[4] Jonassen D. (2008). *Meaningful Learning with Technology*. Upper Saddle River: Pearson.

[5] The New Media Consorium. The HORIZON report 2010 edition. 2013. http://net.educause.edu/ir/library/pdf/hr2010.pdf. [Accessed 1 June 2013]

[6] The New Media Consortium. The HORIZON report 2011 edition. 2013. http://net.educause.edu/ir/library/pdf/hr2011.pdf. Accessed 1 June 2013

[7] Wu H., Lee S., Chang H., Liang J. (2013). Current status, opportunities and challenges of augmented reality in education. *Comput Educ.*;62:41–9.

[8] Hugues O., Fuchs P., Nannipieri O. (2011). New augmented reality taxonomy: technologies and features of augmented environment. In: Furth B, editor. *Handbook of Augmented Reality*. New York: Springer; p. 47–63.

[9] Zhou F., Duh H. B. L., Billinghurst M. (2008 Sept 15–18). Trends in augmented reality tracking, interaction and display: a review of ten years of ISMAR. In: *7th IEEE and ACM International Symposium on Mixed and Augmented Reality (ISMAR 2008)*; Cambridge, UK; 2008.

[10] Milgram P., Takemura H., Utsumi F., Kishino. (1994). Augmented reality: a class of displays on the reality– virtuality continuum. In: *Proceedings of Telemanipulator and Telepresence Technologies*, vol. 2351; p. 282–292.

[11] Yu D., Jin J. S., Luo S., Lai W. (2010). A useful visualization technique: a literature review for augmented reality and its application, limitation & future direction. In: Huang ML, Nguyen QV, Zhang K, editors. *Visual Information Communication.* New York: Springer; p. 311–337

[12] Genc Y., Riedel S., Souvannavong F., Akinlar C., (2002). Navab N. Marker-less tracking for AR: a learningbased approach. In: Proceedings of the international symposium on mixed and augmented reality (ISMAR'02); *Darmstadt: IEEE Computer Society*; p. 295–304; IEEE Xplore.

[13] Frank J. R., (2005). Royal College of Physicians and Surgeons Canada RCPSC. *The CanMEDS 2005 physician competency framework: better standards, better physicians, better care*, vol. vii. Ottawa: Royal College of Physicians and Surgeons Canada; p. 29.

[14] Winkelmann A. (2007). Anatomical dissection as a teaching method in medical school: a review of the evidence. *Med Educ*. 41:15–22

[15] Chien C-H, Chen C-H, Jeng T-S. (2010). An interactive augmented reality system for learning anatomy structure. In: *Proceedings of the International Multiconference of Engineers and Computer Scientists*, IMECS, vol. 1; p. 1–6

[16] Vassilakaki, Evgenia evasilakaki yahoo gr, and Valentini Moniarou-Papaconstantinou. "A Systematic Literature Review Informing Library and Information Professionals' Emerging Roles

[17] Zhu, Egui, Anneliese Lilienthal, Lauren Aquino Shluzas, Italo Masiello, and Nabil Zary. "Design of Mobile Augmented Reality in Health Care Education:

[18] Masters, Ken, Rachel H. Ellaway, David Topps, Douglas Archibald, and Rebecca J. Hogue. "Mobile Technologies in Medical Education: Amee Guide No. 105."

Computational Methods in Science and Technology Volume 1 – Kaur et al. (Eds)
© 2025 the Editor(s), ISBN: 978-1-032-81760-6

Author index